10th Joint Conference on Chemistry 2015

IOP Conference Series: Materials Science and Engineering
Volume 107

Solo, Indonesia
8 - 9 September 2015

ISBN: 978-1-5108-2084-5
ISSN: 1757-8981

10th Joint Conference on Chemistry 2015

IOP Conference Series: Materials Science and Engineering
Volume 107

Solo, Indonesia
8 - 9 September 2015

IOP
Institute of Physics

Printed from e-media with permission by:

Curran Associates, Inc.
57 Morehouse Lane
Red Hook, NY 12571

Printed by Curran Associates, Inc. (2016)

For permission requests, please contact the Institute of Physics at the address below.

Institute of Physics
Dirac House, Temple Back
Bristol BS1 6BE UK

Phone: 44 1 17 929 7481
Fax: 44 1 17 920 0979

techtracking@iop.org

Additional copies of this publication are available from:

Curran Associates, Inc.
57 Morehouse Lane
Red Hook, NY 12571 USA
Phone: 845-758-0400
Fax: 845-758-2634
Email: curran@proceedings.com
Web: www.proceedings.com

Table of contents

Volume 107

10th Joint Conference on Chemistry
8–9 September 2015, Solo, Indonesia

Accepted papers received: 21 December 2015
Published online: 5 February 2016

Preface
011001

10th Joint Conference on Chemistry OPEN ACCESS

011002

Peer review statement OPEN ACCESS

Papers

Catalysis

012001

Photocatalytic removal of phenol over titanium dioxide- reduced graphene oxide photocatalyst OPEN ACCESS Nor Shuhada Alim, Hendrik O. Lintang and Leny Yuliati pg. 1

012002

The shape conversion of silver nanoparticles through heating and its application as homogeneous catalyst in reduction of 4- nitrophenol OPEN ACCESS H A Ariyanta and Y Yulizar pg. 8

012003

Preparation of TiO_2-ZnO and its activity test in sonophotocatalytic degradation of phenol OPEN ACCESS Is Fatimah and Novitasari pg. 16

012004

Chitosan and N-Alkyl chitosan as a heterogeneous based catalyst in the transesterification reaction of used cooking oil OPEN ACCESS T S Julianto and R A Mumpuni pg. 24

012005

High photocatalytic activity of mixed anatase-rutile phases on commercial TiO_2 nanoparticles OPEN ACCESS Wai Ruu Siah, Hendrik O Lintang, Mustaffa Shamsuddin and Leny Yuliati pg. 30

Enviromental Chemistry

012006

Synthesis of α-Bismuth oxide using solution combustion method and its photocatalytic properties OPEN ACCESS Y. Astuti, A. Fauziyah, S. Nurhayati, A.D. Wulansari, R. Andianingrum, A.R. Hakim and G. Bhaduri pg. 38

012007

Aliphatics hydrocarbon content in surface sediment from Jakarta Bay, Indonesia OPEN ACCESS M YAzis, L Asia, A Piram, P Doumenq and A D Syakti pg. 45

012008

Occurrence of aliphatic and polyaromatic hydrocarbons (PAHs) in Mytillus galloprovincialis from the traditional market in Marseille, France, by Gas Chromatography triplequadropole tandem Mass Spectrometry (GC-QQQ/MS) OPEN ACCESS M Y Azis, Yelmiza, L Asia, A Piram, B. Bucharil, P Doumenq and A D Syakti pg. 53

012009

Synthesis of thiazole silica hybrid from waste glass for adsorption of cadmium(II) OPEN ACCESS C. Azmiyawati and TaslimahVirkyanov pg. 62

012010

Alginate cryogel based glucose biosensor OPEN ACCESS Amin Fatoni, Dian Windy Dwiasi and Dadan Hermawan pg. 69

012011

Adsorption of remazol yellow FG from aqueous solution on chitosan-linked P-T-Butylcalix[4]Arene OPEN ACCESS Desi Suci Handayani, Candra Purnawan, Pranoto, Sri Hastuti and Diniyah Hilmiyana pg. 75

012012

Photocatalytic removal of 2,4-dichlorophenoxyacetic acid herbicide on copper oxide/titanium dioxide prepared by co-precipitation method OPEN ACCESS Shu Chin Lee, Norhasnita Hasan, Hendrik O. Lintang, Mustaffa Shamsuddin and Leny Yuliati pg. 86

012013

Analysis of diazinon pesticide using potentiometric biosensor based on enzyme immobilized cellulose acetate membrane in gold electrode OPEN ACCESS Mashuni, L O A N Ramadhan, M Jahiding and Herniati pg. 94

012014

Cross-linking of succinate-grafted chitosan and its effect on the capability to adsorb Pb(II) ion OPEN ACCESS Abu Masykur, Sri Juari Santosa and Dwi Siswanta dan Jumina pg. 101

012015

Development of electrokinetic remediation for caesium: A feasibility study of 2D electrode configuration system OPEN ACCESS Rudy Syah Putra pg. 111

012016

Improving quality of textile wastewater with organic materials as multi soil layering OPEN ACCESS Supriyadi, H Widijanto, Pranoto and AK Dewi pg. 119

012017

Adsorption and desorption of Zn(II) and Cu(II) on Ca- alginate immobilized activated rice bran OPEN ACCESS A Suratman, N Z Kamalia and W A Kusumawati pg. 129

012018

Optimization of biosurfactant production in soybean oil by *rhodococcus rhodochrous* and its utilization in remediation of cadmium-contaminated solution OPEN ACCESS Venty Suryanti, Sri Hastuti and Dewi Andriani pg. 136

012019

Development of inorganic composite material based TiO_2 for environmental application OPEN ACCESS Sayekti Wahyuningsih, Ari Handono Ramelan, Edi Pramono, Candra

Purnawan, Velina Anjani, Puji Estianingsih, Ludfiaastu Rinawati and Khusnan Fadli pg. 142

012020

Modification of chitosan by swelling and crosslinking using epichlorohydrin as heavy metal Cr (VI) adsorbent in batik industry wastes OPEN ACCESS B Hastuti, A Masykur and S Hadi pg. 150

Material Science and Engineering

012021

Solubility dynamic of methyl yellow and carbon black in microemulsions and lamellar liquid crystal of water, non ionic surfactants and cyclohexane system OPEN ACCESS A Amran, R Harfianto, W Y Dewi, D Beri and A Putra pg. 159

012022

Synthesis and characterization of geopolymer from bottom ash and rice husk ash OPEN ACCESS Ufafa Anggarini and Ndaru C Sukmana pg. 165

012023

Electrical conductivity of zirconia and yttrium-doped zirconia from Indonesian local zircon as prospective material for fuel cells OPEN ACCESS Karima Apriany, Ita Permadani, Dani G. Syarif, Syoni Soepriyanto and Fitria Rahmawati pg. 170

012024

Optical simulation of surface textured TCO using FDTD method OPEN ACCESS I L Elviyanti, H Purwanto and Kusumandari pg. 180

012025

Effect of reaction time and (Ca+Mg)/Al molar ratios on crystallinity of Ca-Mg-Al layered double Hydroxide OPEN ACCESS E Heraldy, K D Nugrahaningtyas, F B Sanjaya, A A Darojat, D S Handayani and Y Hidayat pg. 185

012026

Intercalation of diclofenac in modified Zn/Al hydrotalcite-like preparation OPEN ACCESS E Heraldy, R W Suprihatin and Pranoto pg. 192

012027

Detection of nitrite and nitrate ions in water by graphene oxide as a potential fluorescence sensor OPEN ACCESS Ing Hua Tang, Rita Sundari, Hendrik O Lintang and Leny Yuliati pg. 199

012028

The study of FTO surface texturing fabrication using Argon plasma etching technique for DSSC applications OPEN ACCESS Lindha Jayanti, Kusumandari, Tjipto Sujitno and Risa Suryana pg. 205

012029

Degradation of self-compacting concrete (SCC) due to sulfuric acid attack: Experiment investigation on the effect of high volume fly ash content OPEN ACCESS S A Kristiawan, Sunarmasto and G P Tyas pg. 209

012030

Solvothermal and electrochemical synthetic method of HKUST-1 and its methane storage capacity OPEN ACCESS Witri Wahyu Lestari, Marisa Adreane, Candra Purnawan, Hamzah Fansuri, Nurul Widiastuti and Sentot Budi Rahardjo pg. 216

012031

Changes in grammage, tearing resistance, and water vapor transmission rate of active paper incorporated with Cinnamaldehyde during storage at various temperatures OPEN ACCESS G J Manuhara, L U Khasanah and R Utami pg. 228

012032

Design of reflective optical fiber sensor for determining refractive index and sugar concentration of aqueous solutions OPEN ACCESS Ahmad Marzuki, Nila Wulan Sari and Riatun pg. 235

012033

Synthesis of $LiNi_xMn_{2-x}O4$ by low-temperature solid-state reaction and its microstructure OPEN ACCESS D Purwaningsih, R Roto and H Sutrisno pg. 239

012034

Synthesis and characterization of *bis*-(2-cyano-1-methyl-3-{2- {{(5-methylimidazol-4-yl)methyl}thio}ethyl)guanidine copper(II) sulfate tetrahydrate OPEN ACCESS Sentot B Rahardjo, Teguh Endah Saraswati, Edy Pramono and Nur Fitriana pg. 246

012035

The single cell of low temperature solid oxide fuel cell with sodium carbonate-SDC (samarium-doped ceria) as electrolyte and biodiesel as fuel OPEN ACCESS F Rahmawati, A Nuryanto and K D Nugrahaningtyas pg. 256

012036

Physical and optical properties of lead doped tellurite glasses OPEN ACCESS Riyatun, Lita Rahmasari and Ahmad Marzuki pg. 263

012037

The influence of the composition of P3HT:TiO_2 on the characteristics of hybrid polymer solar cell OPEN ACCESS Erlyta Septa Rosa, Shobih, Hanik Munawaroh and Pranoto pg. 268

012038

Effect of cyclical thermal to mechanical properties of Hybrid Laminate Composites (HLC) with skin recycle polypropylene/natural fiber/halloysite and core PP/KF composites OPEN ACCESS Neng Sri Suharty, Hanafi Ismail, Desi Suci Handayani, Kuncoro Diharjo, Fajar Rachman Wibowo and Margaretha Arnita Wuri pg. 274

012039

Physical characteristics of chitosan-silica composite of rice husk ash OPEN ACCESS Woro Sumarni, Retno Sri Iswari, Putut Marwoto and Endah F Rahayu pg. 281

012040

Study of the acidic properties of ZrO_2-pillared bentonite OPEN ACCESS Ahmad Suseno, Priyono, Karna Wijaya and Wega Trisunaryanti pg. 290

012041

Effect of glycerol on mechanical and physical properties of silver-chitosan nanocomposite films OPEN ACCESS E Susilowati, I Kartini, S J Santosa and Triyono pg. 295

012042

Functionalization of C-4-methoxyphenylcalix[4]resorcinarene with several ammonium compounds OPEN ACCESS Suryadi B Utomo, Agung N C Saputro and Yudi Rinanto pg. 302

012043

The Effects of plasticizers and palmitic acid toward the properties of the carrageenan Film OPEN ACCESS Atmanto Heru Wibowo, Oktaviana Listiyawati and Candra Purnawan pg. 312

012044

Synthesis H-Zeolite catalyst by impregnation KI/KIO_3 and performance test catalyst for biodiesel production OPEN ACCESS W. Widayat, Adit Rizky Wicaksono, Lukman Hakim Firdaus and Ndaru Okvitarini pg. 322

012045

Nanocellulose prepared by acid hydrolysis of isolated cellulose from sugarcane bagasse OPEN ACCESS W T Wulandari, A Rochliadi and I M Arcana pg. 331

012046

Laboratory investigation on the role of tubular shaped micro resonators phononic crystal insertion on the absorption coefficient of profiled sound absorber OPEN ACCESS I Yahya, J I Kusuma, Harjana, R Kristiani and R Hanina pg. 338

012047

Preparation and properties of arenga starch-chitosan based edible film OPEN ACCESS Putri Rahmawati, Dwi Ni'maturrohmah, Putri Luthfiana Sari, Putri Puspita Sari, Tetty Nur Indah Sari and Venti Suryanti pg. 343

012048

Synthesis and characterization of free-isocyanate polyurethane as renewable coating materials OPEN ACCESS O A Saputra, K Apriany, F R Putri and M Firdaus pg. 349

012049

Yeast fuel cell: Application for desalination OPEN ACCESS Ummy Mardiana, Christophe Innocent, Marc Cretin, Buchari Buchari and Suryo Gandasasmita pg. 356

012050

Fabrication of organic solar cells with design blend P3HT: PCBM variation of mass ratio OPEN ACCESS Agus Supriyanto, Amrina Mustaqim, Maya Agustin, Ari H Ramelan, Suyitno, Erlyta Septa Rosa, Yofentina and Fahru Nurosyid pg. 369

012051

X-ray Absorption Improvement of Single Wall Carbon Nanotube through Gadolinium Encapsulation OPEN ACCESS Alimin, I Narsito, Kartini and S J Santosa pg. 375

Medicinal Chemistry

012052

HLA-DP, HLA-DQ, and HLA-DR-restricted epitopes in GRA5 of *toxoplasma gondii* strains OPEN ACCESS S Haryati, Y Sari, A APrasetyo and R Sariyatun pg. 384

012053

Virtual screening of Indonesian flavonoid as neuraminidase inhibitor of influenza a subtype H5N1 OPEN ACCESS A A Parikesit, B Ardiansah, D M Handayani, U S F Tambunan and D Kerami pg. 392

012054

In silico modification of Zn^{2+} binding group of suberoylanilide hydroxamic acid (SAHA) by organoselenium compounds as *Homo sapiens* class II HDAC inhibitor of cervical cancer OPEN ACCESS Usman Sumo Friend Tambunan, Ridla Bakri, Arli Aditya Parikesit, Titin Ariyani, Ratih Dyah Puspitasari and Djati Kerami pg. 401

012055

Strain-specific variations in Toxoplasma gondii GRA1, GRA5, GRA6, GRA8, and GRA14 OPEN ACCESS S Haryati, Y Sari, A A Prasetyo and R Sariyatun pg. 411

Natural Product

012056

Screening of Marine Actinomycetes from Segara Anakan for Natural Pigment and Hydrolytic Activities OPEN ACCESS A Asnani, D Ryandini and Suwandri pg. 415

012057

Characterization of active paper packaging incorporated with ginger pulp oleoresin OPEN ACCESS T Wiastuti, L U Khasanah, W Atmaka Kawiji, G J Manuhara and R Utami pg. 424

012058

Biological evaluation of certain substituted hydantoins and benzalhydantoins against microbes OPEN ACCESS Ika-Wiani Hidayat, Yee Yee Thu, David St. C Black and Roger W Read pg. 432

012059

Ethyl-2-(3,5-Dihidroxyfenol): Phloroglucinol derivatives as potential anticancer material OPEN ACCESS Triana Kusumaningsih, Maulidan Firdaus, Muhammad Widyo Wartono, Anif Nur Artanti, Desi Suci Handayani and Angga Eryanto Putro pg. 439

012060

Acidity-controlled selective oxidation of alpha-pinene, isolated from Indonesian pine's turpentine oils (*pinus merkusii*) OPEN ACCESS Masruri, Mohamad Farid Rahman and Bagus Nurkam Ramadhan pg. 448

012061

The co-pigmentation of anthocyanin isolated from mangosteen pericarp (*Garcinia Mangostana L.*) as Natural Dye for Dye- Sensitized Solar Cells (DSSC) OPEN ACCESS H Munawaroh, G F adillah, L N M Z Saputri, Q A Hanif, R Hidayat and S Wahyuningsih pg. 454

012062

Addition reaction of methyl cinnamate with 2-amino-4- nitrophenol OPEN ACCESS Venty Suryanti, Fajar Rakhman Wibowo, Pranoto, Siti Robingatun Isnaeni, Meiyanti Ratna Kumala Sari and Sekar Handayani pg. 460

012063

Isolation and identification of curcumin and bisacurone from rhizome extract of temu glenyeh (*Curcuma soloensis.* Val) OPEN ACCESS Rista A Vitasari, Fajar R Wibowo, Soerya D Marliyana and M Widyo Wartono pg. 466

012064

Physical characteristics of cinnamon oil microcapsule OPEN ACCESS R F Hermanto, L U Khasanah, Kawiji, W Atmaka, G J Manuhara and R Utami pg. 471

012065

O-methylation of natural phenolic compounds based on green chemistry using dimethyl carbonate OPEN ACCESS N I Prakoso, P H Pangestu and T D Wahyuningsih pg. 480

012066

The effect of polyethylene glycol on shellac stability OPEN ACCESS Khairuddin, Edi Pramono, Suryadi Budi Utomo, Viki Wulandari, W A'an Zahrotul and Francis Clegg pg. 487

Theory, Characterization, Modelling

012067

The langmuir isotherm adsorption equation: The monolayer approach OPEN ACCESS Patiha, E Heraldy, Y Hidayat and M Firdaus pg. 495

012068

Theoretical study of adsorption and dissociation of NH_3 on pentanuclear Fe(111) surface OPEN ACCESS G Purwiandono, Triyono and K Wijaya pg. 504

012069

The effects of sintering temperature on dielectric constant of Barium Titanate ($BaTiO_3$) OPEN ACCESS DianisaKhoirum Sandi, Agus Supriyanto, Anif, Jamaluddin and Yofentina Iriani pg. 511

IOP Conf. Series: Materials Science and Engineering **107** (2016) 011001 doi:10.1088/1757-899X/107/1/011001

Editorial Materials (preface and photograph)

PREFACE

The 10th Joint Conference on Chemistry is an international conference organized by 4 chemistry departments of 4 universities in central java, Indonesia. The universities are Sebelas Maret University, Diponegoro University, Semarang State University and Soedirman University. The venue was at Solo, Indonesia, at September 8-9, 2015. The total conference participants are 133 including the invited speakers. The conference emphasized the multidisciplinary chemical issue and impact of today's sustainable chemistry which covering the following topics:

- Material innovation for sustainable goals
- Development of renewable and sustainable energy based on chemistry
- New drug design, experimental and theoretical methods
- Green synthesis and characterization of material (from molecule to functionalized materials)
- Catalysis as core technology in industry
- Natural product isolation and optimization

The conference speakers are:

Prof. Takuji Ogawa

Single molecular electronic devices and their integration, Physical Organic Chemistry, Osaka University

Osaka University, Japan. Research

Prof. Dr. Evamarie. Hey-Hawkins

Catalysis- A powerful means for sustainable developments. Institute of Inorganic Chemistry

Leipzig University, Germany. Research

Prof. Masaaki Nagatsu

Plasma science and technology, Shizuoka University

Shizuoka University, Japan. Research

Prof. Santiago Gómez Ruiz

In search of novel sustainable systems based on nanostructured materials for different catalytic applications.",
Rey Juan Carlos University,

Móstoles (Madrid), Spain
Research

Dr. Henning Störz

Current State and Future Prospects Including Monomer Preparation from Biomass
Institute of Agriculture Technology,
Johann Heinrich Von Thuenen Institute.

Braunschweig Germany.

Dr. Leny Yuliati

Photocatalytic hydrogen production,
Ibnu Sina Institute

University Teknologi Malaysia.
Research

Prof. Effendy, PhD

Molecular Structure Analysis by Single X-ray Diffraction Method,

Dr.rer.nat Hery.Haerudin

Catalytic and sustainable Processes in PERTAMINA,

Dr. Hasanudin Abdurakhman

Advanced Material Research in Toray,
Toray Foundation

Dr. Fitria Rahmawati

Materials for Solid Electrolytes,
Sebelas Maret University

. Research1 Research2

IOP Conf. Series: Materials Science and Engineering **107** (2016) 011002 doi:10.1088/1757-899X/107/1/011002

Peer review statement

All papers published in this volume of *IOP Conference Series: Materials Science and Engineering* have been peer reviewed through processes administered by the proceedings Editors. Reviews were conducted by expert referees to the professional and scientific standards expected of a proceedings journal published by IOP Publishing.

IOP Conf. Series: Materials Science and Engineering **107** (2016) 012001 doi:10.1088/1757-899X/107/1/012001

Photocatalytic removal of phenol over titanium dioxide-reduced graphene oxide photocatalyst

Nor Shuhada Alim[1], Hendrik O. Lintang[2], and Leny Yuliati[2*]
[1]Department of Chemistry, Faculty of Science, Universiti Teknologi Malaysia, 81310 UTM Johor Bahru, Johor, Malaysia.
[2]Centre for Sustainable Nanomaterials, Ibnu Sina Institute for Scientific and Industrial Research, Universiti Teknologi Malaysia, 81310 UTM Johor Bahru, Johor, Malaysia.

[*]E-mail: leny@ibnusina.utm.my

Abstract. Titanium dioxide (TiO_2) has been one of the most investigated semiconductors due to its high activity for the removal of organic pollutants. In order to improve the efficiency of the TiO_2, series of TiO_2-reduced graphene oxide (rGO) composites with various loading amounts of graphene oxide (GO), which were 0.5, 1, 3 and 5 wt% were prepared by UV-assisted photocatalytic reduction method. The X-ray diffraction (XRD) patterns and Fourier transform infrared spectroscopy (FTIR) spectra confirmed that all the TiO_2-rGO composites samples were successfully synthesized without disrupting the structure of the TiO_2. Fluorescence spectroscopy revealed the role of the rGO to reduce the electron-hole recombination on the TiO_2, while the transmission electron microscopy-energy dispersive X-ray spectroscopy (TEM-EDS) confirmed the morphology and the presence of both TiO_2 and rGO. In the photocatalytic removal of phenol, all the TiO_2-rGO composites showed better photocatalytic activities than the TiO_2 under UV light irradiation. The activity of the TiO_2 was enhanced by more than two times with the addition of the GO with the optimum amount (3 wt%). It was proposed that the good photocatalytic performance obtained on the composites were caused by the successful suppression of electron-hole recombination by the rGO on the TiO_2.

1. Introduction

Titanium (IV) oxide (TiO_2) is known to be a good photocatalyst for degradation of environmental contaminants nowadays [1]. Due to its good properties, such as having a wide band gap of (~ 3 eV) and anti-photocorossion, TiO_2 has been reported to give high photocatalytic activities in degradation of organic pollutants and dyes under UV light irradiation. However, TiO_2 still suffers from high electron-hole recombination rates that result in the low photocatalytic activity. Performing a composite between TiO_2 semiconductor and graphene (TiO_2-GR) has been reported to give a higher photocatalytic activity [2-9].

Graphene is a 2-dimensional graphitic material with abundance of sp^2 hybridization. It has tremendous unique properties, such as high transport conductivity of ~ 5000 W m^{-1} K^{-1}, excellent mechanical properties, flexible structure, and large theoretical specific surface area up to 2600 m^2 g^{-1} [10]. Due to its outstanding properties, many researchers have used graphene to design novel graphene-based materials for various photocatalytic applications, such as degradation of organic pollutants [2-5] and photocatalytic water splitting [6-9]. These graphene based materials are commonly synthesized from graphite to graphene oxide (GO) followed by reduction process to produce reduced graphene oxide (rGO). Various methods have been reported in reducing the GO, such as chemical reduction, thermal reduction, and hydrothermal reduction [2,3,6]. However, there are some limitations to these methods, such as the use of hydrazine that is highly toxic and high temperature that will affect the structure of the formed rGO [6]. Recently, Williams *et al.* discovered a clean method by using TiO_2 as a photocatalyst to convert GO to rGO at room temperature [11]. The advantages of the photocatalytic reduction method are low cost and feasible to form partially reduced graphene oxide, which is favorable for the photocatalytic applications. In this study, UV-assisted photocatalytic reduction method was used to prepare the TiO_2-rGO composites by using methanol as the reducing agent. The photocatalytic activities of the TiO_2-rGO composite samples were assessed by examining the degradation of phenol under UV light irradiation.

2. Experimental

2.1 Preparation of TiO_2-rGO composites

GO was prepared via an improved Hummers' method [12] by reacting a commercial graphite powder (Sigma-Aldrich) with a mixture of $KMnO_4$, H_2SO_4 and H_3PO_4. The GO was centrifuged and washed twice with HCl, followed by double distilled water until pH 7. The obtained solid was dispersed in methanol and ultrasonicated for 1 h. Finally, the solution was evaporated to remove excess solvent and dried at room temperature overnight. The TiO_2-rGO composites were prepared by UV-assisted photocatalytic reduction method [11] with GO contents of 0.5, 1, 3, and 5 wt %. The composite samples were labelled as TiO_2-rGO(x), which x shows the GO contents. For a typical synthesis of TiO_2-rGO(*0.5*), TiO_2 (1 g, Sigma-Aldrich) and the prepared GO (0.005 g) was dispersed in methanol (60 mL) and sonicated for 30 min. After sonication, the mixture was transferred into a 100 mL jacketed beaker and placed on a stirring plate with a cooling water system to ensure that the reaction could be carried out at room temperature. The synthesis reaction was conducted in a closed box for 24 h under UV light irradiation (λ = 254 nm, I = 0.32 mw cm^{-2}). After each reaction, the obtained solid was filtered and washed with distilled water and methanol consecutively. The final product was then dried at room temperature. Similar procedures were used to prepare the other TiO_2-rGO(x) composite photocatalysts.

2.2 Characterizations of TiO_2-rGO composites

The prepared composite photocatalysts were characterized using different types of instruments. X-ray diffraction (XRD, D8 Advance, Bruker) was used to record the diffraction patterns of photocatalyst. Fluorescence spectroscopy (FP-8500 JASCO) was used to identify the emission sites in the prepared sample. Fourier transform infrared spectroscopy (FTIR, Nicolet iS50, Thermo Scientific) was used to detect the presence of functional groups in the composite photocatalyst. Transmission electron microscopy equipped with energy dispersive X-ray spectroscopy (TEM-EDS, JEM 2100, JEOL) was used to study the morphology of the selected composite sample.

2.3 Photocatalytic removal of phenol

The photocatalytic removal of phenol was conducted by adding photocatalyst (0.05 g) in 50 mL of 50 ppm phenol aqueous solution in a 100 mL beaker attached to the cooling system. Prior to the reaction, the mixture was placed on a stirring plate in a closed box under dark condition for 30 minutes to reach adsorption equilibrium. The mixture was then put under exposure of UV light irradiation (8 W, λ = 254 nm, I = 0.32 mw cm^{-2}) from top of the beaker for 3 h. After reaction, the concentration of the solution was measured by a gas chromatography equipped with a flame ionization detector (GC-FID, 7820A, Agilent Technologies) to determine the percentage removal of phenol.

3. Result and discussions

3.1 Characterizations of the TiO_2-rGO composite photocatalysts

The XRD patterns of the GO and the TiO_2-rGO composite photocatalysts are shown in Figure 1. As can be seen in Figure 1(a), a weak diffraction peak was observed at a 2θ value of 10.4° (002), which attributed to the interplanar distance of 0.80 nm and was in good agreement with the previous reported literatures [2,12]. This result indicated that the GO was successfully synthesized from oxidation of graphite via the improved Hummers' method. Figure 1(b-f) illustrate the XRD patterns of the TiO_2-rGO composite photocatalysts. It was clear that all the diffraction peaks matched the standard data for pure anatase and rutile phases of TiO_2 (JCPDS 84-1286 and 88-1175, respectively). The diffraction peaks identified at 2θ values of 25.4, 36.8, 37.7, 38.5, 48.1, 53.9, 55.0, 62.8, and 69.0° can be ascribed as (101), (103), (004), (112), (200), (105), (211), (118) and (220) crystal planes of anatase TiO_2. Meanwhile, peaks at 2θ values of 27.6, 36.2, and 41.2° can be indexed as (110), (101), and (111) for rutile phase of TiO_2. Addition of GO did not affect the crystallinity and the structural phases of the TiO_2. Notably, the characteristic peak of the GO was disappeared while the rGO could not be detected in the composites, suggesting the low amount or low diffraction intensity of rGO.

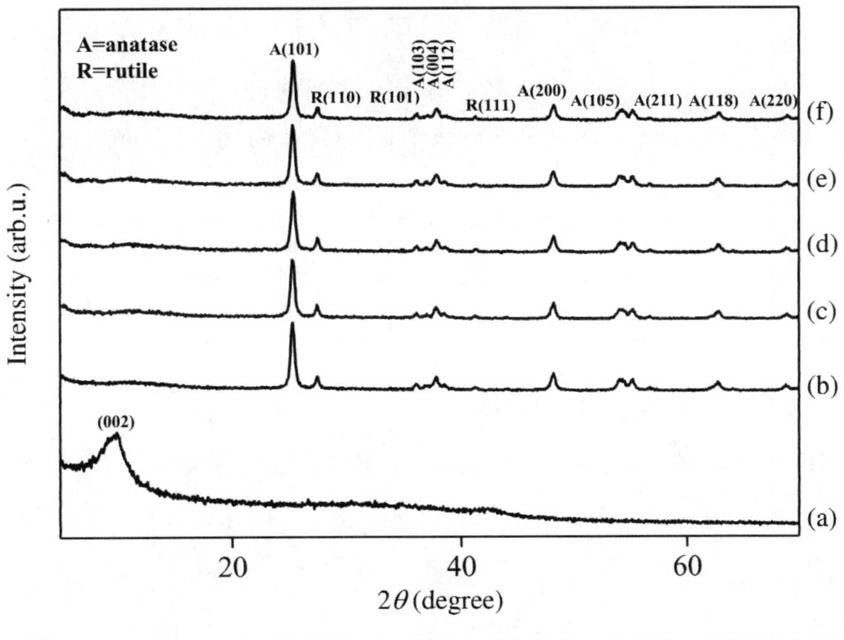

Figure 1. XRD patterns of (a) GO, (b) TiO_2, (c) TiO_2-rGO(*0.5*), (d) TiO_2-rGO(*1*), (e) TiO_2-rGO(*3*), and (f) TiO_2-rGO(*5*).

The molecular structures of the GO and the TiO_2-rGO composite photocatalysts were characterized by FTIR spectroscopy. Figure 2 illustrates the FTIR spectra of the GO, the TiO_2 and the TiO_2-rGO composites. As can be seen from Figure 2(a), the peak of C=C aromatic bonding at around 1630 cm^{-1} and the broad peak at around 3400 cm^{-1} due to physically adsorbed water molecules on the GO were observed [4]. On the other hand, TiO_2 exhibited additional broad peaks below 800 cm^{-1}, which corresponded to Ti-O-Ti groups [4]. All the TiO_2-rGO composites gave characteristic peaks of both GO and TiO_2, in which the peak at around 1630 cm^{-1} became more significant with the increase of GO loading, suggesting the presence of GO in the composite photocatalysts.

IOP Conf. Series: Materials Science and Engineering **107** (2016) 012001 doi:10.1088/1757-899X/107/1/012001

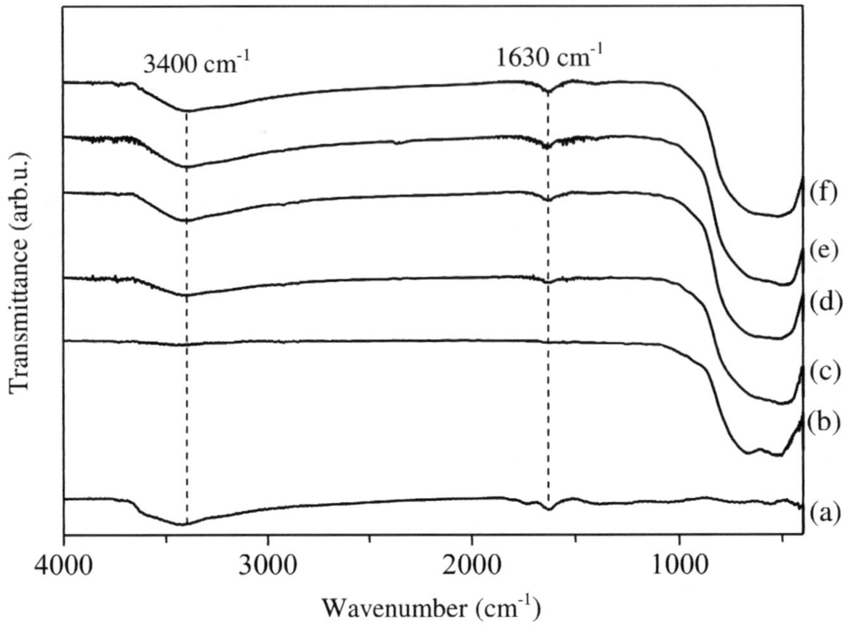

Figure 2. FTIR spectra of (a) GO, (b) TiO_2, (c) TiO_2-rGO(*0.5*), (d) TiO_2-rGO(*1*), (e) TiO_2-rGO(*3*), and (f) TiO_2-rGO(*5*).

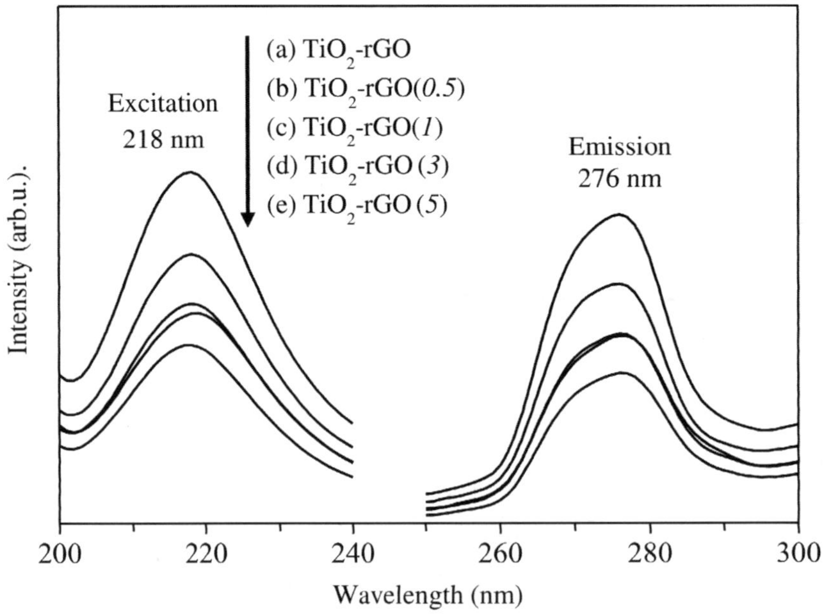

Figure 3. Fluorescence spectra (a) TiO_2, (b) TiO_2-rGO(*0.5*), (c) TiO_2-rGO(*1*), (d) TiO_2-rGO(*3*), and (e) TiO_2-rGO(*5*).

IOP Conf. Series: Materials Science and Engineering **107** (2016) 012001 doi:10.1088/1757-899X/107/1/012001

The emission sites of TiO_2 with different loadings of the GO were studied by fluorescence spectroscopy. Figure 3 illustrates the excitation and emission spectra of TiO_2 and TiO_2-rGO composite photocatalysts. TiO_2 showed strong emission spectrum at 276 nm with excitation wavelength of 218 nm. It was confirmed that the addition of the GO did not give any influences on the excitation and emission wavelengths but the intensity of the spectra was found to be affected by the presence of rGO. The intensity of excitation and emission spectra decreased with the increase of the GO contents (*0.5 - 5* wt%). This result suggested that the rGO was able to reduce the electron-hole recombination in the TiO_2.

The morphology of the selected TiO_2-rGO composite photocatalyst was investigated by TEM. Figure 4 shows the TEM image of TiO_2-rGO(*3*) composite. As can be seen in Figure 4(b), the aggregated TiO_2 was retained and unaffected after addition of rGO. Moreover, the rGO sheets with slight aggregation could also be observed, suggesting the presence of rGO on the surface of the composite photocatalysts. The presence of both TiO_2 and rGO was also confirmed by EDS as shown in Table 1. The amount of carbon (2.2. wt%) was found to be close to the added GO amount (3 wt%).

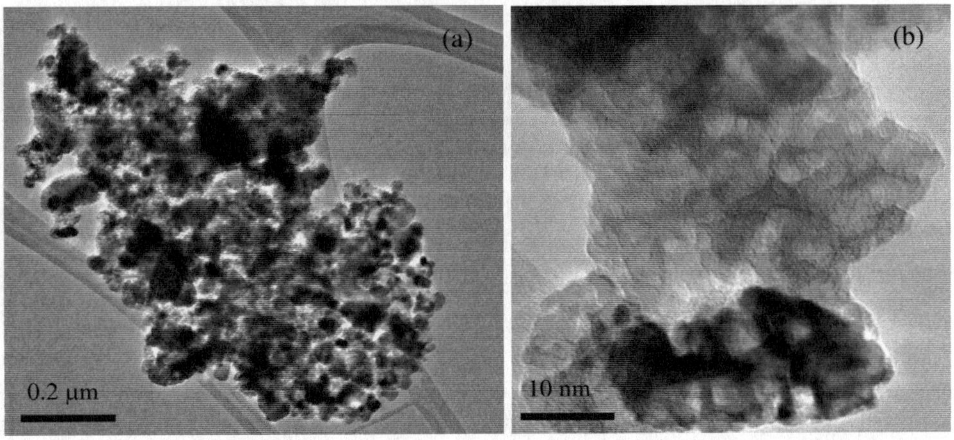

Figure 4. The TEM images of the TiO_2-rGO(*3*) with different magnifications of (a) 10,000 and (b) 250,000 times.

Table 1 The EDS elemental microanalysis of the TiO_2-rGO(*3*)

Element	Weight %[a]	Atomic %[b]
C	2.2	4.9
O	35.2	59.6
Ti	62.7	35.5

[a]Weight % shows the ratio percentage between mass of each element and total mass of all elements. [b]Atomic % shows the ratio percentage between atomic number of each element and total atomic number of all elements.

3.2 Photocatalytic activities of the TiO_2-rGO composite photocatalysts

The photocatalytic performances of the TiO_2-rGO composite photocatalysts were investigated for the removal of phenol under 3 h UV light irradiation at room temperature. Photolysis and adsorption experiments were also conducted to support that the decrease in the concentration of phenol was contributed by the photocatalytic reaction. After 3 h of UV light irradiation without photocatalyst, the concentration of phenol was unchanged. This result suggested that phenol was stable under UV light irradiation, and phenol could not be degraded under UV light irradiation without the photocatalyst. On the other hand, adsorption test under dark condition showed that adsorption equilibrium of phenol was achieved after 30 min by stirring phenol solution in the presence of the photocatalyst. It was confirmed that only *ca.* 3% phenol was adsorbed on the TiO_2 photocatalyst. This result suggested that phenol could not be degraded in the presence of photocatalyst without UV light irradiation. Table 2 shows the percentage removal of phenol on the TiO_2 and the TiO_2-rGO composites with different GO contents. All the TiO_2-rGO composites showed higher photocatalytic activity than the unmodified TiO_2 sample, which only showed 7% removal of phenol. The photocatalytic activity of the TiO_2 was enhanced from 7 to 17% when the amount loading of GO was 3 wt%. This good improvement (2.4 times) would be due to the presence of rGO in the composite that suppressed the electron-hole recombination of TiO_2, as supported by the fluorescence spectra.

Table 2 Photocatalytic activity of the TiO_2 and the TiO_2-rGO composites for phenol removal under UV light irradiation[a]

Sample	Percentage of phenol removal (%)[b]
TiO_2	7
TiO_2-rGO(*0.5*)	9
TiO_2-rGO(*1*)	11
TiO_2-rGO(*3*)	17
TiO_2-rGO(*5*)	15

[a]Photocatalyst used was 0.05 g, phenol was 50 mL (50 ppm). [b]Percentage of phenol removal was calculated as $(C_o-C_t)/C_o \times 100\%$, where C_o is the initial concentration of phenol and C_t is the concentration of phenol after photocatalytic reaction.

4. Conclusions

The TiO_2-rGO composite photocatalysts were successfully prepared via a UV-assisted photocatalytic reduction method. It was revealed that the addition of GO did affect the crystalline structure of TiO_2, which was remained to be a mixture of anatase and rutile in the TiO_2-rGO composite. The presence of rGO was found to decrease the emission intensity of the TiO_2, implying the successful reduced electron-hole recombination. The TiO_2-rGO composite photocatalysts showed significantly enhanced photocatalytic activities for the removal of phenol. The TiO_2-rGO sample with 3 wt% of GO loading exhibited the highest photocatalytic activity, which was 2.4 times higher than pure TiO_2. The efficient suppression of the electron-hole recombination in the TiO_2 led to the improvement of the photocatalytic performance.

Acknowledgements

The research leading to these results has received funding from the European Union Seventh Framework Program (EU-FP7) under 4G-PHOTOCAT grant (agreement no.: 309636). This work has been also financially supported by Ministry of Higher Education (MOHE) and Universiti Teknologi Malaysia (UTM, Malaysia) through the international contract matching grant (cost center code: R.J130000.7626.4C035).

References

[1] Fujishima A, Rao T N and Tryk D A 2000 *J. Photochem. Photobiol. C: Photochem. Rev.* **1** 1
[2] Zhang Y, Tang Z-R, Fu X and Xu Y-J 2010 *ACS Nano* **4** 7303

[3] Sanjaya D P, Ruperto G M, Khiem V, Nijem N, Oliver S, Yves C and Kenneth J B 2012 *ACS Catal.* **2** 949
[4] Wojtoniszak M, Zielinska B, Chen X, Kalenczuk R J, Borowiak-Palen E 2012 *J. Mater. Sci.* **47** 3185
[5] Mohamed R M 2012 *Desalin Water Treat.* **50** 147
[6] Iwase A, Ng Y H, Ishiguro Y, Kudo A and Amal R 2011 *J. Am. Chem. Soc.* **133** 11054
[7] Fan W, Lai Q, Zhang Q and Wang Y 2011 *J. Phys. Chem. C.* **115** 10694
[8] Cheng P, Yang Z, Wang H, Cheng W, Chen M, Wenfeng S and Ding G 2012 *Int. J. Hydrogen Energy* **37** 2224
[9] Zhang X Y, Li H P, Chui X L and Lin Y 2010 *J. Mater. Chem.* **20** 2801
[10] Berger C, Song Z, Li T, Li X, Ogbazghi A Y, Feng R, Dai Z, Marchenkov A N, Conrad E H, First P N and de Heer W A D 2004 *J. Phys. Chem. B.* **108** 19912
[11] Williams G, Seger B and Kamat P V 2008 *ACS Nano* **2** 1487
[12] Marcano D C, Kosynkin D V, Berlin J M, Sinitskii A, Sun Z, Slesarev A, Alemany L B, Lu W and Tour J M 2010 *ACS Nano* **4** 4806

IOP Conf. Series: Materials Science and Engineering **107** (2016) 012002 doi:10.1088/1757-899X/107/1/012002

The shape conversion of silver nanoparticles through heating and its application as homogeneous catalyst in reduction of 4-nitrophenol

H A Ariyanta and Y Yulizar*
Department of Chemistry, Faculty of Mathematics and Natural Sciences, Universitas Indonesia, Kampus UI Depok 16424, Indonesia

*E-mail: yokiy@ui.ac.id

Abstract. The shape conversion of silver nanoparticles (AgNPs) through heating and its application as a homogeneous catalyst in the reduction of 4-nitrophenol is reported here. Synthesis of AgNPs by reduction of $AgNO_3$ using $NaBH_4$ and sodium citrate as reducing agent were successfully conducted. The addition of PVP was used as stabilizing agent. The synthesized AgNPs were heated at 95 °C and observed using UV-Vis spectrophotometer, transmission electron microscopy (TEM), Fourier-transformed infrared (FTIR) spectroscopy and particle size analyzer (PSA). Characteristics of AgNPs before heated were blue with UV-Vis absorbance spectrum at λ_{max} = 786 nm and the shape was pseudo nano prism sized ± 28 nm. During the heating process, the color changed gradually from blue (λ_{max} = 786 nm) to orange (λ_{max} = 486 nm) and also its shape from nano prism to nanodisk. Silver nano prism has a lattice constant, 4.160 Å, larger than the silver nanodisk, 4.081 Å, which was possibly achieved through rearrangement of silver atoms on the surface of AgNPs. Both silver nanodisk and nano prism were tested as a homogeneous catalyst for the reduction of 4-nitrophenol (4-NP) with $NaBH_4$.

1. Introduction

Silver nanoparticles (AgNPs) received much attention from researchers in the world due to its unique. Besides having a good antimicrobial ability [1,2], AgNPs also have unique optical properties, namely LSPR (Localized Surface Plasmon Resonance). The LSPR properties lead AgNPs to have an attractive color and luminescence [3,4]. Therefore, AgNPs have many benefits such as chemical sensors, biological sensors, catalysts, as well as the diagnosis and cure of disease.

Differences synthesis conditions will produce different shapes and sizes of AgNPs. Generally, the synthesis of AgNPs with reduction method produces spherical particles at a maximum wavelength (λ_{max}) 395 nm [7]. However with certain techniques such as photochemical, thermochemical and biochemical [8-10], the resulting nanoparticles can be deformed. Tang et al. [11] reported shape conversion mechanism of AgNPs from triangular to spherical shapes with heating (thermochemical). Shape conversion of AgNPs can be monitored via color change from blue (triangular) to orange (spherical) at λ_{max} 655 nm to 472 nm.

Metal nanoparticles are often used in redox catalysis process because they have a large surface area. Even though silver is not the best redox catalyst, silver can be used as an active component of the catalyst for the partial oxidation [12]. AgNPs are effectively used as a homogeneous catalyst and can be immobilized on a suitable support material such as silica, alumina, carbon, zeolites and metal oxides as heterogeneous catalysts [13]. In this research, we studied the effect of the shape conversion of AgNPs through heating to their catalytic activity for the first time. Reaction model used to test the catalytic activity of AgNPs was the reduction reaction of 4-nitrophenol (4-NP) to 4-aminophenol (4-AP). Previous studies have been made out in order to optimize the 4-NP reduction using AgNPs catalyst with a wide variety of reactant conditions [5]. But until this day, there has never been a report about the effect of the shape conversion of AgNPs to their catalytic activity for 4-NP reduction.

2. Experimental

2.1. Materials

$AgNO_3$, Trisodium citrate, Polivinilpirolidon (PVP), H_2O_2 30 wt%, $NaBH_4$, and 4-nitrofenol were purchased from Merck. All chemicals were of analytic grade and were used without further purification. MilliQ water (18.2 Ω cm^{-1}) was used to make aqueous solutions.

2.2. Heat assisted shape conversion of AgNPs

The initial triangular AgNPs were synthesized according to the Mirkin's method [14]. An aqueous solution of 100 mL $AgNO_3$ 0.1 mM, 1.8 mL trisodium citrate 100 mM, 0.06 g PVP, and 0.24 mL hydrogen peroxide 30 wt% were mixed and vigorously stirred under ambient condition. Then, 1,5 mL $NaBH_4$ solution 100 mM was rapidly injected into the mixtures. After about 30 min, colloidal blue silver nano prism was obtained.

The synthesized silver nano prism was placed in a water bath at 95 °C. The color of colloidal AgNPs changed from blue to purple, red and then orange during the heating process. The AgNPs colloid were monitored using ultraviolet-visible spectrophotometer (Shimadzu 2600), Fourier transmits infrared (FTIR) (Shimadzu Prestige 21), particle size analyzer (PSA) Malvern ZEN 1600 and TEM (JEM 1400).

2.3. Catalytic activity test of 4-NP reduction

The catalytic activity of AgNPs as a homogeneous catalyst was determined by adding 1 mL of $NaBH_4$ 0,1 M into 5 mL of 4-NP 8,6x10^{-5} M in a centrifuge tube. The solution showed a change in color of 4-NP which are colorless into yellow. Blue (AgNPs-blue) and orange (AgNPs-orange) homogeneous catalysts were added to the mixture, vortexed at the variation of time and characterized using UV-Vis spectrophotometer.

3. Results and discussion

3.1. Shape conversion of AgNPs

The AgNPs-Blue colloid gradually changed into purple, maroon and orange during heated at 95 °C as shown in figure 1. The color changes related to the SPR (surface plasmon resonance) properties of AgNPs. Color changes which are accompanied by changes in UV-Vis absorption spectra indicate that heating could lead to the shape conversion of AgNPs [9].

Figure 1. Images of conversion process of AgNPs

The AgNPs-Blue colloidal have an absorbance at λ_{max} 786 nm. The increasing of heating time (every 10 min) caused the blue shift at λ_{max} 534 nm (AgNPs-Purple), 501 nm (AgNPs-Marun), and 486 nm (AgNPs-Orange), which indicates that the nanoparticle size decreases (figure 2).

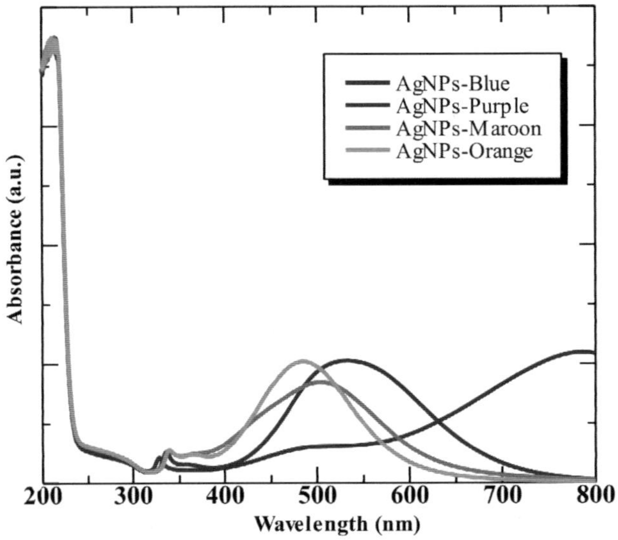

Figure 2. SPR spectra changes of AgNPs during heating.

TEM characterization has been used to clarify the shape conversion of AgNPs through the differences of heating time. Figure 3 and 4 shows the differences in morphology between AgNPs-blue and AgNPs-orange. AgNPs-blue was the initial color after synthesis while AgNPs-orange was the last color change of AgNPs colloidal after heating ± 30 minutes.

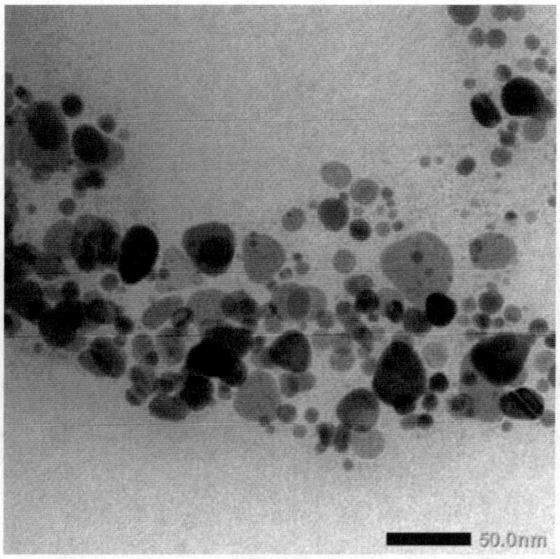

Figure 3. TEM characterization results for the morphological of AgNPs-blue.

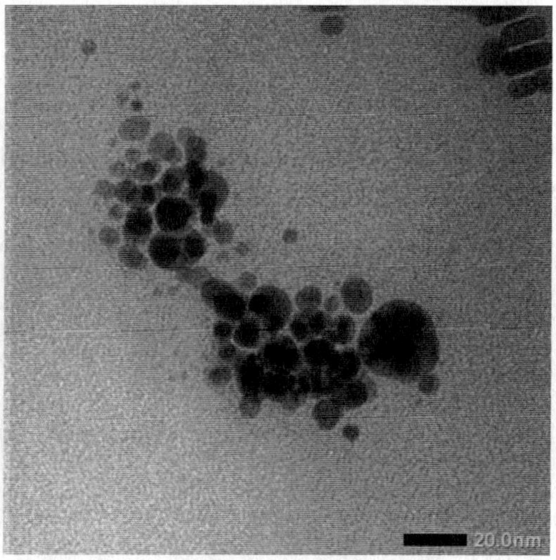

Figure 4. TEM characterization results for the morphological of AgNPs-orange.

The Morphology of AgNPs-blue showed that most AgNPs have pseudo triangular shape with side length ± 25 nm while AgNPs-orange is almost spherical with diameters ± 12 nm. Based on the results, (1) AgNPs morphology can be converted from triangular to spherical by heating at 95 °C, (2) AgNPs size get smaller with increasing heating time. Shape and size conversion process of AgNPs by heating are illustrated in figure 1.

Shape conversion of AgNPs likely to occur because of the migration of silver atoms from the less stable sides to the more stable, thereby causing surface energy difference and selective adsorption of citrate. According to Tang et al. [11], the sharp edges on the triangular AgNPs have a larger surface energy and less stable, so it can lead to changes the shape of triangular become spherical with heating. This raises the idea that spherical AgNPs can be formed because of the displacement of the silver atoms on the less stable side with a high surface energy and the rearrangement of the atoms that have been "dissolved" in the spherical AgNPs colloidal formed by heat-assisted. Displacement of the silver atom from sharp edges on triangular particles to colloids causes the formation of very small and unstable AgNPs to allow the rearrangement of these particles on spherical AgNPs surfaces. Rearrangement of the silver atoms can be proved through the analysis of SAED data from TEM characterization. Based on the diffraction pattern (Figure 5 and 6), the crystal structure of AgNPs can be determined.

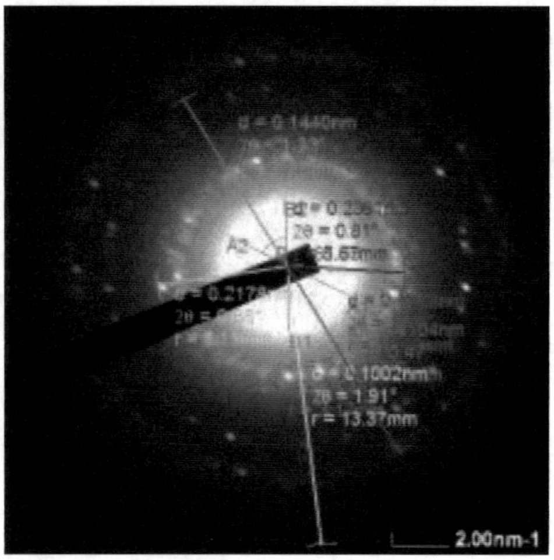

Figure 5. Diffraction patterns of AgNPs-Blue with SAED analysis using TEM

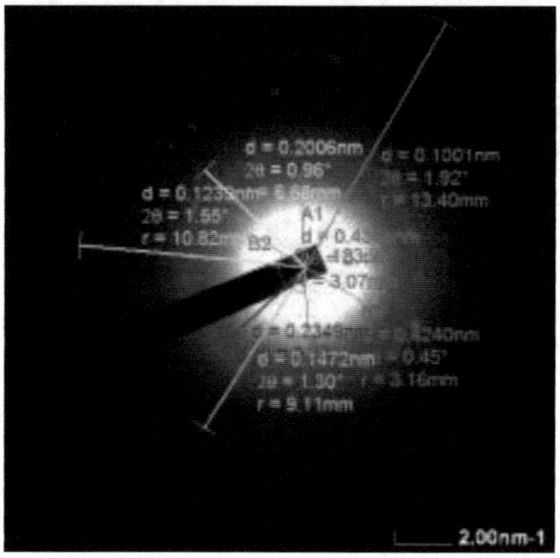

Figure 6. Diffraction patterns of AgNPs-Orange with SAED analysis using TEM

The rings patterns with plane distances from Table 1 are consistent with the plane families of pure face-centered cubic (fcc) silver structure (JCPDS, File No. 4-0787).

Table 1. Diffraction data of AgNPs compared with JCPDSAg

Plane distances (Å)			**Hkl**		
JCPDS Ag	AgNPs-Blue	AgNPs-Orange	**JCPDS Ag**	AgNPs-Blue	AgNPs-Orange
2,32	2,36	2,34	**111**	111	111
2,05	2,17	2,006	**200**	200	200
1,44	1.44	1.47	**220**	220	220
1,22	1.28	1.239	**311**	311	311
1,18	1.002	1.001	**331**	331	331

The Fcc structure of AgNPs-Blue and AgNPs-Orange in this study have lattice parameter values 4,160 Å and 4.081 Å, respectively. Decrease in the value of the lattice parameter of triangular AgNPs-Blue to spherical AgNPs-Orange indicates rearrangement of the AgNPs atoms which was initially shaped triangular then atoms on the edges "detached" and undergo rearrangement into spherical nanoparticles.

3.2. Catalytic activity of AgNPs for 4-NP reduction

The catalytic activity of AgNPs as homogeneous catalysts was tested using reaction reduction model of 4-NP into 4-AP. The reaction model was used because in addition to the interest of reducing the harmful compounds into the medicinal raw materials compounds, also due to the two compounds can be easily detected using UV-Vis spectrophotometer on a laboratory scale. Reduction reaction of 4-NP into 4-AP in the presence of $NaBH_4$ can occur thermodynamically when the value of potential standard (E_0) for 4-NP / 4-AP = -0.76 V and for H_3BO_3 / BH_4 = -1.33 V. But in kinetics, the reaction without the presence of catalyst restricted (no changes absorbance for approximately 2 days) [15]. Therefore, the presence of AgNPs as catalyst expected the reaction runs more effectively.

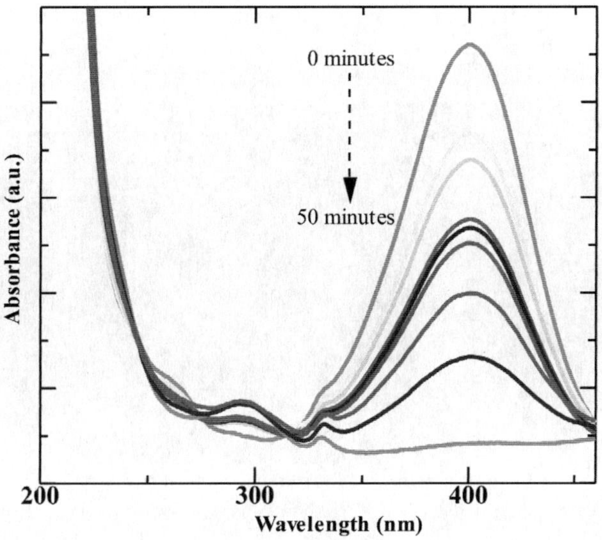

Figure 7. UV-Vis absorption spectra of the reduction of 4-NP in aqueous solution using 1 mL AgNPs-Blue as a catalyst.

4-NP (λ_{max} 317 nm) was reduced to nitrophenolate ions (λ_{max} 400 nm) in the presence of $NaBH_4$. However, with the presence of AgNPs catalyst, absorbance peak decreased, and a new peak appeared at λ_{maks} 300 nm which is the absorbance of 4-AP (Figure 7). The reaction was assumed to be a pseudo-first order reaction because of the excessive amount of $NaBH_4$ in comparison to 4-NP [6]. A graph with a plot of ln [A] versus time, where [A] is the absorbance of 4-NP at *t* time, will give a linear correlation with the rate constant as slope. The kinetic equation can be drawn in equation (1), (2), and (3).

$$\frac{d[A]}{dt} = -k[A] \qquad (1)$$

$$ln\frac{[A]}{[A]_0} = -kt \qquad (2)$$

$$ln[A] = -kt + ln[A]_0 \qquad (3)$$

Figure 8 shows the decrease of 4-NP's concentration against reaction time using AgNPs-Blue and AgNPs-Orange catalyst, in which the rate constant *k* are 0.0458 min^{-1} and 0.0356 min^{-1}, respectively. Based on TEM characterization, the AgNPs-Blue catalyst that showed higher catalytic activity in this study has pseudo triangular particles shape. While AgNPs-Orange catalyst has a spherical shape.

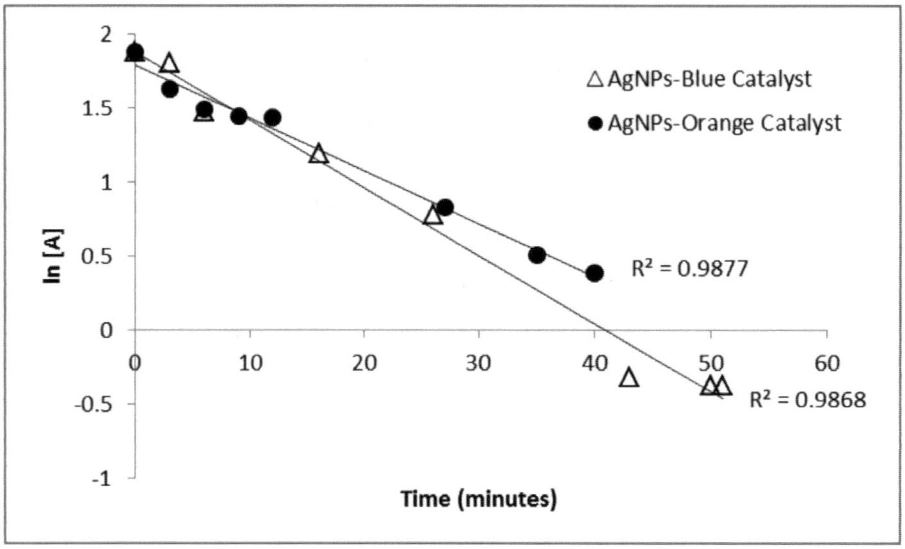

Figure 8. Plots of ln [A] versus reaction time for AgNPs-Blue and AgNPs-Orange Catalyst

Although AgNPs-Blue catalyst has a larger particle size than AgNPs-Orange catalyst, based on its shape, AgNPs-Blue has more active side than AgNPs-Orange catalyst so that the catalytic activity in the electron-transfer process becomes larger.

Mahmood et al. [16] state that the catalytic activity of metal nanoparticles that have sharper edges and corners are more active than the smooth and blunt. That is because particles with sharper edges and corners have more valence-unsatisfied surface atoms, atoms that do not have the complete number of bonds that they can chemically accommodate, which will act as active side than fine and blunt particles

Accordingly, the catalytic activity of AgNPs with triangular shape was better than AgNPs with a spherical shape for 4-NP reduction despite larger particle size. Triangular shape has sharp edges that act as an active site on the catalyst as electron transfer media due to more valency-unsatisfied surface atoms than spherical AgNPs catalysts. The amount of the active side of triangular AgNPs accelerates the 4-NP reduction in the presence of $NaBH_4$.

4. Conclusions

Shape conversion of AgNPs have been done through the heating at 95°C and applied for reduction catalyst of 4NP to 4-AP. TEM characterization results show the occurrence of morphological changes of triangular to spherical AgNPs. The catalytic activity of triangular AgNPs was better than a spherical one.

Acknowledgement
This work was funded by Hibah Penelitian Unggulan Perguruan Tinggi (PUPT) 2015 from the Directorate General of Higher Education, Ministry of Research, Technology and Higher Education of Indonesia through the Directorate of Research and Community Services, Universitas Indonesia (No. 0544/UN2.R12/HKP.05.00/2015).

References and notes

[1] Ahluwalia V, Kumar J, Sisodia R, Shakil N A and Walia S 2014 *Industrial Crops and Products* **54** 202
[2] Kumar D A, Palanichami V and Roopan S M 2014 *Spectrochimica Acta Part A: Molecular and Biomolecular Spectroscopy* **127** 168
[3] Cruz D, Fale P L, Mourato A, Vaz P D, Serralheiro M L and Lino A R L 2010 *Colloids and Surfaces B: Biointerfaces* **81** 67
[4] Ahmad N, Sharma S, Alam Md K, Singh V N, Shamsi S F, Mehta B R and Fatma A 2010 *Colloids, and Surfaces B: Biointerfaces* **81** 81
[5] Yulizar Y, Kadja G T M and Safaat M 2015 *Reac Kinet Mech Cat.* **doi:**10.1007/s11144-015-0916-2
[6] Joseph S and Mathew B 2015 *Spectrochimica Acta Part A: Molecular and Biomolecular Spectroscopy* **136** 1371
[7] Jin R C, Cao Y C, Hao E, Metraux G S, Schatz G C and Mirkin C A 2003 *Nature* **425** 487
[8] Grilet N, Manchon D, Cottancin E, Bertorelle F, Bonnet C, Broyer M, Lerme J and Pellarin M 2013 *The Journal of Physical Chemistry C* **117** 2274
[9] Tang Bin, An J, Zheng X, Xu S, Li D, Ji Z, Zhao B and Xu X 2008 *J. Phys. Chem. C* **112** 18361
[10] Jin R C, Cao Y W, Mirkin C A, Kelly K L, Schatz G C and Zheng J G 2001 *Science* **294** 1901
[11] Tang B, Xu S, Hou X, Li J, Sun L, Xu W and Wang X 2013 *ACS Applied Materials & Interfaces* **5** 646
[12] Raji, Vadakkekara, Chakraborty M and Parikh P A 2012 *I&EC Research* **51** 5691
[13] Narayanan K B, Park H and Sakthivel N 2013 *Spectrochimica Acta Part A: Molecular and Biomolecular Spectroscopy* **116** 485
[14] Jin R C, Cao Y W, Mirkin C A, Kelly K L, Schatz G C and Zheng J G 2001 *Science* **294** 1901
[15] Chi Y, Tu J, Wang M, Li X and Zhao Z 2014 *Journal of Colloid and Interface Science* **423** 54
[16] Mahmoud M A, Narayanan R and El-Sayed M A 2013 *Accounts of Chemical Research* **46** 1795
[17] Banerjee S, Loza K, Zaika M W, Prymak O and Epple 2014 *Chemistry of Materials* **26** 951
[18] Chern JM and Chien YW 2002 *Water Research* **36** 647
[19] Grilet N, Manchon D, Cottancin E, Bertorelle F, Bonnet C, Broyer M, Lerme J, and Pellarin M 2013 *The Journal of Physical Chemistry C* **117** 2274
[20] Kareem, T A, Pragatheeswaran and Kaliani A 2010 *Jurnal Nanosains & Nanoteknologi* **3** 29
[21] Mafune F, Kohno J, Takeda Y and Kondow T 2000 *The Journal of Phyical Chemistry* **104** 8333
[22] Chou and Lu 2008 *Patent Application Publication* **US.2008/0064767 A1**
[22] Panacek A, Prucek R, Hrbac J, Nevecna T, Steffkova J, Zboril R and Kvitek L 2013 *Chemistry of Materials* **26** 1332
[23] Yu A B and Jiang X C 2008 *Langmuir* **24** 4300

IOP Conf. Series: Materials Science and Engineering **107** (2016) 012003 doi:10.1088/1757-899X/107/1/012003

Preparation of TiO_2-ZnO and its activity test in sonophotocatalytic degradation of phenol

Is Fatimah*, Novitasari
Chemistry Department, Islamic University of Indonesia
Kampus Terpadu UII, Jl. Kaliurang Km 14, Sleman, Yogyakarta, Indonesia 55584

*E-mail: isfatimah@uii.ac.id

Abstract. Synthesis of TiO_2-ZnO and its activity test in Sono photocatalysis degradation of phenol has been conducted. The synthesis was performed by the sol-gel mechanism by using titanium isopropoxide and zinc acetate as precursors with the Ti: Zn ratio of 5:1. Characterization of material were conducted by x-ray diffraction analysis, surface area analysis and also diffuse reflectance UV-Visible spectrophotometry. The material obtained from the synthesis was tested in photocatalysis, Sono catalysis and Sono photocatalysis degradation of phenol solution. Results showed that material exhibited the activity of varied mechanism o-phenol degradation. In advance, the Sono photocatalysis degradation produced the synergy index of 1.169 compared to both photocatalysis and Sono catalysis.

1. Introduction

In recent years, photocatalytic degradation processes of the organic contaminant with ultraviolet (UV) radiation over semiconductor photocatalyst like titanium dioxide (TiO_2) and zinc oxide (ZnO) have gained interest as an effective water purification method. The effectivity of the process in decomposing target molecules by mineralizing toxic content is the positive aspect compared to the previous method of water purification such as adsorption. The important goal of these processes was to change the toxic dyes into harmless end products than original compounds. For these purposes, a comprehensive review of the UV assisted heterogeneous photocatalytic oxidation process revealed that the combination of both TiO_2 and ZnO can produce rapid degradation and potentially attain full mineralization of the organic contaminants under mild reaction conditions such as ambient pressure and temperature. Furthermore, due to the ability of semiconductor TiO_2 and ZnO to catch energy from ultrasonic wavelength, Sono catalysis degradation also becomes an opportunity in effort to find the new effective technique of degradation[1]. The recent experimental investigations have showed that toxic compounds can be degraded by coupled advanced oxidation processes via the exciting of powerful like hydroxyl radicals (•OH). Beside of photochemistry, sonochemistry is one of the important processes for this similar scheme. Sonolysis involves the use of ultrasonic waves for the production of cavitations bubbles. The pyrolysis in the cavitation of bubbles, as well as the generated hydroxyl radicals that required to, oxidizes the targeted organic compounds.

The formation of TiO_2 and ZnO in TiO_2-ZnO composite is widely investigated for such application of photocatalysis and also sonocatalysis [2] [3,4]. Several investigation reported the feasibility of TiO_2-ZnO in producing photoactive material for compounds degradation. From previous works the activity of in photocatalysis and Sono catalysis can create the synergistic effect by the combined photocatalytic and Sono catalytic methods called as sonophotocatalysis [5,6]. Phenol degradation was chosen as reaction model to evaluate the effect of each process for the activity of the catalyst in Sono photocatalysis was studied. The work is aimed to evaluate the synthesis of TiO_2-ZnO material and study the effectiveness of Sono photocatalytic compared to Sono catalysis and photocatalysis process by the prepared catalyst.

2. Materials and method

2.1. Materials

The following chemicals were used: titanium (IV)-isopropoxide (TTIP, 97%, Aldrich, Germany), zinc acetate dihydrate, isopropanol, and phenol were purchased from Merck-Millipore, Germany.

2.2. Instrumentation

The physicochemical character of prepared TiO_2-ZnO was studied by x-ray diffraction (XRD) measurement, surface area analysis and scanning electron microscope-energy dispersive x-ray (SEM-EDX0. The x-ray diffraction (XRD Philips) analysis was performed by using Ni-filtered Cu Kα radiation (l = 1.5418 A °) at 40 kV and 40 mA. The diffraction patterns were taken over a 2θ range from 3-65°. Surface morphology was recorded by scanning electron microscope SEM (Zeiss Ultra plus 40) while specific surface area analysis, pore volume, and pore radius determination were obtained by measurement with NOVA 2000e instrument.

2.3. Synthesis

Synthesis of TiO_2-ZnO was prepared by sol-gel method. A solution of titanium isopropoxide in isopropanol: water (1:1) solvent was titrated slowly by zinc acetate solution under stirring until the mole ratio of Ti: Zn = 5:1 with the addition of acetic acid of 0.5N. The mixture was then stirred for a night. A suspension obtained was then evaporated, dried and calcined at 400 °C for 4 h As reference materials, a metal oxide of TiO_2 and ZnO were also prepared from the same precursors. TiO_2 was prepared by diluting titanium isopropoxide solution in isopropanol: water:1:1 and slowly added with acetic acid 0.5 N. The mixture was then stirred for 4 h. The obtained gel was then dried in an oven for a night before calcination at 400 °C for 4 h. Similar method was applied for ZnO synthesis by using zinc acetate dehydrate precursor. Both methods are in the presence of acetic acid as a hydrolyzing agent, and the calcination temperature is also 400 °C for 4 h.

2.4. Activity Test

Activity test experiments were conducted by phenol degradation solution. A solution of phenol with the concentration of 10 mg/L were prepared in distilled water and were mixed with prepared TiO_2-ZnO at the ratio of 500 mL phenol solution:0.25g TiO_2-ZnO taken in the reactor. The degradation experiments consist of varied mechanisms, i.e.,
Adsorption in which the material was mixed with phenol solution under stirring, without UV irradiation, without sonication. Photolysis in which the phenol solution was treated with stirring, with UV irradiation, without sonication. Photocatalysis in which the material was mixed with phenol solution under stirring and UV irradiation, without sonication. Sonocatalysis in which the material was mixed with phenol solution under stirring, without UV irradiation, with sonication. Sonophotocatalysis in which the material was mixed with phenol solution under stirring, with UV irradiation, with sonication

2.3. Experimental Setup

The photocatalytic and Sono photocatalytic experiments were carried out in a photoreactor. An immersion well types photoreactor made of Pyrex glass set with a magnetic bar in the reactor on the magnetic stirrer for continuous agitation. UV light provided by 40 watts bulb (Philips) placed at 30 cm on top of the photoreactor. Ultrasound compartment for Sono catalysis and Sono photocatalysis was assembled by using Neytech sonicator operated at a frequency of 35 kHz and output power of 300 W.

Treated solution was analyzed by gas chromatography and chemical oxygen demand measurement. The COD determination tests were performed according to standard dichromate method using COD digester. The photodegradation efficiency was calculated with equation (1).

$$\eta = \frac{COD_i - COD_f}{COD_i} x\ 100 \qquad (1)$$

where η is the degradation efficiency, COD_i is the initial chemical oxygen demand, COD_f is the chemical oxygen demand of treated solution. A variation of phenol concentration and also a time of treatment were also performed in this study.

3. Result and Discussion

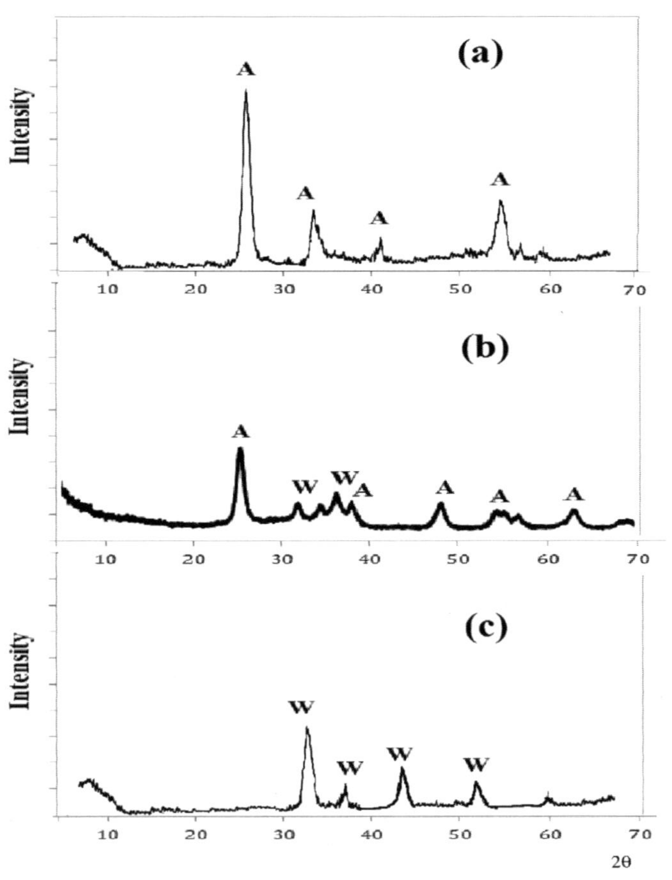

Figure 1. XRD pattern of TiO_2-ZnO

From XRD analysis, it is concluded that composite material of TiO_2-ZnO consists of TiO_2 in anatase and ZnO in wurtzite which those phases are also found in pure TiO_2 and ZnO. Reflection at $2\theta = 31.91^\circ$ and 36.41° are correspond to wurtzite phase (JCPDS 36–1451)., while anatase phase is identified by reflections at : 25.3°, 37.8, 48.0, 54.5 (JCPDS 21-1272). The formation of wurtzite phase is related to that under normal conditions; wurtzite ZnO is thermodynamically stable compared to cubic and another phase. The wurtzite phase also gives higher photoactivity (3.4eV) compared to cubic one (2.7 eV).

Figure 2. SEM images of the prepared materials

The surface profile of materials (Figure 2) exhibits the difference in surface morphology of TiO_2-ZnO compared to TiO_2 and ZnO. ZnO shows the aggregation of granules while TiO_2 has a smaller size of particles and TiO_2-ZnO express the combination of ZnO and TiO_2 morphologies. Furthermore, surface parameters consist of specific surface area, pore volume and pore radius are other parameters required in both Sono catalysis and photocatalysis. The N_2 adsorption-desorption profile of prepared materials is presented in Fig.2, and calculated parameters are listed in Table 1. Adsorption-desorption pattern also indicates that the adsorption-desorption pattern of TiO_2-ZnO lies between the pattern of TiO_2 and ZnO. TiO_2 and TiO_2-ZnO obey type III in IUPAC classification while ZnO fit to type IV correspond to the small adsorbate-absorbent interaction potentials (similar to type III), and are also associated with pores in the 1.5 – 100 nm range.

Table 1. Surface parameter from N_2 adsorption-desorption analysis

Parameter	**TiO_2-ZnO**	**TiO_2**	**ZnO**
BET surface area(m^2/g)	91.05	66.54	36.89
Single point surface area(m^2/g)	88.51	45.46	21.76
t-plot external surface area(m^2/g)	92.15	59.17	32.22
Pore Size (Å)	14.99	11.05	15.09
Pore Volume(cc/g)	0.341	0.289	0.198
Edge wavelength (nm)	389.05	389.05	392.00
Band gap energy(eV)	3.19	3.19	3.16

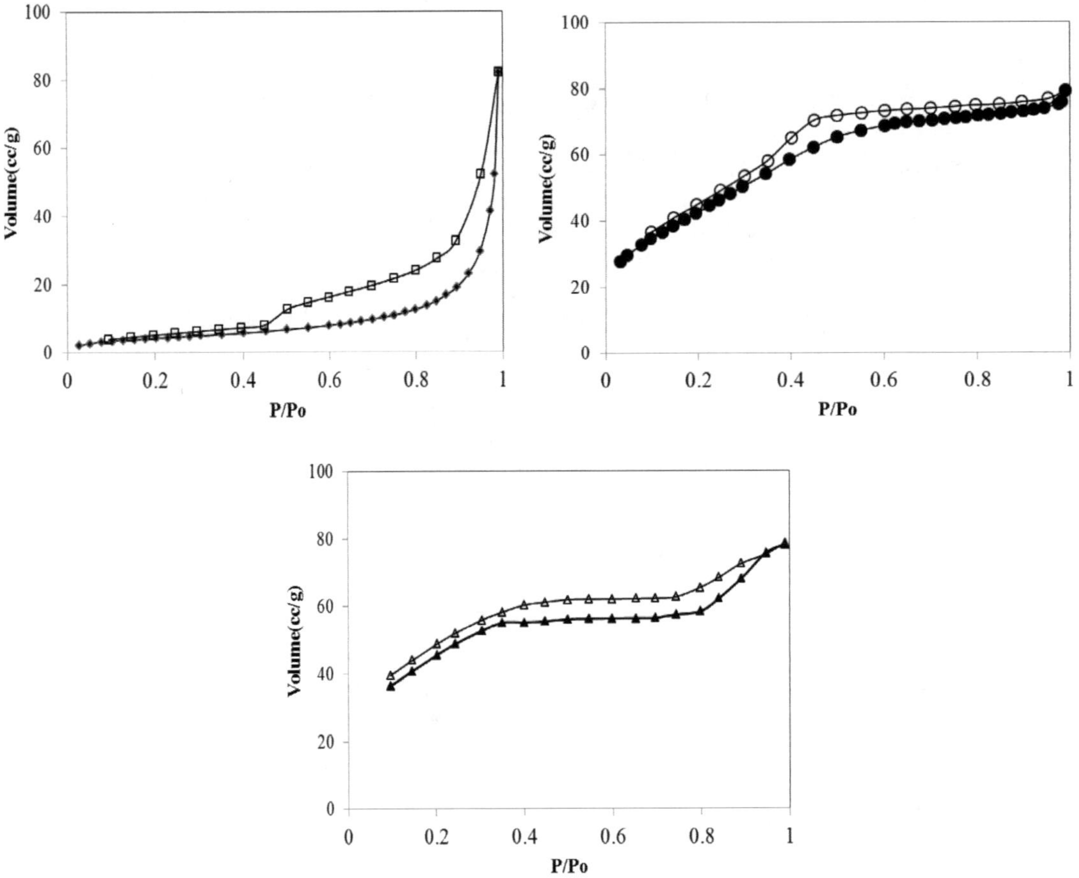

Figure 3. N_2 adsorption-desorption profile of (a) ZnO (b) TiO_2 (c) TiO_2-ZnO

Band gap energy is an important parameter in both Sono catalysis and photocatalysis. The measurement of band gap energy was conducted by DRUV-Vis, and the spectra are presented in Figure 3. The spectra exhibited the edge wavelength correspond to band gap energy values tabulated in Table 1. TiO_2-ZnO expressed the same band gap energy value with TiO_2 material which is 3.19 eV while ZnO shows the value of 3.16 eV. The same values of TiO_2-ZnO with TiO_2 is probably due to the domination of TiO_2 in TiO_2-ZnO in which the mole ratio of Ti: Zn is equal to 5:1.

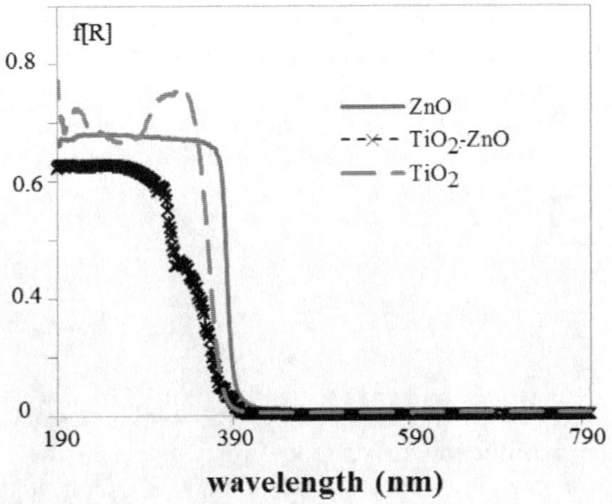

Figure 4. DRUV-Vis spectra of TiO_2-ZnO compared to TiO_2 and ZnO

The catalytic activity of the material in the varied process; photocatalysis, Sono catalysis and Sono photocatalysis of phenol degradation are determined by the reduction of phenol content and chemical oxygen demand (COD). Chromatogram of treated phenol solution by varied treatment is depicted in Figure 4. Phenol in solutions is approximated from the peak at the retention time of 0.456 min. From the chromatogram, it is noted that all treatment give less phenol in solution as showed by the lower peak height/area. Compared to adsorption, Sono catalysis, photocatalysis and also Sono photocatalysis give specific pattern in that new peaks identified. New peaks are the indication of new products from the catalytic reaction which is not from adsorption process. Phenol reduction and η parameters obtained by varied treatment using varied catalysts is depicted in Figure 5.

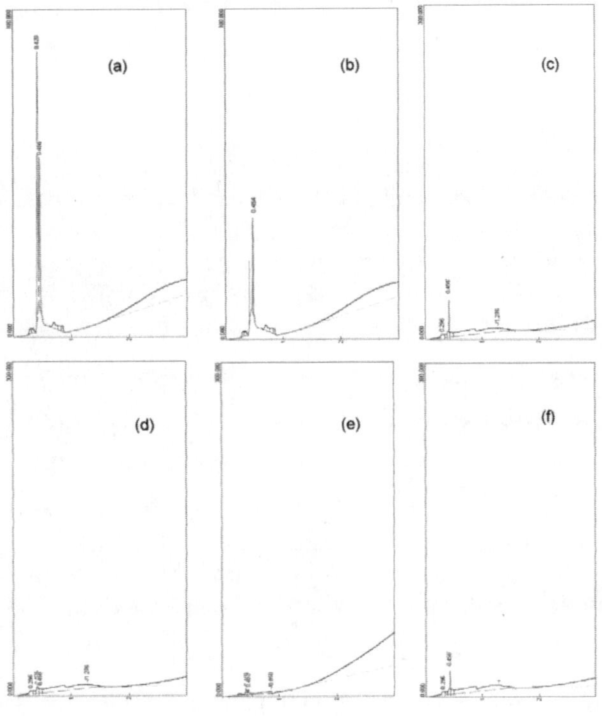

Figure 5. Chromatogram of phenol solution (a) initial (b) after adsorption (b) sonication + UV (c) Sono photocatalysis (d) Sono catalysis (e) photocatalysis

From the data it is concluded that COD reduction is in line with phenol reduction . From various treatment the data imply that COD reduction is related to the less molecules to oxidize in treated solution especially from the Sono photocatalysis treatment. As reported in many works on photocatalytic degradation of phenol, the oxidation occurs by the following mechanism (equation (2-5)).

Photocatalyst —> $h^+ + e-$ (2)
$H_2O + h^+$ —> $\bullet OH + H^+$ (3)
$O_2 + e^-$ —> $\bullet O_2^-$ (4)
$\bullet O_2^- + H+$ —> $\bullet HO_2$ (5)

The results show that in the presence of TiO_2-ZnO the combination of light and sound (Sono photocatalysis) enhances the degradation of phenol significantly. In the case of Sono catalysis, the degradation is 37.66% during 1 h while from photocatalysis, the degradation is 47.82%. However, Sono photocatalysis resulted in this research is approx. 99.31% degradation. From data in Table 2.

Table 2. Phenol reduction and η of various process using varied materials

Process	TiO_2-ZnO		TiO_2		ZnO	
	Phenol reduction (%)	η	Phenol reduction (%)	η	Phenol reduction (%)	η
Adsorption	50.28	49.5	40.28	39.5	20.55	28.89
Photocatalysis	47.82	68.99	47.02	56.89	39.72	44.89
Sonolysis + UV (w.o catalyst)	61.45	72.7	56.45	61.47	46.45	61.47
Sonocatalysis	37.66	67.16	29.66	17.16	24.66	17.16
Sonophotocatalysis	99.99	89.31	90.99	79.91	80.99	79.91

Thus, it is evident that the degradation of phenol under UV-light and ultrasound irradiation is more than the sum of degradation under individual photocatalysis and Sono catalysis process, thereby showing a synergistic effect. Effect of the presence of TiO_2-ZnO as a catalyst within the synergistic process is also indicated by the higher degradation efficiency of Sono photocatalysis than that of the treatment by using UV irradiation and sonication without a catalyst. The synergy index in the case of both catalyzes is calculated from the rate of degradation using the following equation (6).

$$Synergy\ index = \frac{R_{sonophotocatalysis}}{R_{sonocatalysis}+R_{photocatalysis}} \qquad (6)$$

Obtained values for the synergistic index of Sono photocatalysis by using TiO_2-ZnO, TiO_2 and ZnO, are 1.169; 1.040 and 1.24 respectively. The index of the process by using TiO_2-ZnO is lower compared to the use of ZnO. This comparison was similar with was reported by previous research using ZnO [2], however from the compared photocatalysis and Sono catalysis degradation efficiency data by using single ZnO and TiO_2 prepared by a similar process and precursor, prepared TiO_2-ZnO is also contribute to give synergistic effect (Figure 5). With the same initial concentration of phenol and also a time of irradiation, photocatalytic degradation of phenol by using TiO_2 resulted in 25.12% while the Sono catalysis gives 12.67% reduction. In the case of ZnO utilization, the degradation of 17.20% and 12.55% reduction were obtained by photocatalysis and Sono catalysis respectively. By comparing both mechanisms, the synergistic of Sono catalysis process, is higher compared to the photocatalysis. Physico-chemical character data listed in Table 1 suggest that the contribution of band gap energy rather than a specific surface area of the catalyst. From both Sono catalysis and photocatalysis, photocatalysis is more dominant compared to Sono catalysis as indicated by the higher phenol reduction efficiency.

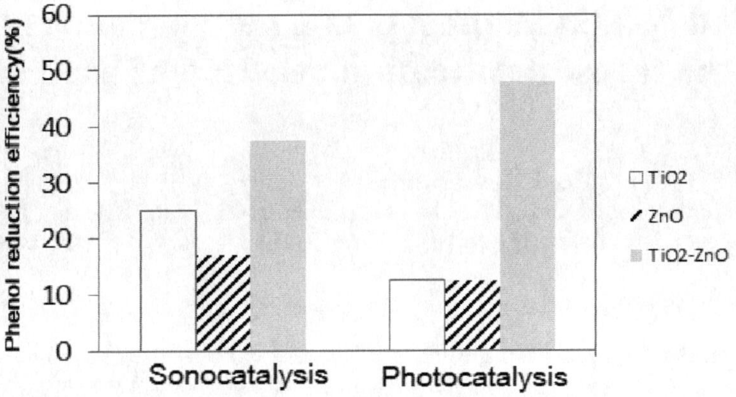

Figure 6. Phenol reduction efficiency over TiO_2, ZnO and TiO_2-ZnO catalyst

4. Conclusion

TiO_2-ZnO catalyst was successfully prepared and found to be active in Sono catalytic, photocatalytic and also Sono photocatalytic degradation of phenol. Synergistic effect of Sono catalysis and photocatalysis is indicated by the synergistic index of Sono photocatalysis equal to 1.169.

References

[1] Kavitha S K and Palanisamy P N 2011 Photocatalytic and Sonophotocatalytic Degradation of Reactive Red 120 using Dye Sensitized TiO 2 under Visible Light *Engineering and Technology* 1–6

[2] Anju S G, Yesodharan S and Yesodharan E P 2012 Zinc oxide mediated Sono photocatalytic degradation of phenol in water *Chemical Engineering Journal* **189-190** 84–93

[3] Ashok C H and Rao K V 2014 Superlattices and Microstructures ZnO / TiO 2 nanocomposite rods synthesized by microwave-assisted method for humidity sensor application *Superlattices and Microstructures* **76** 46–54

[4] Samurai S A M, Rahman M Y A, Umar A and Salleh M 2015 Effect of zinc acetate dihydrate precursor concentration on the properties of TiO 2 – ZnO core – shell nanograms hetero-structure *Journal of Alloys and Compounds* **623** 460–5

[5] Pang Y L, Abdullah A Z and Bhatia S 2011 Review on sonochemical methods in the presence of catalysts and chemical additives for treatment of organic pollutants in wastewater *Desalination* **277** 1–14

[6] Abdul M, Khan N, Siddique M, Wahid F and Khan R 2015 Ultrasonics Sonochemistry Removal of reactive blue 19 dye by song , photo and Sono photocatalytic oxidation using visible light *Ultrasonics - Sonochemistry* **26** 370–7

IOP Conf. Series: Materials Science and Engineering **107** (2016) 012004 doi:10.1088/1757-899X/107/1/012004

Chitosan and *N*-Alkyl chitosan as a heterogeneous based catalyst in the transesterification reaction of used cooking oil

T S Julianto[*] and R A Mumpuni
Department of Chemistry, Faculty of Mathematics of Natural Sciences, Islamic University of Indonesia, JL. Kaliurang KM 14.5, Sleman, Daerah Istimewa Yogyakarta 55584, Indonesia

[*]E-mail: tatang_shabur@staff.uii.ac.id

Abstract. This paper explains the result of research on the use of chitosan and *N*-alkyl chitosan (*N*-methyl chitosan, *N*-isopropyl chitosan) as a heterogeneous base catalyst in the transesterification reaction of used cooking oil. *N*-methyl chitosan and *N*-isopropyl chitosan synthesized by reacting chitosan with formaldehyde and acetone successively forming Schiff bases were then reduced using $NaBH_4$. Then chitosan, *N*-methyl chitosan, *N*-isopropyl chitosan are used as heterogeneous catalysts in the transesterification reaction of used cooking oil. The results show that inclusion of alkyl groups to the chitosan structure leads to increase the amount of base in which chitosan, *N*-methyl chitosan, and *N*-isopropyl chitosan are 0.066 mole.g^{-1}, 0.0068 mole.g^{-1}, 0.0070 mole.g^{-1} respectively. It is also evident from the increasing yield of methyl esters. *N*-isopropyl chitosan gives the greatest content of methyl esters yield.

1. Introduction
Chitosan and the derivatives are rarely used as a heterogeneous catalyst for transesterification reaction. It is usually used as supporting material of metal catalyst [1], enzyme [2] metal oxide catalyst [3].

Chitosan is a polysaccharide contain units of *N*-acetyl-D-glucosamine and D- glucosamine produced from the *N*-deacetylation of natural chitin polymers derived from the shells of marine animals, or fungi [4]. The high reactivity of free amine groups makes chitosan has potential as a Lewis base. The longer chain of chitosan gives more content of free amine groups so that the amount of base increase. Chitosan and the derivatives are expected to replace the homogeneous base catalysts used in transesterification processes such as NaOH and KOH. The amount of base chitosan amine group can be improved by incorporating the electron driving group to the amine group. One of the electron driving group is an alkyl functional group. The electron driving group will increase the stability of cation about the amine group that increases the amount of base [5].

This work reports the use of chitosan and *N*-alkyl chitosan (*N*-methyl chitosan and *N*-isopropyl chitosan) in transesterification reaction of used cooking oil and their amount of base evaluation.

2. Methods
2.1. Chemicals and Analysis Instruments
All chemicals were obtained from Merck, Germany. Chitosan was kindly donated by LIPI Yogyakarta. Used cooking oil was collected from "pecellele" food stalls around the campus of the Islamic University of Indonesia.

Methyl esters and functional group analysis were performed using *Gas Chromatography* (Agilent Technologies 6890 N), *Mass Spectrometer* (Agilent Technologies 5975 B), and Fourier Transform Infrared (FTIR) Spectrophotometer (Thermo Nicolet Avatar 360).

2.2. Synthesis of N-methyl chitosan
Synthesized chitosan derivative refers to the study of Kim and Choi (2001) was made of chitosan solution of 1% w/v by dissolving 5 grams of chitosan into glacial acetic acid 1% v/v of 500 mL. Furthermore, 1.6683 mL of formaldehyde 10% was added to the chitosan solution at room temperature and stirred using a magnetic stirrer at 500 rpm. After one hour, the pH of the solution was conditioned to 4.5 with NaOH 1M solution was added. Then this solution was added to the $NaBH_4$ solution of 10% w/v in 3.17 mL of water and the solution mixture was stirred for 1.5 hours. *N*-methyl chitosan produced subsequently precipitated at the pH 10. Precipitated material was washed with distilled water to neutralize the product. The precipitate was then dried at a temperature of 35 °C for several days to obtain dry *N*-methyl chitosan. Products were then characterized using FTIR spectrophotometer [6].

2.3. Synthesis of N-isopropyl chitosan
Synthesized chitosan derivative refers to the study of Kim and Choi (2001) was made of chitosan solution of 1% w/v by dissolving 5 grams of chitosan into glacial acetic acid 1% v/v of 500 mL. Furthermore, 4.411 mL of acetone 10%was added to the chitosan solution at room temperature and stirred using a magnetic stirrer at 500 rpm. After one hour, the pH of the solution was conditioned to 4.5 with 1M NaOH solution was added. Then this solution was added to the $NaBH_4$ solution of 10% w/v in 3.17 mL of water and the solution mixture was stirred for 1.5 hours. *N*-isopropyl chitosan produced subsequently precipitated by conditioning the pH 10. The material precipitate was washed with distilled water to neutralize the product. The precipitate was then dried at a temperature of 35 °C for several days to obtain dry *N*-isopropyl chitosan. Products were then characterized using FTIR spectrophotometer [6].

2.4. Amount of base determination of chitosan and N-alkyl chitosan
A total of 0.5 grams of chitosan or N-alkyl chitosan dissolved in 60 mL of HCl 0.0782 N. Few drops of phenolphthalein indicators incorporated into the solution. The solution was titrated using 0.108 N NaOH, and the volumes are noted to know the volume of HCl remaining. Moles of HCl determine the level of the amount of base chitosan and *N*-alkyl chitosan.

2.5. Treatment of used cooking oil
Used cooking oil was filtered through filter paper to remove solid particles ballpark. Then the filtrate is cleaned by using adsorbents of activated charcoal

2.6. Transesterification reaction of used cooking oil
Transesterification reaction was carried out in a two-neck round bottom flask of 250 mL capacity that has equipped with a hot plate, thermometer, magnetic stirrer, and condenser. One gram of chitosan inserted into a two-neck round bottom flask and then coupled with the condenser.The amount of filtered used cooking oil and methanol is added to the flask with a mole ratio of oil: methanol is 1:12 and refluxed for 2 hours at a temperature of 60 °C.After 2 hours, the reflux was stopped, and the mixture formed was poured into a test tube, The methyl esters were taken, weighed and calculated the yield of methyl esters. Transesterification process above is repeated for the other catalysts, *N*-methyl chitosan, and *N*-isopropyl chitosan

3. Result and Discussion
3.1. Synthesis and characterization of chitosan and N-alkyl chitosan
Chitosan, *N*-methyl chitosan, and *N*-isopropyl chitosan synthesized by reacting chitosan with formaldehyde and acetone successively forming Schiff bases were then reduced using $NaBH_4$.

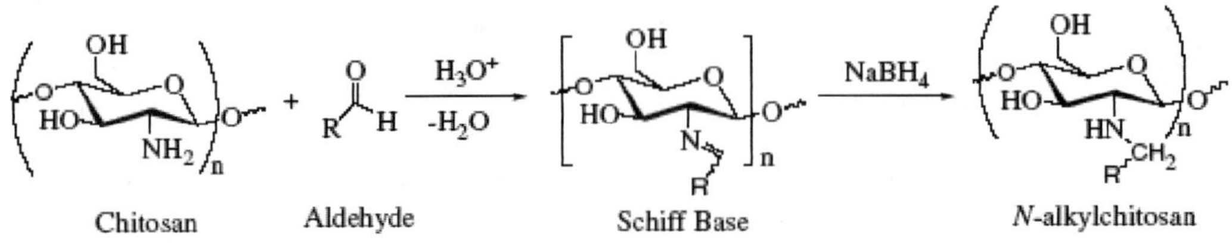

N-methyl chitosan was produced by weighing 4.35 grams of pure chitosan weight used 5 grams. While the N-isopropyl chitosan products produced weighing 4.26 grams of pure chitosan weight of 5 grams. *N*-methyl chitosan and *N*-isopropyl chitosan were then characterized using FTIR and FTIR results were compared with the data of chitosan. FTIR spectra of chitosan, *N*-methyl chitosan, and *N*-isopropyl chitosan is shown in Figure 1.

FTIR spectra of chitosan in Figure 4(A) showed that 3444 cm^{-1} is the absorption of hydrogen bonded to hydroxyl groups. The absorption band at 1020 - 1155cm^{-1} showed the C-O vibration of chitosan. The absorption band at 2919 cm^{-1} shows alkanes and also gives absorption at 1457 cm-1 that indicates the type CH_2- group of alkanes. Absorption in the region 1580 - 1650 cm^{-1} (2 peaks) showed primary amine NH_2.The difference between the absorption of chitosan and chitosan *N*-methyl group is located on NH absorption. NH_2 on chitosan occurred in the 1580 – 1650 cm^{-1} with two peaks while the *N*-methyl chitosan is only formed one peak which means the NH_2 group has been transformed into NH. Furthermore, to know that there has been a methyl group on chitosan (*N*- methyl chitosan) can be identified by the presence of absorption at 1381cm^{-1}. Uptake of NH with -CH_3 CN formed at 1250 cm^{-1}. FTIR spectra of *N*-isopropyl chitosan in Figure 1(C) has an absorption at ~ 1380-1400 cm^{-1}(2 peaks) with a high intensity. It indicates that there is isopropyl group in the *N*-isopropyl chitosan catalysts.

3.2. *Amount of base of chitosan and N-alkyl chitosan*

Determination of amount of base of chitosan and *N*-alkyl chitosan showed that inclusion of alkyl groups to the chitosan structure leads to increase of base that is shown in Table 1.

Table 1. Amount of base of chitosan and *N*-alkyl chitosan

Catalyst	**Amount of base mole.g^{-1}**
Chitosan	0.066
N-methyl chitosan	0.068
N-isopropyl chitosan	0.070

Amine group in D- glucosamine unit of chitosan is the weak base. The amount of base chitosan amine group can be improved by incorporating the electron driving group to the amine group. Alkyl functional group will increase the stability of cation about the amine group, so that increase the amount of base.

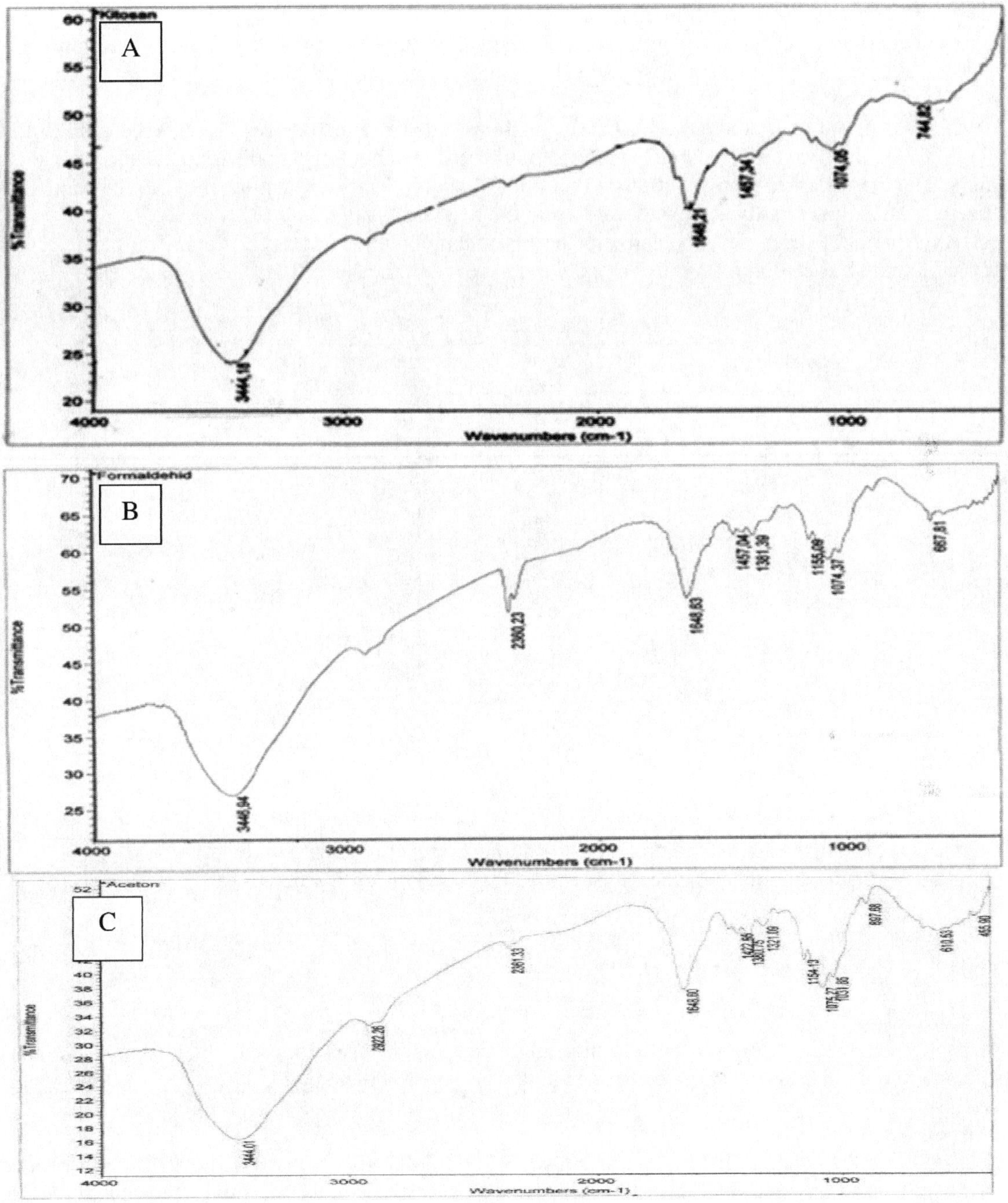

Figure 1. FTIR Spectra of (A) chitosan, (B) N-methyl chitosan, (C) N-isopropyl chitosan

3.3. *Transesterification of used cooking oil*

Transesterification reaction gave unsatisfaction result of methyl ester yield. Transesterification using chitosan catalyst only gave 2.0% of methyl ester. Catalyst *N*-methyl chitosan and *N*-isopropyl chitosan gave 3.1% and 8.6% methyl ester yield, respectively.

When compared with homogeneous catalysts, the results obtained are very low. According to Lee et al. (2009), the synthesis of methyl esters (biodiesel) in laboratory scale is influenced by some factors. The factors are the type of catalyst (acid/base), the mole ratio of methanol to oil, the content of impurities (usually free fatty acids and water) and the reaction temperature [7].

Methyl palmitate and methyl oleate are the main compounds of methyl esters. The peak area of chromatogram showed that methyl esters yield increased with the use of catalysts (Figure 2).

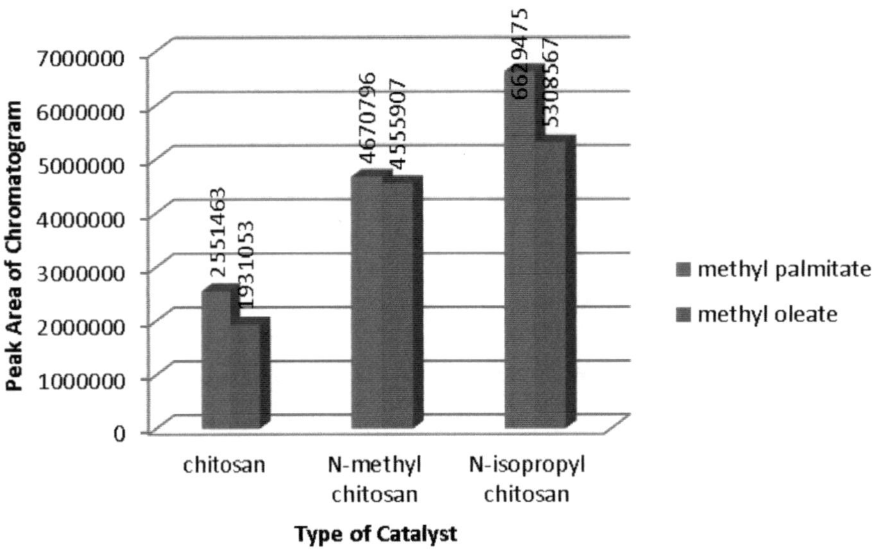

Figure 2. The methyl esters yield of methyl palmitate and methyl oleate with variation of chitosan and N-alkyl chitosan as a catalyst for the transesterification reaction

The results above showed the increasing of the amount of base enhanced the methyl ester yield of methyl palmitate and methyl oleate.

4. Conclusion

The inclusion of alkyl groups to the chitosan structure increases the amount of base in which increasing the methyl ester yield of methyl palmitate and methyl oleate.

References

[1] Da Silva R B, Neto A F L and dos Santos L S 2008 Catalysts of Cu(II) and Co(II) ions adsorbed in chitosan used in transesterification of soybean and babassu oils – A new route for biodiesel syntheses *Bioresource Technology* **99** 6793–6798

[2] Almerindo G I, Probst L F D, Campos C E M, de Almeida R M, Meneghetti S M P, Meneghetti M R, Clacens J M and Fajardo H V 2011 Magnesium oxide prepared via metal–chitosan complexation method:Application as catalyst for transesterification of soybean oil and catalyst deactivation studies *Journal of Power Sources* **196** 8057– 8063

[3] Nasratun M, Said H A, Noraziah A and Alla A N A 2009 Immobilization of Lipase from *Candida rugosa* ChitosanBeads for Transesterification Reaction *American Journal of Applied Sciences* **6 (9)** 1653-1657

[4] Dunn E T, Grandmaison E W and Goosen M F A 1997 *Application and Properties of Chitosan* Technomic Pub Basel p. 3-30
[5] Carey F A 2000 *Organic Chemistry* 4th edition McGraw-Hill Companies Inc New York
[6] Kim C H, Cho J W and Chun H J 1997 Synthesis of Chitosan Derivatives with Quaternary Ammonium Salt and Their Antibacterial Activity *Polym Bull* 38:387–393 in Entsar I Rabea, Mohamed EI Badawy, Tina M Rogge, Christian V Stevens, Walter Steurbaut, Monica H, Oftel and Guy Smagghel 2006 Enhancement of Fungicidal and Insecticidal Activity by Reductive Alkylation of Chitosan *Pest Manag Sci.* **62**, 890–897
[7] Lee D W, Park Y M and Lee K Y 2009 Heterogeneous Base Catalysts for Transesterification in Biodiesel Synthesis *Catal Surv Asia* **13** 63–77

High photocatalytic activity of mixed anatase-rutile phases on commercial TiO_2 nanoparticles

Wai Ruu Siah[1], Hendrik O. Lintang[1], Mustaffa Shamsuddin[2] and Leny Yuliati[1*]

[1] Centre for Sustainable Nanomaterials, Ibnu Sina Institute for Scientific and Industrial Research, Universiti Teknologi Malaysia, 81310 UTM Johor Bahru, Johor, Malaysia
[2] Department of Chemistry, Faculty of Science, Universiti Teknologi Malaysia, 81310 UTM Johor Bahru, Johor, Malaysia

[*]E-mail: leny@ibnusina.utm.my

Abstract. Titanium dioxide (TiO_2) is well-known as an active photocatalyst for degradation of various organic pollutants. Over the years, a wide range of TiO_2 nanoparticles with different phase compositions, crystallinities, and surface areas have been developed. Due to the different methods and conditions used to synthesize these commercial TiO_2 nanoparticles, the properties and photocatalytic performance would also be different from each other. In this study, the photocatalytic removal of 2,4-dichlorophenoxyacetic acid (2,4-D) and 2,4,5-trichlorophenoxyacetic acid (2,4,5-T) was investigated on commercial Evonik P25, Evonik P90, Hombikat UV100 and Hombikat N100 TiO_2 nanoparticles. Upon photocatalytic tests, it was found that overall, the photocatalytic activities of the P25 and the P90 were higher than the N100 and the UV100 for the removal of both 2,4-D and 2,4,5-T. The high activities of the P25 and the P90 could be attributed to their phase compositions, which are made up of a mixture of anatase and rutile phases of TiO_2. Whereas, the UV100 and the N100 are made up of 100% anatase phase of TiO_2. The synergistic effect of the anatase/rutile mixture was reported to slow down the recombination rate of photogenerated electron-hole pairs. Consequently, the photocatalytic activity was increased on these TiO_2 nanoparticles.

1. Introduction

Photocatalytic decomposition of organic pollutants is a promising solution for agricultural, industrial and domestic wastewater treatment. Titanium dioxide (TiO_2) is widely used for the photodegradation of hazardous chemical wastes due to its high activity, stability and non-toxicity [1, 2]. In particular, nanoparticulate TiO_2 is of great interest due to its higher surface area and photocatalytic activity than the bulk TiO_2. Hence, various TiO_2 nanoparticles have been developed and supplied for large scale commercial applications. Different synthesis methods would result in different properties of TiO_2, such as different phase compositions, crystallinities, and surface areas, which would affect the performance of the TiO_2 as a photocatalyst. In this context, it is of great interest to investigate the influence of various properties on the photocatalytic activity of the commercial TiO_2 nanoparticles. In preceding literatures, the photocatalytic activity of some commercial TiO_2 has been investigated for dye degradation and pollutant decomposition [3, 4], which can be considered as models for potential environmental wastewater treatment systems.

In this study, the photocatalytic activities of commercial Evonik P25, Evonik P90, Hombikat UV100 and Hombikat N100 TiO_2 nanoparticles for the removal of 2,4-dichlorophenoxyacetic acid (2,4-D) and 2,4,5-trichlorophenoxyacetic acid (2,4,5-T) were investigated. The P25 TiO_2 is a commercial nanoparticulate titania powder that is widely recognized as the benchmark material for photocatalytic degradation studies, while the P90 TiO_2 is a similar material synthesized by analogous preparation, comprised of smaller sized nanoparticles of mainly anatase phase [5]. Likewise, the UV100 and the N100 are commercially synthesized by an analogous method and used for a variety of applications [6]. While the test pollutants (2,4-D and 2,4,5-T) are widely used as herbicides, they have been categorized as drinking water contaminants as they are toxic, hardly biodegradable and difficult to remove from the environment [7]. This study showed that the commercial TiO_2 photocatalysts gave large variations in the photocatalytic activity and the highest activity was observed on the photocatalyst having mixture of anatase-rutile phases.

2. Materials and methods

2.1. Materials

Commercial P25 TiO_2 and P90 TiO_2 were obtained from Evonik Industries. Commercial Hombikat UV100 TiO_2 and Hombikat N100 TiO_2 were obtained from Sachtleben Chemie. The test pollutants 2,4-dichlorophenoxyacetic acid (2,4-D, $C_8H_6Cl_2O_3$, ≥ 98%) and 2,4,5-trichlorophenoxyacetic acid (2,4,5-T, $C_8H_5Cl_3O_3$, 95%) were purchased from Sigma.

2.2. Methods

2.2.1. Characterizations. Diffraction patterns of the samples were collected in the 2θ range of 20 to 60° by an X-Ray diffraction (XRD; Bruker D8 Advance diffractometer) at room temperature, using Cu-K_α radiation (λ = 1.5406 Å) at 40 kV and 40 mA. The diffuse reflectance ultraviolet-visible (DR UV-Vis) spectra of the samples were measured on a Shimadzu, UV-2600 under ambient conditions using barium sulphate as the reference. The morphology of the catalysts was studied by a transmission electron microscopy (TEM, JEM-2100) at an accelerating voltage of 200 kV. The sample was dispersed in ethanol and a drop of the suspension was placed on a carbon-coated copper grid.

2.2.2. Photocatalytic activity tests. Model pollutants used for photocatalytic tests were 2,4-D and 2,4,5-T herbicides. The commercial TiO_2 sample (50 mg) was first dispersed in 50 ml of the pollutant solution (0.5 mM). The mixture was then stirred in the dark for 1 hour to achieve adsorption-desorption equilibrium. While under constant stirring, the mixture was then exposed for 1 hour to a Xe-Hg UV lamp (200 W, $\Phi_i = 4$ mW cm^{-2}) with an optical filter cutting off wavelength above 400 nm. All reactions were made open to air in order to provide sufficient oxygen for the oxidative degradation of the pollutants. The amount of pollutant removed during the adsorpstion process and after UV lamp exposure was determined by a high performance liquid chromatography (HPLC; Shimadzu Prominence LC-20A). The adsorption capacity of the catalysts was determined from the measurement of pollutant concentration before and after the adsorption process. The percentage of pollutant adsorbed on the catalyst surface was determined from the equation (1).

$$\text{Adsorption of pollutant (\%)} = \frac{C_0 - C_A}{C_0} \times 100 \qquad (1),$$

where C_0 is the initial concentration of pollutant and C_A is the concentration of pollutant after 1-hour adsorption under dark condition. Likewise, the photocatalytic activity of the catalysts was determined from the measurement of pollutant concentration before and after UV light irradiation. The removal of pollutant was determined from the equation (2).

$$\text{Removal of pollutant (\%)} = \frac{C_A - C_R}{C_A} \times 100 \qquad (2),$$

where C_A is the concentration of pollutant after adsorption and C_R is the concentration of pollutant after reaction for 1 hour under UV light irradiation.

3. Results and discussion

3.1. Crystalline structure and optical properties

Figure 1(a) shows the XRD patterns of the P25, P90, UV100 and N100 commercial TiO_2, indexed to anatase and rutile phases. It can be seen that while the P25 and the P90 are made up of a mixture of anatase and rutile phases, the UV100 and the N100 are made up of TiO_2 anatase phase only. The ratio of anatase to rutile was calculated and shown in table 1. The anatase:rutile ratio of the P25 and the P90 was determined to be 80:20 and 92:8, respectively. The crystallite size of the commercial TiO_2 samples was estimated by the Scherrer equation and the results are also shown in table 1. As for the anatase phase, the P25 and the N100 commercial TiO_2 have a similar crystallite size of about 17-18 nm. On the other hand, the P90 has smaller crystallite size of 10.4 nm, followed by the UV100 that has the smallest crystallite size of 9.2 nm. As for the rutile phase, the crystallite sizes were 61.8 and 53.4 nm for the P25 and the P90 TiO_2, respectively. Based on the diffraction intensity, the P25 was more crystalline than the P90, while the N100 has better crystallinity than the UV100.

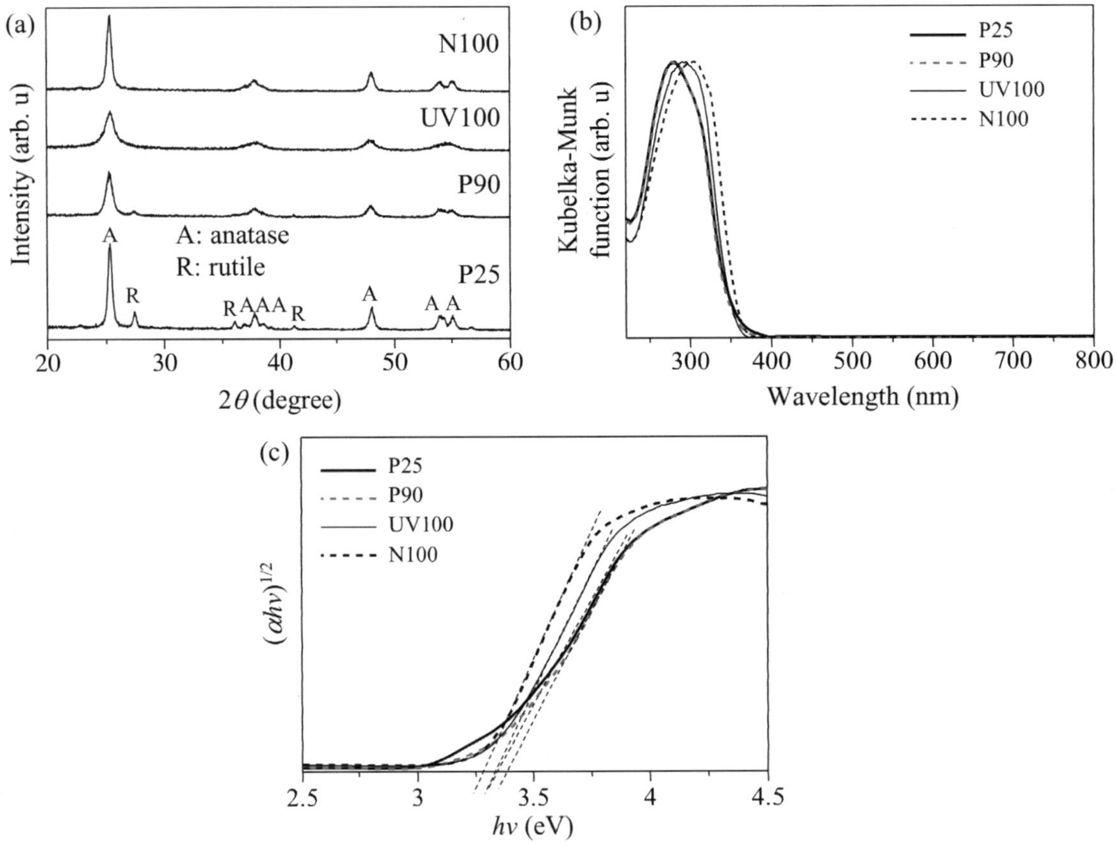

Figure 1. (a) XRD patterns, (b) DR UV-Vis spectra, and (c) Tauc plots of the commercial P25, P90, UV100 and N100 TiO_2 photocatalysts.

Figure 1(b) shows the DR UV-vis spectra of the commercial TiO_2 samples, representing the optical absorption of the samples. It was observed that the absorption edge of the P25, the P90, the UV100 and the N100 samples was similar to each other at around 390 nm, which is typical for absorption edge of TiO_2 [8]. The bandgap energy values of the samples were estimated by a Tauc plot ($(\alpha hv)^{1/2}$ versus hv), whereby the extrapolation of the linear part of the curve gives direct values of bandgap [9]. The results are shown in figure 1(c) and table 1. It was confirmed that the bandgap values of the commercial TiO_2 samples were similar to each other and in the range of 3.30 to 3.40 eV. The specific surface areas are also shown in table 1. The P25 has the lowest surface area while the UV100 has the largest surface area of the samples.

Table 1. Crystallite size, anatase: rutile ratio, bandgap and surface area of the commercial TiO_2 samples.

Commercial TiO_2	Crystallite size[a] (nm)		Anatase:rutile ratio[b]	Bandgap energy[c] (eV)	Specific surface area[d] (m^2/g)
	Anatase	Rutile			
P25	17.9	61.8	80:20	3.37	50 [5, 8, 10]
P90	10.4	53.4	92: 8	3.40	91 [5, 11]
UV100	9.2	-	100:0	3.36	290 [8]
N100	17.1	-	100:0	3.30	104 [12]

[a]Estimated from FWHM of XRD anatase (101) or rutile (110) reflection by a Scherrer equation.

[b]Estimated from XRD by the formula $A(\%) = (100 \times I_A)/(I_A + 1.265 \times I_R)$, whereby A(%) is the percentage of anatase, I_A and I_R correspond to the intensity of the anatase (101) and rutile (110) reflections.

[c]Estimated from Tauc plot of DR UV-Vis data.

[d]Obtained from the previously reported works.

3.2. Morphology and particle size

The commercial P25 and P90 TiO_2 were supplied by Evonik Industries, which have been reported to have similar morphology to each other [8, 11]. Likewise, the UV100 and the N100 both supplied by Sachtleben Chemie were also expected to have similar morphology. Hence, the P25 and the UV100 were selected as the representative samples to study morphology and particle size by TEM. Figure 2(a) shows the TEM image of the P25 TiO_2. The P25 has irregular polyhedra shape, and most of the particles fall within the size range of about 15-25 nm, which would be the anatase phase of the P25 sample. The larger particles in the range of 50-60 nm were also observed due to the rutile phase of the P25. On the other hand, the TEM image of the UV100 sample is shown in figure 2(b). Comparison to the P25, the UV100 has aggregates with a smaller primary particle size of about 7 nm. The particle size observed from the TEM for both the P25 and the UV100 agreed well with the crystallite size results obtained from the XRD mentioned above.

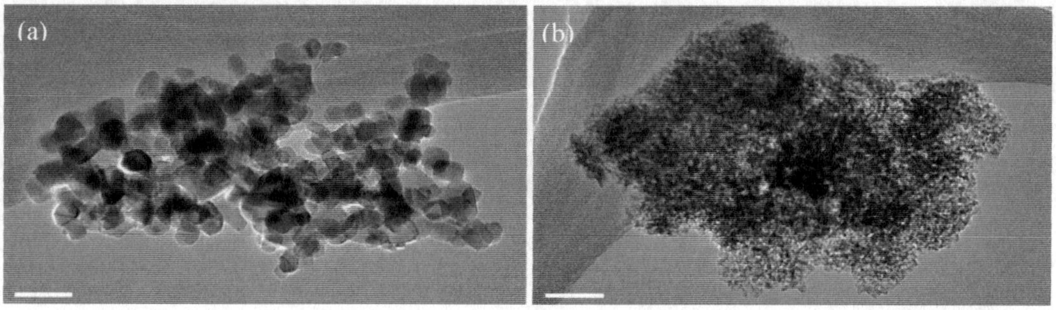

Figure 2. TEM images of commercial (a) P25 TiO_2 and (b) UV100 TiO_2.

3.3. Photocatalytic activity

The photocatalytic activities of the commercial TiO_2 samples for the removal of 2,4-D and 2,4,5-T herbicides are shown in figure 3. It can be clearly seen that after 1-hour exposure to the UV irradiation, regardless of the type of commercial TiO_2 used for the reactions, the amount of the 2,4,5-T removed was higher than that of the 2,4-D. The higher removal of the 2,4,5-T compared to the 2,4-D could be rationalized by considering its adsorption to the commercial TiO_2 samples.

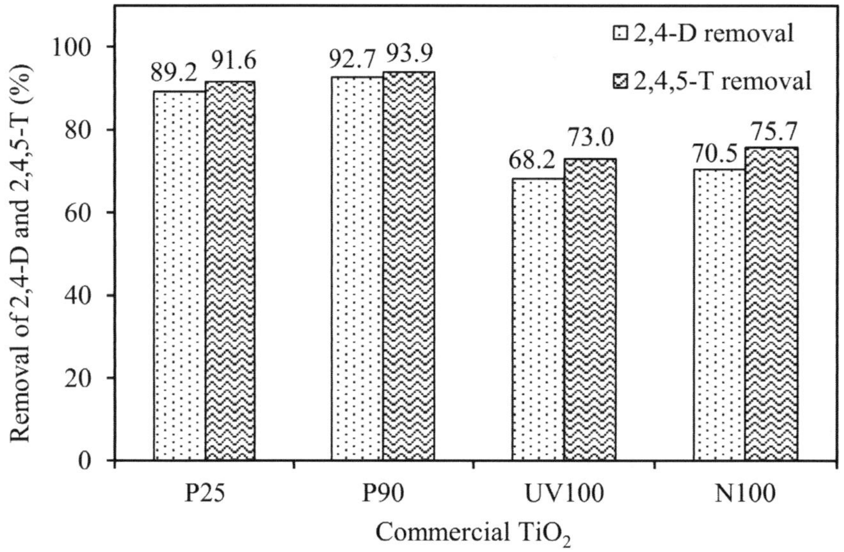

Figure 3. Photocatalytic removal of the 2,4-D and the 2,4,5-T model pollutants in the presence of commercial P25, P90, UV100 and N100 TiO_2 photocatalysts.

Figure 4 shows the percentage of 2,4-D and 2,4,5-T adsorbed onto the surface of the commercial TiO_2 samples when adsorption-desorption equilibrium was achieved. It was shown that for all types of commercial TiO_2 used, a greater amount of 2,4,5-T was adsorbed compared to the 2,4-D. It has been widely accepted that there is a corresponding relation between the surface adsorption of the pollutants with their photocatalytic degradation activity in the presence of TiO_2 [10]. Higher adsorption of the pollutant molecules implies that more of the pollutants are at close proximity to the TiO_2 surface, where active species are generated. Hence, the higher adsorption of the 2,4,5-T led to a higher photocatalytic removal activity compared to the 2,4-D.

The chemical structures of the 2,4-D and the 2,4,5-T are shown in figure 5. It can be seen that the main difference between the 2,4-D and the 2,4,5-T is the extra chlorine substituent at the 5th carbon of 2,4,5-T. The larger and more highly substituted 2,4,5-T is less soluble in water than the 2,4-D. Consequently, it interacts more strongly with the TiO_2 surfaces. The increased absorption of the 2,4,5-T on the surface of TiO_2 could also be explained by considering the electron donor-acceptor (EDA) complex mechanism as has been proposed for adsorption of chlorophenol on mesoporous SBA-15 [13]. The aromatic ring of the 2,4-D and the 2,4,5-T acts as an electron acceptor and the oxygen atoms on TiO_2 surface serves as a donor. Since the chloro group is an electron-withdrawing group, increasing the number of chlorine substituents leads to a decrease of π-electron density in the aromatic ring. As a result, the affinity of the 2,4,5-T for the non-binding electrons at TiO_2 surface oxygen also increased and this, in turn, leads to the increased adsorption of the 2,4,5-T as compared to the 2,4-D.

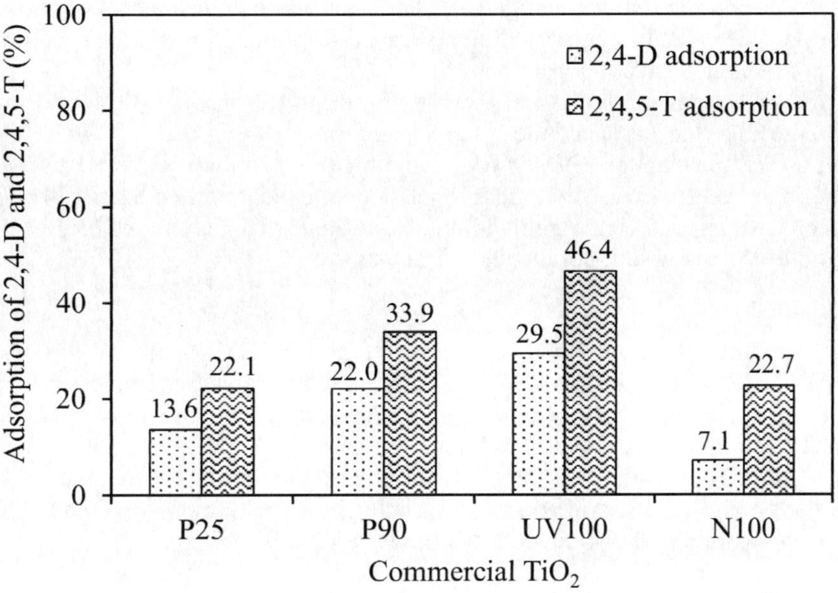

Figure 4. Adsorption of the 2,4-D and the 2,4,5-T model pollutants on the commercial P25, P90, UV100 and N100 TiO_2 photocatalysts.

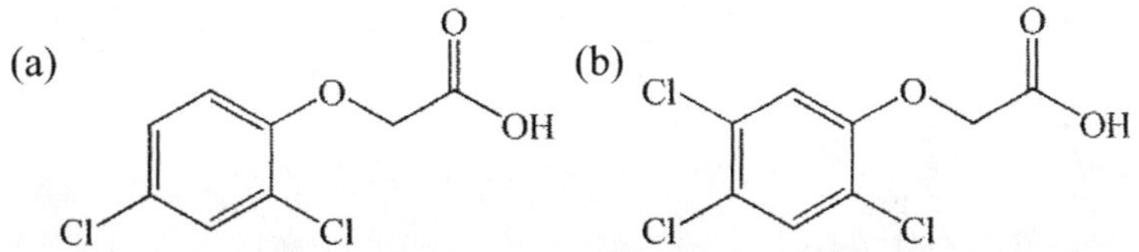

Figure 5. The chemical structures of (a) 2,4-D and (b) 2,4,5-T.

While the removal of the pollutants was affected by the type of the pollutants due to the different adsorption, it was also strongly depended on the types of commercial TiO_2. Among the investigated samples, it was obvious that the P25 and the P90 gave similar photocatalytic activity to each other, and the activity was higher than the UV100 and the N100. However, it appeared that the specific surface area did not play a significant role in the adsorption and photocatalytic reactions. As shown in table 1, the specific surface area was in the order of P25 < P90 < N100 < UV100. On the other hand, the adsorption of pollutants was in the order of N100 < P25 < P90 < UV100. Moreover, the higher adsorption towards the pollutants did not necessarily lead to higher catalytic activity. For example, while having the highest adsorption of the 2,4-D and the 2,4,5-T, UV100 TiO_2 was not the most active catalyst for the removal of the 2,4-D and the 2,4,5-T. Instead, the P25 and the P90 were more active catalysts for the photocatalytic removal of these herbicides. These results suggested that there are other factors influencing the photocatalytic reactions apart from the surface adsorption.

As previously shown, the P25 and the P90 have a mixture of anatase-rutile phase, while the UV100 and the N100 only have anatase phase. This main difference caused the P25 and the P90 have higher photocatalytic activity than the UV100 and the N100. The mixture of anatase and rutile phase would lead to a synergistic electron transfer that improved the charge separation of the photogenerated electron-hole pairs [14-17]. One of the widely accepted mechanism proposed that when anatase and rutile are in direct interfacial contact, the resultant band alignment leads to a favorable electron transfer pathway from rutile to anatase that stabilizes charge separation and prevents rapid recombination [14, 15]. This, in turn, caused the improved photocatalytic activity of the P25 and the P90 TiO_2. Even though the properties of the P25 and the P90 were slightly different, it was demonstrated that for TiO_2 anatase-rutile photocatalysts, the crystallinity, crystallite size, anatase-rutile ratio, and the specific surface area did not much affect the photocatalytic reactions.

Comparing the activity of the N100 and the UV100, the N100 displayed a higher photocatalytic activity for the removal of the 2,4-D and the 2,4,5-T than the UV100 TiO_2. As shown in figure 1(a), the N100 was more crystalline than the UV100. By having higher crystallinity, the N100 also have fewer defect sites, and consequently longer lifetime of photogenerated electron-hole pairs [4]. Therefore, for TiO_2 anatase photocatalyst, one factor that is important in the photocatalytic removal of the 2,4-D and the 2,4,5-T would be the crystallinity of the photocatalyst.

4. Conclusions

In conclusion, the photocatalytic activity of commercial P25, P90, UV100 and N100 TiO_2 was studied using 2,4-D and 2,4,5-T as the model pollutants. The higher photocatalytic removal of the 2,4,5-T than the 2,4-D on all four commercial TiO_2 samples could be associated with the higher adsorption between the 2,4,5-T and TiO_2, which in turn led to the improved pollutant-catalyst interaction. In addition, the higher photocatalytic activity of the P25 and the P90 compared to the UV100 and the N100 TiO_2 could be attributed to the synergistic effect of anatase and rutile phase mixture of the P25 and the P90, which led to the improved separation of photogenerated charge carriers.

Acknowledgements

The research leading to these results has received funding from the European Union Seventh Framework Program (EU-FP7) under 4G-PHOTOCAT grant (agreement no.: 309636). This work has been also financially supported by Ministry of Higher Education (MOHE) and Universiti Teknologi Malaysia (UTM, Malaysia) through the international contract matching grant (cost center code: R.J130000.7626.4C035).

References

[1] Carp O, Huisman C L and Reller A 2004 *Prog. Solid State Chem.* **32** 33
[2] Gaya U I and Abdullah A H 2008 *J. Photochem. Photobiol. C* **9** 1
[3] Doll T E and Frimmel F H 2005 *Water Res.* **39** 403
[4] Ryu J and Choi W 2007 *Environ. Sci. Technol.* **42** 294
[5] Peiró A M, Colombo C, Doyle G, Nelson J, Mills A, and Durrant J R 2006 *J. Phys. Chem. B* **110** 23255
[6] Sachtleben at a glance: the leading TiO_2 specialty company, Sachtleben Chemie, Germany
[7] Kamble S P, Sawant S B, and Pangarkar V G 2006 *Chem. Eng.Res. Des.* **84** 355
[8] Hidalgo M C, Colón G, and Navío J A 2002 *J. Photochem. Photobiol. A* **148** 341
[9] Znad H and Kawase Y 2009 *J. Mol. Catal. A* **314** 55
[10] Yu Y, Yu J-C, Chan C-Y, Che Y-K, Zhao J-C, Ding L, Ge W-K and Wong P-K 2005 *Appl. Catal. B* **61** 1
[11] Chen H-W, Liang C-P, Huang H-S, Chen J-G, Vittal R, Lin C-Y, Wu K C W and Ho K-C 2011 *Chem. Commun.* **47** 8346
[12] Gunnelius K R, Lundin T C, Rosenholm J B and Peltonen J 2014 *Cement Concrete Res.* **65** 1
[13] Qin Q, Liu K, Fu D, and Gao H 2012 *J. Environ. Sci.* **24** 1411
[14] Hurum D C, Agrios A G, Gray K A, Rajh T and Thurnauer M C 2003 *J. Phys. Chem. B* **107** 4545

[15] Scanlon D O, Dunnill C W, Buckeridge J, Shevlin S A, Logsdail A J, Woodley S M, Catlow C R A, Powell M J, Palgrave R G, Parkin I P, Watson G W, Keal T W, Sherwood P, Walsh A and Sokol A A 2013 *Nat. Mater.* **12** 798
[16] Ohno T, Sarukawa K, Tokieda K and Matsumura M 2001 *J. Catal.* **203** 82
[17] Sun B, Vorontsov A V and Smirniotis P G 2003 *Langmuir* **19** 3151

IOP Conf. Series: Materials Science and Engineering **107** (2016) 012006 doi:10.1088/1757-899X/107/1/012006

Synthesis of α-Bismuth oxide using solution combustion method and its photocatalytic properties

Y. Astuti[1*], A. Fauziyah[1], S. Nurhayati[1], A. D. Wulansari[1], R. Andianingrum[1], A. R. Hakim[1], G. Bhaduri[2]

[1]Chemistry Department, Faculty of Science and Mathematics, Diponegoro University, Jl. Ngumpulsari Raya No.12, Tembalang, Semarang, Jawa Tengah 50277
[2]School of Chemical Engineering and Advanced Materials, Newcastle University, Newcastle, UK NE1 7RU

[*]Email: yayuk_astuti@undip.ac.id

Abstract. The monoclinic bismuth oxide was prepared by the solution combustion method using bismuthyl nitrate as the raw material and citric acid as fuel. The synthesis process consisted of the formation of a clear transparent solution and the formation of white powder after heating the mixture at 250 °C for 2 hours. The yellow pale crystalline materials were obtained after calcination of the white powder at 600 °C for 80 minutes. Furthermore, the photocatalytic activity of the product was also studied using methyl orange as a model pollutant. The result showed that the coral reef-like bismuth oxide was able to degrade 50 mL methyl orange (5 ppm) by 37.8 % within 12 hours irradiation using 75-watt tungsten lamp.

1. Introduction

Use of inorganic semiconductor catalysts for the degradation of environmental wastes has gained extensive research interest recently [1, 2]. Bismuth oxide is one of the non-TiO_2 semiconductor catalysts that have shown effective catalytic properties for the degradation of environmental wastes [2-5]. Bismuth oxide (Bi_2O_3) is a pale yellow solid, having a melting point of about 825 °C [6] and very easily reduced to Bi-metal when being heated with carbon or hydrogen [7]. Bismuth oxide (Bi_2O_3) has excellent optical and electrical properties such as wide band-gap, high refractive index, high dielectric permittivity and high photoconductivity [8]. Other than photocatalysis bismuth oxide can also be used in various other applications such as gas sensing [9] and superconductors [10].

Bismuth oxide exists in five different polymorphs, α-Bi_2O_3, β-Bi_2O_3, γ-Bi_2O_3, δ-Bi_2O_3 and ε-Bi_2O_3 respectively, wherein each polymorph has different properties [7]. The variation of the properties among the types closely relates to the stability of the material to the temperature. In the low temperature, the type of α-Bi_2O_3 (monoclinic) is formed, while there are four phases of Bi_2O_3 stable at high temperatures (more than 600 °C) i.e. β-Bi_2O_3 (tetragonal), γ- Bi_2O_3 (body centred cubic), δ-Bi_2O_3 (face centred cubic) and ε-Bi_2O_3 (triclinic) [11]. δ-Bi_2O_3 phase transforms into β-Bi_2O_3 or γ- Bi_2O_3 if being heated at 650 °C or 639 °C, respectively. Both α-Bi_2O_3 and δ-Bi_2O_3 phases are correspondingly stable at low and high temperature. In contrast, the other phases are metastable; however, stabilization of these materials can be conducted by doping with impurities [12]. Various transitional metals have been doped in Bi_2O_3 to observe its catalytic properties [4, 5, 13]. The β-Bi_2O_3 is one of the most commonly used polymorph for photocatalytic applications [1, 4, 13, 14] whereas use of α-bismuth oxide has just been recently reported [15].

Bismuth oxide can be synthesised by several methods that include sol-gel [6], deposition [16], solution combustion [17], hydrothermal [18], and fusion spray method [19]. Solution combustion has several advantages such over other methods i.e. easy processing, use of low temperatures, easy to control the particle size leading to shorter reaction times, high product purity, crystallinity and nanosized powder with high specific surface area [20]. In the current study bismuth oxide is synthesised by solution combustion method using bismuthyl nitrate as a raw material and citric acid as a fuel. The obtained bismuth oxide was characterized for its morphology, crystal structure and surface functional groups using SEM-EDX, XRD and FTIR respectively. Furthermore, the photocatalytic activity of the bismuth oxide particles was demonstrated by degradation of methyl orange dye.

2. Experimental Methods

2.1. Materials

Bismuth(II) subnitrate ($Bi_5O(OH)_9(NO_3)_4$), nitric acid (65%, v/v), NH_4OH (25%, v/v), polyethylene glycol (PEG 20000), citric acid, methyl orange and aquadest were of analytical grade (except for aquadest) were purchased from Merck Index, Indonesia.

2.2. Synthesis of Bismuth Oxide using Solution Combustion Method

The synthesis procedure for Bi_2O_3 follows the solution combustion method reported by La *et al.* [20] with some modifications. 2.91 g $Bi_5O(OH)_9(NO_3)_4$ and 1.471 g citric acid were dissolved in 10 ml HNO_3. The mixture was then added 0.04 g PEG 20000 and stirring 350 rpm for 5 minutes. A clear transparent solution was formed, that was subsequently heated at 250 °C for 2 hours. A turbid white suspension was formed that was air cooled to room temperature. The white precipitant was filtered from solution and washed using distilled water to remove all the acid. The white powder was then dried in the oven for 24 hour at 110 °C. The sample was later calcined at 600 °C for 80 minutes. A yellow powder was obtained after calcination and used for further characterisation.

2.3. Characterisation of the Samples

The starting material bismuth(II) subnitrate and yellow powder were characterized to observe complete conversion. The characterisation techniques used were XRD, SEM-EDS and FTIR. X-ray diffractogram was obtained using XRD Bruker with CuK radiation (λ = 0.15418 nm) at 40 kV and 30 mA, and with 2θ ranging from 10° to 60°. The morphology of the samples were identified using SEM-EDX (JEOL-JSM-G510LV). The FTIR spectra were recorded using FTIR Prestige 21 (Shimadzu) between wavenumber (400-4000 cm^{-1}).

2.4. Photocatalytic Activity on Methyl Orange Degradation

The absorbance peak of methyl orange was determined by observing the absorbance in the visible range of 300-700 nm using a UV-Vis spectrophotometer (Shimadzu UV-1201). The maximum absorbance was observed at a wavelength of 462 nm which further used for monitoring the reaction end point.

Photocatalytic activity of the synthesised Bi_2O_3 particles was conducted using the following procedure. 0.1 g sample bismuth oxide was added into 50 mL methyl orange solution (5 ppm) and stirred 600 rpm. This sample was then irradiated with tungsten lamp 75 watt (1.4-3.6 eV) and analysed after a time interval of 2 hours for maximum reaction time of 12 hours. A control was carried out with the degradation experiment without any Bi_2O_3 particles. The $Bi_5O(OH)_9(NO_3)_4$ was also tested as a control to observe any catalytic activity (only one experiment for 12 hours was carried out for $Bi_5O(OH)_9(NO_3)_4$). All experiments were performed in duplets.

3. Results and Discussion

3.1. Synthesis of bismuth oxide using combustion method

Synthesis of bismuth oxide using bismuth(II) subnitrate as the raw material produced pale white powder after drying it in oven for 24 hour at 110 ^{0}C as shown in Figure 1a. The color of the powder turned to be yellow when calcination at 600 ^{0}C for 80 minutes was applied as shown in Figure 1b. This yellow color powder indicated that bismuth oxide was formed.

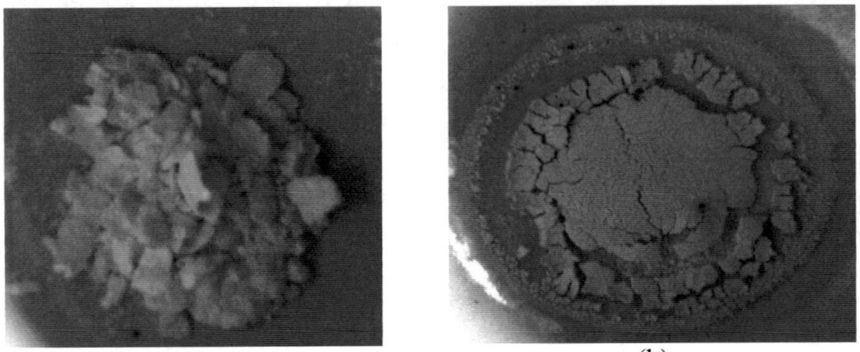
(a) (b)

Figure 1. a) The white powder product after heating at 110 °C for 24 hour in oven, b) The yellow powder product after calcination of the white powder product at 600 °C for 80 minutes.

3.2. X-Ray Diffraction (XRD)
Figure 2 presents the diffractograms of bismuth(II) subnitrate and bismuth oxide, respectively. The XRD pattern in Figure 2a shows three intense peaks at 2θ values of 25.0°, 27.8° and 42.0° respectively, corresponding to bismuth(II) subnitrate [21]. However the pattern in Figure 2b shows intense peaks at 2θ values of 27.5°, 33.2° and 33.4° corresponding to that of α-bismuth oxide [22]. The α-bismuth oxide thus synthesised has a monoclinic structure having the lattice and the angular parameters as $a = 5.8499(3)$, $b = 8.1698(4)$, and $c = 7.5123(3)$ are $\alpha = 90°$, $\beta = 112.988(4)°$ and $\gamma = 90°$ respectively (JCPDS No: 76-1730). These lattices are consistent with the product formed. This results are also in good agreement with previous literature [15]. The absence of peaks at 25.0° in Figure 2b indicates absence of the bismuth(II) subnitrate in the product.

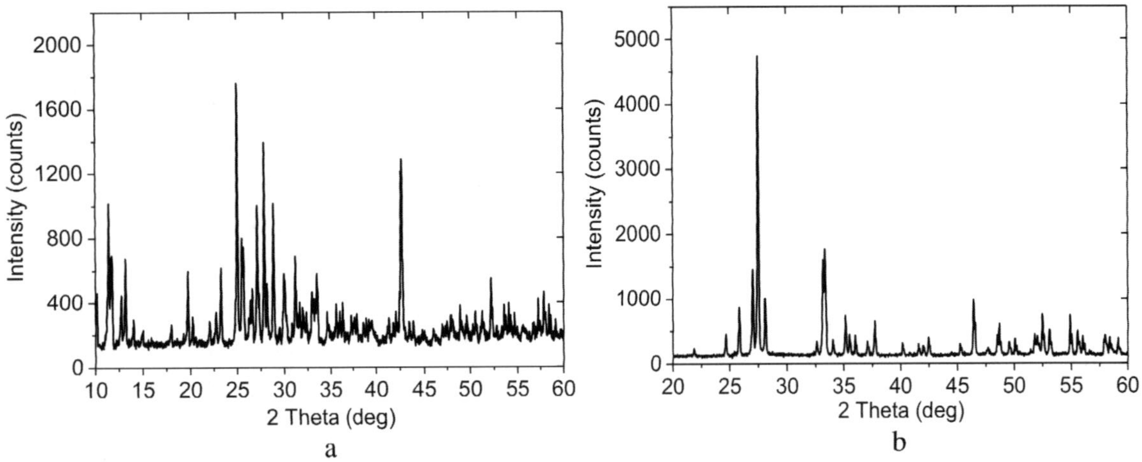

Figure 2. XRD patterns of a) bismuthyl nitrate and b) bismuth oxide synthesised using solution combustion method.

3.3. Fourier Transform Infrared (FTIR)

The IR spectra of the bismuth(II) subnitrate and bismuth oxide are shown in Figure 3. The peak at 3200-3600 cm^{-1} is a characteristic stretch vibration frequency of O-H [23]. This vibration mode is observed in both samples; however, the intensity of this peak decreases in the product. The presence of OH vibration mode observed in the product could be due to the incomplete reaction of raw material became bismuth oxide. Furthermore, significant decrease is also observed in the FTIR spectra of the product for vibration mode at 1200 – 1700 cm^{-1} which is a characteristic for NO_3^- group even barely visible [6]. Meanwhile, the broad peak at 700 – 400 cm^{-1} originating from the metal-oxygen (Bi-O-Bi) vibration [6, 24] is still observed. Therefore, the FTIR results confirm the presence of bismuth oxide in synthesised in the product free of any unreacted bismuth(II) subnitrate.

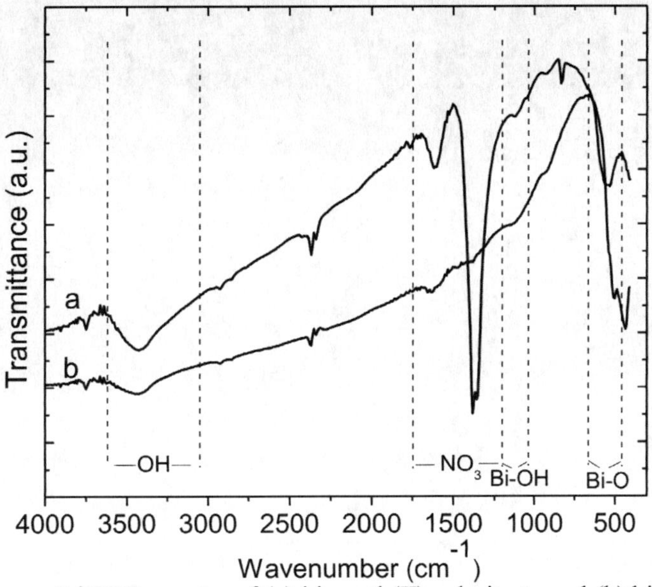

Figure 3. FTIR spectra of (a) bismuth(II) subnitrate and (b) bismuth oxide

3.4. Scanning Electron Microscopy-Energy Dispersive X-Ray Spectroscopy (SEM-EDX)

Figure 4 shows the SEM images of bismuth(II) subnitrate and the resulting product of α-bismuth oxide. It can be seen in Figure 4a that the shape of bismuth(II) subnitrate is briquette with the size of 8 μm. Meanwhile, the morphology of α-bismuth oxide synthesized using solution combustion method shows coral reefs like consisting of agglomerating small materials. Xia *et al* [19] observed similar morphology for α-bismuth oxide synthesised using fusion spraying method. The absence of wire-like structure of bismuth(II) subnitrate in Figure 4b shows complete removal of bismuth(II) subnitrate from the product. The in situ EDX analysis of α-bismuth oxide shows absence of the nitrogen peak confirming the absence of bismuth(II) subnitrate in the product (figure 5).

Figure 4. SEM images of a) bismuth(II) subnitrate and b) bismuth oxide synthesized using solution combustion method.

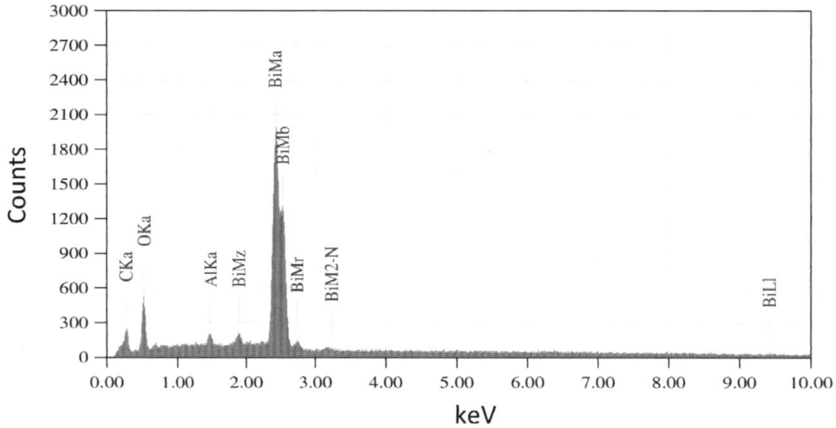

Figure 5. EDS spectrum of bismuth oxide synthesized using solution combustion method

3.5. Photocatalytic Activity

α-Bismuth oxide was tested as a photocatalyst for the degradation of methyl orange under different illumination time interval of 2 hours for reaction time of 12 hours under tungsten lamp irradiation. Photocatalysis was confirmed by studying the dye degradation using α-bismuth oxide without irradiation with tungsten lamp as compared to tungsten lamp irradiation. The result presented in the Figure 6 shows that the degradation (37.83%) of methyl orange using bismuth oxide during 12 hours under light tungsten irradiation. The maximum degradation was obtained after 12 hours irradiation which is the maximum of the time limit working. A linear trend is observed in the degradation of methyl orange in the presence of α-bismuth oxide.

A comparison to observe any catalytic activity of bismuth(II) subnitrate was also conducted. It was observed that the white bismuth(II) subnitrate turned orange in colour when suspended in 5 ppm methyl orange solution. This indicates that the dye is adsorbed on the surface of bismuth(II) subnitrate rather than degradation. No such change in colour of α-bismuth oxide (i.e. from yellow to orange) was observed confirming photocatalytic degradation of the dye. These results proved that the α-bismuth oxide synthesized using the solution combustion method can be used as a photocatalyst. The photocatalytic performance of α-bismuth oxide observed in this study is less than that observed for γ-bismuth oxide prepared by a simple chemical route [25].

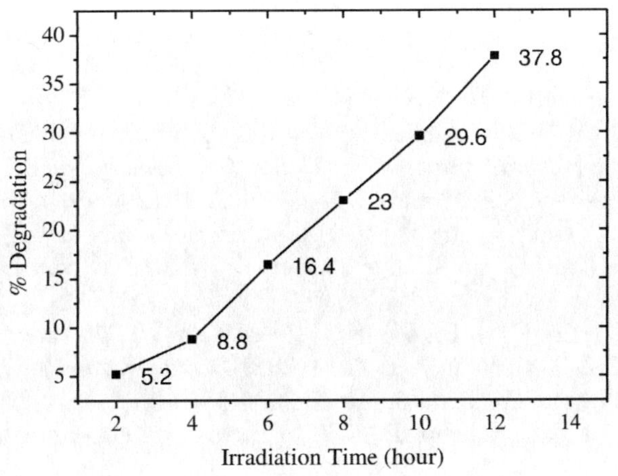

Figure 6. The degradation of methyl orange by bismuth oxide in different time.

4. Conclusion

Synthesis of bismuth oxide using solution combustion method produced α-bismuth oxide material which has a monoclinic structure with irregular shapes (coral reefs like structure) consisting of the small agglomerate particles. The synthesised α-bismuth oxide demonstrated photocatalytic activity in decomposition of methyl orange. The result indicates that this material is potential for photocatalyst for dye degradation.

5. Acknowledgements

The authors would like to thank The Integrated laboratory, Diponegoro University for XRD and SEM-EDS analysis and Laboratory of Organic Chemistry, Gajah Mada University for FTIR analysis.

References

[1] Brezesinski K, Ostermann R, Hartmann P, Perlich J and Brezesinski T 2010 *Chem. Mater.* **22** 3079-3085

[2] Lu Y, Zhao Y, Zhao J, Song Y, Huang Z, Gao F, Li N and Li Y 2015 *Cryst. Growth Des.* **15** 1031-1042

[3] Raza W, Haque M M, Muneer M, Harada T, and Matsumura M 2015 *J. Alloy Compd.* **648** 641-650

[4] Saison T, Chemin N, Chanéac C, Durupthy O, Mariey L, Maugé F, Brezová V and Jolivet J -P 2015 *J. Phys. Chem. C.* **119** 12967-12977

[5] Tucher J, Peuntinger K, Margraf J T, Clark T, Guldi D M and Streb C 2015 *Chem. Eur. J.* **21** 8716-8719

[6] Mallahi M, Shokuhfar A, Vaezi M R, Esmaeilirad A and Mazinani V 2014 *Am. J. Eng. Res.* **03** 162-165

[7] Febrianto E Y and Indayaningsih N 2009 *Jurnal Fisika Himpunan Fisika Indonesia* **1** 0854-3046

[8] Milena C 2012 *Dyna* **176** 139-148

[9] Xinglong G, Rong L, Guoxiu W, Zhixin C and David W 2009 *Nanotechnology* **20** 495501

[10] Koza J A, Bohannan E W and Switzer J A 2013 *ACS Nano* **7** 9940-9946

[11] Yilmaz S, Turkoglu O, Ari M and Belenli I 2011 *Cerâmica* **57** 185-192

[12] Cornei N, Tancret N, Abraham F and Mentré O 2006 *Inorg. Chem.* **45** 4886-4888

IOP Conf. Series: Materials Science and Engineering **107** (2016) 012006 doi:10.1088/1757-899X/107/1/012006

[13] Saison T, Chemin N, Chanéac C, Durupthy O, Ruaux V, Mariey L, Maugé F, Beaunier P and Jolivet J -P 2011 *J. Phys. Chem. C* **115** 5657-5666
[14] Qin F, Li G, Wang R, Wu J, Sun H and Chen R 2012 *Chem-Eur. J.* **18** 16491-16497
[15] Oudghiri-Hassani H, Rakass S, Al Wadaani F T, Al-ghamdi K J, Omer A, Messali M and Abboudi M 2015 *J. Taibah Univ. Sci.* **155** 1-5
[16] Hernandez-Delgadillo R, Velasco-Arias D, Martinez-Sanmiguel J J, Diaz D, Zumeta-Dube I, Arevalo-Niño K and Cabral-Romero C 2013 *Int. J. Nanomedicine* **8** 1645-1652
[17] Aruna S T and Mukasyan A S 2008 *Curr Opin Solid ST M* **12** 44-50
[18] Wu C, Shen L, Huang Q and Zhang Y -C 2011 *Materials Letters* **65** 1134-1136
[19] Xia J -y, Tang M -t, Chen C, Jin S -m and Chen Y -m 2012 *T. Nonferr Metal Soc.* **22** 2289-2294
[20] La J, Huang Y, Luo G, Lai J, Liu C and Chu G 2013 *Particul. Sci. Technol.* **31** 287-290
[21] You Q, Fu Y, Ding Z, Wu L, Wang X and Li Z 2011 *Dalton Trans* **40** 5774-5780
[22] Jabeen Fatima M J, Navaneeth A and Sindhu S 2015 *RSC Adv* **5** 2504-2510
[23] Bartonickova E, Cihlar J and Castkova K 2007 *Processing and Application of Ceramics* **1** 29-33
[24] Li W 2006 *Mater. Chem. Phys.* **99** 174-180
[25] Iyyapushpam S, Nishanthi S T, and Pathinettam Padiyan D 2014 *J. Alloy Compd.* **601** 85-87

IOP Conf. Series: Materials Science and Engineering **107** (2016) 012007 doi:10.1088/1757-899X/107/1/012007

Aliphatics hydrocarbon content in surface sediment from Jakarta Bay, Indonesia

M YAzis[1,2,3*], L Asia[2], A Piram[2],BBuchari[1],P Doumenq[2], A D Syakti[3]
[1]Analytical Chemistry Laboratory, Department of Chemistry, Institute of Technology Bandung, Jl. Ganesha No. 10, Bandung 40132, Indonesia
[2]Environmental Analytical Chemistry Laboratory (LCE), FRE-CNRS 3416 Equipe MPO, Europôle de l'Arbois-Bâtiment Villemin BP 80 Aix en Provence Cedex 4, France
[3]Center for Maritime Bioscience Studies-Institute for Research and Community Service, Jenderal Soedirman University, Jl. Dr. Suparno, Purwokerto 53123, Indonesia

[*]Email: yudis_chempd@yahoo.com

Abstract. Sedimentary aliphatic hydrocarbons content have been studied quantitatively and qualitatively using GC/MS method in eight coastal stations located in the Jakarta Bay, North of Jakarta, Indonesia. The total concentrations n-alkanes have ranged from 480 $\mu g.kg^{-1}$to 1,935 $\mu g.kg^{-1}$sediment dry weight. Several ratios (e.g. CPI_{24-32}, NAR, TAR, Pr/Phy, n-C_{17}/Pr, n-C_{18}/Phyt,n-C_{29}/n-C_{17}, $\sum n$-alkanes/n-C_{16}LMW/HMW, Paq and TMD) were used to evaluate the possible sources of terrestrial-marine inputs of these hydrocarbons in the sediments. The various origins of aliphatic hydrocarbons were generally biogenic, including both terrigenous and marine, with an anthropogenic pyrolytic contribution (petrogenic and biogenic combustion). Two stations (G,H) were thehighest concentration and had potential risk to environment

1. Introduction

Jakarta Bay is located in northern part of Jakarta,Indonesia, which is considered as the most crowded in maritime traffic in Java sea. Jakarta Bay is defined by two flanking delta systems which supply large volumes of sediment from 13 rivers watersheds [1,2, 8]. Several human-related activities may become the pollutant sources i.e. domestic waste, industrial effluent, urban run off, shipping wastes and illegal fishing as well as the oil and gas exploitation activities. The later may occur via oil spills with has a potential hazard to the living biota and ecosystems.We used a group of compound namely aliphatic hydrocarbons i.e. n-alkane, pristine (Pr) and phytane (Phy) from terrestrial plants or their decomposition and metabolic activity of aquatic organisms [3,4]. Aliphatic hydrocarbons are one of the widespread contaminants in coastal, river, bay and lagoons. They are formed by biological (decarboxylation affecting linear and branched fatty acids), chemical or synthesized process. Taking into account that aliphatic hydrocarbons can enter the marine environment by natural (i.e. pyrolysis, diagenesis and biosynthesis) and anthropogenic processes (i.e. industrial activities, combustion of fossil compounds, oil plants and refineries, oil waste, urban waste, etc.) [3,4,9] source of aliphatic hydrocarbons are multiple and difficult to determine. Such saturated hydrocarbons generally have to range from n-C_{12} to n-C_{40} and might have particular predominance number of atoms n-C_{15} and n-C_{17} in the case of marine while the n-C_{29} and n-C_{31}in case of terrestrial apportionment. Hence, the objective of this study was to determine of aliphatic hydrocarbons level and potential source/ origin of contamination of marine sediment in Jakarta Bay, Indonesia.

2. Materials and Methods

2.1 Study area and sampling

This study focused on the detection of aliphatic hydrocarbon condition in Jakarta Bay. A global positioning systems (GPS, Germin Etrex Summit HC, Kansas City USA) and Quantum GIS (Geographic information system) were used in determining different sampling sites. In December 2013, there are 8 stations of surface sediment were collected from Jakarta bay, 4 stations near from urban and industrial area (A, B, G and H). These stations were canalized from several rivers form an integral part of the stormwater and sewage transport systems within Jakarta's urban centre [2,3], and 4 stations were characterized by coralline sand substrates with high traffic of shipping transport (C,D,Eand F) (**Figure 1**). Surface sediment samples were taken using a stainless steel grab sampler, and then placed immediately in an aluminium container and stored in a cool box to minimize microbial degradation during transport to the laboratory. Sediment sample were subsequently freeze-dried (Over on at -55 ^{0}C and 0.0080 mtorr, Korea), crash it in mortar porcelain then sieved (stainless steel sieved) at 200 μm and homogenised.

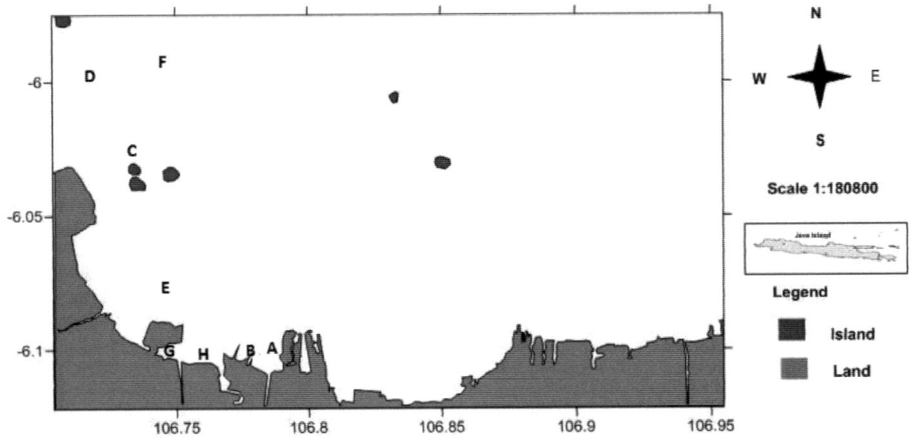

Figure 1. Site description of Jakarta Bay, Indonesia

2.2. Chemicals

All solvents were of Supra Solv grade (Merck, Darmstadt, Germany) such as n-Heptane (HEPT), Dichloromethane (DCM), HCl (37%) was of pure grade (Fisher, Scientific, Leicestershire, UK), Individual deuterated internal standard of aliphatic hydrocarbons [$^2H_{40}$] n-nonadecane, m/z 66 (Sigma-Aldrich, St Louis, MO, USA); 1-eicosene, m/z 71 (Dr. Ehrenstrofer Laboratories, Augsburg, Germany) for n-alkanes standard control. Alumina, Silica gel 60 (200-300 mesh) and Cu powder (>230 mesh) were purchased from VWR international and Merck (Darmstadt, Germany).

2.3. Extraction and Separation

Precleaned thimble cellulose extraction was prepared and filled 10 g accurately weighed of freeze-dried sediment then extracted by six-term apparatus for 3 h with a 150 mL mixture of HEPT/DCM (1/1;v/v). Before extraction, 0.5 mL in ternal standard (15 mg/L) was added into sample. The extract or Extractible Organic Material (EOM) was evaporated under gentle nitrogen stream and weighed (microbalance Perkin-Elmer AD2Z) part of the EOM was dissolved in HEPT and applied to a 50% silica/ 50% alumina (8 g of each, both deactivated with 5% H_2O) with chromatography column adsorption, borosilicate glass column (30 m x 1.0cm). The saturated fraction (*n*-alkanes) was eluted with 30 mL of HEPT and 20 mL of HEPT/ DCM (90/10;v/v) on the column then fraction was evaporated on a rotary evaporator under a stream of nitrogen, and dry residues were quantified gravimetrically (microbalance Perkin-Elmer AD2Z, France). Cu powder which had been washed with diluted HCl (1/4;v/v) was added and mixedwith extract to remove their sulphur interference. Cu powder residue was removed, and extract solution was diluted in HEPT to 1 mL. The n-alkanes fraction were analyzed by Agilent 7000 A GC/ QQQ-MS (triple quadrupole Gas Chromatography-Mass Spectrometry system, Agilent Technologies, USA) with DB-5MS (30 m x 0.25 mm x 0.25µm) capillary column, nitrogen as carrier gas, temperature programmed from 80 ^{0}C (2 min) then raised to 300 ^{0}C (12 ^{0}C min^{-1}). The mass spectrometer was operated in electron ionization (EI) mode (70 eV) and simultaneously scanned in both Full Scan and Selected Ion Monitoring (SIFI mode). We have analyzed variety of n-alkanes (n-C_{13-33}) at m/z 71.The n-alkanes were identified by comparison of mass spectra with NB75 mass spectra library and matching their retention times against corresponding individual standard solution (1-eicosane).

2.4. Quality control

All data were subject to quality and control procedures. Blank HEPT and matrix duplicate samples were analyzed as quality control procedures. One internal standard (n-nonadecane-d_{40}) was added to each sample and matrix blank prior to extraction as surrogates to assess the recovery. n-alkane surrogate recovery was 112%.Individual n-alkanes were quantified using deuterated internal standard (*n*-nonadecane d_{40} m/z 66) with 12 variation concentrations between 0.375-15,000 pg.µL^{-1} (data not shown).

3. Result and Discussion

The gravimetric data for each fraction are mainly composed of saturated hydrocarbons and the extractable organic matter (EOM) content varied from 980-2,380 sed. dry weight mg.kg^{-1}. The fraction masses were ranged from 680-1,440 sed.dry weight mg.kg^{-1}sed dry weight. Station E has a large amount of EOM and aliphatic hydrocarbon fractions because physically it is mud sediment and plant. Gravimetric data the extract of aliphatic hydrocarbons have showedin **Table 1**.

Table.1. Gravimetric data (mg.kg^{-1} sed.dry weight)

Ratio	St.A	St.B	St.C	St.D	St.E	St.F	St.G	St.H
EOM	1.060	1.870	1.630	1.150	2.380	980	1.683	2.240
F1	700	1.280	1.120	680	1.440	620	980	990

EOM = Extractible Organic Materials; F1 = aliphatic hydrocarbons fraction

The distribution spatial of total n-alkanes (n-C_{13}-n-C_{33}) in aliphatic hydrocarbon fraction ranged from 480 to 1934 sed.dry weight µg.kg^{-1}(**Figure 2**, and **Table 2**). The highest concentration was station H, which site near from urban area and from two main watershed inputs (**Figure 1**). Contamination levels found in sediments are lower than those measured in previous studies such as in Gulf of Fos (French Mediterranean Cost), ($\sum n$-C_{14-32}: 548-4794 µg.kg^{-1} sed.dw) [10] Arc River and BerreLagoon,Western Mediteranean Sea ($\sum n$-C_{14-34}: 563-5068 µg.kg^{-1} sed.dw)[4], Huveaune river, France ($\sum n$-C14-34 183.6-236780.4 µg.kg^{-1}sed.dw) [15] Cretan island, Greece, Eastern Mediterranean Sea ($\sum n$-C_{17-41}: 326-3758 µg.kg^{-1} sed.dw) [5]and coast of Aliaga, Turkey ($\sum n$-C_{10-34} 340-56480µg.kg^{-1}sed.dw) [16]. In America, fishing village-Z3. Patos lagoon, Pelotas region, Brazil ($\sum n$-C_{14-34} 337.6-5868.9 µg.kg^{-1} sed.dw) [17] In Asia, yellow sea (Huanghai) Northwestpacific ocean, China ($\sum n$-C_{14-}

$_{33}$:700-15800 μg.kg^{-1} sed.dw) [18] and Segara Anakan lagoon, Cilacap, Indonesia, total of *n*-alkanes ($\sum n$-C_{13-34} ranged from 2786 to 129027 μg.kg^{-1}sed.dw) [9].

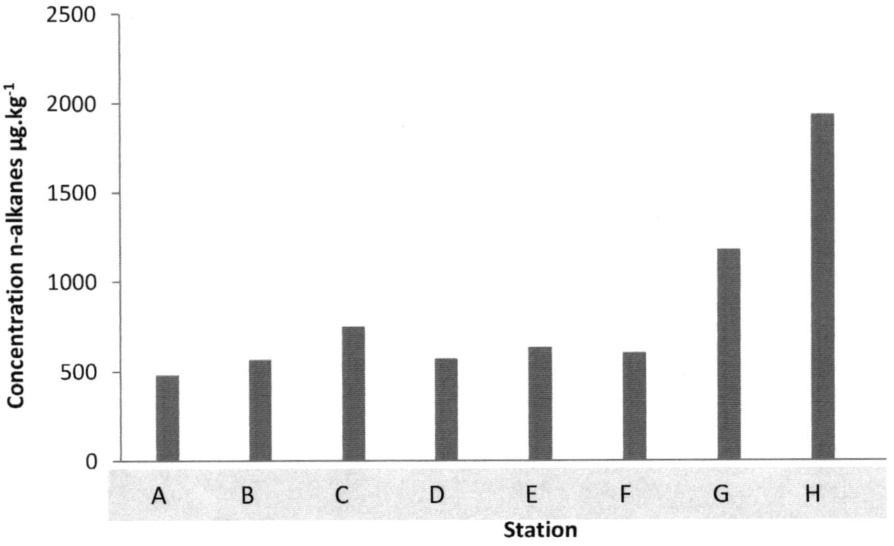

Figure 2.Total concentration of *n*-alkanes in surface sediment of Jakarta Bay, Indonesia

3.1 The possible source

To assess the possible source of saturated hydrocarbons or n-alkanes in the marine environment are complex and of diverse origins. Several criteria are used to determine the probability and to assess the possibility of their origins (anthropogenic or biogenic) in reserve sediment [15 and 17].We calculated 9 ratios: CPI (n-C_{24}-n-C_{33}), n-C_{17}/Pr , n-C_{18}/Phy, Pr/Phy,n-C_{29}/n-C_{17}, Terrigenous/ aquatic ratio (TAR)[3,5], Natural Alkane Ratio (NAR) [3,5,9,10,15 and17], $\sum n$-alkanes/n-C_{16}, TMD (Terrestrial Marine Discriminant) and Proxy ratio (Paq) [9,14]. Proxy ratio (Paq) indicate the presence of macrophyte inputs from terrestrial plants [9,13]. Values $<$ 0.1 proposed non-emergent plant input characteristic. The values are between 0.1 to 0.4 indicate emergent macrophytes. In the other hand, those from 0.4 to 1.0 indicate submerged/ floating macrophytes. The result showed that 5 stations (C,D,E,G,H) were characterized by emergent macrophytes input while 3 other stations (A,B and F) were submerged/ floating macrophyte input. This ratio is characteristics for mangroves vegetation and their association with coastal ecosystems [9].

The Carbon Preference Index (CPI) which indicate a predominance of odd numbered n-alkanes in the range n-C_{24} to n-C_{32} or range n-C_{25} to n-C_{33} are characteristics of terrestrial higher plant debris to the shelf sediments [3,5, 17]. This ratio also was used to assess the relative inputs from biogenic and petroleum-related sources [7]. CPI values close to one also indicate from marine microorganisms and/or recycle organic matter, another hand these values also indicate n-alkanes source from fossil fuels, including petroleum sources n-alkanes from biogenic sources (land plant material) have CPI close to 5-10 (station H). All station except station H were significantly higher than 1 (from 1.26 to 3.24) which were terrigenous inputs [11,12]. Our finding is in accordance with the previous study that showed Jakarta Bay was affected by the contribution of petrogenic sources and these petrogenic inputs might be related to the extensive use of fossil fuels to increase of industry and motor vehicle in Jakarta [1]. The presence of short odd alkanes particularly n-C_{17} is indicative of biogenic marine organisms (algae, plankton, marine animals, etc.) n-C_{17}/ Pr and n-C_{18}/ Phy ratios: these ratios are used to evaluate the presence of oil and the relative biodegradation of n-alkanes [9,10]. Pristane (Pr) and Phytane (Phy)

are indices of biogenic origin and consist of naturally hydrocarbons present in zooplankton, algae or bacteria. However, thepresence of oil can be either biogenic or diagenetic in origin [5,6]. n-C_{18}/ Phy ratio (>1) found in all station that confirm oil input and lower than one confirm such selective biodegradation phenomena [5] n-C_{29}/n-C_{17} ratio: n-C_{29} is abundant in land plants, and n-C_{17} is prominent in marine organisms, the ratio of these two alkanes reflects the relative contributions [4].
The NAR (Natural n-Alkane Ratio) represents the proportions between natural and petroleum long chain n-alkanes [3,4,9,10]. This ratio is close to zero for petroleum hydrocarbons and crude oils terrestrial plants or marine plants source for the value close to one [2,4]. For all stations, this ratio ranges between 1.45 and 1.93 indicate from terrestrial inputs.The TAR (Terrigenous Aquatic Ratio) represented a crude indicator of the relative strength of terrigenous versus aquatic sources [3,5,9] this ratio between the concentration of long-chain n-alkanes (n-C_{27}+n-C_{29}+n-C_{31}) to short-chain n-alkanes (n-C_{15}+n-C_{17}+n-C_{19}). This ratio was found below 1 for most of thestations except for stations G (1.66) and H (6.96) which can be assessed as predominant terrestrial input. (**Table 2**). The LMW/HMW (Low molecular weight (sum of n-C_{14} to n-C_{20}) / High molecular weight (sum of n-C_{21} to n-C_{34}) ratios is indicated n-alkanes from petrogenic and plankton sources, if this value close to 1 and < 1 usually represent biogenic source which produced by plants, marine animals and sedimentary bacteria [4, 14]. This ratio also indicates the presence of fresh oil in sediment for the value >2 [13]. For Jakarta bay, the station A,B,C,D,E,F are higher than 1 confirm essentially from amix between natural and petrogenic source. Then for station G and H confirm the biogenic source.∑n-alkanes/n-C_{16} ratio indicated a biogenic input for value >50 and whereas values < 15 indicate petroleum-contaminated samples [9]. Two stations (Station A and H) characterized mix between biogenic and petroleum input.TMD (Terrestrial Marine Discriminant) is indicated from terrestrial and marine sources with a value 1. A value > 1 a dominant terrestrial input (Stations G and H), whereas a value <0.5 confirm dominant marine contribution (Stations A,B,E and F). A value between 0.5 and 1 indicate mixed terrestrial and marine input (Stations C and D). Moreover, TMD confirmed the n-alkane from vascular plants and phytoplankton, same function like TAR index [9, 14].

Table 2. Concentration ($\mu g.kg^{-1}$ sed.dry weight) of n-alkanes, pristane and phytane in sediment samples from Jakarta Bay (Stations A to H)

Composes	St.A	St.B	St.C	St.D	St.E	St.F	St.G	St.H
∑n-alkanes (n-C_{13-33})	479.97	564.75	752.30	572.35	635.39	602.75	1178.86	1934.52
n-C_{17}/Pr	0.78	1.09	0.72	0.97	1.10	1.04	1.06	1.08
n-C_{18}/Phy	1.09	1.29	1.16	1.50	1.24	1.95	1.24	1.30
CPI_{24-32}	1.78	1.26	2.08	1.87	1.88	1.96	3.24	5.01
n-C_{29}/n-C_{17}	2.74	5.71	1.81	1.93	3.89	2.56	0.82	0.21
Pr/Phy	0.91	0.82	0.81	0.93	0.96	0.83	1.09	0.93
NAR	1.83	1.87	1.78	1.93	1.73	1.82	1.61	1.45
TAR	0.50	0.27	0.77	0.95	0.39	0.47	1.66	6.96
Pr/Ph	0.91	0.82	0.81	0.93	0.96	0.83	1.09	0.93
TMD	0.46	0.30	0.76	0.94	0.42	0.46	1.57	5.78
Paq	0.44	0.51	0.35	0.30	0.29	0.44	0.21	0.12
∑n-alkanes/n-C_{16}	22.20	16.55	14.11	10.30	15.41	11.25	28.61	52.87
LMW/HMW	1.45	1.98	1.18	1.29	1.84	1.71	0.59	0.20

Pr=Pristane, Phy=Phytane, CPI (Carbon Preference Index) as 2*(n-C_{25}+n-C_{27}+n-C_{29})/(n-C_{24}+2*(n-C_{26}+n-C_{28}+n-C_{30}), NAR (Natural n-Alkanes Ratio) calculated as (∑(n-C_{19-33})-(2*∑ (n-C_{20-32})/∑ (n-C_{19-33})), TAR (The terrigenous/aquatic ratio) calculated as (n-C_{27}+n-C_{29}+n-C_{31})/(n-C_{15}+n-C_{17}+n-C_{19}), LMW/HMW (∑(n-C_{14} to n-C_{20})/ ∑ (n-C_{21} to n-C_{34})), TMD (Terrigenous Marine Discriminant) : (n-C_{25}+ n-C_{27}+n-C_{29}+n-C_{31}+n-C_{33}/ (n-C_{15}+n-C_{17}+n-C_{19}+n-C_{21}+n-C_{23}), Paq (Proxy ratio) : (n-C_{23}+n-C_{25})/(n-C_{23}+n-C_{25}+n-C_{29}+n-C_{31}).

GC chromatograms showed the series of resolved compounds as a sign of petroleum input and/or the presence of biodegraded complex mixture of hydrocarbons, mainly *n*-alkanes and some branched alkanes called Unresolved Complex Mixture (UCM). It is amount coelution of structurally complex isomers and homologues of branched and cyclic hydrocarbons. Figure 2 was showed the concentration of *n*-alkanes in surface sediment from Jakarta Bay. The UCM ranging from *n*-C_{13} to *n*-C_{34} were observed in bimodal patternin the most station except for station H. Such bimodal UCMs indicated the presence coexistence between a recent and an old pollution.

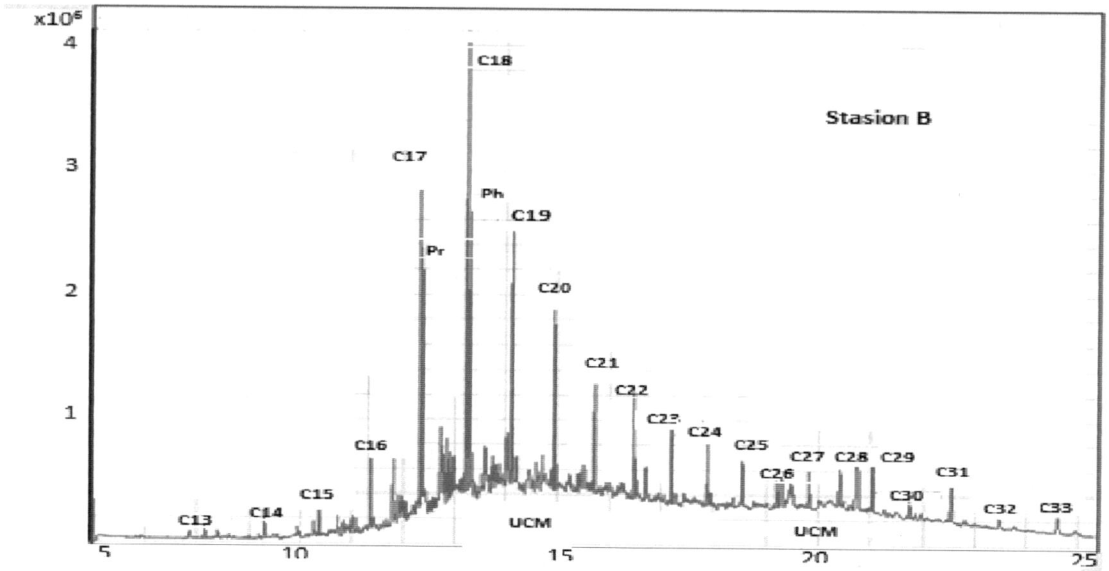

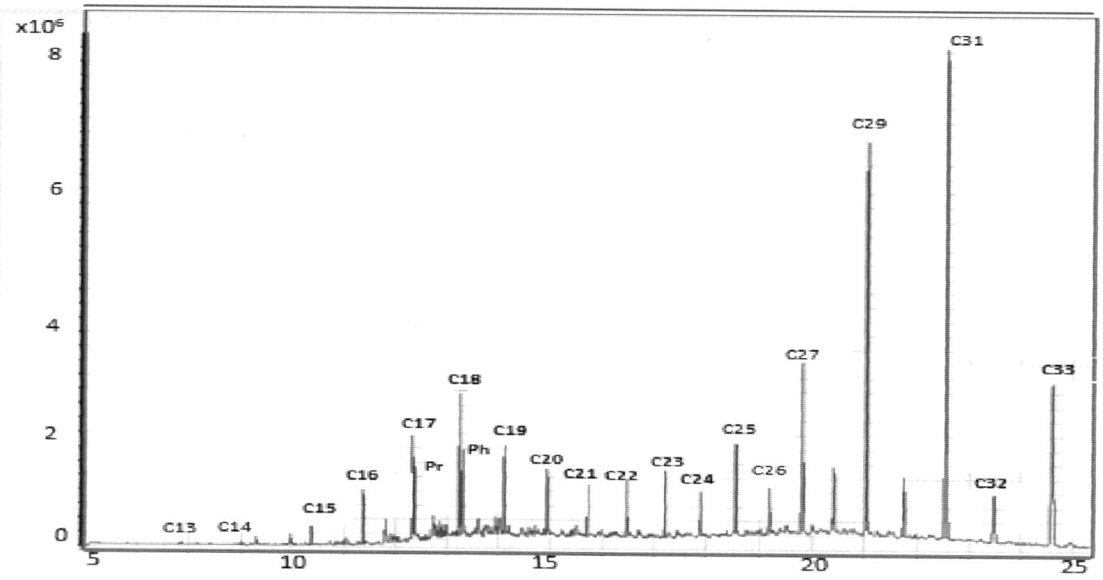

Figure 2. Capillary column gas chromatograms of n-alkanes fractions in station A (above) and Station H (under), Jakarta Bay, Indonesia

Capillary column gas chromatograms profiles (**Figure 2**) showed Station A indicate the profile n-alkanes from petroleum input with dominant compounds found in the whole sediments are n-C_{17}, n-C_{18}, Pr, Phy, n-C_{19} and n-C_{20}. On the other hand, station H has predominant intensity for n-C_{27}, n-C_{29},n-C_{31} and n-C_{33} which can be assessedas terrestrial input. Interestingly, we clearly demonstrated the important extent of n-C_{33} that refers mangroves leaf fingerprint [9].

4. Conclusion

The total n-alkanes (n-C_{13}-nC_{33}) concentrations ranged from 480 to 1,934 μg.kg^{-1}sediment dry weight (dw).The ratios of specific n-alkanes(n-C_{17}/Pr, n-C_{18}/Phy,n-C_{17}/n-C_{29}, Pr/Phy, TAR, ∑n-alcanes/n-C_{16}, NAR, LMW/HMW, CPI_{27-34}, Paq and TMD index) showed that the hydrocarbons source in the sediments were generally petrogenic and terrigenous input(in thecase of G and H). These data can use baseline reference and evaluating recent environmental condition in this site for future efficient environmental management.

References

[1] Rinawati, Koike T, Koike H, Kurumisawa R, Ito M, Sakurai S, Togo A, Saha M, Arifin Z and Takada H 2012 Distribution, source identification and historical trends of organic micropollutants in coastal sediment in Jakarta Bay, Indonesia *J. Hazard.Matter.***217-218** 208-216

[2] Rees JG, Setia permana D, Sharp VA, Weeks JM and Williams TM 1999 Evaluation of the impacts of land-based contaminants on the benthic faunas of Jakarta Bay, Indonesia *Oceano. Act.***22** 627-640

[3] Mille G, Asia L, Guiliano M, Malleret L and Doumenq P 2007 Hydrocarbons in coastal sediments from the Mediterranean sea (Gulf of Fos area, France) *Mar.Pollut.Bull.* **54** 566-575

[4] Kanzari F, SyaktiA D, Asia L, Malleret L, Mille G, Jamoussi B, Abderrabba M and Doumenq P 2012 Aliphatic hydrocarbons, polycyclic aromatic hydrocarbons, polychlorinated biphenyls, organochlorine and organophosphorus pesticides in surface sediments from Arc river and the Berre lagoon, France. *Environ.Sci.Pollut.Res.***19** 559-576

[5] Mandalakis M, Polymenakou P N, TselepidesA and Lampadariou N 2014 Distribution of aliphatic hydrocarbons, polycyclic aromatic hydrocarbons and organochlorinated pollutants in deep-sea sediments of the southern Cretan margin, eastern Meditteranean Sea; a baseline assessment *Chromosphere* **106** 28-35

[6] Zrafi I, Hizem L, Chalgmi H, Ghrabi A, Rouabhia M and Saidane-Moushabi D 2013 Aliphatic and aromatic biomarkers for petroleum hydrocarbon investigation in marine sediment. *J. Petroleum. Sci.Res.* **2** 145-155

[7] Jeng W L 2006 Higher plant n-alkane average chain length as an indicator of petrogenic hydrocarbon contamination in marine sediments. *Mar.Chem.***102** 242-251

[8] Williams T M, Rees J G and Setiapermana D 2000 Metals and trace organic compounds in sediments and waters of Jakarta Bay and the Pulau Seribu Complex, Indonesia *Mar. Pollut. Bull.* **40** 277-285

[9] Syakti A D, Hidayati N V, Hilmi E, Piram A and Doumenq P 2013 Source apportionment of sedimentary hydrocarbons in the Segara Anakan Nature Reserve, Indonesia *Mar.Pollut.Bull* **74** 141-148

[10] Asia L, Mazouz S, Guilliano M, Doumenq P and Mille G 2009 Occurence and distribution of hydrocarbons in surface sediments from Marseille Bay (France) *Mar.Pollut.Bull.*, **58**, 443-451

[11] UNEP/IOC/IAEA. 1992. Determination of petroleum hydrocarbons in sediments. Reference Methods for Marine Pollution Studies No 20 UNEP Nairobi pp 75

[12] Commendatore M G, Nievas M L, Amin O and Esteves J L 2012 Source and distribution of aliphatic and polyaromatic hydrocarbons in coastal sediments from Ushuaia Bay (Tierra del Fuego, Patagonia, Argentina) *Mar. Environ. Res.***74** 20-31
[13] Oyo-ita O E, Ekpo B O, Oros D R and Simoneit B R T 2010 Distributions and sources of aliphatic hydrocarbons and ketones in surface sediments from the Cross River estuary.S.E. Niger Delta Nigeria *J.Appl.Sci.Environ.Sanit.* **5** 1-11
[14] Meyers P A 2003 Application of organic geochemistry to paleolimnological reconstruction: a summary of examples from the Laurentian Great Lake *Org. Geochem.* **34** 261-289
[15] Kanzari F, Syakti A D, Asia L, Malleret L, Piram A, Mille G and Doumenq P 2014 Distribution and sources of persistent organic pollutants (aliphatic hydrocarbons, PAHs, PCBs and pesticides) in surface sediments of an industrialized urban river (Huveaune), France *Sci.Tot.Environ.* **478** 141-151
[16] Neser G, Kontas A, Ünsalan D, Altay O, Darilmaz E, Uluturhan E, Küçüksezgin F, Tekogul N and Yercan F 2012 Polycyclic aromatic and aliphatic hydrocarbons pollution at the coast of Aliaga (Turkey) ship recycling zone *Mar.Pollut.Bull* .**64** 1055-1059
[17] Filho S P J, Luz L P, Betemps G R, Silva M D G R and Caramao E B 2013 Studies of n-alkanes in the sediments of colony Z3 (Pelotas-RS-Brazil) *Braz. J.Aquat. Sci.Technol.* **17**(1) 27-33
[18] Wu Y, Zhang Y, Mie T-Z and Li B 2001 Occurence of n-alkanes and polycylic aromatic hydrocarbons in the core sediments of the yellow sea *Mar.Chem.* **76**1-15

IOP Conf. Series: Materials Science and Engineering **107** (2016) 012008 doi:10.1088/1757-899X/107/1/012008

Occurrence of aliphatic and polyaromatic hydrocarbons (PAHs) in Mytillus galloprovincialis from the traditional market in Marseille, France, by Gas Chromatography triplequadropole tandem Mass Spectrometry (GC-QQQ/MS)

M Y Azis[1,2,3*], Yelmiza[1,2], L Asia[2], A Piram[2], B. Bucharil, P Doumenq[2] and A D Syakti[3]

[1]Analytical Chemistry Laboratory, Department of Chemistry, Institute of Technology Bandung, Jl. Ganesha No. 10, Bandung 40132, Indonesia
[2]Environmental Analytical Chemistry Laboratory (LCE), FRE-CNRS 3416 Equipe MPO, Europôle de l'Arbois-Bátiment Villemin BP 80 Aix en Provence Cedex 4, France
[3]Center for Maritime Bioscience Studies-Institute for Research and Community Service, Jenderal Soedirman University, Jl. Dr. Suparno, Purwokerto 53123, Indonesia

[*]E-mail: yudis_chempd@yahoo.com

Abstract. Mediterranean mussel, *Mytillusgalloprovincialis* collected from the traditional market in Marseille, France,have been analysed using GC-QQQ/MS for their hydrocarbons (n-alkanes and polyaromatic hydrocarbons (PAHs)) extentwith two different solvent extraction, such as heptane:dichloromethane (HEP:DCM;1:1) and heptane:acetone (HEP:ACE; 1:1). The results showed hydrocarbons yielded from heptane:acetone extractionwere 28335μg.kg^{-1}mussels dw (Σ n-alkanes C_{15-34}) and 202 μg.kg^{-1}mussels dw(ΣPAHs) while the yield from heptane:DCM extract was lower ca. 27026 μg.kg^{-1}musselsdw and 133 μg.kg^{-1} mussels dw respectively from the Σn-alkanes$C_{15-34}$and ΣPAHs.High hydrocarbon levels can be affected by the presence of lipids or other metabolites in mussels that have the same polarity with hydrocarbon compounds which has interferred the measurement. Several ratio parameter of n-alcanes and PAHs source in the mussels were evaluated to asses the origins of their hydrocarbons in mussels from which we suggested origins of hydrocarbons were pyrolytic and biogenic rather than petrogenic.

1. Introduction

Petroleum hydrocarbons compounds *i.e.* aliphatic hydrocarbons and polyaromatic PAHs that known as toxic pollutants to the environment *e.g.*rivers, lagoons, and marine, and some of them have mutagenic and carcinogenic properties [1,2,6,7]. Those compounds may be derived from petroleum or oil spills rafinery product (petrogenic sources), incomplete combustion of organic matter, and biomass fossile fuels (pyrolytic source) [1,3,6,7,10]. For that reason, petroleum hydrocarbons analysis within the organims are important to monitoring activity of pollutant in environment as one integrative approach [14].This study used Mediteranean mussel (*M.galloprovincialis*) to monitor marine ecosystems quality. Hydrocarbons can absorb into sediment via particulate or colloid and potentially ingested by marine benthic organisms in marine environment [2,9][2,4,5,8,9]. Bivalves such as Mussels had used as sentinel organisms of marine pollution and have the capability to reflect the environment quality which related to their habitat, food chains, and pollutant hydrophobicity properties that have the tendency to be absorbed into their living tissues [2,9]. *Mytillus* families widely used since 1990's and become one of the most successful model organisms for time-integrated responses to complex mixture of pollutants [14], also as bioindicator reference of hydrocarbons level [8,13]. This study focus on the hydrocarbons content in mussel from traditional market that have direct impact to human, because as we know that the traditional market sell unknown source of mussels which can be from pollutant site or propre site for human consumption.

The significance of this study related to the extraction, fractination, and analysis method using GC-QQQ/MS [1,2,6,7, 10]. Quadropole can be used in scanning or filtering mode. Triple quadropole system contains of three quadropoles. Quadropole 1 and 3 as mass filter while quadropole 2 as collision cell [15].Several studies used heptane: dichloromethane (1:1 v/v)[1,6,7,9,10] and heptane:acetone (1:1, v/v) to extract the hydrocarbons[4,13]. Thematrice of mussels tissues are more complex than sediment and can influence the interference effect to separate the hydrocarbons content from their matrices.This study aimed to analyze the content of hydrocarbons in *M.galloprovincialis* mussel species that derived from traditional markets in Marseille, France, by using protocol analysis of hydrocarbon sediments and to compare the different extraction solution to extract hydrocarbons that can be optimized using GC-QQQ/MS.

2. Material and Methods

2.1. Sampling
Approximately 3 kg wet mussels, *M. galloprovincialis* have taken from Noailles traditional market, Marseille, France. Fresh seafood market is located about 300 meters from Mediteranean port. After purchased, the wet mussel stored in *freezer* -20^{0}C.

2.2. Chemicals
n-heptane (HEP), acetone (ACE), and dichloromethane(DCM) for organic solvent (Suprasolve, GC analysis grade (Merck, Pessac, France)). Individual deuterated standards for PAHs ([$^2H_{10}$]Anthracene, m/z 188 [$^2H_{12}$] Chrysene, m/z 240 (CIL Cluzeau, Andover, MA, USA) and individual labeled standards of n-alkane ([$^2H_{40}$] n-nonadecane, m/z 66 (PESTANAL Sigma-Aldrich, St Quentin, Fallavier, France) and 1-eicosane (n-C_{20}) (Dr.Ehrenstrofer Laboratories, Augsburg, Germany). 16 PAHs US EPA (PAH mix 25) standar mix solutions(Dr. Ehrenstonfer Laboratories, Augsburg, Germany). Silice gel-60 and Alumine gel-60 (Merck, Darmstatd, Germany, size 200-300 mesh for each of them.

2.3. Sampel Preparation
The tissues taken and placed in buchner. Water content in wet tissues removed by gravitation for 30 minutes [3], then wetmass weighed with microbalance (Perkin Elmer AD2Z, Marseille France, weighed approximately ± 250 gr). The tissues were *freeze-dried* under pressure 0.080 bar and freeze at -50^{0}C for 48 h. Then it cutted and crashed by mortar to small size and the dry mass weighed and stored under temperature ambient before the extraction [8,13].

2.4. Extraction and Purification
Hydrocarbons extracted from mussel tissues by using protocole extraction of hydrocarbons for sediments [1,6,7,11,12]. 10 gram freeze-dried mussels tissues accurately weighed and placed into pre-cleaned cellulose extraction thimble by soxhterm apparatus for 3.5 h with 150mL HEP:DCM (1:1 v.v) solvent extraction. Beside that, same extraction method has been done, with another solvent extraction, 150mL HEPT:ACE (1:1. v:v). 0.5 mL each internal surrogate deuterated standard solutions of hydrocarbons (n-alkane (nonadecane-d_{40} (15 mg/L)) and PAH (anthracene-d_{10} (20mg/L); chrysene-d_{12} (20mg/L)) added into sample before the extraction. Then, the extract concentrated to 2 mL with

rotary evaporator (Heindolph, Laborata 4000 efficient, Krackeler Scientific, Pessac, France) and evaporated under a gentle nitrogen stream to obtain the total extractable organic matter (EOM). The extract (EOM) weighed on a microbalance (Perkin-Elmer AD2Z) and part of EOM dissolved in a minimum of HEP. The fractination of aliphatic and aromatic hydrocarbons using 1.0 x 3.0 cm borosilicate glass column with 8 gr of silica gel and 8 gr of alumina (bottom and top, respectively), both adsorbents activated with 5% H_2O. Saturated hydrocarbon fraction or n-alkanes fractions eluted with 30 mL HEP and 20 mL HEP :DCM (90:10, v:v). For PAHs fraction or fraction 2 eluted with 40mL HEP:DCM (80:20, v/v). Both of fraction evaporated on rotary evaporator at 0.5mL. Then, dry fraction (F1 and F2) weighed and obtained the total hydrocarbon content. Fraction 1 dilluted with 300μL HEP and for fraction 2 dilluted in 300μL DCM.

2.5. Analysis by capillary gas chromatography coupled to mass spectrometry(GC/MS)

Both of fractions (F1 and F2) analyzed by GC-QQQ/MS (Autosystem XL GC and TurboMass triple quadropole mass spectrometer, Perkin Elmer, USA). Chromatographic conditions follows: splitless injection (30 s), Elite 5MS capillary column (30 m x 0.25 mm i.d. x 0.25 μm). The GC oven temperature- programmed from 40 ^{0}C (isotherm 2 min) then raised to 120 ^{0}C (45 ^{0}C.min^{-1}) and then raised to 310 ^{0}C (5 ^{0}C.min^{-1}) and finally held isothermally for 20 min. Helium (1mL.min-^{1}) used as carrier gas in constant flow mode. For injector temperature 50 ^{0}C (isotherm 0.1 min) to 250 ^{0}C (20 ^{0}C/min) for reduce solvant peaks. The mass spectrometer operated in the electron ionization (EI) positive ion mode (70 eV) and simultaneously scanned in both Full Scan and Selected Ion Monitoring (SIFI mode). MS scanned with comparing with NB library reference to confirm and matching their retention times and fragmentation profiles againts corresponding standards. Detection of n-alcanes (m/z 71),), 8 PAH target ions (Acenaphtene, m/z 152; Acenaphtylene, m/z 154; Phenanthrene, m/z 178; Anthracene, m/z 178; Fluoranthene, m/z 202; Pyrene, m/z 202; Benzo(a)anthracene, m/z 228; Chrysene, m/z 228; Indeno(123cdi) pyrene, m/z 276; Benzo(ghi)perylene, m/z 276). internal labeled standard of PAHs (Anthracene-d_{10}, m/z 188 and Chrysene-d_{12}, m/z 240) and n-alkanes (nanodecane-d40, m/z 66) to evaluate hydrocarbons analysis.

2.6. Quality Control

Calibration Curves have been performance with external standards methods at different concentrations levels usingdilluted individual labeled hydrocarbons standards (*i.e.* 0.05; 0.2; 0.5; 1.5; 2; 3;3.5 mg/L). In duplicate samples have been calculated that the mean recovery are 97% (HEP:ACE ; 1:1, v/v) and 93% (HEP:DCM, 1:1,v/v). For PAHs are anthracene-d_{10} and chrysene-d_{12} respectively (82%; 89%(HEP:DCM ; 1:1, v/v) and 90%; 101% HEP:ACE ; 1:1,v/v).The coefficient of correlation was 0.99 forall labeled standard compounds.

3. Result and discussion

Extractable Organic Matter (EOM) concentrations, Total Hydrocarbons Content (THC), quantities of F1 and F2, comparing between THC and MOE, F1 and F2 from two different solvant extraction are given gravimetrically in **Table 1**. The solvant extraction(HEP:ACE ; 1:1 ; v/v) solvant has a lot EOM, THC and fractions content then HEP:DCM (1:1, v/v).

Total hydrocarbon content from both extract are < 1 mg.g^{-1} generally. The proportions of THC in EOM are 18 % for HEP:DCM (1:1),v/v) and 53% for (HEP:ACET (1:1), v/v). Fraction saturated (F1) has more quantity then F2. Comparing of F1/F2 with solvant extraction HEP :ACE (1:1), v/v) more abundance than HEPT :DCM (1:1), v/v).

Generally, n-alkanes are nonpolar compounds and semipolar for PAHs, that can extracted and separated using fractionation method with comparing different polarity of eluent. GC/MS analysis with Single Ion Monitoring (SIM)mode has found 20 peaks n-alkanes (n-C_{15-34}) and 8 peaks identic of PAHs (**Figure 1** and **Figure 2**).

Table 1. Gravimetric data (mg.g^{-1}tissuesdry weight)

Extraction	MOE (mg/g)	THC (mg/g)	F1 (mg/g)	F2 (mg/g)	THC/MOE %	F1/F2 (mg/g)	F1/THC %	F2/THC%
HEP:DCM	55.67	0.10	0.075	0.025	18	3	75	25
HEP:ACE	72.90	0.39	0.325	0.065	53	5	83	17

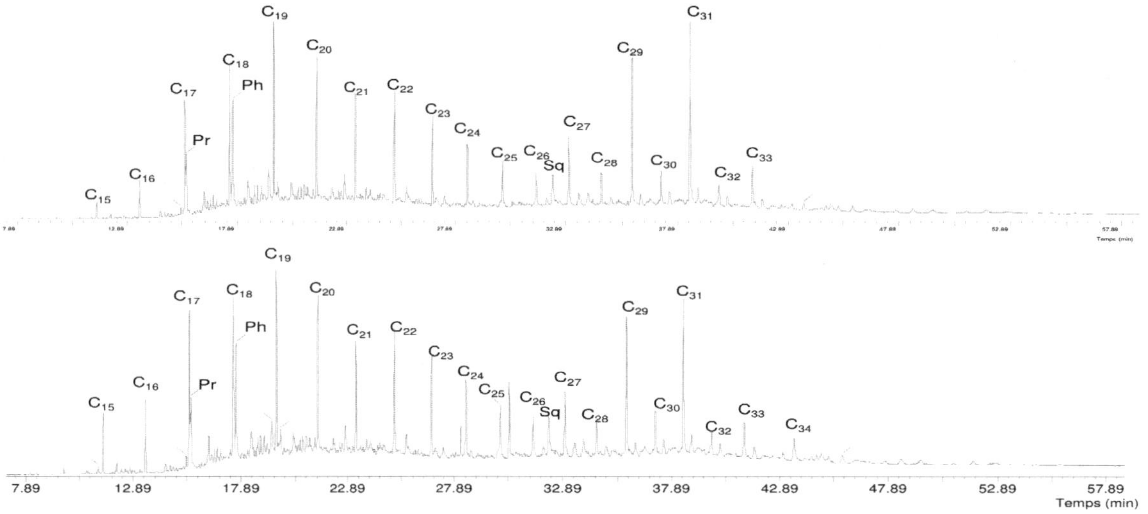

Figure 1. Capillary column gas chromatograms of the n-alkanes fraction with HEP:DCM (1:1, v/v) (under) and HEP:ACE (1:1,v/v) (above) as solvant extraction; Sq: Squalene; Pr: Pristane; Ph: Phytane.

According from chromatograms, both of chromatograms have similar profile. Besides *n*-alkanes derivatives, isomer found from n-alkanes such Pristane, Phytane which indicate the presence of crude oils that might be from natural (biogenic or diagenetic) and isomer from *n*-alkenes, squalene which derived from microbial degradation [1].Figure 1 showed that there are two high peaks (n-C_{29} and n-C_{31}) in the last chromatograms and n-C_{17}, n-C_{18}, n-C_{19} and n-C_{20} have high peaks in the beginning.

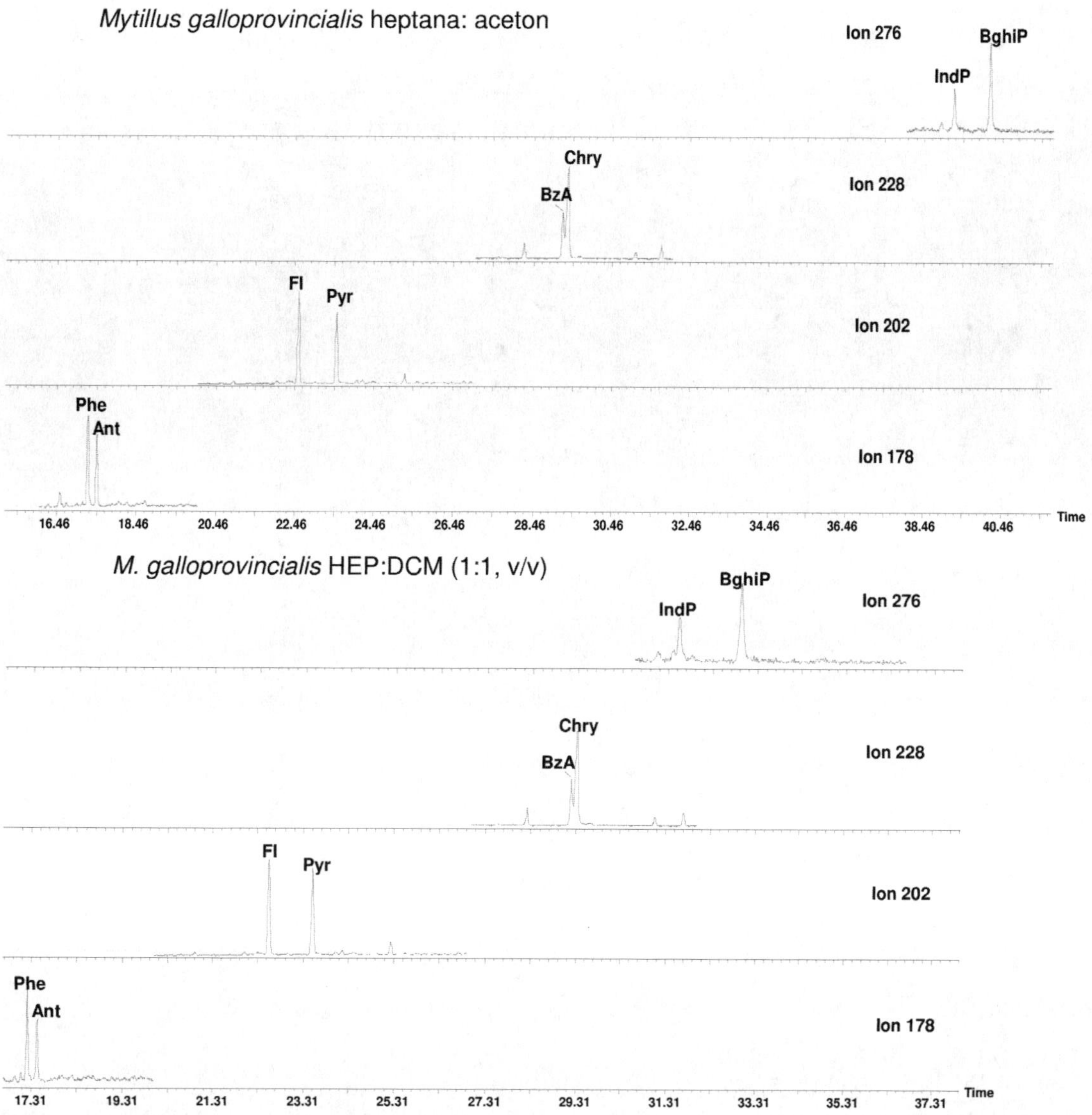

Figure 2. Capillary column gas chromatograms of the polycyclic aromatic hydrocarbon fraction with HEP:DCM (1:1, v/v) (above) and HEP:ACE (1:1, v/v) (under) as solvant extraction. Phe: Phenanthrene; An: Anthracene; Fl: Fluoranthene; Pyr: Pyrene; BzA: Benzo(a)Anthracene; Chry: Chrysene; IndP: Indeno (123 cdi) pyrene; BghiP: Benzo(ghi)Perylene.

Total of *n*-alkanes (n-C_{15-34}) and PAHs respectively were 27026 and 133μg.kg^{-1}tissues.dw for those extracted with HEP:DCM (1:1, v/v) and for HEP:ACE (1:1, v/v), $\sum n$-$C_{15-34}$and ∑PAHs respectively have value 28335 and 202μg.kg^{-1}tissues.dw (see **Table 2**). The high level of *n*-alkanes content have the possibilities that originated from disrupting of lipid and metabolites, or these mussels have high contamination in mussels tissues from Mediteranean sea.

Table 2. Concentrations of *n*-alkanes, pristane and phytane and PAHs in *M.galloprovincialis* from traditional market, Noailles, Marseille, France (μg.kg^{-1}tissues. dw)

Compounds	(HEP:ACE)	(HEP:DCM)
n-C_{15}	442	654
n-C_{16}	663	712
n-C_{17}	1775	1737
Pristane	940	1082
n-C_{18}	1794	1682
Phytane	2023	1541
n-C_{19}	2288	2244
n-C_{20}	2039	1885
n-C_{21}	1541	1412
n-C_{22}	1725	1568
n-C_{23}	1424	1289
n-C_{24}	1075	953
n-C_{25}	971	634
n-C_{26}	574	375
n-C_{27}	1103	1126
n-C_{28}	535	525
n-C_{29}	2009	2400
n-C_{30}	589	521
n-C_{31}	3332	3352
n-C_{32}	449	398
n-C_{33}	699	714
n-C_{34}	346	222
∑*n*-C_{15-34}	28335	27026
Phenanthrene	24	13
Anthracene	9	8
Fluoranthene	93	52
Pyrene	62	45
Benz(a)anthracene	3	4
Chrysene	8	9
Indeno(1233cdi)pyr	1	1
Benz(ghi)perylene	2	1
∑PAH	202	133

The mixed solvents HEP:ACE (1:1, v/v) known to be more polar than HEP:DCM (1:1, v/v).From this result, hydrocarbons content from *M.galloprovincialis* were determined. The properties of mussels tissues are presumably more complex because of their macromolecules content such as lipid, protein, and other metabolite [4]. Such compounds are capable to interference while separating process and these metabolites macromolecules can affect the extraction yield since they have the same polarity with hydrocarbons compounds. For that reason, hydrocarbons can associated with this macromolecul and have the same ion fragmentation with ion target [4]. The method of fractionation need another method to separate these macromolecules before the hydrocarbons fractination process. One of method for macromolecules separation is permeable gel chromatography, which is the macromolecule can be removed from mussel tissues before the fractionation process. Our finding showed aliphatic and PAHs extraction yield within the same greater magnitude that compared to previous studies using sediment as environmental matrices. Partially concluded thatthe mussels from traditional market were potentially polluted by hydrocarbons [1,6,7 and 10]. Obviously, further research is needed to re-validate and to find out the effective method to separate another molecules prior for the environmental analysis. In addition, high concentration of the hydrocarbons in *M. Galloprovincialis* has been observed, compare to another previous studies concerning PAHsin *Mytillus edulis* (27,6-442 μg.kg^{-1}mussels dw) [12]; *M.galloprovincialis* from coastal of Saronikos, Gulf, Greece, (17PAHs: 219-1487 μg.kg^{-1} mussels tissues dw) [13] and for *n*-alkanes in M.*galloprovincialis* from Galicia coast (n-$C_{8\text{-}35}$: 89.46-5098.01 μg.kg^{-1}mussels tissues dw) [16].

7 ratios for n-alkanes used to assess the potential sources of hydrocarbons contamination: CPI (*Carbon Preference Index*), NAR (*Natural alkane ratio*, nC18/phyt, nC17/phyt, prist/phyt,TAR (Terrigenous Aquatic Ratio), and LMW/HMW (Low Molecular Weight/High Molecular Weight), while PAHsindices are derived from comparison identical molecular weight such as (anthracene/∑m/z 178; Fluorene/∑m/z 202; Bz(a) anthracene/∑m/z 228; indeno (123 cd)pyrene/∑m/z 176). Generally, there are two main sources of hydrocarbons, origin from anthropogenic activities and from biogenic process (natural process) or biological activites. Concern tobiogenic source, it may be derived from plant or animal residualor both of them.

Table 3. Characteristic of *n*-alkanes from different sources [1,6 and 7]

Parameters	Origin		
	Oil	Aquatic environment	Terrestrial environment
CPI $_{(24\text{-}32)}$	≈ 1		>1
n-C_{29}/n-C_{17}	<1	<1	>1
n-C_{17}/prist	>1	<1	>1
n-C_{18}/Phyt	>1	≈ 1	≈ 1
Prist/Phyt	<2	>2	>2
NAR	≈ 0		>0.5-1
TAR		<1	>1
LMW/HMW	>2	≈ 1	<1

CPI: Carbon Preference index; TAR: Terrigenous/Aquatic Ratio; NAR: Natural n-alkane ratio; LMW/HMW (Low Molecular Weight/High Molecular Weight); Pr: Pristane; Phy: Phytane

Parameters index and their values of hydrocarbons are given at **Table 3** for n-alkanes and **Table 4** for PAHs.

Table 4. Characteristics of PAH source from different source (μg.kg^{-1} mussels tissues dw)

	Origin from oil	Origin from mix between oil and combustion product	Origin from combustion of fossil and plant
An/∑178	< 0.1		> 0.1
Fl/∑202	< 0.4	0.4 – 0.5 Mixed source from combustion of fossil	> 0.5 Cerosen combustion, charbone or biomass (herbal, wood…)
BzA/∑228	< 0.2	Mixed	> 0.35
IndP/∑276	< 0.2	0.2 – 0.5 Combustion result or flammable fossil liquid	> 0.5 combustion of charbone or from biomass (wood, herbal)

Table 5. Selected *n*-alkanes ratios in *M galloprovincialis* (μg.kg^{-1} mussels tissues dw)

Mussels	CPI	n-C_{17}/Pr	n-C_{18}/Ph	Pr/Ph	n-C_{29}/n-C_{17}	NAR	TAR	LMW/HMW
M.galloprovinsialis (HEP:DCM)	3.0	1.6	1.1	0.7	1.4	0.4	1.5	0.7
M.galloprovincialis (HEP:ACE)	2.6	1.9	0.9	0.5	1.1	0.3	1.4	0.8

CPI: Carbon Preference Index = $2*(n\text{-}C_{25}+n\text{-}C_{27}+n\text{-}C_{29}+n\text{-}C_{31})/(n\text{-}C_{24}+2*(n\text{-}C_{26}+n\text{-}C_{28}+n\text{-}C_{30})+n\text{-}C_{32})$; NAR: Natural n-alkane ratio = $((\sum n\text{-}C_{19\text{-}32})\text{-}\ 2\sum \text{even}\ n\text{-}C_{20\text{-}32})/(\sum n\text{-}C_{19\text{-}32})$; TAR: Terrigenous/Aquatic Ratios = $(n\text{-}C_{27}+n\text{-}C_{29}+n\text{-}C_{31})/(n\text{-}C_{15}+n\text{-}C_{17}+n\text{-}C_{19})$; Pris: Pristane; Phy: Phytane; LMW/HMW (Low Molecular Weight/High Molecular Weight) = $(n\text{-}C_{15\text{-}21})/(n\text{-}C_{22\text{-}34})$ [1,6,7,10, 16]

The results showed an apportionment of the n-alkanes level from biogenic sources, probable from aquatic organism or terrestrial environment with index (CPI, n-C17/Pr, n-C18/Phy, and TAR) >1, for NAR is close to zero < 1, indicates for petroleum hydrocarbons, LMW/HMW< 1 usually represent n-alkanes that are produced by higher plants, marine animals and sedimentary bacteria.there are found parameters ratio from oil which origin from n-alkanes isomer (Pristane/Phytane) < 2. The source n-alkanes from this experiment are shown in Table 5.

Source of PAHs Fluoranthene/ (Fluoranthene+pyrene) (Fl/∑202) is often used to discriminate PAH sources (1, 10). Ratio<0.5 is generally characteristic as petrogenic sources and ratio > 0.5 founded in kerosene, grass, coal, and wood combustion samples. The ratio of anthraceene/(Anthracene+Phenanthrene) ratio (An/∑178) < 0.1 indicates a petrogenic source while a ratio > 0.1 reflects a combustion source [1,2,6,10]. Benz (a) anthracene/chrysene (Benzo(a)Anthracene/Benzo(a) anthracene+Chrysene) ratio (BzA/∑228) < 0.2 indicates petroleum sources whileratio > 0.35 indicates combustion sources and values between 0.2-0.35 cannot used to discriminate petroleum and/or combustions sources [1,6].

The index indicates the source of PAHs in mussels from traditional market derived as natural or biogenic sources which origin from the burning of fossil or living creatures biosynthesis results with given indices < 0.5 (anthracene / Σm/z 178, benzo (a) anthracene / Σm/z 228; indeno (123cdi) pyrene / Σ m/z 176 and it also has the probability that those indices came from oil with indices > 0.5 (Fluorene / Σ m/z 202) parameters source of PAH resulting from the two solutions of different extracts that shown in Table 6.

Table 6. Characteristic of PAH sources fromM. galloprovincialis (μg.kg^{-1} mussels tissues dw)

	An/∑178[a]	Fl/∑202[b]	BzA/∑228[c]	IndP/∑276[d]
***M.galloprovincialis* (HEP:ACE)**	0.27	0.53	0.32	0.29
***M.galloprovincialis* (HEP:DCM)**	0.38	0.54	0.31	0.31

[a] An/∑178 (Anthracene/ (Anthracene+ Phenanthrene)
[b] Fl/∑202 (Fluoranthene/(Fluoranthene+Pyrene)
[c] BzA/∑228 (Benzo(a)anthracene/Bz(a)anthracene+chrysene)
[d] IndP/∑276(Indeno(123cdi)pyrene/Indeno(123cdi)pyrene+Benzo(ghi)perylene)

The profile of capillary column gas chromatograms of n-alkanes fraction in mussels extract between HEP:ACE (1:1, v/v) and HEP: DCM (1:1, v/v) as solvent extraction have same profile. $n\text{-}C_{15\text{-}34}$ were found with different yield.

4. Conclusion

The extract solution of HEP:ACE (Σn-C15-34: 28 335 μg.kg^{-1}tissues dw and Σ PAHs: 202 μg.kg^{-1} tissues dw) better than HEP:DCM (Σn-C15-34: 27026 μg.kg^{-1}tissues dw and Σ PAHs: 133 μg.kg^{-}

[1]tissues dw). High hydrocarbons extract yield might be come from two possibilities (i).direct disruption of lipids and metabolites during analysis; (ii). the apportionment as source of n-alkanes and PAHs from the use of diagnostic molecular indices that indicated pyrolytic and biogenic, also petrogenic sources.

References

[1] Asia L, Mazouz S, Guiliano M, Doumenq P and Mille G 2009 Occurence and distribution of hydrocarbons in surface sediments from Marseille Bay (France) *Mar. Pollut.Bull.* **58** 443-451

[2] Baumard P, Budzinski H, Michon Q,Garrigues P, Narbonne J F, Burgeot T, Michel X and Bellocq T 1999 Polycyclic aromatic hydrocarbon (PAH) burden of mussel (Mytillus sp) in different marine environments in relation with sediment PAH contamination and bioavailability *Mar.Environ.Res.* **47** 415-439

[3] UNEP/RAMOGE 1999 Manual on the Biomarkers Recommended for the MEDPOL Biomonitoring Programme *UNEP* Athens

[4] Jaouen-Madoulet A, Abarnou A, Le-Guellec A M Loizeau V and Lebouleger F 2000 Validation of an analytical procedure for polychlorinated biphenyls, coplanar polychlorinated biphenyls and polycyclic aromatic hydrocarbons in environmental samples *J.Chrom A.* **886** 153-173

[5] Lima I, Moreira S M, Osten J R V, Soares A M V M and Guilhermino L 2007 Biochemical responses of the marine mussel *Mytillus galloprovincialis* to petrochemical environmental contamination along the North-western coast of Portugal *Chemosphere* **66** 1230-1242

[6] Mille G, Asia L, Guiliano M, Malleret L and Doumenq P 2007 Hydrocarbons in coastal sediments from the Mediterranean sea (Gulf of Fos area, France) *Mar.Pollut.Bull.* **54** 566-575

[7] Mille G, Guiliano M, Asia L, Malleret L and Jalaluddin N 2006 Sources of hydocarbons in sediments of the Bay of Fort de France (Martinique) *Chemomsphere* **64** 1062-1073

[8] Nesto N, Romanon S, Moschino M, Mauri M and Da Ros L 2007 Bioaccumulation and biomarker responses of trace metals and micro-organic pollutants in mussels and fish from the Lagoonof Venice, Italy*Mar.Pollut.Bull* **55** 469-489

[9] Wetzel D W and Van Vleet V S 2004 Accumulation and distribution of petroleum hydrocarbons found in mussels (*Mytilusgalloprovincialis*) in the canals of Vence Italy *Mar.Pollut. Bull* **48** 927-936

[10] Syakti A D, Hidayati N V, HilmiE, Piram A and Doumenq P 2013 Source apportionment of sedimentary hydrocarbons in the Segara Anakan Nature Reserve Indonesia*Mar.Pollut. Bull* **74** 141-148

[11] Syakti A D, Hidayati N V and Siregar A S 2012 Agen Pencemar Laut *IPB press Bogor* **1** 100-101

[12] Wise S A 2012 Certificat of Analysis Standard Reference Material 1974 C Organics in Mussel Tissue (*Mytilus edulis*) *NIST* MD 20899 USA 1-10

[13] Valvanidis A, Vlachogiann T, Triantafillaki S, Dassenaki M, Androutsos F and Scoullos M 2008Polycyclic aromatic hydrocarbons in surface seawater and in indegenous mussels (*Mytillus galloprovincialis*) from coastal areas of the Saronikos Gulf (Greece) *Estuar.Coast. Shelf Sci.* **79** 733-739

[14] Sundt R C, Pampanin D M, Grung M, Barsiene J and Ruus A 2011 PAH body burden and biomarker responses in mussels (*Mytillus edulis*) exposed to produced water from a North Sea oil field: Laboratory an field assessment *Mar.Pollut.Bull.* **62** 1498-1505

[15] Schreiber A 2010 Advantages of using triple quadropole over single quadropole mass spectrometry to quantify and identify the presence of pesticides in water and soil samples Food and Chemisrtry AB *Sciex Concord Ontarion* **1** 1-6

Synthesis of thiazole silica hybrid from waste glass for adsorption of cadmium(II)

C. Azmiyawati,* Taslimah, and Virkyanov

Chemistry Department, Diponegoro University, Semarang, Indonesia

*E-mail: choiril.azmi@gmail.com or choiril_azmiyawati@undip.ac.id

Abstract. Synthesis of thiazole silica hybrid from waste glass to adsorption of cadmium (II) metal ion has been performed. The synthesis was done by attaching thiazole group through liaison compound γ-glycidoxy propyl tri-methoxy silane with silica gel obtained from waste glass. In this study, the effect of adsorption contact time and the concentration of cadmium (II) was studied to determine the reaction rate and the amount of adsorption thermodynamics. The existence of the cluster thiazole on silica gel indicated by IR spectra at wavelengths around 2576 cm^{-1} of mercaptan groups that previously did not appear on silica gel without modification. The synthesized TSH showed a high adsorption capacity of 9.363 mmol/g of Cd(II). The adsorption isotherm obtained with Langmuir isotherm model gives the negative values of ΔG^o, i.e. -15.488 kJ/mol for Cd(II), indicating the spontaneous process of adsorption. Kinetic studies showed that the adsorption of Cd(II) ion into TSH follows the pseudo-second-order kinetics.

1. Introduction

Glass has the main composition in the form of silica. Extraction of silica from the waste glass is one way to increase the economic value of the waste glass [1]. The extraction results can be used as a source of silica in the silica-based synthetic such as thiazole silica hybrid adsorbent. The investigation of adsorbent has been focused on synthesis and application. Thiazole silica hybrid can be formed by grafting and sol-gel methods [2-3] Researches about sol-gel method besides the grafting one have been reported, i.e. preparing sulfonated-silica (osmosis) hydrogels [4], polyacrylate silica nanocomposite [5], and the hexagonal mesoporous silica modified with 2-mercaptothiazoline [6]. The sol-gel method has special quality because the formation of bonding at sol process. Therefore, in this study, the formation of thiazole silica was done using the sol-gel method.Thiazole-silica hybrid (TSH) is a product of silica and organic compounds containing thiazole groups. One advantage of the synthesis of silica materials via sol-gel process is the possibility to synthesize organic-silica hybrid materials in mild process condition and a low temperature. The paper reports the synthesis of TSH through the sol-gel process using silica precursor of sodium silicate solution extracted from waste glass, a γ-glisidoxypropyltrimethoxysilane bridge and mercaptobenzothiazole (MBT) as the active group source. The composition of TSH is elucidated based on FTIR spectra. A batch system was used in this experiment to evaluate parameters of the adsorption of Cadmium(II).

2. Experimental

2.1. Chemicals

The thiazole-silica hybrid was synthesized using *Mercaptobenzothiazole (MBT)*, sodium hydroxide (NaOH), Potasium hydroxide, γ-glysidoxyprophyltrimethoxysilane (GPTMS), *hydrochloric acid* (HCl) 37%, *waste glass,* aquabidest. The metal cations were from cadmium chloride hydrate ($CdCl_2.H_2O$), *dimethylformamide* (DMF), asam sulfat p.a (H_2SO_4), etanol 96% (C_2H_5OH), sodium thiosulfat ($Na_2S_2O_3$). All other reagents unless silica, were of analytical reagent grade (Merck).

2.2. Procedure
The preparation of silica precursor was performed using a procedure reported by Kamath [7] with modification.

2.2.1. The Synthesis of thiazole silica hybrid. Synthesis TSH was performed by two-step processes, extraction of silica from WG and synthesis TSH. The sodium silicate were prepared by suspending 25 grams of fine glass powder in 125 mL of 5M sodium hydroxide (NaOH), stirred and heated on a sand bath with a Teflon container to boil for 4 hours. Subsequently, 150 ml of sodium silicate solution coupled with 6 M HCl to form silicate acid sol. Into the silicate sol is added to a solution of mercaptobenzothiazole (0.37 g MBT in 93 mL dimethylformamide (DMF) and 12.88 mL GPTMS) with stirring. The results obtained was aged for 24 hours, filtered, neutralized with distilled water, dried in an oven at a temperature of 100 °C, and was analyzed by FTIR spectrophotometer.

2.2.2. Adsorption Process. The metal salt of $CdCl_2.H_2O$ was used as the source of metal ions. Cd(II) solutions (1000 mg/L) was prepared by dissolving appropriate amounts of metal salt in a double distilled water. The working solution was prepared by diluting the stock solution to appropriate volumes. All the adsorption experiments were conducted batch wise [8]; the sorption equilibrium was obtained by shaking 0.1 g of TSH in 20 mL of aqueous solution containing cadmium ion for 2 hours. Adsorption experiments were carried out in an incubator at a temperature of 25°C. After 2 h was shaken, the mixture was aged for 24 h, the solution was filtered, and the cadmium ion concentrations in the aqueous solutions were measured. Initial and equilibrium cadmium ion concentrations in the aqueous solutions were determined by using flame atomic absorption spectrometer. Initial pH of the solutions was adjusted to the desired pH by adding nitric acid to the medium to maintain a constant pH. Experiments were repeated three times in each case. The amount of adsorbed cadmium ion was calculated from the difference between metal concentration in the aqueous solution before and after equilibrium and the certain weight of the dry TSH.

3. Result and Discussion
Silica hybrid material is a material modification of silica gel with organic compounds through chemical bonds. This material has advantages in applications, such as adsorbents, catalysts, and ion exchange. The ability of silica hybrid materials depends on the success of the synthesis of these materials. One indication was done by the characterization of the products [9]. The synthetic procedure was based on the production of the new sylilant agent named TSH (Figure 1) from the reaction between mercaptobenzothiazole (MBT) and γ-glysidoxipropyltrimethoxysilane (GPTMS) according to Eq. (1). This reaction is analog with the synthetic procedure of the reaction between 2-mercaptothiazoline (MTZ) and 3-chloropropyltrimethoxysilane (SiCl) to produce SiMTZ has reported by Evangelista et al. [6]. The final silica modified with MBT was produced by ordered polymerization of Si-thiazole according to equation (2a and 2b).

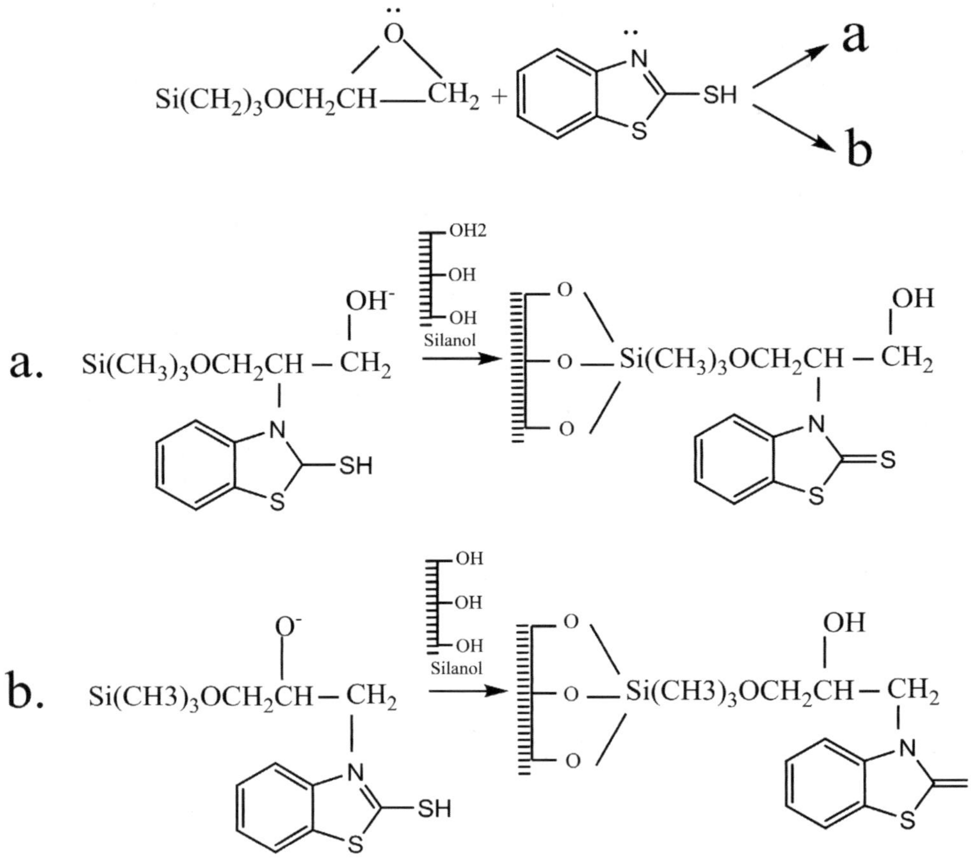

Figure 1. The reaction equation (1) between and γ-glysidoxipropyltrimethoxysilane (GPTMS) to produce Si-SH and (2) polymerization of Si-SH, sodium silicate, and hydrochloric acid to produce TSH.

From infrared spectra (Figure 2) obtained for solid samples, it was possible to conclude the successful of the thiazole attachment onto silica gel due to the presence of three well-defined peaks at 2576 (S-H stretching), 1050 (C-H aromatic), and 910 cm^{-1} (for asymmetric C-S stretching) in TSH spectrum [10-11]. However, Si-O stretching asymmetric and symmetry of Si-O-Si that appears in silica gel and TSH at about 1080 to 1100 cm^{-1} and 750 to 790 cm^{-1}, respectively. The binding of organic compounds on the silica gel is also characterized by the emergence of the peak at about 750, 1050 and 3040~3060 cm^{-1} of TSH in a CH bonds aromatic bending , CH bending aromatic , and C-H stretching aromatic, respectively [12-13]. While sequentially it can be said that peaks at ~910 and 1450~1490 cm^{-1} for C-C stretching aromatic asymmetric C-S stretching. However, the silica hybrid, TSH, presented a characteristic peak at 2962 cm^{-1} is assigned to C-H stretching of sp^3 carbon and confirm the presence of the organic group attached onto silica surface and peak at 3450 cm^{-1} was characteristic for –OH.

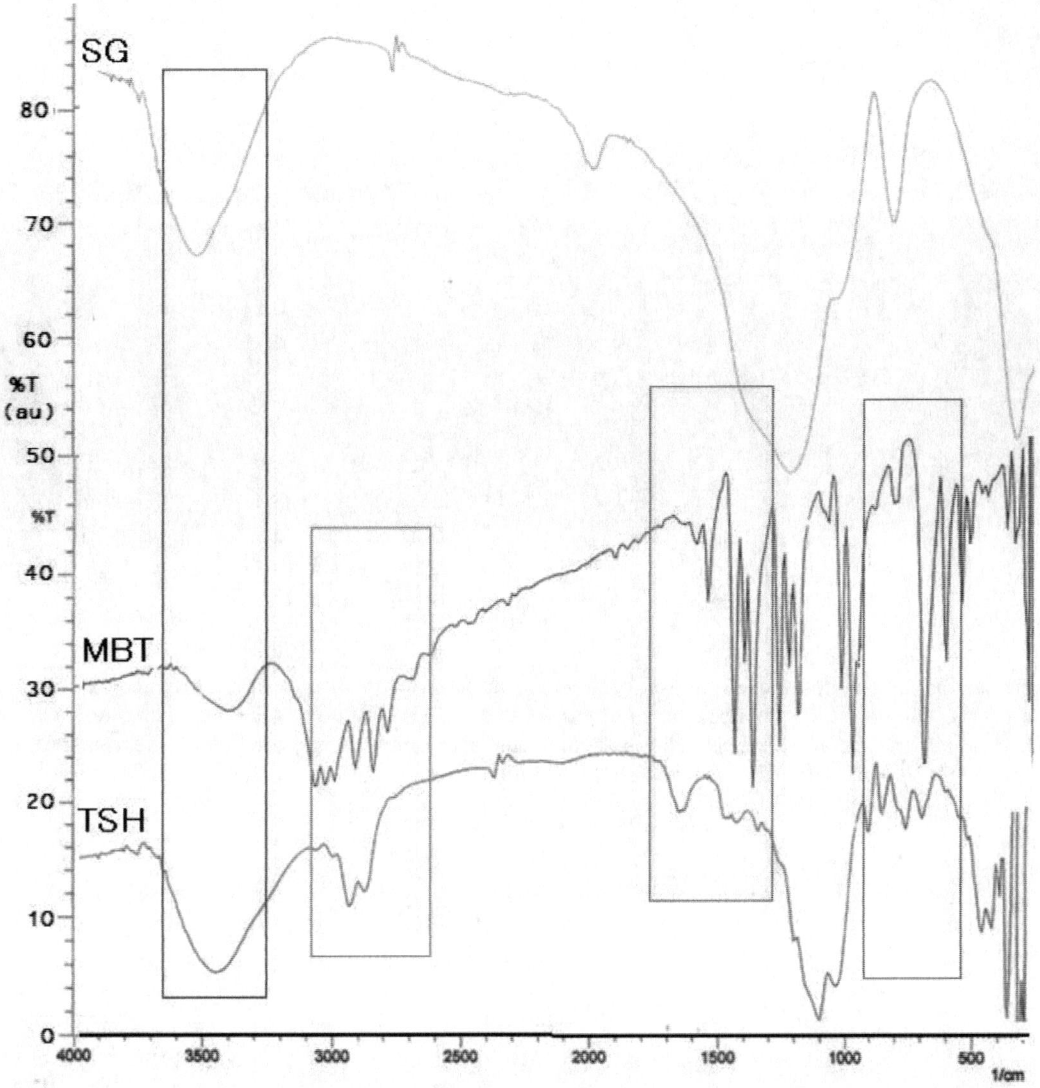

Figure 2. FT-IR spectra of Silica gel (SG), Mercaptobenzothiazole (MBT) and Thiazole Silica Hybrid (TSH).

3.1. Kinetic studies

Chemical kinetics gives information about reaction pathways and times to reach equilibrium. Adsorption kinetics shows a large dependence on the physical and/or chemical characteristics of the sorbent material. The study of adsorption dynamic describes the solute uptake rate, and this rate controls the habitation time of adsorbate uptake at the solid-solution interface (Figure 3).

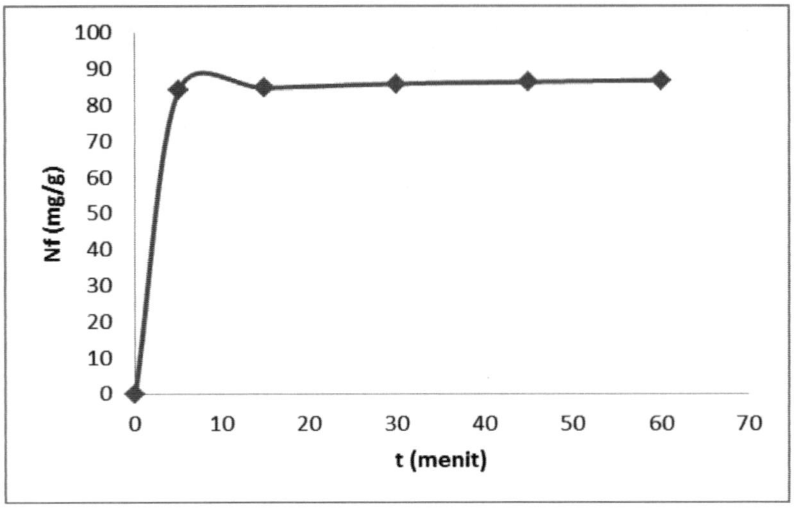

Figure 3. The sorption of Cd(II) ion onto TSH.

Various models have been described reaction sequence kinetic adsorption system is based on the concentration of the solution [6]. In this study, an absorption kinetics batch Cd (II) ion with TSH has been studied in terms of first-order kinetics, the second-order kinetic, pseudo-first-order kinetics, and the pseudo-second-order models (Table 1). Pseudo-second order kinetics models are given in the equation (1) [9][14][15].

$$\frac{t}{q_t} = \frac{1}{k_2 q_e^2} + \frac{t}{q_e} \quad (1)$$

Where k_2 is the rate constant of pseudo-second-order reaction ($g.mg^{-1}min^{-1}$), q_e and q_t are the amounts of solute adsorbed at equilibrium and at any time (mgg^{-1}), respectively. Plot the straight line of t/q_t against t is used to obtain constant pseudo-second-order reaction. Here, the initial absorption rate follows equation (2).

$$k_o = k_2 q_e^2 \quad (2)$$

The greatest value of the correlation factor R^2, obtained from the field of pseudo-second-order kinetics are given in Table 1. These results indicate that the adsorption of Cd (II) ion to the following TSH also pseudo-second-order kinetics model. Expression levels of pseudo-second order is used to describe physisorptions involve valence forces through the sharing or exchange of electrons between the adsorbent and adsorbate. In recent years, the expression level of pseudo-second-order has been widely applied to adsorption of pollutants from aqueous solutions [9]

Table 1. Adsorption kinetics models for Cd(II) ion onto TSH*.

Kinetics models	Cd(II)	
	Rate constant (k)	R^2
First order	-0.003 min^{-1}	0.958
Second order	0.00003 min^{-1}	0.966
Pseudo-first order	-0.001 min^{-1}	0.958
Pseudo-second order	0.076 min^{-1}	0.999

*at C_0=25 mg/L, pH 6.0; TSH= 0.1 g; temperature 25°C

3.2. *Thermodynamic studies*

Isotherm data analysis is important to develop an accurate equation is the result and can be used for design purposes. In order to investigate the adsorption isotherms, two equilibrium isotherms were analyzed: Langmuir and Freundlich. Constant parameter equation for this process of adsorption isotherms calculated by linear regression of isotherm equation. Constant parameters and correlation coefficients (R^2) are summarized in Table 2.

Table 2. Adsorption isotherm models for Cd(II) ion onto TSH*

Parameter		Cd(II)
Langmuir parameter	q_m (mmol/g)	9.363
	K_a (x10^4 mg/g)	1052.434
	ΔG°(kJ/mol)	-15.488
	R^2	0.991
Freundlich parameter	N (mmol/g)	2.242
	K_F (mg/g)	21.184
	R^2	0.970

*at C_0=25 mg/L, pH 6.0; TSH= 0.1 g; temperature 25°C

Langmuir adsorption isotherm has been successfully applied to many real absorption process [9][16-18]. It predicts the maximum adsorption capacity adsorbent monolayer and also determine whether the adsorption is profitable or not. Langmuir (Ce/qe vs Ce) field of Cd (II) were found linear for the entire concentration range studied and the correlation coefficient is very high. The maximum adsorption capacity was determined as 9.363 mmol/g for Cd(II (Table 2).

The Freundlich isotherm is known to describe the adsorption isotherm initial relationship [19]. Satisfactory empirical isotherm can be used in adsorption from aqueous solution. The Langmuir and Freundlich isotherms constants are shown in Table 2. Thermodynamic considerations adsorption process necessary to conclude whether or not this process is spontaneous. The Gibbs free energy change, ΔG°, an indication of the spontaneity of chemical reactions and, therefore, is an important criterion for spontaneity. The reaction occurs spontaneously at a certain temperature if ΔG° is a negative quantity. Free energy of adsorption, calculates based on adsorption equilibrium constant K_a is given by the following equation Gibbs (3)[16]

$$\Delta G^o = -RT\ln K_a \qquad (3)$$

The ΔG° which is the standard free-energy change (J / mol), R the universal gas constant (8.314 J / Molk), and T is the absolute temperature (K). The ΔG° parameters for adsorption process using the K_a of the Langmuir isotherm shown in Table 1. The calculated ΔG° give a negative value for the adsorption of Cd(II) onto TSH, confirmed the feasibility of the process and the spontaneous nature of adsorption.

4. Conclusion

Synthesis of thiazole silica hybrid can be done through the sol-gel process. The presence of the thiazole modifier on the surface of silica gel was evidence using FTIR. The TSH can be used to absorb cadmium ion from the aqueous solution. The kinetics of adsorption processes fit a pseudo-second-order kinetic model. The thermodynamic parameter, ΔG° has been evaluated from adsorption equilibrium constant based on a Langmuir, isotherm model. The negative values of ΔG° indicate the spontaneous nature of adsorption Cd(II) onto TSH.

References

[1] Zhuravlev L T 2006 *The Surface Chemistry of Silica-The Zhuravlef Model in book* Colloidal Silica Fundamentals and Applications CRC-Press New York **261**
[2] Azmiyawati, C Nuryono and Narsito 2005 *Indo. J.Chem.* **5** (1) March. p. 31-35
[3] Hofen K, Weber S, Chan C P C and Majewski P 2011 *Ap. Surf. Sci.* **257** p. 2576-2580
[4] Aylward W M and Pickup P G 2008 *El. Chim. Act.,* **53** p. 3897-3902
[5] Ma J Z, Hu J and Zhang Z J 2007 *Euro. Poly. J.,* **43** p. 4169-4177
[6] Evangelista S M, Deoliviera E Castro, G R Zara and L F Prado 2007 *Surf. Sci. Science Direct Elsevier* **601** p. 2194-2202
[7] Kamath S R and Proctor A 1997 *Arkansas Rice Research Studies*
[8] Ünlu N and Ersoz M 2006 *J. of Hazard. Mat.* B **136** p. 272-280
[9] Azmiyawati C, Nuryono and Narsito November 2012 Adsorption of Mg(II) and Ca(II) on Disulfonate-silica Hybrid *Indo. J.Chem.* **12** (3) pp. 223-228
[10] Oh Y-K, Hong L Y, Asthana Y and Kim D P 2006 *J. Ind. Eng. Chem.* **12** (6) p. 911-917
[11] Troitsky V, Berzina T, Shchukin D, Sukhorukov G, Erokhin V and Fontana M P 2004 *Colloids Surf. A: Physicochem. Eng. Asp.* 245 163

[12] Prado A G S, Sales J A A, Carvalho R M, Rubim J C and Airoldi C 2004 *Journal of Non-Crystalline Solids* **333** p. 61-67
[13] Gunzler H and Gremlich H-U 2002 *IR Spectroscopy An Introduction* Wiley-VCH Weinheim
[14] Ho Y-S 2006 *J. of Hazard. Mat.* **B** (136) p 681-689
[15] Hameed B H, Din A T M and Ahmad A L 2007 *J. of Hazard. Mat.* **141** p. 819-825
[16] Langmuir I 1918 *J. Am. Chem. Soc.* **40** p.1361-1403
[17] Ho Y-S, Huang C-T and Huang H W 2002 *Proc. Biochem*. **37** p. 1421-1430
[18] Kul A R and Koyuncu H 2010 *J. of Hazard. Mat*. **179** p. 332-339
[19] Freundlich H M F 1906 *Z. Phys. Chem.* **57 A** p. 385-470.

IOP Conf. Series: Materials Science and Engineering **107** (2016) 012010 doi:10.1088/1757-899X/107/1/012010

Alginate cryogel based glucose biosensor

Amin Fatoni*, Dian Windy Dwiasi and Dadan Hermawan

Department of Chemistry, Faculty of Mathematics and Natural Sciences, Universitas Jenderal Soedirman, Jl. Dr. Soeparno, Purwokerto 53123, Indonesia

*E-mail: aminfatoni@unsoed.ac.id

Abstract. Cryogel is macroporous structure provides a large surface area for biomolecule immobilization. In this work, an alginate cryogel based biosensor was developed to detect glucose. The cryogel was prepared using alginate cross-linked by calcium chloride under sub-zero temperature. This porous structure was growth in a 100 µL micropipette tip with a glucose oxidase enzyme entrapped inside the cryogel. The glucose detection was based on the colour change of redox indicator, potassium permanganate, by the hydrogen peroxide resulted from the conversion of glucose. The result showed a porous structure of alginate cryogel with pores diameter of 20-50 µm. The developed glucose biosensor was showed a linear response in the glucose detection from 1.0 to 5.0 mM with a regression of $y = 0.01x+0.02$ and R^2 of 0.994. Furthermore, the glucose biosensor was showed a high operational stability up to 10 times of uninterrupted glucose detections.

1. Introduction

The biosensor is an analytical device consist of the biological sensing element and a transducer which convert the biological recognition into a measurable output signal [1]. The advantages of the biosensor such as high selectivity, low-cost, high sensitivity, miniaturization and real-time measurement ability, stimulates researchers to develop this method continuously. One of the strategies reported to improve the biosensor performances is increase the sensitivity and stability, such as the use of cryogel, porous materials, for supporting material in the biosensor development [2].

Chitosan cryogel based biosensors have been reported in order to improve the biosensor performances mainly their sensitivity and stability in the biosensor development to detect glucose [3], sialic acid [4] and microalbumin [5]. However, those previous cryogel based glucose biosensor developments were used electrochemical detection with some disadvantages such as complex electrode preparation and relatively use an expensive instrument. The use of lower cost detection systems has been reported such as cheap potentiostat [6], pocket camera, mobile phone camera [7] and commercial scanner [8].The use of color change detection system provide some advantages such as simple instrument, easy to prepare a standard solution and the color change sometimes can be easily detected by naked eyes [9].

In this work, the advantages of cryogel were used to prepare glucose biosensor using an alginate as polymer backbone. The biosensor was built by developing cryogel in a micropipette tip, thus, easy to operate by sucking and pumping the sample solution. Glucose oxidase enzyme was used as the biological sensing element and a redox indicator was use to record the resulting hydrogen peroxide, which was finally its color change detected by a spectrophotometer.

2. Materials and Methods

2.1. Materials
Glucose oxidase (GOD) (EC 1.1.3.4, Type II-S, 15-50 unit mg^{-1}) were procured from Sigma (Steinheim, Germany). Commercial sodium alginate, calcium chloride and D-(+)-glucose anhydrous (≥ 98.0%), potassium permanganate, hydrogen peroxide, sodium dihydrogen o-phosphate and disodium hydrogen o-phosphate were from Merck (Germany).

2.2. Apparatus and measurements
Scanning electron microscopy (SEM) Table Top TM3000 (Hitachi, Japan) was used to characterize the morphology of the alginate cryogel. The sample colour change was measured using a Shimadzu Biospec 1601 UV-Vis spectrophotometer (Shimadzu, Japan). The alginate cryogel biosensor preparation and measurement were used a 10-100 μL micropipette (Eppendorf, Germany).

2.3. Cryogel Preparation
Alginate solution was prepared by dissolving 2.0 g sodium alginate with sodium acetate buffer (50 mM, pH of 4.5) to make 100 mL solutions. The alginate solution of 100 μL was then placed in the microcentrifuge tube to prepare the cryogel. The glucose oxidase (GOD) enzyme of 35 U was then added to the alginate solution, added 12 μL of calcium chloride solution (5% w/v in distilled water) and mixed immediately. A 100 μL micropipette tip was used to support this biosensor system by filling this tip using the alginate-GOD-calcium chloride solution of 100 μL and a 0.5 mm diameter stainless rod as core template. This micropipette tip was then kept in the freezer (-20 ^{0}C) for 6 h to allow cryogelation. The alginate-GOD cryogel was thawed in the refrigerator (4 ^{0}C) for 1 h. The brief of the biosensor fabrication was described in Figure 1. The porous structure of alginate cryogel was then characterized using a scanning electron microscope (SEM). The pore size of the cryogel was measured using a freeware of ImageJ (http://imagej.nih.gov/ij/).

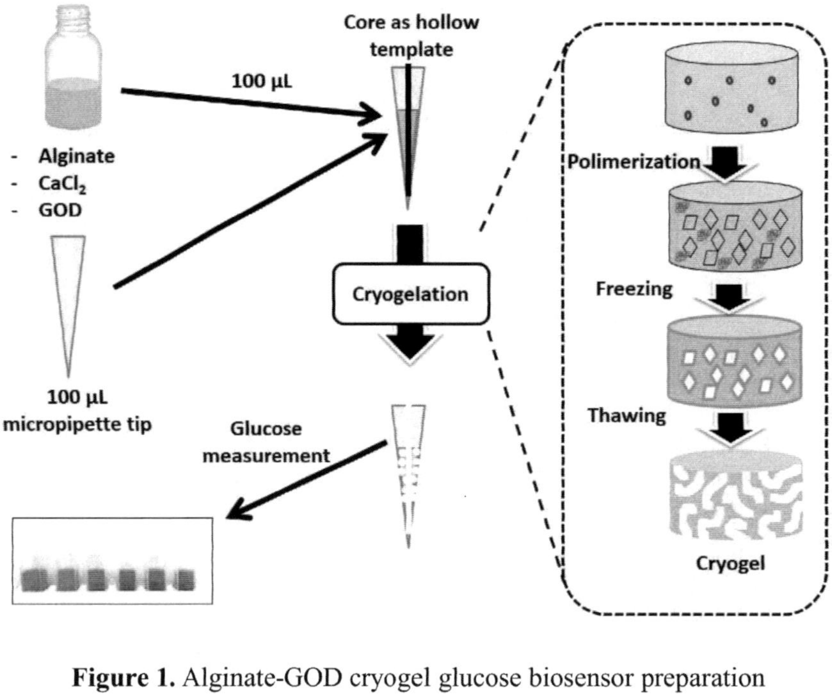

Figure 1. Alginate-GOD cryogel glucose biosensor preparation

IOP Conf. Series: Materials Science and Engineering **107** (2016) 012010 doi:10.1088/1757-899X/107/1/012010

2.4. Glucose detection

The glucose detection was based on the redox indicator colour change by the resulted hydrogen peroxide as an enzymatic product of glucose catalysed with glucose oxidase. The redox indicator used was potassium permanganate in appropriate concentration (0.1 mM). The indicator of 2.5 mL was placed in a test tube and then added a 100 µL of sample. The first indicator colour change was studied using a standard solution of 1.0 to 5.0 hydrogen peroxide. Then the biosensor measurement was performed by filling the in-tip cryogel biosensor with 100 µL of glucose sample, allowed the enzymatic reaction for 2 minutes and the reaction product was dropped to the 2.5 ml of redox indicator. The filling and releasing of the sample – product in the biosensor was controlled by micropipette sucking and pumping. The redox indicator colour change by hydrogen peroxide or enzymatic glucose product was recorded using a spectrophotometer UV-Vis. The relation between glucose concentration and the colour change was then plotted in the regression line to get the linear equation of $y = ax + b$.

2.5. Stability study

The stability study was performed by measuring glucose solution at certain concentration uninterrupted, similar to the glucose detection procedure above. The stability was determined by the biosensor response given for more than 90% in a series measurements.

3. Results and Discussion

3.1. Alginate cryogel

Cryogel is one of the interesting porous material, especially due to its large surface area, thus, enhance the biosensor performance when it was used as supporting material in the biosensor development [2-4]. The various polymer can be used to prepare cryogel such as polyvinyl alcohol, chitosan and alginate. In this study, the cryogel as biosensor supporting material was prepared using alginate as polymer backbone. The crosslinking of sodium alginate was by divalent calcium cations of calcium chloride. Native sodium alginate has a functional group of –COONa, which in the aqueous solution became $-COO^-$ and Na^+. The calcium ions of calcium chloride replace the sodium ion in the polymer, each calcium ion can attach two polymer strands. This cross-linking process allows to prepare alginate hydrogel in the room temperature. The cryogel preparation or cryogelation was prepared by crosslinking of the sodium alginate at subzero temperatures, which allowed create hydrogels with large interconnected pores [10]. In this process, the reactants remain in the unfrozen phases and form a cross-linked network upon polymerization, while the ice crystals nucleated from the aqueous phase during freezing, act as pyrogens. The interconnected macroporous networks were formed when the ice crystal melt at a temperature above the freezing temperature of water. Scanning electron microscope image showed a porous structure of the cryogel (Figure 2) with pores size diameter of 20-50 µm.

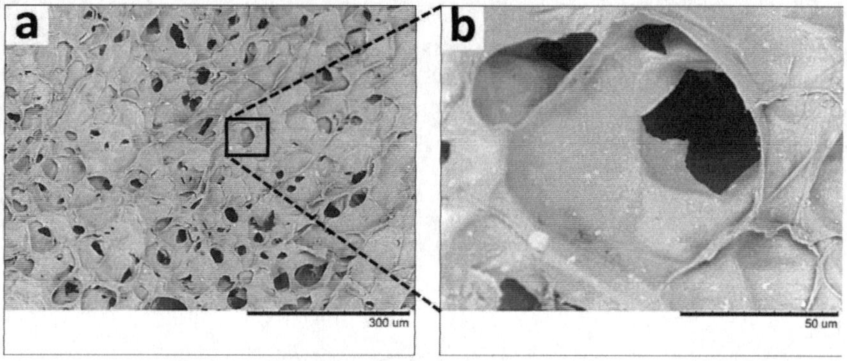

Figure 2. SEM of alginate cryogel surface at x250 (a) and x1500 (b).

3.2. Glucose detection

Glucose detection of the fabricated biosensor was based on potassium permanganate color change by resulted hydrogen peroxide according to the equation (1).

$$H_2O_2 + Mn^{7+}{}_{(purple)} \rightarrow H_2O + O_2 + Mn^{2+}{}_{(pale\ yellow)} \quad (1)$$

Furthermore, the color change with the increasing glucose concentration was studied in the range of 1.0 to 5.0 mM. The result showed a good relation (Figure 3) between the color changes of the indicator with the glucose concentration with R^2 of 0.994. The use of potassium permanganate as an indicator was used for the model in this preliminary study to verify the work of the fabricated glucose biosensor since the potassium permanganate was not specific reduce by hydrogen peroxide only. In the real glucose biosensor application, such for glucose detection in the blood, where there are many biological interferences, the specific hydrogen peroxide indicator should be used such as silver nano prism [11], 4-nitrophenyl boronic acid [12] and titanium oxy sulphate [13].

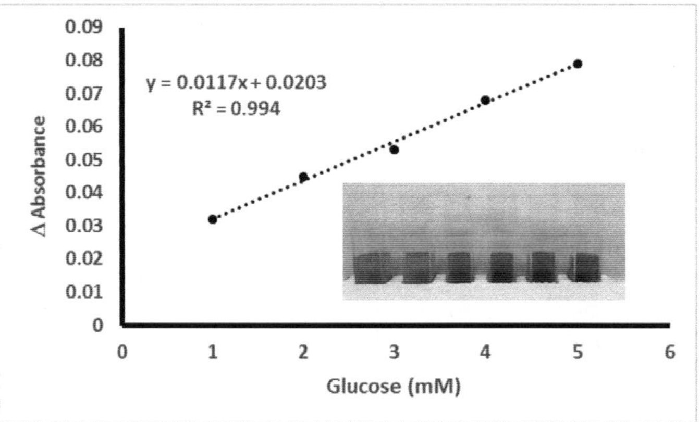

Figure 3. Calibration curve of fabricated alginate cryogel based glucose biosensor at a series glucose concentrations(1.0 – 5.0 mM). Inset, the example of redox indicator colour change by the resulted hydrogen peroxide as glucose conversion by glucose oxidase.

3.3. Operational stability study

The advantages of cryogel in the biosensor development were its ability to hold the enzyme activity, thus, resulted in a high stable biosensor [3, 4]. This was the main purpose of the use of cryogel as biological sensing element supporting material in the biosensor fabrication. However, different polymer material to build the cryogel may result in various stability profile to hold the biological element, due to their nature characteristic, for example, biomaterials usually more biocompatible compare to synthetic material. In this work, the alginate cryogel performance to support the glucose oxidase immobilization was studied by uninterrupted measurement of 3.0 mM glucose solution. The results showed the fabricated biosensor was stable up to 10 times analysis (Figure 4) without significantly lost its activity.

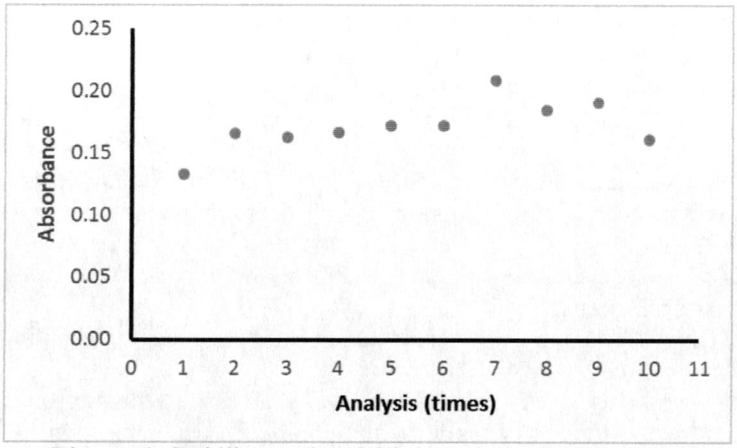

Figure 4. Stability study of fabricated glucose biosensor by measuring uninterrupted analysis of 3.0 mM glucose solution.

4. Conclusion

The alginate cryogel as a supporting material in the developed glucose biosensor showed a porous structure, thus, provide a large surface area for glucose oxidase immobilization. The developed glucose biosensor based on alginate cryogel showed a good linear range in the glucose detection, with an operational stability up to 10 times uninterrupted analysis without significant responses decreasing. This simple alginate cryogel based glucose biosensor with colorimetric detection would be an excellent model for other biosensor application.

Acknowledgement

We would like to thank the Directorate General of Higher Education (DGHE, DIKTI) and the Jenderal Soedirman University for supporting this research through "Riset Unggulan" grant. We also like to thank Fauziyah Sri P and Makhmudita Eka S for their helping in this research.

References

[1] Collings A F and Caruso F 1997 *Biosensors: recent advances* Reports on Progress in Physics **60** 1397-445

[2] Hedström M, Plieva F, Galaev I and Mattiasson B 2008 Monolithic macroporous albumin/chitosan cryogel structure: a new matrix for enzyme immobilization *Analytical and Bioanalytical Chemistry* **390** 907-12

[3] Fatoni A, Numnuam A, Kanatharana P, Limbut W, Thammakhet C and Thavarungkul P 2013 A highly stable oxygen-independent glucose biosensor based on a chitosan-albumin cryogel incorporated with carbon nanotubes and ferrocene *Sensors and Actuators B: Chemical* **185** 725-34

[4] Fatoni A, Numnuam A, Kanatharana P, Limbut W and Thavarungkul P 2014 A conductive porous structured chitosan-grafted polyaniline cryogel for use as a sialic acid biosensor *Electrochimica Acta* **130** 296-304

[5] Fatoni A, Numnuam A, Kanatharana P, Limbut W and Thavarungkul P 2014 A novel molecularly imprinted chitosan-acrylamide, graphene, ferrocene composite cryogel biosensor used to detect microalbumin *Analyst* **139** 6160-7

[6] Rowe A A, Bonham A J, White R J, Zimmer M P, Yadgar R J, Hobza T M, Honea J W, Yaaco I B and Plaxco K W 2011 Cheapstat: an open-source,"do-it-yourself" potentiostat for analytical and educational applications *PloS one* **6** e23783

[7] Oncescu V, O'Dell D and Erickson D 2013 Smartphone based health accessory for colorimetric detection of biomarkers in sweat and saliva, *Lab on a Chip*, **13** 3232-8

[8] Kompany-Zareh M, Mansourian M and Ravaee F 2002 Simple method for colorimetric spot-test quantitative analysis of fe(iii) using a computer controlled hand-scanner, *Analytica Chimica Acta*, **471** 97-104

[9] Puiu P D 2012 Color Sensors and Their Applications *Optical Nano-and Microsystems for Bioanalytics* Springer pp. 3-45

[10] Plieva F M, Galaev I Y, Noppe W and Mattiasson B 2008 Cryogel applications in microbiology *Trends in microbiology* **16** 543-51

[11] Nitinaivinij K, Parnklang T, Thammacharoen C, Ekgasit S and Wongravee K 2014 Colorimetric determination of hydrogen peroxide by morphological decomposition of silver nanoprisms coupled with chromaticity analysis *Analytical Methods* **6** 9816-24

[12] Su G, Wei Y and Guo M 2011 Direct colorimetric detection of hydrogen peroxide using 4-nitrophenyl boronic acid or its pinacol ester *American Journal of Analytical Chemistry* **2** 879

[13] Eisenberg G 1943 Colorimetric determination of hydrogen peroxide *Industrial & Engineering Chemistry Analytical Edition* **15** 327-8

IOP Conf. Series: Materials Science and Engineering **107** (2016) 012011 doi:10.1088/1757-899X/107/1/012011

Adsorption of remazol yellow FG from aqueous solution on chitosan-linked P-T-Butylcalix[4]Arene

Desi Suci Handayani*, Candra Purnawan, Pranoto, Sri Hastuti and Diniyah Hilmiyana

Department of Chemistry, Faculty of Mathematics and Natural Sciences, Sebelas Maret University, Jl. Ir Sutami 36A Surakarta Indonesia

*E-mail: dsuci72@gmail.com

Abstract. Chitosan-liked p-t-butylcalix[4]arene can be applied as an adsorbent of dye Remazol Yellow FG because it has an active group of hydroxy (-OH) and amine ($-NH_2$). Adsorption had done by comparing to chitosan. Adsorption was carried out in a batch system. Several variables including pH, contact time, and initial concentration of dye Remazol Yellow FG were determined. The optimum adsorption conditions of dye Remazol Yellow FG by chitosan dan chitosan-liked p-t-butylcalix[4]arene occurred at pH 4 and a contact time of 135 minutes. The adsorption kinetics of dye Remazol Yellow FG using chitosan and chitosan-liked p-t-butylcalix[4]arene followed Ho kinetics model, pseudo-second order and the rate constant of adsorption was 2.42×10^{-4} g/mg.menit and 3.63×10^{-4} g/mg.menit. The isothermal study showed that adsorption of dye Remazol Yellow FG using chitosan and chitosan-liked p-t-butylcalix[4]arene tends to follow the Langmuir isotherm, with adsorption energy was 24.92 kJ/mol and 32.21 kJ/mol.

1. Introduction

The development of the textile industry has a positive impact that can improve the welfare of the community. In another hand, development of the textile industry could also have a negative impact that environmental pollution when the wastewater from the textile dyeing process settled into rivers or sewers without treatment processing. Batik industrial waste water is originating from the dyeing, washing, printing and finishing the process. Waste staining of batik industry contains several components such as a residual dye, salt and additive materials such as urea, sodium alginate, sodium bicarbonate and water (residual staining and washing) [1].

Residual dye was the most dominant component of the waste dyed batik industry. Classification of dyes based on their use that was an acid dye, basic, direct, mordant, metal complexes, azoat, sulphur, vessel, dispersions and reactive [2]. Reactive dyes are often used in the batik industry include Procion, Cibracon, Drimaren, and Lavafix, which can hold a substitution reaction with the fiber, and azo dyes are Remazol, forecasts allowing, and Primaxin, which can hold an addition reaction [3].

Remazol Yellow FG is an azo dye which is a kind of synthesis of dyes are quite important and widely used. More than 27.2% of the textile industry uses Remazol Yellow FG dye. Azo dye has a chromophore system of the azo group (-N=N-), which binds to the aromatic group. Reactive dye chromophore is usually the azo and anthraquinone systems with relatively small molecular weight. Absorption of the fiber was not great so that the dye does not react with the fibers easily removed [4].

Many methods has been developed for the treatment of waste water, including precipitation, extraction, separation with membranes [5], ion exchange [6], and adsorption. Precipitation method was the most economical method, but it was not efficient to the aqueous solution, ion exchange method, and reverse osmosis are generally effective, but requires equipment and operating costs are relatively high. Adsorption method was one potential alternative method because the process is relatively simple, can work at low concentrations, can be recycled, and the costs involved are relatively inexpensive [7].

Adsorption method has been proven effective to reduce the concentration of the dye in water as reported by various studies, including the adsorption of the dye textile waste types BR Red HE 7B using rice straw [8]. Adsorption used the fly ash [9]. The use of natural adsorbent which cheap and abundant certainly very profitable, but to know the active compound which acts as the adsorbent of such materials was difficult, it can be difficult to explain the phenomenon of adsorption occurs. It was, therefore, necessary to find another adsorbent whose structure was known with certainty, and this can be obtained from the synthesis.

One group of synthetic compounds that have great potential to be developed as an adsorbent was chitosan. Chitosan is the result of deacetylation of chitin which has undergone the removal of acetyl groups. Acetyl group was missing from the process of deacetylation leaving a free amino group that causes the chitosan was polycationic [10]. Research on the application of chitosan as an adsorbent has been widely studied, including Rakhmawati (2007) [4] which successfully applying chitosan as an adsorbent dye. Mahatmanti (2003) [11] was using chitosan as an adsorbent for the adsorption of methyl orange dye. Chitosan which has specific properties such as biocompatibility, a non-toxic and renewable biopolymer made of chitosan as serious attention to continue to be developed and researched. Improving the ability of chitosan can be crosslinked on macromolecular compounds namely crown ether p-t-butylcalix[4]arene. p-t-butylcalix[4]arene has a geometry such as baskets and hollow, can be used in guest-host systems (guest host) [12]. Modification of chitosan has to improve as an adsorbent that is by p-t-butylcalix[4]arene attached to the chitosan. In this research, dye adsorption Remazol Yellow FG by using research results Hand (2014) [13] p-t-butylcalix[4]arene tied to generate chitosan, chitosan-pt-butylcalix[4]arene. Chitosan-p-t-butylcalix[4]arene was expected to effectively and efficiently so it can be useful to reduce environmental pollution, especially in the textile industry.

2. Research Methods

2.1. Effect of solution pH Remazol Yellow FG

A total of 25 mL of Remazol Yellow FG at a concentration of 100 ppm respectively set the initial pH to 1.5; 2; 4; 6; 8; 10 with HNO_3 and NaOH. Then into each solution was added 20 mg of chitosan-p-t-butylcalix[4]arene. The chitosan-p-t-butylcalix[4]arene was synthesized based on the previous research [4]. The mixture was stirred at a speed of 200 rpm for 2 hours. Afterward, the solution was filtered, and the filtrate was analyzed by UV-Vis spectroscopy. Equal treatment performed on chitosan as a comparison.

2.2. Effect of Stirring Time

A total of 25 mL of Remazol Yellow FG each with a variation of time at 5, 15, 45, 135, 405 minutes was set at optimum pH. Then into a solution, the the 20 mg of chitosan-p-t- butylcalix[4]arene adsorbent was added and stirred for optimum time with a speed of 200 rpm. The solution is then filtered and analyzed by UV-Vis spectroscopy (*Optima SP-300*). Equal treatment performed on chitosan as a comparison.

2.3. Influence of Solution Concentration Remazol Yellow FG

A total of 25 mL of Remazol Yellow FG each with varying concentrations of 100, 150, 200, 250, 300 ppm was set at optimum pH. Then into a solution, 20 mg of chitosan-p-t- butylcalix[4]arenes adsorbent was added and stirred for optimum time with a speed of 200 rpm. The solution is then filtered and analyzed by UV-Vis spectroscopy. The same treatment was done for chitosan as a comparison. Morphological analysis was conducted by Scanning Electron Analysis (SEM) JEOL JSM 6360 equipped with an elemental analysis of EDX (JED-2200). Meanwhile, the functional group analysis was conducted by FTIR analysis (Prestige-21 Shimadzu).

3. Results And Discussion

3.1. Determination of Optimum pH

Effect of pH was very important in the process of adsorption, adsorption processes depending on the type of adsorbent, the adsorbed species and the type of solvent. Variations initial pH of the solution at pH 1.5, 2, 4, 6, 8 and 10, get results as shown in Figure 1.

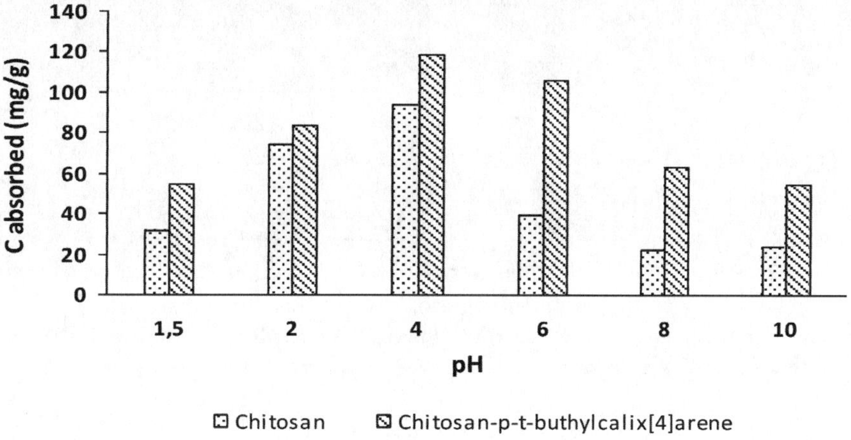

Figure 1. Concentration of Dye Absorbed Remazol Yellow FG on the variation of pH

Based on the data in Figure 1 shows that both adsorbent experience optimum absorption at pH 4, so that it can be said that the optimum pH-absorption occurs at pH 4. The phenomenon that occurs possibly because, at pH 4, the atmosphere is not strongly acidic, proton (H^+) was used to protonated D-$SO_2CH_2CH_2OSO_3$-Na^+ group into vinyl sulphone (D-SO_2-CH=CH_2) groups did not take place completely. Vinyl sulphone group formed perfectly leads to reduced interaction with the chitosan and chitosan-p-t-butylcalix[4]arene as an adsorbent.

Sakkayawong (2005) [5] suggested that at acidic conditions the hydrogen atoms (H^+) in solution can protonate primary amine group (R-NH_2) of chitosan becomes (R-NH_3^+). Similarly, on the chitosan-p-t-butylcalix[4]arene adsorbent, hydrogen atoms (H^+) in solution can protonate secondary amine group (R-R'-NH). In aqueous solution, the dye will dissolve, and sulphonate groups on reactive dye Remazol Yellow FG dissociated and transformed into ions. The adsorption process was then generated from the interaction between the dye with an adsorbent.

Adsorption process which takes place at pH 1.5 and pH 2 has a low adsorption efficiency value; it was possible proton (H^+) are used for protonated D-$SO_2CH_2CH_2OSO_3Na$ group into vinyl sulphone group (-SO_2-CH=CH_2) has complete, so that the interaction between chitosan and chitosan-p-t-butylcalix[4]arene adsorbent was reduced.

Adsorption of Remazol Yellow FG dye by chitosan and chitosan-p-t-butylcalix[4]arene at pH 8 and pH 10 had a small adsorption value. This was because the highly alkaline pH condition, D-$SO_2CH_2CH_2OSO_3Na$ group into vinyl sulphone group (-SO_2-CH=CH_2) was complete. As at pH 6, the atmosphere was not too alkaline vinyl sulphone group formation is not running perfectly, so still have a sulphonate group in the reactive dye Remazol Yellow FG, which dissociates and transformed into ions, then interact with the adsorbent.

3.2. Effect of Contact Time and Parameter Kinetics

The contact time was the time required adsorbent to adsorb dye Remazol Yellow FG. Variation of the contacts time in this study was 5, 15, 45, 135 and 405 minutes. Effect of contact time on the amount of dye adsorbed can be seen in Figure 2.

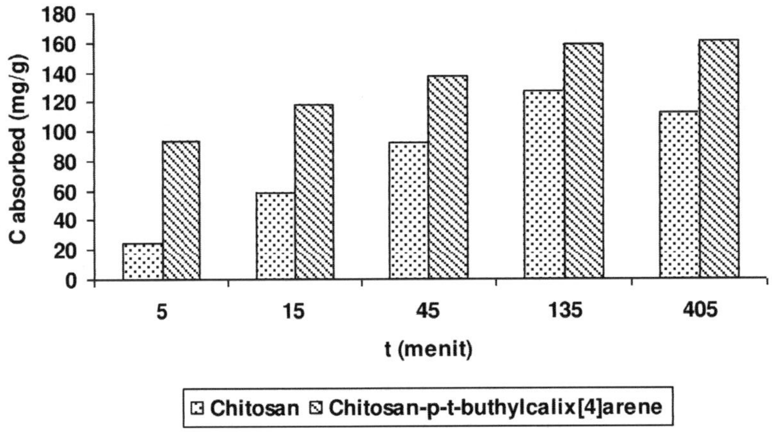

Figure 2. Concentration of Dye Absorbed Remazol Yellow FG on the variation of contact time

Based on data from the influence of contact time, then it can be studied adsorption kinetics. Results of the adsorption kinetics study of both adsorbents shown in Table 1.

Table 1. Assessment of the dye adsorption kinetics of Remazol Yellow FG

adsorption kinetics	Adsorbent	
	Chitosan	chitosan-p-t-butylcalix[4]arene
Lagergren:		
K (min^{-1})	0.012	0.015
R^2	0.979	0.791
Ho:		
K ($g\ mg^{-1}\ min^{-1}$)	$2.42 x 10^{-4}$	$3.63 x 10^{-4}$
R^2	0.999	0.966

Based on the linearity calculation, it can be concluded that kinetics equation Lagergren pseudo first-order less suitable to be applied as adsorption kinetics model for both adsorbents, while the second order equation pseudo-Ho had high linear coefficient so that both of adsorbent was likely to follow a pseudo second order Ho.

Rate constants adsorption value of second order Ho equation for both chitosan and chitosan-p-t-butylcalix[4]arene adsorbent was 2.42 x 10^{-4} g/mg.min and 3.63 x 10^{-4} g/mg.min, respectively. Based on the adsorption rate constant value, capabilities interaction of chitosan-p-t-butylcalix[4]arene faster than with chitosan.

3.3. Effect of Initial Concentration Dyes and Parameter Isotherm

The influence of the initial concentration of Remazol Yellow FG against the adsorbent was shown in Figure 3.

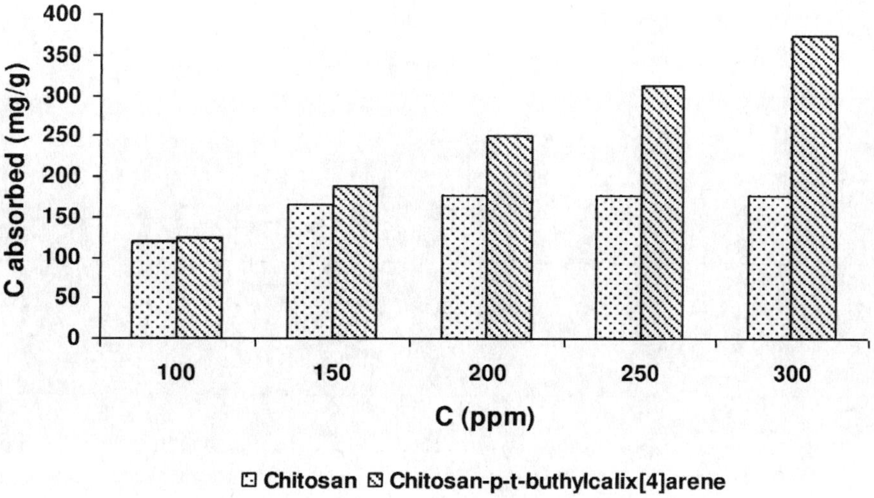

Figure 3. Initial concentration of Dyes variation Remazol Yellow FG on the initial concentration of dye

Figure 3 shown the concentration of adsorbed solution tends to increase with increasing adsorbate concentration. This was because an increase in competition between the adsorbate with the active adsorbent side. The adsorption process was also influenced by the surface area of the adsorbent for adsorption availability. The surface area of the adsorbent was determined by the size and magnitude of the adsorbent pores, the greater of the surface area, the greater on the adsorption had done. Handayani [14] states that the chitosan-p-t-butylcalix[4]arene adsorbent had an outer surface of 16.352 m^2/g, while chitosan according to Cahyaningrum et al. [16] mentions that the surface area is 0.847270 m^2/g, and Kelesorgu [17] states that the specific surface area of chitosan was 4.45 m^2/g. From these results it can be seen that the surface area of chitosan-p-t-butylcalix[4]arene adsorbent was larger than the specific surface area of chitosan, this causes the adsorption of chitosan-p-t-butylcalix[4]arene more than chitosan.

Study of adsorption isotherms was tested to determine whether the dye adsorption with adsorbent followed Langmuir isotherm or Freundlich isotherm using absorbance calculation data. The isotherm study results for adsorption Remazol Yellow FG dye using adsorbents chitosan and chitosan-p-t-butylcalix[4]arene were presented in Table 2.

Based on Table 2, the linearity graph adsorption level of Langmuir isotherm models was higher than Freundlich isotherm for both chitosan and chitosan-p-t-butylcalix[4]arene, with a linearity value of 0.975 and 0.910 for chitosan adsorbent and 0.998 and 0.746 for the chitosan-p-t-butylcalix[4]arene adsorbent. Thus the dye adsorption isotherm Remazol Yellow FG on chitosan and chitosan-p-t-butylcalix[4]arene more appropriated and tend to follow the model of the Langmuir adsorption isotherm, which showed that both adsorbent chitosan and chitosan-p-t-butylcalix[4]arene occurred in monolayer, which assumes that the maximum adsorption occurs when all of the active sites of the adsorbent filled by adsorbate forming monolayer. In addition Langmuir models indicated that the adsorption process happening was chemisorption. Adsorption process which influenced by the bond between the active group of dyes that react with a hydroxyl group (OH) on the adsorbent.

Table 2. The dye adsorption isotherms Remazol Yellow FG with adsorbent

adsorption isotherms		Adsorbent	
		Chitosan	chitosan-p-t-butylcalix[4]arene
Langmuir:			
–	q_m (mg/g)	188.31	187.28
–	K (L/mol)	21897.37	406114.46
–	$E_{adsorpsi}$(kJ/mol)	24.92	32.21
–	R^2	0.975	0.998
Freundlich:			
–	n	2.39	6.41
–	K (mol/gr)	110.92	790.68
–	R^2	0.910	0.746

Comparison of optimum adsorption capacity from adsorbent that has been used to adsorb Remazol Yellow FG dye, chitosan and chitosan-p-t-butylcalix[4]arene adsorbent have optimum adsorption capacity greater than other adsorbents, as shown in Table 3.

Table 3. Adsorption Remazol Yellow FG dye by several adsorbent

Adsorbent	**qm (mg/gr)**
Active water hyacinth [18]	4.43
Reeds activated NaOH	7.85
Copoly(Eugenol-DVB)-Calix[4]resorsinarene	12.26
Chitosan	188.31
Chitosan-p-t-butylcalix[4]arene	187.28

The optimum adsorption capacity value between the chitosan and chitosan-p-t-butylcalix[4]arene adsorbent had a great optimum capacity value at chitosan adsorbent of 188.31 mg/g and 187.28 mg/g for chitosan and chitosan-p-t-butylcalix[4]arene. This phenomenon allegedly because chitosan-p-t-butylcalix[4]arene adsorbent had bonding process occurs with chitosan causes the adsorbent was increasingly full or bulky, so the adsorption capacity tends less than chitosan without bound. In addition to the chitosan adsorbent adsorption mechanism not only chemisorption but accompanied by physisorption, can be seen from the determination coefficient value on chitosan which was not too far from Langmuir and Freundlich models. Phyisisorpsi adsorption process involves intermolecular forces (Van der Waals force), and the adsorbed substances were relatively easily separated which allows the desorption.

The determination coefficient (R^2) in the chitosan-p-t-butylcalix[4]arene adsorbent was 0.998 which indicated that the adsorption process takes place in Chemisorption, the chemical bonding that occurs between chitosan adsorbent chitosan-p-t-butylcalix[4]arene with Remazol Yellow FG dye. This was supported by K value that Langmuir provision was a provision related to the free energy or adsorption enthalpy. Low K value indicates that low adsorption energy and the desorption process rate is larger the rate of the sorption processes. The K values of chitosan are 21897.37 L/mol, and the K value of chitosan-p-t-butylcalix[4]arene is 406114.46 L/mol. It indicates that the desorption of chitosan was larger than the desorption from the chitosan-p-t-butylcalix[4]arene.

Determination of adsorption isotherms can also be done by calculating the adsorption energy. In which, if the energy of adsorption is less than 20 kJ/mol, the adsorption follows Freundlich isotherm, while adsorption energy at $>$ 20 kJ/mol follows the Langmuir isotherm [18]. Based on calculations, the adsorption energy of chitosan adsorbent was 24.92 kJ/mol, while the adsorption energy of chitosan-p-t-butylcalix[4]arene adsorbent was 32.21 kJ/mol, which states that chitosan adsorbent and chitosan-p-t-butylcalix[4]arene follows the Langmuir isotherm. Adsorption energy value can be determined by the Gibbs free energy equation. Gibbs, free energy was negative indicates the adsorption process takes place spontaneously [13], from the calculation of the energy obtained adsorption for the both adsorbent was negative. Thus, adsorption process of chitosan and chitosan-p-t-butylcalix[4]arene adsorbents were spontaneous.

3.4. Adsorbent characterization With FTIR

Characterization of adsorbent was using an infrared spectrophotometer has been conducted to identify the functional groups contained in the adsorbent. FTIR spectra results of Remazol Yellow FG dye as well as a combination of both chitosan before and after being used to adsorb dye Remazol Yellow FG can be seen in Figure 4.

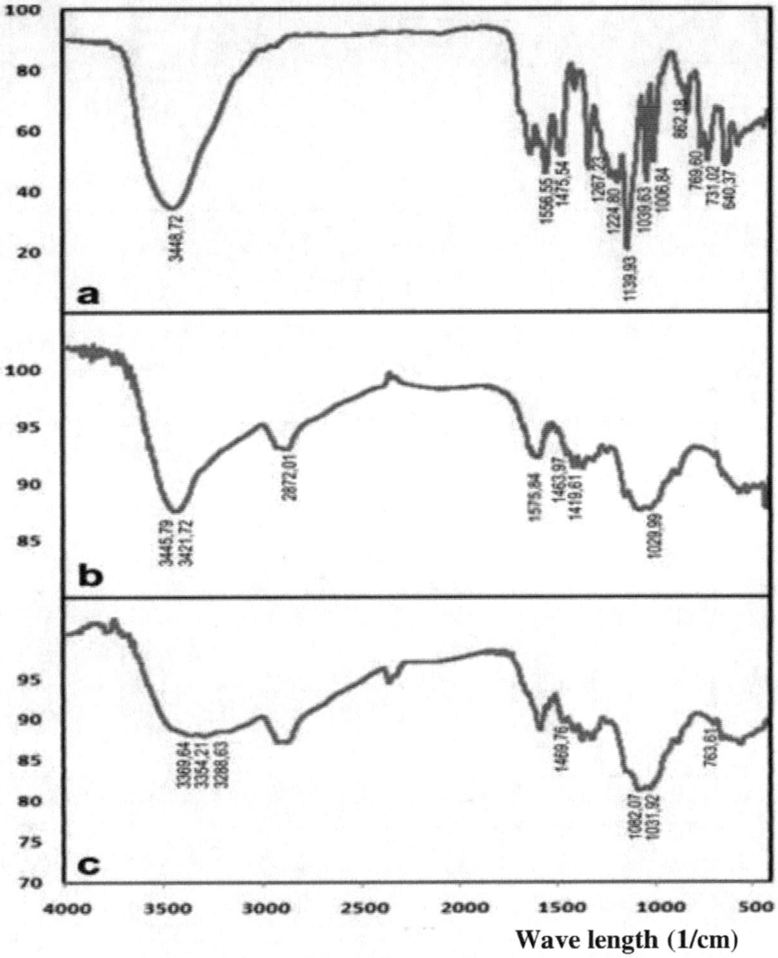

Figure 4. FTIR spectrum of Remazol Yellow FG dye (a), chitosan before (b) and after (c) adsorption Remazol Yellow FG dye.

Based on FTIR spectra can be obtained information for functional groups contained within the adsorbent. FTIR spectra in Figure 4 can be interpreted that after being used in the adsorption, chitosan provides a wide absorption at 3369.64 cm^{-1} region that showed the hydrogen bonds. That hydrogen bonding occurs between chitosan adsorbent and Remazol Yellow FG dye.

Adsorption peak on the wave number 1082.07 and 1031.92 cm^{-1} indicated a C-O bond, the wavelength of 1253.73 cm^{-1} indicated a S=O bond and a wavelength of 763.61 cm^{-1} indicated the presence of the -Cl bond on adsorbent which has adsorbed with Remazol Yellow FG dye [20].
The groups contained in Remazol Yellow FG dye. This shows that the chitosan adsorbent can absorb the Remazol Yellow FG dye. The other peaks that are in the dye structure do not appear in clearly due to its low intensity.

Remazol Yellow FG FTIR analysis of chitosan-p-t-butylcalix[4]arene before and after used to adsorb Remazol Yellow FG dye can be seen in Figure 5.

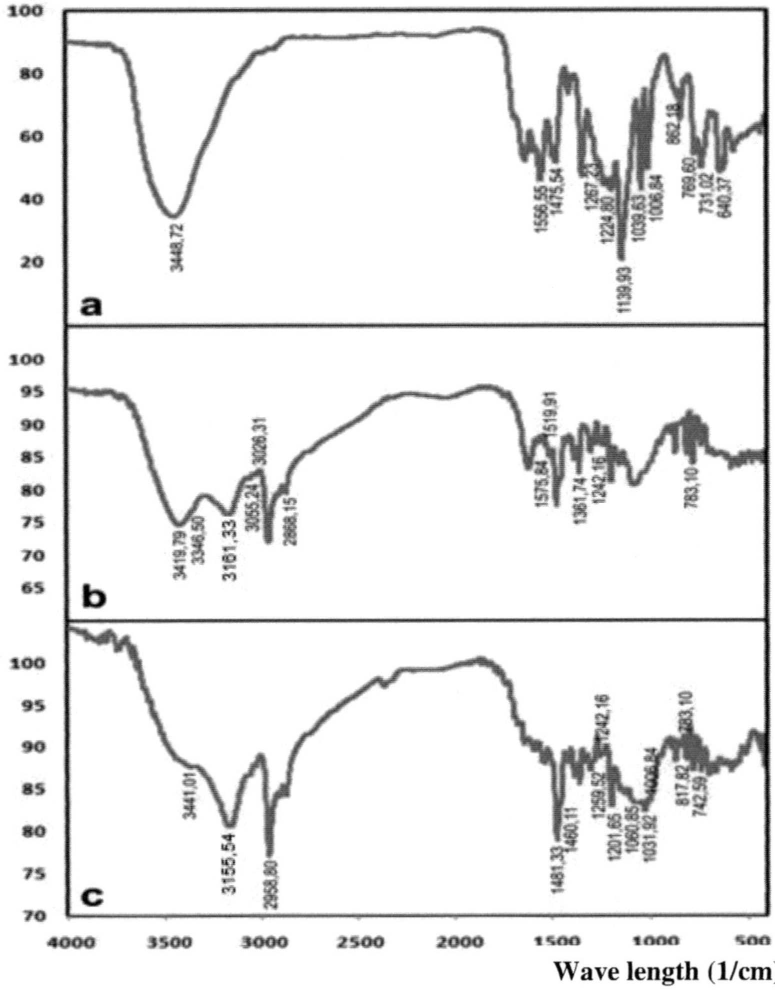

Figure 5. FTIR spectra of the dye Remazol Yellow FG (a), chitosan-pt-butylcalix[4]arene before adsorption, and (b) after adsorption. The adsorbate is the Remazol Yellow FG (c)

FTIR spectrum in Figure 5 shows that the chitosan-p-t-butylcalix[4]arene adsorbent after adsorption showed broad absorption at 3441.01 area. This indicated the presence of hydrogen bonding that occurs between the adsorbent with Remazol Yellow FG dye, besides the difference in the pattern with the advent of peaks at wave number 1060.85; 1031.92; 1006.84 is the group C=S, peak at around the wave number 1201.65 is a group S=O and peak at wave number around 783.10; 742.59 was a -Cl groups [20].

The groups contained in Remazol Yellow FG dye there was also appeared in the adsorbent which had adsorbed. This shows that the chitosan-p-t-butylcalix[4]arene adsorbent can absorb the Remazol Yellow FG dye. Other peaks that are in the dye structure does not appear in clearly because it had a low intensity thus covered by the peaks of chitosan-p-t-butylcalix[4]arene.

3.5. Adsorbent characterization By SEM - EDX

Chitosan adsorbent morphology before and after adsorption were characterized using Scanning Electron Microscopy (SEM) as shown in Figure 6. Based on Figure 6, visible differences between the structure of chitosan adsorbent before and after adsorbed. Chitosan adsorbent which has not been adsorbed was shown in Figure 6a, shows a slotted structure which allows the adsorbate dye can meet the gap on the surface of chitosan. Chitosan adsorbent which has adsorbed shown in Figure 6b, where the visible surface structure has been covered by Remazol Yellow FG dye. This is supported by the results which showed an increase EDX atoms Na, S, and Cl in chitosan adsorbent.

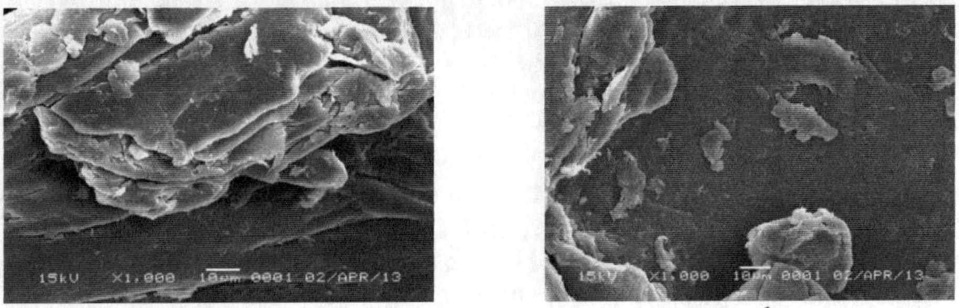

a b

Figure 6. SEM Results of chitosan Before Adsorption (a), after Adsorption (b)

SEM analysis result of chitosan-p-t-butylcalix[4]arene adsorbent before and after the adsorption was shown in Figure 7. Based on Figure 7 Visible differences between the structure of adsorbent chitosan-p-t-butylcalix[4]arene before and after adsorbed. Chitosan-p-t-butylcalix[4]arene adsorbent that has not adsorbed shown in Figure 7a shows the structure more porous compared to chitosan, which allows adsorbate dye can fulfill porous more than chitosan. chitosan-p-t-butylcalix[4]arene adsorbent that has been adsorbed shown in Figure 15b, where the visible surface structure has been coated with Remazol Yellow FG dye. This was also supported by EDX results which showed an increase of Na, S and Cl atoms in the chitosan-p-t-butylcalix[4]arene adsorbent.

a b

Figure 7. SEM Results of chitosan-p-t-butylcalix[4]arene before Adsorption (a), after Adsorption (b).

4. Conclusion

Adsorption of the dye Remazol Yellow FG using chitosan and chitosan-p-t-butylcalix[4]arene following the adsorption pseudo second order kinetics of the Ho equation, while the isotherm adsorption test for both adsorbents tends to follow the Langmuir isotherm.

Acknowledgement

The financial support from Indonesian Directorate General of Higher Education through the research project of Penelitian Unggulan Perguruan Tinggi (PUPT) 2015 is greatly appreciated.

References

[1] Atmaji P, Wahyu P and Edi P 1999 Daur Ulang Limbah Hasil Pewarnaan industri Tekstil *Jurnal Sains dan Teknologi Indonesia* **1** 9-15
[2] Isminingsih G, L Djufri and Rassid 1978 *Pengantar Kimia Zat Warna* Institut Teknologi Tekstil Bandung
[3] Rasjid D, G A Kasoenarno, Astini S and Arifin L 1976 *Teknologi Pengelantangan Pencelupan dan Pencapan* Institut Teknologi Tekstil Bandung
[4] Rakhmawati E 2007 *Pemanfaatan Kitosan Hasil Deasetilasi Kitin Cangkang Bekicot sebagai Adsorben Zat Warna Remazol Yellow* Skripsi FMIPA Universitas Sebelas Maret Surakarta.
[5] Jain V K, Pillai S G, Pandya R A, Agrawal Y K and Shrivastav P S 2005 Selective Extraction, Preconcentration, and Transport Studies of Thorium(IV) Using Octa-Functionalized Calix[4]resorcinarene-Hydroxamic Acid *Anal. Sci.* **21** 129-135
[6] Sivaiah M V, Venkatesan K A, Sasidhar P, Krishna R M and Murthy G S 2004 Ion Exchange Studies of Cerium(III) on Uranium antimonate *J. Nucl. Radiochem. Sci.* **5** 7-10
[7] Sardjono Ratningsih Eko 2007 *Sintesis dan Penggunaan Tetramer Siklis Seri Kaliksresorsinarena, Alkoksialkilsarena, dan Alkenilkaliksarena untuk Adsorpsi Kation Logam Berat* Disertasi FMIPA Universitas Gajah Mada Yogyakarta
[8] Suwarsa S 1998 Penyerapan Zat Warna Tekstil BR Red HE 7B oleh Jerami Padi *Jurnal JMS* **3** 32–40
[9] Mufrodi Z, Widiastuti N and Kardika R C 2008 Adsorpsi Zat Warna Tekstil Dengan Menggunakan Abu Terbang*(Fly Ash)* Untuk Variasi Massa Adsorben Dan Suhu Operasi, *Prosiding Seminar Nasional Teknoin Bidang Teknik Kimia dan Tekstil,* ISBN : 978-979-3980-15-7
[10] Emma R M T, Suhartono, Dahrul S and Sugiyono 2004 Karakterisasi Kitosan Hasil Deasetilasi *Enzimatis oleh Kitin Deasetilasi Isolat Bacilus Papandayan K29-14*
[11] Mahatmanti, F.W, Woro S., 2003, Kajian Termodinamika Penyerapan Zat Warna Indikator Metil Orange (MO) Dalam Larutan Air Oleh Adsorben Kitosan. *JSKA*. vol. **VI**, page. 1-19.
[12] Linane P and Shinkai S 1994 Calixarenes: Adaptable Hosts Par Exellence *Chem. Ind.* **32** 811-814
[13] Han Xiulin, Wang Wei, and Ma Xiaojian 2011 Adsorption characteristics of methylene blue onto low-cost biomass material lotus leaf *Journal Chemical Engineering Elsevier* **171** 1-8
[14] Handayani D S, Kusumaningsih T and Restuti A 2014 Modification and Characterization a New Polymer Chitosan-p-t-b-calix[4]arenes *Prosiding ICOPIA part of series: Advance in Physic Research* Atlantis Press 21-25
[15] Sakkayawong N, Thiravetyan P and Nakbanpote W 2005 Adsorption Mechanism of Synthetic Dye Wastewater By Chitosan *Journal of Colloid and Interface Science* **286** 36-42
[16] Cahyaningrum S E, Narsito Sri J S and Rudiana S 2008 Pemanfaatan Kitosan Limbah Udang Windu (Penaus monodon) Sebagai Adsorben Ion Logam Ca(II) dalam Medium Air *J. Kimia Lingkungan* **10** 59-65
[17] Kolesorgu S 2006 *Comparativ Adsorption Studies of Heavy Metal Ions on Chitin and Chitosan Biopolymers* Tesis Izmir Institute of Technology Turki

[18] Sadewo Gigih 2013 *Adsorpsi Zat Warna Remazol Yellow FG Menggunakan Adsorben Kopoli(Eugenol-Dvb) Terimpregnasi C-Heksil kaliks[4]Resorsinarena* Skripsi FMIPA Universitas Sebelas Maret Surakarta
[19] Utomo S B, Jumina and Wahyuningsih T D 2009 The Adsorption Of Pb(II) And Cr(III) By Polypropylcalix[4]Arene Polymer *Indo. J. Chem* 2009 **9** 437–444
[20] Silverstein D F, Atkins P W and Morill T C 1984 *Penyelidikan Spektometrik Senyawa Organik* Edisi Keempat Erlangga Jakarta

Photocatalytic removal of 2,4-dichlorophenoxyacetic acid herbicide on copper oxide/titanium dioxide prepared by co-precipitation method

Shu Chin Lee[1], Norhasnita Hasan[2], Hendrik O. Lintang[1], Mustaffa Shamsuddin[2] and LenyYuliati[1*]
[1]Centre for Sustainable Nanomaterials, Ibnu Sina Institute for Scientific and Industrial Research, Universiti Teknologi Malaysia, 81310UTM Johor Bahru, Johor, Malaysia
[2]Department of Chemistry, Faculty of Science, Universiti Teknologi Malaysia, 81310 UTM, Johor Bahru, Johor, Malaysia

[*]E-mail: leny@ibnusina.utm.my

Abstract. In this work, suppression of the charge recombination on the titanium dioxide (TiO_2) was reported by the addition of copper oxide (CuO), which led to a higher activity of TiO_2 for removal of 2,4-dichlorophenoxyacetic acid (2,4-D) herbicide. A series of CuO/TiO_2 with CuO loadings of 0.1-1 wt% was prepared through a co-precipitation method. X-ray diffraction patterns revealed that the presence of CuO could not be detected as the low loading amount of CuO might have good dispersion on the surface of TiO_2. Diffuse reflectance UV-visible spectra suggested that low loading amount of CuO did not influence the optical property of TiO_2. Fluorescence spectroscopy revealed that TiO_2 possessed a dominant emission peak of 407 nm at an excitation wavelength of 218 nm. The increasing loading amount of CuO decreased the emission intensity of TiO_2, suggesting the successful reduction of charge recombination.After irradiation under UV light for 1 h, CuO(0.1 wt%)/TiO_2 gave the highest percentage removal of the herbicide among the samples. The optimum loading amount of CuOmight improve the charge separation and reduce the electron-hole recombination on TiO_2 without blocking the active sites, thus leading to the improved photocatalytic activity. This work showed that CuO/TiO_2 is a potential photocatalyst for environmental remediation.

1. Introduction

Water contamination is a great concern that gains attentions due to the adverse effects it will bring to the ecosystem. The herbicide is one of the contaminants that can be detected from water source because of its wide usage for weeds control to increase production of crops [1, 2]. In order to have a clean water source, efficient water treatment approaches must be developed.

Photocatalysis has been considered as one green approach for environmental remediation since organic pollutants can be mineralized to non-harmful end products such as carbon dioxide, water and inorganic ions. Titanium dioxide (TiO_2) photocatalyst has been established for degradation of various organic pollutants due to several advantages such as low toxicity, low cost, good stability and good photocatalytic activity. Although TiO_2 possesses high photocatalytic activity as compared to many other photocatalysts, its photocatalytic efficiency is still limited by the high recombination rate of charge carriers. Various modifications have been done on TiO_2 in order to further improve its photocatalytic performance, such as using noble metals, metal oxides, non-metal ions and dyes [3, 4].

Recently, roles of cocatalyst in enhancing the photocatalytic efficiency of a semiconductor photocatalyst have been highlighted due to its capability in promoting charge separation and suppressing photo corrosion of a semiconductor photocatalyst [5-9]. Among types of cocatalyst, transition metal compounds are especially the topic of interest because they have advantages of earth abundance and low cost [5]. One of the most studied transition compounds is copper species. It has been reported that copper species-modified TiO_2 showed enhanced activity up to 3.4 and 1.6 times higher than the unmodified TiO_2 for photocatalytic degradation of organic pollutants such as 2-naphthol [6] and herbicides [7, 8], respectively.

In the present work, CuO modified TiO_2 photocatalysts prepared via a co-precipitation method was evaluated for removal of 2,4-dichlorophenoxyacetic acid (2,4-D) herbicide, which is one of the commonly used herbicides. The effect of CuO cocatalyst on the properties and photocatalytic activity of TiO_2 was investigated.

2. Experimental

2.1. Preparation of samples

1 g of the commercial Hombikat UV100 TiO_2 (UV100, Sachtleben, Germany) without any treatments was weighed and transferred into a clean beaker. After addition of 20 mL double distilled water, the mixture was ultrasonicated for 5 minutes. The solution of copper(II) nitrate trihydrate (Sigma-Aldrich, 99-104%)with different weight percentages (wt%), which were 0.1, 0.25, 0.5, and 1.0 wt%, was added into the beaker while stirring, followed by the addition of ammonium hydroxide (Sigma-Aldrich, 28%) (dropwise) to adjust the pH to 9. After stirring for 1 hour, the mixture was dried at 55 °C until dried. The dried powder of TiO_2 was then calcined at a rate of 10 °C min^{-1} to reach a temperature of 300 °C and was tempered for 4 hours. The prepared photocatalysts were denoted as CuO(x)/TiO_2, where x showed the loading of CuO (in wt%).

2.2. Characterizations of samples

The TiO_2 and the prepared CuO/TiO_2 samples were characterized using various types of instruments. A Bruker D8 Advance diffractometer was used to record X-ray diffraction (XRD) patterns of the prepared samples, which was analysed at the 2θ range of 20-70° with a scan rate of 0.05° s^{-1}. A Shimadzu UV-2600 diffuse reflectance UV-Visible (DR UV-Vis) spectrophotometer was used to measure the absorption spectra of samples using barium sulfate ($BaSO_4$) as a reference. A JASCO FP-8500 spectrofluorometer was used to identify the emission sites of the samples, which the emission of TiO_2 was monitored at an excitation wavelength of 218 nm. For the determination of charge separation efficiency, electrochemical impedance spectra (EIS) were recorded on a Gamry Interface 1000 potentiostat/galvamostat/ZRA. For the preparation of working electrode for the measurements of EIS, 10 mg photocatalyst and 10 μL nafion (Aldrich, 5%) were dispersed in 1 mL water and sonicated for 10 min to form a homogeneous colloidal suspension. 20 μL photocatalyst colloid was deposited onto the working electrode of a screen printed electrodes (SPE, Dropsens DRP-110). The SPE was immersed in 6 mL electrolyte containing 0.1 M sodium sulfate (Fisher Chemical) and 2.5 mM potassium ferricyanide (Riedel-de Haën). The EIS was recorded at 10 mV over a frequency range of 1 MHz to 100 mHz.

2.3. Photocatalytic tests

The photocatalytic activities of the TiO_2 and the prepared CuO/TiO_2 samples were evaluated for the removal of 2,4-D herbicides under UV light irradiation. Typically,0.5mM 2,4-D was prepared by dissolving an appropriate amount of 2,4-D (Sigma, 98%) in double distilled water. 28 mL of 2,4-D solution was then transferred into a photocatalytic reactor containing 28 mg photocatalyst. The suspension was stirred under the dark condition for 1 hour to achieve adsorption-desorption equilibrium. After 1 hour, 3 mL of the suspension was taken and filtered for the measurement of absorbance monitored at 283 nm using a Shimadzu UV-2600 UV-Vis spectrophotometer, which the

value was taken as initial absorbance, A_0. The 25 mL suspension in an open glass reactor was then irradiated by UV light from the top (200 W Hg-Xe lamp, the dominant wavelength of UV light at 365 nm with light intensity of 15 mW/cm^2) for 1 hour at room temperature. After the reaction, the suspension was filtered and the absorbance of 2,4-D, A_t was measured.
Percentage of 2,4-D removal was calculated according to the equation (1).

$$\frac{A_0 - A_t}{A_0} \times 100\% \qquad (1)$$

where A_0 showed the initial absorbance of 2,4-D after adsorption in dark condition, while A_t showed the 2,4-D absorbance after the reaction.A high-performance liquid chromatography (HPLC, Shimadzu Prominence LC-20A) equipped with a Hypersil GOLD PFP column was also used to analyse the intermediate of the photocatalytic reaction.

3. Results and Discussion

Figure 1 shows the XRD patterns of the TiO_2 and the prepared CuO/TiO_2 samples. The TiO_2 showed diffraction peaks at 2θ of 25.35° (101), 38.10° (004), 48.05° (200), 54.55° (105), and 62.60° (204) corresponding to the anatase phase [8, 10, 11], which are in good agreement with the standard values reported by International Centre for Diffraction Data (ICDD) number 211272. As for the CuO/TiO_2 series, the addition of CuO did not affect the diffraction patterns of TiO_2, in which the characteristic diffraction peaks of anatase phase of TiO_2 remained in all samples. On the other hand, there were no characteristic peaks of CuO found on the XRD patterns according to the ICDD number 801916, as has been reported on the CuO nanoparticles [12, 13]. The absence of CuO peaks on the CuO/TiO_2 series suggested that addition of 0.1 to 1.0 wt% could not be detected by XRD. This might be due to the low amount of CuO added to TiO_2 or the low amount of CuO added might be dispersed well on the surface of the TiO_2, which the result was in a good agreement with the ones reported elsewhere [7, 14].

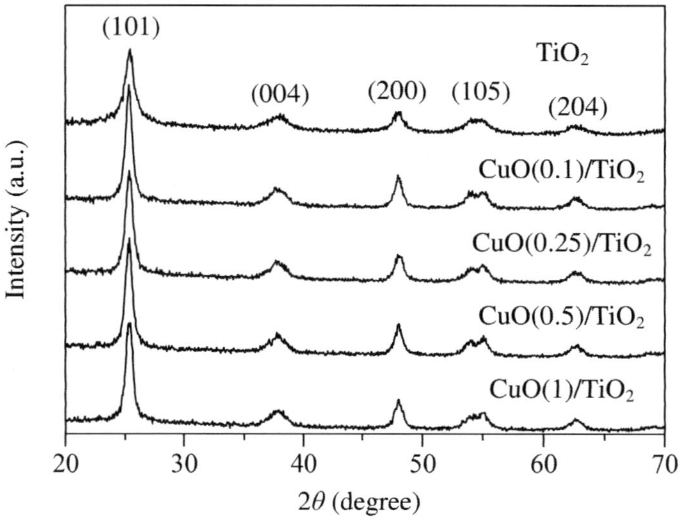

Figure 1. XRD patterns of the TiO_2 and the CuO/TiO_2.

In order to investigate the effect of CuO on the crystallite size of the TiO_2, the crystallite size of all the samples were calculated using Scherrer equation as shown in equation (2) based on peaks obtained from XRD patterns.

$$t = \frac{K\lambda}{\beta\cos\theta} \qquad (2)$$

IOP Conf. Series: Materials Science and Engineering **107** (2016) 012012 doi:10.1088/1757-899X/107/1/012012

where t is the average crystallite size in nm, while K is dimensionless shapes factor generally taken as 0.89, λ is the wavelength of the X-ray equals to 1.540, β is full width at half maximum (FWHM) at 2θ of 25.35° and θ is Bragg's angle. Table 1 shows the crystallite size of the TiO_2 and CuO/TiO_2 series. It was found that the crystallite size of the TiO_2 was 9.3 nm. Crystallinity of the TiO_2 was slightly improved after addition of CuO, which could be due to the thermal treatment during the synthesis process [3]. No significant changes of the crystallite size were obtained on the TiO_2 with the increase loading amount of CuO, in which the CuO/TiO_2 series have a crystallite size in the range of 11.3-12.1 nm.

Table 1. Crystallite size and bandgap energy of the $TiO_2$and the CuO/TiO_2 series.

Sample	Crystallite size (nm)[a]	E_g (eV)[b]
TiO_2	9.3	3.30
$CuO(0.1)/TiO_2$	12.1	3.28
$CuO(0.25)/TiO_2$	12.1	3.29
$CuO(0.5)/TiO_2$	11.3	3.29
$CuO(1)/TiO_2$	11.3	3.28

[a] Crystallite size was calculated using the Scherrer equation.
[b] Bandgap energy (E_g) was determined from the Tauc plot.

Figure 2 shows the DR UV-Vis spectra of the TiO_2 and CuO modified-TiO_2 samples. The TiO_2 exhibited an absorption peak at around 300 nm, as also reported elsewhere [6, 10]. With the addition of CuO, the optical property of TiO_2 was not much altered. This might be due to the low amount of CuO added to TiO_2.

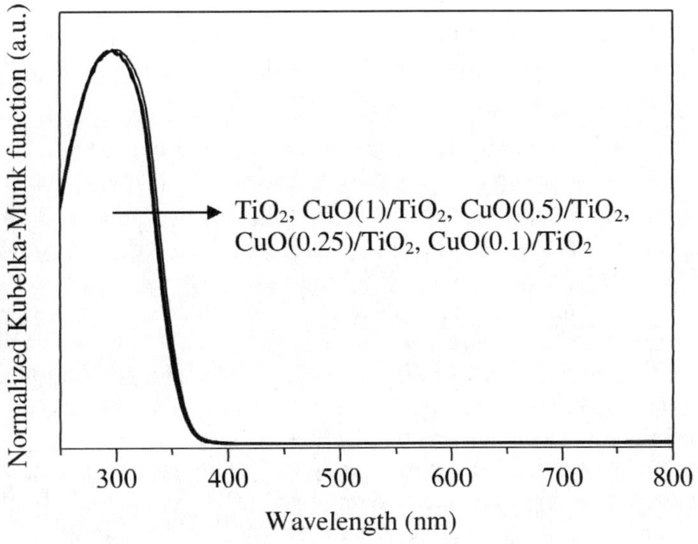

Figure 2. DR UV-Vis spectra of the TiO_2 and CuO/TiO_2.

Besides UV absorption (Figure 2), the effect of CuO on the bandgap energy (E_g) of the CuO/TiO_2 series was also determined. The E_g can be estimated from Tauc plot by plotting $(\alpha h\nu)^{1/2}$ versus hν using equation (3).

$$\alpha h\nu = A(h\nu - E_g)^{n/2} \quad (3),$$

where α is absorption coefficient, h is a Plank's constant, ν is the frequency of vibrant, A is a proportional constant, E_g is the bandgap energy and $n/2$ denotes the type of transition. It has been reported that anatase gives indirect transition [10]. Since the TiO_2 studied was consisted of anatase phase that prefers indirect transition, the n value should be 4. The E_g was determined from the value in x-intercept by taking the linear extrapolation in the plot of $(\alpha h\nu)^{1/2}$ versus $h\nu$, as shown in Figure 3. The E_g of the samples is summarized in Table 1. The TiO_2 gave E_g of 3.30 eV. With the addition of CuO, the E_g of TiO_2 was not much affected, which the value of CuO/TiO_2 were in a range of 3.28-3.29 eV. Since the added amount of CuO was low, no clear correlation was observed between the crystallite size and the band gap energy.

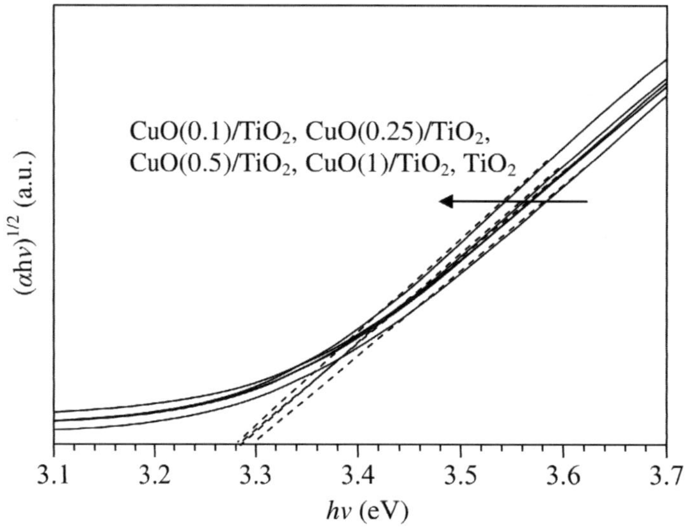

Figure 3. Tauc plots of the TiO_2 and CuO/TiO_2.

Figure 4 demonstrated the fluorescence spectra of the TiO_2 and CuO/TiO_2 series that were monitored at an excitation wavelength of 218 nm. There were three emission peaks obtained at 407, 466 and 562 nm, which the former one was the emission of the bandgap transition while the latter two were emission signals originating from the charge transfer transition of an oxygen vacancy trapped electron. The results were in good agreement with the reported TiO_2 [11]. As shown in Figure 4, the unmodified TiO_2 showed the highest emission intensity. It was found that addition of CuO reduced the emission intensity of the TiO_2, which the emission intensity of TiO_2 decreased gradually with the increasing of CuO loading amount, suggesting that CuO was loaded on the TiO_2.

In order to investigate the photocatalytic efficiency of the prepared photocatalysts, the TiO_2 and CuO/TiO_2 series were evaluated for photocatalytic removal of 2,4-D under UV light irradiation for one hour at room temperature. It was confirmed that the 2,4-D was stable and not degraded under UV light irradiation in the absence of photocatalyst. On the other hand, TiO_2 gave 2,4-D adsorption of about 30%, while the CuO/TiO_2 series gave 13-18% after one hour of the adsorption process. As shown in the equation (1), the photocatalytic activity was calculated by taking account of the remained 2,4-D after adsorption process as the initial concentration. Therefore, the obtained photocatalytic activities were solely caused by the photocatalytic reaction. As shown in Figure 5, the unmodified TiO_2 showed

21% of 2,4-D removal. An appropriate loading amount of CuO was found to improve the photocatalytic activity of the TiO_2 for removal of 2,4-D. It was obtained that addition of 0.1 wt% CuO enhanced the photocatalytic activity of the TiO_2up to 38%, which the activity was 1.8 times higher than the unmodified TiO_2. Further addition of CuO from 0.25-1 wt% decreased the activity of the TiO_2, in which 1 wt% CuO showed activity similar to the one of the unmodified TiO_2. These results clearly suggested that small amount of CuO (0.1 wt%) was the optimum loading to achieve the improved activity of the TiO_2.The formation of the intermediate product, which was 2,4-dichlorophenol (2,4-DCP), was confirmed by HPLC on all photocatalysts. This is in good agreement with previously reported literature [15,16], suggesting that the 2,4-DCP was a crucial intermediate in the photocatalytic removal of 2,4-D.

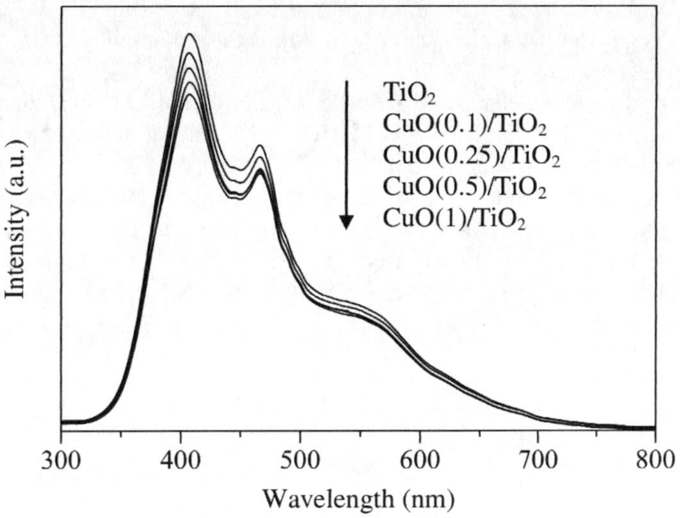

Figure 4. Fluorescence spectra of the TiO_2 and CuO/TiO_2 series monitored at an excitation wavelength of 218 nm.

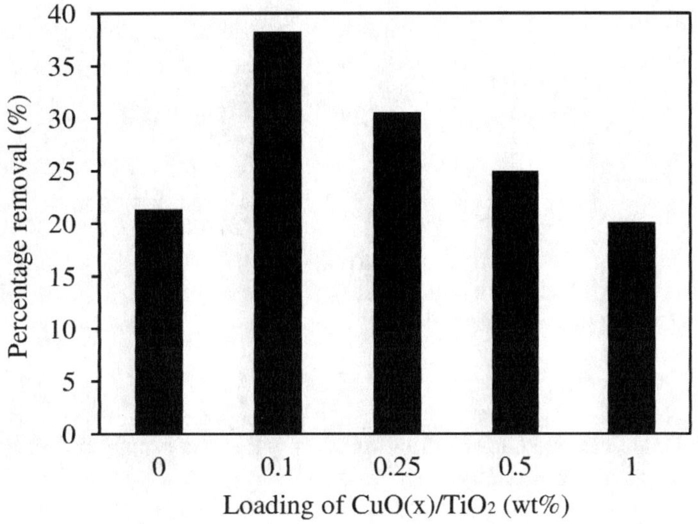

Figure 5. Photocatalytic removal of 2,4-D on the TiO_2 and CuO/TiO_2 series under UV light irradiation for 1 hour.

The photocatalytic activities obtained on the CuO/TiO_2 revealed the importance of CuO cocatalyst loading amount to improve the photocatalytic performance of TiO_2. A suitable amount of cocatalyst loaded on a semiconductor could trap photogenerated electrons efficiently to suppress electron-hole recombination, which improves the photocatalytic efficiency of a semiconductor. Since photoluminescence study has been used to investigate and associate recombination of electron-hole pairs of semiconductors [11], the high photocatalytic activity obtained on the $CuO(0.1)/TiO_2$ could be related to the suppressed electron-hole recombination, as supported by the fluorescence results (Figure 4). The reduced emission intensity of the TiO_2 with the increased loading of CuOsuggested that lower amount of CuO (0.1 wt%) acted to trap electrons and suppress the electron-hole recombination, leading to high photocatalytic activity. However, further reduced emission intensity with the increasing loading of CuO (above 0.1 wt%) might be due to the covering of TiO_2 emission sites by the loaded CuO, as can be suggested from the decreased photocatalytic activity of TiO_2 with CuO loading of higher than 0.1 wt% (Figure 5).

In order to support the good charge separation of the $CuO(0.1)/TiO_2$ that lead to the high photocatalytic activity, EIS of TiO_2 and CuO/TiO_2 with CuO loading amount of 0.1 and 1 wt% were recorded as shown in Figure 6. Impedance arc radius of the $CuO(0.1)/TiO_2$ was much smaller than the unmodified TiO_2, which indicated that conductivity of the sample was enhanced when 0.1 wt% CuO was loaded, revealing the better charge separation. When 1 wt% CuO was loaded, the impedance arc radius was similar to the one of the unmodified TiO_2. It is therefore suggested that only suitable loading of CuO would show good charge separation on the TiO_2 that would improve the photocatalytic activity.

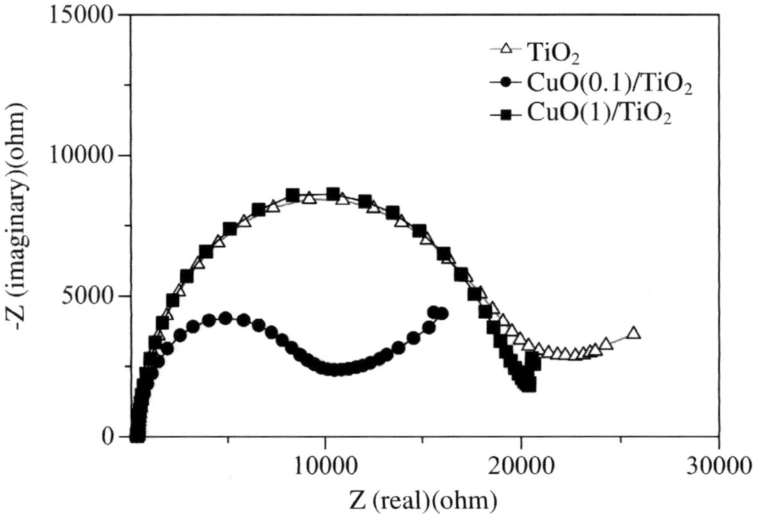

Figure 6. EIS measurements of the TiO_2 and CuO/TiO_2 with CuO loading amount of 0.1 and 1 wt%.

4. Conclusions

A series of CuO/TiO_2 with various CuO loading of 0.1-1 wt% was prepared through a co-precipitation method. All samples were UV light-active and possessed E_g of about 3.3 eV. The CuO with 0.1 wt% was the optimum loading amount for the modification on the TiO_2, which showed 1.8 times higher photocatalytic activity than the unmodified TiO_2 for the removal of 2,4-D under 1 hour of UV light irradiation. The high photocatalytic activity obtained on the $CuO(0.1\ wt\%)/TiO_2$ could be due to the good dispersion of CuO on the surface of the TiO_2 as well as efficient charge separation and suppressed electron-hole recombination on the TiO_2 without blocking the active sites.

Acknowledgements
The research leading to these results has received funding from the European Union Seventh Framework Program (EU-FP7) under a 4G-PHOTOCAT grant (agreement no.: 309636). This work has been also financially supported by Ministry of Higher Education (MOHE) and Universiti Teknologi Malaysia (UTM, Malaysia) through the international contract matching grant (cost center code: R.J130000.7626.4C035).

References
[1] Majzik E S, Tóth F, Benke L and Kiss Z 2006 SPE-LC-MS-MS determination of phenoxy acid herbicides in surface and ground water *Chroma.***63** S105-9
[2] Barbash J, Thelin G, Kolpin D and Gilliom R 2001 Major herbicides in ground water: results from the National Water-Quality Assessment *J. Environ. Qual.***30** 831-45
[3] Gupta S and Tripathi M 2011 A review of TiO_2 nanoparticles *Chin. Sci. Bull.***56** 1639-57
[4] Lazar M, Varghese S and Nair S 2012 Photocatalytic water treatment by titanium dioxide: Recent updates *Catalysts***2** 572-601
[5] Ran J, Zhang J, Yu J, Jaroniec M and Qiao S Z 2014 Earth-abundant cocatalysts for semiconductor-based photocatalytic water splitting *Chem. Soc. Rev.***43** 7787-812
[6] Jin Q, Fujishima M, Iwaszuk A, Nolan M and Tada H 2013 Loading effect in copper(II) oxide cluster-surface-modified titanium(IV) oxide on visible- and UV-light activities *J. Phys. Chem. C***117** 23848-57
[7] An X, Liu H, Qu J, Moniz S J A and Tang J 2015 Photocatalytic mineralisation of herbicide 2,4,5-trichlorophenoxyacetic acid: enhanced performance by triple junction Cu-TiO_2-Cu_2O and the underlying reaction mechanism *New. J. Chem.***39** 314-20
[8] Roslan N A, Lintang H O and Yuliati L 2015 Enhanced photocatalytic performance of copper-modified titanium dioxide prepared by UV reduction method *Adv. Mater. Res.***1112** 180-3
[9] Yang J, Wang D, Han H and Li C 2013 Roles of cocatalysts in photocatalysis and photoelectrocatalysis *Acc. Chem. Res.***46** 1900-9
[10] Valencia S, Marín J M and Restrepo G 2010 Study of the bandgap of synthesized titanium dioxide nanoparticules using the sol-gel method and a hydrothermal treatment *Open Mater. Sci. J***4** 9-14
[11] Cong Y, Zhang J, Chen F and Anpo M 2007 Synthesis and characterization of nitrogen-doped TiO_2 nanophotocatalyst with high visible light activity *J. Phys. Chem. C***111** 6976-82
[12] Kumar P P N V, Shameem U, Kollu P, Kalyani R L and Pammi S V N 2015 Green synthesis of copper oxide nanoparticles using aloe vera leaf extract and its antibacterial activity against fish bacterial pathogens *BioNanoSci.*5 135-9
[13] Srivastava S, Kumar M, Agrawal A and Dwivedi S K 2013 Synthesis and characterisation of copper oxide nanoparticles *IOSR J. Appl. Phys.***5** 61-5
[14] Xu B, Dong L and Chen Y 1998 Influence of CuO loading on dispersion and reduction behavior of CuO/TiO_2 (anatase) system *J. Chem. Soc., Faraday Trans.***94** 1905-9
[15] Terashima Y, Ozaki H, Giri R R, Tano T, Nakatsuji S, Takanami R and Taniguchi S 2006 Photocatalytic oxidation of low concentration 2,4-D solution with new TiO_2 fiber catalyst in a continuous flow reactor *Water Sci. Technol.* **54** 55-63
[16] Kim J and Choi W 2011 TiO_2 modified with both phosphate and platinum and its photocatalytic activities *Appl. Catal. B: Environ.* **106** 39-45

Analysis of diazinon pesticide using potentiometric biosensor based on enzyme immobilized cellulose acetate membrane in gold electrode

Mashuni[1*], L O A N Ramadhan[1], M Jahiding[2], Herniati[1]
[1]Department of Chemistry, Faculty of Mathematics and Natural Sciences, Halu Oleo University (UHO), Kendari, Sulawesi Tenggara, 93231
[2]Department of Physics, Faculty of Mathematics and Natural Sciences, Halu Oleo University (UHO), Kendari, Sulawesi Tenggara, 93231

[*]E-mail: mashuni2696@yahoo.com

Abstract. Biosensor for analysis of diazinon pesticide using Potentiometric transducer has been developed. The basic element of this biosensor was a gold electrode modified with an immobilized acetylcholinesterase enzyme layer formed by entrapment with glutaraldehyde crosslinked-cellulose acetate. The aim of the research is to determine the composition of glutaraldehyde crosslinked-cellulose acetate in the gold electrode which provide optimum performance of biosensors of diazinon pesticide analysis on characterization include a range of working concentration, sensitivity, and detection limit. The results showed the composition of the cellulose acetate 15% and glutaraldehyde 25% that obtain optimum performance in the measurement of diazinon pesticide with a range of working concentration of 10-6 ppm to 1 ppm, the value of sensitivity 20.275 mV/decade and detection limit 10-6 ppm. The use of cellulose acetate provides highly sensitive devices allowing the efficient analysis of pesticides. The response time of electrode is on the measurement of pesticide diazinon with concentration variation of 10-6 ppm to 1 ppm with response time is about 5 minutes.

1. Introduction

In the last few decades, the use of pesticide was dominated by two types of insecticide, i.e. carbamate and organophosphate compound. Both compounds were used a lot the agricultural process because of its low stability in the environment, compared to organochlorine compound [1]. The pesticide is a chemical material which is used to control the growth of pests, diseases, and weeds. The intensive use of pesticides in agriculture, especially organophosphate, has left a little residue in plants and been a problem for environment and health of living creatures, especially human [2]. Pesticide of diazinon is one of the insecticides from the group of organophosphate, used a lot in an agricultural business to eradicate the pests. Along with the intensive use of pesticide, the analysis of pesticide residue is very important to do because of its high toxicity and its serious risk for the health of human and environment [3].

The analytical method of pesticide residue used a lot is by using Gas Chromatography (GC) and High-Pressure Liquid Chromatography (HPLC). When GC or HPLC are used to detect the samples, many pretreatment procedures including extraction, purification, and concentration are required before analysis and thus, GC or HPLC method shows some disadvantages, such as being troublesome, time-consuming, and costly. Moreover, it needs requirement of solvent that has to be used in extraction. Therefore to overcome the weaknesses, a new method in the analysis of pesticide residue is used, i.e. by using a biosensor, whose application is more simple and cheap [4]. However, the biosensor method possesses some advantages, such as the simple operation of the sample preparation, low cost, and it needs only 25 min to complete the analytical process. Therefore, the potentiometric enzymatic membrane biosensor is likely to play an important role in OPs detection for agricultural products in the future [5].

A biosensor is an analytical tool which is a combination of biological component and physical transducer to detect a targetted compound. Biosensor consists of three elements, i.e. biological element, transducer, and electronic system processing a signal. One of the biological elements commonly used in designing is enzyme immobilized in a transducer. One of the enzymes that can be used in the biosensor is acetylcholinesterase [6]. The efficiency of carbamate and organophosphate as a pesticide and their toxicity to human and animals is caused by the ability of inhibiting enzyme group of a hydrolase, called esterase. The enzyme of acetylcholinesterase (AChE) is very important for central nerve system in human and insects [7].

In this research, the biosensor of pesticide of diazinon by immobilizing enzyme of acetylcholinesterase (AChE) in the material of cellulose acetate (CA) membrane and glutaraldehyde (GA) membrane functioning as a bifunctional reactant in gold wire. The success of a biosensor, few of them, depends on the ability of an enzyme of being bound in the surface of biosensor and still being active during the desired application, as well as the role of glutaraldehyde (GA) solution as an agent of bifunctional reactant between enzyme and membrane. Its mechanical strength and stability of membrane cellulose acetate are the reason of the user in this research.

2. Methods

2.1. Chemicals

The materials used in this research are: active material of pure diazinon pesticide, acetylcholinesterase (AChE) enzyme from Electrophorus electricus (electric eel) (Type C3389), acetylthiocholine chloride (ATCh) substrate (Type A5626), glutaraldehyde 25% from Aldrich-Sigma, cellulose acetate with Mr = 61000 g/mol, $Na_2HPO_4.12H_2O$, $NaH_2PO_4.H_2O$, KCl, acetone p.a (E. Merck), ethanol p.a (E. Merck), Gold wire, bronze wire, platinum wire, silver wire, tin wire, silica gel, parafilm and distilled water.

2.2. Procedure

Preparation of the phosphate buffer solutions of NaH_2PO_4. H_2O (A) and Na_2HPO_4. $12H_2O$ (B). Next, A and B were each made into phosphate buffer solution pH 8.0. A standard substrate solution of acetylthiocholine chloride was made in a solvent with phosphate buffer solution pH 8.0 at the concentration of 1 x 10-3 M. After that, diazinon pesticide solution 1 ppm -10-6 ppm was prepared by diazinon pesticide which then be diluted using ethanol in a flask. KCl 0.1 M solution was prepared by diluting 0.7445 gram of KCl in 100 ml quads. AChE enzyme from Electrophorus electricus Sigma 1.17 mg with the activity of 500 UN was diluted in a solvent with 9.0 ml buffer phosphate pH 8.0 and 1 ml KCl 0.1 M [8]. Cellulose acetate (CA) 15% were made by scaling 1.5 g CA, and then being diluted in 10 ml acetone. Glutaraldehyde solution (GA) 25% used in this study was produced from Aldrich-Sigma.

2.2.1. Design of Electrode Biosensor of Layered Wire

The design of biosensor based on acetylcholinesterase for analysis of diazinon is based on electrochemical analysis by using Potentiometer. Electrode biosensor consists of two electrodes, i.e. standard and working electrodes. Electrode of Ag/AgCl as the standard electrode is made through electrolysis of Ag wire as anode connected to positive pole of battery and platinum (Pt) wire as cathode connected to negative pole of the battery where silver (Ag) wire was sunk in a solution of KCl 0.1 M for 30 minutes.

The enzymatic biosensor or working electrode bodies were made from a 7 cm bronze wire of 0.5 cm in diameter which was connected with a 2.0 cm Gold (Au) wire of 0.4 mm in diameter, which was then soldered with tin (Sn) wire. It was then put into a blue tape with the Gold wire (Au) sticking out at 1.5 cm which was used as the working electrode bodies. At the bodies of all electrode were the parafilm coiled around as Cu and Au wire. The edge of the electrode, which was Au wire, was immersed in homogeneous cellulose acetate (CA). Working electrode of diazinon pesticide biosensor is made of layering membrane (CA) with a concentration of 15% in the surface of the gold wire. After the membrane electrode was formed, the electrodes were washed using a distilled water at 3 times. Next, the part of Au wire layered with membrane CA was immersed in glutaraldehyde 25% (GA) for 8 hours. After that, the electrodes were washed with phosphate buffer pH 8.0. This formed membrane electrode which was then immersed in phosphate buffer pH 8.0 containing acetylcholinesterase enzyme for 48 hours (enzymatic biosensor). The enzymatic biosensor or working electrode which had not been used remained immersed in the phosphate buffer solution pH 8.0 at the temperature of 4 °C [9].

2.2.2. Measurement of Range of Working Concentration, Sensitivity, Detection Limit and Response Time of the Diazinon Pesticide Biosensor.

The prepared biosensor was immersed into acetylthiocholine chloride (ATCh) substrate solution with the concentration of 10-3 M, and the initial potential was recorded. The potential of ATCh substrate with diazinon pesticide inhibitor was measured by way of immersing into varying concentration between 10-6 - 1 ppm of the diazinon pesticide solution for 30 minutes, then immersed into ATCh solution, and the potential response values were finally measured under the optimal measurement conditions. Determining the response time is done in the substrate concentration of 10-3 M, i.e. when the measured potential is already constant. To determine response time potential of pesticides, analyte solutions are measured every interval of 1 minute until the constant value is gained.

3. Results And Discussion

The Working principle of the biosensor to analyze the diazinon pesticide is based on the enzyme of AChE. The biological component (enzyme) was immobilized in a matrix of polymer membrane of CA which was connected to the gold wire integrated with transducer signal of Potentiometer. Component of AChE enzyme in the membrane of working electrode of biosensor functions as an electroactive sensor. The potentiometric method is based on electricity of a solution substance analyzed in an electrochemical cell. The potentiometric method used in working electrode in the type of layered wire. The use of working electrode of layered wire to detect diazinon pesticide is because of having some advantages, i.e. quick analysis process, simple, usable for routine analysis, and having a small size so it is easier to carry. Work of biosensor is very influenced from the thickness of membrane as supporting material in working electrode of the biosensor. The process of making membrane is conducted by using phase inversion method of instantaneous liquid-liquid demixing, i.e. change of polymer form from liquid to solid by removing solvent using a solvent having a different character. In this research, the solvent of working electrode of biosensor that has been layered by a membrane of cellulose acetate is then removed by using a distilled water and enzyme of AChE with the cross-linking method using 25% of glutaraldehyde is immobilized. Glutaraldehyde can form a bound connecting enzyme and cellulose acetate 15% as supporting material. The success of work of pesticide biosensor depends not only on the potentiometric transducer and membrane as supporting material but also on the existence of enzyme immobilized in electrode membrane of biosensor functioning as substrate catalysator being a product measurable in potentiometer transducer in the form of potential signal.

IOP Conf. Series: Materials Science and Engineering **107** (2016) 012013 doi:10.1088/1757-899X/107/1/012013

3.1. Performance of Diazinon Pesticide Biosensor

3.1.1. Range of Working Concentration, Sensitivity and Detection Limit.

Work test of pesticide biosensor is based on the mechanism of inhibiting diazinon pesticide towards enzyme of AChE in the hydrolyzing substrate of acetylthiocholine into thiocholine and acetic acid as shown in Figure 1.

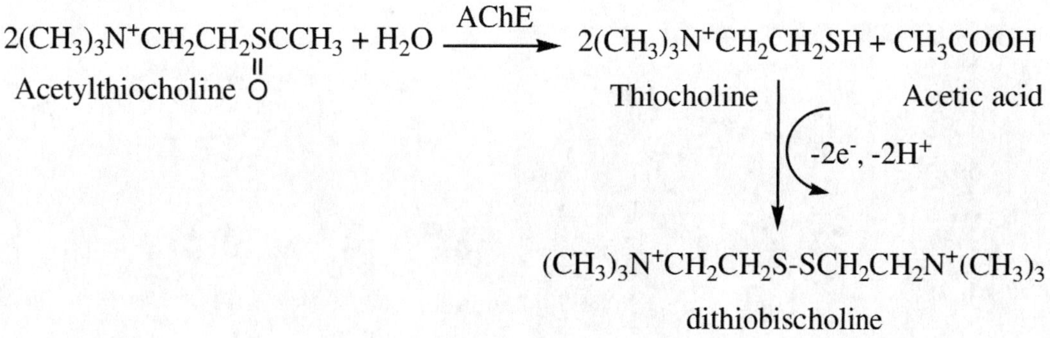

Figure 1. Hydrolysis reaction of acetylthiocholine substrate with AChE enzyme.

Interaction happening between acetylthiocholine substrate and AChE enzyme in the reaction of hydrolysis produces thiocholine and acetic acid (Figure 1), with inhibition from diazinon pesticide so it causes phosphorylation i.e. insertion of the phosphate group in serine amino acid inactive side of enzyme which can decrease the working activity of AChE enzyme in hydrolyzing substrate [10]. AChE catalyzes the cleavage of acetylthiocholine chloride substrate to thiocholine, which was oxidized to give a dithiobischoline compound. All the experiments were carried out in phosphate buffer solution at pH 8.0 at room temperature. The response of the biosensor showed a good linearity range with an incubation time of 30 min for diazinon (Table 1). Work test of diazinon pesticide biosensor conducted covers: range of working concentration, sensitivity, and detection limit.

Table 1. The range of working concentration and Nernst factor (sensitivity) and detection limit of the biosensor of diazinon pesticide for the composition of electrode membrane CA 15%, GA 25%.

Substrate concentration [ATCh] M	Inhibitor concentration [diazinon] ppm	Potential (mV)
10^{-3}	1	30.7
10^{-3}	10^{-1}	42.1
10^{-3}	10^{-2}	58.1
10^{-3}	10^{-3}	78.6
10^{-3}	10^{-4}	109.1
10^{-3}	10^{-5}	137.4
10^{-3}	10^{-6}	139.4
Nernst factor (sensitivity)	20.275 mV	
R^2 (coefficient of correlation)	0.9722	
Range of working concentration	10-6 - 1 ppm	

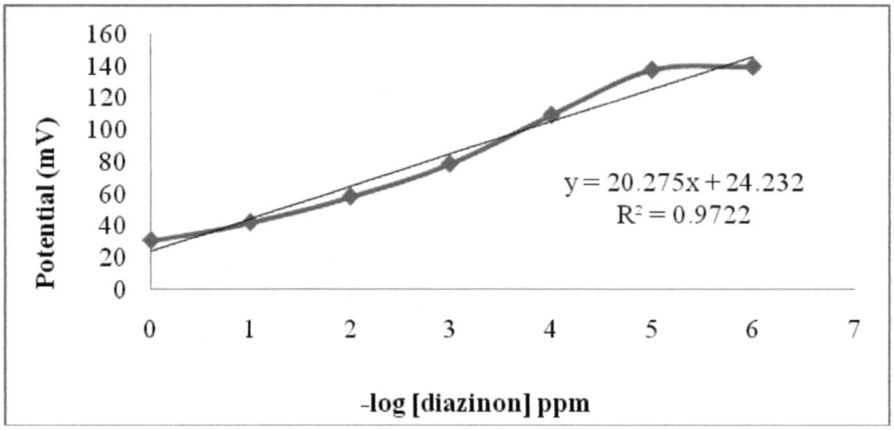

Figure 2. The curve of potential towards–log [diazinon] ppm for electrode membrane composition of CA 15%, GA 25%.

The value of sensitivity can be seen from Nernst factor value of 29.6 mV/decade in temperature of 24 °C ± 1 °C. It happened because oxidation of acetylthiocholine involved two electrons. Figure 2 shows the curve of potential towards –log [diazinon] ppm for membrane composition of CA 15% towards GA 25% in substrate concentration of 10-3 M and value of Nernst factor as well as measurement range. Electrode of the biosensor with the membrane of CA 15% and GA 25% has measurement range in diazinon pesticide concentration of 10-6 ppm to 1 ppm. Based on the obtained data, it shows that value of regression approaches 1, i.e. 0.9722 and value of Nernst factor (sensitivity) is 20.275 mV/decade.

The detection limit is the smallest concentration of an analyte that is still able to be determined or detected by a biosensor after inhibition by the pesticide. Detection limit shows that the smallest concentration that still gives linear in the range of working concentration in this research is 10-6 ppm.

3.1.2. Response Time.

Pesticide biosensor is an enzyme-containing biosensor immobilized on the surface of the electrode that is able to give a specific response to the substrate [11]. The working principal of enzyme-based biosensor that was produced from immobilization of biological component (enzyme, bacteria, etc) on matrix membrane polymer integrated with transducer signal in the analyte. The biological components on the membrane of working electrode of biosensor function as an electroactive sensor. Response time is the time needed by a biosensor electrode to reach the constant potential. Determining the response time is done in the substrate concentration of 10-3 M, i.e. when the measured potential is already constant. To determine response time potential of pesticides, analyte solutions are measured every interval of 1 minute until the constant value is gained. The faster the time needed to reach the constant value, the better the quality of biosensor electrode.

Table 2. The response time of biosensor of diazinon pesticide for the composition of electrode membrane CA 15%, GA 25%.

Minute	Potential (mV)						
	1 ppm	10^{-1} ppm	10^{-2} ppm	10^{-3} ppm	10^{-4} ppm	10^{-5} ppm	10^{-6} ppm
1	39.7	48.7	64.5	83.4	109.3	144.6	146
2	38.4	46.7	63.2	81	108.9	143.4	144.9
3	37.2	45.5	61.9	80.3	107.9	142.5	143.6
4	36.5	44.5	60.7	79.1	107.6	141.7	142.2
5	34.9	43.9	59.6	78.1	107.9	140.2	141.4
6	33.1	43.3	58.7	77.9	108.1	139.6	140.4
7	32	42.8	58.2	78.2	108.3	138.1	139.7
8	31.2	42.5	58.1	78.5	108.7	137.7	139.5
9	30.9	42.1	58.3	78.7	109.1	137.4	139.4
10	30.7	42.2	58.2	78.6	109.1	137.4	139.4

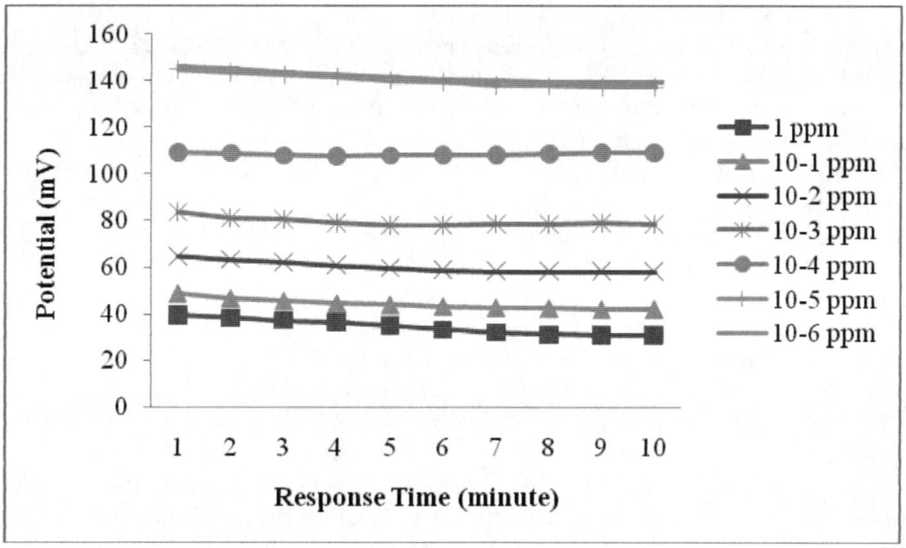

Figure 3. The curve of the response time of biosensor in membrane composition of CA 15% and GA 25%.

The value of response time is determined from the duration of every electrode to reach a constant potential in analyzing pesticide in a variation of diazinon pesticide concentration, each has10^{-6} ppm, 10^{-5} ppm, 10^{-4} ppm, 10^{-3} ppm, 10^{-2} ppm, 10^{-1} ppm dan 1 ppm. The faster the time needed to reach the constant value, the better the quality of biosensor electrode. The analysis result of response time value for biosensor electrode membrane composition is CA 15% and GA 25% is shown in table 2 and figure 3. It shows about 5 to 10 minutes to produce constant electrode potential value in analyzing diazinon pesticide.

4. Conclusion

The biosensor of diazinon pesticide was developed by using gold electrode in the type of layered wire with immobilization of acetylcholinesterase (AChE) enzyme in the membrane of cellulose acetate (CA) 15% and glutaraldehyde (GA) 25% used as working electrode and Ag/AgCl as a standard electrode. The research result shows that biosensor of diazinon pesticide gives optimal work in the analysis of diazinon pesticide using acetylthiocholine chloride substrate 10-3 M and potentiometer transducer with a range of working concentration of 10-6 ppm to 1 ppm and sensitivity of 20.275 mV/decade. The limit of detection value was found to be 10-6 ppm. The response time is the time needed by a biosensor electrode to reach the constant potential. The response time of electrode is on the measurement of pesticide diazinon with concentration variation of 10-6 ppm to 1 ppm is about 5 minutes.

References

[1] Pogacnik L and Franko M 2003 Detection of Organophosphate and Carbamate Pesticides in Vegetable Samples by a Photothermal Biosensor *Biosensors and Bioelectronics* **18** 109

[2] Ciucu A A, Negulescu C and Baldwin R P 2003 Detection of Pesticides Using anAmperometric Biosensors Based on Ferophthalocyanine Chemically Modified Carbon Paste Electrode and Immobilized Bienzymatic System *Biosensors and Bioelectronics* **18** 303-310

[3] Raghu P, Madhusudana Reddy T, Reddaiah K, Kumara Swamy B E and Sreedhar M 2014 Acetylcholinesterase based biosensor for monitoring of Malathion and Acephate in food samples: A voltammetric study *Anal. Methods* **142** 188-196

[4] Mehrvar M and Abdi M 2004 Recent Developments Characteristics and Potential Applications of Electrochemical Biosensors *Analytical Science* **20** 1113-1126

[5] Zhang J, Luo A, Liu P, Wei S, Wangs G and Wei S 2009 Detection of Organophosphorus Pesticides Using Potentiometric Enzymatic Membrane Biosensor Based on Methylcellulose Immobilization *Anal. Sciences* **25** 511

[6] Guilbault G G, Pravda M and Kreuzer M 2004 Biosensors-42 Years and Counting *Anal. Lett.* **37** 14481- 14496

[7] Prieto-Simon B, Campas M, Andreescu S and Marty J 2006 Trends in Flow-based Biosensing Systems for Pesticide Assessment *Sensors* **6** 1161-1186

[8] Mashuni, Syahrul M, Ahyar A and Wahab A W 2010 Pesticides Using A Biosensor-Based Enzyme Acetylcholinesterase Determination of Carbamate and Cholin Oxidase on Platinum Electrode *Indonesian Journal of Chemistry* 10 **3** 290-294

[9] Mashuni, Ramadhan L O A N, Jahiding M and Marlina Y 2015 Optimation of Cellulose Acetate Membrane in Biosensor Electrode of Profenofos Pesticide *Proc. Int. Conf.* ISSN: 2338-0128 Vol. **5** 241-244

[10] Giacoppo J O S, Lima W E A, Kamil K, Franca T C C, Da Cunca E F F and Ramalho T C 2014 Guerra Quimica: Perspectives No Estudo De Reactivators Da Enzima Acetylcholinesterase Inibida For Organofosforados *Rev. Virtual Quim.* ISSN: 1984-6835 **6** 656-670

[11] Wang K, Xu J J, Sun D C, Wei H and Xia X H 2005 Selective glucose detection based on the concept of electrochemical depletion of electroactive species in diffusion layer *Biosensors and Bioelectronics* **20** 1366-1372

Cross-linking of succinate-grafted chitosan and its effect on the capability to adsorb Pb(II) ion

Abu Masykur[1*], Sri Juari Santosa[2], Dwi Siswanta[2] dan Jumina[2]

[1]Department of Chemistry, Faculty of Mathematics and Natural Sciences, Sebelas Maret University, Jl. Ir.Sutami 36A Surakarta 57126, Indonesia

[2]Department of Chemistry, Faculty of Mathematics and Natural Sciences, Gadjah Mada University, Sekip Utara Kotak Pos BLS 21, Yogyakarta Indonesia

[*]Email:abu.masykur@gmail.com

Abstract. The aim of this research was to improve the adsorption capacity of chitosan by modification of the chitosan using various cross-linking agents and followed by grafting using succinate anhydride. Succinate anhydride was grafted into chitosan that had been cross-linked using ethylene glycol di-glycidyl ether (EGDE), diethylene glycol diglycidyl ether (DEGDE) andbisphenolAdiglycidyl ether (BADGE) on the hydroxyl group of chitosan to yield Chit-EGDE-Suc, Chit-DEGDE-Suc, and Chit-BADGE-Suc, respectively. Modified chitosans were analyzed using FTIR and TG-DTA and then applied as adsorbents for Pb(II) ion. Adsorption was carried out in batch condition with a variation of solution pH, contact time, and concentration of Pb(II) in the solution. Adsorption ofPb(II) ion reached optimum condition at pH 5 and contact time of 120 minutes. Adsorption of Pb(II) ion on all of the adsorbents fit well the pseudo-second order kinetic equation. Adsorption capacities of Pb(II) on Chit-EGDE-Suc, Chit-DEGDE-SucdanChit-BADGE-Suc were 0.333, 0.388 and 0.898 $mmolg^{-1}$, respectively, which mean that the adsorption of Chit-BADGE-Suc was the highest and followed by Chit-DEGDE-Suc andChit-EGDE-Suc.

1. Introduction

Contamination of water resources by heavy metals ion becomes serious problem for public health and the environment. Heavy metals ion are considered toxic owing to their non-biodegradability and accumulation in living organism [1]. Lead (Pb(II)) is one of the industrial pollutants. The sources of lead pollution are the wastewater from processing industries engaged in metal electroplating, paints, lead storage battery, pigment and electronic. Lead is toxic metal that is harmful if inhaled or swallowed. It has both acute and chronic effects on human being, as it may cause anemia, nervous system problems, high blood pressure, kidney damage, and headache. Hence, it necessary to remove this element completely from industrial wastewater before discharge into natural water systems [2, 3].

There are many methods to remove lead from wastewater include ion exchange, reverse osmosis, chemical precipitation, solvent extraction, electroflotation, coagulation, and adsorption. Biosorption of lead is promising alternative method due to high removal ability, high selectivity, easy handling and low operating cost [4]. Biomaterial such as agricultural waste, fungal or bacterial biomass, alginate, chitin, and chitosan are prospective adsorbent due to biodegradability, non-toxic and low cost [5]

Chitosan is a product of the partial N-deacetylation of chitin, which has several desirable properties like biodegradability, hydrophilicity, antibacterial and non-toxicity [6-7]. The presence of a large number of functional groups such as amino (NH_2) and hydroxyl groups (OH) allow chitosan to have a high adsorption capacity for metal ions[8-9]. Chitosan is poor acidic resistance and mechanical strength. Therefore, it is necessary to improve the chemical resistance of chitosan to apply it as an adsorbent [1]. Attempts were made to improve the chemical stabilities chitosan in acid condition through chemical cross-linking with cross-linking agents such as glutaraldehyde (GLA), epichlorohydrin (ECH), and ethylene glycol diglycidyl ether (EGDE) [1,4,10-13].

Cross-linking agents prefer to react with the amine groups instead of the hydroxyl groups in chitosan and consequently the adsorption capacity of the cross-linked chitosan is reduced. The primary amine groups of chitosan are known to be the main chelating sites for many types of heavy metal ions [14]. Chitosan cross-linked enhance the resistance of chitosan against acid and alkali medium [15]. Li and Bai (2005) [16] reported a new cross-linking method that protected the amine groups in chitosan using formaldehyde before the cross-linking reaction and subsequently releases the protected amine groups after the cross-linking reaction. Oshita et al. (2008) [10] used benzaldehyde as protecting agent.

Adsorption capacity and selectivity of modified chitosan can be increased by grafting functional groups on chitosan cross-linked. Amine groups can be reacted with an aldehyde, alkyl halide, epoxide and anhydrate groups. Many works were done to graft other functional groups to chitosan such as phenylarsonic acid [18], EDTA[19], 3,4-dihydroxy benzoic acid [20] and carboxylic acid [21].

In this paper, chitosan was modified by cross-linking using three cross-linkers namely: EGDE, DEGDE, and BADGE. All of the cross-linkers are bifunctional reagent having two epoxide groups. Epoxide groups can be reacted with amine and hydroxyl groups. The structure of the cross-linkers is shown on figure 1.

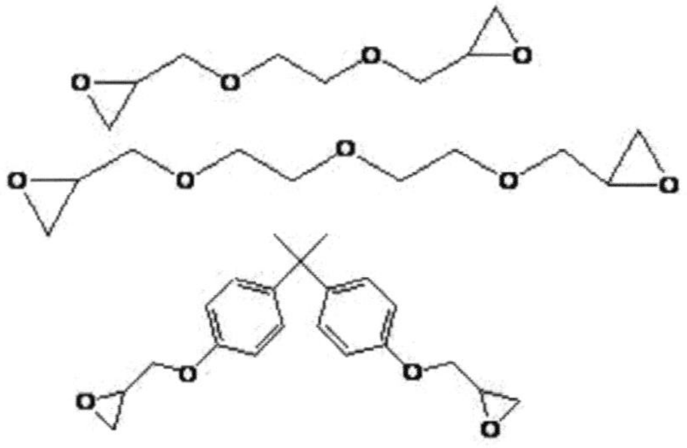

Figure 1. The structure of cross-linkers were used: EGDE (A), DEGDE (B) and BADGE (C)

Cross-linked chitosans were then grafted using succinate anhydride on their amine groups. Grafting succinate anhydride will produce carbonyl and carboxylate groups in addition to hydroxyl and amine groups. These succinate-grafted materials were used as adsorbed for Pb(II) ions. The effect of pH solution, adsorption kinetic and adsorption isotherm have been studied

2. Methods

2.1. Material.

Chitosan flakes (DD>75), EGDE, DEGDE, and BADGE, were obtained from Sigma-Aldrich Germany. Dioxane, benzaldehyde, ethanol absolute, NaOH pellet, HCl 37%, dimethyl sulfoxide, succinate anhydride, HNO_3 65% and $Pb(NO_3)_2$ were obtained from E Merck Germany and distilled water obtained from Gadjah Mada University.

2.2. Instrument.
Instruments used in this study are consisted of reflux apparatus, analytical balance (Metler AE 100), hot plate and stirrer, filter Buchner, oven (Heraeus), thermometer, magnetic bar, shaker (MarusInstrument), pH meters (TOA-DKK HM-30R), FTIR(Shimadzu,Prestige-21), Thermoanalysis TG-DTA (ThermoplusEvoRigaku, T 8120), and Atomic absorption spectrophotometer (AAS) (Perkin Elmer Model 3110).

2.3. Adsorbent preparation.
Cross-linked chitosan was synthesized using a procedure that was reported by Oshita et al. (2008)[10]. Chitosan (5,0 g) was suspended in 50 mL of ethanol, added 20 mL benzaldehyde and stirred for overnight. The solution was filtered using Whatman filter paper and washed using ethanol and water to remove unreacted benzaldehyde to obtain benzaldehyde-protected chitosan (Chit-Ben). The obtained Chit-Ben was put in a round flask containing 75 mL of dioxane and then added 10 mL of 1 M NaOH and 5.3 g EGDE or DEGDE (6,7 g) or BADGE (5,3 g). The mixture was refluxed for 6 hours add the existing solid was separated and then washed with ethanol and distilled water until neutral pH to obtain cross-linked chitosans. Benzaldehyde was released from the amine group of cross-linked chitosan by stirring it in 0.5 M HCl (200 mL) for 15 hours, and this process was done twice. Cross-linked chitosans were washed with ethanol and water to remove any remaining benzaldehyde and cross-linkers. Cross-linked chitosans were grafted using succinate anhydride according to the procedure reported by Yan et al. (2006) [22]. Cross-linked chitosans we resuspended in 100 mL DMSO, then added 7.7250 g succinate anhydride and stirred for 24 hours in 60 °C. The mixture was neutralized using acetic acid and washed with water and ethanol until neutral. The mixture was dried in the oven to obtain Chit-EGDE-Suc, Chit-DEGDE-Suc, and Chit-BADGE-Suc. Chitosan and its derivative were analyzed by FTIR usingKBr pellet method on wave numbers 400-4000 cm^{-1}. TG-DTA was done at a temperature between 30-600 °C in an air atmosphere with increasing temperature rate of 10°C per second.

2.4. Influence of pH solution.
Batch experiments were carried out using 15,0 mg adsorbent, added bottle containing 25 mL of Pb(II) solution (150 mg L^{-1}). The solution was agitated using shaker at 40 rpm for 1200 min. The Ph solution was adjusted with a solution of 0.1 M HNO_3 or 0.1 M NaOH. The solution was filtered using Whatman filter paper and the concentration of Pb(II) was analyzed using AAS. The adsorbed Pb(II) on adsorbent was calculated using equation (1).

$$Q = [(C_o - C_e)V]/(m) \tag{1}$$

where Q is the amount of Pb(II) adsorbed on a gram adsorbent (mmol g^{-1}). $C_o$danC$_e$ is the initial and final concentration Pb(II) in the solution (mmol L^{-1}), V is the volume of metal solution (mL), and m is the weight of the adsorbent used (g).

2.5. Kinetic study.
The kinetic study was carried out using 15.0 mg adsorbent in 25 mL of Pb(II) at pH 5 with initial concentration 150 mg L^{-1}. Solutions were agitated using shaker at 40 rpm at predetermined time intervals 5-1200 min. The solution was filtered using Whatman filter paper to separate adsorbent and filtrate. The concentration of Pb(II) infiltrate was analyzed using AAS.

2.6. Adsorption isotherms.
Adsorption isotherms were studied using 15.0 mg adsorbent in 25 mL of Pb(II) at pH 5 in variations initial concentration 25-500 mg L^{-1}. Solutions were agitated using shaker at 40 rpm for 20 hours. Solutions were filtered using Whatman filter paper and the concentration of Pb(II) in the filtrate was analyzed using AAS.

3. Result and Discussion

The cross-linking grafted-chitosan with succinate anhydride.Chitosan cross-linked were done on hydroxyl groups using EGDE, DEGDE and BADGE as cross-linker. On the other hand, grafting Grafted succinate anhydride on chitosan cross-linked were done on amine groups. Chitosan cross-linked and grafted were analyzed using FTIR and TG-DTA.

3.1. FTIR spectroscopy.
FTIR spectra of chitosan, Chit-EGDE-Suc, Chit-DEGDE-Suc and Chit-BADGE-Suc was shown in figure 2. The FTIR spectrum of chitosan exhibits a strong peak at 3335 cm^{-1} which can be assigned to stretching vibration of O-H and N-H groups. The peak at 2868 cm^{-1} indicates C-H stretching vibration of –CH and CH_2. The peak at 1635 cm^{-1} is the characteristic vibration of C=O from amide that is overlap with N-H vibration from amine group at 1592 cm^{-1}. The band at 1381 cm^{-1} indicates deformation of asymmetry $-CH_3$ and bending vibration CH. The broadband at 1080 is skeletal vibration involving C-O stretching that is characteristic of chitosan's saccharide structure [23, 24]. The FTIR spectra of Chit-EGDE-Suc and Chit-DEGDE-Suc showed new peaks at 1726 cm^{-1} and 1558 cm^{-1} from grafting succinate (C=O vibration from carboxylic and C=O amide). The peak at 1101 cm^{-1} become stronger due to increasing amount of C-O groups from EGDE or DEGDE. Spectrum Chit-BADGE-Suc exhibits new peaks at 835 cm^{-1} and 1510 cm^{-1} because of benzene groups in BADGE [25-26].

3.2. Thermal analysis.
Thermal properties of Chit-EGDE-Suc, Chit-DEGDE-Suc, and Chit-BADGE-Suc were studied thermal properties at temperature 30 to 600 °C and the result of TG-DTA thermogram shown in figure 3. TG-DTA of modified chitosans showed the fourth stages of degradation. For thermogram of Chit-EGDE-Suc, the first stage was at temperature 40 to 110 °C with a weight reduction of 8%. This peak was an endothermic reaction due to releasing adsorbed water. The second stages were an exothermic reaction at 190 to 274 °C with a weight reduction of 31% due to initial degradation of adsorbent, especially functional groups attached to the main chain of chitosan such as hydroxyl, amine and carboxylate groups. The third stages were an exothermic reaction at 274 to 356 °C and weight loss of 33% due to decomposition of main chains of chitosan and cross-linked. The last stages were exothermic degradation and weight loss of 27% due to continued degradation at a temperature above 400 °C and produced water, carbon dioxide, and other small gas compounds [27, 15].

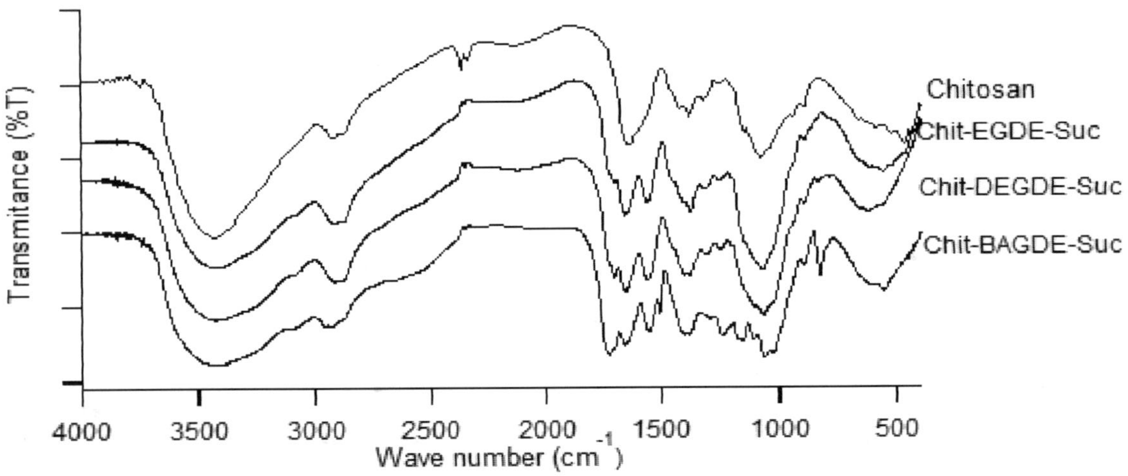

Figure 2. Spectrum FTIR of Chitosan, Chit-EGDE-Suk, Chit-DEGDE-Sucdan Chit- BADGE-Suc.

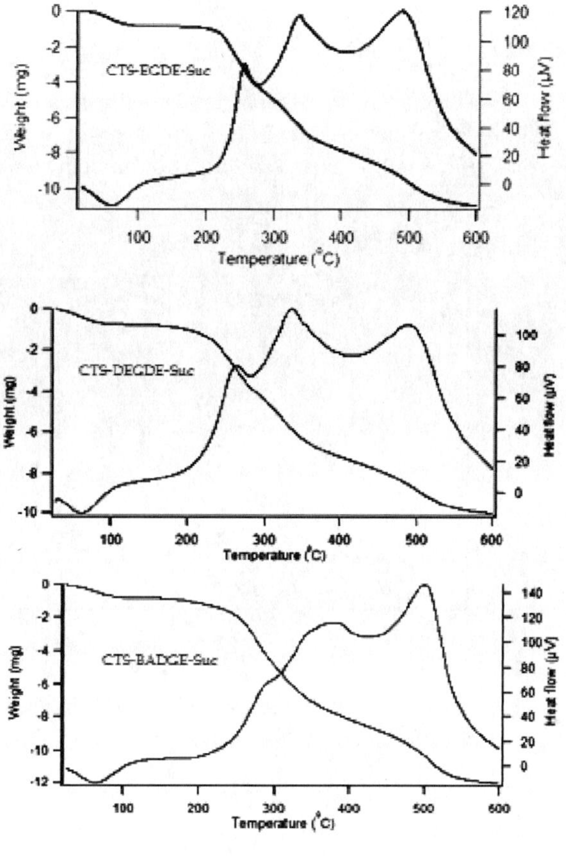

Figure 3. Thermogram TG-DTA of Chit-EGDE-Suc, Chit-DEGDE-Suc and Chit- BADGE-Suc.

Thermogram of Chit-DEGDE-Suc showed similarly curve as Chit-EGDE-Suc but in the third stages, the energy that released higher because DEGDE structure heavier than EGDE. Thermogram of Chit-BADGE-Suc showed the high exothermic reaction at 500 ^{o}C due to degradation of cross-linked. BADGE have two benzene rings that its decomposition occurs at the high temperature.

3.3. Effect of pH on sorption of Pb(II).

The pH of the sorption medium is an important parameter in adsorption processes. In this study, the effect of pH was carried out in the pH range 1.0-6.0 to ensure to the solubility of metal ions. The effect of pH on adsorption of Pb(II) on Chit-EGDE-Suc, Chit-DEGDE-Suc and Chit-BADGE-Suc were shown on figure 4. The adsorption of Pb(II) ion was highly dependent on the pH because pH can affect the solubility of the metal ions and at the same time influence the ionization state of the functional groups on the adsorbent. In the lower pH, adsorption ability of adsorbent is very low since more proton are available, and functional groups are protonated. Active sites on adsorbent are amine group (pKa 6.2) and carboxylate group (pKa 4.6) [6,17,28]. The positive charge of functional groups exerts a high electrostatic repulsive force, which prevent Pb(II) ion from approaching the adsorbent surface. At around pH 4 and 5, the inhibitory effect of H^+ decrease, which led to an increase in the Pb(II) adsorption. When pH > 6, precipitation of lead hydroxide occurs. As a result, the maximum uptake of Pb(II) ions takes place at pH 5. Adsorption ability of Chit-BADGE-Suc at each pH was higher than Chit-EGDEE-Suc and Chit-DEGDE-Suc, this is may be due to the presence of benzene ring which has a high affinity toward Pb(II) by π bounding [28].

IOP Conf. Series: Materials Science and Engineering **107** (2016) 012014 doi:10.1088/1757-899X/107/1/012014

3.4. Adsorption is kinetic.

To investigate the mechanism of Pb(II) adsorption and to identify the rate-determining step, the pseudo first-order, and pseudo second-order equations were applied to analyze the kinetic data. The profile of time contact effect on adsorption of Pb(II)was displayed in figure 5. In the initial, adsorption capacity increased quickly due to all of active site in the adsorbent fully empty. Equilibrium times was reached 80 min, and beyond, this time, no change in adsorption occurred.

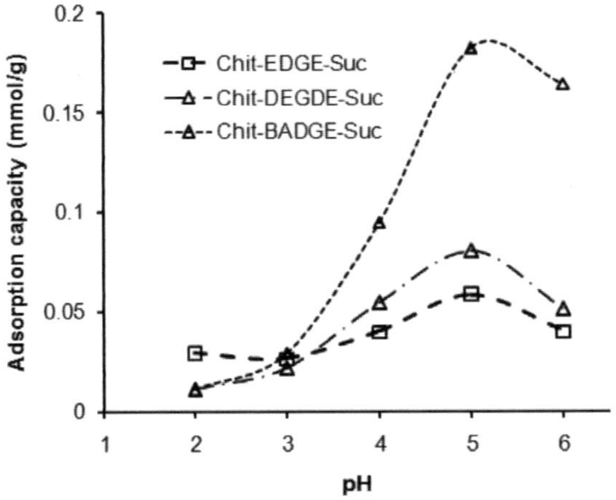

Figure 4. Effect of pH on the adsorption capacity of Chit-EGDE-Suc, Chit-DEGDE-Suc and Chit- BADGE-Suc for Pb(II) ions.

The pseudo-first-order equation (Ho) is given by eq.(2) [30]

$$\ln(q_e\text{-}q_t) = \ln(q_e) - k_1\, t \qquad (2)$$

where q_e and q_t are the adsorption capacities (mmol g^{-1}) at equilibrium and at time t (min), respectively, and k_1 is the rate constant of the first-order adsorption (min^{-1}).

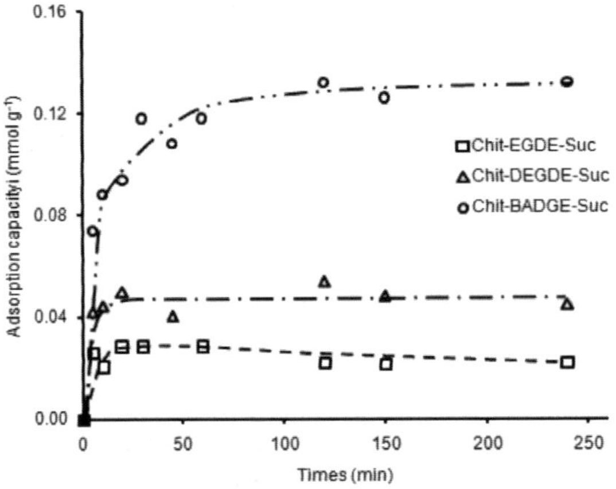

Figure 5. Effect of contact time on the adsorption capacity of Chit-EGDE-Suc, Chit-DEGDE-Suc and Chit- BADGE-Suc for Pb(II) ions.

IOP Conf. Series: Materials Science and Engineering **107** (2016) 012014 doi:10.1088/1757-899X/107/1/012014

The pseudo-second-order kinetic models (Ho) was illustrated as eq.(3) [30]

$$\frac{t}{q_t} = \frac{1}{k_2 q_e^2} + \frac{1}{q_e} \quad (3)$$

Where k_2 (g $mmol^{-1}$ min^{-1}) is the rate constant of adsorption.

The parameters obtained from the pseudo first-order model, and pseudo second order model is shown in Table 1. The adsorption Pb(II) using Chit-EGDE-Suc, Chit-DEGDE-Suc, and Chit-BADGE-Suc did not fit the pseudo-first-order equation, owing to its low correlation coefficient value (R^2): 0.356, 0.127 and 0.961, respectively. The third modified chitosan follows the pseudo second-order equation, given by its high R^2 value: 0.944, 0.996 and 0.998, respectively. The pseudo-second-order assumes that chemisorption is the rate-determining step of the adsorption process. It is suggested that the overall of Pb(II) adsorption process be controlled by the chemical attachment and not by mass transport. The adsorption mechanism may involve valency forces through sharing of an electron between Pb(II) and adsorbent [29].

3.5. Adsorption Isotherm.

The effect of concentration of Pb(II) ions on the adsorption by Chit-EGDE-Suc, Chit-DEGDE-Suc, and Chit-BADGE-Suc, is shown in figure 6. The result indicated that the adsorption of Pb(II) ions increased with the increasing concentration of Pb(II) in the initial stage and then relatively constant with the increasing concentration. This was due to the fixed total active sites in the adsorbent, and the adsorption amounts would reach a state of saturation.

The experimental data were analyzed by Langmuir and Freundlich adsorption isotherm. Langmuir adsorption isotherm based on the monolayer adsorption on the active sites. The Langmuir adsorption isotherm can be expressed in the linear form as follows [31]:

$$\frac{C_e}{q_e} = \frac{1}{bK} + \frac{1}{b} C_e \quad (4)$$

Where q_e is capacity adsorption in the equilibrium ($mmolg^{-1}$), C_e is the equilibrium concentration of metal in the solution ($mmolL^{-1}$), bis adsorption capacity maximum ($mmolg^{-1}$), K is Langmuir constant (L mol^{-1}).

Freundlich isotherm assumed multiple layer adsorption with uniform energy and can be expressed as eq.(5) [30]

$$\log m = \log K_f + \frac{1}{n} \log Ce \quad (5)$$

where m is amount of Pb(II) ion adsorbed (mmol g^{-1}), K_f (L mol^{-1}) is the Freundlich constant related to adsorption capacity, and n is the Freundlich constant related to adsorption intensity.

Table 1. Kinetic parameters for Pb(II) on Chit-EGDE-Suc, Chit-DEGDE-Suc, and Chit-BADGE-Suc

Adsorbent	Pseudo-first-order		Pseudo-second-order	
	K(min^{-1})	R^2	K (g $mmol^{-1}$ min^{-1})	R^2
Chit-EGDE-Suc	1.7×10^{-2}	0.356	3.1×10^{-3}	0.944
Chit-DEGDE-Suc	2.3×10^{-4}	0.127	3.8×10^{-2}	0.996
Chit-BADGE-Suc	1.6×10^{-2}	0.961	4.3×10^{-3}	0.998

Table 2.Langmuir and Freundlich parameter for adsorption isotherm Pb(II) on Chit-EGDE-Suc, Chit-DEGDE-Suc and Chit-BADGE-Suc

Adsorbent	Adsorption Isotherm						
	Langmuir				Freundlich		
	b (mmol g^{-1})	$K\times10^3$ (L $mmol^{-1}$)	R^2	R_L	N	K_f (mmol g^{-1})	R^2
Chit-EGDE-Suc	0.333	1.442	0.957	0.841-0.221	2.042	0.185	0.921
Chit-DEGDE-Suc	0.388	2.363	0.974	0.777-0.217	2.488	0.251	0.983
Chit-BADGE-Suc	0.898	2.634	0.962	0.907-0.159	2.053	0.621	0.933

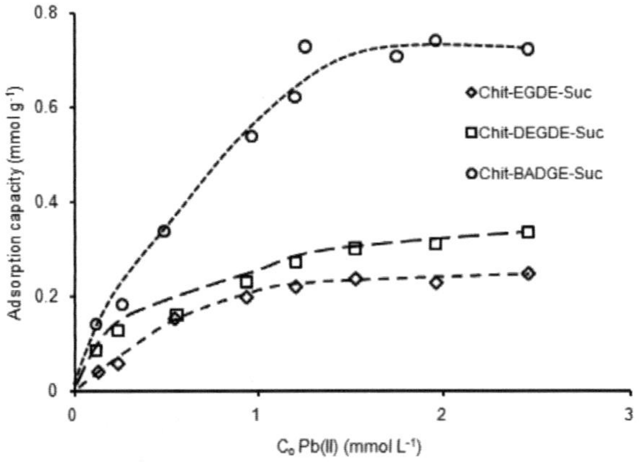

Figure 6. Effect of initial concentration on the adsorption of Chit-EGDE-Suc, Chit-DEGDE-Suc, and Chit- BADGE-Suc for Pb(II) ions.

The parameters of Langmuir adsorption isotherm and Freundlich adsorption isotherm of Chit-EGDE-Suk, Chit-DEGDE-Suc, and Chit-BADGE-Suc for Pb(II) are shown in Table 2. The coefficient correlation (R^2) value of Langmuir and Freundlich adsorption isotherm are similar. Pb(II) ion can be bound on hydroxyl, amine or carboxylate simultaneously.

The adsorption capacity of Pb(II) on Chit-EGDE-Suc, Chit-DEGDE-Suc and Chit-BADGE-Suc are 0.333, 0.388 and 0.898 mmol g^{-1}, respectively. Adsorbent Chit-BADGE-Suc has the highest capacity of Pb(II) ions because cross-linker BADGE has benzene groups that are rich with an electron. The benzene ring binds Pb(II) ion through π electrons on its ring. The adsorption capacity of Pb(II) ions on Chit-DEGDE-Suc is higher than Chit-EGDE-Suc because cross-linker DEGDE is longer than EGDE. The longer cross-linker made space between chains of chitosan increase, so it is easier for the active site to bind Pb(II) ion

Langmuir parameter (R_L) is used to predict the favorability of adsorption process. The (R_L) was calculated from the following equation using Langmuir constant (6) [32]:

$$R_L = \frac{1}{(1+KC_o)} \quad (6)$$

where K is the Langmuir constant (L $mmol^{-1}$) and C_o is the initial concentration Pb(II) ion (mmol L^{-1}). It is considered to be unfavorable (R_L>1), linear (R_L=1.0), favorable (1>R_L>0) and unfavorable (R_L=0). The R_L values for adsorption Pb(II) on Chit-EGDE-Suc, Chit-DEGDE-Suc, and Chit-BADGE-Suc were in the range 1>R_L>0, indicating that the adsorption process was favorable.

4. Conclusion

In this study, we observed that cross-linking and grafting chitosan produced material that was chemically stable at low pH value, allowing its use in the acidic solution. Type of cross-linker influenced capacity adsorption of Pb(II). Chit-BADGE-Suc showed the highest capacity adsorption of Pb(II) due to its benzene ring that bind Pb(II) ion effectively. The adsorption was influenced by some parameters such as the pH of solution, contact time and initial Pb(II) concentration

Acknowledgement

This research was funded by the Competitive Research Grants Program of the Directorate General of Higher Education of the Republic of Indonesia. Thank for Prof.Dr. Shogo Shimazu from Chiba University, Japan for analysis of TG-DTA and Ageng Trisna Surya Pradana and Adhani Fajar Sari Rahmayanti, who helped conduct this research.

References

[1] Meneghetti E, Baroni P, Vieira R S, da SilvaMGC, and BeppuM 2010 *Mater. Res.***13**89.
[2] Hussain M A, Salleh A, and Milow P 2009 *Am. J. Biochem& Biotech.* **5** 75.
[3] Jiang W, Su H, Huo H and Tan T 2010 *Appl. Biochem. Biotechnol.* **160** 467.
[4] Vieira R S, Guibal E, Silva E A and Beppu M 2007 *Adsorption***13** 603.
[5] Copello G J, Varela F, Vivot M and Diaz L E 2008 *Bioresour. Technol.* **99** 6538.
[6] Guibal E 2004 *Sep. Purif. Technol.***38** 43
[7] Hasan S, Krishnaiah A, Ghosh T K, Viswanath D S, Boddu V M, and Smith E D 2006 *Ind. Eng. Chem. Res.***45** 5066.
[8] MelloKGPC, Bernusso L C, Pitombo R N M and Polakiwicz B 2006 *Braz. Arch. Biol. Tech.***49** 665.
[9] Miretzky P, and Cirelli A F 2009 *J. Hazard. Matter.***167** 10.
[10] Oshita K, Takayanagi T,Oshima M and Motomizu S 2007 *Anal. Sci.***23** 1431.
[11] Vieira R S and Beppu M 2005 *Adsorption* **11** 731.
[12] Sigh A,NarviS S,Dutta P K and Pandey N D 2006 *Bull. Mater. Sci.***29** 233.
[13] Oliveira B F, Santana M H A and Re M E 2005 *Braz. J. Chem. Eng.* **22** 353.
[14] Ngah W S W, Endud C S and Mayanar M 2002 *React. Funct. Polym.***50** 181.
[15] Leach J B,Wolinsky J B, Stone P J and Wong J Y, 2005 *Acta Biomater.***1** 155.
[16] Li N and Bai R 2005 *Ind. Eng. Chem. Res.* **44**, 669.
[17] Mourya V K, Inamdar N N and Tiwari A, 2010 *Adv. Mat. Lett.***1** 11.
[18] Oshita K, Seo K, Sabarudin A, Oshima M, Takayanagi T and Motomizu S 2008 *Anal. Bioanal. Chem.***390** 1927
[19] Katarina R S,Takayanagi T, Oshita K,Oshima M, and Motomizu S 2008 *Anal. Sci.***24** 1537.
[20] Sabarudin A, Noguchi O, Oshima M, Higuchi K, and Motomizu S 2007 *Microchim. Acta*, **159** 341.
[21] Ladenslager M J, Schiffman J D and Schauer C L 2008 *Biomacromol.***9**2682.
[22] Yan C, Chen D, Gu D, Hu H, Zhao X and Qiao M 2006 *Yakugaku Zasshi* **126** 789
[23] Dong Y, Xu C, Wang J, Wang M, Wu Y and Ruan Y 2001 *Sci. In China* **B. 44**216.

IOP Conf. Series: Materials Science and Engineering **107** (2016) 012014 doi:10.1088/1757-899X/107/1/012014

[24] Van de Velde Kand Kiekens P 2004 *Carbohydr.Polym.***58** 409.
[25] Zhang J, Zhang Y, Li R and Pan Q 2012 *Ind J., of Chem. Technol.***19**161.
[26] Chan Z, Ping Q, Zhang H and Shen J 2003 *Eur. Polym.J.***39**1629.
[27] Jiao T F,Zhou J, Zhou J X,Gao L H,Xing Y Y and Li X H 2011 *Iranian Polym. J.***20** 123
[28] Reddy A S and Sastry G N 2005 *J. Phys. Chem. A.***109**8893.
[29] Bunting J W and Thong K 1971 *Canadian J. Chem.***48** 1654
[30] Ho Y S and McKay G 1999 *Process. Biochem.***34** 451.
[31] Erosa M S D, Medina T T S, Mendoza R N, Rodriguez M A and Guibal E 2011 *Hydrometallurgy* **61** 157
[32] Dalida P M I, Mariano A F V, Futalan C M, Kan C,Tsai W and Wan M 2011 *Desalination* **275** 154.

IOP Conf. Series: Materials Science and Engineering **107** (2016) 012015 doi:10.1088/1757-899X/107/1/012015

Development of electrokinetic remediation for caesium: A feasibility study of 2D electrode configuration system

Rudy Syah Putra

Department of Chemistry, Faculty of Mathematic and Natural Sciences, Universitas Islam Indonesia, Jl. Kaliurang KM. 14.5, Sleman, Daerah Istimewa Yogyakarta 55584, Indonesia

E-mail: rudy.syahputra@uii.ac.id

Abstract. Agar matrix was artificially contaminated with caesium and subjected to rapid assessment of electrokinetic treatment on the basis of the *2D* electrode configuration. The effect of caesium concentration on the process was investigated using different electrode configuration (i.e. rectangular, hexagonal and triangular). During treatment the in situ pH distribution, the current flow, and the potential distribution were monitored. At the end of the treatment, the caesium concentration distribution was measured. The results of these experiments showed that for caesium contamination, pH control is essential in order to create a suitable environment throughout the agar matrix to enable contaminant removal. It was found that the type of electrode configuration used to control the pH affected the rate of caesium accumulation. All of the electrode configurations tested was effective, but the highest caesium extraction was achieved when the hexagonal pattern was used to control the pH. After 72 h of treatment at 50 mA, the concentration of caesium decreased gradually from the second and first layer of agar matrix throughout the cell, suggesting that most of the caesium was concentrated on the cathode part.

1. Introduction

Now day Japan has faced a crisis on the radionuclide spreading after the accident in Fukushima Nuclear Power plant after hits by a tsunami following the M8.7 earthquake on 11 March 2011. Based on data released by MEXT and other resources, the radionuclide contaminants (Cs-137) are spreading in the soil and also water around 30 km from the accident location[1]. The deposition of ^{137}Cs occurs primarily in association with precipitation [2], and its initial depth distribution in the soil layer is affected by infiltration of precipitation into soil and adsorption of radionuclides by the soil particles of the surface horizon [3]. On one hand, shallow penetration and slow downward migration rate of ^{137}Cs result in its long residence time in plant root zone. On the other hand, deeper penetration and fast migration in soil may increase the risk of water contamination by the radionuclide. Even though today the radioactive concentration is still far below from the indexes of drinking water and food safety limit reported by the Nuclear Safety Commission of Japan, but it was still necessary to restore these contaminants location including around the nuclear reactor.

Recently there has been considerable interest in the international community in a range of in-situ and ex-situ remediation technologies for radioactively contaminated soil and groundwater, including a low expense and large scale application of phytoremediation [4, 6]. While these technologies have been applied in a range of environments, their application in clay and silt-rich soils may have raised the problem, due largely to the low hydraulic conductivities of these soils. Also, a major factor which influences the mobility of contaminants in soil is the soil pH. One emerging technology that has, however, received much attention as a practical in situ remediation technology for low permeability clay-rich soil is electrokinetic remediation, which applying a dc current to the soil results in the generation of acid at the anode and alkali at the cathode [5, 7, 8]. This leads to an acid front at the anode and a basic front at the cathode. In the presence of the electric field, the acid front tends to migrate towards the cathode and the basic front advances towards the anode by electromigration. This method has been successfully employed at laboratory scale on a range of radioactively contaminated soil and groundwater [9]. A common problem with electrokinetic removal of heavy metals is precipitation of the metal ions which has been found to occur close to the cathode, hindering the extraction process [10, 11]. The generation of alkali conditions at the cathode tends to immobilise heavy metal contaminants by hydroxide precipitation. A low pH environment can promote the removal of the metallic contaminant from fine grain soils as a low pH promotes desorption (by ion exchange) of most heavy metals and cations from the soil surface. The latest results has confirmed that the combined process of electrokinetic with other methods, such as permeable reactive barrier (PRB) material using drinking water sludge as an entrapping zone (i.e. entrapping zone electrokinetic remediation–EZ-EKR) and phytoremediation using Kentucky bluegrass as an accumulator species (i.e. electro-assisted phytoremediation–EAPR), has a promising result to apply for the removing heavy metal from contaminated soil [12, 13] and water [14].

An open electrode configuration allowing the exchange of solution between the electrode and the subsurface environment is essential for proper functioning of the electrokinetic process. The electrodes can be installed horizontally or vertically. Most bench-scale and large-scale laboratory and pilot-scale field studies on electrokinetic remediation and electro-assisted phytoremediation (EAPR) system performed to date have been horizontally one-dimensional (1*D*) [15, 16, 17]. Limited research has been conducted to study the effect of the vertically 1*D* electrode configuration on the efficiency of electrokinetic remediation and EAPR system [18, 19, 20]. Limited numerical sensitivity analyses have been performed on two-dimensional (2*D*) or axisymmetrical configurations (Renaud and Probstein 1987; Jin and Sharma 1991) and also on EAPR system [22]. In that configuration, the cathode is placed at the center, and the anodes are placed on the perimeter to maximize the spread of the acidic environment generated by the anodes and to minimize the extent of the basic environment generated by the cathode. These configurations of electrodes generate 2*D* non-linear electric fields. In a 1*D* configuration, the electric current density, i.e. current per unit area, is independent of location. In 2*D* configurations, however, the electric current density increases linearly with distance towards the cathode [23]. Therefore, the electric field strength also increases linearly with distance towards the cathode.

The purpose of this study was to evaluate a designed 2*D* electrode configuration on the electrokinetic system for removal of Cs from soil for further application in the field at Fukushima area, Japan. The study was conducted using an agar matrix for rapid assessment on the basis of pH changes as an indicator the effectiveness of the electrokinetic process.

2. Methodology

2.1. Designed electrokinetic cell

The study was focused on the development of modified electrokinetic system using 2*D* electrode configuration as shown in Figure 1. This reactor was provided to evaluate the application of vertical upward current, which is characterized by the removal of radionuclide from a lower subsurface soil (i.e. depth 5 ~10 cm). Furthermore, a low cost/free waste recycling and re-utilization material, such as dewatering drinking water sludge (DWS), will be applied as an adsorbent material for the combined electrokinetic and entrapping zone system as shown in Figure 2. This material in Japan is now a day produced in large quantity and dumped for the landfill as aluminum drinking water treatment residuals (Al-WTRs). Through the introduction of a ready cheap and available material, the method can be applied in low-cost operation and environmentally friendly technology.

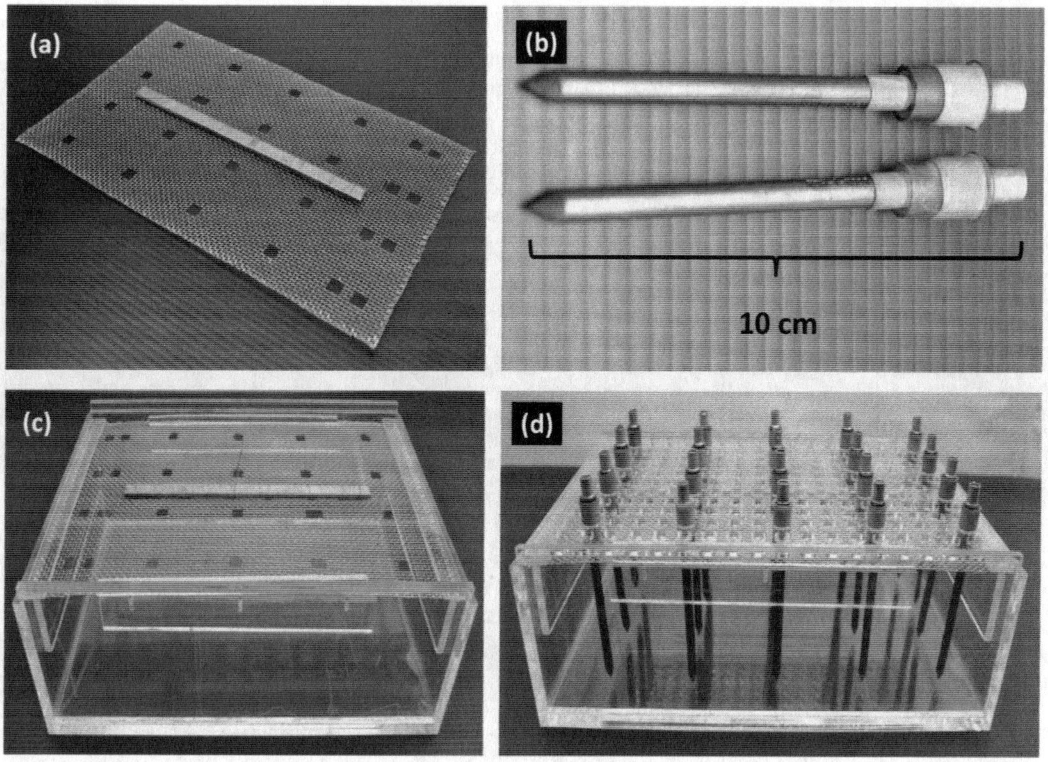

Figure 1. The electrokinetic apparatus used in this study. (a) the cathode electrode made from stainless still bar (10 mm (W) x 200 mm (L), 3 mm thickness) and net wire (150 mm (W) x 200 mm (L), 10 mesh); (b) anode electrode made from graphite (100 mm (L), Ø 58 mm); (c) dimensional electrokinetic cell (200 mm (W) x 300 mm (L) x 100 mm (H)); (d) lateral images of full scale *2D* electrode configuration, respectively.

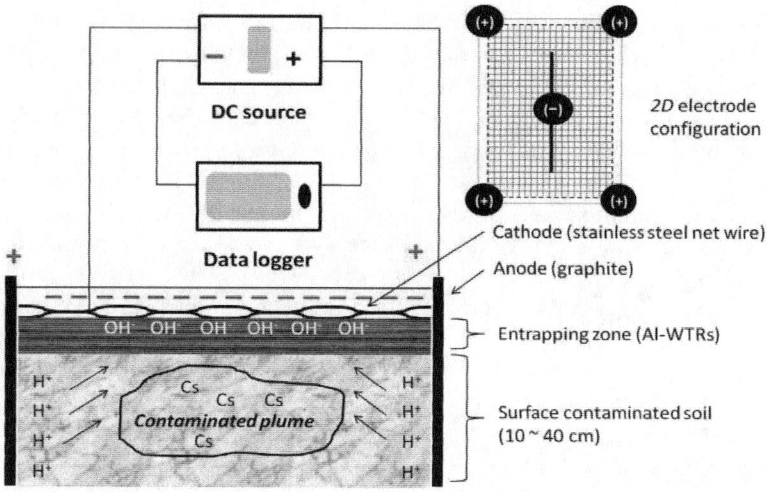

Figure 2. Schematic diagram of modified 2*D* reactor for EK/EZ system.

2.2 Evaluation of 2D electrode configuration of electrokinetic cell

The aim of the study was to evaluate whether the construction of 2*D* electrode configuration could be used efficiently to enhance the electromigration of caesium ion from contaminated plume during the electrokinetic process. The test was preceded in the agar matrixes representing as a liquid media, where rapid assessment of the caesium ions movement in a short time (i.e. 72 h) could be easily evaluated. The contaminated part was set-up in the bottom tray; meanwhile another part in the tray was filled with clean agar.Each layer of agar matrix was prepared by pouring of 30 g agar in 4 L of 0.01 M KCl solution. The contaminated agar matrix was prepared by addition of 100 mg/L of CsCl. The results from this step were further used to design the effectiveness of electrode configuration in the soil matrix for the electrokinetic process.The studies in agar matrix were carried out in three different electrode configurations such as rectangular, hexagonal and triangular patterns as shown in Figure 3. All studies were run in 50 mA of constant current.

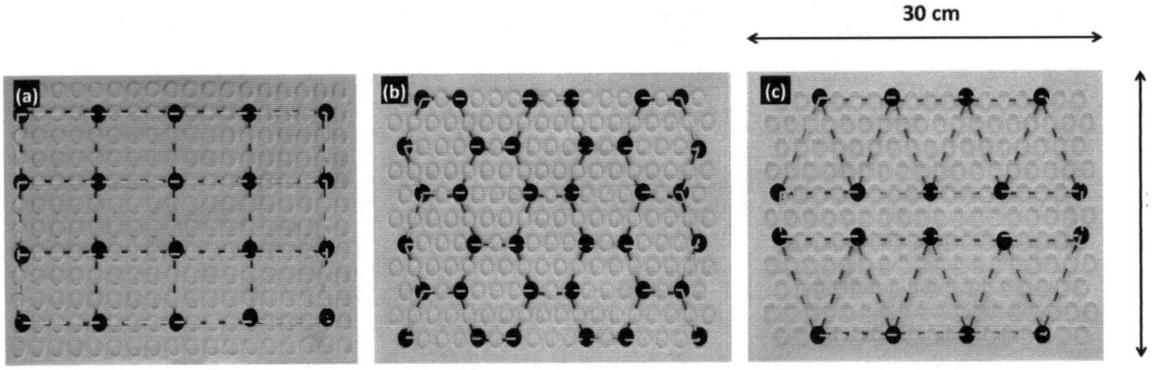

Figure 3. 2D electrode configuration for the electrokinetic system. (a) rectangular configuration needs for 20 electrodes, (b) hexagonal configuration needs for 36 electrodes, and (c) triangular configuration needs for 18 electrodes, respectively. The black dot representing the electrode position and the connection of electric wire between points were represented by dashed yellow line.

3. Discussions And Implications

A series of experiments were carried out by using different *2D* electrode configuration (i.e. rectangular, hexagonal and triangular) for 72 h electrokinetic process. Figure 4 shows the pH profiles from contaminated agar for rectangular electrode configuration. The decreasing agar pH which was indicated by purple color would increase the solubility of Cs ions. At first, of 2 h electrokinetic process acid pH gradually changed into mild acid as indicating with dark red color (i.e. pH range 4.2 ~ 6.2). After 72 h of the electrokinetic process, totally the agar media changed the acidic environment as indicating with red color. However, the pink color in the upper layer of agar matrix representing the alkaline environment (i.e. pH range 7.8 ~ 10.0). Also, a similar result has been obtained when hexagonal electrode configuration was applied in the system as shown in Figure 6. In this study, the anode position changed after 24 h electrokinetic process to spread the current for another half part of the electrokinetic cell.

Figure 5. shows the average caesium concentration in the upper layer of the electrokinetic cell was much increase after the test. The concentration of caesium much decreased from the contaminated layer (i.e., second layer) toward the first layer and an upper layer of agar matrix, suggesting that most of the caesium was concentrated on the cathode part. Furthermore, the distribution of caesium at upper layer much concentrated in the left part (S8 ~ S12) compared to the right part of the cell (S1 ~ S7) indicating the electromigration of Cs ion in the cell was not normally spreading due to unbalance a current distribution. However, for the hexagonal electrode configuration (figure 7), the electromigration of caesium at upper layer was grouped into two part of the hexagonal cluster. For example, the Cs ion has a high concentration at the right hexagonal group (i.e. S1 ~ S8) and left the hexagonal group (i.e. S9 ~ S15) indicating the current distribution in the cell tends to concentrate on the hexagonal structure.These results suggested that the electrokinetic process could be enhanced by *2D* electrode configuration system.

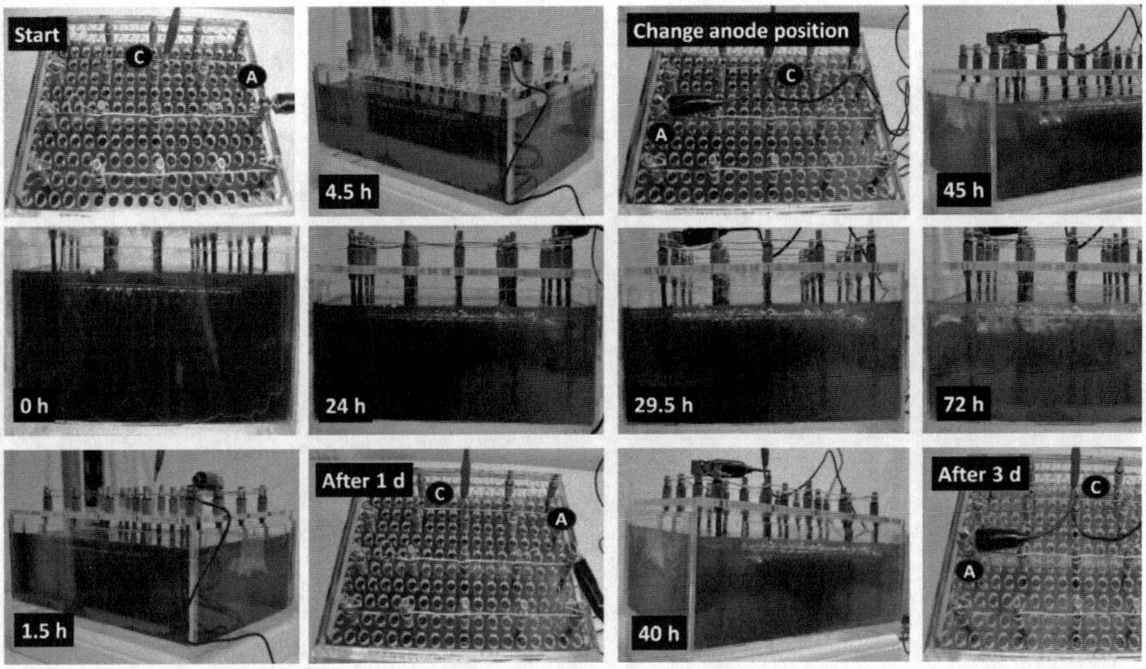

Figure 4. pH profiles from contaminated agar for rectangular electrode configuration in the electrokinetic process. The Color degradation is representing of the pH profiles in the electrokinetic cell, i.e. purple for bromophenol blue at pH range 3.0 ~ 4.6; dark red for methyl red at pH range 4.2 ~ 6.2; and pink for phenolphthalein at pH range 7.8 ~ 10.0.

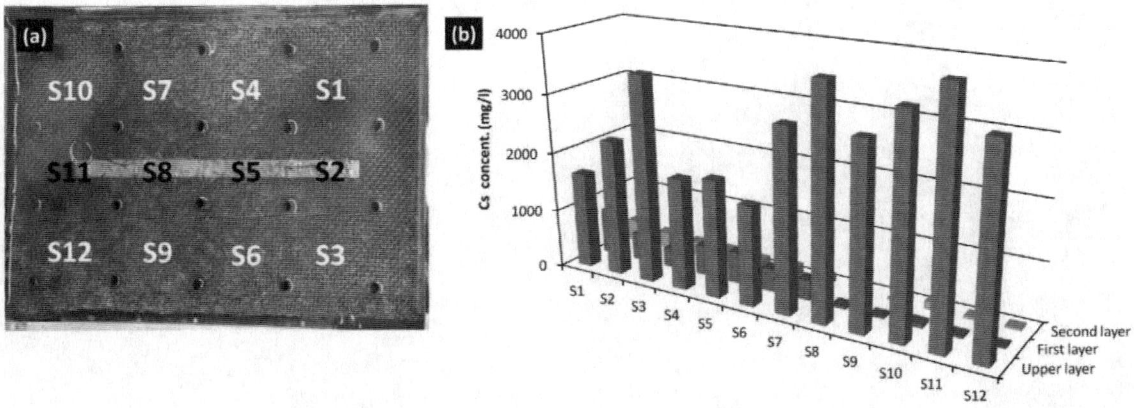

Figure 5. Caesium migration profiles from contaminated agar matrix toward cathode rectangular electrode configuration. (a) Sampling points of agar in the electrokinetic cell; (b) Concentration of Cs in each sectional of agar at sampling points.

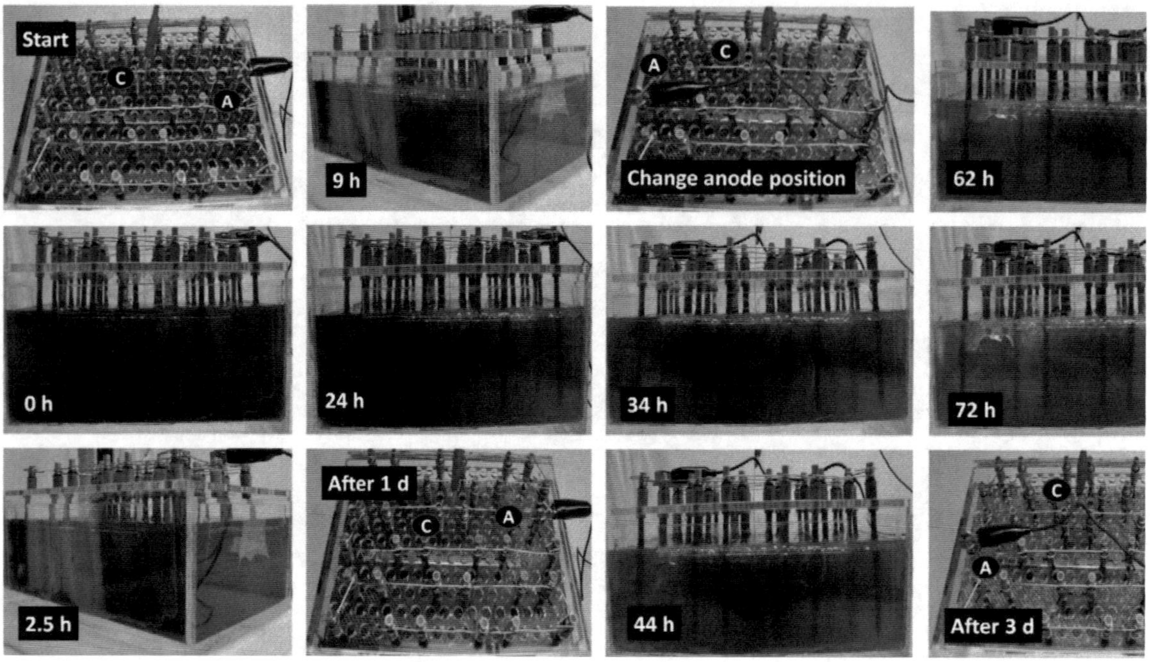

Figure 6. pH profiles from contaminated agar for hexagonal electrode configuration in the electrokinetic process. The Color degradation is representing of the pH profiles in the electrokinetic cell, i.e. purple for bromophenol blue at pH range 3.0 ~ 4.6; dark red for methyl red at pH range 4.2 ~ 6.2; and pink for phenolphthalein at pH range 7.8 ~ 10.0.

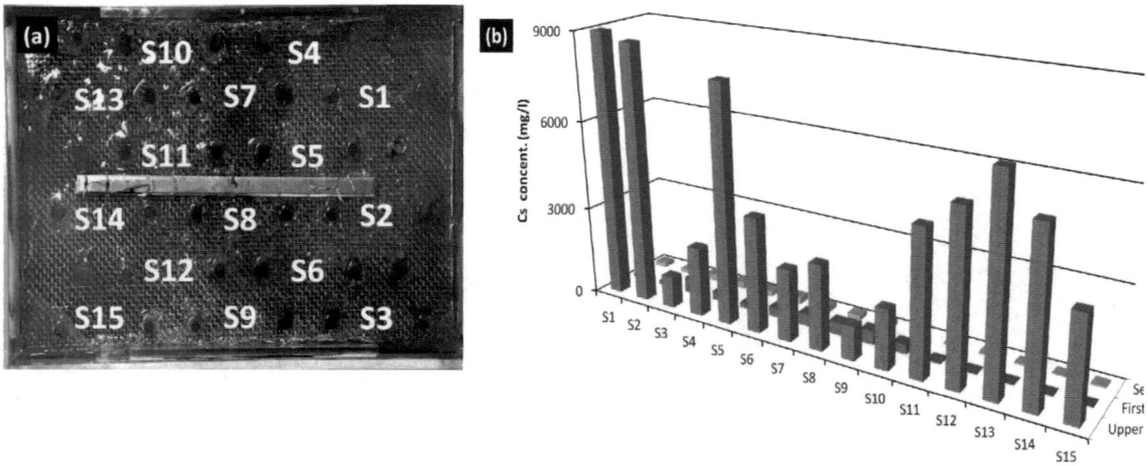

Figure 7. Caesium migration profiles from contaminated agar matrix toward cathode hexagonal electrode configuration. (a) Sampling points of agar in the electrokinetic cell; (b) Concentration of Cs in each sectional of agar at sampling points.

4. Conclusions

The shape of the caesium concentration profile would depend in particular the pH of agar matrix. Regions with low pH would be expected to show higher rates of extraction, as the caesium is more readily dissolved in the pore fluid as indicating with red color (i.e. pH range 4.2 ~ 6.2). However, the pink color in the upper layer of agar matrix representing the alkaline environment (i.e. pH range 7.8 ~

10.0) which indicating of heavy metal precipitation. The electrokinetic process could be enhanced using both of 2*D* electrode configurations (e.g. rectangular and hexagonal).

Acknowledgement(s)
The author would like to thank the SAME program 2013 of the Directorate of Higher Education of the Republic of Indonesia for academic exchange grand and also greatly appreciate to Prof. Shun Tanaka laboratory for providing the experimental equipment.

References

[1] Endo S, Kimura S, Takatsuji T, Nagasawa K, Imanaka T and Shizuma K 2012 Measurement of soil contamination by radionuclides due to the Fukushima Dai-ichi Nuclear Power Plant accident and associated estimated cumulative external dose estimation *J. EnviRadioact* **111** 18-27

[2] Tamura T 1964 Selective sorption reaction of caesium with mineral soils *Nuclear Safety* **5** 262-268

[3] Ritchie J C and McHenry J R 1990 Application of radioactive fallout Caesium-137 for measuring soil erosion and sediment accumulation rates and patterns a review. *J. EnviQual* **19** 215-233

[4] Campbel L S l and Davies B E 1997 Experimental investigation of plant uptake of caesium from soils amended with clinoptilolite and calcium carbonate *Plant and Soil* **18** 65–74

[5] J Trombly 1994 Electrochemical remediation takes to the field *Environ. Sci. Technol.* **28** 289–291

[6] Uchida S, Tagami K, Shang Z R and Choi Y H 2009 Uptake of radionuclides and stable elements from paddy soil to rice: a review *J. Enviro. Radio.* **100** 739–745

[7] Acar Y B and Alshawabkeh A N 1993 Principles of electrokinetic remediation *Environ. Sci. Technol.* **27** 2638–2647

[8] Acar Y B, Gale R J, Alshawabkeh A N, Marks R E, Puppala S, Bricka M and Parker R 1995 Electrokinetic remediation: basics and technology status *J. Hazard. Mater.* **40** 117–137.

[9] Al-Shahrani S S and Roberststs E P L 2005 Electrokinetic removal of caesium from kaolin *J. Hazard.Mater.* B**122** 91 – 101

[10] Lageman R, Pool W and Seffinga G 1989 Electro-reclamation: theory and practice *Chem. Ind.* 585–590

[11] Eykholt G R and Daniel D E 1994 Impact of system chemistry on electro osmosis in contaminated soil *J. Geotech. Eng.* **120** 797–815

[12] Putra R S and Tanaka S 2011 Aluminum drinking water treatment residuals (Al-WTRs) as an entrapping zone for the lead in soil by electrokinetic remediation *Sep.Pur Techno* **79** 208-215

[13] Putra R S, Ohkawa Y and Tanaka S 2013 Application of EAPR system on the removal lead from sandy soil and uptake by Kentucky bluegrass (*Poapratensis* L.*) Sep.Pur Techno.* **102** 34–42

[14] Putra R S, Cahyana F and Novarita D 2015 Removal of lead and copper from contaminated water using EAPR system and uptake by water lettuce (*PistiaStratiotes* L.) *Procedia Chemistry* **14** 381-386

[15] O`Connor C S, Lepp N W, Edwards R and Sunderland G 2003 The combined use of electrokinetic remediation and phytoremediation to decontaminate metal-polluted soils: a laboratory-scale feasibility study *Enviro. Monit.Assess.* **84** 141-158

[16] Aboughalma H, Bi R and Schlaak M 2008 Electrokinetic enhancement on phytoremediation in Zn, Pb, Cu and Cd-contaminated soil using potato plants *J. Enviro. Sci. Health* Part A **43** 926-933

[17] Lim J M, Salido A L and Butcher D J 2004 Phytoremediation of lead using Indian mustard (Brassica juncea) with EDTA and electronics, *Microchem.* J., **76**, 3-9.

[18] Wang J Y, Huang X J and Kao J C M 2006 Stabnikova, Removal of heavy metals from kaolin using an upward electrokinetic soil remediation (UESR) technology, *J. Hazard. Mater.*, B**136** 532-541.
[19] Zhou D M, Chen H F, Cang L and Wang Y C 2007 Ryegrass uptake of soil Cu/Zn-induced by EDTA/EDDS together with a vertical direct-current electrical field, *Chemosphere*, **67**, 1671-1676.
[20] Cang L, Wang Q Y, Zhou D M and Xu H 2010 Effect of electrokinetic-assisted phytoremediation of multiple metal contaminated soil on soil metal bioavailability and uptake by Indian mustard, *Separ., Pur., Techno.*, **79**, 246-253.
[21] Renaud P C and Probstein R F 1987 Electroosmotic control of hazardous wastes, PCH *PhysicoChemical Hydrodynamics*, **9(1/2)** 345-360.
[22] Hodko D, Hyfte J V, Denvir A and Magnuson J W 2000 Methods for enhancing phytoextraction of contaminants from porous media using electrokinetic phenomena, U.S. Patent No. **6**,145,244, Nov. 14, 2000.
[23] Alshawabkeh A N, Yeung A T and Bricka M R 1999 Practical aspects of in-situ electrokinetic extraction, *J. Environ. Eng.,* ASCE, **125 (1)** 27-35.

IOP Conf. Series: Materials Science and Engineering **107** (2016) 012016 doi:10.1088/1757-899X/107/1/012016

Improving quality of textile wastewater with organic materials as multi soil layering

Supriyadi[1*], H Widijanto[1], Pranoto[2], AK Dewi
[1]Department of Agrotechnology, Agriculture Faculty, Sebelas Maret University, Jl.Ir. Sutami 36A, Surakarta 57126, Indonesia
[2]Department of Chemistry, Science and Mathematics Faculty, Sebelas Maret University, Jl.Ir. Sutami 36A, Surakarta 57126, Indonesia

[*]E-mail: supriyadi_uns@yahoo.com

Abstract. On agricultural land, fresh water is needed especially for irrigation. Alternative ways to fulfill needs of fresh water is by utilizing wastewater from industry. Wastewater that produced in the industry in Surakarta is over flowing especially textile wastewater. Wastewater that produced from industry has many pollutants that affected decreasing fresh water quality for irrigation. Multi Soil Layering (MSL) is one of method that utilize the soil ability as main media by increasing its function of soil structure to purify wastewater, so it does not contaminate the environment and reusable. This research was purposed to know affectivity of organic materials (such as rice straw, baggase, sawdust, coconut fibre, and corncob) and dosage (5%, 10% and 25%) in MSL, also get alternative purification ways with easy and cheaper price as natural adsorbent. This study using field and laboratory experiment. The result shows that MSL can be an alternative method of purification of wastewater. The appropriate composition of organic materials that can be used as adsorbent is MSL with wood sawdust 10% dosage because it can increase pH, decrease the number of Cr, ammonia, and phosphate but less effective to decrease BOD and COD.

1. Introduction

Needs for fresh water is increasing every year, it also gives an impact on the agricultural sector. Wastewater utilization is alternative ways to appropriately irrigation needs. According to Supriyadi (2013) waste is the usually outcome of the production process, both industrial and domestic, can also be produced by nature or classified as having no economic value [1].

Wastewater generated from the industry in the region of Surakarta is large in amount, especially the textile waste. Industrial wastewater usually contains many contaminants including heavy metals that can pollute main waterways receiver. The pollution gives impact on the quality of water, especially for irrigation.

Wastewater treatment systems in various countries around the world have been using physical, chemical, and biological properties of soil. Multi Soil Layering (MSL) applied might do purification of waste from natural materials. MSL is a kind of method that utilizes the ability of the soil as the main medium to enhance its function through the structure of soil to fresh wastewater so can decrease pollutant in the environment. MSL method is formed in a coating construction ground with rocks, organic minerals and other materials as bricks compiled with zeolite for treating wastewater by flow through the wastewater into the structure of the layer. A layer of soil mixture act as part of anaerobic and rock layers act as part of aerobic [2-4].

Andisols in Indonesia still rarely used as water purification, actually it has a very good ability to adsorb pollutant. Andisols is expected to help the purification of wastewater if used in the MSL, so that wastewater can be fully utilized. Andisols is the kind of soil that has many short-range order minerals such as allophane, imogolit, ferrihydrite, or Al-humus complexes [5]. Allophones are aluminosilicate minerals such as clay contained in the soil has a high adsorption for heavy metals [6].

According to Pranoto et al. (2013), Andisols in some volcanic mountains in Indonesia such as Papandayan, Arjuna, and Willis contain allophane [7]. Allophane has great ability to adsorb heavy metals. Activation of allophanes by NaOH is needed to increase heavy metals adsorption. Heavy metal level such as Fe, Mn, Cr, Cd, Cu, and Pb proved to be lowered percentages. Utilization of allophane in Andisols soil can be expanded on Multi Soil Layering. Based on these problems then conducted an experiment to determine effectiveness and efficiency of the organic materials that used in the MSL and compose the appropriate dosage for application in liquid industrial waste treatment.

2. Materials and Methods

2.1. Materials and Tools

Andisols used for MSL was taken from Lawu Mountain, which had altitude about 900 meters above the sea. Organic materials used include: rice straw, corncob, sawdust, coconut fibre and baggase. Other materials, which support this MSL, are zeolite, iron sands, and adhesives. Textile industry waste water sample obtained from batik industry factory in Sukoharjo, with sampling in the waste outlet reservoirs tub that circulated.

2.2 Research Method

This research through several phases including: soil sampling, experimental in greenhouse and laboratory analysis. This study uses the basic design of completely randomized design (CRD) with type 2 factorial experiment. The first factor is the type of organic material used and the second factor is the dosage of organic materials. The first factor: the type of organic materials, including: B1 (rice straw), B2 (baggase), B3 (sawdust), B4 (coconut fibre), B5 (corn cob). The second factor is dosage of organic materials, with 3 levels: P1 (5% dosage), P2 (10% dosage), P3 (25% dosage).

Research data were analyzed with SPSS 16.0 software applications and Excel 2007. Research data were analyzed using an ANOVA F test with a 5% level to determine differences between the treatments of the data calculation. If there is a significant difference in treatment then followed by Duncan's Multiple Range Test (DMRT) at the level of 95%. Another test that related to this research maybe applied, such as correlation test used to determine the relationship between variables, and Principal Component Analysis (PCA) used to determine the variables that influence.

2.3 Soil Sampling

Soil samples used in this study came from the area around an altitude of 900 m above sea level, from Lawu, sampling is done by using a drill ground at some point in a different location by first separating about 60 cm layer of soil over the top of the darken colour.

2.4 Making Multi-Soil Layering

Early stages of MSL are to prepare bricks made from a mixture of organic materials according to treatment, with a brick size 15 cm x 10 cm x 3 cm. Furthermore, container / jar that is used for the manufacture of MSL. The size of container/plastic jars used were 22 cm x 22 cm x 22 cm. The containers then fitted with a hole on one side near the base of it, then it fitted with a hose to the outlet of the wastewater.

MSL is formed by arranging the bricks that were produced alternately with zeolite. Given base layer of zeolite ± 2 cm, then placed the bricks that have been previously prepared. The next layer is a zeolite with a thickness of ± 1 cm and continued laying a second layer of bricks. Provision is also made for the zeolite third layer with a thickness of ± 1 cm and then topped with a hose to aerator. Giving hose laying bricks followed by a third layer and the top layer was closed using a zeolite with a thickness of ± 2 cm. MSL layout can be viewed in Figure 1.

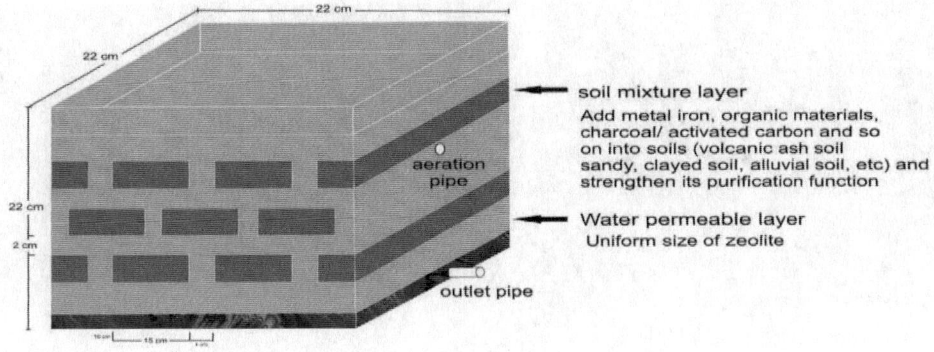

Figure 1. Multi Soil Layering Layout.

Application of wastewater to the MSL, first, wastewater is fed to the MSL ± 2 L in each container. Second, put 2.5 L of wastewater derived from the pre-experiment conducted in order to produce the amount of output ± 600 ml of waste water which will be used in the analysis. Then, output of wastewater put in bottle to make easier when saving the samples.

2.5 Testing the performance of Multi Soil Layering
The results of the analysis of each application variables in the MSL as compared with that previously contained in the wastewater. The performance of the MSL can also be seen from the comparison between the control treatments with a variety of other treatments in MSL testing.

3. Result and Discussion
The presence of the waste usually cam e from the production process, both industrial and domestic (household, for example: garbage), it can also have a negative impact on the environment and human health. Textile industry such as "*Batik*" is also generating the waste from the washing. The process raises a large wastewater and containing the rest of the colours, high BOD, high oil and toxic content. Required treatment of the wastewater is by recycling.

The recycling process can be divided into three phases: physics, chemistry, and biology. One method that can be used decrease toxicity content in wastewater is used Multi-Soil Layering (MSL). The performance of the MSL can be known through the results of the analysis of pH, ammonia content, content of dissolved phosphate, chromium content, BOD and COD level, and physical testing (colour and odour).

3.1 pH effluent on the MSL
Preliminary analysis of the wastewater samples showed that the pH value of the sample is 10.7 pH of laboratory test results presented in Figure 2 shows the average of pH value on the three replicates on each treatment.

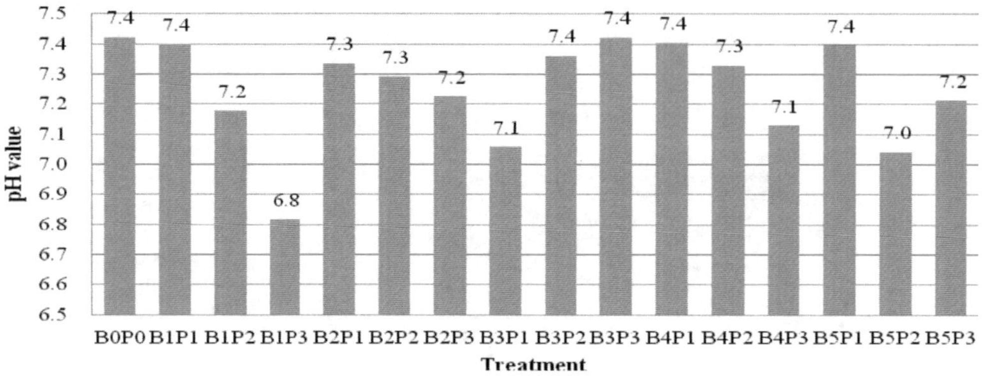

Figure 2. Average of pH value in MSL.

The lowest pH value was 6.8 contained in the MSL with 25% dosage of rice straw organic material (B1P3), while B0P0 and B3P3 have the highest pH value (7.4). B0P0 is a control and B3P3 is a combination of residual sawdust with 25% dosage. B0P0 (controls) have a tendency to high pH because although it does not use a mixture of organic materials but are in MSL zeolite. According to Yulianto et al. (2009) the addition of rice straw can reduce the pH value [8]. The pH value can also affect the process of removal of chromium (Cr) in wastewater. The pH value was effective in lowering serum Cr levels was 6.0 to 9.5 [9].

3.2 Chromium (Cr) value on the MSL
Based on Cr laboratory test result, which presented in Figure 3 shows average value of the three replicates of Cr on each treatment. The highest value of 1.58 ppm Cr contained in B0P0 (control), while the lowest value of 0.09 ppm found in the MSL with corncob organic materials at 10% dosage (B5P2). Maximum allowable levels of chromium for the textile industry waste is 2.0 mg/l based on Badan Pengendalian Dampak Lingkungan (BAPEDAL) [10]. Although it appropriates quality standards, Cr has an accumulative character in nature, so it should be removed from the water. Chromium is a dangerous element found in the earth's surface and the condition of the oxide between Cr (II) to Cr (VI) [11]. Chromium is one example of heavy metals that are carcinogenic.

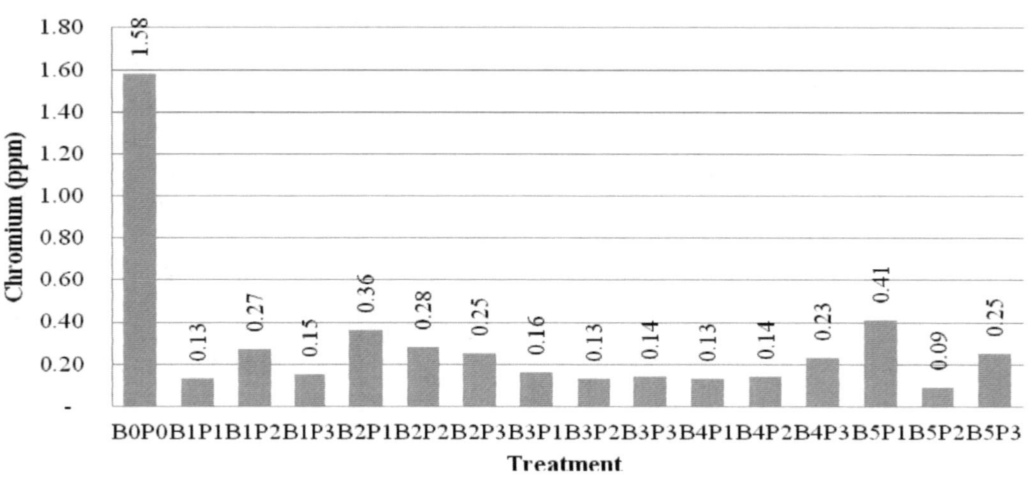

Figure 3. Average of Chromium value on MSL.

Adsorption capacity for Cr (VI) ion removal is depend on temperature, pH effluent, effect of concentration variation, and contact time. The most appropriate pH condition for chromium removal on ranges 6.0-8.0. Application of adsorption with patented hybrid surfaces, organofunctionalization with alkylaminesis necessary for separation of solid/liquid for chromium removal in industrial and environmental [12].

3.3 Ammonia value on the MSL

Based on ammonia laboratory test results that presented in Figure 4 shows the average value of the three replicates of ammonia on each treatment.

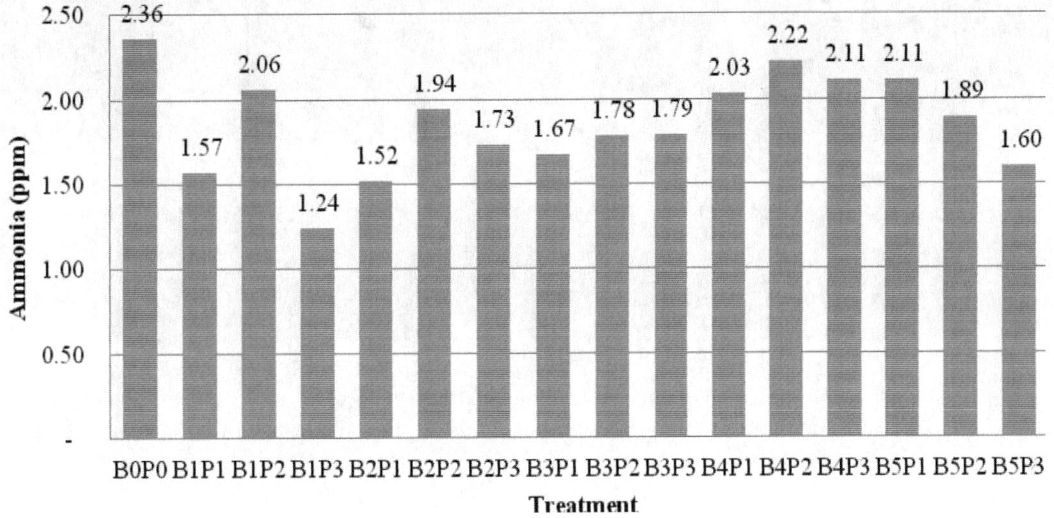

Figure 4. Average of Ammonia value on MSL.

The average of ammonia content in test showed that the lowest value of 1.24 ppm found in the MSL with rice straw organic materials at a 25% dosage (B1P3). The highest value (2.36 ppm) produced in controls (B0P0). A maximum allowable level of ammonia in the textile industry waste is 8.0 ppm based on [10]. All over the treatments has an average value of ammonia <8.0 ppm which indicate that ammonia levels after treatment MSL still at the threshold of the permissible standard. Research from Sato et al. (2010) suggest that aeration is needed in upper parts of MSL for nitrification, and extra organic materials in the lower part of MSL for denitrification [13]. It was because ammonia adsorption and nitrification were almost completed up to 3rd layer and denitrification not much proceeds after 4th layer.

3.4 Dissolved phosphate value in MSL

Dissolved phosphate in laboratory test results presented in Figure 5 shows the average value of the three replicates phosphate on each treatment.

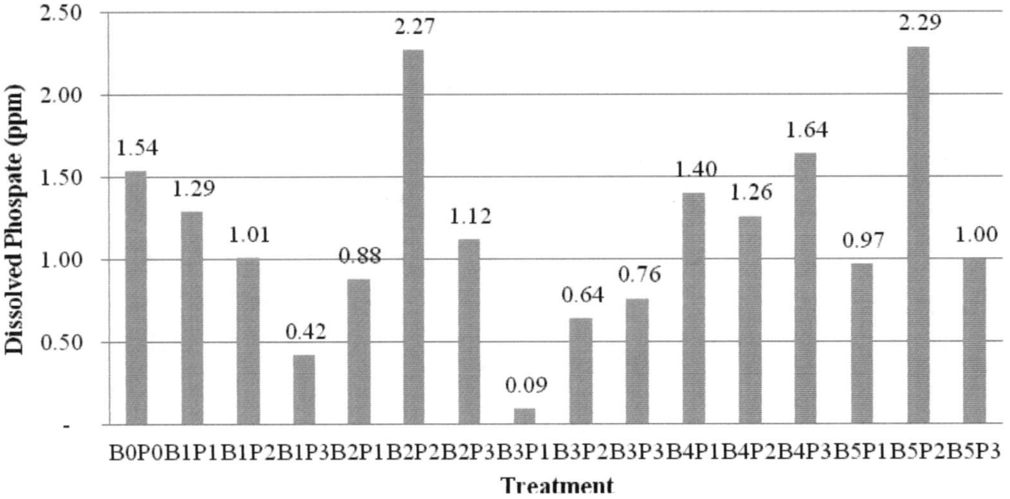

Figure 5. Average of Dissolved Phosphate value on MSL.

The average of dissolved phosphate based on histogram shows that the highest value of phosphate is 2.29 ppm contained in the MSL with corncob organic materials at 10% dosage (B5P2). Phosphate lowest value (0.09 ppm) contained in the MSL with sawdust organic materials of 5% dosage (B3P1) by 0.09 pm. Based on the regional regulation of Central Java No. 10 years 2004, total of phosphate for raw threshold is 5.0 ppm. The average of each treatment showed ppm phosphate 5.0 ppm, which indicates that, the ppm phosphate in wastewater after the application MSL still on the threshold allowed. Permeability of MSL was also very important for phosphorus removal in MSL system. Phosphorus can be removed by ferric hydroxides on soil mixture. The adsorption of phosphorus on decreased or the adsorbed phosphorus leached from the MSL due to establish an anaerobic condition [13]. Andisols basically have amorphous materials such as mineral allophane, imogolit and Al or Feoxide minerals that have a high phosphate adsorption capacity [14]. According to Yunarsih (2013) reduced levels of dissolved phosphate can be through adsorption by chitosan or materials containing high lignin [15]. Research of Wikanta et al. (2013) shows that wood sawdust has cellulose (48.89%), lignin (28.90%), ash (2.09%), water content (6.02%) and pentose (14.10 %) [16].

3.5 Biological Oxygen Demand (BOD) in wastewater

Based on BOD test results are presented in Figure 6 shows the average of the BOD value of the three replicates on each treatment. Based on the available data in the histogram, it is known that the average lowest value (149.53) contained in the MSL with bagasse organic material at 5% dosage (B2P1). The highest average value (393.83 ppm) contained in the MSL with organic material in the rice straw at a 25% dosage (B1P3). Maximum of allowable levels of BOD for the textile industry waste is 85 mg/l based on BAPEDAL [10].

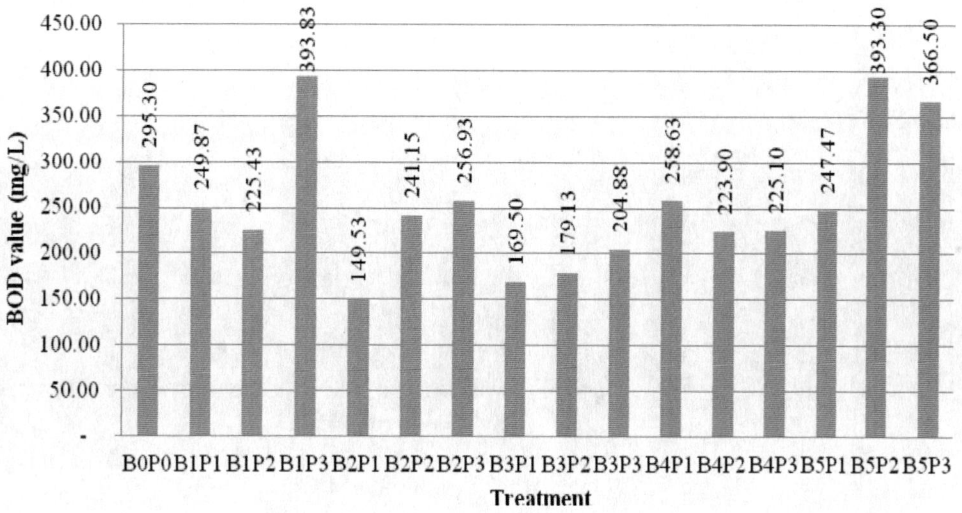

Figure 6. Average of BOD value on MSL.

Although BOD value after application of MSL still above threshold standard but has decreased from the initial BOD value (780 ppm). BOD indicates the amount of biological oxygen dissolved. The average value of BOD of MSL treatment outcomes when compared with standard threshold is 85 ppm, all of the treatment does not appropriate the threshold standard. The average BOD value of each treatment showed a high value is >100 ppm, which indicates that the MSL still not effective in lowering the levels of BOD.

According Sato et al. [13] the water flow rate in MSL was an essential factor for removal of organic materials, phosphorus and ammonia. There are some factors that can keep the proper flow rate in MSL, that selection of soil with proper particle size, mixing of high permeable materials with MSL and the setting of aeration regime. Proper aeration is effective in the prevention of clogging.

3.6 Chemical Oxygen Demand (COD) in wastewater

COD laboratory test results are presented in Figure 7 shows the average of the COD value of the three replicates on each treatment. Average value of COD on MSL treatment outcomes when compared with standard threshold of 150 ppm, all of the treatment does not appropriate the threshold standard. The average of the COD value of each treatment showed a high value is >300 ppm, which indicates that the MSL still not effective in lowering the levels of COD.

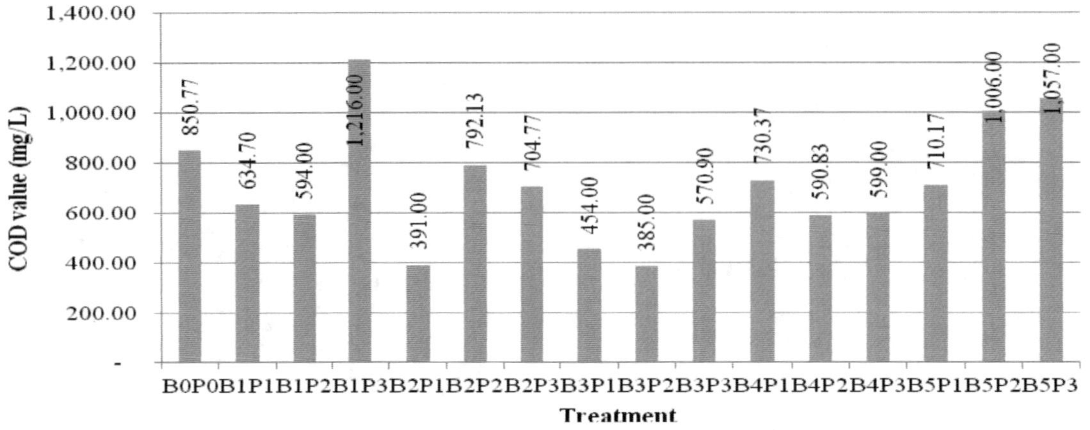

Figure 7. Average of COD value on MSL.

The highest average of COD value (1216 ppm) based on the histogram shown by MSL rice straw organic materials at 25% dosage (B1P3), while the lowest average COD value (385 ppm) contained in the MSL with organic material sawdust at 10% dosage (B3P2). Maximum allowable levels of COD for the textile industry waste is 250 mg/l based on BAPEDAL (1995). The COD value at the beginning of the sample is equal to 1520 ppm, after MSL applications tend to decrease even though its value is not significant.

Impairment of COD based on the research of Georgiou et al. (2003) was also influenced by the pH value of the environment. Higher pH values can inhibit a decrease in COD values, whereas low pH can help reduce the COD by 60%. In COD and BOD removal, COD needed number of layers than BOD. It was because COD includes easily and slowly decomposable organic materials, while BOD represents easily decomposable organic materials [13].

3.7 Testing Odor and Colour Changes

Based on Table 1 known that any samples taken have dark colour and pungent odour. The first, second and third samples show that wastewater plant output produces a pungent odour (detergents and dyes) of various treatment processes. It shows that the wastewater does not appropriate to threshold of fresh water ready to be disposed of into the water reservoir canal.

Table 1. The results of physical observation (colour and odor) in the sample of wastewater

Sample	Odor	Colour
1	Detergents and dyes pungent	Dark blue, slightly cloudy
2	Detergents and dyes pungent	Livid, slightly cloudy
3	Detergents and dyes pungent	Dark blue, slightly cloudy

The results of physical observation (colour and odour) in the sample after application of MSL on each repetition are also not different from the initial waste samples. Some samples also showed changes in colour though not completely - can completely eliminate the existing colour in the liquid waste. MSL has not been effective in removing odours and colours might be due to the application of MSL has not used the system on a regular circulation causes the incubation time in MSL less waste, and environmental conditions (pH) which does not support the purification so that the absorption of odours and colours at less than the maximum waste. pH value can affect the change of colour in textile wastewater. Low pH conditions can improve discolouration than when the pH is likely to be high. The reason is alkaline environment can make H_2O_2 is unstable and tends to decompose the formation of oxygen and water than hydroxyl radicals [18].

3.8 MSL Performance Effectiveness

MSL proven effective to raise the pH, lowering ppm Cr, lowering dissolved ammonia and phosphate. Application of MSL is less effective to reduce the BOD and COD. The correlation test shows the correlation of the factors used. The results of the correlation test showed that the correlation found between BOD to COD, pH dissolved phosphate and ammonia with Cr levels. Additional testing is also done to determine the factors that really affect the performance of MSL.

Based on PCA test, known that the factors that influence them are COD, soluble phosphate, and Cr levels. Treatment is best seen from the factors that most influence in the PCA test is B3P2 that MSL with sawdust organic material with 10% dosage. Treatment B3P2 COD value a low of 385.00 ppm, 0.64 ppm of dissolved phosphate levels, and Cr content is 0.13 ppm. Sudarja & Novi (2012) research shows that sawdust can be used as activated charcoal, which has the capability of adsorbing heavy metals in high level such as Cd, Pb, and Cr [19]. However, in high saturation of sawdust uses will impact to decreasing of adsorption. Smell and colour test of the waste showed textile wastewater quality still does not appropriate for threshold of safety standard for distributed into a major waterways receiver.

4. Conclusion

The results showed the organic material can be used to raise the pH, Cr lower level, ammonia and phosphate lower level less effective to reduce the BOD and COD. Multi Layering Soil composition is suitable as a natural adsorbent is MSL with organic material sawdust with a dosage of 10%. Multi-Soil Layering can be used as effective and efficient alternative method to dissolve Cr metal.

Acknowledgement

The experiment was supported by Indonesia Higher Ministry of Education of Sebelas Maret University Surakarta.

References

[1] Supriyadi 2013 *Environmental Toxicology* Study Case Surakarta UNS Press

[2] Masunaga T, Sato K, Mori J, Shirahama M, Kudo H and Wakatsuki T 2007 Characteristics of Wastewater Treatment Using A Multi-Soil-Layering System in Relation to Wastewater Contamination Levels and Hydraulic Oading Rates. *Japanese Journal of Soil Science and Plant Nutrition* **53** 215–223

[3] Sato K, Masunaga T and Wakatsuki T 2002 Treatment of Bio-generic Wastewater by Multi-Soil-Layering Method Effect of Various Positions of Aeration Using Small Model Systems. *Journal of Japan Society on Water Environtment* **25** 234–241

[4] Sato K, Masunaga T and Wakatsuki T 2005 Water Movement Characteristics in a Multi-Soil-Layering System *Soil Science and Plant Nutrition* **51**(1) 75–82 doi:10.1111/j.1747-0765.2005.tb00009.x

[5] Miller R W 1990 *Soils: An Introduction to Soils and Plant Growth* USA John Willey and Son

[6] Jara A A, Goldberg S and Mora M L 2005 Studies of the Surface Charge of Amorphous Aluminosilicates using Surface Complexation Models *Journal of Colloid and Interface Science* **292** 160–170

[7] Pranoto, Suranto, Sugiyarto K H and Ashadi 2013 Javanese Volcanic Allophane Adsorbent as Heavy Metal Adsorber to Improve the Quality of Drinking Water in Surakarta *Journal of Environment and Earth Science* **3**(5) 48–56 Retrieved from www.iiste.org

[8] Yulianto M E, Diyono I, Hartati I , N R S and J F P 2009 Development of Biomass Enzymatic Hydrolysis of Rice Straw to Ethanol Production For Bioethanol Production In *National Symposium RAPI VIII* (pp. 66–73)

[9] Liu T, Tsang D C W and Lo I M C 2008 *Environ. Sci. Technol* **42** 2092–2098

[10] Environmental Impact Control Agency (in Indonesia: BAPEDAL) 1995 *Hazardous and Toxic Materials* Jakarta Environmental Impact Control Agency

[11] Suhendrayatna 2001 Bioremoval Heavy Metals by Using Microorganisms: A References Study Tokyo Institute for Science and Techology Study

[12] Guerra D J L, Mello I, Freitas L R, Resende R and Silva R a R 2014 Equilibrium, Thermodynamic, And Kinetic of Cr(VI) Adsorption Using A Modified And Unmodified Bentonite Clay *International Journal of Mining Science and Technology* **24** (4) 525–535 doi:10.1016/j.ijmst.2014.05.017

[13] Sato K, Iwashima N, Matsumoto T, Wakatsuki T and Masunaga T 2010 Wastewater treatment processes and mechanisms of organic matter , phosphorus , and nitrogen removal in a multi-soil-layering system. In *World Congress of Soil Science, Soil Solutions for a Changing World* (pp. 85–88)

[14] Ilyas, Syekhfani and Sugeng Prijono 2000 Analysis of Agricultural Waste Giving Abu husk as a source of silicate in Andisol and Oxisol to release the adsorbed phosphorus Peruntut Engineering pp 103 - 110. *Meeting of Scientific Research and Development Isotopes and Radiation Technology*

[15] Yunarsih N M 2013 Chitosan Membrane effectiveness of Leather Galah Shrimp (*Macrobrachium rosenbergii*) to Reduce Phosphate in Wastewater Laundry. *Thesis*. Graduate Program of Universitas Udayana Denpasar Unpublished

[16] Wikanta D , Arifan F, Azmi L, Devi O, Septriani R, Rekyan P and Siti A 2013 Optimalisasi Limbah Serbuk Kayu Menjadi Bioetanol Sebagai Energi Alternatif Terbarukan Menggunakan distilasi Gelombang Mikro pp 6 – 11 *Proceeding SNST ke-4* Semarang Study Program of Chemistry PSD III University of Diponegoro

[17] Georgiou D, Aivazidis A, Hatiras J and Gimouhopoulos K 2003 Treatment of Cotton Textile Wastewater Using Lime and Ferrous Sulfate. *Water Research* **37** 2248–2250 Retrieved from www.elsevier.com/locate/watres

[18] Deshannavar U B, Murgod A A, Golangade M S, Koli P B, Samyak B and Naik N M 2012 Photo-Oxidation Process – Application for Removal of Color from Textile Industry Effluent *Research Journal of Chemical Sciences* **2** (10) 75–79

[19] Sudarja and Novi Caroko 2012 Assess the Effectiveness Absorption Experimental Liquid Waste of Batik Taman Sari Using Active Charcoal Mesh 80 of the Waste Saws Teak *Journal of Ilmiah Semesta Teknika* Vol **14** (1) 50 – 58

IOP Conf. Series: Materials Science and Engineering **107** (2016) 012017 doi:10.1088/1757-899X/107/1/012017

Adsorption and desorption of Zn(II) and Cu(II) on Ca-alginate immobilized activated rice bran

A Suratman*, N Z Kamalia and W A Kusumawati
Department of Chemistry, Faculty of Mathematics and Natural Sciences
Gadjah Mada University, Sekip Utara, Yogyakarta 55281

*E-mail: adhitasari@ugm.ac.id

Abstract. Ca-alginate immobilized activated rice bran has been used for adsorption of Zn(II) and Cu(II) from aqueous solution. The effect of the pH, kinetics model, adsorption isotherm and desorption on the adsorption performance was investigated. Activated rice bran was immobilized by the entrapment in alginate beads. The adsorption strength of Ca-alginate immobilized activated rice bran was compared to Ca-alginate and non-immobilized activated rice bran. The concentrations of adsorbed ions were analyzed using Atomic Absorption Spectrophotometer (AAS).The result showed that pH of 4.0 and the contact time of 120 min are the optimum condition for adsorption of Zn(II) and Cu(II). The adsorption kinetic of Zn(II) and Cu(II) followed the pseudo-second-order model with adsorption rate constant 4.9 x 10^{-2} and 3.14 $g.mg^{-1}.min^{-1}$, respectively. The both adsorption processes obeyed Langmuir isotherm with adsorption capacity of 2.03 and 2.42 $mg.g^{-1}$ of adsorbent, respectively. The strength of Zn adsorption on Ca-alginate immobilized activated rice bran (86.63%) was more effective compared to Ca-alginate beads (60.96%) and activated rice bran (43.85%). The strength of Cu adsorption was 80.00%, 61.50% and 22.10%, respectively. The desorption of Zn(II) and Cu(II) showed that recovery percentage of the adsorption was 76.56% and 57.80% with the condition of using HCl 0.1 M as desorption agent for 1 hour.

1. Introduction

The heavy metal pollution in natural water arises from mainly acid mine drainage and waste water released by industries, agricultural effluents, and various sources. These heavy metals such as Cu, Pb, Hg, Ni, Zn and Cd are toxic to the living system that is very important for environmental quality.

Some methods such as ion exchange, adsorption, membrane separation, solvent extraction, chemical precipitation, reverse osmosis, evaporation and electrodialysis have been used for metal removing from wastewater [1]. Since trace heavy metals should be removed in wastewater, adsorption may provide an effective alternative method for a low cost for the removal of heavy metals from wastewater [2].

Biosorption of heavy metals is proven to be quite effective for the removal of metal ions at low concentration from wastewater with a low cost treatment, potential for sorbent regeneration, high velocity of adsorption and desorption, and environmental friendly manner [3, 4]. Algae, fungi, yeast and bacteria are some of the microorganisms capable on sorption of toxic heavy metal ions from dilute aqueous solution [4, 5].

Rice bran is a by-product of the rice milling industry and found in a big amount that cause disposal problem. The rice bran has been successfully used to remove metal from waste water with the relatively low-cost process [6, 7].

Alginate is one of the most common biopolymers used for immobilization of material like kaolin/zeolite, microorganism or bacteria or biomass [5, 8, 9]. Alginate is the ammonium or alkali salt of chains of 1,4 linked β-D-manuronic and α-L-guluronic acid residues. Alginate becomes a negatively charged anionic polymer on acidification with the formation of –COO- groups. After reaction with calcium salts, alginate can form gels so that it can retain its shape and size unchanged. Calcium ions can be bound to the polymer by simple ionic bridging with two carboxyl groups. These –COO- groups have interaction with metal ions in waste water and calcium ions can be readily exchanged with the metal ions in the waste water [10]. In this experiment, alginate is used as immobilized activated rice bran to remove Cu(II) and Zn(II) from water.

The process of biosorption can be economical if sorbent can be used in a continuous adsorption-desorption cycle. Biosorbent should be cheap, reusable, both uptake and release of metals ions should be efficient and rapid, and desorption process should be metal-selective [5].

2. Materials and methods

2.1. Materials

Rice bran, Na-alginate from Sigma-Aldrich, $CaCl_2.2H_2O$, $ZnSO_4.7H_2O$, $CuCl_2.2H_2O$, HNO_3, KNO_3, Na_2EDTA, NaOH, HCl, CH_3COOH, H_3PO_4, Na_2HPO_4 were supplied by Merck. All chemicals were analytical reagent grade. Distilled water was used in the preparation of standard and all solution.

2.2. Methods

2.2.1. Preparation of activated Rice Bran. The rice bran was weighed and washed with 30 mL of HCl 0.1 M two times, centrifused at 2800 rpm for 15 min, then washed with distilled water. Activation of rice bran was treated by addition of 150 mL NaOH 0.1 M were and soaked for 24 h. The solution was then centrifuged at 2800 rpm for 15 min again and washed with distilled water for three times. Water was removed by filteration. Biomass thus was dried at 50 °C and used for adsorption investigation.

2.2.2. Immobilization of Activated Rice Bran Biomass. Na-alginate was prepared for immobilization of activated rice bran. Na-alginate solution 3% (w/v) was prepared with distilled water. Then, 1.6 g of activated rice bran was added to the alginate solution. The solution was drown through needles and introduced into 0.1M $CaCl_2.2H_2O$. The alginate beads were allowed to treat for 30 min, then were dried and stored at 4 °C for the next use. The characterization of Ca-alginate immobilized activated rice bran was performed by FT-IR.

2.2.3. Adsorption and desorption studies. The adsorption of Cu(II) and Zn(II) on the Ca-alginate immobilized activated rice bran was investigated. The effect of pH (range 3-5), contact time (10-240 min), initial concentrations of Cu(II) and Zn(II) (20 to 300 mg/L) on the adsorption rate and capacity were studied. The comparation between activated rice bran, Ca-alginate and Ca-alginate immobilized activated rice bran were carried out under the optimal condition. In order to study metal desorption, the adsorbed Cu(II) and Zn(II) in Ca-alginate immobilized activated rice bran were rinsed by sequential desorption of H_2O, 1M KNO_3, 0.5 M HNO_3, 0.1 M Na_2EDTA, respectively. The adsorbed Cu(II) and Zn(II) were rinsed by direct desorption of 0.5 M HNO_3 and 0.1 M HCl for 1 h. The concentration of Cu(II) and Zn(II) were analyzed by 3110-perkin Elmer atomic absorption spectrophotometer.

3. Results and discussion

3.1. Characterization of Ca-alginate immobilized activated rice bran

FTIR spectra of Ca-alginate immobilized activated rice bran are shown in Figure 1. The combination of Ca-alginate and rice bran FTIR spectrum showed the characteristic peaks at 3448 cm^{-1} (OH^-

stretching), 1635 and 1427 cm^{-1} (COO^- asymmetric and symmetric stretching), 1088-1034 cm^{-1} (C-O-C antisymetric stretching), and carboxyl and carboxylate at about 1427and 1635 cm^{-1}. From rice bran, -Si-O from silanol group at 879 and 439cm^{-1}, siloxane (Si-O-Si) at 1000-1100 $cm^{-1}$was found. Based on the FTIR analysis, the immobilization of rice bran in Ca-alginate will increase the adsorption of active group of Ca-alginat and free rice bran on Cu(II) and Zn(II).

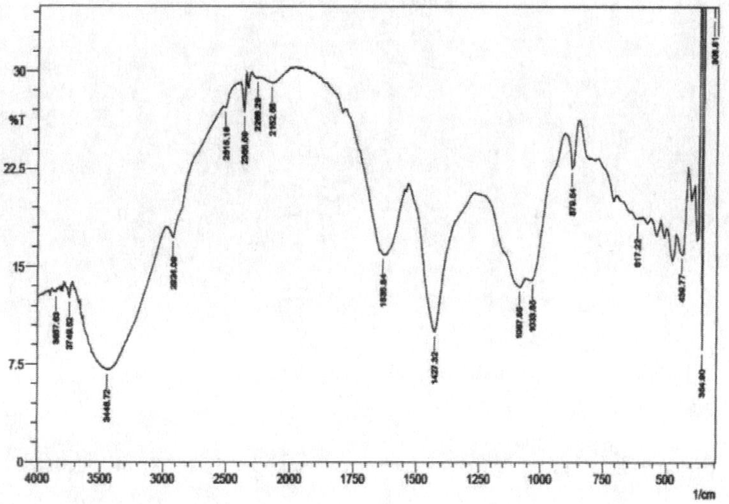

Figure 1. FTIR spectra of ca-alginate immobilized activated rice bran

3.2. Effect of pH on Cu(II) and Zn(II)adsorption

The effect of pH on the Cu(II) and Zn(II) adsorption by Ca-alginate immobilized activated rice bran was studied and the observed results are shown in Figure 2. Test was undertaken with different initial pH values of Cu (II) and Zn(II) solutions in water. There was an increase in adsorption capacity with increasing pH from 3.0 to 4.0 and a decrease in pH 4.0. The pH 4.0 is found to be highest efficiency for the Cu(II) and Zn(II) adsorption. At acidic pH, protonation of the adsorbent influenced the adsorption capacity. The proton concentration in solution is high and competes with free Cu(II) and Zn(II) for adsorption sites on surface of adsorbent. This effect becomes minor with increasing pH in the solution.

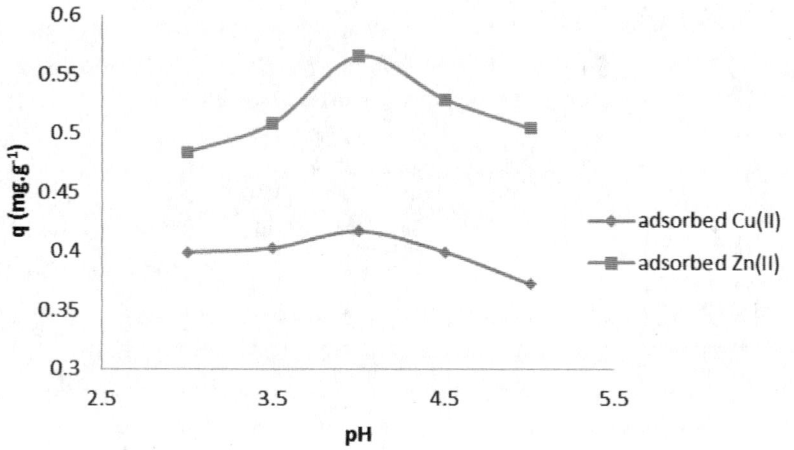

Figure 2. Effect of pH on Cu(II) and Zn(II) adsorption

3.3. Adsorption kinetic study

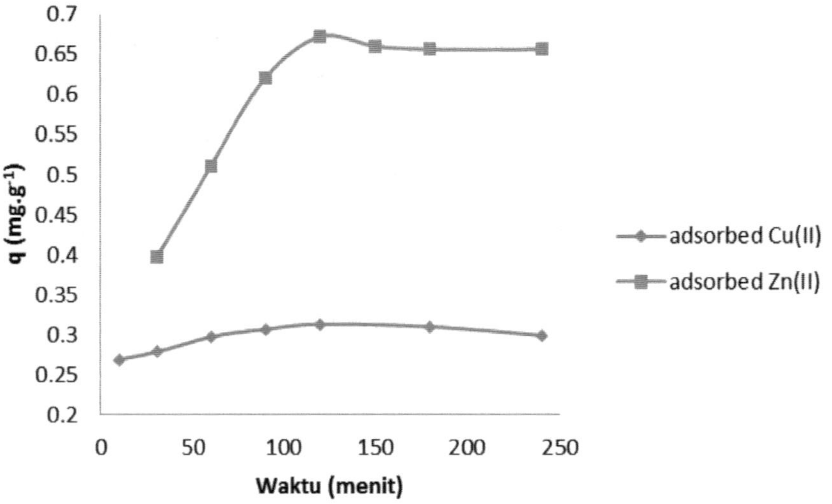

Figure 3. Effect of contact time on Cu(II) and Zn(II) adsorption

The contact time on Cu(II) and Zn(II) adsorption by Ca-alginate immobilized activated rice bran was investigated and the experimental results are showed in Figure 3. In order to find out the optimum contact time, adsorption of Cu(II) and Zn(II) have been carried out for 240 minutes. The equilibrium of both metal adsorptions is achieved after 120 min. The adsorption rate was initially fast due to availability of binding sites and after that the change of rate became constant.

In order to explain the adsorption kinetics, four common kinetic models, which are the first order, second order, pseudo-first order and pseudo-second-order kinetic model were tested. All the parameters were determined and listed in Table 1. The best kinetic model was selected based on the linear regression correlation coefficient (R^2). Furthermore, the correlation coefficient (R^2) for pseudo-second-order model was much closer to 1 than other models. These results confirmed that the adsorption of Cd (II) and Zn (II) ions onto alginate immobilized activated rice bran can be represented by pseudo-second-order kinetic model.

Table 1. The kinetic parameters evaluated for Cu(II) and Zn(II) adsorption

Kinetic Model	Equation	Cu(II) adsorption		Zn(II) adsorption	
		k	R^2	k	R^2
First order	$\ln C_a = -k_1t + \ln C_0$	3.3×10^{-3} min^{-1}	0.973	5.03×10^{-3} min^{-1}	0.782
Second order	$(1/C_a) = k_2t + (1/C_0)$	1.0×10^{-4} M.min^{-1}	0.984	1.3×10^{-4} M.min^{-1}	0.783
Pseudo-first order	$\log (q_e-q_t) = \log q_e - (k_3/2.303)t$	3.22×10^{-3} min^{-1}	0.974	5.02×10^{-3} min^{-1}	0.782
Pseudo-second order	$t/q_t = 1/(k_4q_e^2) + (1/q_e)t$	3.14 g.mg^{-1}.min^{-1}	0.999	4.9 g.mg^{-1}.min^{-1}	0.999

3.4. Equilibrium Adsorption Study

Adsorption of Cu(II) and Zn(II) by Ca-alginate immobilized activated rice bran is due to surface phenomena that would follow either Freundlich or Langmuir isotherm models. The Langmuir model makes several assumptions, the metal adsorption is a process occurring on homogenous surface and form monolayer coverage, while Freundlich model is based on physicochemical adsorption on heterogeneous surfaces. The Langmuir model gives the best representation and showed in Table 2. Regression correlation coefficients for Cu(II) and Zn(II) adsorption by Ca-alginate immobilized activated rice bran are very high. According to the Langmuir isotherm model, the value of adsorption capacity (Q_{max}) appears to be higher for the Cu(II) than Zn(II). Ca-alginate immobilized activated rice bran shows good performance of Cu(II) and Zn(II) adsorption. Both Ca-alginate and rice bran have negatively charge on the functional groups that attract the cationic metals. This contributed to high sorption capacity for alginate immobilized rice bran.

Table 2. Langmuir isotherm for Cu(II) and Zn(II) adsorption

Parameter	**Q_{max} (mg.g^{-1})**	**R^2**
Cu(II)	2.42	0.969
Zn(II)	2.03	0.967

3.5. The effectivity of Ca-Alginate/rice bran to adsorb Cu(II) and Zn(II)

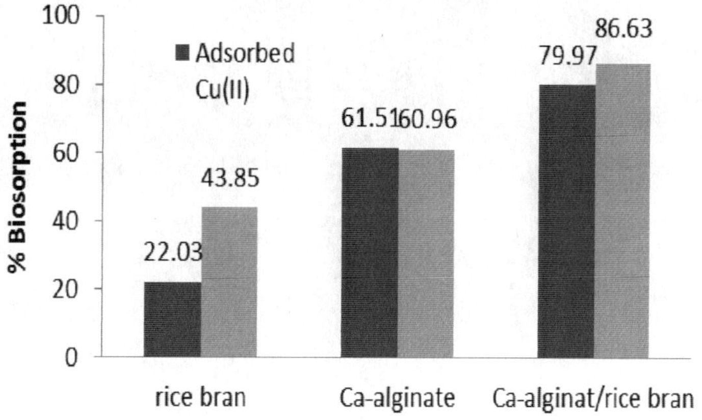

Figure 4. The strength of Ca-Alginate/rice bran to adsorb Cu(II) and Zn(II) compare to rice bran and Ca-alginate

The effectivity of Cu(II) and Zn(II) adsorption between activated rice bran, Ca-alginate, and Ca-alginate immobilization activated rice bran were carried out under the optimal condition. Figure 4 showed that the strength of Zn(II) and Cu(II) adsorption on Ca-alginate immobilized activated rice bran (79.97% for Cu(II), 86.63% for Zn(II)) were more effective compared to Ca-alginate beads (61.51%, 60.96%) and activated rice bran (43.85%, 22.03%). It is because of alginate and rice bran have an active site that can be interact with Cu(II) or Zn(II). The combination of both can increase the active site so that the Cu(II) and Zn(II) adsorption will be more effective.

3.5. Desorption efficiency of Cu(II) and Zn(II) absorbed by Ca-alginate immobilized activated rice bran

Adsorbent generation by desorption of adsorbed metal is necessary in order to minimize the process cost. Optimum desorption was achieved by a proper selection of pollutants, which depends on the type of adsorbent and the mechanism of adsorption. The adsorbed Cu(II) and Zn(II) in Ca-alginate immobilized activated rice bran were rinsed by sequential desorption of H_2O, 1M KNO_3, 0.5 M HNO_3, 0.1 M Na_2EDTA, respectively. Those desorption agents are responsible for adsorption of Cu(II) and Zn(II) through entrapment, ion exchange and complexation mechanism [5]. During desorption studies, maximum desorption of adsorbed Cu(II) was recorded. The results shown in Figure 5 indicates that high recoveries were possible for the Cu(II) and Zn(II) by using HNO_3. It indicates that adsorption

IOP Conf. Series: Materials Science and Engineering **107** (2016) 012017 doi:10.1088/1757-899X/107/1/012017

process involves electrostatic interaction between negative charges of the adsorbent and cationic metal.

In this study, the desorption efficiency was tested by 0.1 M HCl compare to 0.5 M HNO_3 for 1 h. The results that showed in Figure 6 indicate that HCl effective to desorb Cu(II), and Zn(II) compare to HNO_3. The recovery of Cu(II) and Zn(II) were found to be 57.89% and 76.56%, respectively.

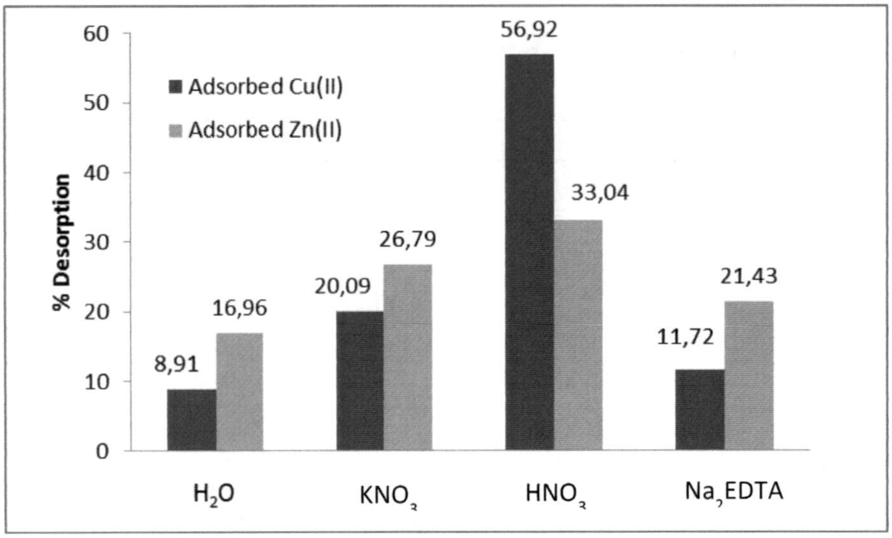

Figure 5. Sequential desorption of adsorbed Cu(II) and Zn(II)

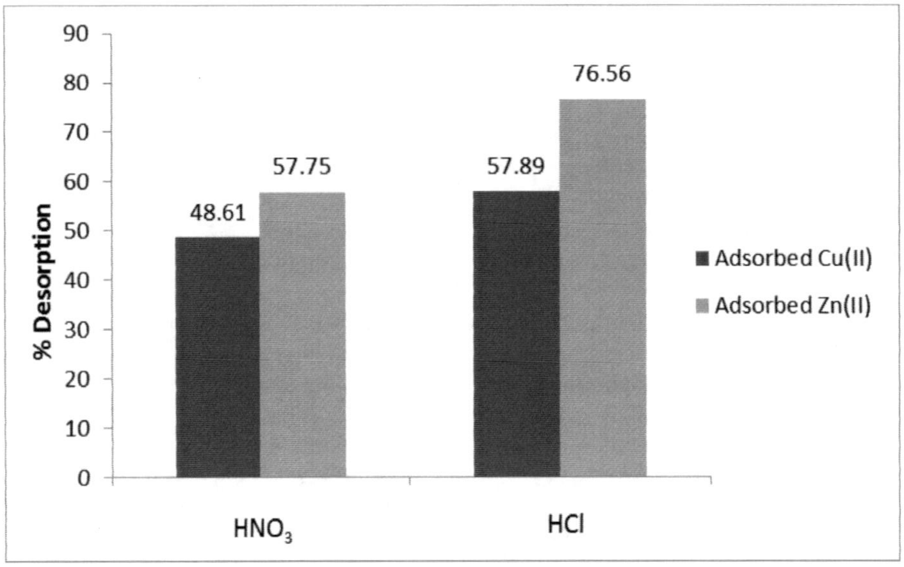

Figure 6.Effectivity desorption of adsorbed Cu(II) and Zn(II) by HNO_3 and HCl

4. Conclusions

Ca-alginate Immobilized activated rice bran can be used as adsorbents for the removal of Cu(II) and Zn(II) from the aqueous medium. The maximum adsorption capacity of adsorbent was achieved at pH 4 and contact time 120 min. Kinetic adsorption of Zn(II) and Cu(II) can be described by the pseudo-second-order kinetic model with adsorption rate constant 4.9 x 10^{-2} and 3.14 $g.mg^{-1}.min^{-1}$. The Zn(II) and Cu(II) adsorption process obeyed Langmuir isotherm with adsorption capacity of 2.03 and 2.42 $mg.g^{-1}$ of adsorbent. Compared with free activated rice bran and Ca-alginate only, the Ca-alginate immobilized activated rice bran showed improved performance to adsorb Cu(II) and Zn(II) and could be used as an efficient adsorbent system for the wastewater treatment. The desorption of Zn(II) and Cu(II) showed that recovery percentage of the adsorption was 76.56% and 57.89% with the condition of using HCl 0.1 M.

References

[1] Fu F and Wang Q 2011 Removal of heavy metal ions from wastewaters: a review, *J Environ Manage* **92**(3) 407-18
[2] Ghomri F, Lahsini A, Laajeb A and Addaou 2013 The removal of heavy metal ions (copper, zinc, nickel and cobalt) by natural bentonite, *Larhyss Journal* **12** 37-54
[3] Volesky B 2001 Detoxification of Metal Bearing Effluents: Biosorption for the Next Century, *Hydrometallurgy* **59**(2-3) 203-216
[4] Mudhoo A, Garg V K and Wang S 2012 Removal of heavy metals by biosorption, *Environment Chem Lett* **10**(2) 109-117
[5] Mishra S P 2014 Review: adsorption-desorption of heavy metal ions *Current Science* **107**(4) 601-612
[6] Ye H and Yu Z 2010 Adsorption of Pb(II) onto modified rice bran, *Natural resources*, **1** 104-109
[7] Ye H P, Zhu Q and Du D Y 2010 Adsorptive Removal of Cd(II) from Aqueous Solution Using Natural and Modified Rice Husk, Bioresource Technology **101** (14) 5175-5179
[8] Tzu T W, Tsuritani T and Sato K 2013 Sorption of Pb(II), Cd(II), and Ni(II) Toxic Metal Ions by Alginate-Bentonite, *Journal of Environmental Protection* **4** 51-55
[9] Mishra S P 2013 Adsorption of Cu and Zn on calcium alginate immobilized Penicillium sp., *Indian Journal of Chemical Technology* **20** 21-25
[10] Singh L, Asalapuram R P, Ramnath L and Gunaratna K R 2012 Effective Removal of Cu^{2+} Ions from Aqueous Medium Using Alginate as Biosorbent, *The Journal of Ecotechnology* **38**(1) 119-124

IOP Conf. Series: Materials Science and Engineering **107** (2016) 012018 doi:10.1088/1757-899X/107/1/012018

Optimization of biosurfactant production in soybean oil by *rhodococcus rhodochrous* and its utilization in remediation of cadmium-contaminated solution

Venty Suryanti*, Sri Hastuti and Dewi Andriani
Department of Chemistry, Faculty of Mathematics and Natural Sciences, Sebelas Maret University, Jl. Ir. Sutami 36A Surakarta, Central Java 57126 Indonesia

*E-mail: venty@mipa.uns.ac.id

Abstract. Biosurfactant production by *Rhodococcus rhodochrous* in soybean oil was developed, where the effect of medium composition and fermentation time were evaluated. The optimum condition for biosurfactant production was achieved when a medium containing 30 g/L TSB (tryptic soy broth) and 20% v/v soybean oil was used as media with 7 days of fermentation. Biosurfactant was identified as glycolipids type biosurfactant which had critical micelle concentration (CMC) value of 896 mg/L. The biosurfactant had oil in water emulsion type and was able to reduce the surface tension of palm oil about 52% which could stabilize the emulsion up to 12 days. The batch removal of cadmium metal ion by crude and partially purified biosurfactants have been examined from synthetic aqueous solution at pH 6. The results exhibited that the crude biosurfactant had a much better adsorption ability of Cd(II) than that of partially purified biosurfactant. However, it was found that there was no significant difference in the adsorption of Cd(II) with 5 and 10 minutes of contact time. The results indicated that the biosurfactant could be used in remediation of heavy metals from contaminated aqueous solution.

1. Introduction

Biosurfactants are surface-active compounds which are produced by a variety of bacteria, fungi, and yeast [1-2]. They have better environmental compatibility than that of the chemical surfactants, e.g., biodegradable and lower toxicity [3]. In recent years, research has expanded a lot in developing the method of biosurfactant production and the application of biosurfactants [4-5].

Fermentation conditions and nutrient availability affects the production of biosurfactant. Different substrates have been used to produce biosurfactants [6-7]. We have reported that vegetable oil (soybean oil) could be used for rhamnolipids biosurfactants production by *P. aeruginosin* [8]. We also demonstrated that industrial waste (cassava flour wastewater) could be applied to the production of rhamnolipids biosurfactants by *P. fluorescence* [9].

Biosurfactants have potential use in numerous areas, such as in pharmaceutics, agriculture, food industry, the oil industry and the paper and pulp industry [5,10]. Application of biosurfactants in remediation technologies for organic and inorganic pollutants, such as hydrocarbons and metals, from contaminated soil and solutions, has shown an increasing interest [11-14]. The removal of metals from contaminated soil and solutions is mostly based on the ability of biosurfactants to form complexes with metals [15-17].

It is paramount importance in optimizing factors that influence growth in biosurfactant producing organisms for commercial application. As an extension of our work, we are interested in the production of biosurfactants by *Rhodococcus rhodochrous* using soybean oil as a substrate and the use of these biosurfactants in the removal of cadmium from contaminated aqueous solution.

2. Materials and Methods

2.1. Materials
All chemical were used are analytical grade from e-Merck, whereas soybean oil from Sunbeam and palm oil from Bimoli. The strain used in this work, *R. rhodochrous* FNCC 0066, was purchased from Pusat Antar Universitas Gadjah Mada, Indonesia.

2.2. Media Used and Growth Condition

Cultures of bacteria were maintained on tryptic soy agar. Biosurfactant production optimization were performed using media composed of tryptic soy broth (30 g/l), and soybean oil (0, 5, 10 and 20% v/v). Fermentation was carried out on a rotary shaker (150 rpm) at room temperature for 12 days. The optical density of the cultures was monitored by UV-VIS Spectrophotometer, Shimadzu UV-160 IPC.

2.3. Biosurfactant recovery
Culture liquid of *R. rhodochrous* was centrifuged at 12,500 *g* for 20 minutes to obtain a crude preparation of biosurfactant (CPB). Partially purified biosurfactant (PPB) was obtained by extraction of culture supernatant with n-hexane and subsequently with chloroform. The extract was then evaporated to get biosurfactants free of the solvent. The PPB was analyzed by FT-IR spectrophotometer using Shimadzu FTIR-8201 PC spectrometer.

2.4. Biosurfactants Characterization
Emulsification index (E24) was determined by adding 2 mL of oil and the same amount of PPB, mixed by a vortex for 2 mins and left to stand for 24 h. The E24 is given as a percentage of the height of emulsified layer (mm) divided by the total height of the liquid column (mm). The surface tension and interfacial tensions were measured by the capillary rise method. The CMC value was obtained by dissolving of PPB in distilled water, and the surface tension was measured with various concentrations of the PPB. The sudden change in the surface tension was represented as the CMC, which was determined by plotting the surface tension as a function of PPB concentration.

2.5. Emulsion Type of Biosurfactants
Conductivity test was used to identify the emulsion type of PPB. Sodium chloride (1% w/w) as an electrolyte was added to the emulsion and the conductivity was measured. If the conductance increases, the emulsion is oil in water type. On the contrary, if there is no significant change, it is water in oil type.

2.6. Emulsification Properties of Biosurfactants
The E24 of the PPB were tested by addition of 0.1 mg of PPB to a crew-capped tube containing 1 mL of distilled water and 1 mL of palm oil. The E24 of the formed emulsions were monitored for 15 days.

2.7. Removal of Cd(II) Ion by Biosurfactants
The protocol of samples were prepared by dissolving cadmium ($Cd(NO_3)_2$. $4H_2O$ in 1 ml of 0.1 M HNO_3 to produce a metal solution (1000 ppm). The ph solution was adjusted to 6.0 before the biosurfactants were added. Batch adsorption technique was performed, where 10 mL of 2.0 ppm Cd(II) solution was mixed with either CPB (2 mL) or PPB (10 g) in a flask and stirred. The samples were kept at room temperature in a shaker at 150 rpm for 5 and 10 minutes. The solution was then filtered through Whatman filter paper No. 42. Residual metal ion concentration in the filtered samples was measured by Atomic Absorption Spectroscopy (AAS), Shimadzu type AA-6650. Experiments were made in triplicate, and the average value was calculated and presented as a percentage of metal removed.

2.8. Calculation
The adsorption capacity (q_e) of Cd(II) ion was calculated using the following equation

$$q_e = (C_i - C_f)\ V / m \quad (1)$$

where q_e is the amount of Cd(II) adsorbed per mass unit of biosurfactants (mg/g), C_i is the Cd(II) ion initial concentration (mg/L) and C_f is the Cd(II) ion final concentration (after treatment with PPB or CPB) (mg/L), V is the volume of the aqueous phase (L) and m is the amount of the CPB, or PPB used (g). The results were statistically examined using one-way analysis of variance (ANOVA), and significant differences were set at $p<0.05$.

3. Results and Discussion

3.1. Biosurfactants Production Optimization
Optimization of biosurfactant production was carried out in media containing 5, 10 and 20% v/v of soybean oil which was incubated for 12 days. The media with no addition of soybean oil was also performed as a negative control. The optical density, surface tension, and E24 were monitored during fermentation.

As presented in Figures 1(a)-(c), the addition of soybean oil in all concentrations significantly reduce the surface tension and enhance the optical density and E24 ($p<0.05$). Statistically, the addition of 20% v/v soybean oil at 7 days of fermentation gave the highest optical density ($p<0.05$). The addition of 20% v/v soybean oil provided notable significant different from each other where the highest E24 and the lowest surface tension were obtained. Although the optimum fermentation time could not be evaluated statistically based on the E24 and surface tension data ($p>0.05$), the figures showed that the optimum fermentation time was achieved at 7 days of incubation. Overall, the addition of 20% v/v of soybean oil and 7 days of fermentation time were elected for the optimum condition for the biosurfactant production.

3.2. Biosurfactants Characterization
The obtained FT-IR spectra revealed a strong and broadband at 3400 cm^{-1} which might be related to the O-H stretching vibration. The –CH stretch was confirmed by the band at 2923 and 2854 cm^{-1}. The asymmetric stretching vibration of C=O was identified by the band at 1747 cm^{-1} for the carboxylate and 1651 cm^{-1} for the ketone. The weak band at 1461 and 1377 cm^{-1} were obtained as a result of deformation and bending vibrations of aliphatic chains (-CH_3- and CH_2-), reflecting the presence of alkyl chains in the compounds. The stretching of C-O group was found by the presence of the intense and strong band at 1242 cm^{-1}. The obtained peaks of biosurfactants suggested that glycolipids biosurfactants were produced by *R. rhodochrous* as reported by other researchers that glycolipids (mostly trehalolipids) biosurfactants are produced by *Rhodococcus* sp. Trehalose tetra ester biosurfactants have been produced by *R. erythropolis* 51T7 [18], and a surface-active complex of acidic polysaccharides and lipids has been produced by *R. rhodochrous* CF222 [19]. Trehalolipids from rhodococci occur as complex mixtures with high structural diversities where their compositions vary depending on strain physiology and growth condition [20].

The produced biosurfactants presented the CMC value of 896 mg/L and the surface tension value of 37 mN/m. The produced biosurfactants were possibly ionic surfactant types as indicated by the higher value of CMC and the presence of carboxylate and hydroxyl groups. The addition of sodium chloride increased the conductance significantly, indicating that the emulsion type of the biosurfactants was oil in water (o/w). The biosurfactants were found to be effective emulsifiers. When palm oil was used as an immiscible compound, reducing the surface tension of water of about 52% from 62 mN/m to 32 mN/m. Furthermore, the biosurfactant enhanced the E24 of about 55% from 43% to 96% and the stable emulsion was reached up to 12 days.

3.3. Cadmium Removal from Synthetic Solutions by Biosurfactants
The cadmium removal from aqueous solution was examined for crude preparation biosurfactants (CPB) and partially purified biosurfactant (PPB). In this study, the pH value of 6.0 was chosen. At higher pH, the precipitation of the metal was observed whereas at lower pH, the excess of hydrogen

ions competed with the metal ion for binding with the active sites available in the sorbent [21]. Since the removal of cadmium was not correlated with the metabolism process, the absorption experiments were conducted with short contact time, e.g. 5 and 10 minutes.

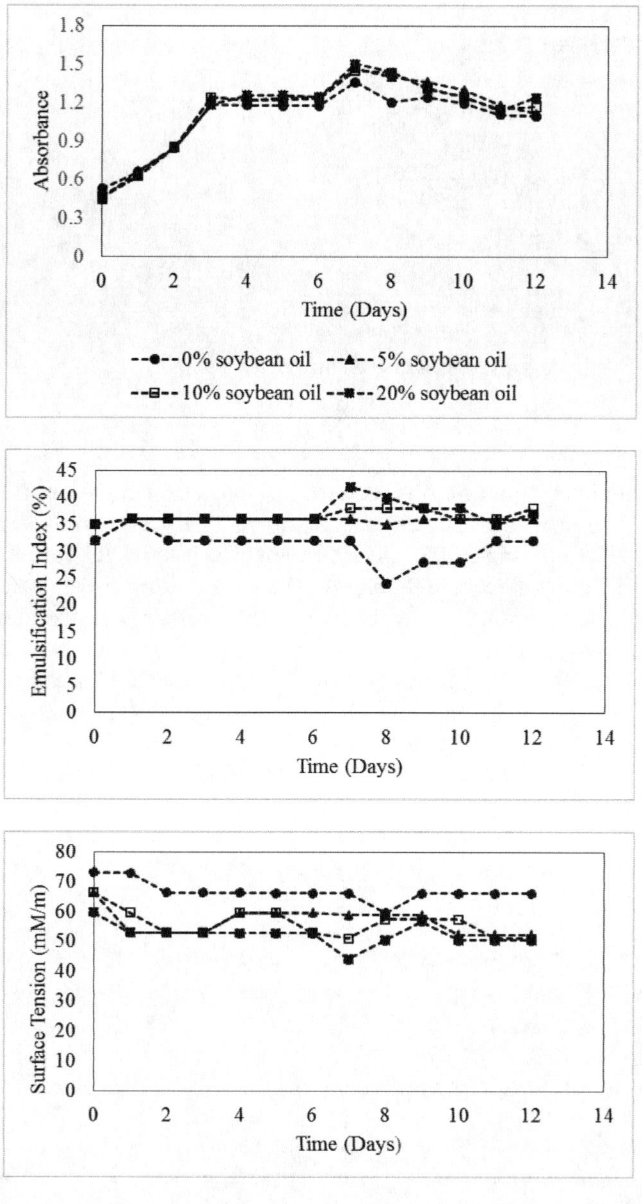

Figure 1. Optimization of biosurfactants production by *R. rhodochrous* grown in different amounts of soybean oil: (a) cell growth (optical density), (b) emulsification index and (c) surface tension.

The adsorption capacities of Cd(II) by CPB with 5 and 10 minutes of contact time were noted to be 1.659 and 1.766 mg/g. respectively. The adsorption capacities of Cd(II) by PPB with 5 and 10 minutes of contact time were found to be 0.344 and 0.558 mg/g. respectively (Figure 2). The results confirmed that the CPB has a much better absorption ability of Cd(II) removals than that of the PPB. Reduction in the number of functional groups of adsorbent due to purification occurred [15]. Accordingly, the amount of Cd(II) bound to the active adsorbent sites decreased, resulting in the adsorption capacity reduction. However, the results established that there was no significantly different in the removal of Cd(II) by CPB with the contact time of 5 and 10 minutes. Biosurfactants have been reported to be employed in the remediation of heavy metals, however, this potential for *Rhodococcus* biosurfactants still needs to be determined [20] and these results demonstrated that the biosurfactant produced by *R. rhodochrous* could be used for heavy metals removal from contaminated aqueous solution.

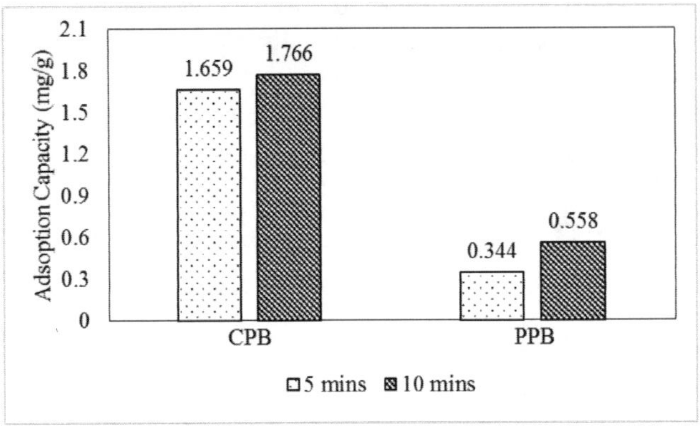

Figure 2. The adsorption capacities of Cd(II) by Crude Preparation Biosurfactants (CPB) and Partially Purified Biosurfactants (PPB) with 5 and 10 minutes of contact time.

4. Conclusion

Extracellular glycolipids biosurfactants could be produced by *R. rhodochrous* with soybean oil as a substrate. The optimum condition for biosurfactant production was reached at 7 days of incubation in a medium containing of 20% v/v soybean oil. The produced biosurfactants showed effective emulsifier properties. Biosurfactant exhibited CMC value of 896 mg/L and reduced the surface tension of water of about 52% when palm oil was used as an immiscible compound. Additionally, the biosurfactant enhanced the E24 of about 55%, and the emulsion was stable up to 12 days. The results suggested that the biosurfactant could be applied in remediation of heavy metals from aqueous solution. The crude biosurfactants performed a much better adsorption ability of Cd(II) than that of partially purified biosurfactants. Nevertheless, there was no considerable difference in the adsorption of Cd(II) by crude biosurfactants with 5 and 10 minutes of contact time.

Acknowledgments

The authors acknowledge their thanks to the Indonesian Toray Science Foundation (ITSF) for the financial support of this research in the form of Science and Technology Research Grant.

References

[1] Amara P F, Coelho M A, Marrucho I M and Coutinho J A 2010 *Adv Exp Med Biol* **672** 236-244
[2] Muthusamy K, Gopalakrishnan S, Ravi T K and Sivachidambaram P 2008 *Curr Sci* **94** 736-747
[3] Mukherjee S, Das P and Sen R 2006 *Trends Biotechnol* **24 11** 509-515
[4] Gautam, K and Tyagi V K 2006 *J Oleo Sci* **55 4** 155-166
[5] Banat I, Franzetti A, Gandolfi I, Bestetti G, Martinotti M, Fracchia L, Smyth T J and Merchant R 2010 *Appl. Microbiol Biotechnol* **87 2** 427-444
[6] Makkar R S and Cameotra S 2002 *Appl Microbiol Biotechnol* **58** 428-434
[7] Rodrigues L R, Teixeira J A and Oliveira R 2006 *Biochem Eng J* **32** 135-142
[8] Suryanti V, Hastuti S, Wahyuningsih T W, Mudasir and Muliawati D I 2009 *Indo J Chem* **9 1** 107-112
[9] Suryanti V, Marliyana S D, Handayani D S and Ratnaningrum D 2013 *Indo J Chem* **13 3** 229-235
[10] Banat I M, Makkar R S and Cameotra S S 2000 P *Appl Microbiol Biotechnol* **53 5** 495-508
[11] Cameotra S S and Makkar R S 2010 *Pure Appl Chem* **82 1** 97-116
[12] Garcia-Junco M, Gomez-Lahoz C, Niqui-Arroyo J L and Ortega-Calvo J J 2003 *Environ Sci Technol* **37 13** 2988-2996
[13] Zouboulis A, Matis K, Lazaridis N and Golyshin P 2003 *Miner Eng* **16 11** 1231-1236
[14] Sivakami R, Saravanan S and Premkishore G 2015 *Int. J. Curr Res Aca Rev* **3 4** 272-281
[15] Pacwa-Płociniczak M, Płaza G A, Piotrowska-Seget Z and Cameotra S S 2011 *Int J Mol Sci* **12**

633-654
[16] Wang S and Mulligan C N 2004 *Water Air Soil Poll* **157** 315-330
[17] Ochoa-Loza F J, Artiola J F and Maier R M 2001 *J Environ Qua* **30** 479-485
[18] Marques A M, Pinazo Farfan A M, Aranda F J, Teruel J A, Ortiz A, Manreza A and Espuny M J 2009 *Chem Phys Lipids* **158** 110-117
[19] Iwabuchi N, Sunairi M, Anzai H, Nakajima M and Harayama S 2000 *Appl Environ Microbiol* **66** 5073-5077
[20] Kuyukina, M S and Ivshina I B 2010 *Rhodococcus* biosurfactants: Biosynthesis, properties and potential applications in Biology of *Rhodococcus* edited by Alvarez H M *Springer-Verlag Berlin Heidelberg* 291-314
[21] Abdel-Ghani N T and El-Chaghaby G A 2014 *Int J Latest Res Sci Technol* **3 1** 24-42

IOP Conf. Series: Materials Science and Engineering **107** (2016) 012019 doi:10.1088/1757-899X/107/1/012019

Development of inorganic composite material based TiO_2 for environmental application

Sayekti Wahyuningsih*, Ari Handono Ramelan, Edi Pramono, Candra Purnawan, Velina Anjani, Puji Estianingsih, Ludfiaastu Rinawati, Khusnan Fadli

Inorganic Material Research Group, Faculty of Mathematic and Natural Science, Sebelas Maret University, Ir. Sutami street No. 36A Surakarta 57126 Indonesia

*E-mail: sayekti@mipa.uns.ac.id

Abstract. Syntheses of various materials, for green energy nanotechnology applications have special attention to develop emerging areas, such as environmental as well as energy materials. Various approaches for preparing nanostructured photocatalysts, such as titanium dioxide, nickel oxide, lead oxide and their composites, was introduced. The use of nanomaterials as photocatalysts water detoxification by visible light photocatalyst of an inorganic composite as well as dye-sensitized photoreduction was also discussed. The enhancement of selective photocatalyst system was gain by the use of photocatalyst composite materials and applied potential bias on the system. The photoelectrocatalytic degradation of rhodamine B (RB) and Remazol Yellow FG (RY) as water contaminant using the thin film of modified TiO_2 as the electrode was investigated via a series of potentials, and various pH. The result showed that the anodic potential bias influenced the degradation rate of water contaminant and exhibited better performance by the positive anodic bias was applied. The pH conditions influence the active dye structure whereas it will interact with inorganic semiconductor photocatalyst. Using dye-sensitized TiO_2 system (DSTs), we have applied this system to build water decolorization as a novelty environmental remediation system.

1. Introduction

Environmental pollution has become a serious problem for developing and developed countries in the world. Therefore, there is a need for techniques to remediate the polluted environment. Photocatalyst oxidation using a semiconductor as a catalyst has become an attractive candidate as a green technology for environmental application such as water cleaning [1], air purification [2] and soil remediation [3]. Photocatalyst has been intensively studied over the last two decades as a clean and green process to decompose pollutants in the environment. Recently, nanomaterial has received significant attention for environmental and energy applications because their shape, crystallinity, and the surface can provide unique physical and chemical properties, such as high surface area, well-defined structure, high dispersibility, and high reactivity [4-5].

Intensive research efforts have been focused on the development of novel nanostructured materials. Many nanostructure materials, such as titanium dioxide, zinc oxide, iron oxide metal sulfide have been synthesized with difference techniques, including sol-gel method, hydrothermal method, and mechanochemical method [6-9]. Green nanotechnology has the potential to help us to obtain desirable materials with low cost, low toxicity, high chemical and thermal stability, and high degradation activity for environmental detoxification.

IOP Conf. Series: Materials Science and Engineering **107** (2016) 012019 doi:10.1088/1757-899X/107/1/012019

Interest in application of titanium dioxide in photocatalytic processes has increased significantly in recent years [10, 11, 12, 13]. The use of titanium dioxide as a photocatalyst was first reported in 1972 [14]. Since then, TiO_2 nanoparticles have been used as photocatalysts in many fields as environmental applications of various pollutants [15, 16, 17, 18, 19, 20, 21]. Recently, dye sensitization technique has found its application in solar cells [22]. Dye sensitization can further be applied for water treatment and sacrificial hydrogen generation using photocatalysis [23, 24]. The success of the process depends on the choice of suitable dye, semiconductor material, electron donor, and light sources. The sensitized-anatase TiO_2 has also been shown to be a good photocatalyst under visible light irradiation. Hence, there has been much research into the photoelectrocatalytic properties of the sensitized TiO_2 thin film to investigate its activity in the photodegradation of dye and study the related mechanism under visible radiation [25, 26]. Attachment of dye molecule on photocatalyst surface leads to subsequent electron transfer into the conduction band of semiconductor, therefore strongly bound anchoring groups are preferred. Fixation of dye molecule on semiconductor surface also improves the electron transfer process. Incorporation of noble metals on semiconductor surface enhances their photocatalytic activity by reducing the electron/hole recombination rate. Ruthenium-based dyes in solar cells and dye-sensitized photocatalysis are the best reported so far, however, researchers are gradually switching towards inexpensive and environment-friendly organic dyes and/or natural dyes from vegetable sources [27].

2. Methodology

2.1. Chemicals

Titanium isopropoxide (TTIP, $Ti(O^iPr)_4$) (Aldrich, Singapore), Sodium Chloride (NaCl) and acetic acid (HAc) (Merck, Indonesia) were used in the titania nanocrystals synthesis procedure. The sensitizer dyes for DSTs preparation use N_3 (di(thiocyanate)bis(2,2'-bipyridyl-4,4'-dicarboxylate)ruthenium(II)) (Dye Sol)) and anthocyanin. Anthocyanin were extracted by alcoholic solution from Mangoeteed rind (Garcinia mangostana L.). Whereas the rhodamine B (RB) and Remazol Yellow FG (RY) used in this study is a local material. All glassware was washed with soap, rinsed with nitric acid, and then washed with deionized water prior to use.

2.2. Preparation of nanostructured TiO_2 photocatalyst

Preparation of TiO_2 nanostructure by sol-gel method refers to Wahyuningsih et al. methods [28]. A 10 mL solution of TTIP was hydrolyzed in the 100 mL acetic acid glacial solution and mixed under vigorous stirring in an ice water bath (10–15 ºC) until a clear yellow solution of TiO_2 nanocrystals was formed. The solution was heated at 90 °C for 2 h when it became a gel, which was then placed in an oven at 150 ºC for 24 h to undergo an aging process. Next, the TiO_2 xerogel was ground and pulverized into a fine powder and annealed in a muffle furnace at 400 °C for 2 h at a heating rate of 10 $°C.min^{-1}$. Characterization of the TiO_2 nanoparticle obtained was carried out using X-ray diffraction.

2.3. Preparation of nanostructured based TiO_2 composite electrodes

Preparation of Ti/TiO_2 electrode was done by mixing TiO_2 powder into ethanol then it was stirred and sonicated for 60 minutes. TiO_2 suspension then was sprayed onto titanium plate and heated at a temperature of 200 °C for 15 minutes. While preparation of Ti/TiO_2-PbO electrode was done by mixing TiO_2-PbO composite (12.5 g) in ethanol (50 mL) then it was stirred and sonicated for 60 minutes. The TiO_2-PbO solution then was sprayed onto titanium plate and heated at a temperature of 200 °C for 15 minutes. The Ti/TiO_2-NiO electrode has been synthesized by procedures that have been published previously [29]. 1.145 grams of synthesized xerogel TiO_2 was added in 0.81 grams of $NiNO_3.6H_2O$ that has dissolved in 25 mL of distilled water. The mixed solution was stirred. Then, the uniform mixture was dried at 110 ˚C for 3 h. To improve the crystalline TiO_2-NiO after the green suspension was formed; it was calcinated at 400 ˚C for 4 hours at 10 ˚C/min. The crystalline TiO_2-NiO composite was superimposed onto a Ti plate electrode and was dried at 110°C for 3 h. The coating and drying treatments were repeated three times.

The working electrode of sensitized TiO_2 thin film was prepared in three steps including (1) thin films FTO/TiO_2 preparation, (2) dye sensitizer synthesis of anthocyanin and N_3 complexes and (3) sensitization process of the TiO_2 thin film. Sensitized TiO_2 thin films on FTO glass were prepared from the anatase TiO_2 powder prepared previously [28]. A certain amount of TiO_2 powder (3.5 g) was added to 15 mL of ethanol in the beaker under continuous magnetic stirring. The colloidal TiO_2 was then transferred to a sonicator for 10 min until the gel was fully formed. The gel was applied onto FTO plate (2×□2 cm^2, sheet resistance, 15 Ohm per square; Dye Sol Australia). The glass plates were dried at 200°C in an oven to evaporate the ethanol. The TiO_2 thin film was dipped into dye sensitizer solution for 12 h to obtain the sensitized TiO_2 thin film (DSTs).

2.4. *The use of nanostructure based-TiO_2 as photocatalysts water decolorization*

The photoelectrocatalytic degradation of RY was conducted by Ti/TiO_2-PbO as working electrode, and Ti/TiO_2 plate as the counter electrode. The photoelectrocatalytic process of RY (10 ppm) has been carried out at various pH (pH = 3, 5, 7, and 10). The process was arranged under 7.5 volts positive potential bias for 10 minutes. Meanwhile, the photoelectrocatalytic degradation of RB were conducted by Ti/TiO_2-NiO as working electrode, and Ti/TiO_2 plate as the counter electrode. 10 ppm of RB solution at optimum pH (pH = 2, 7, 8, and 9) was performed by the batch method under 6 volts positive potential bias for 10 min. The front of the working electrode was irradiated with a 300 W visible light lamp (Halogen lamp, Osram) and the progress of the photoelectrocatalytic degradation was recorded at defined intervals time by ultraviolet – visible spectroscopy.

2.5. *The use nanostructure based-TiO_2 as a dye-sensitized photocatalytic oxidation*

In order to investigate the photoelectrocatalytic (PEC) activity of the prepared sensitized TiO_2 thin film (DSTs), a series of degradation experiments of RB in aqueous solutions were performed. The initial concentration of RB was 5 $mg.L^{-1}$, with 0.05 $mol.L^{-1}$ of NaCl used as the supporting electrolyte. A copper wire was used as the counter electrode. The PEC degradation of RB was performed at voltages of +1.0 V and +2.0 V under 300 watts visible light irradiation. The progress of the photoelectrocatalytic degradation was recorded at defined intervals time by Ultra violet-Visible spectroscopy.

3. Result and discussion

3.1. *Preparation and Characterization of nanostructure based TiO_2 photocatalyst*

In a typical sol-gel process, a colloidal suspension, or sol, is formed from the hydrolysis and polymerization reactions of the precursors, which are usually inorganic metal salts or metal organic compounds such as metal alkoxides. Complete polymerization and loss of solvent lead to the transition from a liquid sol into a solid gel phase. Titanium dioxide has been reported synthesized by sol-gel method using titanium tetrachloride, titanium alkoxide, and titanyl sulfate as well as titanyl chloride precursors [28, 29, 30]. If TiO_2 is reacted with a strong base it will hydrolyze, this is splitting the polymer chains of TiO_2 (depolymerization) by the presence of OH- groups. Depolymerization occurs because the strength of OH- ions that break the Ti-O-Ti in polymeric structure. A result of subsequent alkaline hydrolysis is neutralized by the addition of HCl. At pH 7, it will obtain the most optimal condensation process.

The synthesized anatase TiO_2 from titanium tetra isopropoxide precursor was already obtained after an annealing process at 400 °C (annealing rate of 10 °C min). The fully anatase type of TiO_2 was obtained by synthesis using acetic acid solution media (Figure 1). Nanostructure-based TiO_2 photocatalyst was carried out to prepare TiO_2-based electrode.

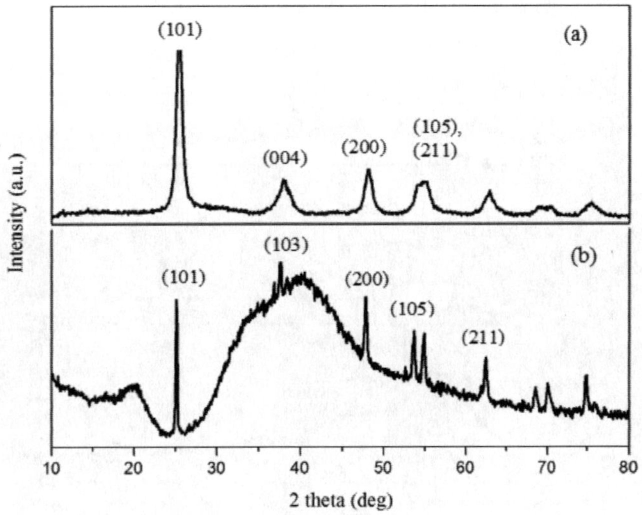

Figure 1. XRD pattern of (a) TiO_2 Powder and (b) Ti/TiO_2 electrode.

The XRD pattern of Ti/TiO_2 electrode shows the diffraction pattern of anatase TiO_2 (Figure 1). The diffractogram peaks represented of TiO_2 content were shown at 2θ = 25.37° (d101 = 3.5165 Å), 2θ = 36.95° (d103 = 2.4307 Å), 2θ = 48.04° (d200 = 1.8922 Å), 2θ = 53.88° (d105 = 1.7000 Å), and 2θ = 55.07° (d211 = 1.6663 Å). The diffractogram characteristics of anatase TiO_2 were according to the standard JCPDS No. 78-2486.

The XRD pattern of Ti/TiO_2-PbO electrode (Figure 2) also shows the diffraction pattern of anatase TiO_2 at 2θ = 25.18° (d101 = 3.5334 Å), 2θ = 37.71° (d004 = 2.3848 Å), 2θ = 40.1° (d112 = 2.2457 Å), 2θ = 53.71° (d105 = 1.7042 Å), 2θ = 62.61° (d110 = 1.4842 Å), according to the standard JCPDS No. 78-2486. Whereas, the diffractogram pattern at 2θ = 28.73° (d111 = 3.1060 Å), 2θ = 48.06° (d022 = 1.8926 Å), and 2θ = 54.73° (d311 = 1.6782 Å) were characteristic diffractograms of PbO according to the standard JCPDS (No. 78-1666).

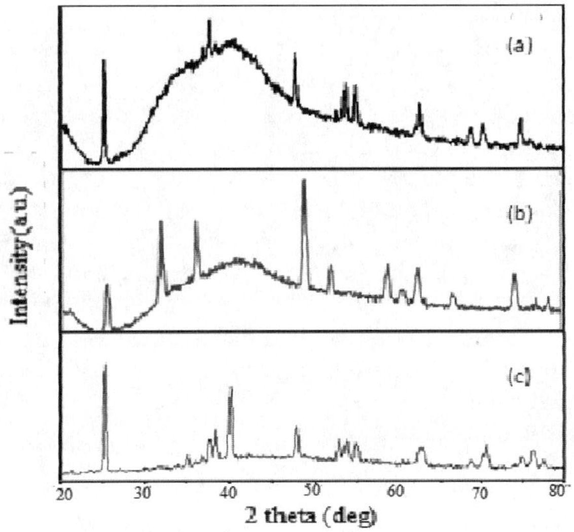

Figure 2. XRD pattern of electrode materials (a) Ti/TiO_2 (b)TiO_2-PbO (c)Ti/TiO_2-PbO.

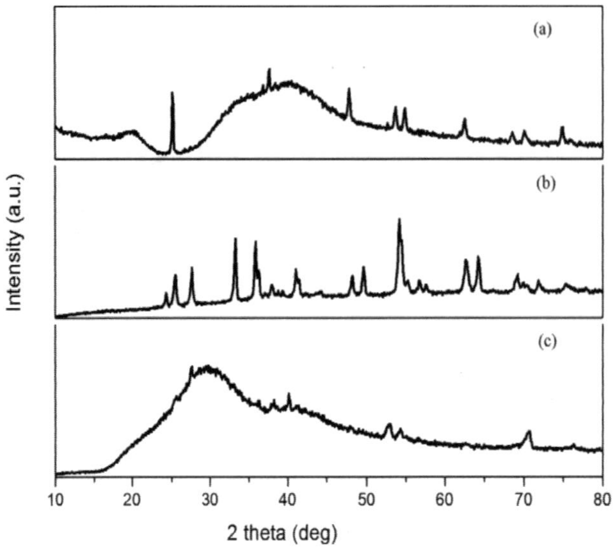

Figure 3. XRD pattern of Ti/TiO_2-NiO electrode materials (a) Ti/TiO_2 (b) TiO_2-NiO (c) TiO_2-PbO.

The composite of TiO_2-NiO separately, has been obtained after an annealing process at 700 °C (annealing rate of 10 °C min) (Figure 3). Futhermore, The XRD pattern of Ti/TiO_2-NiO electrode also shows the diffraction pattern of anatase TiO_2 at 2θ = 25.18° (d101 = 3.5334 Å), 2θ = 37.71° (d004 = 2.3848 Å), 2θ = 40.1° (d112 = 2.2457 Å), 2θ = 53.71° (d105 = 1.7042 Å), 2θ = 62.61° (d110 = 1.4842 Å), according to the standard JCPDS No. 78-2486. The characteristic difractograms of NiO in electrode were appear at 2θ = 24° (d012 = 3.6584 Å), 2θ = 33° (d110 = 2.6911 Å), 2θ = 49° (d024 = 1.8356 Å), and 2θ = 57° (d018 = 1.5969 Å) (Figure 3) All peaks are in good agreement with the standard spectrums JCPDS No. 753-75.

Modification of TiO_2 by sensitization using dye adsorption does not change the crystal structure indicated there are not differences in diffractogram X-ray result.

3.2. Photoelectrodegradation of dyes on TiO_2 based photocatalyst nanostructure

The optimum degradation occurs in the pH 3 RY, about 72.756%. The result in Fig. 4 shows that the pH of RY solution influence degradation yield. This is due to, acidic RY solution already protonated to make a positive charge, thus affinity to the reaction of hydroxyl radicals ($OH^{\bullet}$) and superoxide anion radicals ($O^{\bullet-}$) is higher. The addition of HCl also affects the degradation RY. The Cl can be oxidized to generate the ClO^{3-} species. These species turn out to be powerful oxidizing agents, which together with the hydroxyl radicals can degrade the dye thus at pH 3 the degradation RY becomes higher. When the potential was 3.0 V, the rate of photoelectron transport process was lower and controlled the overall PEC oxidation. Thus, the photocurrent increased within the potential of 3.0– 5.0 V. Once the applied bias exceeded 5.0 V, the interfacial oxidation, which was slower than the photoelectron transport, became the rate-determining step of the overall process. Under this condition, the photocurrent saturated gradually.

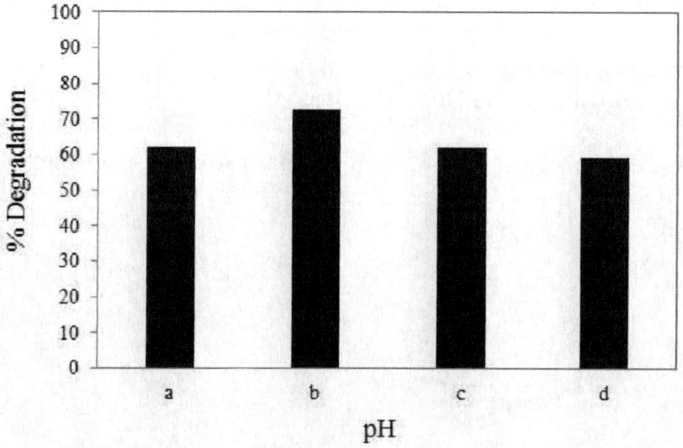

Figure 4. Photoelectrodegradation of RY with applied bias potential 7.5 V for 10 min at various pH (a) RY initial (at pH 6,8); (b) RY at pH 3; (c) RY at pH 7; and (d) RY at pH 10.

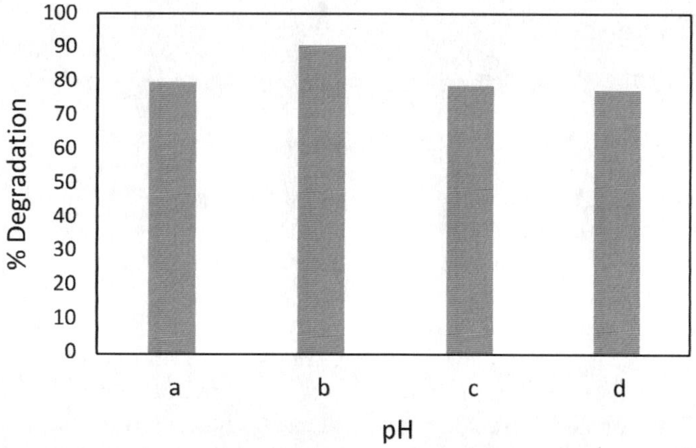

Figure 5. Photoelectrodegradation of RB with applied bias potential 6 V for 10 min at various pH (a) RB neutral (at pH 6,5); (b) RB at pH 2; (c) RB at pH 8; and (d) RB at pH 9.

Figure 5 showed optimum photoelectrodegradation activity under acidic conditions (pH 2) of RB with percentage degradation of RB reached 90.76%. This is due to the cationic structure of RB capable of interacting with $OH^{\bullet}$ and $O_2^{\bullet-}$ formed by photoanode. The zwitter ionic structure of RB less degraded than the cationic form [28].

The anodic potential was an important parameter in the process of photoelectrocatalytic degradation of RB. The effect of applied potential has been determined by a series of potentials bias. To study the key factors affecting the photodegradation of RB, a series of tests were executed under different experimental conditions in which the deduction of visible absorbance was estimated. After 5 min degradation (batch system), about 96.3% of RB was removed. PEC under halogen lamp irradiation using the Ti/TiO_2-NiO photoelectrode with a potential voltage of +10.0V do the electrode surface damage rapidly. Furthermore, the continuous systems were done with applied potential bias of 4, 5, and 6 volts [29].

We also have been investigated water detoxification system using dye sensitized TiO_2 photocatalyst. Sensitisation of titanium dioxide was performed using a dye, i.e., anthocyanin and N_3 complexes. The photoelectrocatalytic degradation of rhodamine B (RB) using ITO/TiO_2/dye as electrode was investigated at +1.0 V to +2.0 V. The decreasing in the absorbance of RB with various potentials applied indicated that the RB concentration decreased with increasing anodic potential bias. The degradation cell exhibited better performance when the enough positive anodic bias was applied. The pH values of RB in solution systems also influence the photoelectrodegradation process because of the different RB species present. NaCl concentration also affects the activity of RB photoelectrocatalytic degradation due to changes in the ionic strength character of the electrolyte [28]. Using dye-sensitized TiO_2 system; we have applied that system to build green energy production as a novelty environmental remediation system.

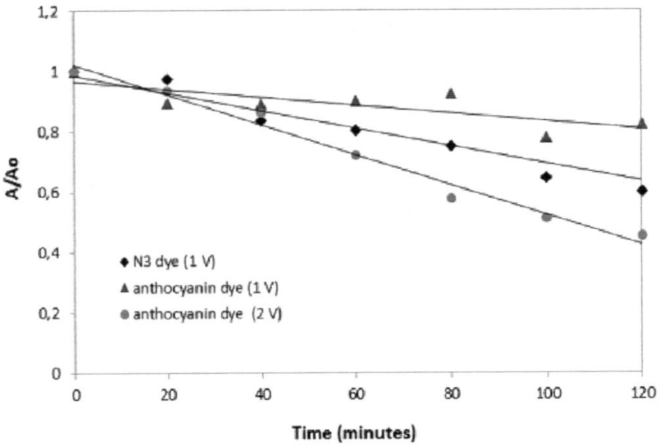

Figure 6. The photoelectrocatalytic degradation of RB using dye sensitized TiO_2 photocatalyst.

4. Conclusion

Photocatalysis based TiO_2 nanomaterial can be used for water purification. The decolorization of RB and RY by photoelectrocatalytic degradation system highly was affected by the structure of each RB and RY, hydroxyl radicals ($OH^{\bullet}$) and superoxide anion radicals ($O_2^{\bullet-}$) formed by photoanode. Decolorization of RB by photoelectrocatalytic degradation has optimum activity at RB initial pH 2, this is due to cationic structure of RB that formed by acidic condition and able to interact with $OH^{\bullet}$ and $O_2^{\bullet-}$. Whereas the decolorization of RY by photoelectrocatalytic degradation has optimum activity at RY initial pH 3, this is due to acid solution RY already protonated form a positive charge, thus affinity to the reaction of $OH^{\bullet}$ and $O_2^{\bullet-}$ is higher. The RB concentration decreased rapidly in the dye-sensitized TiO_2 photoanode system by anodic potential bias applied.

Acknowledgement.
The researchers wish to express their gratitude to Sebelas Maret University Kemenristekdikti Republic of Indonesia, for supporting this research and to the Integrated Mathematics and Natural Science Laboratory of Sebelas Maret University for supporting and providing the facilities for this research.

References

[1] Vítor R, Vilar J P, Gomes A I E, Ramos V M, Maldonado M and Boaventura R A R 2009 *Photochem. Photobiol. Sci.* **8** 691–698
[2] Pichat P 2010 *Appl. Catal.* B. **99** 428–434
[3] Higarashi M M and Jardim W F 2002 *Catal. Today.* ***76*** 201–207
[4] Dimitrov D 2006 *Colloids Surf. A.* ***282*** 8–10
[5] Dastjeridi R and Montazer M A 2010 *Colloids Surf. B.* **79** 5–18
[6] Gonzales-Oliver C J R, James P F and Rawson H 1982 *Journal of Non-Crystalline Solids* **48**

129-152
[7] Kuznetzov V A 1968 *Journal of Crystal Growth* **4** 405-410
[8] Lindh A 1972 *Lithos* **5** 325-343
[9] Razumovskii S D, Podmaster'ev V V and Zelenetskii A N 2011 *Biocatalysis* **3** 23-27
[10] Wilhelm P and Stephan D 2007 *Journal of Photochemistry and Photobiology A: Chemistry*. **185** 19–25
[11] Qu P, Zhao J P, Shen T and Hidaka H 1998 *Journal of Molecular Catalysis A: Chemical* **129** 257–268
[12] Liu G M and Zhao J P 2000 *New Journal of Chemistry* **24** 411–417
[13] Hashimoto K, Irie H and Fujishima A 2005 *Japanese Journal of Applied Physics* **44** 8269–8285
[14] Fujishima A and Honda K 1972 *Nature* **238** 37–39
[15] Chatterjee D and Mahata A 2001 *Applied Catalysis B: Environmental* **33** 119–125
[16] Chatterjee D and Mahata 2002 *Journal of Photochemistry and Photobiology A: Chemistry* **153** 199–204
[17] Chatterjee D and Dasgupta S 2005 *Journal of Photochemistry and Photobiology C: Photochemistry Reviews* **6** 186–205
[18] Chen C C, Li X Z, Mao W H, Zhao J C, Hidaka H and Serpone N 2002 *The Journal of Physical Chemistry B*. **106** 318–324
[19] Zhong H, Yang S G, Ju Y and Sun C 2009 Journal *of Environmental Sciences*. **21** 268–272
[20] Chen X B and Mao S S, 2007 *Chemical Reviews*. **107** 2891–2959
[21] Chen C C, Maa W H and Zhao J C 2010 *Chemical Society Reviews* **39** 4206– 4219
[22] O'Regan B and Grätzel M, 1991 *Nature*. **353** 737–740
[23] Tariq M, Chuncheng C, Lili L, Dan Z, Wanghong M, Jun L and Jincai Z 2009 *Journal of Environmental Sciences* **21** 263–267
[24] Cho Y M, Choi W Y, Lee C H, T, Hyeon T Y, and Lee H I 2001 *Environmental Science & Technology* **35** 966–970
[25] Kim W, Tachikawa T, Majima T, and Choi W 2009 *J. Phys. Chem. C.* **113** 10603-10609
[26] Chowdhury P, Moreira J, Gomaa H, and Ray AK 2012 *Industrial & Engineering Chemistry Research*. **51** 4523-4532
[27] Lee C H, LuengShie J, Tsai C Y, Yang Y T, and Chang C Y 2013 *Journal of Clean Energy Technologies* **1** 2
[28] Wahyuningsih S, Purnawan C, Kartikasari PA, and Praistia N 2014 *Chemical paper* DOI: 10.2478/s11696-013-0476-8
[29] Wahyuningsih S, Purnawan C, Saraswati T E, Pramono E, Ramelan A H, Pramono S, and Wisnugroho A 2014 *Journal of Environmental Protection* 1630-1640
[30] Wahyuningsih S, Ramelan A H, Juliana G, Khoirunisa A G, Rahardjo S B, Pramono E, Suharyana S, Suryana R, and Supriyanto A 2014 *Proc. of SPIE*. **9137** 913719-1

Modification of chitosan by swelling and crosslinking using epichlorohydrin as heavy metal Cr (VI) adsorbent in batik industry wastes

B Hastuti[1*], A Masykur[2] and S Hadi[3]
[1]Chemistry Education Department, Faculty of Teacher Training and Education, UNS
[2]Chemistry Department, Faculty of Mathematics and Natural Sciences, UNS
3Pharmaceutical Department, Faculty of Mathematics and Natural Sciences, UNS

*E-mail: Bhastuti.uns@gmail.com

Abstract. Study on chitosan modification by swelling and crosslinking and its application as a selective adsorbent for heavy metals Cr (VI) in batik industry wastes was done. Swelling is intended to improve chitosan porosity, whereas crosslinking is to increase the resistance of chitosan against acid. Natural samples are generally acidic, thus limiting chitosan application as an adsorbent. Modification of chitosan by combining swelling and crosslinking is expected to increase its adsorption capacity in binding heavy metal ions in water. The modified chitosan was later contacted with Cr (VI) to test its adsorption capacity with a variation of pH and contact time. Finally, application of modified chitosan was done in batik industry waste containing Cr (IV). Based on the results, chitosan-ECH 25% ($^{v}/_{v}$) was the optimum concentration of crosslinker to adsorb Cr (VI) ions. Modified chitosan has a solubility resistance to acids, even though a strong acid. Modification of chitosan also improved its adsorption capacity to Cr (VI) from 74% (pure chitosan) to 89% with contact time 30 min at pH 3. On the application to the batik wastes, the modified chitosan were able to adsorb Cr (IV) up to the level of 5 ppm. Thus, the modified chitosan has a potential to be applied to as an adsorbent of Cr (VI) in batik industry wastes.

1. Introduction

The process of industrialization can not be separated from the negative effects, the presence of residual materials industry either solid or liquid form can affect the surrounding environment. When the remains are released into free waters, there will be changes in the value of the waters both the quality and quantity that can be considered polluted water. The dynamic relationship of the balance of the composition of the nutrient components, organic matter, and biomass are very important for the stability of aquatic ecosystems, but the stability relations will be soon interrupted when occurs input materials which are toxic. The entry of pollutants will be able to reduce the potential of biological resources.

Pollution by industrial materials which contains hazardous materials, such as pesticides, heavy metals, toxic metals (such as Hg, Cr, Pb, Zn, Cu, Ni, Cd, As, Co, Sn, etc.), a concerning issue related to threatening human health and ecosystems [1]. Heavy metals, harmful because of the heavy metals tend to accumulate in the body in sensitive organs and tissues. In humans, heavy metal poisoning will cause severe renal dysfunction, disrupt the reproductive system, damaging internal organs such as liver, brain damage, and central nervous system [2].

Type of heavy metal pollution comes from industrial waste that goes into the environment. So that, for treating wastewater that has been generated, requires an environmentally friendly process. Thus, it is necessary to do research to reduce the heavy metal content in the waste industry. As one of the alternative method that is easy and inexpensive in the reduction of heavy metals by using chitosan as an adsorbent [3-4].

Chitosan was selected as the basis of an adsorbent material for its characteristic that can be attributed as a natural polymer, nontoxicity, biodegradability, biocompatibility, bioactivity and attractive physical and mechanical performance[5,6]. In addition, the cost of production is low, does not generate new waste, and effective at low ionic concentration. Chitosan also has a selectivity and high adsorption capacity. Chitosan is an excellent bio-sorbent of the heavy metal ion. It contains a large number of $-NH_2$ and $-OH$ groups and its able to remove these stuff in near-neural solutions. The ability of chitosan can be attributed to 1. its high hydrophilicity generated by a large number of hydroxyl groups of glucose unit; 2. the presence of a large number of functional groups; 3. the high chemical reactivity of these groups and 4. the flexible structure of polymer chain [7]. The reactive amino groups selectively bind to virtually al group III transition metal ion but not a group I and II ion [8]. Chitosan shows a cationic behavior in acidic media. The protonation of amine groups leads to adsorption of metal anions by ion exchange [9,10].

Chitosan can absorb heavy metal contaminants in the water due to the presence of amine and hydroxyl groups which are highly reactive and alkaline. Chitosan will exchange protons owned by metal contaminants with electrons owned by nitrogen (N). However, chitosan is easily dissolved in acetic acid and also partially dissolve in dilute acids, such as HNO_3, HCl, $HClO_4$, and others, so that direct use of chitosan as an adsorbent will be less effective. Chitosan adsorption process will be interrupted by the acidic conditions in the solution. In acidic conditions will dissolve chitosan, so that will limit the properties of chitosan as an adsorbent. This is due to acidic will cause a reduction in the amine group which is the active group that is able to bind to metal ions. For that it is necessary to crosslinking chitosan to improve resistance to acid, increasing the stability of chitosan to form intermediate compounds.

Crosslinking chitosan use of aldehydes as a crosslinking agent that makes chitosan not dissolve in acidic media [11-13]. This crosslinking, cross-linking between the polymer chitosan at $-NH_2$ groups. This can reduce the absorption of chitosan and metal ions. Crosslinking chitosan using epoxy compound classes such as ethylene glycol diglycidyl ether (EDGE) [14-15], Epichlorohydrin (ECH) [16-17].

To increase the speed of chitosan adsorption to adsorb heavy metal ions, as well as to optimize further the adsorption capacity of chitosan, before crosslinking, swelling the chitosan first to improve the chitosan axis value. Thus, modification of chitosan by making it into a composite by combining the two methods of activation of the swelling method chitosan followed by crosslinking chitosan is expected to increase the adsorption capacity of chitosan to bind heavy metal ions in water.

2. Methodology

2.1. Equipment and material

2.1.1. Equipment used. Equipment used includes laboratory glassware, a set of reflux, Buchi evaporator, electronic scales, heater and a magnetic stirrer, Buchner filter, oven, thermometer, magnetic stirrer bar, pH meters, FTIR spectrometer (Shimadzu 8201 PC), AAS (Shimadzu) and SEM (JSM-35C).

2.1.2. Materials used. Materials used in this study include chitosan (local), NaOH (E-Merck), HCl (E-Merk), $NaHCO_3$ (E-Merck), CH_3COOH (E-Merck), HNO_3 (E-Merck), H_2SO_4 (E-Merck), Epichlorohydrin (ECH) (Aldrich), Dioxane (E-Merck), Benzaldehyde (E-Merck), Ethanol (95%), Aqua bidest (local), P*aper filter (Whatman 21),* $K_2Cr_2O_7$ *(E-Merck), buffer pH and distilled water.*

2.2. Research Methods

2.2.1. Preparation of Chitosan Epichlorohydrin adsorbent bead. 2 grams of chitosan was dissolved in 100 ml of 1% acetic acid (v/v) and stirred continuously for 2 hours. Then $NaHCO_3$ was added to the chitosan solution at room temperature with a composition variation of $NaHCO_3$: chitosan = 0,2:2,0 (w/w) and stirred for about 1 hour. After the solution was sprayed with a 5% NaOH solution of 300 mL (w/v) using chitosan gel injection formed, then washed with distilled water until neutral. Then, the formed chitosan beads added ECH crosslinker agent respectively 8%, 10%, 20% and 25% through the following procedure:

2.2.1.1. Protection with benzaldehyde. Two grams of chitosan beads was added 10 ml benzaldehyde and then stirred for 2 hours at room temperature. It was then filtered and washed using ethanol: water= 1:1, and rinsed with distilled water.

2.2.1.2. Crosslink process of chitosan with ECH. Two grams of chitosan suspended in 20 ml of dioxane in the round bottom flask and stirred. 1M NaOH solution of 4 ml was added slowly to a stirred solution and then was added with varying ECH 8%, 10%, 20%, and 25%. The solution was refluxed for 6 hours. The precipitate that has been obtained was then filtered using Whatman 42 filter paper with a Buchner filter and washed with ethanol and excess water. It was dried for 3 hours at a temperature of 1100°C (to dry). Chitosan that has been crosslinked with ECH crosslinker that has been protected with benzaldehyde then analyzed by infrared spectrometry.

2.2.1.3. Deprotection Benzaldehyde on chitosan that has crosslinked with ECH. Chitosan that has crosslinked with treatment as mentioned above, then as much as 2 grams suspended in a solution of 10 ml 0,5M HCl and allowed to stand for 12 hours. The precipitate obtained was filtered with Whatman 42 filter paper using a Buchner filter and washed with ethanol and water. The powder obtained was dried in an oven at a temperature 1100 C for 3 hours. The results were analyzed by infrared spectrometry.

2.2.2. Adsorption of chitosan modified

2.2.2.1. Effect of pH on the adsorption of Cr (VI) using chitosan adsorbent. Adsorption was performed using a batch system at room temperature. A solution of Cr (VI) 5ppm, set the pH (2, 3, 5, and pH 7) and then added 1 ml of pH buffer. The pH was measured using pH meter TOA by adding HCl or NaOH. 20 ml of Cr (VI) 5 ppm that has set its pH is added to 50 ml beaker glass and added 10 mg of adsorbent (chitosan and chitosan-ECH beads) and then stirred with a magnetic stirrer for 30 minutes. The solution was filtered and then measured the concentration using AAS. A standard solution prepared from mother liquor of Cr (VI) of 100 ppm and diluted with distilled water.

2.2.2.2. Effect of contact time adsorption of Cr (VI) using chitosan adsorbent. Adsorption was performed using a batch system at room temperature. A solution of Cr (VI) 5ppm, pH 3 was measured using pH meter TOA by adding HCl or NaOH, then added 1 ml of pH buffer. 20 ml of Cr (VI) 5 ppm that has set its pH was added to 50 ml beaker glass and added 10 mg of adsorbent (chitosan and chitosan-ECH beads) and then stirred with a magnetic stirrer, contact time was set for 15,30,60, and 90 minutes. The solution was filtered and then measured the concentration using AAS. A standard solution prepared from the mother liquor of Cr (VI) of 100 ppm and diluted with distilled water.

2.2.2.3. Effect of Epichlorohydrin crosslinker concentration on adsorption of Cr (VI). Adsorption was performed using a batch system at room temperature. A solution of Cr (VI) 5 ppm, pH 3 was measured using pH meter TOA by adding HCl or NaOH, then added 1 ml of pH buffer. 20 ml of Cr (VI) 5 ppm that has set its pH was added to 50 ml beaker glass and added 10 mg of adsorbent (chitosan and chitosan-ECH bead). Chitosan-ECH bead was made by varying the concentration of epichlorohydrin 8, 10, 20 and 25%. Furthermore, a solution of Cr(VI) was contacted with each chitosan modified with stirring using magnetic stirrer for 30 minutes at pH 3 which measured using pH meter TOA by adding HCl or NaOH, then add 1 ml of pH buffer. 20 ml of Cr(VI) 5 ppm which has set its pH was inserted into 50 ml beaker glass and added 10 mg of adsorbent (chitosan and chitosan-ECH bead).

3. Results and Discussion

3.1. Preparation of Chitosan-ECH bead adsorbent

The process of making a modified chitosan begins with swelling process by dissolving chitosan in acetic acid until form chitosan gel. After that, the gel formed, sprayed into the NaOH solution until form chitosan bead. The swelling process intended to increase the surface area and to increase the chitosan axis value. After swelling process, then performed chitosan crosslinking process with ECH. The crosslinking process performed by reacting ECH with chitosan that has protected its amine group with benzaldehyde. The purpose of this protection is to prevent the reaction between ECH with the amine group so that the adsorption capacity of chitosan that has crosslinked not decreased since the amine group is dominant binding to metal ions. IR spectra of Chitosan crosslinked by ECH shown in figure 1.

Figure 1 shows that the protection process of the amine group with benzaldehyde successfully performed, indicated by the appearance of imine peak at 1558 cm^{-1}, and reinforced their new peak at 756 cm^{-1}, and 684 cm^{-1}, which is the bending vibration of benzaldehyde. After deprotection process, imine peak decreased significantly, and the peak of 756 cm^{-1} and 684 cm^{-1} disappeared. It can be concluded that almost all benzaldehyde has been separated from the chitosan. The occurrence of chitosan crosslink with ECH is not so apparent when viewed from the infrared spectra, but the physical properties, crosslinked chitosan is insoluble in concentrated acid, indicates there has been a crosslink between chitosan with ECH.

Crosslink reaction occurred between ECH with the hydroxyl group, and after crosslinking, benzaldehyde released from the amine group by soaking in a solution of 0.5 M HCl to decide imine bond. The scheme of crosslinking reaction with Epichlorohydrin is described in figure 2.

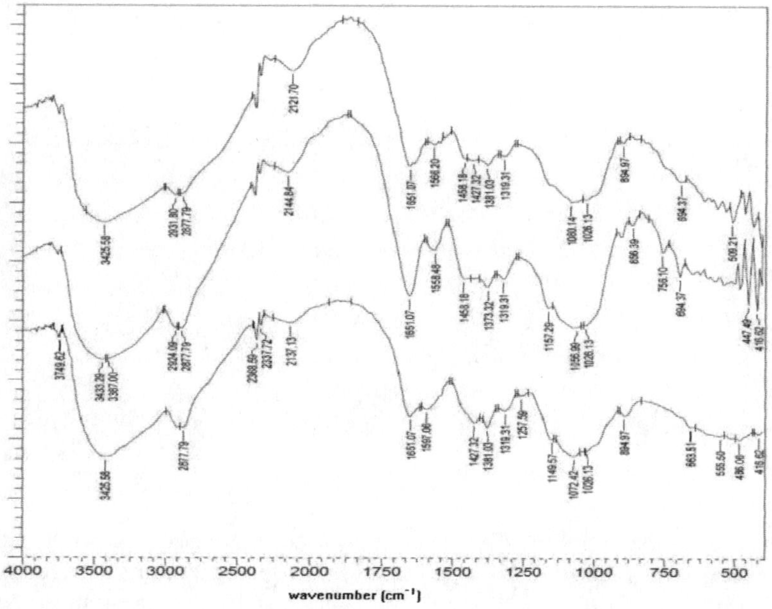

Figure 1. IR Spectra of chitosan-chitosan ECH, benzaldehyde, and chitosan

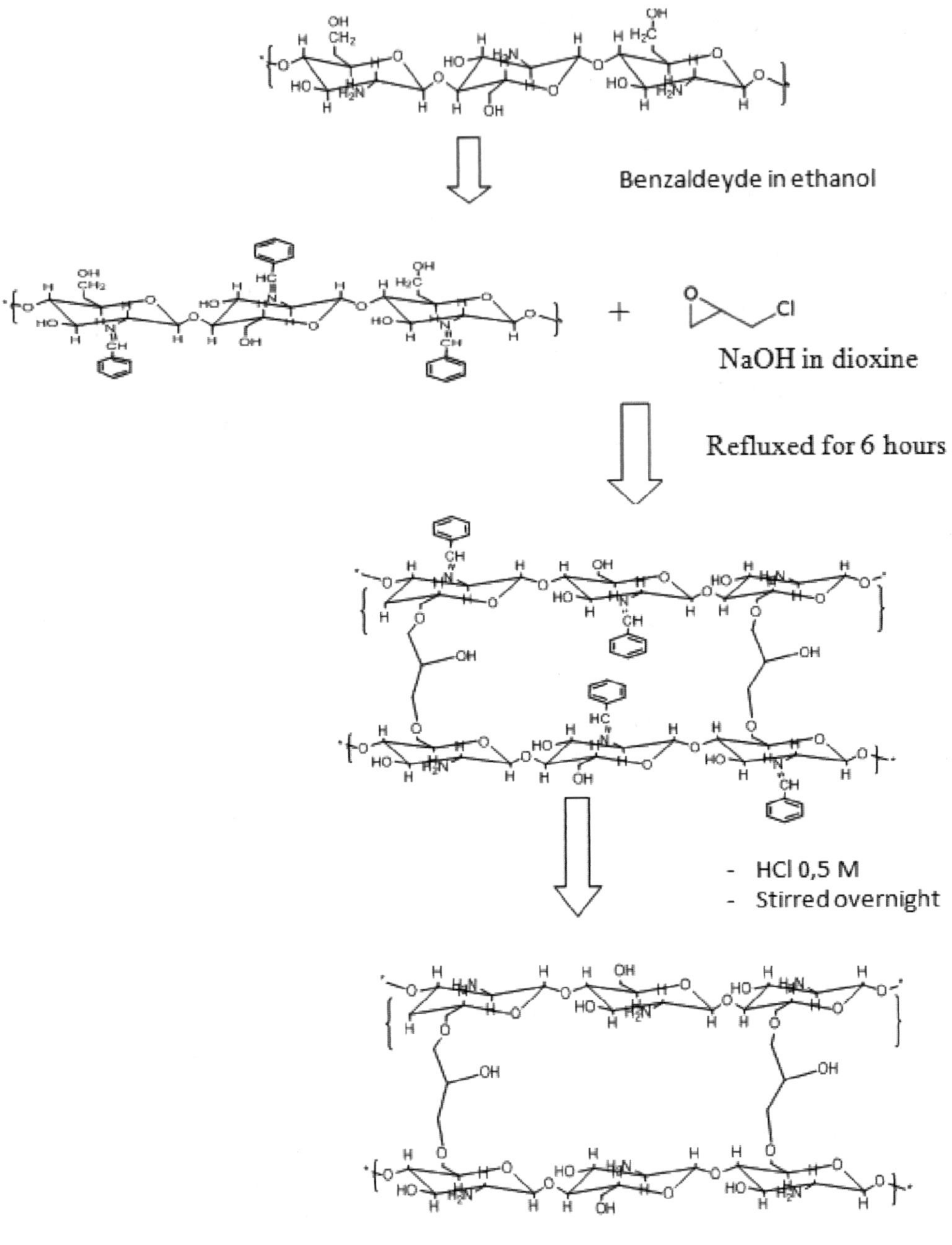

Figure 2. Scheme of crosslink reaction with Epichlorohydrin

3.2. Adsorption process

3.2.1. Determination of optimum pH. Determination of optimum pH was conducted to determine the pH of the interaction in which the adsorbent absorbs to the maximum. Results of research on the effect of pH on an amount of Cr(VI) metal ions adsorbed by the chitosan adsorbent or modified chitosan can be seen in figure 3.

Absorption of the Cr (VI) metal ions by chitosan adsorbent is more stable at various pH compared to modified chitosan, unfortunately, it have lower absorption. It is due to the amine group on the positively charged of chitosan binds the negatively charged of Cr (VI) ions. The greatest absorptive capacity exists at pH 3. At pH 3, the absorption of both chitosan adsorbent has stabilized, thus, pH 3 as a reference for future research. It is in a good agreement with that proposed by Sakkayawong et.al.

(2005), that under acidic condition, the atoms of hydrogen (H^+) in solution can do protonation to an amine group ($-NH_2$) of chitosan, as defined in equation (1).

$$R'{-}NH_2 + H^+ \qquad R'{-}NH_3^+ \qquad (1)$$

The protonated amine group that will be used to adsorb Cr(VI) through the mechanism of exchange of bonds. Not only that, but ion exchange also occurs in the H atoms in the OH groups of chitosan, with Cr metal from solution. With this ion exchange, chitosan complexes formed with Cr(VI) metal ions.

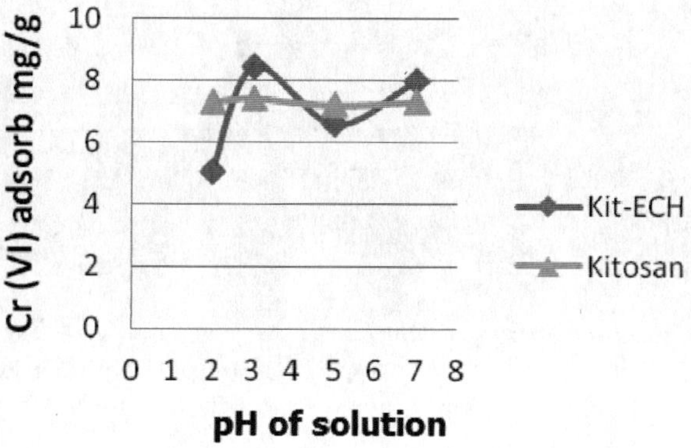

Figure 3. Variations in pH of chitosan adsorption process to Cr(VI)

3.2.2. Determination of contact time. The contact time is the time required adsorbent to adsorb substance of Cr(VI) metal. To determine the optimum contact time, 5 mg of adsorbent in 20 ml of 5 ppm Cr(VI) solution at pH 3 shaken by the variation of contact time of 15, 30, 60, and 90 minutes. Yield curve adsorption of Cr(VI) by chitosan and chitosan modified the contact time variation can be seen in figure 4.

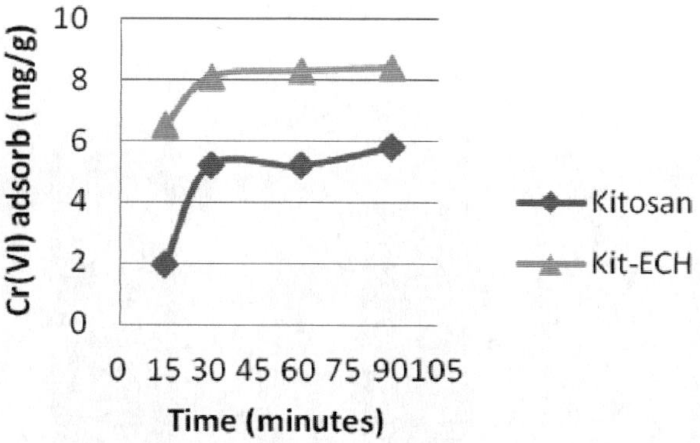

Figure 4. Contact time variation of chitosan adsorbent to Cr(VI)

In figure 4 it can be seen that the adsorption of Cr(VI) show a pattern similar to the adsorption of chitosan and chitosan – ECH adsorbent. On the 15 minutes early the adsorption increase sharply and the addition of the next time experiencing the stability of adsorption. The optimum adsorption of chitosan and chitosan modified adsorb Cr(VI) on the same time at 30 minutes. At 30 minutes, both adsorbents of chitosan were able to absorb optimally, thus, the contact time of 30 minutes serve as a reference for determining the contact time on the adsorption of chitosan and chitosan modified to Cr(VI) in industrial batik waste.

3.2.3. Variation of crosslinker concentration. By varying the crosslinker concentration, epichlorohydrin able to increase chitosan adsorption on the Cr(VI) in solution. figure 5 shows the optimum of crosslinker concentration to adsorb Cr(VI) metal.

Based on figure 5, the ECH concentration of 25% is an optimal concentration to adsorb Cr(VI) metal ion, so it used as a concentration references of crosslinked the chitosan to produce a selective adsorbent that can adsorb heavy metals, especially Cr(VI) with a high capacity.

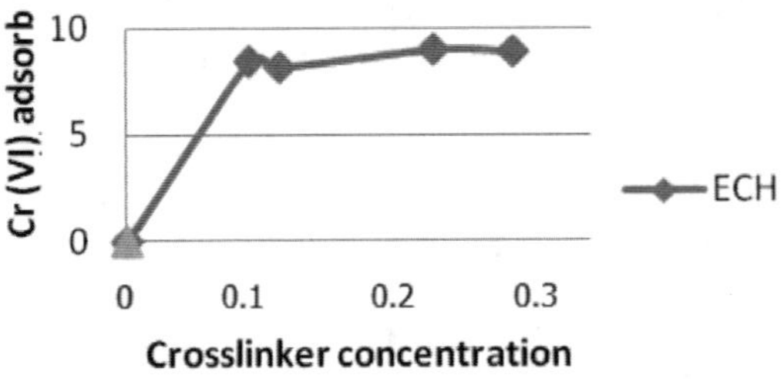

Figure 5. The variation of Epichlorohydrin (ECH) crosslinker concentration to adsorb Cr(VI) metal

3.2.4. Comparison of the adsorption of the Cr(VI) in optimum condition. Adsorption was done at the optimum condition with ECH concentration of 25%, pH 3 and a contact time of 30 minutes. Comparison of adsorption by chitosan and chitosan modified to adsorb 5 ppm Cr(VI) at the optimum condition can be seen in table 1 and figure 6.

Table 1. Comparison of adsorption capacity of chitosan adsorbent

Type of Adsorbent	Concentration of Cr(VI) after adsorption	Adsorption capacity	
	ppm	mg/g	%
Chitosan	1.32	7.4	74
chi-ECH	0.55	8.9	89

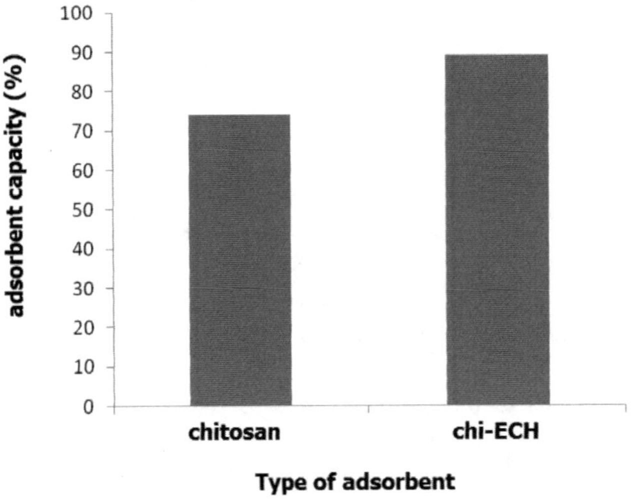

Figure 6. adsorption capacity of chitosan adsorbent

Table 1 and figure 6 show that the modified chitosan and chitosan has a high adsorption capacity due to chitosan has the amine group, there is an element of N that is highly reactive, and alkaline and the hydroxyl groups can bind to the metal through ion exchange mechanism. However, the adsorption capacity of modified chitosan is higher when compared to chitosan, it due to the modified chitosan has through the swelling process and the crosslinking process. The both process seem to further increase the adsorption ability and the resistance to acidic conditions. The presence of larger cavities on the modified chitosan compared to the chitosan can be seen in figure 7. The scanning electron microscope (SEM) micrograph of the modified chitosan (figure 7b) shows larger cavity on their surface morphology. It proved a great increase in its surface porosity compared to the corresponding non-modified chitosan(figure 7a). It proves that the combination of swelling and crosslink process could increase the adsorption capacity on the Cr(VI) metal. The adsorption capacity of Chit-ECH in adsorbing Cr(VI) ions is 89%; it indicates that the modified chitosan can be used as a good adsorbent for Cr(VI) as well as for Cr(VI) metal in the batik waste.

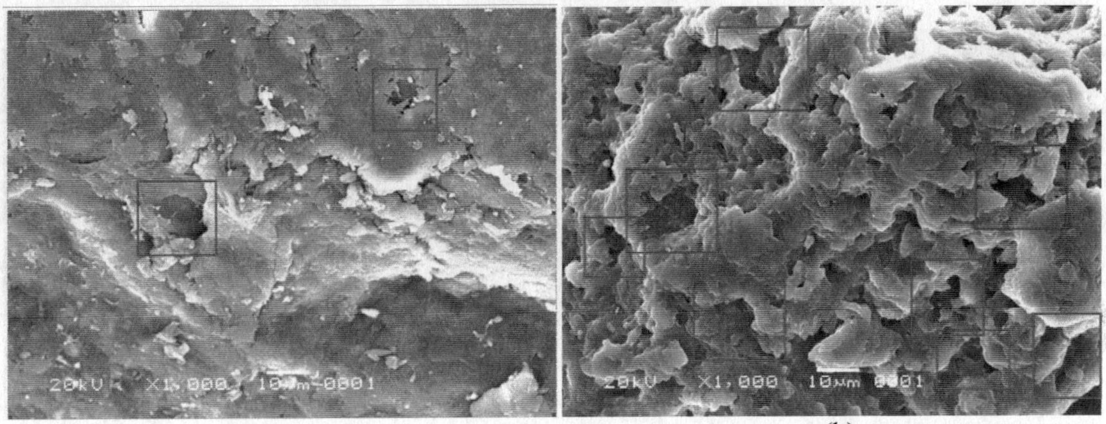

(a) (b)

Figure 7. (a) The surface of chitosan and (b) The surface of Chitosan-ECH beads.

3.3. Application of adsorbent on Liquid Batik Waste
Chitosan modified contacted with batik waste is chitosan that is swelled first then crosslinked using epichlorohydrin 25%, with a contact time of 30 minutes at pH 3, subsequently measured the concentration using AAS. AAS test result is shown in table 2.

Table 2. The adsorption of Cr(VI) metal in batik waste by chitosan and chitosan-ECH measured by AAS

Initial concentration of Cr(VI) in batik sample (ppm)	Final concentration after adsorption (ppm)	
	Chitosan	Chitosan-ECH
0.037	0	0

The data in table 2 shows that both of Chitosan and Chitosan-ECH can adsorb Cr(VI) metal in batik waste up to the maximum. Thus, it can be concluded that the Chitosan and Chitosan-ECH are very effective to be used as a super adsorbent to adsorb Cr(VI) heavy metal in the batik waste.

4. Conclusions
The modification of chitosan through swelling process that was combined with crosslinking process produced the modified-chitosan. The modified chitosan then has been used to absorp the Cr (VI) metal. Modification of chitosan with Epichlorohydrin (ECH) 25% crosslinker agent produce an adsorbent that able to adsorb Cr(VI) metal with the adsorption capacity of 89%. Meanwhile, the non-modified chitosan only provide the adsorption capacity of 74%. The modified chitosan also can absorb the Cr(VI) metal in the batik waste. It is proved by the elimination of Cr(VI) content from 0.037 ppm into 0 ppm.

References
[1] Wang J L and Chen C 2006 a review *Biotechnol Adv.* **24** 427–51
[2] Manahan S E 1994 *Environmental Chemistry* (Boca Raton, FL, USA: Lewis)
[3] Varma A J, Deshpande S V and Kennedy J F 2004 A review *Carbohydrate Polymers.* **55** 77-93
[4] Crini G 2006 Non-conventional low-cost adsorbents for dye removal A review *Bioresource*

Tech **97** 1061–1085
[5] Liu B J, Wang D F, Li Y Xu, H Y and Zhang L 2011 *Desalination* **272** 286-292
[6] Liu B J, Wang D F, Yu G and Meng X 2013 *J. Ocean Univ Cina.* **12** 500-508
[7] Crini G 2005 Progress in *Polymer Science* **30** 38-70
[8] Muzzarelli R A A 1973 Natural chelating polymers Pergamon Press Oxford UK p1
[9] Guibal E, Saucedo I, Jansson M, Charrier Delanghe B and Le Cloirec P 1994 *J. Water Sci. Technol.* **30** 9 183-190
[10] Kunkoro E P, Roussy J and Guibal E 2005 *Separation Sci and Tech* **40** 659-684
[11] Chang Y, Huang C, Hsu W and Chang F 2006 *J. Appl. Polym. Sci.* **104** 2866-2905
[12] Chen X, Liu J, Feng Z and Shao 2005 *J. Apply. Polym. Sci* **96** 1267-1274
[13] Wang A, Wang L and Sun S 2005 *J. Hazard Mater.* **136** 930-937
[14] Katarina R K, Takayanagi T, Oshita K, Oshima M and Motomizu S 2008 *J. Anal. Sci.* **24** 1537-1544
[15] Oshita K, Takayanagi T, Oshima M and Motomizu S 2008 *J. Anal. Sci.* **24** 665-668
[16] Masykur A, Santosa S J, Siswanta D and Jumina 2014 *Indo. J. Chem.* **14** (1) 63-70
[17] Chen C Y, Yang C Y and Chen A H 2011 *J. of Environmental Management* **92** 796-802

Solubility dynamic of methyl yellow and carbon black in microemulsions and lamellar liquid crystal of water, non ionic surfactants and cyclohexane system

A Amran*, R Harfianto, W Y Dewi, D Beri and A Putra
Department of Chemistry, Faculty of Mathematics and Sciences, Universitas Negeri Padang, Jl. Prof. Dr. Hamka, Air Tawar Barat, Padang, Sumatera Barat, Indonesia 25131

*E-mail: amrana.unp@gmail.com

Abstract. Solubility dynamics of methyl yellow and carbon black in microemulsions and liquid crystals of water, non-ionic surfactants and cyclohexane system, have been investigated. Actually, solubility dynamics of these dyes both in microemulsion (w/o microemulsions) and the lamellar liquid crystal (LLC) were strongly related to the chemical composition, nature and characteristics of microemulsions and the lamellar liquid crystals.

1. Introduction

Interaction among organic dyes and surfactants became an interested and technological research areas [1-4], since organic dyes can be able to probe of association colloid [5-6]. Solubility area of organic dyes has been investigated and focus on micellar solution, whereas solubility area has been limited studied [7, 8]. Additionally, it has been evaluated the hydrophilic-liphophilic equilibrium. The goal of this research at getting the interface interaction that can be able to compare more specific interface study [9, 10]. In fact, this fundamental knowledge will be useful and will be able to understand the thermodynamic and general solubility rules [11]. This information is also important to explain and establish the relationship between colloidal structures in microemulsion system and the organic dyes solubilities [12, 13]

Solubility of organic pigments in microemulsions and liquid crystals systems have importantly scientific and technological values which can be able to apply spontaneously in industries of paints [14], inks and pharmacies [15], cosmetics and some other manufactures [16]. Information about physical and chemical properties can be got from mapping of diagram phases in some needed physical and chemical systems [17, 18]. By doing so, it can be able to make improvisation or to more increased application values of non-ionic surfactants.

In fact, dynamics solubility in water, non-ionic surfactants, and cyclohexane system became more interesting area, because the system, at least, performing four surfactant association structures i.e. normal micel, and inversed micel (oil in water, o/w and water in oil, w/o microemulsions, in general term), lamellar liquid crystal, and hexagonal liquid crystal. The four surfactant association structures have different physical and chemical properties, as well [17].

This study aimed at knowing solubility dynamics of methyl yellow and carbon black in microemulsions and lamellar liquid crystal regions. Methyl yellow was able to get: "a permanently red colour in the water of pH = 1, and a permanently yellow colour in the water pH =

5. Whereas, fine divided carbon black was able to get a black colour. Ultimately, permanently red, yellow, and black colours of microemulsions and LLC will be used to fulfil ballpoint ink and cartridge printer.

2. Experimental section

2.1. Materials and Method
The following chemicals were used without further furification. Non-ionic surfactants (Tween-20, Tween 40, Tween 60, and Tween 80) and cyclohexane, were purchased from Merck, Germany. Methyl yellow and fine divided carbon black were purchased from Wako Pure Chemical Industries Ltd. Double distilled water was purchased from Rafi Medika. Nitric acid (70%, ACS certified, Baxter Hearthcare Corp, Scientific Product Division) was used as acid medium to adjust the water to a pH = 1, and pH = 5.

2.2. Phase diagram preparation
The phase regions of microemulsions and liquid crystals were obtained by titration of cyclohexane/non-ionic surfactant with the water (pH = 1, and pH = 5). The phase region limits were determined by visual observation between crossed polarizers with parafilms and was confirmed by the results from Hund-Wetzlar® optical polarizing microscope.

2.2.1. Solubility of Methyl Yellow and Carbon Black in Microemulsions and Lamellar Liquid Crystals. Solubility of methyl yellow in w/o microemulsions and lamellar liquid crystals was determined for system of water (pH = 1, and pH = 5), Tweens (Tween-20, Tween-40, Tween-60, and Tween-80), and cyclohexane system. Solubility of methyl yellow and fine divided carbon black in hexagonal liquid crystal was not determined. The procedure used: a small quantity of methyl yellow and fine divided carbon black were added gradually into a tube which already filled with given compositions of microemulsions and/or lamellar liquid crystals phases of interested samples. Then, the mixture of samples was mixed using Vertexer. Hence, the homogeneity of dyes solubility was observed visually using Red laser light and the Hund Weslar. The addition of dyes was stopped when a precipitation of dyes appeared.

3. Results and Discussion
Figure 1 showed the structure of Tween 40 as an example of Tween compounds that can be widely used as emulsifier, solubilizers, and wetting agents. The solubility of Tween aqueous solutions increased with the degree of ethoxylation. For a fixed degree of ethoxylation, aqueous solubility decreased as the number of ester groupings increased. For fixed degree of ethoxylation and esterification, aqueous solubility decreased as the molecular weight of the fatty acid increased.

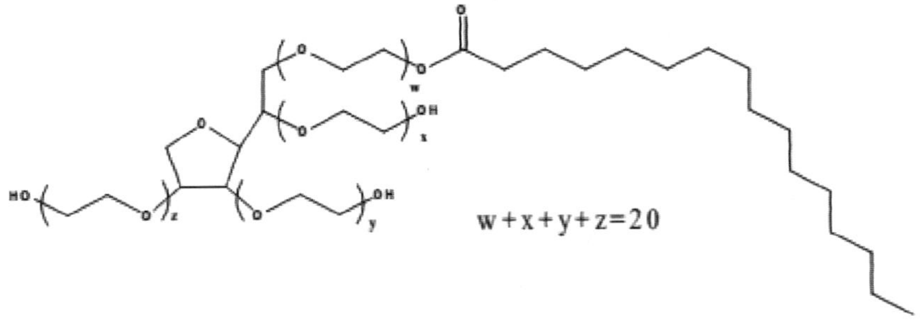

Figure 1. Example of Tween structure (Tween-40)

In this research, methyl yellow and carbon black were used as dyes. Methyl yellow was a chemical compound which might be used as pH indicator. In aqueous solution at low pH, methyl yellow appeared red. Between pH = 2.9 and pH = 4.0 methyl yellow underwent a transition, to became yellow above pH = 4. So, methyl yellow was able to get red colour in water of pH = 1, and yellow colour in water of pH = 5. Figure 2 showed the structure of methyl yellow. Furthermore, carbon black was able to get a black colour in water (pH = 1 and pH = 5).

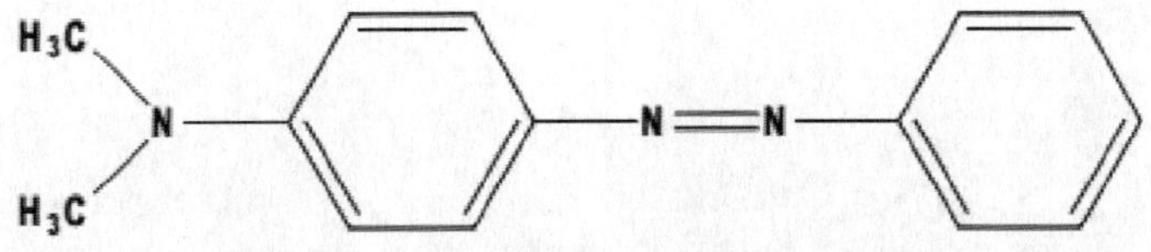

Figure 2. Structure of methyl yellow

3.1. Phase diagram

The results were interesting phase diagrams of water (pH = 1), non-ionic surfactants (Tween-20, Tween-40, Tween-60, and Tween-80), and cyclohexane system, was performed in figure 3. It was shown from this figure that the performance of surfactant association structures, i.e w/o microemulsions, lamellar liquid crystals (LLC) areas formed larger by increasing the total number of moles of ethylene oxide (20 to 80) of Tweens. Except for Tween-60 and Tween-80, the surfactant association structures of hexagonal liquid crystals were also formed.

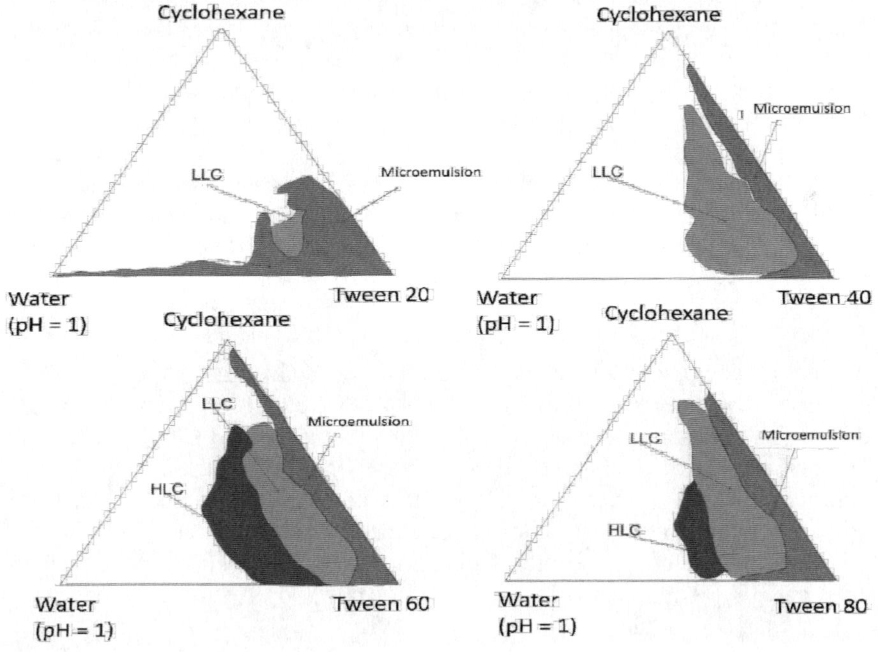

Figure 3. Phase diagram of water (pH=1), Tweens (Tween-20, Tween-40, Tween-60, and Tween-80), and Cyclohexane System.

The phase diagrams of water (pH = 5), non-ionic surfactants (Tween-20, Tween-40, Tween-60, and Tween-80), and cyclohexane system, was performed in figure 4 which relatively similar to the phase diagrams of water (pH = 1) non-ionic surfactant (Tween-20, Tween-40, Tween-60, and Tween-80), and cyclohexane system. It meant that the exchanging of pH (from 1 to 5) of water would not exchange the formation of the surfactant association structures.

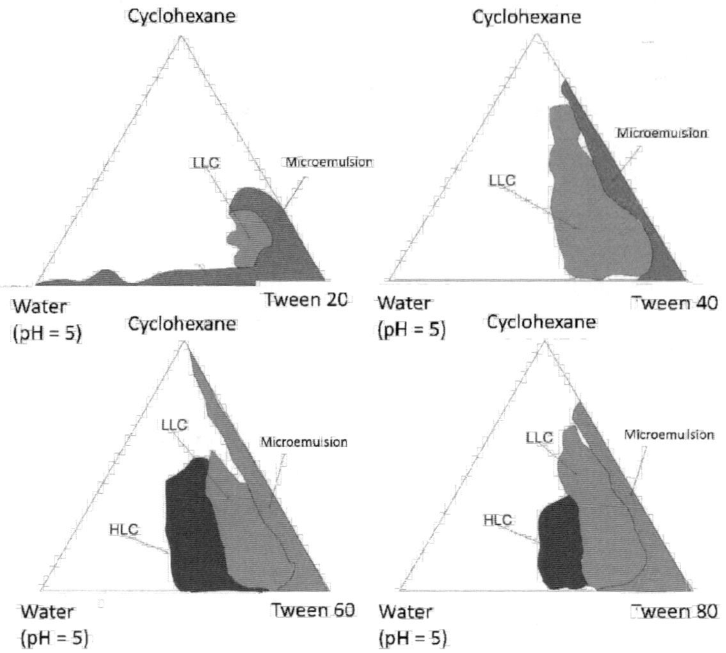

Figure 4. Phase diagram of water (pH=5), Tween (Tween-20, Tween-40, Tween-60, and Tween-80), and Cyclohexane System.

3.2. Solubility of methyl yellow and carbon black in microemulsion and lamellar liquid crystals

Table 1 showed the solubilities of methyl yellow and carbon black in the water (pH = 1, and pH = 5), non ionic surfactant (Tweens) and cyclohexane system. The solubilities of both methyl yellow and carbon black were not exchanged by changing the pH of water from 1 to 5 in the region of w/o microemulsions. It just maintained colour stability of methyl yellow and carbon black. The solubilities of methyl yellow in the w/o microemulson regions which involved Tween 20 was higher than the other Tweens, because in that region the surfactant contents were higher.

On the other hand, the solubilities of methyl yellow in the region of LLC were lower than in that of w/o microemulsion regions, because the physical structure of LLC was more compact than physical structure of w/o microemulsion as shown in the figure 5. In the system with the water of pH=1, the solubilities of methyl yellow were lower than that of the system with the pH = 5. These might water of be due to the physical structure of LLC became less compact.

The solubilities of carbon black in both association structures' regions, w/o microemulsion and LCC, and both water pH (pH = 1 and pH = 5), were too low (less than 0.2 mg/g). It indicated that the solvent medium was not suitable.

Table 1. The solubilities of methyl yellow and carbon black in the water (pH = 1, and pH = 5), non ionic surfactant (Tweens) and cyclohexane system

Surfactants	Regions	Dyes (pH= 1)		Dyes (pH= 5)	
		Methyl yellow (mg/g)	Carbon black (mg/g)	Methyl yellow (mg/g)	Carbon black (mg/g)
Tween-20	w/o microemulsion	1.6	<0.2	1.6	<0.2
	LLC	<0.2	<0.2	1.2	<0.2
Tween-40	w/o microemulsion	0.4	<0.2	0.466	<0.2
	LLC	<0.2	<0.2	0.284	<0.2
Tween-60	w/o microemulsion	<0.2	<0.2	<0.2	<0.2
	LLC	0.4	<0.2	0.6	<0.2
Tween-80	w/o microemulsion	0.266	<0.2	0.666	<0.2
	LLC	0.17	<0.2	0.638	<0.2

The solubility of methyl yellow in w/o microemulsion, o/w microemulsion, and lamellar liquid crystal, was proposed as shown in Figure 5. It was seen that the polar head of methyl yellow would dissolved in polar medium, like water. The non-polar ring (as a tail) of methyl yellow, on the other hand, would dissolved in oil phase.

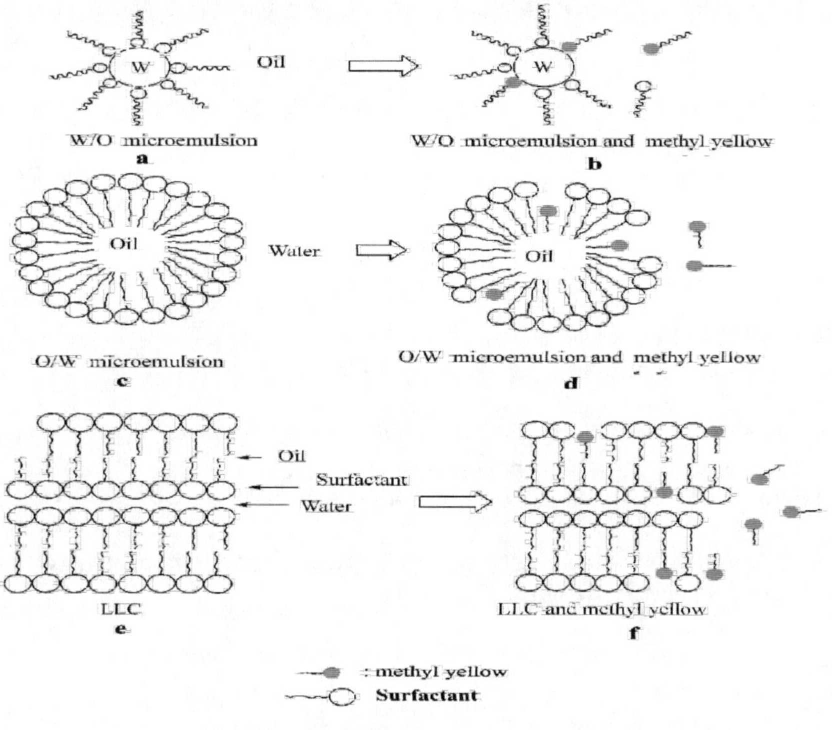

Figure 5. Solubility illustration of methyl yellow

4. Conclusions

It concluded that solubility dynamics of methyl yellow and carbon black, both in w/o microemulsions and the lamellar liquid crystal (LLC), were strongly related to the chemical composition, nature and characteristics of microemulsions and the lamellar liquid crystals. In the system with the water of pH = 1, the solubilities of methyl yellow were lower than that of the system with the water of pH = 5. These might be due to the physical structure of LLC became less compact. Whereas, the solubilities of carbon black in both association structures' regions, w/o microemulsion and LCC, and both water pH (pH = 1 and pH = 5), were too low. It indicated that the solvent medium was not suitable, yet.

Acknowledgement

This research was founded by the Ministry of Research and Technology and Higher Education of The Republic of Indonesia through Decentralized Competitive Research Fund Scheme (2013-2015), contract no. 298.a.54/UN35.2/PG/2014.

References

[1] Xue C H, Shi M N, Chen H Z, Wu G, Wang N 2006 *Col. Surf. A: Physicochemical and Engineering Aspects* **287** 1–3 147–152
[2] Löwen H 2012 *J. Phys: Condensed Matter* **24** 46 460201
[3] Jensen K E, Pennachio D, Weitz D A and Spaepen F 2012 *Soft Matter* **9** 1 320–328
[4] Niemann B, Rauscher F, Adityawarman D, Voigt A and Sundmacher K 2006 *Chem. Eng. and Proc.: Process Intensification* **45** 10 917–935
[5] Wongwailikhit K and Horwongsakul S 2011 *Mat. Let* **65** 17–18 2820–2822
[6] Jing C and Hanbing S X 2007 *Dyes and Pigments* **75** 3 766–769
[7] Amar I, Aserin A and Garti N 2004 *Coll. Surf. B: Biointerfaces* **33** 3–4 143–150
[8] Lin C C, Lin H Y, Chen H-C, Yu M W and Lee M H 2009 *Food Chem.* **116** 4 923–928
[9] Banerjee S, Sutanto S, Klejin M J, Rooamalen M J E, Witkamp G-J and Struart M A C 2012 *Adv. Coll. Interface Sci.* **175** 11–24
[10] Zeuner B, Kontogeorgis G M, Riisager A and Meyer A S 2012 *New Biotech.* **29** 3 255–270
[11] Gaucher J F, Ries-Kautt M, Reiss-Husson F and Ducruix A 1997 *FEBS Letters* **401** 113–116
[12] Shah D O, Bansal V K, Chan K and Hsieh W C 1977 The Structure, Formation and Phase–Inversion of Microemulsions *Improved Oil Recovery by Surfactant and Polymer Flooding* 293–337
[13] García Vior M C, Monteagudo E, Dicelio L E and Awruch J 2011 *Dyes and Pigments* **91** 2 208–214
[14] Hota G, Jain S and Khilar K C 2004 *Colloids and Surfaces A: Physicachemical and Engineering Aspects* **149** (1-3) 263-277
[15] Malcolmson C and Lawrence M J 1995 *Colloids and Surfaces B: Biointerfaces* **4** (2) 97-109
[16] Sahle F F, Metz H, Wohlrab J and Neubert R H 2012 *Eropean Journal of Pharmaceutics and Biopharmaceutics* **82** (1), 139-150
[17] Friberg S E and Bendiksen B 1979 *J. Chem. Educ.* **56** 8 553
[18] Kumoro A C, Retnowati D S and Budiyati C S 2010 *J. Chem. Eng.* Data, vol. **55** 7 2603 – 2606

Synthesis and characterization of geopolymer from bottom ash and rice husk ash

Ufafa Anggarini[1*] and Ndaru C Sukmana[2]
[1]Department of Chemical Engineering, Universitas Internasional Semen Indonesia
[2]Department of Engineering Management, Universitas Internasional Semen Indonesia

[*]E-mail: ufafa.anggarini@uisi.ac.id

Abstract. All Geopolymer (GP) has been synthesized from bottom ash and rice husk ash. This research aims to determine the effect of Si/Al ratio on geopolymer synthesis. Geopolymer was synthesized with various Si/Al ratio of 2, 3 and 4. The characterization result using XRD and SEM indicated that by using a different ratio of Si/A, it will produce geopolymer with varied structure and morphology. Diffractogram result shows that polymerization has been done for all samples (GP2, GP3, GP4) with the presence of hump peak at 2θ = 27-35°. In GP4, no peak at 2θ = 18° indicating sodalite phase forming. Besides that, the morphology of geopolymer with a varied ratio of Si/Al shows that higher ratio will produce geopolymer with higher particle size. The highest compressive strength of geopolymer was obtained at a ratio of Si/Al = 4, with a maximum load of 12866 kgf.

1. Introduction

The contamination of heavy metals become attention because it has a negative impact on the environment. Heavy metals waste are produced by several industrial activities such as electroplating, battery, printed circuit board (PCB), metal coating treatment [1], mining and smelting of metalliferous ores, municipal wastes, fertilizers, pesticides, and the use of pigment [2]. Rochayatun (2006) states that the content of Pb, Ni, Cu, Zn and Cd in Cisadane river are high, and even exceed the limit that published by KepMen LH/51/1995 about the standard of heavy metals waste.

The method that can be used for heavy metal treatment namely waste treatment using high pressure and temperature, the use of aerobic bacteria for degradation, combustion of the waste under high pressure and catalytic oxidation [4]. Yet, those methods have disadvantages, such as the limits ability to decrease the high concentration of waste, requiring extreme conditions, and limiting as each catalyst.

The use of porous materials is currently noted as an adsorbent. The use of porous material focuses on zeolite, carbon, geopolymer and the other materials. Geopolymer is amorphous, three-dimensional aluminosilicate binder material, that can be synthesized by mixing aluminosilicate reactive materials and strong alkaline solutions, followed by curing it at room temperature [5-6]. Geopolymer has potential applications in immobilization of heavy metal, reduce the emissions of CO_2 associated with cement manufacturing, high mechanical strength, flexibility, and durability. Based on that information, in this work, we report the synthesis of geopolymer using bottom ash and rice husk ash as aluminosilicate source. The compressive strength of geopolymer was measured after 28 days. The structure and morphology of this material were observed with XRD and SEM, respectively.

2. Experimental Section

2.1 Material and Instrumentation
The materials used in this experiment were rice husk ash from Madiun, bottom ash from PLTU Paiton, Na_2SiO_3, $NaAlO_2$, NaOH pellet (99% pa) as alkali activator, and distilled water. Powder X-ray diffraction (XRD) patterns were collected using CuKα radiation to identify crystal phase. Scanning Electron Microscope images were collected to identify the morphology of geopolymer. The compressive strength of geopolymer was investigated after 28 days in Physical Laboratorium of PT Semen Indonesia.

2.2 Procedure
Geopolymer samples were synthesized with mole ratio composition of NaOH/rice husk ash. All procedures were conducted at room temperature to minimize the effect of it. Geopolymer samples were synthesized by mixing bottom ash with NaOH as activating solutions for 5 minutes. Rice husk ash was added to the mixture and stirred for 5 minutes. The final composition of each geopolymer sample was cast in 5x5x5 cm^3 cube mold. Final geopolymer was dried for 24 hours at room temperature. After that, this geopolymer was dried at 75 °C for 48 hours and cooled to room temperature. The compressive strength of each geopolymer was measured after 28 days.

3. Result and Discussion

3.1 Effect of Si/Al Ratio Towards the Structure of Geopolymer
The XRD characterization was used to determine the phase of geopolymer. It is because, the polymerization reaction can form multiple crystal phases. Crystalline phase were characterized by the presence of sharp diffractogram peaks, whereas the amorphous phase were characterized by the presence of hump with irregular intensity [11]. XRD characterization was performed on three samples of geopolymer (GP 2, GP 3 and GP 4) as seen in Figure 1. X-ray diffraction pattern of all geopolymer samples with Si/Al ratio variation showed the formation of some phase such as quartz, mullite, magnetite, and sodalite. Kim (2012) suggests that the presence of hump peak at 2θ = 27-35° indicates the formation of geopolymer has been done.

GP 4 has high intensity peak at 2θ = 27° that confirms as quartz phase formation. In addition, GP 2 and GP 3 have apeak at 2θ = 18° that confirm as sodalite phase formation. According to the Kusumastuti, (2012) experiment, sodalite phase causes the product of geopolymer become fragile and the compressive strength of this material becomes low.

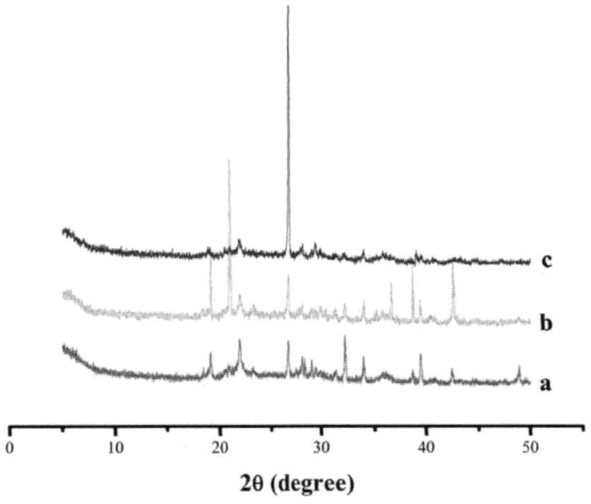

Figure 1. The XRD pattern of (a) GP 2 (Geopolymer with Si/Al ratio = 2), (b)GP 3 (Geopolymer with Si/Al ratio = 3), GP 4 (Geopolymer with Si/Al ratio = 4)

3.2 Effect of Si/Al Ratio Towards the Morphology of Geopolymer

Xu et al. (2000) state that geopolymer product does not have a stoichiometric composition and the mixture of amorphous to semi-crystalline structure and crystalline Al-Si particles through nano-crystalline that disperse in an alumino-silicate gel. Geopolymerization process includes leaching, diffusion, condensation and hardening step. Figure 2 showed the morphology of geopolymer with Si/Al ratio of 2, 3 and 4. A higher Si/Al atomic ratio corresponded to the more crystalline aluminosilicate gel and the increase in crystalline indicated increasing in compressive strength [7]. Figure 2 showed that the higher Si/Al ratio can produce higher crystallinity of geopolymer particle.

(a) GP 2

(b) GP 3

(c) GP 4

Figure 2. The morphology of (a) GP 2 (x500 and x1000); (b)GP 3 (x500 and x1000); (c) GP 4 (x500 and x1000)

3.3 Effect of Si/Al Ratio Towards the Compressive Strength of Geopolymer

Geopolymerization process occurs under an alkaline condition with rapid reaction. Minerals containing silica and alumina will react to produce chain and ring that contain three-dimensional polymers with Si-O-Al-O bond [9]. The source of silica and alumina for geopolymer synthesis can be derived from a natural source or commercial source. Chindaprasirt (2009) has been synthesis geopolymer from fly ash and bottom ash because these materials contain high silica content.

The compressive strength of geopolymer was investigated after 28 days. Tavor et al. (2007) state that the compressive strength of geopolymer will increase by time, up to 28 days. This means that polymerization process requires 28 days. The result of compressive strength test was showed in Figure 3. When the Si/Al ratio were higher, the compressive strength of geopolymer increases. This result occurs while higher ratio affects the increase of geopolymer crystallinity. This result is appropriate to the XRD result. Figure 3 shows that the higher Si/Al ratio of geopolymer sample, the higher the maximum load that can be detained. Davidovits (1991) states that there is a relationship between mechanical strength and silica content. While silica content in geopolymer increases, it will improve the amount of Si-O-Si bond. This type of bond is stronger than Si-O-Al and Al-O-Al bonds. In other cases, if the ratio of Si/Al too high, will decrease the mechanical strength of geopolymer, due to the mixture of geopolymer producing the material that cannot react [12].

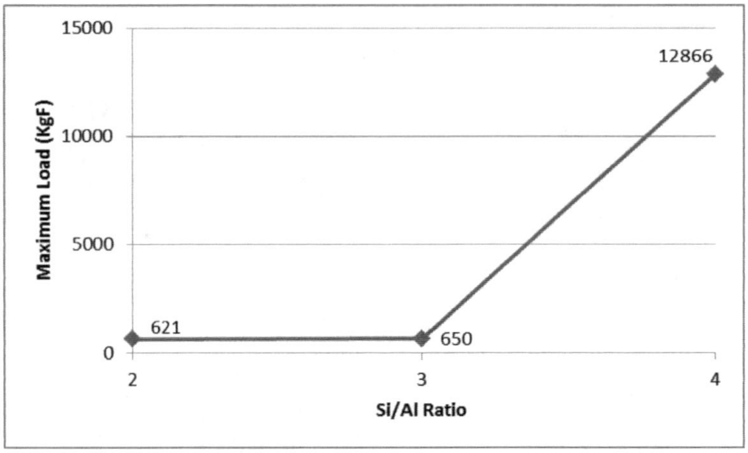

Figure 3. Maximum load that can be detained by geopolymer with Si/Al ratio variation

4. Conclusion

Geopolymer has been synthesized from bottom ash and rice husk ash with various ratio of Si/Al at 2, 3 and 4. The XRD and SEM result shows higher Si/Al ratio producing geopolymer with better structure and higher particle size, respectively.

Acknowledgements
The authors wish to thank LPPM STiMSI for providing *Hibah Penelitian Dosen STiMSI* , contract number 003/PL/LPPM/12/2014 date December 8th, 2014

References

[1] Cheng T W, Lee M L, Ko M S, Ueng T H and Yang S F 2012 The heavy metal adsorption characteristics on metakaolin-based geopolymer: *Applied Clay Science*. Vol. **56**, p. 90-96

[2] Leyval C, Turnau K, Haselwandter K 1997 Effect of heavy metal pollution on mycorrhizal colonization and function: physiological, ecological and applied aspects: *Mycorrhiza*. Vol. **7** p. 139-153

[3] Rochyatun, Endang, Kisupy, M., Taufik, Rozak, Abdul 2006 Distribusi logam berat dalam muara air dan sediment di perairanmuarasungai Cisadane: *Journal of Makara Sains*. Vol.**10**, p. 35-40

[4] Tavor, Dorith, Wolfson, Adi, Shamaev, Anat 2007 Recycling of industrial wastewater by its immobilization in geopolymer cement: Ind.Eng. Chem. Res. Vol. **46**, p. 6801-6805

[5] Davidovits J 1991 Geopolymer: Inorganic Polymeric New Materials, *J.Therm*. Anal. Vol. **37**, p. 1633

[6] Xu H, van Deventer 2000 The polymerization of Aluminosilicate minerals: *Int.J.Miner*. Process. Vol. **59**, p. 247

[7] Rattanasak U, Chindaprasirt P 2009 Influence of NaOH solution on the synthesis of fly ash geopolymer, *Min. Engine*. Vol **22**, p. 1073-1078

[8] Tavor, Dorith, Wolfson, Adi, Shamaev, Anat 2007 Recycling of Industrial Wastewater by Its Immobilization in Geopolymer Cement, Ind. *Eng. Chem*. Vol **46**, p. 6801-6805

[9] Abdullah M M A, Hussin K, Bnhussain M, Ismail K N and Ibrahim W M W 2011 Review Paper: Mechanism and Chemical Reaction of Fly Ash Geopolymer Cement, *Int. Journal of Pure and App. Sci and Tech*, Vol. **6**, p. 35-44

[10] Chindaprasirt, Prinya, Jaturapitakkul, Chai, Chalee, Wichian, Rattanasak and Ubolluk 2009 Comparative Study on The Characteristics of Fly Ash and Bottom Ash Geopolymers, *Journal of Waste Management*, Vol. **29**, p. 539-543

[11] Kusumastuti E 2012 Pemanfaatan Abu Vulkanik Gunung Merapi sebagai Geopolimer (Suatu Geopolymer Anorganik Aluminosilicate), *Jurnal MIPA* **35**

[12] Kim E H 2012 Understanding Effects of Silicon/Aluminium Ratio and Calcium Hydroxide on Chemical Composition Nanostructure and Compressive, theUniversity of Illinois at Urbana-Champaign

Electrical conductivity of zirconia and yttrium-doped zirconia from Indonesian local zircon as prospective material for fuel cells

Karima Apriany[1], Ita Permadani[1], Dani G. Syarif[2], Syoni Soepriyanto[3], Fitria Rahmawati[1*]

[1]Research Group of Solid State Chemistry & Catalysis, Chemistry Department, Sebelas Maret University, Jl.Ir. Sutami 36 A Kentingan Surakarta 57126
[2]Lab. of Material Physics, Badan Tenaga Atom Nasional, Jl. Taman Sari Bandung
[3]Department of Metallurgy, Faculty of Mining and Petroleum Engineering, Institut Teknologi Bandung, Jl. Ganesha 10 Bandung

[*]E-mail: fitria@mipa.uns.ac.id

Abstract. In this research, zirconium dioxide, ZrO_2, was synthesized from high-grade zircon sand that was founded from Bangka Island, Sumatra, Indonesia. The zircon sand is a side product of Tin mining plant industry. The synthesis was conducted by caustic fusion method with considering definite stoichiometric mole at every reaction step. Yttrium has been doped into the prepared zirconia by solid state reaction. The prepared materials were then being analyzed by X-ray diffraction equipped with Le Bail refinement to study its crystal structure and cell parameters. Electrical conductivity was studied through impedance measurement at a frequency range of 20 Hz- 5 MHz. Morphological analysis was conducted through Scanning Electron Microscopy (SEM) equipped with Energy Dispersive X-ray (EDX) for elemental analysis. The results show that the prepared yttrium stabilized zirconia, YSZ, was crystallized in the cubic structure with a space group of *P42/NMC*. The sintered zirconia and yttrium stabilized zirconia at 8 mol% of yttrium ions (8YSZ) show dense surface morphology with a grain size less than 10 μm. Elemental analysis on the sintered zirconia and 8YSZ show that sintering at 1500 °C could eliminate the impurities, and the purity became 81.30%. Impedance analysis shows that ZrO_2 provide grain and grain boundary conductivity meanwhile 8YSZ only provide grain mechanism. The yttrium doping enhanced the conductivity up to 1.5 orders. The ionic conductivity of the prepared 8YSZ is categorized as a good material with conductivity reach 7.01×10^{-3} at 700 °C. The ionic conductivities are still lower than commercial 8YSZ at various temperature. It indicates that purity of raw material might significantly contribute to the electrical conductivity.

1. Introduction

Zirconia is metal oxide for multifunctional applications such as for ceramics, due to the ability to reflect light and a good ceramics coating [1], for catalysts, due to its good thermal stability [2] and its high surface area [3]. It is also a potential material for electrolyte and electrode of fuel cells [4-7]. Zircon sand is a potential raw material for zirconia production. Zircon sand in Bangka Island Indonesia is siding product of tin mining plant that has low economic value. Therefore, some researchers such as Soepriyanto and Hidayat [8] did the effort to increase zircon ($ZrSiO_4$) concentration through flowsheet modification in separation method. It can increase zircon concentration from 20% to 98%. The commercial zircon usually contains of 66wt% min. ZrO_2 + HfO_2, 32-33 wt% (weight%) of SiO_2, 0.1 wt% TiO_2 dan 0.1 wt% max. Fe_2O_3 [9]. Zircon is very stable chemical compound due to the strong bond between zirconia and silica in its molecule. Therefore, the zircon extraction must be initiated by the chemical bonds break down. Alkali fusion method is a more favorable technique than traditional chlorination breakdown [10].

A variety of synthesis method is also being explored and developed for the production of stabilized and nonstabilized nanoscale zirconia particles. The methods are thermal decomposition [11-12], sol–gel [13], precipitation [13, 14], emulsion precipitation [15], hydrothermal [16], reverse microemulsion [17], and mechanical milling [18]. Since 1960, the caustic fusion method showed that zircon ore could be fused at 600 °C with the mixture of NaOH and water. The product is a hydrolyzed zirconate, which was further treated using HCl to form a zirconium oxychloride before precipitation [19]. Manivasan et al. [19] used a slight variation on this caustic fusion process and used nitric acid to form a zirconyl nitrate, which can be precipitated. The use of ball milling as a mean for breaking down loose agglomerates is also explored. Rahmawati et al. [4] have applied caustic fusion method as it was conducted by Soepriyanto and Hidayat [8] to produce ZrO_2 from zircon sand from Bangka Island, Indonesia. However, the synthesis did not follow the stoichiometric amount of each reagent, such as the zircon sand was crushed with solid NaOH at the same weight of the mixture. The produced ZrO_2 then was doped with yttrium and combined it with LSGM-8282. It found that the prepared YSZ is still non homogeneous and consist of high silica content. Even though the caustic fusion method still provides low purity of zirconia, however caustic fusion or alkali fusion is a simple non-expensive method to produce zirconia from zircon concentrate, $ZrSiO_4$ [20]. Therefore in this research, the caustic fusion method is conducted by applying a definite mole ratio at each reaction step based on the stoichiometric reaction. The prepared materials were sintered at a different temperature to study the effect of sintering temperature to its crystal structure and conductivity. Comparison to the commercial product was also conducted to study the prospect ability of the prepared materials.

2. Experiment

ZrO_2 was synthesized by caustic fusion method. NaOH (99%) 30 g solid and 33 g of High-Grade Zircon were grounded in mortar and pestle for 2 hours. Weight ratio was determined from mol calculation based on the chemical reaction in equation (1).

$$ZrSiO_{4(s)} + 4\ NaOH_{(s)} \rightarrow Na_2ZrO_{3(s)} + Na_2SiO_3 + 2\ H_2O_{(g)} \quad (1)$$

The mixture is then dissolved in 120 mL of 5 M HCl solution during the acid leaching process to produce $ZrOCl_2$. Acid leaching was conducted at 85 °C for 30 minutes. The produced zirconium oxychloride then being reacted with ammonium hydroxide 1 M until pH of the solution reaches 9. This step was conducted to produce zirconium hydroxide, $Zr(OH)_4$. The addition of ammonium hydroxide was conducted under stirring at 80 °C. After 24 hours aging, zirconium hydroxide deposit was filtered and washed with hot water until reach neutral pH (pH=7) condition. The residue then was heated at 150 °C for 3 hours and being calcined at 800 °C for 5 hours to produce zirconia powder.

Doping yttrium ions into zirconia were conducted by solid state reaction between Y_2O_3 and the prepared ZrO_2 at definite mol amount to produce $Y_{0.08}Zr_{0.92}O_{2-\delta}$ (8YSZ) and followed by calcination at 1000 °C for 5 hours. The powder was analyzed by X-ray Diffraction, XRD (Bruker D8) equipped with Le Bail refinement to study its crystallinity, crystal structure and its cell parameters. The X-ray fluorescence, XRF (BRUKER S2 Ranger) analysis was conducted for elemental analysis. Surface morphology and elemental content were analyzed by Scanning Electron Microscopy, SEM/EDX (FEI Type Inspect S50).

Sintering of the prepared materials was conducted at 1500 °C for 5 hours. The conductivity analysis was conducted through calculation of conductivity from impedance data that was measured by LCR meter (GW Instek LCR-8105G, frequency range 20 Hz – 5 MHz) at 400 – 700 °C. Conductivity value was calculated by using equation (2). Capacitance value was used to determined the conductivity type, whether grain (~ 0.1 pF) , grain boundary (~1.4 nF) or electrode-electrolyte interface (> 1 uF) [21].

$$\frac{1}{\sigma} = \frac{1}{R}\frac{l}{A} \tag{2}$$

σ is specific conductivity (S cm^{-1}), R is resistance (Ω), l is the thickness of pellet (cm), and A is area of active electrode (cm^2).

The Arrhenius plot of conductivity to temperature was provided to analyze the activation energy of ionic conduction and the pre-exponential factor that correlates with the number of vacancies which potentially contribute to the ionic conductivity of the material. The plot followed Arrhenius equation (3).

$$\sigma = \sigma_o \exp^{-Ea/kT} \tag{3}$$

The σ_o is the pre-exponential factor, the Ea is the activation energy, k is Boltzmann constant, and T is temperature.

3. Results and discussion

The caustic fusion of zircon occurs within each step as listed in reaction equations (4) – (6). The decomposition reaction with sodium hydroxide follows equation (1) [10]. Acid leaching with hydrochloride acid was proceeded as described in equation (4) to form $ZrOCl_2$. Then $ZrOCl_2$ was reacted with ammonium hydroxide to form zirconium hydroxide (equation (5)).

$$nNa_2ZrO_3\Downarrow + 4n\ HCl \rightarrow nZrOCl_2.nH_2O + 2n\ NaCl + nH_2O \tag{4}$$
$$ZrOCl_2.nH_2O_{(aq)} + 2NH_4OH_{(aq)} \rightarrow Zr(OH)_{4(s)} + 2NH_4Cl_{(aq)} + (n-1)H_2O_{(l)} \tag{5}$$

Zirconium oxide was formed during calcination at 800°C for 5 hours, as it is described in equation (6)

$$2Zr(OH)_4 \rightarrow t\text{-}ZrO_2 + m\text{-}ZrO_2 + 4H_2O \tag{6}$$

X-ray diffraction analysis of the initial HG Zircon and the prepared ZrO_2 produced diffraction equipped with the Le Bail refinement have been studied in our previous paper [22]. The Le Bail refinement shows that the prepared ZrO_2 was crystallized in tetragonal structure with a space group of *P42/NMMC* with cell parameters a=b= 3.592(1) Å and c= 5.192(2) Å. The cell volume is 67.00(4) $Å^3$ with an angle α=β=γ= 90°. The refinement proceeded well at % residual of Rp 5.51% and Rwp of 6.75 %. XRF analysis shows that the concentration of ZrO_2 in the prepared powder is 72.17%. Some impurities are still present such as sodium oxide and silica [22].

Yttrium doping was conducted by solid state reaction of Y_2O_3 with the prepared ZrO_2. The XRD pattern is depicted in Figure 3 in comparison with standard 8YSZ ICSD #90885. XRD patterns show that the mixture of Y_2O_3 with ZrO_2 without calcination has many peaks with low crystallinity. After being calcined at 1000 °C for 2 hours, some peaks were disappeared and the diffraction has many sharp peaks that indicate high crystallinity. However, comparison of the calcined sample with standard diffraction of 8YSZ ICSD #90885 shows four peaks that are not in agreement with the standard. The presence of sodium oxide and silica as it was shown by elemental analysis may confirm that the unidentified peaks are related to sodium oxide and silicon dioxide as it is described in Figure 1.

The morphological analysis shows that the sintered ZrO_2 has dense surface morphology with a grain size less than 10 μm as it is described in Figure 2. EDX analysis confirms that the sintered ZrO_2 has a high content of Zr i.e. 81.30% weight (Figure 4(b)), and there are no impurities detected. It indicates that some impurities might be released while sintering at 1500 °C for 5 hours, such as the sodium oxide was melted at 1132 °C. Meanwhile, the silicon dioxide is still present due to the compound has high melting point, i.e., 1600 °C.

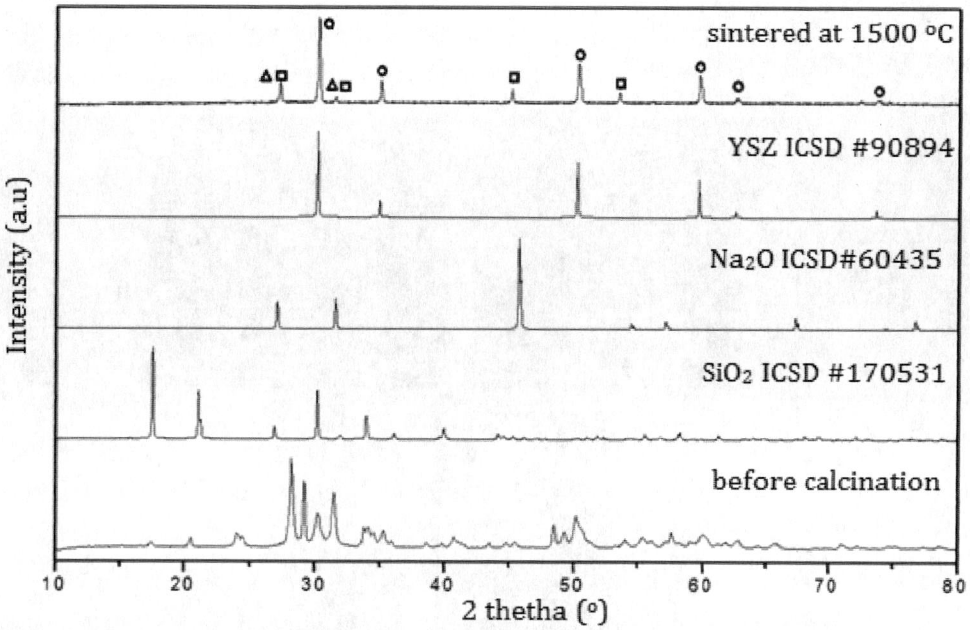

Figure 1. XRD pattern of the mixture of Y_2O_3 and ZrO_2 without calcination, the prepared 8YSZ after being sintered at 1500 °C, and the XRD pattern of standard 8YSZ from ICSD #90885, standard diffraction of SiO_2 and standard diffraction of Na_2O. Sign o refers to the YSZ peaks, □ refers to the disodium oxide and Δ refers to the silicone dioxide.

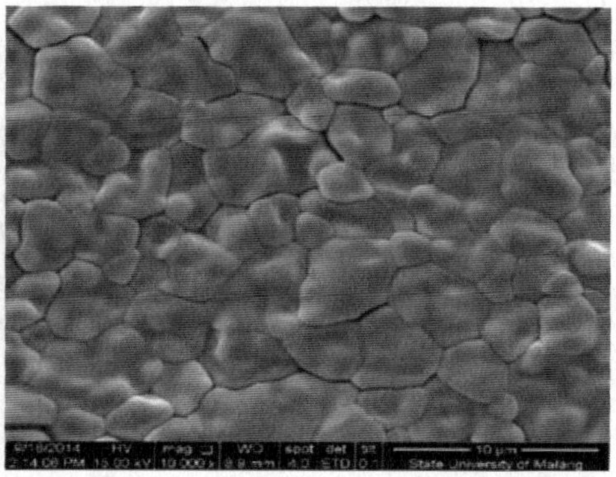

Figure 2. (a) SEM image of the prepared zirconia that was sintered for 5 hours at 1500 °C

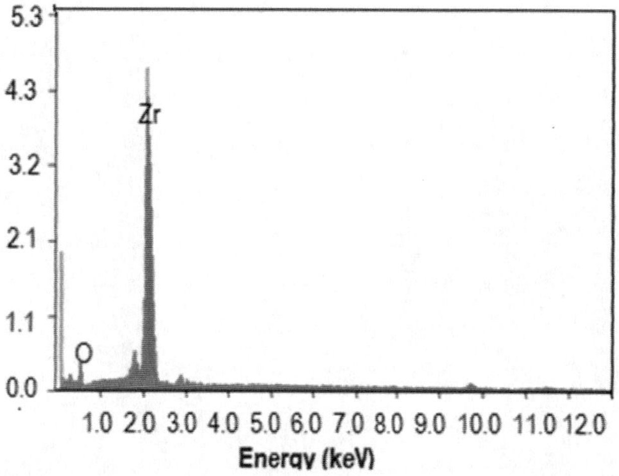

Figure 2. (b) EDX result of the prepared zirconia that was sintered for 5 hours at 1500 °C

IOP Conf. Series: Materials Science and Engineering **107** (2016) 012023 doi:10.1088/1757-899X/107/1/012023

Meanwhile, 8YSZ shows more dense surface morphology with grain size around 10 μm. However, the morphology is less dense than the commercial 8YSZ. SEM images are described in Figure 3. EDX analysis (Figure 4) shows that the prepared 8YSZ content of 68.25% weight and yttrium of 11. 02% weight.

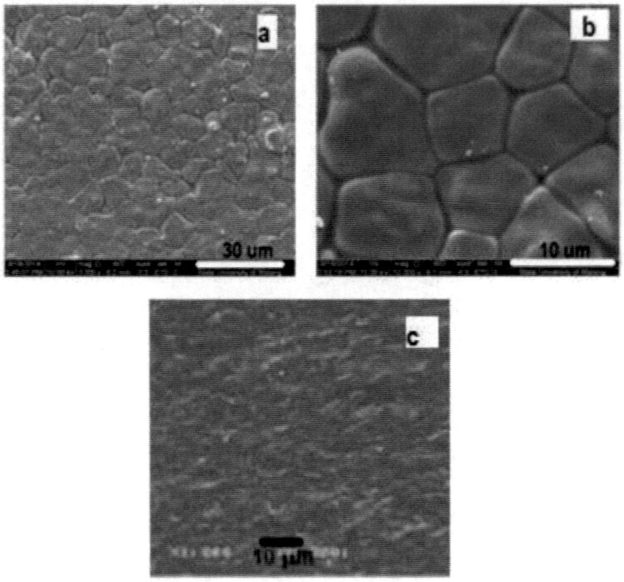

Figure 3. SEM images of the prepared 8YSZ (a) and (b) at different magnification and SEM image of commercial 8YSZ (c)

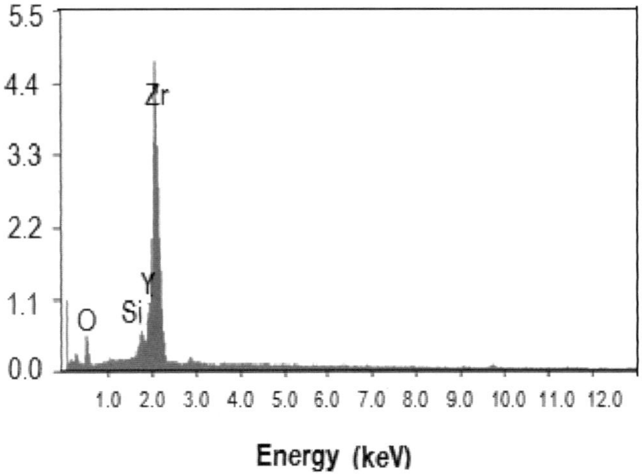

Element	Wt%	At%
OK	18.26	54.31
SiK	02.47	04.18
YL	11.02	05.90
ZrL	68.25	35.60

Figure 4. EDX result of the prepared 8YSZ from zircon concentrate

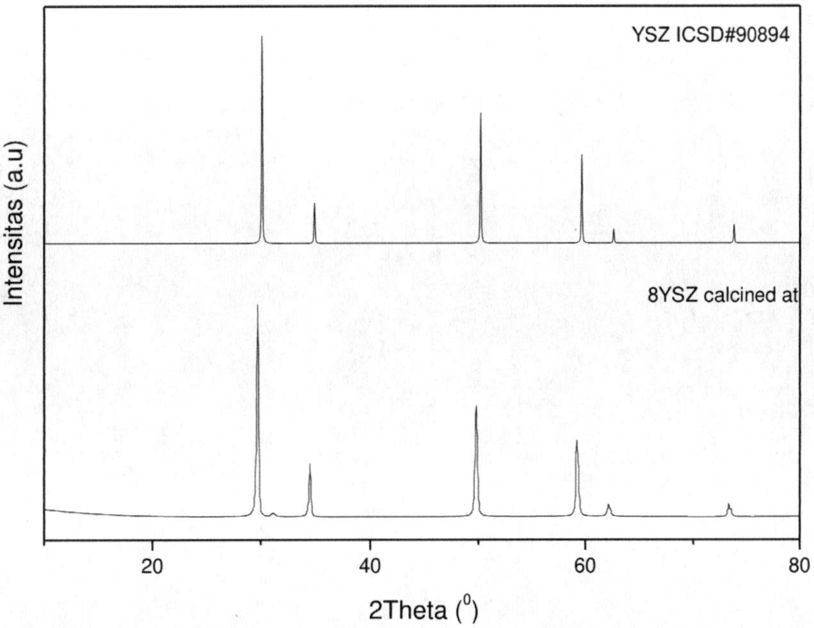

Figure 5. The diffraction patterns of 8YSZ after sintering at 1500 °C

The prepared 8YSZ that was sintered at 1500 °C has diffraction peaks that in agreement with the standard diffraction of ICSD #90894 (Figure 5). The refinement results that the prepared 8YSZ has a cubic structure with cell parameters of a=b=c= 5.1442 (5) A° with a cell volume of 136.14(2)A^{o3}. The residual factors are Rp 6.27% and Rwp 13.5%.

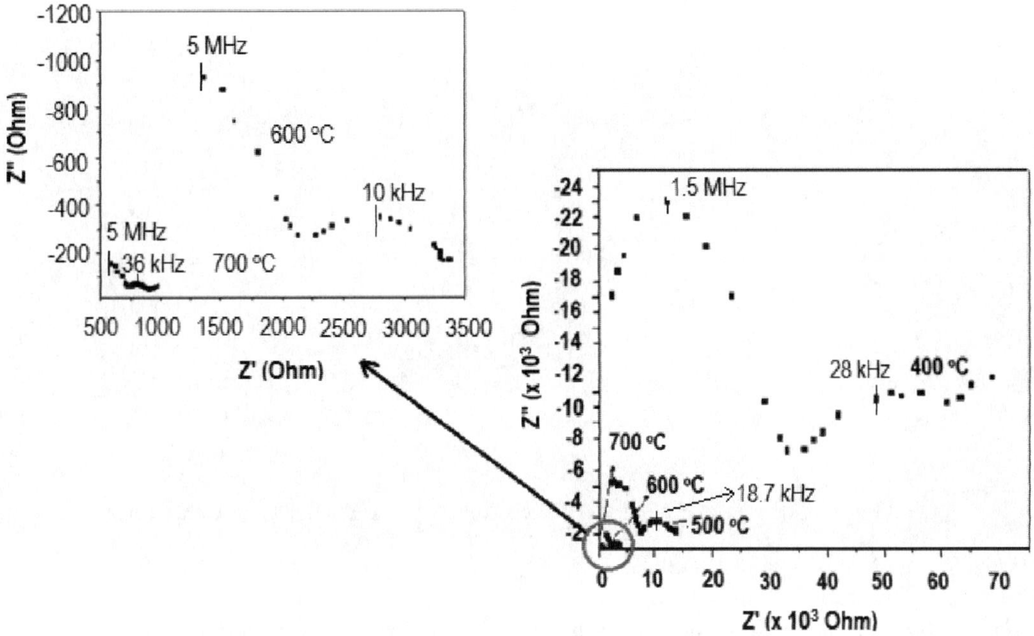

Figure 6. The Nyquist plots of imaginary impedance (Z”) to the real impedance (Z’) of the prepared ZrO_2 at various temperature.

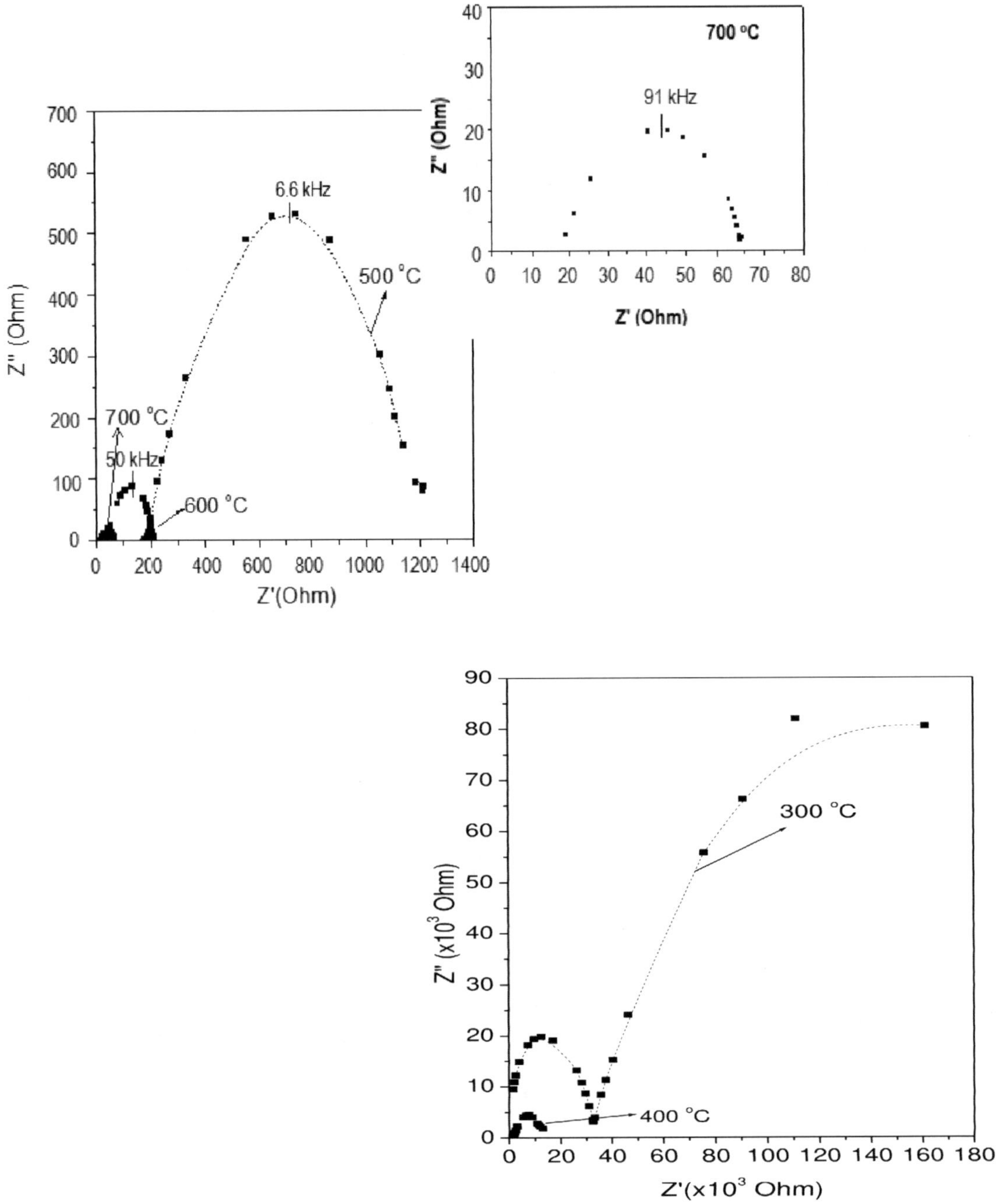

Figure 7 . Impedance plots of the prepared 8YSZ at various temperature.

Impedance analysis at 400 – 700 °C for ZrO_2 and 8YSZ produce Nyquist plots as it is described in Figure 6 and 7. Zirconia shows two semicircles that indicate migration of charge carriers might occur through different mechanisms such as grain and grain boundary conduction or electrolyte-electrode interface. Meanwhile, the prepared 8YSZ shows two semicircles at 300 °C and 400 °C, and the curve became single semicircle at 500 °C and above. Fitting on impedance plot confirms that the resistance value decreased as the temperature increase. Capacitance values confirm that zirconia provides grain resistance only at 300 °C and provide grain and grain boundary resistance at temperature 400 C and above. These were confirmed by capacitance values that listed in Table 2. Calculation of conductivity with equation (1) shows that the prepared zirconia shows good ionic conductivity that is higher than 1×10^{-4} $S.cm^{-1}$ [23] at a temperature higher than 500 °C. Meanwhile, at 300 – 400 °C the ionic conductivity values are lower than 1×10^{-4} $S.cm^{-1}$. Therefore, it is not suitable to be applied as an electrolyte for the fuel cell.

Doping of 8% mol yttrium into the prepared zirconia could increase ionic conductivity values at about two orders. Therefore, the material become suitable for the electrolyte in fuel cell even at 400 °C, due to the ionic conductivity is 2.92×10^{-4} $S.cm^{-1}$, which categorized as good electrolyte [23]. The conductivity values are listed in Table 2. At temperature above 400 °C, the ionic conductivity was dominated by grain boundary conductivity, as confirmed by their capacitance values (Table 2). At 300 – 500 °C the ionic conductivity values are lower than the commercial 8YSZ. The ionic conductivity of the prepared 8YSZ become higher than the commercial 8YSZ at above 500 °C. The Arrhenius plot in Figure 8 shows clearly the conductivity trendline of the materials. It indicates that the 8YSZ that was prepared from Indonesian zircon sand might provide better performance for the high-temperature solid oxide fuel cell than the commercial one.

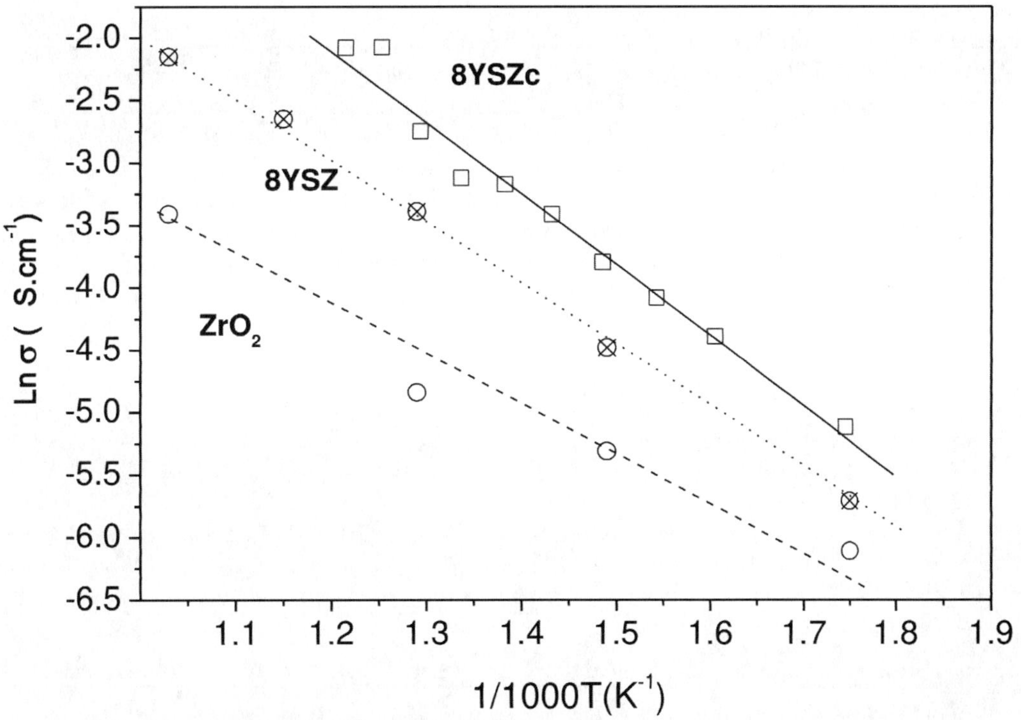

Figure 8. Plot Arrhenius of the prepared ZrO_2, YSZ and the commercial YSZ (YSZc)

Arrhenius plot also can be used to study the energetic of ionic migration inside crystal structure. The conductivity plot of each material shows only a single slope. The slope might consist of activation energy of ionic hoping and the activation energy of ionic migration. Rahmawati et al. [4] also found the only single slope of 8YSZ. Meanwhile, Zhang et al. [24] found two slopes of Arrhenius plot of nanosize 8YSZ. A slope at 600-700 °C with greater value than slope at 700-1000 °C. The temperature of conductivity measurement that only until 700 °C, might become a factor that cause the prepared 8YSZ only shows one slope. It means that if the measurement proceeds up to higher than 700 °C, the Arrhenius plot might consist of two slopes.

The activation energies of the prepared 8YSZ and commercial 8YSZ are 1.002 eV and 1.094 eV, respectively. These are consistent with the result of Madani et al. [25] who found activation energy of 1.08 eV for 9.5% mol YSZ and activation energy of 1.14 eV [24]. Meanwhile, Gong et al. [26] found 0.91 eV at measurement temperature higher than 650 °C. The activation energy of the prepared 8YSZ is lower than the commercial 8YSZ. It indicates that ions might migrate easier in the crystal structure of the prepared 8YSZ. However, the pre-exponential factor of commercial 8YSZ is three times higher than 8YSZ from local zircon. Pre-exponential factor is representing the number of vacancies, as it can be seen in equation (7) and (8) for Frenkel and Schottky defect, respectively [27]. It indicates that commercial 8YSZ has high vacancies concentration than the 8YSZ from zircon sand. It probably of the presence of impurities inhibits the formation of oxygen vacancies.

$$\sigma_o \equiv \frac{n_f}{\exp(-g_f / 2kT)}(a^2 q^2 / kT)\upsilon_o \tag{7}$$

$$\sigma_o \equiv \frac{n_s}{\exp(-g_s / 2kT)}(a^2 q^2 / kT)\upsilon_o \tag{8}$$

Table 2. Electrical conductivity of the prepared ZrO_2, 8YSZ and the commercial YSZc

Materials	Temperature (°C)	σ_{grain} (S.cm^{-1})	$\sigma_{grain\ boundary}$ (S.cm^{-1})	CPE1 (F)	CPE2 (F)
ZrO_2	300	7.74×10^{-7}	-	1.63×10^{-12}	-
	400	9.15×10^{-6}	4.91×10^{-6}	3.44×10^{-12}	1.02×10^{-9}
	500	2.07×10^{-5}	1.45×10^{-5}	2.71×10^{-12}	1.54×10^{-9}
	600	1.90×10^{-4}	1.81×10^{-4}	2.19×10^{-11}	1.04×10^{-8}
	700	9.74×10^{-4}	6.59×10^{-4}	1.03×10^{-10}	1.05×10^{-8}
8YSZ	300	1.96×10^{-5}	2.143×10^{-6}	2.39×10^{-11}	1.11×10^{-10}
	400	2.92×10^{-4}	3.747×10^{-5}	3.77×10^{-11}	3.12×10^{-9}
	500	-	4.04×10^{-4}	-	3.35×10^{-8}
	600	-	2.22×10^{-3}	-	2.46×10^{-8}
	700	-	7.01×10^{-3}	-	4.22×10^{-8}
YSZc	300	3.07×10^{-5}	1.02×10^{-5}	2.857×10^{-11}	8.461×10^{-10}
	400	1.90×10^{-4}	9.95×10^{-4}	1.871×10^{-11}	1.650×10^{-9}
	500	1.79×10^{-3}	-	3.948×10^{-10}	

Table 3. The activation energy and pre-exponential factor of the prepared materials.

Materials	Ea (eV)	pre exponential
ZrO_2	0.732	0.202
8YSZ	1.002	3.082
YSZc	1.094	10.348

4. Conclusion

Doping of yttrium ions into the prepared zirconia from zircon sand produced 8YSZ with the presence of small amount of silica, i.e., 2.47%. The ionic migration in zirconia structure contains grain and grain boundary, meanwhile after being doped with yttrium ions, the mechanism of migration at above 400 °C is dominated by grain boundary conductivity. The ionic conductivity of 8YSZ is higher than 1×10^{-4} S.cm^{-1}, which is categorized as a good electrolyte for the fuel cell.

Acknowledgement
Authors thank Ministry of Research and Technology for funding this research through INSINas Research Project 2015.

References

[1] Bhattacharya S K and Tummala R R 2000 *J Mat Sci: Materials in Electronics* **II** 253-268
[2] Tanabe K 1996 Surface and Catalytic properties of ZrO_2 *Material Chemistry and Physics*. **13** 347-364
[3] Bangi U K H, Sun-Park C, Baek S and Ho-Park H 2013 Powder Tech. **239** 314-318
[4] Rahmawati F, Prijamboedi B, Soepriyanto S and Ismunandar 2012 *Int. J. Min.* Met. Mat. **19**(9) 863-871
[5] Hansch R, Lavergnat D, Menzler N H and Stover D 2005 *Adv Eng Mins* **7** (3) 142-144
[6] Brahim C, Ringuede C, Cassir M, Putkonen M and Niinisto L 2007 App Surf Sci **253,** 3962-3968
[7] Kiratzis N, Holtappels P, Hatchwell C E, Mogensen M and Irvine J T S 2001 Fuel Cells **1** (3-4) 211-218
[8] Soepriyanto S and Hidayat T 2006 Synthesis of nanocrystalline zirconia from by-product of tin processing, in *Proc. Int. Conf. Mathematics and Natural Sciences* (ICMNS) (Bandung, Indonesia
[9] Sakorn minerals Co., Ltd., 2011. Standard Grade Zircon [online]. Available from: http://www.sakornminerals.com/zircon.htm.[accessed: Feb 22, 2015].
[10] Abdelkader A M, Daher A and El-Kashef E 2008 J Alloys and Comp. **460** 577-580
[11] Jiang S, Stangle G C, Amarakoon V R W and Schulze W A 1996 *J Mater Res* **11**(9) 2318-2324
[12] Pang G, Chen S, Zhu Y, Palchik O, Koltypin Y, Zaban A and Gedanken A 2001 *J Phys Chem* B, **105** 4647-4652
[13] Tyagi B, Sidhpuria K, Shaik B and Jasra R V 2006 *Ind. Eng Chem Res* **45** 8643-8650
[14] Hu M Z C, Hunt R D E, Payzant A and Hubbard C R 1999 *J Am Ceram.Soc* **82**(9) 2313-2320
[15] Woudenberg F C M, Sager W F C, Ten Elshof J E and Verweij H 2004 *J Am Ceram Soc* **87**(8) 1430-1435
[16] Agli G D and Mascolo G 2000 *J Eur Ceram. Soc* **20**(2) 139-145
[17] Althues H and Kasel S 2002 Langmuir **18** 7428-7435
[18] Adam J, Drumm R, Klein G and Veithw M 2008 *J Am Ceram.Soc* **91**(9) 2836-2843
[19] Manivasakan P, Rajendran V, Rauta P R,Sahu B B, Panda B K 2011 *J Am Ceram.Soc* **94**(5) 1410-1420
[20] da Silva R J, Dutra A J V and Afonso J C 2012 *Hydrometallurgy* **117-118** 93-100
[21] Martin P, Lopez M L, Pico C, Veiga M L 2007 *Solid State Sciences* **9** 521
[22] Rahmawati F, Permadani I, Soepriyanto S, Syarif D G, Heraldy E 2014 7th International Conference on Physics and Its Applications (ICOPIA 2014).
[23] Agrawal R C and Gupta R K 1999 *J Mater Sci* **34** 1131
[24] Zhang C, Li C J, Zhang G, Ning X J, Li C X, Liao H and Coddet C 2007 *Mat Sci and Eng B* **137** 24 – 30
[25] Madani A, Cheikh-Amdouni A, Touati A, Labidi M, Boussetta H and Monty C 2005 *Sens Actuators B* **109** 107-111
[26] Gong J H, Li , Tang Z L, Xie Y S and Zhang Z T 2002 *Mater.Chem.Phys* **76** 212-216
[27] West A R 1999 *Basic Solid State Chemistry* 2nd Ed (John Wiley & Sons Ltd New York)

IOP Conf. Series: Materials Science and Engineering **107** (2016) 012024 doi:10.1088/1757-899X/107/1/012024

Optical simulation of surface textured TCO using FDTD method

I L Elviyanti[*], H Purwanto and Kusumandari

Department of Physics, Faculty of Mathematics and Natural Sciences, Sebelas Maret University, Jl. Ir. Sutami 36A Kentingan Surakarta 57126, Indonesia

[*]E-mail: isna_elviyanti@yahoo.co.id

Abstract. The purpose of this research is simulating the transmittance of surface textured transparent conducting oxide (TCO) for Dye-Sensitized Solar Cell (DSSC) application. The simulation based on finite difference time domain (FDTD) was performed using the MatLab software for flat and pyramid surface textured TCO. Fluorine-doped tin oxide (FTO) and indium tin oxide (ITO) were used as TCO material. The transmittance simulation of flat TCO was compared to UV-Vis spectrophotometer measurement of real TCO to ensure the accuracy of the simulation. Then, the transmittance simulation of pyramid surface textures of TCO is higher than a flat one. It suggested that surface texturing enhance the path of light through dispersion and reflectance light by the pattern of the surface. This result indicates that surface textured increasing the transmittance of TCO through a complex light trapping mechanism which might be used to increase the light harvesting for DSSC application.

1. Introduction

Dye-Sensitized Solar Cell (DSSC) is one of type solar cell which based on photoelectrochemical process. The DSSC has a sandwich structure with transparent conductive oxide (TCO) as a glass substrate, TiO_2 as working electrode, dye as an absorbent of light, electrolytes as a medium of exchange electron and Pt or carbon as counter electrode [1].

Recently, the DSSC has an efficiency of 11-12.3%. It smaller than conventional Si solar cell which has an efficiency of 25% [2]. Many methods for increasing the efficiency of DSSC such as modification of TiO_2 and variation of dye have been done. Surface texturing is another method to increase the efficiency by scattering or diffract the incident light so that the optical path length and light absorption within the active layer of the solar cell is increased. Recently, conventional solar cell prepared on periodic surface textures have shown conversion efficiency exceeding of 10% [3]. However, the experimental of surface texturing on DSSC is still rare and need more investigation.

Optical simulation is often used to analyze wave propagation and transmittance within solar cell structure to derive optimal textures [4,5]. In this study, simulation uses finite difference time domain (FDTD) two dimension (2D) method computing by Matlab software. The purpose of this study is comparing transmittance and reflectance of flat and pyramid surface textured of TCO.

2. Method

FDTD is a method to solve Maxwell`s equations problem [6]. In this study, FDTD simulation uses Transverse Magnetic (TM) type. TM has component vector of $\hat{E}_z$, H_x and H_y. With $\hat{E}_z$, H_x and H_y is

electric field in x direction, magnetic field in x and y direction, respectively. Two dimension Maxwell's equation at z = 0 depend on media that used as perfect matched layer (PML) [7,8]:
Where S is PML equation, m is vector constant, and L is the length of PML.

$$\nabla x\vec{E}(\omega) = -j\omega\mu_0[\mu_r][S]\,\vec{H}(\omega) \quad (1)$$

$$\nabla x\vec{H}(\omega) = \sigma\vec{E}(\omega) + j\omega[S]\vec{D}(\omega) \quad (2)$$

$$\vec{D}(\omega) = \varepsilon_0[\varepsilon_r]\vec{E}(\omega) \quad (3)$$

The simulation is dividing into four steps. First, we define parameters of object, such as object constant which used in this simulation as shown in Table 1, velocity of light (c_0), maximum refractive index($nmax$), frequency, wave length (λ), resolution grid and size of object. The source frequency are $375 - 857\ THz$. The object have size of width and length of 6000 nm and 2000 nm, respectively.

Table 1. Object constant of FDTD 2D simulation

Object	N	μ_r	ε_r
Glass	1.46	0.57	3.75
FTO	1.47	1.01	2.14
ITO	1.86	0.99	3.45
Air	1	1	1

Second, calculation of modeling of object, source of Gaussian pulse, PML, and field of the coefficient. Third, we start the simulation for 26182 steps of electric and magnetic field using Maxwell [7,8]. And, the last step is analyzing of an electric field to obtain the reflectance and the transmittance. Figure 1 shows optical layout of a TCO structure and incident of the light source. In this study, we use Indium doped Tin Oxide (ITO), and Fluorine doped Tin Oxide (FTO) as TCO.

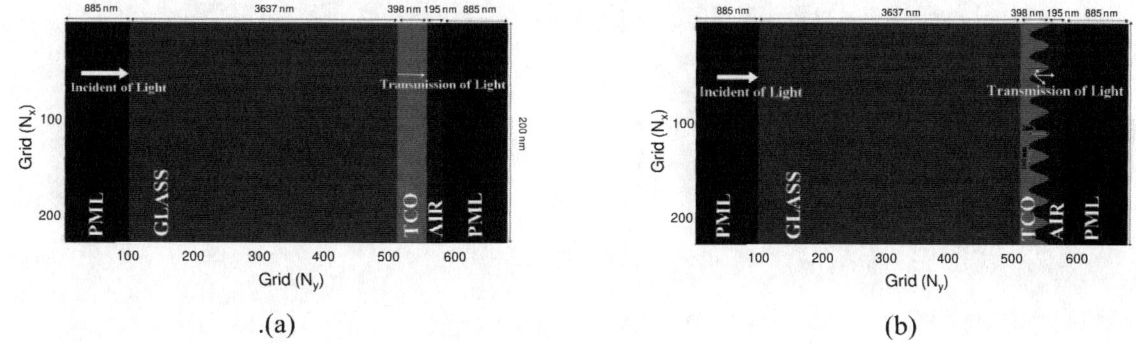

Figure 1. Optical layout structure and incident of light source in (a) flat and (b) pyramid surface textured of TCO

3. **Result and Discussion**

The transmittance of the flat surface of TCO was measured using UV-Vis spectrophotometer. The frequency of incident of light is $375 - 857$ THz. Then, FDTD *2D* simulation uses Gaussian pulse as a source, which is consists of an electric and magnetic field for 26182 steps of propagation.

IOP Conf. Series: Materials Science and Engineering **107** (2016) 012024 doi:10.1088/1757-899X/107/1/012024

The source of Gaussian pulse appears from glass since we consider that glass is not absorbing the incident of light. Gaussian pulse propagates from glass toward to TCO as shown in Figure 1(a) for flat and 1(b) for pyramid surface textured. Pyramid surface textured, showed in Figure 1, have ten pyramids with 200 nm and 292 nm of width and length, respectively. Figure 2(a) and 2(b) show comparison the transmittance of a flat surface as a result of UV-Vis spectrophotometer measurement and simulation of FDTD 2D.

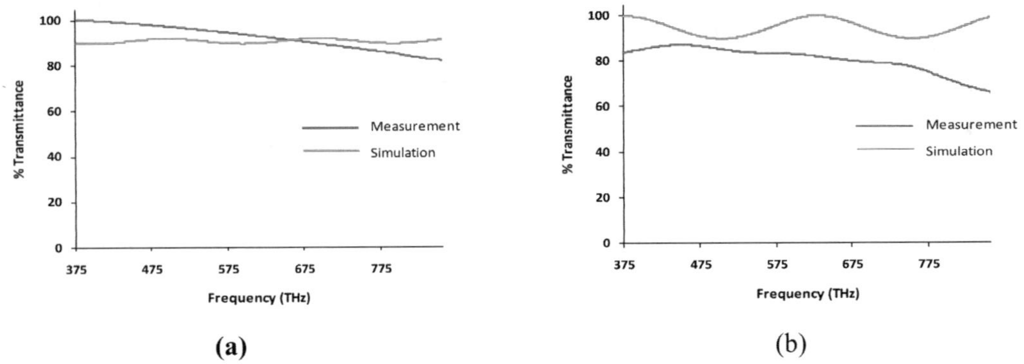

Figure 2. Measurement and simulation transmittance of (a) ITO and (b) FTO

As shown in Figure 2(a) and 2(b), the simulation transmittance not exactly agree with measurement one. In the simulation, we assume that the surface of TCO as the perfectly flat surface. However, in the real TCO, it suggests there is soft roughness caused by a fabrication process which will affect the transmittance.

Then, we compared flat and pyramid surface textured of ITO and FTO. Figure 3(a) and 3(b) show the propagation of light and corresponding transmittance of flat and surface textured ITO, respectively, at 26182 of simulation steps. As a result, the transmittance of pyramid surface textured is higher than flat surface as shown in Figure 4.

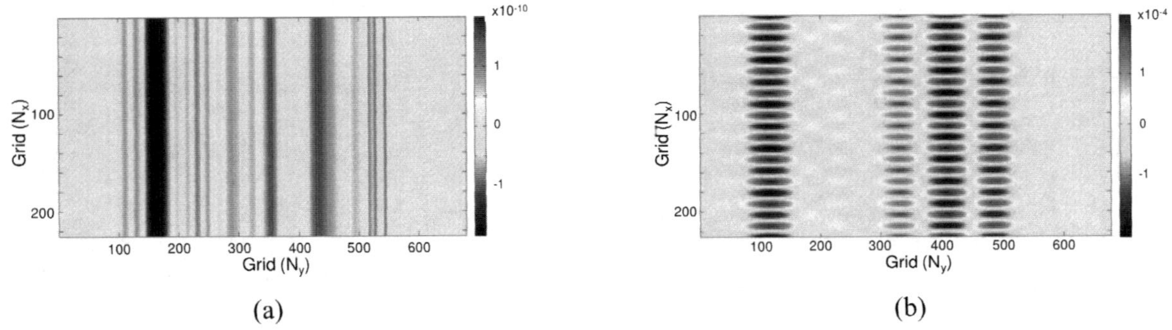

Figure 3. Light propagation in (a) flat and (b) pyramid surface textured of ITO.

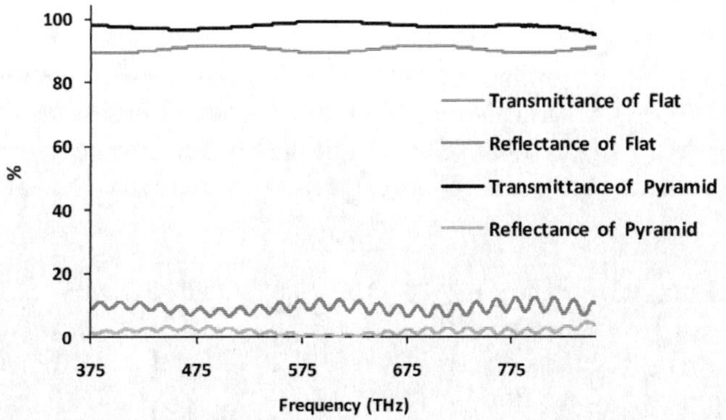

Figure 4. Transmittance and reflectance of flat and pyramid surface textured of ITO

Then, Figure 5(a) and 5(b) show the propagation of light and corresponding transmittance of flat and surface textured FTO, respectively, at 26724 of simulation steps. Similar with ITO, the transmittance of pyramid surface textured of FTO is higher than flat surface as shown in Figure 6. This result indicates that surface texturing on TCO will increase the transmittance which might be caused by scattering in the edge of the texture. The increasing of transmittance is a benefit to enhance the efficiency of DSSC.

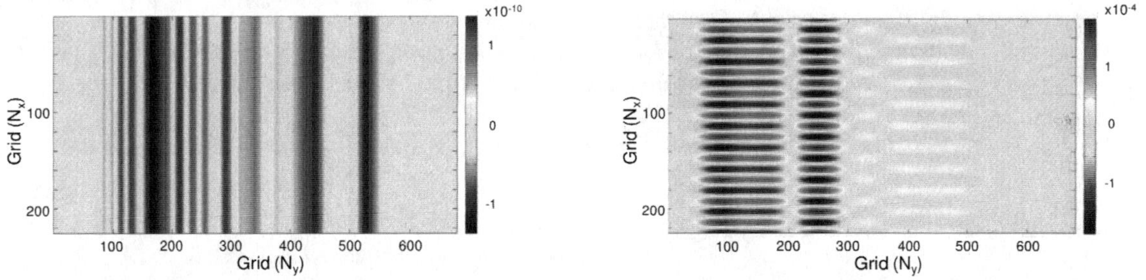

(a) **(b)**

Figure 5. Light propagation of (a) flat and (b) pyramid surface textured of FTO.

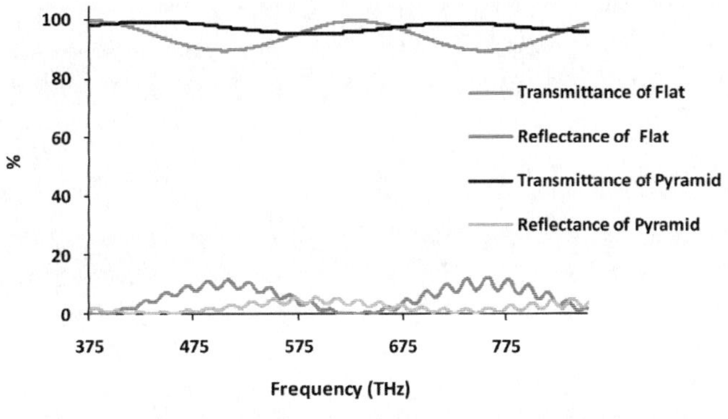

Figure 6. Transmittance and reflectance of flat and pyramid surface textured of FTO

IOP Conf. Series: Materials Science and Engineering **107** (2016) 012024 doi:10.1088/1757-899X/107/1/012024

4. Conclusion

This research investigated the transmittance of flat and pyramid surface textured of TCO using finite difference time domain (FDTD) 2 Dimension (*2D*) method computing by Matlab software. Our result revealed that pyramid surface textured on TCO exhibit higher transmittance than a flat one. Based on the result, pyramid surface textured might be used as a method to increase the efficiency of DSSC.

References

[1] Mona B, Siamak N, Kenneth K S L, Daeyeon L and Masoud S 2013 Theoretical and Experimental Study of a Dye-Sensitized Solar Cell *American Chemical Society*

[2] Green M A, Emery K, Hishikawa Y and Warta W 2011 Solar cell efficiency tables (version 37) *Prog. Photovolt: Res. Appl.* **19** 84–92

[3] Battaglia C, Hsu C M, Soderstrom K, Escarre J, Haug F J, Charriere M, Boccard M, Despeisse M, Alexander D T L,Cantoni M, Cui Y, and Ballif C 2012 Light Trapping Solar Cell: can periodic beat random? ACS Nano **6** (3) 2790-2797

[4] Jäger K, Isabella O, Zhao L and Zeman M 2010 Light scattering properties of surface textured substrates *Phys. Status Solidi C* **7** 945– 948

[5] Kusumandari K and Risa S 2014 Transmittance Simulation of Surface Textured Fluorine Doped Tin Oxide (FTO) for Dye-Sensitized Solar Cell (DSSC) Application *Advances in Physics Research,* 7th International Conference on Physics and Its Applications 2014 (ICOPIA 2014) 86-89

[6] Mohapatra N B and Mohapatra K R 2013 FDTD Optical Simulation To Enhance Light Trapping. *International Journal A Scientific & Engineering Research* **6** 2890-2897

[7] Rathi V, Shrivastava K P and Pokhariya H S 2012 1-D implementation of maxwell's equations in Matlab to study the effect of absorption using pml. *International Journal of Electronics Signals and Systems* (IJESS) **1** 2231-5969

[8] Sullivan M D 2000 *Electromagnetic simulation using the finite difference time domain method.* IEEE Press Inc. New York

Effect of reaction time and (Ca+Mg)/Al molar ratios on crystallinity of Ca-Mg-Al layered double Hydroxide

E Heraldy[1,2*], K D Nugrahaningtyas[1,2], F B Sanjaya[1], A A Darojat[1], D S Handayani[1], Y Hidayat[1,2]
[1]Chemistry Department, Faculty of Mathematics and Natural Sciences, Sebelas Maret University, Jl. Ir. Sutami 36A Kentingan, Surakarta, Indonesia, 57126
[2]Solid State Chemistry & Catalysis, Chemistry Department, Sebelas Maret University, Jl. Ir. Sutami 36A Kentingan, Surakarta, Indonesia, 57126

[*]E-mail: eheraldy@mipa.uns.ac.id

Abstract. Ca-Mg-Al Layered Double Hydroxides (Ca-Mg-Al-LDH) compounds were successfully synthesized from brine water and $AlCl_3.6H_2O$ as the starting materials by coprecipitation method. The product result was characterized by X-ray powder diffraction (XRD) and Fourier transform infrared (FT-IR). The effects of the reaction time and the molar ratios of the raw material on the crystallinity of Ca-Mg-Al-LDH were examining. Results show that increasing reaction time (30; 60 and 90 min.) could improve the crystallinity and monodispersity of layered double hydroxide compounds particles. The well-defined Ca-Mg-Al-LDH could be prepared with (Ca+Mg)/Al molar ratios 0.5.

Keywords: Ca-Mg-Al layered double hydroxide, brine water, reaction time, (Ca+Mg)/Al molar ratios, crystallinity

1. Introduction

Recently, considerable interest has been focused on the synthesis of layered double hydroxides (LDH) or hydrotalcite (HT) is known as one of the minerals of prospective and promising, mainly because it can be synthesized easily and are useful in various applications [1-5].A number of binary, ternary or quaternary LDH compounds could be synthesized via isomorphic substitution of Al or Mg with divalent or trivalent transition metal cations with a similar radius, and it was reported in literature that hydrotalcite-like materials with catalytic properties had been synthesized by introducing various kinds of transition metals or other active metal ions into the brucite-like layer of Mg and Al hydrotalcite. The commonly used methods of preparation of layered double hydroxide compounds at present includes: co-precipitation method, a sol-gel method, hydrothermal method, roasting and recovery method as well as ion exchange method [6].

At present, Mg-Al hydrotalcite has successfully synthesized from brine water as in [7-8]. Meanwhile, Ca-Mg-Al hydrotalcite-like compounds could be prepared by co-precipitation method [9]. With the success as Gao et al. [9] who has the preparation of Ca-Mg-Al-hydrotalcite by combining calcium, magnesium and aluminum ions, brine water utilization (without Ca removal) should be considering for further synthesis.

As one of the most important LDH, Ca-Mg-Al-LDH is a promising material for potential applications in many fields, such as catalyst [9]; fire retardant; and adsorbent [10]. In the present work, we

IOP Conf. Series: Materials Science and Engineering **107** (2016) 012025 doi:10.1088/1757-899X/107/1/012025

employed co-precipitation method to synthesize Ca-Mg-Al-LDH compounds by adjusting (Ca+Mg)/Al molar ratios and variation of reaction temperature.

2. Experiment

2.1. Synthesis procedure

All of the chemicals used in this experiment were analytical grade and were as received without further purification. Brine water sample were taken from Tanjung Jati B CFPP, Jepara, Indonesia. Quantitative analysis of Mg and Ca in brine water using Atomic Absorption Spectrophotometer (AAS) Shimadzu AA 630-12. In the variation of (Ca+Mg)/Al molar ratios, the $AlCl_3.6H_2O$ was added into the water brine solution until (Ca+Mg)/Al molar ratios are 0.5; 1.0 and 2.0 under stirring and then heated at 70 °C for 1 h with the pH of the solution were adjusted to 10. While, in the variation of temperature, the $AlCl_3.6H_2O$ was added into the water brine solution under stirring until (Ca+Mg)/Al molar ratios are 2.0 with the pH of solution pH was adjusted to 10 with 0.1 M Na_2CO_3 solution and then heated to 70 °C for 30; 60 and 90 min. The product was centrifuged to recover the white solid at a speed of 2000 rpm for 20 min. The wet cake was washed to remove chloride ions with aquadest until free of ion Chloride ($AgNO_3$ test). The treated cake then was dried overnight in the oven. To investigate the effects of different temperature on the phase composition of the products, the samples was characterized by X-ray diffraction (XRD). Meanwhile, the Fourier Transform Infrared Spectroscopy (FTIR) was used to analyze functional group.

3. Results and Discussion

3.1. XRD analysis

Figure 1 showed the XRD pattern of the prepared sample at various time at 30; 60 and 90 min.; (Ca+Mg)/Al molar ratios 2.0; pH 10 and temperature at 70 °C; while Figure 2 shows the XRD pattern of as-prepared samples at different (Ca+Mg)/Al molar ratios 0.5; 1.0 and 2.0 ; pH 10 at 70 °C for 1 h. As we could see from Figure 1 and Figure 2, all samples appeared strong diffraction peaks at low angles (2θ = 11°, 23°, and 35° corresponding to the *(003)*, *(006)* and *(009)* reflections) [8]. While the samples appeared weak diffraction peaks at high angles (2θ = 39°, 47°, and 61° corresponding to the *(105)*, *(108)*, and *(110)* reflection. The XRD pattern of the Ca-Mg-Al hydrotalcite consists of both sharp and symmetrical peaks with some asymmetrical peaks at a high angle, indicating good crystallinity [11]. The diffraction peak near 61.0° corresponds to the *(110)* crystal plane and also the existence of diffraction peaks at angles between 61.0 to 62.0° indicates that the interlayer of hydrotalcite is carbonate anionic. It was confirmed by the results of the study as in [12]-[15]. Compared to the JCPDS standard pattern, the synthetic samples had hydrotalcite-like structure. All reflection peaks of different products can be easily indexed as d_{003}, d_{006} and d_{009}that are consistent with the literature value Mg/Al hydrotalcite (JCPDS Card No. 89-0460) and d_{003}, d_{006} and d_{132} Ca/Al hydrotalcite (JCPDS Card No. 87-0493).The basal spacing of all samples is listing in Table 1.

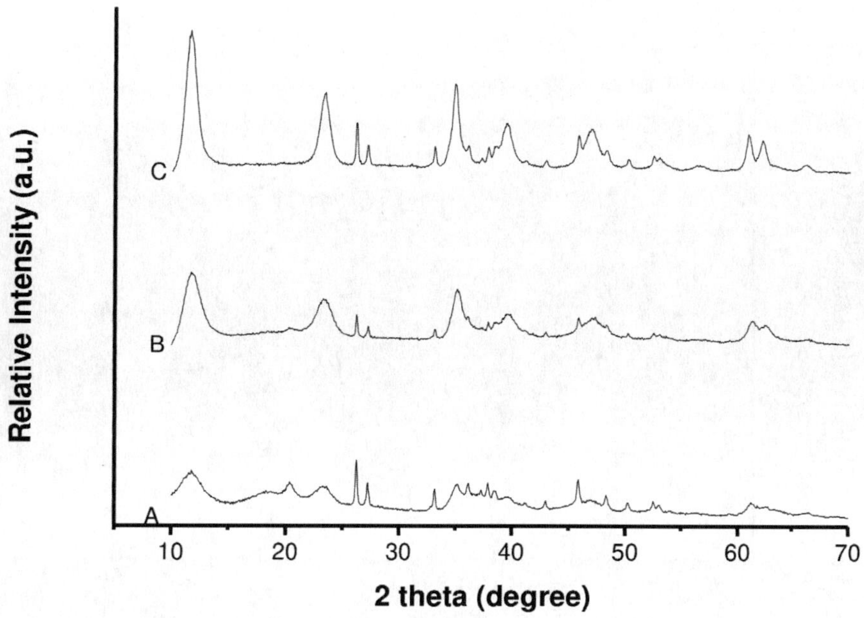

Figure 1. XRD patterns of the samples Ca-Mg-Al-LDH with different reaction time (A) 30; (B) 60 and (C) 90 min. at pH 10, (Ca+Mg)/Al molar ratios 2.0 and temperature 70 °C.

Table 1. The basal spacing of all samples.

Basal	**Reaction time (min.)**			**(Ca+Mg)/Al molar ratio**			**Mg/Al**	**Ca/Al**
spacing	**30**	**60**	**90**	**0.5**	**1.0**	**2.0**	**hydrotalcite**	**hydrotalcite**
d_{003}	7.52 Å	7.59 Å	7.63 Å	7.63 Å	7.63 Å	7.59 Å	7.63 Å	7.60 Å
(2θ)	(11.8°)	(11.7°)	(11.7°)	(11.6°)	(11.6°)	(11.7°)	(11.6°)	(11.7°)
d_{006}	3.89 Å	3.90 Å	3.88 Å	3.87 Å	3.87 Å	3.90 Å	3.87 Å	3.86 Å
(2θ)	(23.3°)	(23.3°)	(23.4°)	(23.4°)	(23.4°)	(23.2°)	(23.4°)	(23.4°)
d_{009}	2.67 Å	2.68 Å	2.70 Å	2.67 Å	2.68 Å	2.68 Å	2.67 Å	-
(2θ)	(35.2°)	(35.1°)	(34.9°)	(35.1°)	(35.0°)	(35.0°)	(35.4°)	-
d_{132}	2.48 Å	2.48 Å	2.43 Å	2.48 Å	2.48 Å	2.48 Å	-	2.48 Å
(2θ)	(38.5°)	(38.4°)	(39.4°)	(38.5°)	(38.5°)	(38.5°)	-	(38.5°)

The characteristic diffraction peaks in Figure 1 became stronger and narrower with the prolonging of reaction time, indicating that crystalline structure of the samples became more perfect. Also, Figure 2 showed that the diffraction peaks were broadened gradually and weakened with an increase of aluminum content, which implying that the structure regularity worsened and the crystallinity decreased. These may be attributing to the trivalent metal cation in a layered structure having a relatively large polarization, which resulted in difficulties in directional alignment. Both the reaction time 60 and 90 min.and(Ca+Mg)/Al molar ratio 0.5 and 2.0 agrees well with the empirical phase composition suggested by the XRD analysis that was confirming according to the references [12]-[13]. However the reaction time 30 °C and (Ca+Mg)/Al molar ratio 1.0, a little of the other hydrotalcite phase generated in the product. It is suspected to be $Mg(OH)_2$ and $Al(OH)_3$.

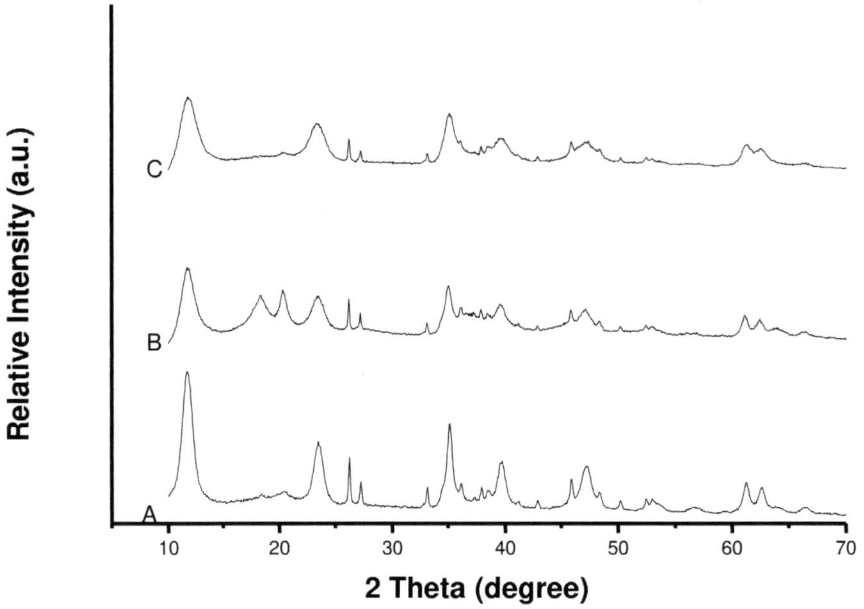

Figure 2. XRD patterns of the samples Ca-Mg-Al-LDH with different (Ca+Mg)/Al molar ratios (A) 0.5; (B) 1.0 and (C) 2.0 at pH 10, temperature 70 °C for 1-hour reaction.

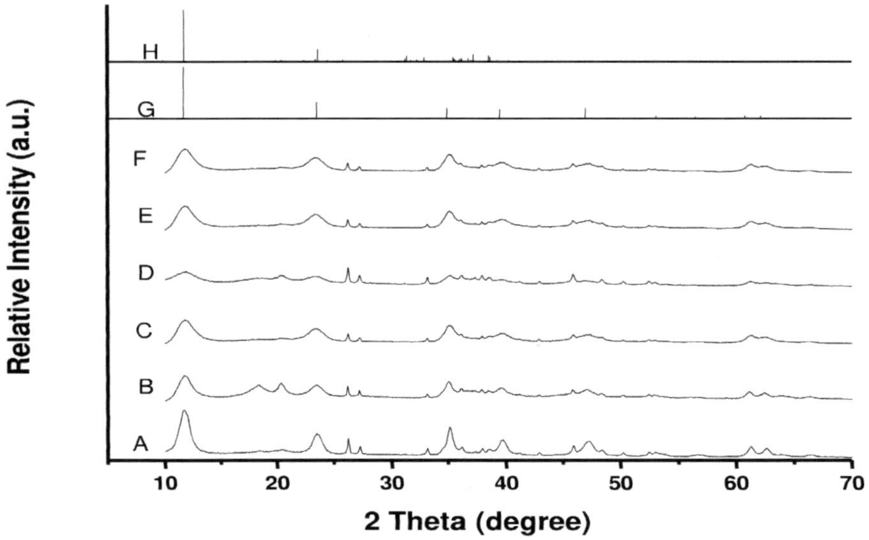

Figure 3. XRD patterns of the samples Ca-Mg-Al-LDH with (Ca+Mg)/Al molar ratios (A) 0.5; (B) 1.0; (C) 2.0 at pH 10, temperature 70 °C for 1 hour reaction and different reaction time (D) 30; (E) 60; (F) 90 min.in comparison with (G) Mg/Al hydrotalcite and (H) Ca/Al hydrotalcite.

3.2. FTIR analysis

Figure 4 and Figure 5 illustrates the FT-IR spectra of Ca-Mg-Al hydrotalcite with carbonate anion interlayer and Table 2 show a comparison of the functional group the hydrotalcite products. As seen from Figure 4 and Table 2 shows the broad absorption bands around 3500-3000 cm^{-1} were derived from the –OH stretching vibration of the hydroxyl groups in layer structures, which shifted towards low wave number in comparison with –OH of the free state, implying the presence of the hydrogen bonds between interlayer water and layer –OH [14,16-19]. The absorption bands at near 1627-1643 cm^{-1} were originated by the bending mode of an interlayer water molecule with anion interlayer [7,16,17,19-21]

All FT-IR spectra of LDH display the bands at 1361-1364 and 1633-1650 cm^{-1} are due to the antisymmetric C-O and C=O stretching vibration mode of interlayer carbonate that also confirmed according to the references [7,14,16,21]. Furthermore, in the low wavenumber region between 400 cm^{-1} and 1000 cm^{-1} at around 677 cm^{-1} show the O=C-O carbonate bending vibration. Peak around 551 cm^{-1} which is associated with stretching vibration Al-O [6,16] and around 448 cm^{-1} as stretching vibration Mg-O. While the appearance of peaks around 449-466 cm^{-1} and 538-561 cm^{-1} that are suspected as stretching vibration Mg-O, Ca-O and Al-O of products. It is corresponding as in Zhou et al. [19] that mention peaks at around 428-433 cm^{-1} are stretching vibration M-O (where M=Mg, Ca and Al). Also, the appearance of the new peaks at 848-858 indicated Ca-O bonding that are consistent according to [19] and [22] who have varied Ca and Mg in synthesis hydrotalcite. The more Ca ions in the synthesis due to new peaks will appear around 800 cm^{-1}. Therefore, from spectra analysis in Figure 4 was shown Mg-O, Ca-O, Al-O bonding, hydroxyl, C=O, and C-O indicated that the product is Ca-Mg-Al hydrotalcite with carbonate anion interlayer.

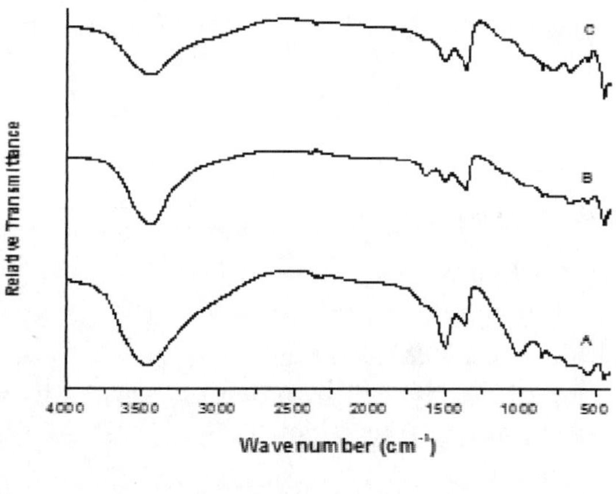

Figure 4. The FT-IR spectra of hydrotalcite with variation reaction time (a) 30; (b) 60 and (c) 90 min.

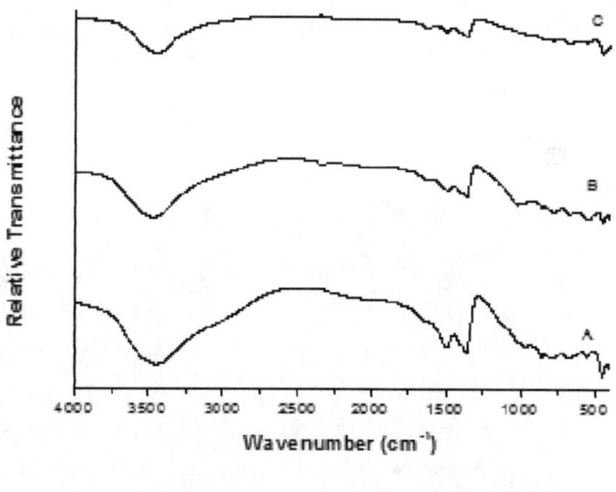

Figure 5. The FT-IR spectra of hydrotalcite with (Ca+Mg)/Al molar ratios (a) 0.5; (b) 1.0 and (c) 2.0.

Table 2. The functional group of all samples.

Functional group	Reference	Reaction time (min.)			(Ca+Mg)/Al molar ratio		
		30	60	90	0.5	1.0	2.0
OH stretching	3300-3700 a,b,c,d,e	3468	3442	3442	3446	3469	3442
OH/C=O bending	1633-1650 a,b,c,d,e,f	1631	1627	1641	1782	1633	1627
C-O stretching	1350-1384 a,b,c,d,e,f	1365	1367	1359	1361	1361	1363
O=C-O bending	650-690 b,c,e,f	673	677	667	677	667	667
M-O stretching	415-565 a,b,c,d,e,f	445	451	449	451	449	451
		551	561	557	555	551	561

Reference: [a][8], [b][14], [c][16], [d][17], [e][20], [f][21]

4. Conclusions

In summary, Ca-Mg-Al-LDH has been synthesized using brine water and $AlCl_3.6H_2O$ as the starting materials by coprecipitation method, and some factors on the synthesis were examined. The results show that the prolonging reaction time at a pH 10 and temperature 70 °C is improving the crystallinity and monodispersity of layered double hydroxide compounds.Moreover, well-defined Ca-Mg-Al-LDH could be synthesized under the (Ca+Mg)/Al molar ratios 0.5 and 2.0.

Acknowledgments

We would like to thank Directorate General of Higher Education Ministry of National Education Indonesia for MP3EI research grants.

References

[1] Kameda T, Yoshioka T, Uchida M, and Okuwaki A 2000 *Mol. Cryst. Liq. Cryst.* **341** 407
[2] Orthman J, Zhu H Y, and Lu G Q 2003 *Sep. Purif. Technol.* **31** 53
[3] Lazaridis N K 2003 *Water Air Soil Poll.* **146** 127
[4] Tong Z, Shichi T, and Takagi K 2003 *Mater. Lett.* **57** 2258
[5] Santosa S J, Kunarti E S and Karmanto 2008 *Appl. Surf. Sci.* **254** 7612
[6] Zhang M, Gao Y, Li L 2011 *Advanced Materials Research* **287-290** 569
[7] Heraldy E, Triyono, Santosa S J, Wijaya K, Prasasti D, Savitri G, Hermawan D, Iwa R, and Suryo E H 2009 *Proc. Inter. Conf. Chem. Bio. and Env. Eng.* (Singapore: Published World Scientific Publishing Co. Pte. Ltd) p. 246
[8] Heraldy E, Triyono, Santosa S J and Wijaya K 2011 *Makara Sains* **15**(1) 9
[9] Gao L, Teng G, Lv J, and Xiao G 2010 *Energy Fuel* **24** 645
[10] Zhou J Z, Xu Z P, Qiao S, Liu J, Liu Q, Xu Y, Zhang J and Qian G 2011 *Appl. Clay Sci.* **54** 196
[11] Ulibarri M A, Pavlovic I, Barriga C, Hermosin M C and Cornejo J 2001 *Appl. Clay Sci.* **18** 17
[12] Johnson C A and Glasser F P 2003 *Clay. Clay Miner.* **51**(1) 1
[13] Kameda T, Yoshioka T, Yabuuchi F, Uchida M and Okuwaki A 2007 *J. Mat. Sci.* **42** 2194
[14] Yang Z, Choi K, Jiang N, and Park S 2007 *Bull.Korean Chem.* **28** 11
[15] Lal J, Sharma M, Gupta S, Parashar P, Sahu P, and Garwal D D 2012 *J. Mol. Catal. A: Chem.* **352** 31
[16] Hickey L, Kloprogge J T and Frost R L 2000 *J. Mater. Sci.* **35** 4347
[17] Sharma S K, Kushwaha P K, Srivastava V K, Bhatt S D, and Jasra R V 2007 *Ind. Eng. Chem. Res.* **46** 4856

[18] Zeng H Y, Wang Y-J, Feng Z, You K-Y, Zhao C, Sun J W and Liu P-L 2010 *Catal. Lett*, **137** 94
[19] Zhou J, Xu Z-X, Qiao S, Liu Q, Xu Y and Qian G 2011 *J. Hazard. Mater.* **189** 586
[20] Kloprogge J T, Hickey L, and Frost R L 2002 *J. Mat. Sci. Lett.* **21**(8) 603
[21] Wang S, Huang J and Chen F 2012 *BioResources* **7**(1) 997
[22] Gupta S, Agarwal D, and Banerjee S 2008 *Indian J. Chem.* **47A** (7) 1004

IOP Conf. Series: Materials Science and Engineering **107** (2016) 012026 doi:10.1088/1757-899X/107/1/012026

Intercalation of diclofenac in modified Zn/Al hydrotalcite-like preparation

E Heraldy[1,2*], R W Suprihatin[1], Pranoto[1]
[1]Chemistry Department, Faculty of Mathematics and Natural Sciences, SebelasMaret University, Jl. Ir. Sutami 36A Kentingan, Surakarta, Indonesia, 57126
[2]Solid State Chemistry & Catalysis, Chemistry Department, SebelasMaret University, Jl. Ir. Sutami 36A Kentingan, Surakarta, Indonesia, 57126

[*]E-mail: eheraldy@mipa.uns.ac.id

Abstract. The intercalation of a pharmaceutically active material diclofenac into modified Zn/Al Hydrotalcite-like (Zn/Al HTlc) preparation has been investigated by the coprecipitation and ion exchange method, respectively. The synthetic materials were characterized using X-Ray Diffraction (XRD); Fourier transforms infrared spectroscopy (FTIR); Scanning Electron Microscope (SEM); X-Ray Fluorescence (XRF) and surface area analyzer. The results show that the basal spacing of the product was expanded to 11.03 Å for direct synthesis and 10.68 Å for indirect synthesis, suggesting that diclofenac anion was intercalated into Zn/Al HTlc and arranged in a tilted bilayer fashion and the specific surface area of material increased after the intercalation of diclofenac.

1. Introduction

Recently, sustained release of drug for more effective delivery systems has received intense attention from the pharmaceutical industry due to the advantages they offer over conventional forms [1]. These materials can include colloids, gels, copolymers, porous media, and inorganic layered materials such as clays. These latter inorganic materials can exist in great variety and possess well-defined, ordered inter lamellae space potentially accessible by foreign species. This ability allows them to act as matrices or hosts for organic molecules, yielding interesting hybrid nanocomposite materials [2]. Among host–guest hybrid materials, synthetic hydrotalcite-like compounds (HTlcs) have emerged as promising delivery systems for efficient transport and release of pharmaceuticals [3]-[4].There have been many studies focusing on the application of hydrotalcite-like (HTlc) as a host for intercalation chemistry [5]-[8]. In addition, HTlc has demonstrated their suitability for different processes in the pharmaceutical and cosmetics industries, as they can absorb or host a large number of active species [9]-[10]. For instance, Carretero et al. [11] and Frunza et al. [12] have reported the intercalation of salicylate in ZnAl-hydrotalcite, and this matrix has been also used in the cosmetic field to prepare new sunscreen products [13]-[16]. Furthermore, MgAl hydrotalcite-like compounds have found application in medicinal chemistry as antacids [17]-[19], and as hosts for intercalation of molecular drug anions with the aim to study these intercalation compounds for storage, transport and ultimately controlled release of drugs [9]; [15]-[16]; [20]-[23]; [24]-[30];.

Layered double hydroxides (LDHs), which are referred to as hydrotalcite-like compounds or as anionic clays, are an important class of lamellar ionic solids [31]. They can be described by the general formula: $[M_{1-x}{}^{2+} M_x{}^{3+} (OH)_2]^{x+} (A^{m-})_{x/m}.nH_2O$, where M^{II} are divalent cations, M^{III} are trivalent cations and A^{m-} is an exchangeable anion with charge (m^-). They have been widely investigated owing to their potential applications as ion exchangers, catalyst supports and so on [32]-[33]. Recently, LDHs have been investigated for storage and delivery of some drugs, such as NSAIDs (nonsteroidal anti-inflammatory drugs) widely used in rheumatism treatment. Similar to another NSAIDs, Diclofenac (DIK) use is associated with rare, but severe and sometimes fatal, gastrointestinal (GI) side effects, including ulceration, and hemorrhage, so it is an ideal candidate for incorporation in a controlled release device to diminish its adverse effects after oral administration.

Although there are a few studies on intercalated Diclofenac into Mg–Al–Cl–LDHs [21], Mg–Al–CO_3–LDHs [24], and Zn-Al-CO_3-LDHs by coprecipitation method [34], little is known in the literature on diclofenac- Zn/Al HTlc which is made by coprecipitation and anion exchange method. In this paper, we present and characterization of intercalation of diclofenac in modified Zn/Al hydrotalcite-like preparation.

2. Materials and methods

2.1. Materials
Zinc nitrate nonahydrate and aluminum nitrate tetrahydrate were purchased from Merck (Germany). The Diclofenac sodium salt (Na-DIK), was supplied by IFARS Company (Karanganyar, Indonesia).

2.2. Synthesis of Zn/Al Hydrotalcite-like (Zn/Al HTlc)
The Zn/Al HTlc precursor was synthesized by the coprecipitation method from a mixed metal nitrate salt solutions with Zn/Al ratio of 2:1 at pH 7. The solution was stirred for 30 min, heated 120 °C for 5 h in the oven. The resulting white slurry was filtered and washed with water (CO_2 free). The product was dried at 80 °C and reserved for characterization.

2.3. Diclofenac intercalation to Zn/Al HTlc
Direct synthetic by coprecipitation

To distilled water (CO_2 free) were added $Zn(NO_3)_2.4H_2O$ and $Al(NO_3)_3.9H_2O$. The pH system was maintained at pH 7 by the addition of NaOH solution slowly and then added diclofenac sodium 0.1 M. The resulting system (HTlc-DIK) was stirred and was aged at 70 °C for 18 h. After aging, the particles were recovered by centrifuging (3000 rpm for 10min). The product was washed and dried at 70 °C for 5 h and reserved for characterization.

Indirect synthetic by anion exchange

In this method, 78 mL of a 50% v/v methanol solution (50% v/v) was transferred to a reaction flask containing Na-DIK. Intercalation reactions were performed after adding 1 g of Zn/Al hydrotalcite-like into the mixture at 60 °C for three days. After cooling, the mixture was centrifuged at 3000 rpm for 10 min. The residue was washed three times with carbon dioxide-free water and finally dried at room temperature and reserved for characterization.

2.4. Characterization
Powder X-ray diffraction (PXRD) patterns were obtained on a Bruker D8 Advance diffractometer, using CuKα radiation. The instrument was set at operating Voltage 35 KV and a current of 40 mA in 2Θ scale 2-70° range. Identification of the crystalline phases was made by comparison with JCPDS files and literature data. The Fourier-transform infrared (FT-IR) spectra were obtained using a Shimadzu IR Prestige-21 by the standard KBr disc method, in the 4000 to 400 cm^{-1} range. Elemental chemical analysis for Zn and Al were carried out in a Bruker X-Ray Fluorescence Bruker S2 Ranger. A Scanning Electron Microscope (SEM-EDX) FEI type Inspect S50 was used to study the surface morphology of the samples. The gas surface area analyzer was performed by Quantachrome Corporation NOVA Win 1200e.

3. Results and Discussion

3.1. XRD analysis

XRD patterns of the products demonstrate successful intercalation of diclofenac into both Zn/Al HTlc hosts by coprecipitation and anionic exchange (figure 1A, B, C respectively). They show the formation of, at least, a well crystallized hydrotalcite-like, with basal diffractions due to (*00l*) basal planes, thus confirming a lamellar structure. Before intercalation, the *d*- spacing of inorganic layer Zn/Al HTlc, is termed as initial basal spacing (8.81Å at 2Θ = 10.04). After intercalation, the basal spacing was expanded. An increase of the basal spacing leads to a shift of the diffraction peak at the lower angles. This expansion is reflected by the *d*-spacing values of the XRD patterns, which are 16.14; 11.63; 8.75 and 7.60 Å, respectively for intercalation diclofenac by coprecipitation method (HTlc-DIK-DS), and 12.35; 11.30; 10.21 and 8.84 Å, respectively for intercalation diclofenac by anion exchange method (HTlc-DIK-AE). As a result, a basal spacing value of 11.03 Å for HTlc-DIK-DS and 10.68 Å for HTlc-DIK-AE was obtained. In both products, the XRD patterns show some overlapping peaks. This result indicates the presence of some unreacted Zn/Al-NO_3-HTlc remains in the products. Previous studies have indicated that it is hard to exchange nitrate in HTlc with the incoming anions, due to the parallel orientation of nitrate ions with respect to the hydroxide layers. As a result, some nitrate anions remained after the intercalation process of diclofenac into Zn/Al HTlc, and this explained why some unreacted Zn/Al-NO_3-HTlc remained [35]. A slight discrepancy in the d_{003} value between HTlc-DIK-DS and HTlc-DIK-AE is probably mainly due to the content of water in the interlayer galleries and the presence of nitrate ions in the interlayer region [36].

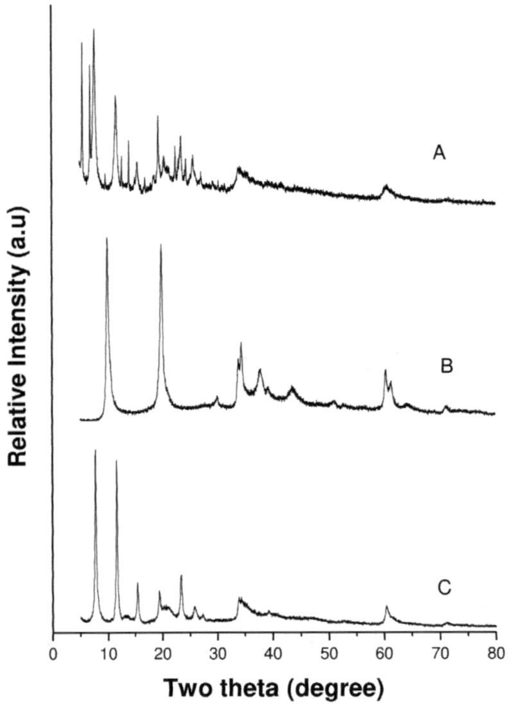

Figure 1. Powder XRD patterns of (A) HTlc-DIK DS, (B) Zn/Al HTlc, (C) HTlc-DIK AE

3.2. FTIR analysis

The FT-IR spectra of the drug-intercalated products are similar for all series. figure 2 show FTIR spectra of the Zn/Al HTlc, HTlc-DIK-DS, and HTlc-DIK-AE. The FTIR spectra show characteristic bands of diclofenac. The spectra show the absorption bands characteristic of the functional groups existing in the intercalated molecule. The spectra shows typical bands of the layered components, composed of absorption bands due to normal vibration modes corresponding to layer hydroxyl groups, water molecules, M–O and M–O–M′ stretching vibrations. The broadband around 3430 cm^{-1} is due to the stretching mode of the hydroxyl groups existing in the layers; broadening of this band is due to the formation of hydrogen bonds. The aromatic ring is responsible for the very weak C–H stretching mode bands just above 3000 cm^{-1}. The aliphatic C–H groups give rise to the weak absorption at 2950 cm^{-1} [37]. It indicates that the diclofenac anions have been intercalated into the interlayer galleries of the Zn/Al HTlc. However, some of the bands are slightly shifted in position, due to the interaction between diclofenac anions and the inorganic HTlc host interlayers. Table 1 shows the FTIR assignments for HTlc-DIK-DS and HTlc-DIK-AE. An intense bands 1575-1578; 1504-1505; 1451-1452 and 1385-1400 cm^{-1} of the HTlc-DIK-DS and HTlc-DIK-AE are attributed to carboxylate stretching of the diclofenac anions, respectively. Peaks at 426-430 $cm^{-1}$are associated with M-O stretching modes in the Zn/Al HTlc; HTlc-DIK-DS and HTlc-DIK-AE [38].

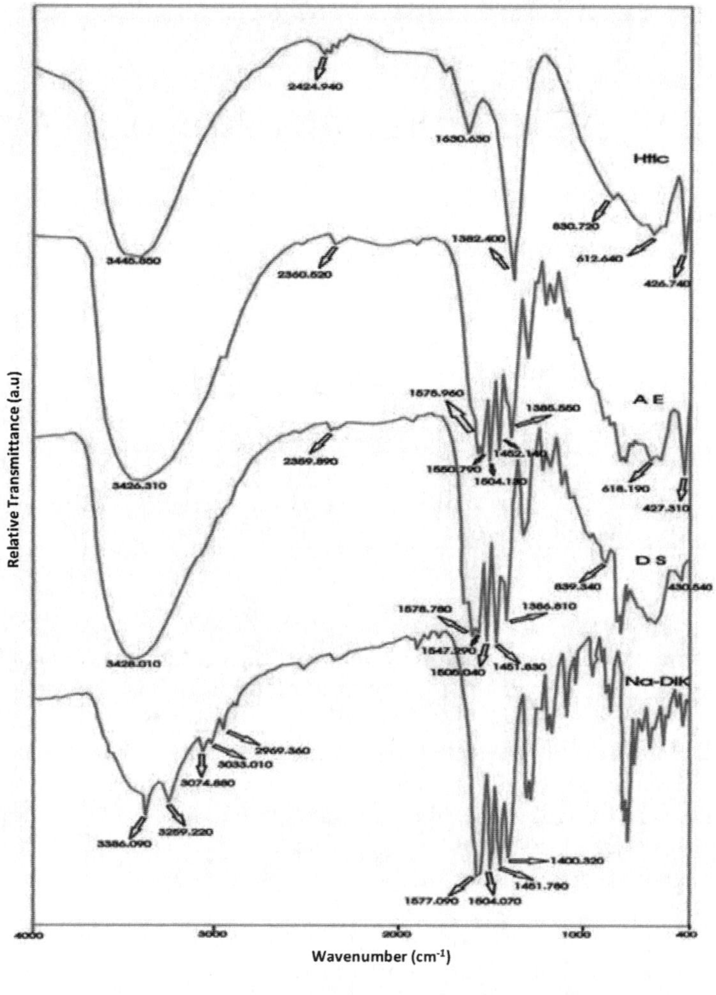

Figure 2. FTIR spectra of Zn/Al HTlc (A); HTlc-DIK-AE (B); HTlc-DIK-DS (C) and Na-DIK (D)

Table 1. The FTIR assignments of HTlc-DIK-AE and HTlc-DIK-DS

Functional group	Wavenumber (cm^{-1})			
	Zn/Al HTlc	HTlc-DIK-AE	HTlc-DIK-DS	Na-DIK
OH stretching	3445.85	3426.31	3428.01	3386.09
NO_3	1382.40; 612-830	618.19	839.34	-
Carboxylate stretching		1451-1452; 1385-1400	1575-1578; 1504-1505	1504-1577; 1400-1451
M-O stretching	426.74	427.31	430.54	-

3.3. Elemental Analysis

Elemental analysis was done for the inorganic composition of HTlc-DIK-DS and HTlc-DIK-AE, as shown in Table 2. As expected, the HTlc-DIK-DS and HTlc-DIK-AE contained inorganic constituents from diclofenac. Both elemental analysis and XRD study indicate that intercalation occurred in which diclofenac was intercalated into the HTlc inorganic interlayers.

Table 2. An elemental analysis of the samples

Samples	ZnO (%)	Cl (%)	Al_2O_3 (%)	Na_2O (%)	Trace Total (%)
Zn/Al HTlc	78.71	0.38	17.30	-	3.28
HTlc-DIK-DS	51.45	29.86	16.26	-	2.33
HTlc-DIK-AE	43.79	13.93	13.28	13.22	15.46

3.4. Specific Surface Area Analysis

The specific surface area of Zn/Al HTlc; HTlc-DIK-DS and HTlc-DIK-AE were also characterized. The calculated B.E.T. surface areas of this sample are shown in Table 3.

Table 3. The surface area of samples

Sample	S_{BET} ($m^2.g^{-1}$)	average (Å)	T_{otal} ($10^{-3}.m^3.g^{-1}$)
Zn/Al HTlc	0.752	90.528	3.406
HTlc-DIK-DS	15.427	18.174	14.019
HTlc-DIK-AE	6.268	15.347	4.810

Table 3 shows the comparison of the surface properties of the material Zn/Al HTlc, which are intercalated with both methods. Besides an intercalation of diclofenac into the area between the layers, it is possible there is also the mechanism of adsorption on metal hydroxide layers (brucite-like) Zn/Al hydrotalcite-like.

In addition, the diclofenac will enter into the pores, causing the inside of the pores was covered by diclofenac molecule. By the closing of the pores of the brucite-like due to a decrease in the average pore radius.It is demonstrated by the data of the average pore radius Zn/Al HTlc amounted to 90.528 Å while intercalation decreased to 18.174 Å on the direct synthesis method by coprecipitation and 15.347 Å on anion exchange methods.

Besides the effect on the decrease in the average pore radius, intercalation of diclofenac will increase the total pore volume. It is evidenced by the increase in value the total pore volume of 3.406 into 14.019×10^{-3} $cm^3.g^{-1}$ on the direct synthesis method and 4.810×10^{-3} $cm^3.g^{-1}$ anion exchange methods. In addition, increasing the value of the total pore volume will affect the brucite-like surface area. The intercalation by directly synthesis method due to increased surface area of 0.752 $m^2.g^{-1}$ to 15.427 $m^2.g^{-1}$ while the intercalation with anion exchange methods increases to 6.268 $m^2.g^{-1}$. A greater increase of pore surface area in the direct synthesis method, suggesting that blending the diclofenac into the Zn/Al HTlc is preferred in this method rather than anion exchange methods.

3.5. Surface Characterization

The samples prepared have been also studied by SEM; some micrographs are included Zn/Al HTlc; HTlc-DIK-DS and HTlc-DIK-AE are shown in figure 3. A similar morphology is observed for both samples. Aggregated lamellar particles can be observed, although somewhat loose. All samples show non-uniform, irregular agglomerates of compact and non-porous plate-like structure. It is the usual morphology of hydrotalcite, described as a result of stacking of hexagonal layers.

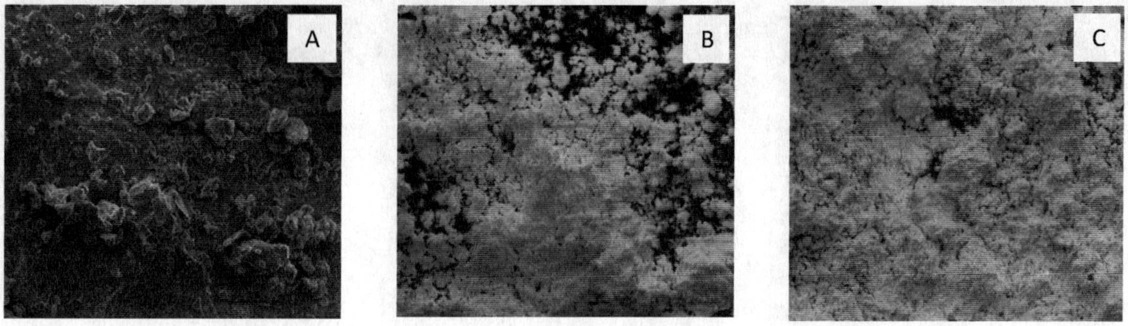

Figure 3. Scanning electron micrographs of Zn/Al HTlc (A); HTlc-DIK-DS (B) and HTlc-DIK-AE (C)

4. Conclusions

Intercalation of diclofenac into modified Zn/Al hydrotalcite-like preparation using coprecipitation and anion exchange method to obtain HTlc-DIK-DS and HTlc-DIK-AE was accomplished well-crystallized solids, with basal spacings between 10 and 12Å. The XRD results confirmed the intercalation process in which diclofenac is arranged between the interlayers as tilted bilayers. Intercalation is further confirmed by FT-IR spectroscopy. FTIR study shows that the bands of the resulting product samples correspond to the characteristic functional groups of the diclofenac and HTlc structures, which support the intercalation process, although slightly shifted from their original positions due to the interactions between the layers.

References

[1] Kumar M 2000 *J. Pharm. Pharm. Sci.* **3** 234
[2] Choy J-H, Oh J M and Biswick T T 2009 *J. Mater. Chem.* **19** 2553
[3] Choi S J, Oh J M, and Choy J H 2009 *J. Inorg. Biochem.* **103** 463
[4] Oh J M, Choi S J, Lee G E, Kim J-E and Choy J H 2009 *Chem. Asian. J.*, **4** 67
[5] Williams G R, Dunbar T G, Beer A J, Fogg A M and O'Hare D 2006 *J. Mater. Chem.* **16** 1222
[6] Marangoni R, Ramos L P and Wypych F 2009 *J. Coll. Interf. Sci.* **330** 303
[7] Hayashi A, Yoshikawa Y, Ryu N, Nakayama H, Tsuhak M and Eguchi T 2008 *Phosphorous Research Bulletin* **22** 48
[8] Gasser, M S 2009 *Colloids and Surfaces B: Biointerfaces* **73** 103
[9] Del Arco M, Gutiérrez S, Martín C, Rives V and Rocha J 2004 *J. Solid State Chem.* **177** 3954
[10] Del Hoyo C 2007 *Appl. Clay Sci.* **36** 103
[11] Carretero M I, Lagaly G and Showalter M 2007 *Appl. Clay Sci.* **36** 1
[12] Frunza M, Popa M I and Hulea V 2007 *J. Optoelectron. Adv. Mater.* **9** 3376

[13] Perioli L, Ambrogi V, Bertini B, Ricci M, Nocchetti M, Latterini L and Rossi C 2006a *Eur. J. Pharm. Biopharm.* **62** 185
[14] Perioli L, Ambrogi V, Rossi C, Latterini L, Nocchetti M, Costantino U 2006b *J. Phys. Chem. Solids* **67** 1079
[15] Perioli L, Nocchetti M, Ambrogi V, Latterini L, Rossi C and Costantino U 2008 *Microporous Mesoporous Mater.* **107** 180
[16] Costantino U, Nocchetti M, Ambrogi V and Perioli L 2008 *Microporous Mesoporous Mater.* **107** 149
[17] Lin M S, Sun P and Yu H Y 1998 *J. Formos.Med. Assoc.* **97** 704
[18] Linares C F, Solano S and Infante G 2004 *Microporous Mesoporous Mater.* **74** 105
[19] Konturek J W, Beneke M, Koppermann R, Petersen-Braun M and Weingartner U 2007*Med. Sci. Monit.***13** 44
[20] Ambrogi V, Fardella G, Grandolini G and Perioli L 2001 *Int. J. Pharm.* **220** 23
[21] Ambrogi V, Fardella G, Grandolini G, Perioli L and Tiralti M C 2002 *AAPS Pharm. Sci.Technol.* **3**(3) article 26
[22] Ambrogi V, Fardella G, Perioli L, Nocchetti M and Grandolini G 2003 *J. Pharm. Sci* .**92** 1407
[23] Choy J H, Hwang S H and HanY S 2001 *Bull. Korean Chem. Soc.* **22** 1019
[24] Dupin J C, Martinez H, Guimon C, Dumitriu E and Fechete I 2004 *Appl. Clay Sci.* **27** 95
[25] Li B, He J, Evans D G and Duan X 2004 *Appl. Clay Sci.* **27** 199
[26] Mohanambe L and Vasudevan S 2005 *J. Phys. Chem. B* **109** 15651
[27] Williams G R and O'Hare D 2006 *J. Mater. Chem.* **16** 3065
[28] Rives V, Del Arco M, Fernández A and Martín C 2009 *Appl. Clay Sci.* **42** 538
[29] Costantino U, Nocchetti M, Sisani M and Vivani R 2009a *Z. Kristallogr.* **224** 273
[30] Costantino U, Nocchetti M, Tammaro L and Vittoria V 2009b *Appl. Clay Sci.* **43** 350
[31] Cavani F, Trifirro F and Vaccari A 1991 *Catal. Today* **11**(2) 173
[32] Carpani I, Berrettoni M, Ballarin B, Giorgetti M, Scavetta E and Tonelli D 2004 *Solid State Ionics* **168** 167
[33] Terry P A 2004 *Chemosphere* **57** 541
[34] Cao F, Wang Y, Ping Q and Liao Z 2011 *Int. J. Pharm.* **404** (1–2) 250
[35] Chao Y F, Chen P C and Wang S L 2008 *Appl. Clay Sci.* **40** 193
[36] Wei M, Yuan Q, Evans D G, Wang Z and Duan X 2005 *J. Mater.Chem.* **15** 1197
[37] Frunza L, Frunza S, Poterasu M, Beica T, Kosslick H and Stoenescu D 2009 *Spectrochimica Acta Part A: Molecular and Biomolecular Spectroscopy* **72** 248.
[38] Wang Z, Wang E, Gao L and Xu L 2005 *J. Solid State Chem* 2005 **178** 736

Detection of nitrite and nitrate ions in water by graphene oxide as a potential fluorescence sensor

Ing Hua Tang[1], Rita Sundari[2], Hendrik O Lintang[3], and Leny Yuliati[3*]
[1]Department of Chemistry, Faculty of Science, Universiti Teknologi Malaysia, 81310 UTM Johor Bahru, Malaysia
[2]The Research Center of Private Universities Coordination, 13360 Jakarta, Indonesia
[3]Centre for Sustainable Nanomaterials, Ibnu Sina Institute for Scientific and Industrial Research, Universiti Teknologi Malaysia, 81310 UTM Johor Bahru, Malaysia

[*]E-mail: leny@ibnusina.utm.my

Abstract. In this study, graphene oxide (GO) was used as a new fluorescence sensor for detection of nitrite (NO_2^-) and nitrate (NO_3^-) ions. The GO was synthesized via an improved Hummers' method, and the properties of GO were examined by X-ray diffraction (XRD), Fourier transform infrared spectroscopy (FTIR), and transmission electron microscope (TEM). The XRD pattern showed the presence of (002) plane at 2θ of 9.5° while the FTIR spectrum showed the presence of C-OH, C=O, C=C, C-O chemical bonds in the GO. The layer structure of the GO was confirmed from the TEM image. The sensing performance of the GO for NO_2^- and NO_3^- was evaluated by monitoring the emission sites of the GO at 567 nm, which was corresponded to the oxygen functional groups. Fluorescence quenching was observed, suggesting that the GO interacted well with both NO_2^- and NO_3^- ions. The linear Stern-Volmer plots were obtained in the concentration range of 1-10 mM, indicating the potential ability of the GO as the fluorescence sensor. The quenching constants for the detections of NO_2^- and NO_3^- were 4.8×10^{-2} and 1.2×10^{-2} mM^{-1}, respectively, suggesting that the GO has greater sensitivity towards the NO_2^- than the NO_3^- ion.

1. Introduction

The negatively charged nitrogen oxide compounds, NO_2^- and NO_3^- ions can exist together in the groundwater of agricultural land due to the excessive use of fertilizers. The presence of high level of toxic NO_2^- and NO_3^- ions in the lake results in a groundwater contamination. This contamination has been identified to bring the occurrences of methemoglobinemia (blue baby syndrome) in infants, which can be fatal for the affected infants. On the other hand, it also leads to eutrophication and algae bloom in the lake of agriculture land [1].

Due to these negative effects, various techniques have been employed for the detection of NO_2^- and NO_3^- ions, such as high-performance liquid chromatography (HPLC), ion chromatography (IC) and Griess colorimetric assay. However, these techniques come with drawbacks. For HPLC method, several tedious purification steps are required to remove the interfering such as chloride [2] and biogenic amines [3]. While, for the IC, it needs extensive maintenance and sample pre-treatment prior to sample measurement [4]. For Griess assay, it is a lack of sensitivity as all the NO_3^- ions are converted to NO_2^- which later are transformed to colored azo dye [5]. Therefore, the sensitive and simple detection technique is highly required. In this study, fluorescence spectroscopy was used as a sensitive and simple detection method for the NO_2^- and NO_3^- ions.

On the other hand, the two-dimensional single-atom-thick graphene oxide (GO) has been one of the most investigated materials in the recent years due to its superior electronic, optical, thermal, and mechanical properties [6-10]. Besides, GO also has a large surface to volume ratio, good chemical stability, biocompatibility, facile surface modification and low manufacturing cost, which make it an interesting material in biotechnology and biosensor [7-9]. In this work, the GO was synthesized using an improved Hummers' method and used as a new potential fluorescence sensor for detection of the NO_2^- and the NO_3^- ions.

2. Experimental

2.1. Materials

All chemicals were purchased and used without further purification. Deionized distilled water was used throughout the experiments. Sodium nitrite (Merck, 99.5%), sodium nitrate (Merck, 99.5%), graphite flakes (Merck), potassium permanganate (Fisher Scientific, ≥ 99%), sulphuric acid (Fisher Scientific, >95%), phosphoric acid (Merck, 85%), hydrochloric acid (QRëc, 37%), hydrogen peroxide (Fisher Scientific, >30%), methanol (Fisher Scientific, 99.99%), and absolute ethanol (HmbG, 99.98%) were used in this study.

2.2. Methods

2.2.1. Synthesis of GO. The GO was synthesized through an improved Hummers' method, similar to the reported literature [11]. A mixture of concentrated sulfuric and phosphoric acids in the volume ratio of 9:1 was added to a mixture of graphite flake and potassium permanganate. Then, the mixture solution was put under 50 °C with stirring for 24 h in a water bath for homogeneous heating. The solution was subsequently cooled to room temperature and poured onto ice. Deionized water was added to the mixture, followed by dropwise addition of hydrogen peroxide. The mixture was centrifuged and washed with hydrochloric acid. Deionized water was then used to wash the solid until neutral. Lastly, the obtained solid was ultrasonicated in methanol for 1 h and vacuum-dried overnight at room temperature. The obtained product was ground to get the GO powder.

2.2.2. Characterizations. The prepared GO was characterized using the following instruments. The structural property of GO was investigated by X-Ray diffractometer (XRD, Bruker Advance D8 diffractometer) with Cu Kα radiation (λ = 1.5406 Å). The functional group of the GO was determined by Fourier transform infrared spectrometer (FTIR, Thermo Scientific Nicolet iS50) with potassium bromide as the reference sample. The morphology of the GO was examined by transmission electron microscope (TEM, JEM-2100) at an accelerating voltage of 200 kV. Prior to the TEM measurement, the sample was dispersed in absolute ethanol and dropped on a carbon-coated copper grid. The fluorescence property of the GO was investigated by using a fluorescence spectroscopy (JASCO, FP-8500).

2.2.3. Sensing capability tests. The sensing capability of the GO for the detection of NO_2^- and NO_3^- ions was studied using the fluorescence spectroscopy (JASCO, FP-8500) at room temperature. The prepared GO was dispersed in deionized water (100 ppm, 250 mL) by ultrasonication for 30 min. The emission spectrum of GO solution (3 mL) was then monitored at an excitation wavelength of 367 nm. Certain amounts of analytes (NO_2^- and NO_3^- ions) were then added to the GO solution with various concentrations (1-10 mM). The interactions between the GO and the analytes were then studied by monitoring the quenching of the emission intensity at a wavelength of 567 nm.

3. Results and discussion

3.1. Structure of the GO

The structural property of the obtained GO was investigated by the XRD. As shown in Figure 1, the prepared GO exhibited a sharp basal diffraction peak (002) at 2θ of 9.50°, with *d*-spacing of 0.93 nm. Here, the obtained interspacing could be attributed to its oxygenated functional groups intercalated in the interlayer of the GO during the oxidation process of graphite flake. The obtained result agrees well with the reported literature [11], where it indicated the presence of the (002) plane in the prepared GO.

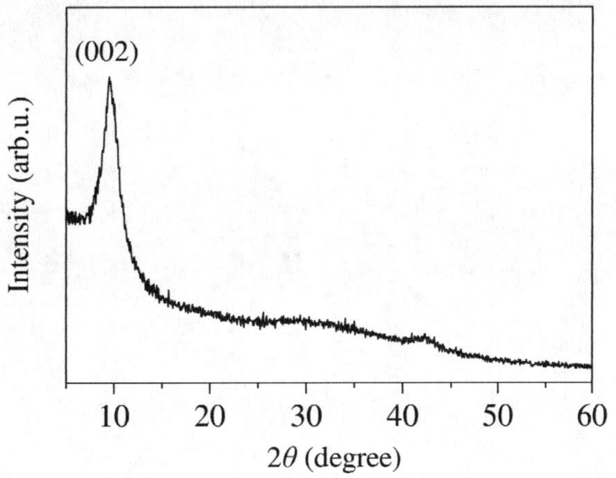

Figure 1. XRD pattern of the prepared GO.

3.2. Functional groups of the GO

The functional groups of the prepared GO were determined by an FTIR spectroscopy. As shown in Figure 2, the prepared GO showed the presence of O-H stretching vibration of hydroxyl at 3416 cm^{-1}, C=O stretching vibration at 1729 cm^{-1}, C=C bonds at 1623 cm^{-1}, C-O stretching vibration for carboxylic acid and epoxide at 1396 cm^{-1} and 1249 cm^{-1}, respectively, and C-OH of alcohol at 1056 cm^{-1}, as has been also reported elsewhere [11-13]. The presence of the mentioned functional groups demonstrated that the graphite flakes has been oxidized to GO.

3.3. Morphology of the GO

The morphology of the prepared GO was examined by the TEM under different magnifications. As can be observed in Figure 3 (a) and (b), when the GO was exfoliated in water under ultrasonic treatment, it was exfoliated into monolayer or very thin layers, where the typical wrinkle morphology of GO could be observed [6, 10].

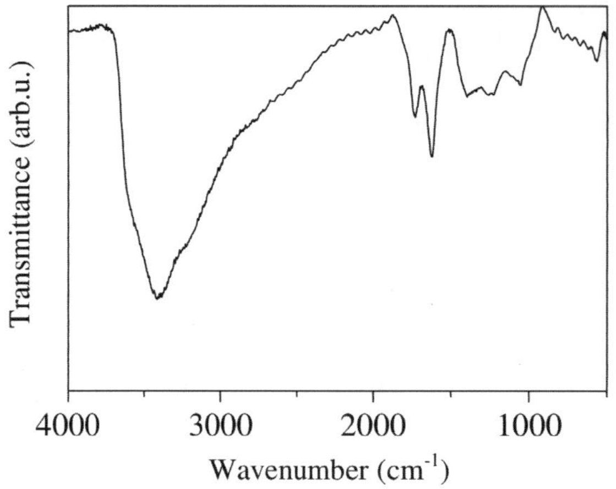

Figure 2. FTIR spectrum of the prepared GO.

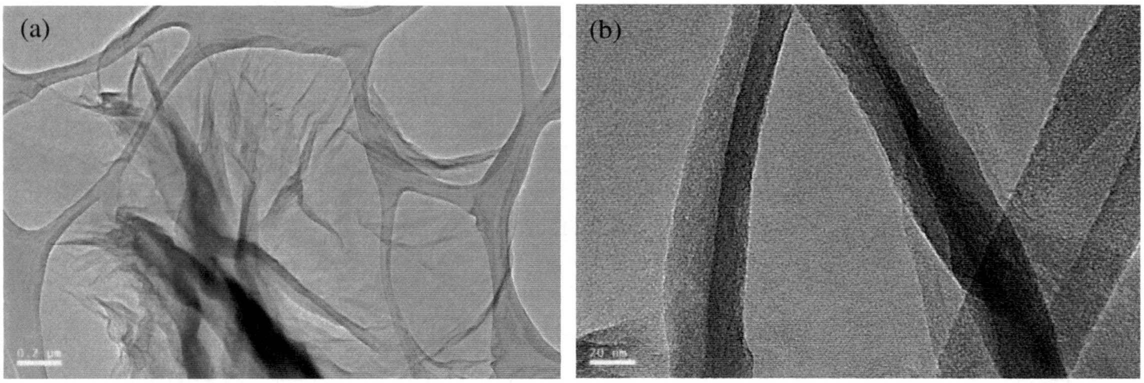

Figure 3. TEM image of the prepared GO at magnification of (a) 10,000 and (b) 100,000 times.

3.4. Sensing capability of the GO

The sensing capability of the prepared GO for NO_2^- and NO_3^- ions was evaluated using the fluorescence spectroscopy, where the implemented concentrations of the analyte ions were in the range of 1-10 mM. It was confirmed that the GO gave maximum excitation peak at 367 nm. When the GO was excited at 367 nm, three emission peaks were observed at 422, 468, and 567 nm. The peak at 422 nm can be assigned to the presence of crystalline graphene [14], while emission peaks at 468 and 567 nm can be assigned to the π to π^* transition in the GO and the oxygen functional groups such as CO, C=O, and O=C-OH [12]. The emission intensity of the GO was monitored in the absence and presence of the analytes. The emission peaks at 422 and 468 nm did not give clear trends in the interactions with the analytes, and only emission peak monitored at 567 nm showed the trend of quenching, suggesting that the oxygen functional groups in the GO would be the possible sites to interact with the analytes. As shown in Figure 4 (a), the emission intensity of the GO at 567 nm decreased with the increase of the NO_2^- concentration from 1 to 10 mM. The same phenomenon was observed when using the NO_3^- ions as the analyte (Figure 4 (b)). The observed quenching strongly indicated that there was a certain interaction between NO_2^- and NO_3^- ions with the emission sites of the GO.

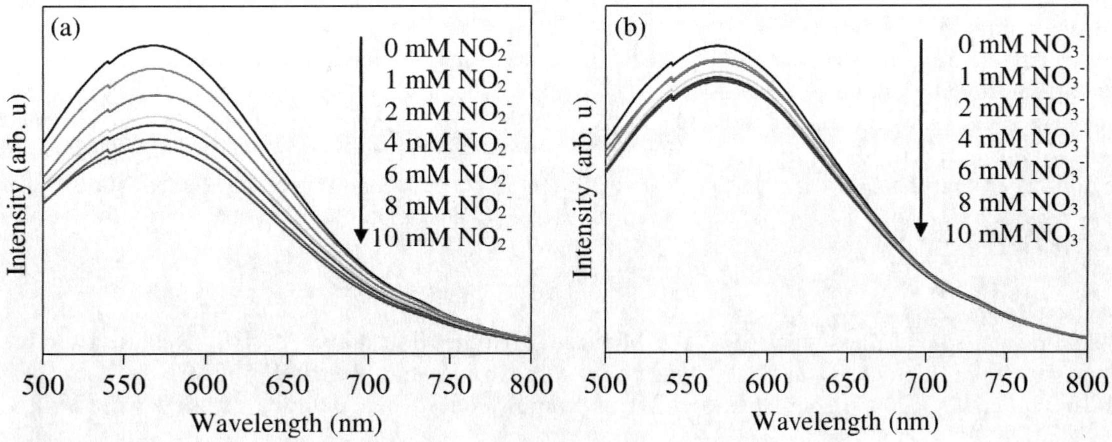

Figure 4. Emission spectra of the GO in the absence and presence of (a) NO_2^- and (b) NO_3^- ions monitored at excitation wavelength of 367 nm.

The Stern-Volmer plot was used to investigate further the sensitivity of the prepared GO towards the detection of the NO_2^- and the NO_3^- ions. The Stern-Volmer equation is depicted in Equation (1),

$$I_o/I = K_{sv}\,[Q] + 1 \qquad (1)$$

where I_o and I are fluorescence intensities in the absence and presence of analyte ion, $[Q]$ is the concentration of analyte ion, and K_{sv} is the Stern-Volmer quenching constant. The obtained gradient in the plot indicated the sensitivity of the sensor to the analyte. As can be observed in Figure 5, the GO showed *ca.* 4 times higher K_{sv} value to NO_2^- ($K_{sv} = 4.8 \times 10^{-2}$ mM^{-1}) than the NO_3^- ions ($K_{sv} = 1.2 \times 10^{-2}$ mM^{-1}), suggesting that the GO was more sensitive for detection of the NO_2^- than the NO_3^- ions. The higher sensitivity of the GO towards the NO_2^- would be possibly caused by the fact that the NO_2^- is smaller than NO_3^-, thus, the NO_2^- would be interacted with the emission sites easier than the NO_3^-. Moreover, the NO_2^- has higher polarity compared to NO_3^-. These factors caused the NO_2^- would have more interactions with the oxygen functional groups of the GO.

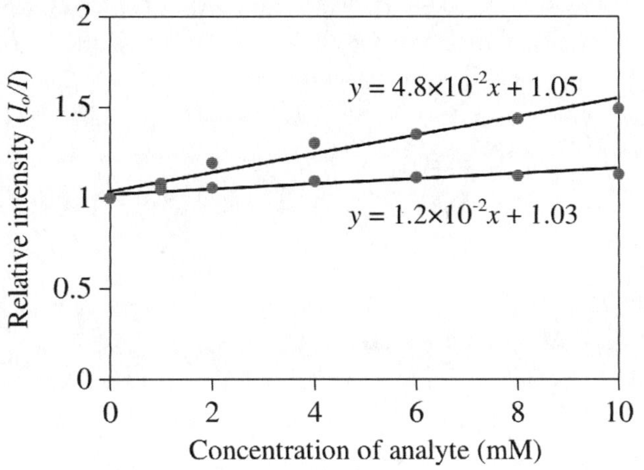

Figure 5. Linear Stern-Volmer plots of the GO for the detection of the NO_2^- (•) and the NO_3^- ions (•).

4. Conclusions

The characterization results from XRD and FTIR showed that GO has been successfully synthesized via the improved Hummers' method. The TEM image showed the layered wrinkle surface of the synthesized GO. The fluorescence results showed that the emission intensity of the GO was quenched linearly with the NO_2^- and the NO_3^- ions in the range concentrations of 1-10 mM. This study demonstrated that the GO was able to act as the potential fluorescence sensor for the detection of the NO_2^- and the NO_3^- ions. The prepared GO was shown to give a greater sensing sensitivity towards the NO_2^- than the NO_3^- ions.

Acknowledgements

The work was financially supported by the Ministry of Higher Education (MOHE, Malaysia) and the Universiti Teknologi Malaysia (UTM, Malaysia) through a Tier-1 Research University Grant (cost center code: Q.J130000.2509.06H66). I.H.T acknowledges the support of MyBrain MyPhd scholarship.

References

[1] Taylor C J, Bain L A, Richardson D J, Spiro S and Russell D A 2004 *Anal. Biochem.* **328** 60
[2] Stratford M R L, Dennis M-F, Cochrane R, Parkins C S and Everett S A 1997 *J. Chromatogr. A* **770** 151
[3] El Menyawi I, Looareesuwan S, Knapp S, Thalhammer F, Stoiser B and Burgmann H 1998 *J. Chromatogr. B* **706** 347
[4] Moorcroft M J, Davis J and Compton R G 2001 *Atlanta* **54** 785
[5] Badea M, Amine A, Palleschi G, Moscone D, Volpe G and Cerulli A 2001 *J. Electroanal. Chem.* **509** 66
[6] Zhu Y, Murali S, Cai W, Li X, Suk J W, Potts J R and Ruoff R S 2010 *Adv. Mater.* **22** 3906
[7] Banning F, Linjie G, Lihua W, Fan L, Jianxin L, Jim G, Chunhai F, and Qing H 2013 *Anal. Chem.* **85** 7732
[8] Feng D, Zhang Y, Feng T, Shi W, Li X and Ma H 2011 *Chem. Commun.* **47** 10680
[9] Cui L, Pu T, Liu Y and He X 2013 *Electrochim. Acta* **88** 559
[10] Shao Y, Wang J, Engelhard M, Wang C and Lin Y 2010 *J. Mater. Chem.* **20** 743
[11] Marcano D C, Kosynkin D V, Berlin J M, Sinitskii A, Sun Z, Slesarev A, Alemany L B, Lu W and Tour J M 2010 *ACS Nano* **4** 4806
[12] Shahriar L and Athawale A A 2014 *Int. J. Renew. Energy Environ. Eng.* **2** 58
[13] Zhang Y, Chi H, Zhang W, Sun Y, Liang Q, Gu Y and Jing R 2014 *Nano Micro Lett* **6** 80
[14] Englert J M, Dotzer C, Yang G, Schmid M, Papp C, Gottfried J M, Steinrück H-P, Spiecker E, Hauke F and Hirsch A 2011 *Nat Chem* **3** 279

The study of FTO surface texturing fabrication using Argon plasma etching technique for DSSC applications

Lindha Jayanti[1], Kusumandari[1*], Tjipto Sujitno[2] and Risa Suryana[1]

[1] Department of Physics, Faculty of Mathematics and Natural Sciences, Sebelas Maret University, Jl. Ir. Sutami 36A Kentingan Surakarta 57126, Indonesia
[2] Center for Accelerator and Material Process Technology, National Nuclear Energy Agency Jl. Babarsari, Yogyakarta 55281, Indonesia

*E-mail: kndari@gmail.com

Abstract. This paper is aimed to investigate the fabrication of the fluorine-doped tin oxide (FTO) texturing by using Argon (Ar) plasma etching. The pressure and temperature of Ar gas during plasma etching were 1.6 mbar and 240-285°C, respectively. The plasma etching time was varied from 3 and 10 min. We also prepared without etching samples as reference. UV-Vis spectrophotometer showed that the transmittances of etching samples are higher than the without etching samples. The root mean square roughness (Rq) of etching samples are lower than the without etching samples. It is considered that the Ar ions bombardment can modify the FTO surfaces. However, the etching time does not significantly affect the FTO surfaces for 3 min and 10 min. The Rq of the without etching sample, the etching sample for 3 min, and the etching sample for 10 min are 11.697 nm, 9.859 nm, and 9.777 nm, respectively. These results are good agreement with the four point probe measurement that indicated that the sheet resistance (RS) for each the without sample, the etching sample for 3 min, and the etching sample for 10 min are 16.817 Ω/sq, 16.067 Ω/sq, and 15.990 Ω/sq. In addition, the optical transmittance of the etching sample for 3 min and the etching sample for 10 min at wavelengths of 350 – 850 nm are almost similar. This is evidence that the etching time below 10 min cannot significantly change the morphology, optical and electrical properties.

1. Introduction

Dye-sensitized solar cell (DSSC) is one of energy alternatives to overcome the energy crisis in the world. The DSSC as a candidate of renewable generation solar cells has advantages relatively to silicon-based solar cells such as it can be fabricated at low cost and it does not require material with high purity [1-2]. The difference between conventional solar cells and DSSC is located at its working principle. For a conventional solar cell, the light absorption process carried out by silicon, while DSSC is performed by its dye. The DSSC efficiency is still lower than the conventional solar cell. There are some efforts to improve the DSSC efficiency i.e. modification of TiO_2 or ZnO layers, variation of dyes, electrolytes in solid or gel shapes, and using light trapping method of transparent conducting oxide (TCO) [3-4]. One of efforts to trap light or photon is using a patterned substrate surface or surface texturing [3].

In our best knowledge, the surface texturing onto the FTO surfaces were not much investigated in the DSSC applications. In this research, the surface texturing of the FTO surfaces are formed using argon plasma etching [5]. In order to know the morphology, the optical, and the electrical properties of the textured-FTO surfaces, it is characterized using atomic force microscope (AFM), ultra violet-visible (UV-Vis) spectrophotometer, and four point probes, respectively.

2. Method

The FTO samples were cut in 2x2 cm^2 size. Samples were cleaned in acetone using an ultrasonic cleaner for 10 min. Then, it is dried at room temperature. The sample was put on the sample holder in the chamber then the vacuum process was performed. The argon gas was introduced into the chamber with the power supply of voltage flux in the range of 0.296 – 0.376 kV and the argon pressure remained at 1.6 mbar. In addition, the temperature was about 240 – 285 °C during the plasma etching. The time of plasma etching was 3 and 10 min for each sample. The textured-FTO samples were characterized by the Atomic Force Microscopy (AFM) to know the roughness and morphology of surfaces, the UV-Vis spectrophotometer to determine the transmittance and the four point probes to confirm the sheet resistances. The samples without the plasma etching were also prepared as a reference.

3. Results and Discussion

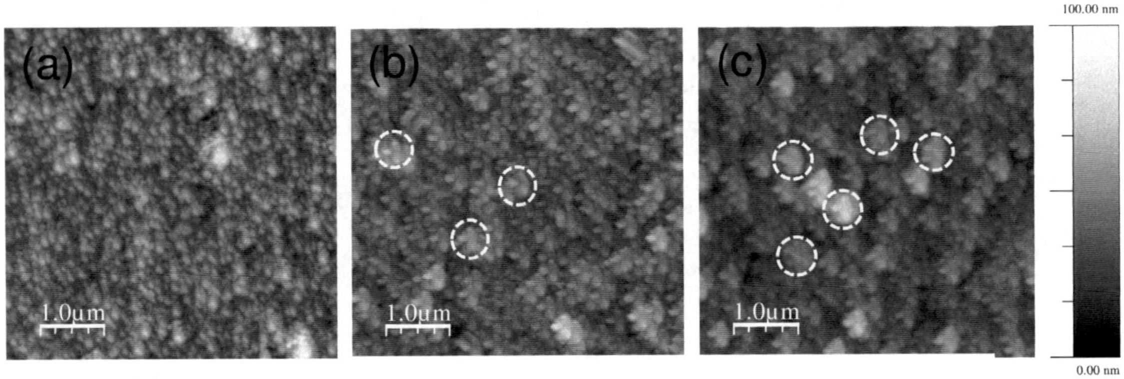

Figure 1. AFM images of FTO surfaces (5 μm x 5 μm) prepared by (a) without argon plasma etching, (b) argon plasma etching for 3 min and (c) argon plasma etching for 10 min.

Figure 1 shows the FTO surfaces change after the argon plasma etching. FTO surface is relatively smooth and uniform as shown Figure 1(a) while it is shown several domains (white circles) in Figure 1(b) after argon plasma etching for 3 min and many domains formed in Figure 1(c) after argon plasma etching for 10 min. It is indicated that argon plasma etching can modify the FTO surfaces. The energized-argon gas bombarded the FTO surfaces, and FTO molecule bonds will be broken then it go out the surface. The root mean square roughness (Rq) of FTO surfaces in area of 5 μm x 5 μm as shown in fig. 1 are 11.697 nm for the without etching sample, 9.859 nm for the etching sample (3 min) and 9.777 nm for the etching sample (10 min). It is considered that etching time from 3 min to 10 min cannot significantly change the FTO surfaces. These results are good agreement to the four point probe measurement as shown in Table 1 that confirmed that the sheet resistance (RS) for each the without sample, the etching sample for 3 min, and the etching sample for 10 min are 16.817 Ω/sq, 16.067 Ω/sq, and 15.990 Ω/sq. The RS can be calculated using equation $RS = \rho/t$ where RS is sheet resistance of thin film (Ω/sq), ρ is resistivity (Ω), and t is film thickness (nm) [6]. It is known that resistivity is related to mobility of electron which depends on several scattering mechanisms including lattice scattering, neutral impurity scattering, and grain boundary scattering [6]. It was reported that large grain size and lower surface roughness will result in small sheet resistance correlated to the dependence of the electron mobility of grain boundary scattering [7]. It is agree with our result.

IOP Conf. Series: Materials Science and Engineering **107** (2016) 012028 doi:10.1088/1757-899X/107/1/012028

Tabel 1. Sheet resistance and Rq of FTO thin films

Sample	Sheet resistance, RS (Ω/sq)	Root mean square roughness, Rq (nm)
Without etching	16.817	11.697
Etching 3 min	16.067	9.859
Etching 10 min	15.990	9.777

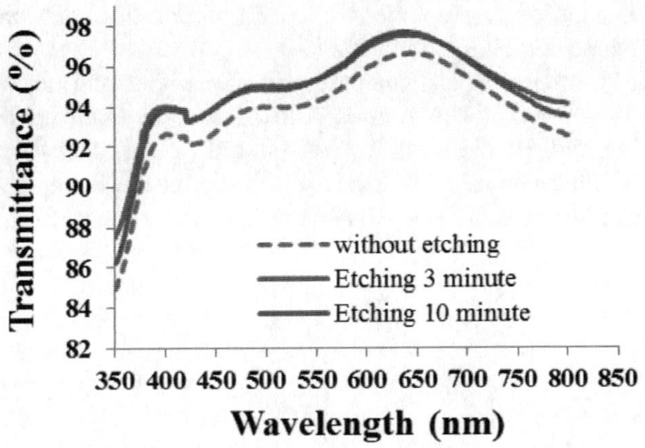

Figure 2. Transmittance of FTO samples of the without etching, the etching for 3 min, and the etching for 10 min

The optical transmittance of FTO samples of the without etching, the etching for 3 min, and the etching for 10 min are shown in Figure 2. For all samples, transmittance peaks occur in wavelength of 400-800 nm with peaks maximum at about 400 nm, 500 nm, and 650 nm. An etching treatment will increase transmittance due to domain formation of FTO will leave much spaces and will be passed the visible lights of UV-Vis spectrophotometer. However, the transmittance values between the etching sample for 3 min and the etching sample for 10 min are almost similar. It is indicated that the etching time below 10 min does not change the optical property. The highest optical transmittance of the FTO films is beneficial for efficient transport of light in DSSC applications [8]. In addition, the transmittance of FTO can affect the efficiency of DSSC because a light trapping structure contributes to increasing the light path length by scattering the incident light at the different angles [9].

4. Conclusion

Textured-FTO surfaces can be formed using an Ar plasma etching. The etching samples can improve the morphological, electrical and optical properties compare to the without etching samples. However, the etching time below 10 min cannot significantly change these properties. Based on the sheet resistance and the optical transmittance values, it is considered that the textured-FTO surfaces will improve the DSSC efficiency.

References

[1] Gratzel M and O'Regan B 1991 A low-cost, high-efficiency solar cell based on dye-sensitized colloidal TiO_2 films *Nature* **353** 737-40

[2] Smestad G P and Gratzel M 1998 Demonstrating electron transfer and nanotechnology: A natural dye-sensitized nanocrystalline energy converter *J Chem Educ* **75** 752-56

[3] Markvart T and Castener L 2005 *Practical Handbook* of *Photovoltaics: Fundamentals and Applications* Oxford: Elsevier advanced Technology

[4] Chen L, Alkaisi M and Liao M 2010 Light trapping technique for dye-sensitized solar cell *J Nonlinear Optical Physics & Materials* **19** **4** 761-66

[5] Hussain S Q, Gi D K, Shihyun A, Snbo K, Hyeongsik P, Anh H T, Chonghoon S, Sangho K, Shahbaz K, Jayapal R, Nagarajan B S, Velumani, Dindier P and Junsin Y 2015 SF6/Ar plasma textured periodic glass surface morphologies with high transmittance and haze ratio of ITO:Zr films for amorphous silicon thin film solar cells *Vacuum* **117** 91-7

[6] Sze S M and Kwok K N 2007 *Physics of Semiconductor Devices Third Edition* New Jersey: Wiley-Interscience

[7] Kulkarnia A K, Schulz K H, Lim T S and Khan M 1999 Dependence of the sheet resistance of indium-tin-oxide thin film on grain size and grain orientation determined from X-ray diffaction technique *Thin Solid Film* **345** 273-277

[8] Shi X-L, Wang J-T and Wang J-N 2014 Roughness improvement of fluorine-doped tin oxide thin films by using different alcohol solvents *J. Alloys and Compounds* **611** 297-302

[9] Tsai C, Sui-Ying H, Tsung-Wei H, Yu-Tang T, Yan-Fang C, Yuang H-J, Lun H, Chung-Ching W, Chieh-Wei C and Chung-Chun Li 2011 Influences of textures in fluorine-doped tin oxide on characteristics of dye-sensitized solar cells *Organic Electronics* **12** 2003-11

Degradation of self-compacting concrete (SCC) due to sulfuric acid attack: Experiment investigation on the effect of high volume fly ash content

S A Kristiawan[1*], Sunarmasto[1] and G P Tyas[2]
[1]SMART Crete Research Group, Civil Engineering Department, Sebelas Maret University, Indonesia
[2]Civil Engineering Department, Sebelas Maret University, Indonesia

[*]E-mail: s.a.kristiawan@ft.uns.ac.id

Abstract. Concrete is susceptible to a variety of chemical attacks. In the sulfuric acid environment, concrete is subjected to a combination of sulfuric and acid attack.This research is aimed to investigate the degradation of self-compacting concrete (SCC) due to sulfuric acid attack based on measurement of compressive strength loss and diameter change. Since the proportion of SCC contains higher cement than that of normal concrete, the vulnerability of this concrete to sulfuric acid attack could be reduced by partial replacement of cement with fly ash at high volume level. The effect of high volume fly ash at 50-70% cement replacement levels on the extent of degradation owing to sulfuric acid will be assessed in this study. It can be shown that an increase in the utilization of fly ash to partially replace cement tends to reduce the degradation as confirmed by less compressive strength loss and diameter change. The effect of fly ash to reduce the degradation of SCC is more pronounced at a later age.

1. Introduction

Concrete is a construction material that has been applied in the development of many types of infrastructures. These infrastructures have been designed, built and maintained to meet certain service life. However, many of these infrastructures have no longer in use before reaching their respective service life. The root of the problem could be due to the degradation of concrete material which eventually affects serviceability and capacity of structural concrete. Therefore, identifying the cause and rate of concrete degradation will aid the engineers to design, build and maintain a better structural concrete for a sustainable infrastructure.

Deterioration or degradation of concrete is any adverse change in normal, mechanical, physical dan chemical properties either on the surface or in the body of concrete, generally due to the disintegration of its components. The effect of deterioration may or may not be manifested visually. The following are three basic visual symptoms of distress in structural concrete: cracking, spalling and disintegration [1]. Aggressive agents from the environment could initiate a chemical attack on the concrete leading to the distress. The present of sulfuric acid produced from either sewage or sulfur dioxide in the atmosphere of industrial cities are examples of an aggressive agent that could be the source of the chemical attack on concrete. The attack is due to the high alkalinity of portland cement, which can react with the sulfuric acid [2]. Calcium hydroxide of the hydrated portland cement first reacts with sulfuric acid to produce gypsum in which the formation of gypsum is accompanied by volume expansion. The expansion initiates distress in the brittle hardened concrete leading to cracking. The second reaction is a reaction between gypsum and calcium aluminate of hydrated portland cement. This second chemical reaction forms ettringite which instigates even greater expansion. At a high alkalinity or lower pH (less than 11.5-12) ettringite is destabilized and decomposed into softer gypsum. At this high alkalinity, concrete is also susceptible to acid attack in which calcium silicate hydrate is decalcified leading to the disintegration of concrete [3]. The deterioration process leads to physical, mechanical and chemical changes on concrete on which sulfuric acid attack may be quantified in the laboratory. Most common practices to measure a sulfuric acid attack on concrete are by quantifying mass loss, dimension change, strength loss, and sulfur concentration in the attacked concrete [2].

The resistance of concrete to sulfuric acid attack is subject to concrete composition especially that of cement type. This is due to the fact that reactions with respect to sulfuric acid attack involve hydrated portland cement. An attempt had been made to reduce the C_3A content of portland cement to obtain sulphate-resistant portland cement (SRPC). ASTM C 150 Type II and ASTM C 150 Type V is an example of such cement for moderate and severe exposure to sulphate environment, respectively. These cement may be used to substitute ordinary portland cement (OPC) in order to reduce the degradation of concrete in the sulfuric acid environment. However, as shown by Attiogbe and Rizkalla [2] prolonged exposure periods does not appear to provide a better resistance to sulfuric acid attack. This is because sulphate attack is only one aspect of sulfuric acid on concrete. The reaction of the acid with calcium silicate hydrate is another aspect of sulfuric acid attack that should be of concern. Another attempts to increase concrete resistance against sulfuric acid attack are by incorporating mineral admixtures such as fly ash, ground granulated blast furnace slag (GBS), silica fume and metakaolin into the concrete composition. The inclusion of mineral admixtures could be that of used as partial cement replacement, mineral addition, binary blended cement or ternary blended cement[4-6]. The effect of mineral admixtures on the deterioration process of concrete due to sulfuric acid attack provide an opportunity for researchers to create or modify concrete composition which can be durable in the sulfuric acid environment.

In the construction industry, a special type of concrete i.e. self-compacting concrete (SCC) has been introduced to overcome the problem of placing and compacting fresh concrete in highly-reinforced members and another area that can not be accomplished using a conventional concreting method. Fresh properties of SCC should meet the following criteria: flowability, filling ability and passing ability under its own weight without a tendency to segregation. With these fresh properties, once it is discharged from the mixer into formwork, it will flow and fill all recesses, reinforcement spaces and voids under its own weight and be self-compacted without any external aid of vibration. The key parameter to obtain SCC is moderate viscosity which can be achieved by a higher composition of fine components, the use of superplasticizer and sometime viscosity agent. SCC tends to have a higher cement content than conventional concrete. For this reason, mineral admixtures may be used as partial cement replacement. This research uses fly ash as partial cement replacement in producing SCC. The replacement level is in the range of 50-70%. Hence, the concrete investigated in this research fall into the category of high volume fly ash-self compacting concrete (HVFA-SCC).

Partial substitution of cement with mineral admixtures in SCC is beneficial with regard to improving the durability of this concrete. It has been noted by other investigators [4] that lowering cement content by incorporating fly ash as partial replacement tends to increase its resistant to sulfuric acid as measured by the mass loss. The increase in the resistance could be due to two factors i.e. a decrease in the actual cement content in the concrete composition and an increase in the formation of C-S-H as a result of the pozzolanic reaction. The first factor is evident from the fact that there is a lower mass loss of SCC with a lower strength (or with a lesser cement content) compared to that of SCC with a higher strength (or with a greater cement content). The sulfuric acid attack is primarily related to chemical reactions between a sulfuric acid and hydrated cement. Hence, a higher cement content in the concrete composition means more hydrated cement (calcium hydroxide and calcium aluminate) are available for sulphate reactions converting the solid products of cement hydration into softer gypsum.For SCC mixture with a pozzolanic addition such ash fly ash, the increased formation of C-S-H promotes refinement of porosity and additional binding property. These could be responsible to the more resistant of SCC containing fly ash to the acidic reaction when this concrete exposed to sulfuric acid. Other investigators [3] show different results in term of the effect of fly ash on the resistance of SCC to sulfuric acid attack. The refinement of capillary pores and eventually a reduction in the total porosity of SCC containing fly ash has negative consequences to sulfuric acid attack. This is due to pores densification creates fewer spaces to accommodate stresses induced by expansion of gypsum as a product of sulphate reaction. Hence, there is a competition between the deterioration process in SCC containing fly ash owing to sulphate attack and acid attack with regard to the effect of refinement porosity.

It should be noted that pozzolanic reaction in concrete containing fly ash takes place at a later age (more than 90 days). Thus, refinement of porosity is not gained at an early age. This is in line with the results of Sunamasto and Kristiawan [7] who confirm that inclusion of high volume fly ash at an early age tends to increase porosity. At an early age, this type of concrete turns out to be more resistant to sulfuric acid attack compared to that of SCC with a lower amount of fly ash [8]. The effect of high volume fly ash in SCC with respect to sulfuric acid attack is further investigated in this research. The varying parameter subjected to this investigation is fly ash content. The effect of sulfuric acid is observed in term of strength loss and dimension (diameter) change.

2. Materials and Method

2.1. Materials

Coarse aggregate (CA) with a maximum size of 10 mm was used in the production of SCC. The aggregate has been subjected to various tests following ASTM C127, ASTM C131 and ASTM C136 to determine its specific gravity, resistance to degradation and gradation, respectively. It was confirmed that this aggregate conforms to the requirements for grading and quality of aggregate as specified by ASTM C33. The fine aggregate or sand (S) was also subjected to various tests which include specific gravity, gradation, organic impurities and clay particles content according to ASTM C127, ASTM C136, ASTM C40 and ASTM C117, respectively. The S conforms to all requirements for grading and quality of aggregate as specified by ASTM C33. Cement used in this research was ordinary portland cement (OPC) or type I cement. Fly ash (FA) was obtained from a power plant in Cilacap. This fly ash is that of Class C as specified by ASTM C618.

All the ingredients for producing SCC was proportioned following the recommendation by Ozawa and Okamura [8]. The key factors to obtain SCC as stated in the recommendation are: the volume of CA is limited to 50% of the total solid volume; S used is no more than 40% of the total mortar volume; volume ratio of water/binder (W/B) is in the range of 0.9-1; and finally, the superplasticizer (Sp) dosage is determined to meet the self-compacting concrete. Based on these guidelines, after few trials the following initial proportion was obtained as per m^3 of SCC: 738 kg of OPC, 579 and 703 kg of S and CA, respectively, 211 kg of W and finally 7.27 kg of Sp. The amount of OPC was then reduced by FA substitution at 50-70% by weight of OPC. The final mixtures proportion is given in Table 1.

Table 1. Proportion of concrete mixtures

ID	OPC(kg)	S(kg)	CA (kg)	W (kg)	Sp (kg)	FA (kg)
C50%	369	579	703	211	7.37	369
C55%	332	579	703	211	7.37	405
C60%	295	579	703	211	7.37	442
C65%	258	579	703	211	7.37	478
C70%	221	579	703	211	7.37	516

2.2. Methods

Cylinder specimens of 150x300 mm were cast. For each mix proportion, 12 cylinder specimens were prepared. On the following day, these cylinders were removed from their casting molds, and immediately a half of these cylinders were immersed in a tank containing 5% concentration of the sulfuric acid solution and the other half were immersed in water. To maintain the 5% concentration of sulfuric acid solution for the period of test, the solution was renewed every week.The immersion was carried out continuously until the cylinder specimens were tested in compression following ASTM C39 at the age of 28 and 90 days. The compressive strength of SCC was determined as average strength of three specimens. The strength loss of SCC due to sulfuric acid attack was determined using equation (1).

$$C_l(\%) = \frac{(C_w - C_{sa})}{C_w} 100 \quad (1)$$

where C_l is strength loss (%), C_w is the compressive strength of concrete immersed in water and C_{sa} is the corresponding compressive strength of concrete immersed in sulfuric acid.

The diameter of cylinder specimens that have been immersed in sulfuric acid solution were measured at 28and 90 days to observed the physical degradation of SCC due to sulfuric acid attack. The physical degradation was quantified in term of changes in diameter of cylinder specimens which were determined using equation (2).

$$D_c(\%) = \frac{(D_0 - D_{sa})}{D_0} 100 \quad (2)$$

where D_c is a percentage of change in diameter after immersion in the sulfuric acid solution, D_0 is the original diameter of cylinder specimens before immersion i.e. 150 mm and D_{sa} is the diameter of cylinder specimens after immersion.

3. Results and discussion

3.1. Compressive strength

The compressive strengths of SCC containing various fly ash concentrated under immersion in 5% sulfuric acid solution for a period of 28 and 90 days are presented in figure 1. The corresponding compressive strengths of the SCCs immersed in water are given in the same figure for comparison and assessment of the strength loss. Comparison between compressive strength of SCC immersed in water and sulfuric acid solution reveals the effect of sulfuric acid attack to deteriorate concrete. It is clearly shown that reduction in compressive strength of SCC due to sulfuric acid attack is greater at a later age. The extent of reduction is also influenced by fly ash content. A higher fly ash content seems to be beneficial to increase the resistance of SCC to sulfuric acid attack as indicated by a lesser compressive strength loss (see figure 2). At an early age, however, the influence of fly ash content appears to be inconsistent. The different tendency of the effect of fly ash content on compressive strength loss due to sulfuric acid attack may be attributed to the pozzolanic reaction.

It is recognized that pozzolanic reaction takes places in the hardened state, and the beneficial effect of this reaction effectively observes at a later age (90 days). The conversion of CaOH into C-S-H as a result of pozzolanic reaction contributes to an increase in strength and reduction in porosity [9-10]. The effect of pozzolanic reaction to increase the compressive strength of SCC investigated in this study may be observed by comparing the strength of SCC immersed in water for 28 and 90 days. It is indicated that prolonging period of immersion in water increases compressive strength especially on SCC with higher fly ash content. On the other hand, prolonging immersion of SCC in sulfuric acid solution tends to decrease the compressive strength of lower fly ash content [2,11]. At higher fly ash content, the reduction in compressive strength due to sulfuric acid is surpassed by the pozzolanic reaction effect. Thus, the final compressive strength of SCC with higher fly ash contents turns to be increased when the comparison is made between those immersed in sulfuric acid for 28 and 90 days. All of these lead to the following findings: at low fly ash content prolonging period of immersion in water and in sulfuric acid causes an increase and decrease of compressive strength, respectively. The different in compressive strength between SCC containing low fly ash immersed in water and sulfuric acid for a longer period becomes greater than that immersed for a shorter period. Thus, a greater compressive strength loss of SCC containing low fly ash content is significantly observed at longer immersion period. At high fly ash content prolonging period of immersion both in water and sulfuric acid tends to increase the compressive strength compared to the corresponding SCC immersed for a shorter period. Hence, at longer immersion period the observed strength loss of SCC at higher fly ash content is lesser than at lower fly ash content. The overall effect of fly ash content may be summarized as follows: at a longer period of immersions, compressive strength loss is reduced when higher fly ash content is incorporated in SCC.

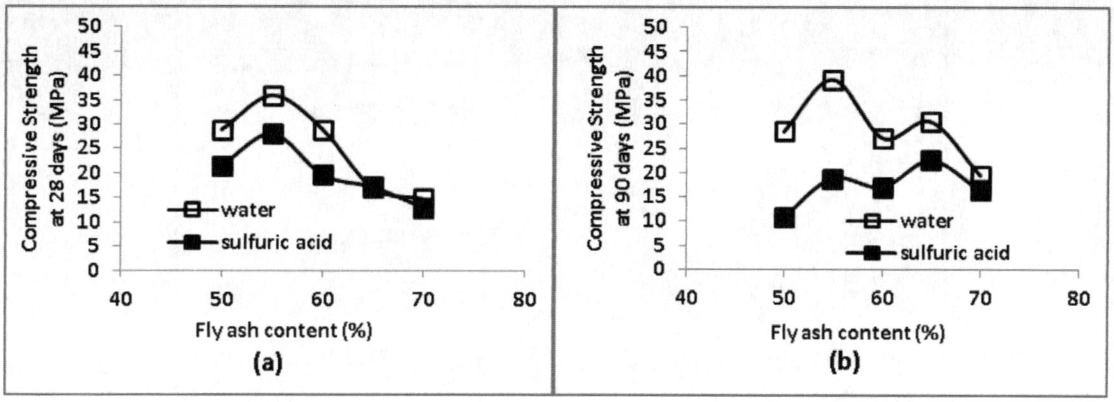

Figure 1. The compressive strength of SCC with various fly ash content determined after 28 (a) and 90 (b) days immersion in water and 5% sulfuric acid solution.

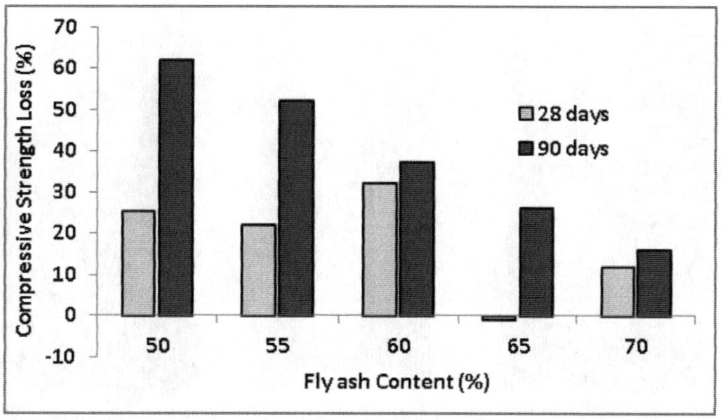

Figure 2. Percentage of compressive strength loss due to sulfuric acid attack

3.2. Diameter change

The effect of the sulfuric acid attack on concrete can also be observed on the basis of changes in dimensions. The changes in dimensions indicate the extent of physical degradation owing to sulfuric acid attack. In this investigation, the physical degradation of SCC with various fly ash contents is quantified on the basis of diameter changes. The result is presented in figure 3. There is a distinct magnitude of diameter changes between SCC immersed in the sulfuric acid solution for 28 days period than that immersed for 90 days period. At shorter period the changes in the diameter are in the range of 1-3% while at the longer period the changes are in the range of 7-11%. The changes in diameter of SCC specimens exposed to sulfuric acid suggest that disintegration of the solid component of SCC has occurred. The disintegration is started in the outer zone of SCC specimens where there is direct contact with the sulfuric acid solution (see figure 4). The disintegration is continuous and goes inwards long as the specimens kept immersed in the sulfuric acid solution. The result is a reduction in diameter of specimens with time. It has been shown by other investigators [4,11] that beyond the depth of attacked zone concrete tends to maintain its solid properties since no degradation occurs in this unattacked zone.

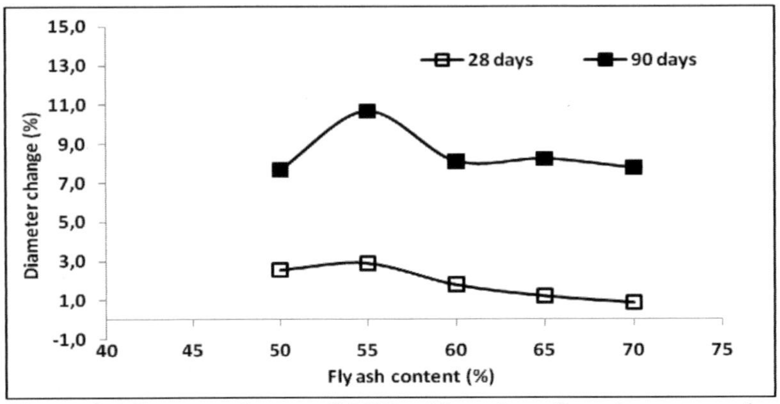

Figure 3. Diameter change of SCC with various fly ash contents due to sulfuric acid attack

(a) 28 days (b) 90 days

Figure 4. Disintegration of specimen at outer zone due to direct contact with sulfuric acid after immersed for 28 and 90 days

An increase in fly ash content has slightly reduced the degradation of SCC as observed from the diameter change with the exception for SCC containing 55% fly ash. Higher cement content of SCC (or lower fly ash content) could be responsible for this behaviour. Since more hydrated cement is available in SCC with higher cement (or lower fly ash) content, there is a higher probability of this concrete to be severely attacked by sulfuric acid. For SCC containing 55% fly ash, the highest strength of this concrete may be due to minimum interparticle spaces at this replacement level (consequently minimum porosity) [7] in addition to the contribution of relatively still higher cement content for hydration to proceed. Hence, when this concrete is attacked by sulfuric acid, the pores system can not accommodate the crystal gypsum leading to induced stresses which eventually deteriorate this concrete more severely [3]. The result is the highest decrease in the diameter of SCC with 55% fly ash content.

4. Conclusions

Based on the experimental results it is concluded that the degradation of SCC due to sulfuric acid attack as measured by compressive strength loss and diameter change can be reduced with a higher inclusion of fly ash to replace partially cement. The extent of the reduced degradation is more pronounced at a later age. The effect of fly ash content to modify SCC resistance against sulfuric acid could be explained by the combination of lower cement content, pozzolanic reaction and refinement of interparticle spaces. Less cement content gives fewer hydrated cement that will be attacked by sulfuric acid. The pozzolanic reaction will improve the strength of concrete which eventually decreases the tendency of concrete to disintegrate owing to sulfuric acid attack. Refinement of interparticle spaces may cause detrimental effect as they can not accommodate stresses induced by expansion of gypsum.

Acknowledgement

This work is part of Hibah Mandatory Research Project which financially supported by Sebelas Maret University for the year of 2015 (Contract No.624/UN27.11/PL/2015).

References

[1] Ferreira R M 2004 *Probability-Based Durability Analysis of Concrete Structures in Marine Environments* (Guimaraes: University of Minho)

[2] Attiogbe E K and Rizkalla S H 1988 *ACI Mat J* **Nov-Dec** 481-8

[3] Size O, Elsen J, Fey D, Herman G, Vandewalle L, Van Gemert D, De Schutter G, Desmet B and Vantomme J 2011 *Int Cong on the Chemistry of Cement (Madrid)* p 436

[4] Siad H, Mesbah H A, Bernard S K, Khelafi H and Mouli M 2010 *The Arabian J for Sci and Eng* **35** 183-195

[5] Drimalas T, Clement J C, Folliard K J, Dhole R and Thomas M D A 2011 *Research Report FHWA/TX-11/0-4889-1* (Austin: FHWA)

[6] Hewayde E, Nehdi M L, Allouche E and Nakhla G 2007 *Proc. Inst. Civ Eng: Constr Mat* **160** 25-35

[7] Sunarmasto and Kristiawan S A 2015 *App Mech Mat* **754-755** 447-451

[8] Okamura H and Ozawa K 1996 *Struc Eng Int* **6** 269-270

[9] Kosmatka S H, Kherkoff B and Panarese W C 2003 *Engineering Bulletin 001* (Illinois: Portland Cement Association)

[10] Douglas R P 2004 *PCA R&D Serial No 2619* (Illinois: Portland Cement Association)

[11] Kristiawan S A, Fatkulloh and K Adrianingtyas 2013 *Konferensi Nasional Teknik Sipil 7* (Surakarta: Universitas Sebelas Maret) p M-297

Solvothermal and electrochemical synthetic method of HKUST-1 and its methane storage capacity

Witri Wahyu Lestari[1*], Marisa Adreane[1], Candra Purnawan[2], Hamzah Fansuri[3], Nurul Widiastuti[3] and Sentot Budi Rahardjo[1]

[1]Inorganic and organo metallic complexes research group, Department of Chemistry, Faculty of Mathematics and Natural Sciences, Sebelas Maret University, Jl. Ir Sutami No.36A, Kentingan, Jebres Surakarta, Indonesia, 57126
[2]Analytical and Environmental Chemistry Research Group, Department of Chemistry, Faculty of Mathematics and Natural Sciences, Sebelas Maret University
[3]Chemistry Department, Institut Teknologi Sepuluh Nopember, Kampus ITS Sukolilo, Surabaya 60111, East Java, Indonesia

[*]E-mail: witri@mipa.uns.ac.id

Abstract. A comparison synthetic strategy of Metal-Organic Frameworks, namely, Hongkong University of Techhnology-1 {HKUST-1[$Cu_3(BTC)]_2$} (BTC = 1,3,5-benzene-tri-carboxylate) through solvothermal and electrochemical method in ethanol:water (1:1) has been conducted. The obtained material was analyzed using powder X-ray diffraction, Scanning Electron Microscopy (SEM), Fourier Transform Infrared Spectroscopy (FTIR), Thermo-Gravimetric Analysis (TGA) and Surface Area Analysis (SAA). While the voltage in the electrochemical method are varied, ranging from 12 to 15 Volt. The results show that at 15 V the texture of the material has the best degree of crystallinity and comparable with solvothermal product. This indicated from XRD data and supported by the SEM image to view the morphology. The thermal stability of the synthesized compounds is up to 320 °C. The shape of the nitrogen sorption isotherm of the compound corresponds to type I of the IUPAC adsorption isotherm classification for microporous materials with BET surface area of 629.2 and 324.3 m^2/g (for solvothermal and electrochemical product respectively) and promising for gas storage application. Herein, the methane storage capacities of these compounds are also tested.

1. Introduction

Methane is an abundant and renewable fuel potentially developed in Indonesia [1]. In addition, it is environmentally friendly due to the cleaner burning than gasoline [2]. This gas can also be locally produced from biogas and biomass. The development and usage of methane is also based on sustainability and social economy to support the governmental programs towards energy self-sufficient community. Meanwhile, the success in the discovery of renewable energy should be supported by efficient energy storage systems [3].

Very recently the U.S. Dept. of Energy has started a new methane storage program [4] with the following targets: 0.5 g (CH_4) per g (sorbent) for gravimetric capacity and r = 0.188 g/cc (11.741 mmol/cc) for volumetric capacity, which corresponds to the density of compressed natural gas (CNG) at 250 bar and 298 K. The new volumetric target is equal to 263 cc (STP: 273.15 K, 1 atm) per cc, which is significantly higher than the previous target of 180 cc (STP) per cc at 35 bar [5].

So far, methane is typically compressed in a pressure containment vessel at ca. 250 bar and room temperature for on-board vehicle applications. At such high pressures, cost issues and volume effectiveness of the fuel tank remain as major concerns. To address these technical challenges, intense research efforts have been carried out to find suitable porous materials to help store methane. However, two major classes of porous materials that are available commercially, such as activated carbon and zeolites, do not have sufficiently high storage capacities and, to date, no suitable material has been commercialized [6]. In response to this important technological challenge, the scientists have recently reported many novel porous materials especially for use in natural gas vehicles (NGVs) with ever increasing methane storage capacities [7, 8, 9]. Many of these materials belong to a versatile class of porous hybrid materials known as metal–organic frameworks (MOFs) that are synthesized by the self-assembly of modular molecular "building blocks" into stable, highly porous crystals [10, 11].

On the other hand, the importance of discovering "green" methods for synthesizing compounds and materials have highlighted due to the rising concerns about the energy efficiency, toxicity and environmental impact of chemical processes. The "green" perspective of research and development in inorganic and metal-organic synthesis has traditionally focused on developing materials for applications recognized as green and environmentally-friendly, such as energy storage, for specific as methane storage. In this contribution, we aim to present an innovative electrochemical synthetic approach, as applied to the synthesis of HKUST-1 [12], the most explored owing the high stability of metal-organic frameworks (MOFs). With this electrochemical method, MOFs have opened a route to metal-organic materials and their syntheses on large industrial scales [13]; environmentally friendly, sustainable and commercially viable synthesis. As comparison, the solvothermal synthetic approach [14] will also be discussed and the methane storage capacity via gravimetric method will be presented.

2. Experimental

2.1. Materials and methods

Cu-plates with 99.9% purity are used as electrode material. 1, 3, 5- benzene tricarboxylic acid (BTC) and tetrabutyl-ammonium-tetrafluoroborate (TBATFB) were purchased from ABCR, Germany and used without further purification. Analytical grade, ethanol HPLC grade was used as received from Merck. Before used, the electrode was activated by immersing in 100 mL HNO_3 (1 M) and scrub with sand-paper until shinny. Methane gas (purity > 98%) which used during gravimetric measurement was supplied from PT. Samator Gas Industry, Tangerang, Indonesia.

2.2. Preparation of HKUST-1 [$Cu_3(BTC)_2$]

The electrochemical method in synthesizing HKUST-1 [$Cu_3(BTC)_2$] was conducted according to literature procedure [15] with slightly modification in the used solvent (water:ethanol, ratio 1:1, with totally 50 ml). The solution was then placed in the electrolysis cell and stirred for 15 min for complete dispersion. Electrolysis was carried out in an electrochemical cell under constant voltage electrolysis (with keeping current varying to make voltage constant) for ca. 2 h to complete the reaction. Finally the sky blue color precipitate of [$Cu_3(BTC)_2$] was collected from the electrolysis cell and allowed to dry at room temperature and then activated at 200 °C for 2 h. The sky blue color which appeared initially was changed to dark blue after activation, i.e., change in color indicates that the co-ordination number of Cu in the complex state is changed from six to four [16]. Further optimization of experimental parameters like, applied voltage ranging from 12 to 15 Volt and reaction temperature were carried out in order to get good yield and highly crystalline [$Cu_3(BTC)_2$]. As comparison, the solvothermal synthetic method of HKUST-1 in solvent mixture water: ethanol (1:1) was also carried out according to Schlichte et al. [14b] The yield of the product corresponding to the amount of weight loss in the copper anode (for electrochemical method) is ranging from 16 to 52%, while the solvothermal method reached until 99% as shown in Table 1.

Table 1. The yield of HKUST-1 synthesized by electrochemical (with voltage and temperature variations) compared to solvothermal method

	Yield in (%) at various temperature condition			
Voltage variation (V)	RT	40 °C	60 °C	80 °C
12	23.8	18.911	26.952	28.962
13	46.4	16.068	12.894	30.284
14	20.0	19.506	17.787	17.919
15	52.4	19.047	31.673	30.416
Solvothermal (120 °C)			99,2	

2.3. Characterization
The synthesized product was characterized by Powder X-Ray diffraction using Cu–Kα radiation (λ = 1.5406 Å), with the voltage and current were held at 40 kV and 30 mA (2θ = 5–50°) at a scan rate of 1°/min. SEM data of samples were collected using FEI type Inspect S50 to confirm their surface morphology. Fourier Transform Infrared spectrum has been recorded on a FT-IR-8201 PC using KBr pellet in the range of 400–4000 cm^{-1}. The surface area and pore volume of the $[Cu_3(BTC)_2]$ are determined from BET adsorption isotherms of nitrogen at 77 K using a static volumetric apparatus Micrometrics, NOVA 1200e. Thermogravimetric analysis was performed using a Q 500 instrument manufactured by TA Instruments and experiments were conducted with a constant heating rate of 10 °C/min under nitrogen atmosphere.

2.4. Gravimetric measurements
The methane uptake experiment was performed at room temperature condition (25-30 °C) and at relatively low pressure (20 psi). Prior to the measurement, the sample HKUST-1 (ca 5 gram) was degassed at 200 °C to remove undesired gases. The sample was then loaded into a sample holder which it was connected to an analytical balance's hook. The mass of the sample after degassing was recorded as the initial mass before methane uptake measurement.

Methane was then flown through the sample at 15 ml/min after which its methane storage capacity was measured using the gravimetric method by observing changes of its weight. Methane storage capacity test was performed until mass of the sample saturated. The mass of the sample was observed every 5-10 minutes and collected by Mettler Toledo MS 4002 SDR analytical balance. The amount of methane uptake capacity was evaluated using Equation (1)

$$\%CH_4 = \left(\frac{m_1 - m_0}{m_0}\right) x\ 100\% \quad (1)$$

where m_0 is the initial mass of the sample, m_1 is the mass of the sample after the methane uptake test and % CH_4 is the amount of the CH_4 adsorbed on HKUST-1 (%wt.).

3. Results and discussion

3.1. Structural study of HKUST-1
HKUST-1 is one of the MOFs material that are widely studied due to its structural stability and various potential applications, such as in catalysis [14], gas separation [18] and gas storage [19, 20]. The specific advantage of HKUST-1 is having structural features such as an open metal coordination sites [3]. The compound has a bimodal pore size distribution, the larger pores are approximately 9 Å in diameter and the smaller pockets have openings approximately 3.5 Å in diameter [21].

Single X-ray analysis revealed that HKUST-1 has dimeric unit of copper-tetracarboxylic (Figure 1) with Cu-Cu bond length of 2.628(2) Å. This framework is electrically neutral. Twelve carboxylate oxygen atoms of two ligands 1,3,5-benzene tri-carboxylate (BTC) bind to the four coordination site for each of the three Cu^{2+} ions of the formula unit. Each metal completes its pseudo-octahedral coordination sphere with an axial aqua ligand opposite to the Cu-Cu vector [12]. The interconnection of Cu-paddle wheel unit with the tridentate BTC linker bond is repeated so that infinitely cubic frameworks are formed.

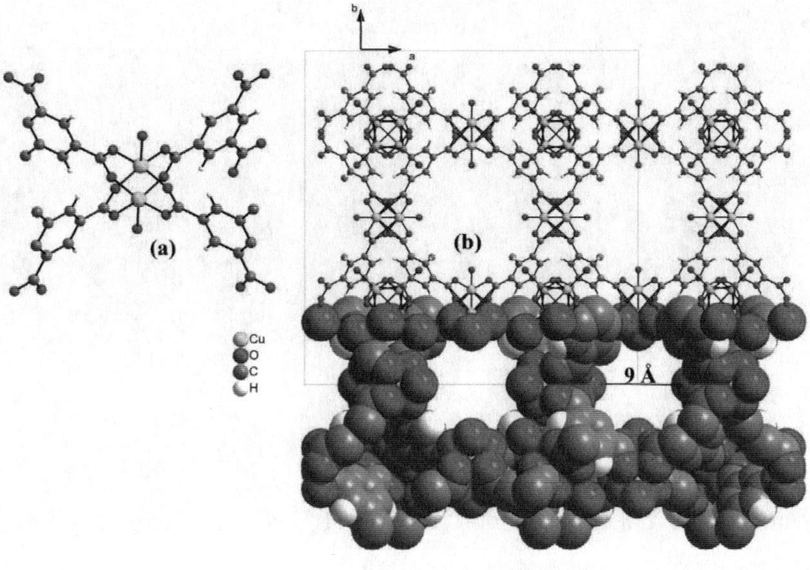

Figure 1. (a) Molecular structure of HKUST-1 with dicopper(II) tetracarboxylate as the basic framework (b) Frameworks of HKUST-1 $[Cu_3(BTC)_2{\cdot}(H_2O)_3]_n$ with orientation [100] and shows nanometer-sized channel.

3.2. Comparison of solvothermal and electrochemical methods

Synthesis of HKUST-1 was firstly reported by Cui et al. [12] through solvothermal for 12 hours at a temperature of 180 °C. However, these reaction conditions lead to the formation of impurities in the form of CuO due to the influence of a high reaction temperature. Schlichte et al. [14b] had optimized the solvothermal reaction in order to get purer crystal of HKUST-1 from Cu_2O impurity by conducting in the solvent mixture water: ethanol (1:1) with a source of Cu^{2+} in the form of salts, for example $Cu(NO_3)_2.\ 3H_2O$ or $CuCl_2.4H_2O$ but using a lower reaction temperature of about 120 °C. The result is a bluish-green crystal which was then washed with ethanol and dried at room temperature with stable frameworks after the drying process and activation. However, the used Cu^{2+} salt led to the formation of acidic by-product that is certainly not environmentally friendly. Reaction time of 12 hours is also considered ineffective, therefore efficient and environmentally friendly synthetic method that is required. In this paper we have tried to use innovation and green synthetic method by electrolysis and then we have compared to solvothermal product. In the electrochemical method, the source of metal salt is from the metal electrode itself; in this case is the copper-plate. The reaction is supported by the presence of the electrolyte, TBATFB. The general reaction scheme of the electrolysis process to synthesize HKUST-1 is depicted in Scheme 1.

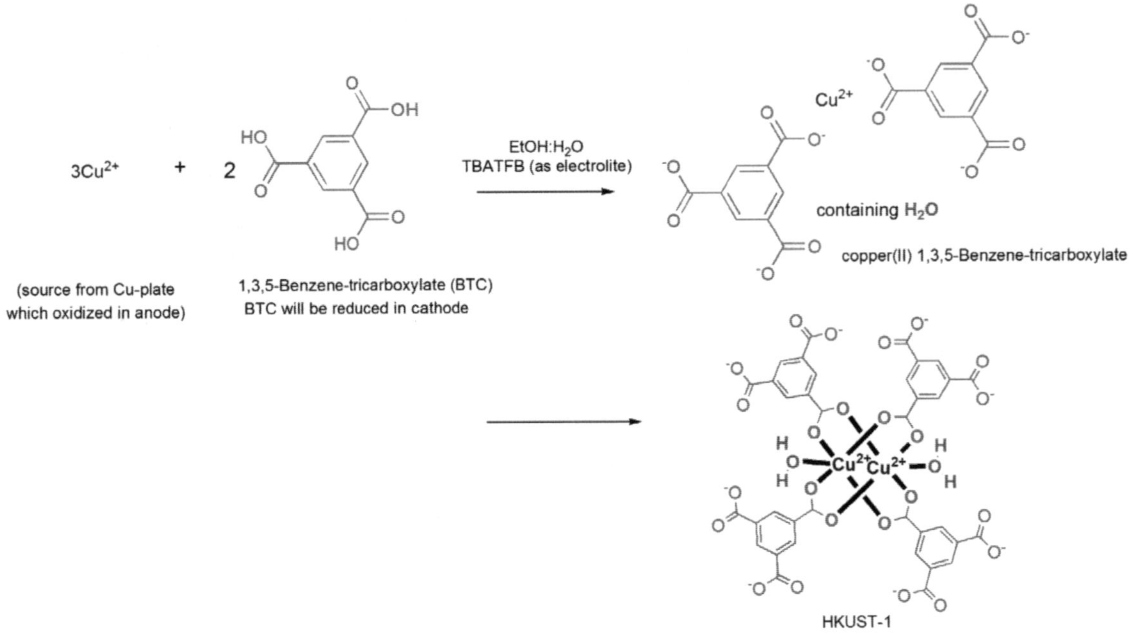

Scheme 1. Synthetic reaction to generate HKUST-1 by electrolysis method.

3.3. Material Characterization

The bulk product either from electrolysis and solvothermal product were characterized by powder X-ray diffraction in order to ensure a high purity of the crystalline phases. The X-ray diffractograms are identical to those calculated from single crystal analysis data reported in literature [12,14b] and the amount of side products is below the limit of detection by means of powder X-ray diffraction (Figure 2).The final results obtained greenish-blue powder which is dried at room temperature. Activation till 200 °C for 2 h lead to the purer and sharper peak of intensity and change to the color from light blue to dark blue indicates the release of coordinated water from copper(II). Temperature variations from RT to 80 °C, however, tend to produce lower yield. The sharper peaks in the range area of 2θ from 20 to 30° assumed represent the peak of the coordinated and or uncoordinated solvent (Figure 3). This can be observed after the activation at 200 °C these peaks are decreased as observed in Figure 2. However, these kind of peaks do not imply the significant impurities and found also in some referred literatures [15,22]. Additionally, it is remarkable that the synthesized $[Cu_3(BTC)_2]$ does not exhibit any peaks corresponding to CuO (2θ = 35.5° and 38.7°) or Cu_2O (2θ = 36.43°) and also in particular no significant impurities peak at 2θ = 11.0° as observed by Hartmann et al. [23] who synthesized $[Cu_3(BTC)_2]$ by hydrothermal method which confirms that higher purity $[Cu_3(BTC)_2]$ was obtained from this electrochemical method.

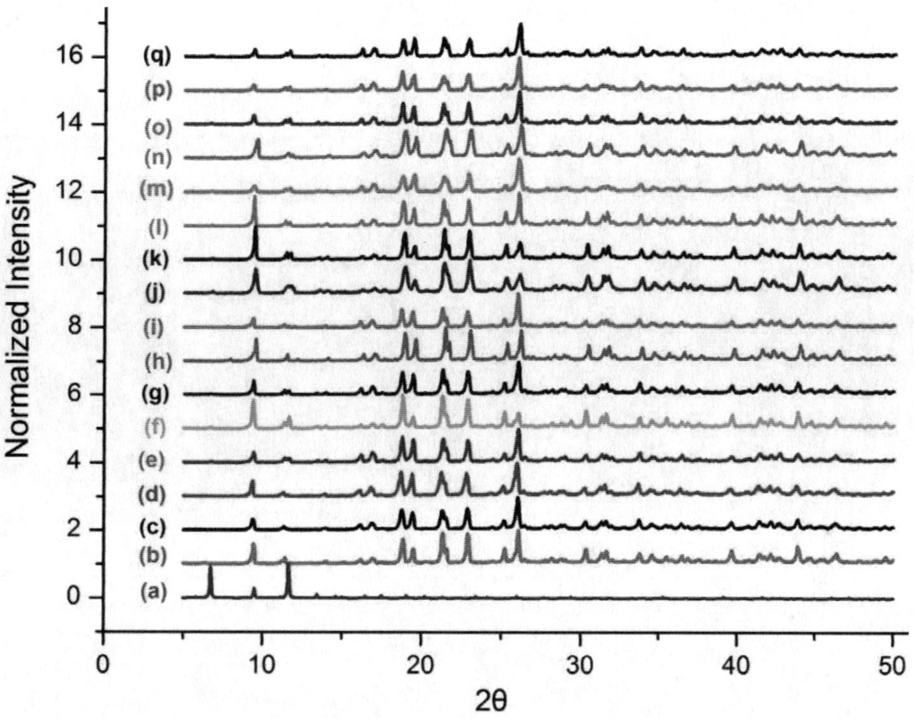

Figure 3. PXRD diffractograms of synthesized HKUST-1 by electrochemical method at 15 V (RT, 40, 60 and 80 °C ; b, c, d, e), 14 V (RT, 40, 60 and 80 °C ; f, g, h, i) , 13 V (RT, 40, 60 and 80 °C ; j, k, l, m) and 12 V (RT, 40, 60 and 80 °C ; n, o, p, q) compared to simulated pattern (a).

3.4. Thermal stability

The thermal stability of the resulting material was tested by thermo-gravimetric analysis (TGA). The electrochemically synthesized HKUST-1 at 15 V that only dried at room temperature indicates a change in the mass of nearly 20% at a temperature of 100-150 °C. This peak indicates the loss of water that fills the pores of HKUST-1. The next mass change of 5% at a temperature of 150-300 °C, these changes indicate a loss of water coordinated to Cu which serves as a node (node) at HKUST-1. Naturally, the higher temperatures required to break bonds of coordinated water with copper(II) rather than to release the water from the pore of HKUST-1. Then frameworks and ligand began to collapse at a temperature of 320 to 400 °C (Figure 4) to a perfect transformed into CO_2 and Cu_2O or metal Cu° [14b].

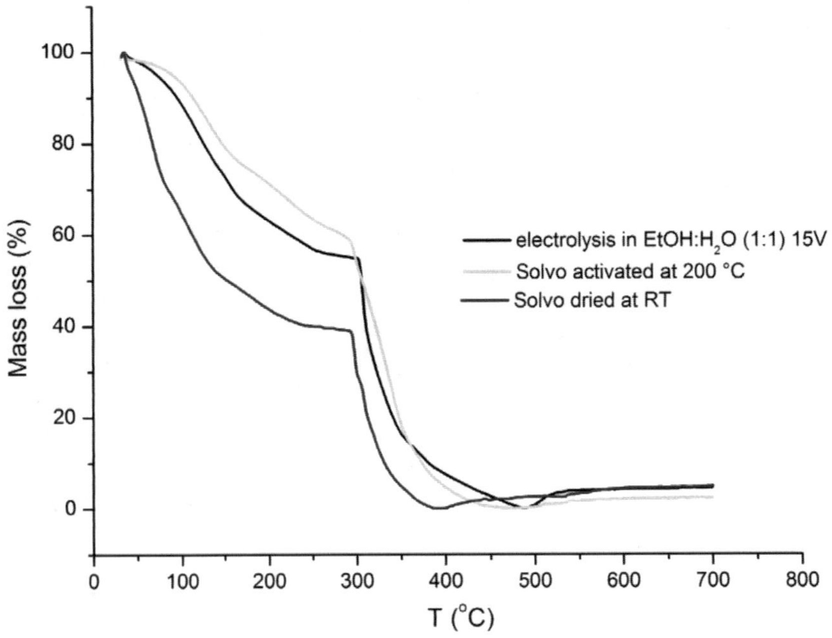

Figure 4. Thermogravimetric analysis of synthesized HKUST-1 either through solvothermal and electrolysis method.

As has been previously mentioned, the chemical and physical bonding of water molecules in HKUST-1 easily removed by heating the material. Dehydration makes part of the coordination of Cu can be accessed by other molecules that may play a role in the process of catalysis and gas storage [14]. The release of water molecules can be seen from the color of HKUST-1 after a synthesis that is turquoise and after drying to dark blue as shown in Figure 5.

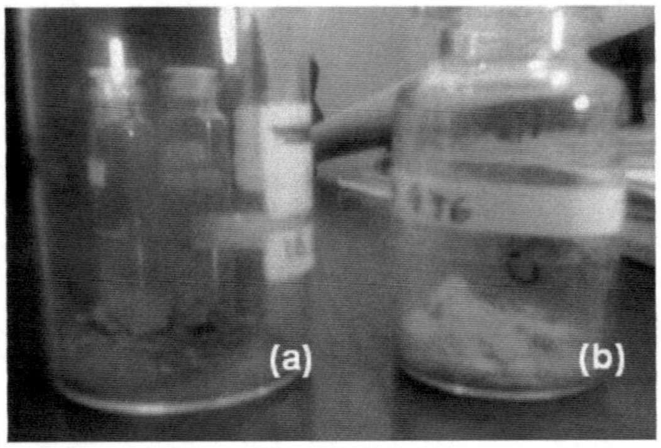

Figure 5. (a) HKUST-1 after being dried at a temperature of 200 °C, (b) HKUST-1 was dried at room temperature.

3.5. Porosity and Surface Area Analysis

To determine the surface area and porosity of HKUST-1 analysis using surface area analysis (SAA) and pore analysis at a temperature of 77 K were performed. The samples were previously activated at

200 °C for 12 hours. Based on this analysis, the synthesized HKUST-1 either by solvothermal or by electrolysis method at a voltage of 15 Volt shows the type I of the nitrogen adsorption isotherms which typical for MOFs and can be categorized as micro-porous material (Figure 6) according to IUPAC (1985) [24]. The synthesized HKUST-1

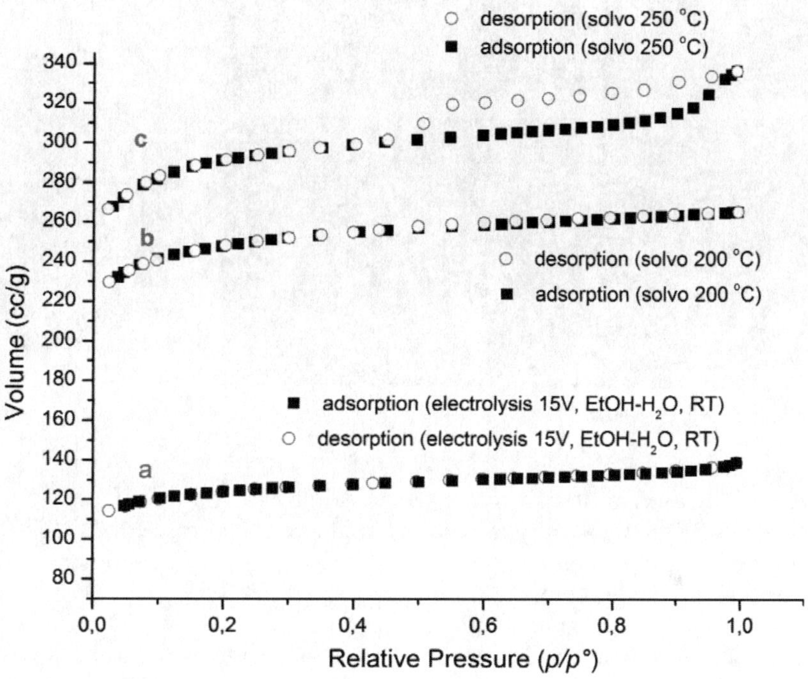

Figure 6. Nitrogen adsorption isotherm of the obtained HKUST-1 which synthesized through electrolysis (a) and solvothermal (b and c)

HKUST-1 by electrolysis at 15 Volt (RT) was obtained BET surface area of 324.331 m^2/g with a pore volume of 0.1902 cc/g and a pore diameter average size of 11.73 Å. While the model of Horvath and Kawazoe (HK method) defined pore size of 1.838 Å. Solvothermally synthesized HKUST-1 showed BET surface area of 635.637 m^2/g with a total size of the pore volume of 0.3531 cc/g and an average size of 11.11 Å pore diameter measured by the HK method. Thus, solvothermal method is capable of producing HKUST-1 with a surface area and pore volume two times larger than the electrolysis method. Nitrogen adsorption isotherm measurement synthesized HKUST-1 via solvothermal method after activation at a temperature of 250 °C showed the BET surface area and pore volume greater than at the time of activation at 200 °C, respectively, the value was 629.2 m^2/g and 0.3777 cc/g with average pore size diameter of about 12 Å (Figure 6).

3.6. Morphological analysis of HKUST-1

The morphology of HKUST-1 was identified by scanning electron microscopic analysis (SEM). Based on Figure 7, it can be seen that the results of the synthesized of HKUST-1 through solvothermal method produce larger material size than the synthesis using electrolysis at a voltage of 15V in solvent combination ethanol: water (1:1). The HKUST-1 morphology with these two methods is solid dark blue with octahedral crystals form with the size of 0.5-5 μm. These results are comparable with the results Mueller et al. (2006).

Figure 7. The morphology of HKUST-1 corresponding SEM analysis (a, b). Synthesis results under solvothermal condition; and (c, d) under electrochemical synthetic method at 15 Volt.

3.7. Infra-Red Measurement

Infrared (IR) analysis is used to determine the functional groups in the BTC ligand and the changes that occur when a BTC ligand coordinated with the copper ion and formed HKUST-1 which mainly changes in absorption of carboxyl groups possessed by BTC ligand. Absorptions around 1715 cm^{-1} corresponds to stretching vibration of C=O acid ligands shown in BTC which after forming HKUST-1 shifted to 1665 cm-1, indicating the deprotonation process occurred in C=O bond. This shift proved that the carboxylate ion participate in the complex formation. Absorption peaks at 410, 500, 610 and 615 cm^{-1} indicating that the synthesis product free of CuO and Cu_2O [25] during the formation of HKUST-1. Moreover, the vibration characteristics of the 714 cm^{-1} may be a Cu-O stretching vibration in which the oxygen atom coordinated with Cu^{2+}. Wide peak at 2500-3300 cm^{-1} correspond to OH stretching absorption of carboxyl group and shifted from 3100 to 3600 at HKUST-1, which indicates the presence of water molecules bond loss at HKUST-1. FTIR comparisons between pure ligand BTC, HKUST-1 were synthesized solvothermal or by electrolysis can be seen in Figure 8.

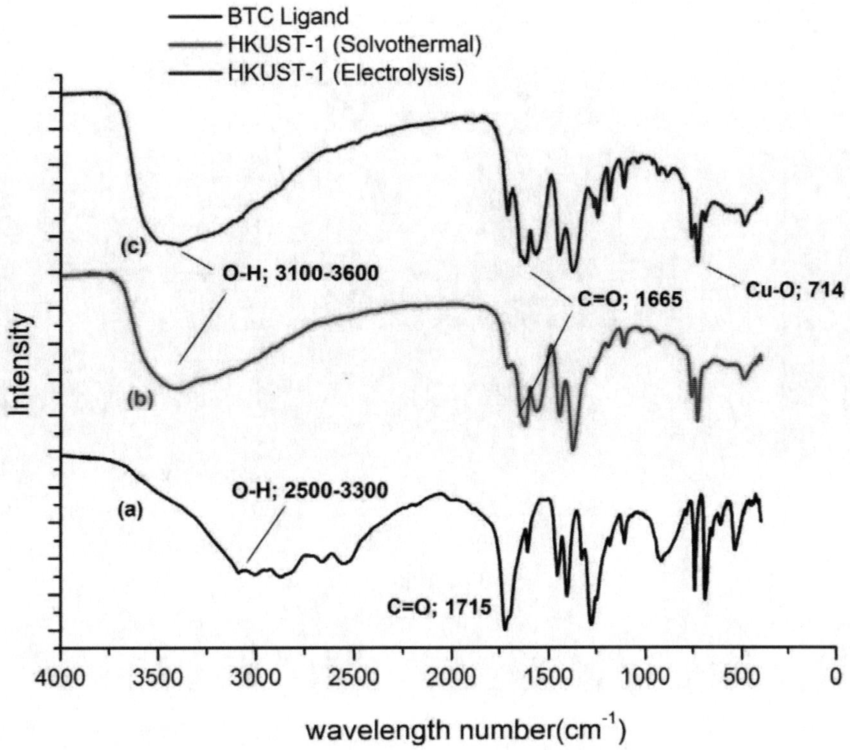

Figure 8. FT-IR spectra of (a) 1,3,5-benzene tricarboxylic acid and (b and c) $[Cu_3(BTC)_2]$ synthesized under solvothermal and electrolysis method.

3.8. Methane Storage Measurement

The synthesized HKUST-1 either by solvothermal or electrochemical method especially the optimum yield produced at 15 V at room temperature were examined for its methane uptake capacities using gravimetric method at relatively low pressure (20 psi ~ 1.38 Bar) following equation (1). Figure 9 shows the methane storage capacities of the HKUST-1. The capacity of electrochemically synthesized HKUST-1 continuously increases even up to 250 minutes with the average ultra high pure (UHP) methane uptake capacities of the HKUST-1 after several measurements was ranging from 7 to 11 %wt. This result is significantly higher than the reported methane storage over HKUST-1 synthesized by solvthermal method by Kaskel and co-worker [14a]. In contrast, the gravimetric measurement of the sample toward mixed gases containing only 5% methane shows lower capacity, although the trend of storage measurement tend to increase till 120 minutes then decrease afterward. This condition indicate. This condition indicates that the presence of gases other than methane (for instance nitrogen) can disturb and compete the methane storage itself. The gravimetric methane storage measurement over the solvothermal product in this case also tend to decrease and only show maximum capacity ca. 6 %wt., which however this condition still need to be optimized.

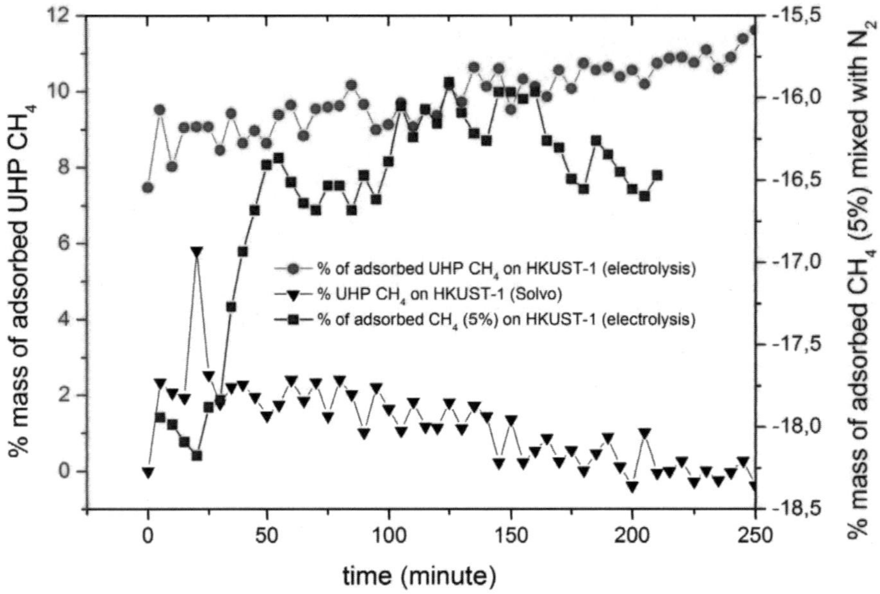

Figure 9. Methane storage by gravimetric analysis of HKUST-1 synthesized by electrolysis and solvothermal method.

The larger capacity of HKUST-1 synthesized by electrochemical method as compared to the solvothermal product can be attributed to the presence of the HKUST-1 synthesized by electrochemical method have relatively high surface area and well-developed pore characteristics. Besides that the method is also suitable to produce large scale of HKUST-1 or MOFs material which applied to industrial scale due to time and cost efficient, environmentally friendly synthesis and support the sustainable development toward the renewable energy.

4. Conclusions

HKUST-1 has been successfully synthesized through an efficient and environmentally friendly electrochemical method. Methane uptake experiment of the corresponding product (at 15 V, RT in the solvent mixture EtOH: H_2O /1:1) showed higher capacity than the solvothermal product and reached up to 11 %wt at 20 Psi (1,38 Bar) according to gravimetric analysis.

Acknowledgement

We gratefully acknowledge financial support by the research grant Doktor Baru and Hibah Mandatory PNBP Sebelas Maret University (UNS). We thank MSc. Wahyu Prasetyo Utomo from ITS Surabaya for conducting the gravimetric measurements.

References

[1] http://finance.detik.com/read/2011/10/11/124232/1741380/1034/ri-simpan-gas-metana-batubara-terbesar-ke-6-di-dunia
[2] http://www.afdc.energy.gov/afdc/fuels/natural_gas_benefi ts.html
[3] Zhou W 2010, The Chemical Record, 10 200
[4] See DOE MOVE program at https://arpa-e-foa.energy.gov/
[5] Burchell T and Rogers M 2000 SAE Tech. Pap. Ser. 2001
[6] Whyatt G A 2010 PNNL-19745
[7] Makal T A, Li J-R. Lu W and Zhou H.-C 2012 Chem. Soc. Rev., **41**, 7761

[8] He Y, Zhou W, Krishna R and Chen B 2012 *Chem Commun* **48** 11813
[9] Guo Z, Wu H, Srinivas G, Zhou Y, Xiang S, Chen Z, Yang Y, Zhou W, O'Keeffe M and Chen B 2011 *Angew Chem Int* Ed **50** 3178
[10] Farha O K and Hupp J T 2010 *Acc Chem* Res. **43**, 1166
[11] Eddaoudi M, Moler D B, Li H, Chen B, Reineke T M , O'Keeffe M and Yaghi O M 2001 *Acc Chem* Res. **34**,319
[12] Chui S.S -Y, Lo S M -F, Charmant J P H, Orpen A G and Williams D 1999 *Science* **283** 1148
[13] a) Müller U, Schubert M, Teich F, Puetter H, Schierle-Arndt K and Pastre J 2006 *J Mater Chem* **16** 626 b) Müller U, Puetter H, Hesse M, Wessel H, Schubert M, Huff J, Guzmann M 2005 WO2005049892-A1; EP1687462-A1
[14] a) Senkovska I and Kaskel S 2008 Microporous Mesoporous Mater. **112**, 108; b) Schlichte K, Kratzke T and Kaskel S 2004 *Microporous and Mesoporo* us Mater **73**, 81
[15] Kumar R S, Kumar S S and Kulandainathan M A 2013 *Microporous and Mesoporous Materials* **168**, 57
[16] Cheng Y, Kondo A, Noguchi H, Kajiro H, Urita K,, Ohba T, Kaneko K and Kanoh H 2009 *Langmuir* **25** 4510
[17] Widiastuti N, Hidayah M. Z. N, Praseytoko D, Fansuri H 2014 *Adv Mat* Lett. **5**(8) 453
[18] Mao Y, Shi L, Huang H, Cao W, Li J, Sun L, Jin X and Peng X 2013 *Chem Commun.* **49** 5666
[19] Mason J A, Veenstra M and Long J R 2014 *Chem* Sci **5** 32
[20] Gensterblum Y 2011 *J of Surface Engineered Materials and Advanced Technology* **1** 23
[21] Vishnyakov A, Ravikovitch P I, Neimark A V, Buelow M and Wang Q M 2003 *Nano Lett.* **3** 713
[22] Joaristi A M, Juan-Alcañiz J, Serra-Crespo P, Kapteijn F and Gascon J 2012 *Cryst. Growth* Des. **12** 3489
[23] Hartmann M, Kunz S, Himsl D and Tangermann O 2008 Langmuir **24** 8634
[24] IUPAC Recommandations 1985 *Pure Appl Chem.* **75** 603
[25] Ardelean I and Cora S 2008 *J Mat Sci Mat Electron* **19** 584

IOP Conf. Series: Materials Science and Engineering **107** (2016) 012031 doi:10.1088/1757-899X/107/1/012031

Changes in grammage, tearing resistance, and water vapor transmission rate of active paper incorporated with Cinnamaldehyde during storage at various temperatures

G J Manuhara*, L U Khasanah, R Utami
Department of Food Science and Technology, Faculty of Agriculture, Sebelas Maret University, Jl. Ir. Sutami no.36A, Surakarta, Central Java, Indonesia

*E-mail: godrasjati@yahoo.com

Abstract. Antimicrobial properties of active paper packaging incorporated with cinnamaldehyde and its application in the storage of agricultural products had been studied. However, changes in grammage, tear resistance and water vapor transmission rate (WVTR) of the active paper during storage is not yet known, whereas it is important to provide consideration in application of the active paper. This study aims to determine the changes in those physical properties during storage (20 days) at various temperatures (10, 20, 25, 30, and 40 °C). The grammage and WVTR of the active paper decreased as increase in storage time and temperature, while the tearing resistance increased as storage time. Higher temperature caused slower increase in tearing resistance, but the results showed fluctuation. The results of Arrhenius plot indicated the activation energy (in kJ/mol) of those physical properties, sorted from the highest to the lowest as follows: 53.6 (grammage), 14.8 (WVTR) and 13.8 (tearing resistance).

1. Introduction

Paper has undergone many developments of its raw materials and application. Fibers from papyrus stems were used as raw material for paper in the kingdom of Ancient Egypt, then the Chinese people used paper for food packaging in the first or second century BC. Modern society has used paper container that is flexible or rigid, as well as coated or mixed with other packaging materials such as plastic and metal foils to improve its function and properties [1, 2]. Paper made from pulp, that was coated or mixed it with a chitosan to enhance its barrier properties against oxygen, carbon dioxide, nitrogen, air and water had also been widely studied [2, 3, 4]. Moreover, recently, the paper with addition of active materials such as cinnamon bark essential oil or cinnamaldehyde (a major compound in cinnamon bark essential oil) had been developed to produce active paper with effective inhibition to microbial growth and deterioration of fresh fruit such as tomatoes, cherries, strawberries, and rambutans [5, 6, 7, 8]. Paper-based active packaging was also reportedly more effective in growth inhibition of bacteria rather than film-based active packaging [9].

However, research on active packaging is still focused on the effect of active material concentration to the antimicrobial activity, and physical properties such as tensile resistance, elastic modulus, elongation, microtopography surface as well as controlled release mechanism of the active materials [10, 11, 12, 13, 14]. Degradation or changes in physical properties of paper-based active packaging during storage have never been investigated.

So far, degradation of ordinary paper (not active paper) during storage had been widely studied, but the studies treated the paper at very high temperature (≥80 °C). The experiment at lower temperatures or close to normal storage temperature was proposed to predict, more accurately, the changes of the paper. This method employed Arrhenius-plot with the slope giving an apparent activation energy which represent minimal energy required for a reaction to take place [15].

Thus, this study was designed to investigate changes in physical properties (grammage, water vapor transmission rate, and tearing resistance) of active paper made from pulp and chitosan incorporated with cinnamaldehyde during storage at various temperatures (10, 20, 25, 30 and 40 °C). It is necessary to study those properties because grammage affects other physical properties, while the water vapor transmission rate (WVTR) and tearing resistance affect active paper application, especially when it is used in agricultural product storage. Finally, the activation energy of each physical properties is analyzed so that the most susceptible property of active paper to change during storage can be determined.

2. Materials and Methods

2.1. Materials

Materials used for the preparation of the active paper is cinnamaldehyde with 99.99% purity (supplied from Aldrich), kraft paper, chitosan (from local supplier/ Chemix Pratama), and acetic acid (from Merck).

2.2. Methods

2.2.1. Preparation of active paper. The active paper was made from the mixture of pulp and chitosan incorporated with cinnamaldehyde. At first, three kinds of ingredients was prepared. They were pulp, solution of chitosan in acetic acid, and cinnamaldehyde emulsion in distilled water. Pieces of kraft paper (dimension about 2 mm x 2 mm; 6%w/v) were soaked in distilled water for 24 hours then beaten in a blender for 5 minutes into pulp. Chitosan powder (0.45%w/v) was dissolved into a solution of 1% acetic acid. In a separate beaker glass, tween-80 (0.45%w/v) was dissolved in distilled water. Cinnamaldehyde (2.5%w/w of kraft paper) was added to the emulsion of tween-80, then it was constantly stirred for 5 minutes to produce cinnamaldehyde emulsion. These three ingredients were slowly mixed in blender for 5 minutes, then evenly poured onto the filter surface in the cast (20 cm x 20 cm) until rough sheet of wet paper was formed. The sheet on filter was then compressed under glass plate with 2.0 kg load for 10 minutes to remove residual water. The sheets were dried at 30 °C for 48 hours and flipped after the first 24 hours of drying.

2.2.2. Storage of active paper. The paper sheets (20 cm x 30 cm) were stored by hanging on a hanger rod made of glass (to avoid migration of compounds from the hanger into the paper) in incubator box with temperature variation of 10, 20, 25, 30, and 40 °C for 20 days. Every 5 days, samples of each temperature were taken to be analyzed for its grammage, water vapor transmission rate (WVTR), and tearing resistance.

2.2.3. Grammage. Grammage test method of paper and paperboard according Indonesia National Standard test (SNI 14-0439-1989) was employed as follows. Paper samples were taken with the size of 10 cm x 10 cm. Grammage (in g/m^2) was then determined by weighing the sample material and dividing the weight by the area.

2.2.4. Water vapor transmission rate (WVTR). The measurement of water vapor transmission rate was performed based on ASTM E096-95 as follows. Paper samples were cut in the shape of a circle. Three test dishes were filled with distilled water up to a level 19 ± 6 mm from the paper samples. The specimens were then attached to the dish. The dishes were then assembled and placed on a true horizontal surface in a chamber. The chamber humidity and temperature must be controlled at 90% relative humidity and 38 °C. The dishes were weighed periodically (every 3 hours) for 24 hours and recorded by automatic device attached to the horizontal surface. The weight then was plotted into a graph against elapsed time, in order to obtain a curve which tends to become straight. WVTR (in $g/m^2.h$) was then calculated by dividing the slope of straight line by the test area (cup mouth area).

2.2.5. Tearing resistance. Tearing resistance measurement was performed based on Elmendorf-type method from SNI (Indonesia National Standard) 0436:2009 which refer to TAPPI (Technical Association for the Pulp and Paper Industry) 414 om-04 and ISO 1974:1990. Samples of paper were clipped vertically on the tool. The test was performed by inducing a crack in the test piece and applying a load perpendicular to the face to pull the paper apart. The pendulum was released to swing freely and tear the paper, then the value indicated by the pointer was recorded as tearing resistance (in mN).

2.2.6. Activation energy. The activation energy was determined by using the Arrhenius-type plot based on the results of grammage, WVTR, and tearing resistance test. Paper storage time (0, 5, 10, 15, 20) was plotted on the horizontal axis, while the physical properties (grammage, WVTR, or tear resistance) was plotted on the vertical axis for the first order or natural logarithmic of the physical properties on the vertical axis, so that slope obtained for each storage temperature (10, 20, 25, 30, and 40 °C). The next curve is made using natural logarithmic slope of the slope on the vertical axis and inversion of the absolute temperature (K^{-1}) on the horizontal axis to produce a slope (Ea/ R) giving an activation energy (Ea).

3. Result and Discussion

3.1. Rammage

The active paper grammage decreased during 20 days of storage (Figure 1). Higher temperature resulted in more rapid decrease of the grammage. Thus, the most rapid decrease in grammage was demonstrated by the active paper stored at 40 °C, while the slowest one was demonstrated by the active paper stored at 10 °C. The decrease of grammage might be associated with moisture content of the active paper that also decreased more rapid as the storage temperature (the data not shown).

The grammage of the active paper was five to six times higher than commercial kraft paper grammage and three to four times higher than the paper made from pulp-chitosan resulted from other studies [2, 16]. Additional material (such as chitosan, tween-80, and cinnamaldehyde) and higher amount of total material used resulted in the striking difference. The active paper grammage met ISO standards for paperboard or slightly below the standard. Grammage affected tensile strength as well as dewatering and rewetting of paper based on the other studies [17, 18].

3.2. Water vapor transmission rate

Water vapor transmission rate (WVTR) of active paper decreased during 20 days of storage (Figure 2). Higher storage temperature resulted in more rapid decrease of WVTR. The most rapid decrease in WVTR was demonstrated by the active paper stored at 40 °C, while the slowest one was demonstrated by the active paper stored at 10 °C. Lower WVTR of indicates that the paper or film layer is more difficult to be passed through by water vapor.

The decrease of WVTR may be associated with moisture content of the active paper that also decreased more rapid as the temperature increase. Other study indicated that the moisture content affected the formation of more porous polymer network structure in the film [19]. The loss of water molecules from the active paper during storage or due to higher temperature might result in more dense structure of polymer chains (both cellulose and chitosan) and more intermolecular interactions, eg hydrogen bonds. The other studies also reported the formation of more solid structure due to higher temperatures employed in the film preparation [20, 21]. Higher density was also associated with lower air permeance of paper [16]. Thus, increased density of the active paper as the increased temperature and time of storage might result in the decrease of WVTR.

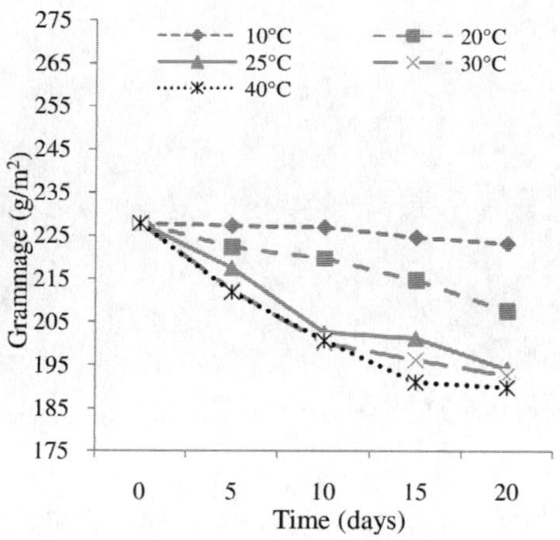

Figure 1. Grammage of paper

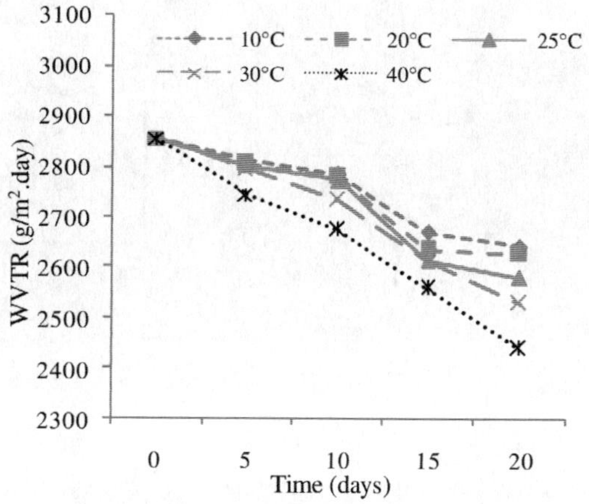

Figure 2. WVTR of paper

Table 1. Activation energy of active paper physical properties

Physical properties	Activation energy (kJ/mol)
Grammage	53.6
WVTR	14.8
Tearing resistance	13.8

3.3. Tearing resistance

Tearing resistance of the active paper increased during storage of 20 days (Figure 3). The result confirmed the increased density of polymer network in the active paper due to the loss of water molecules during storage as indicated by WVTR test result. However, the temperature effect on the rate of increase in the tearing resistance seemed contrary to the temperature effect on the rate of increase in WVTR. Higher temperature resulted in slower increase in tearing resistance, so that the most rapid increase in tearing resistance was demonstrated by the active paper stored at 10 °C, while the slowest increase in tearing resistance was demonstrated by the active paper stored at 40 °C. The result led to speculation that other factor might affect tearing resistance in addition to the density of the polymer network structure or paper moisture content.

The slower increase in tearing resistance at the higher temperature might be associated with the negative effect of paper degradation which was more intense at the higher temperature. Initially, paper degradation begin with polymer chain scission in the amorphous region and continued with the formation of crosslinks between polymers, which led to increased brittleness and decreased tearing resistance [16, 22]. Thus, this result raised other speculation that the paper polymers degradation due to higher temperature only affected tearing resistance, but had no effect on WVTR. It would need further study and comparison. So far, the studies on the temperature effect on the paper degradation focused on folding endurance and tensile resistance [23, 24], while tearing resistance and WVTR never been studied.

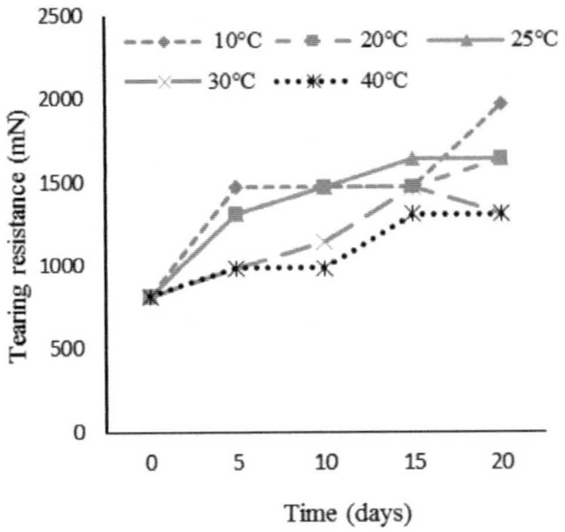

Figure 3. The tearing resistance of paper.

3.4. Activation energy

The activation energy indicates the amount of energy required to initiate the reaction or changes in particular material. The study of the activation energy has been commonly used for food deterioration and cooking as well as chemical reactions based on Arrhenius plot [25, 26]. However, so far, the studies of paper degradation during storage employed one or two high temperature (≥80 °C) at various humidity. Lately, the method is doubtful because the experimental design does not reflect normal storage conditions so that it might lead to wrong conclusion [15, 22]. Therefore, this study employed storage temperature of 10, 20, 25, 30, and 40 °C to provide the better representation of the physical changes that might occur in the active paper. The changes might involve particular chemical reaction in the active paper. The activation energy for the physical properties of the active paper presented in (Table 1).

The result indicated that the paper strength was sensitive to changes during storage. The tearing resistance was the most susceptible properties to be affected during storage. However, the tearing resistance demonstrated increased trend, while the decrease was noted in grammage and WVTR. As discussed earlier, increased tearing resistance due to temperature and storage might be associated with decreased moisture content that result in increased density of the paper materials. It means that the effect of density or moisture content of the paper was greater than the paper degradation. Nevertheless, further research on changes in other physical properties of the active paper such as folding endurance and tensile resistance is needed to confirm this result.

4. Conclusion

The grammage and WVTR of active paper decreased as increased storage time and temperature, while the tearing resistance increased during storage. Higher temperature caused more rapid decrease in the grammage and WVTR, but slower increase in the tearing resistance during storage. The activation energy sorted from the highest to the lowest as follows: grammage, WVTR, and tearing resistance. Therefore, the tearing resistance was the most sensitive properties of the active paper.

References

[1] Raheem D 2012 Application of plastics and paper as food packaging materials – An overview *Emir. J. Food Agric.* **25 (3)** 181

[2] Gallstedt M and Hendqvist M S 2006 Packaging-related mechanical and barrier properties of pulp–fiber–chitosan sheets *Carbohydrate Polymers* **63** 46–53

[3] Kjellgren H, Gallstedt M, Engstrom G and Jarnstrom L 2006 Barrier and surface properties of chitosan-coated greaseproof paper *Carbohydrate Polymers* **65** 453–60

[4] Nada A M A, El-Sakhawy M, Kamel S, Eid M A M and Adel A M 2006 Mechanical and electrical properties of paper sheets treated with chitosan and its derivatives *Carbohydrate Polymer* **63** 113 – 121
[5] Wrona M, Bentayeb K and Nerin C 2015 A novel active packaging for extending the shelf-life of fresh mushrooms (Agaricus bisporus) *Food Control* **54** 200-7
[6] Rodriguez-Lafuente A, Nerin C and Battle R 2010 Active paraffin-based paper packaging for extending the shelf life of cherry tomatoes *J Agric Food Chem.* **58 (11)** 6780-6
[7] Rodriguez A, Batlle R and Nerin C 2007 The use of natural essential oils as antimicrobial solutions in paper packaging - part II *Progress in Organic Coatings* **60** 33–8
[8] Sivakumar D, Wijeratnam R S W, Wijesundera R L C and Abeyesekere M 2002 Control of postharvest diseases of rambutan using cinnamaldehyde *Crop Protection* **21** 847–52
[9] Morelli C L, Mahrous M, Belgacem M N, Branciforti M C, Bretas R E S and Bras J 2015 Natural copaiba oil as antibacterial agent for bio-based *Industrial Crop and Products* **70** 134-41
[10] Goueva D M, Mendonca R C S, Soto M L and Cruz R S 2015 Acetate cellulose film with bacteriophages for potential antimicrobial use in food packaging *Food Science and Technology* **63** 85 – 91
[11] Nostro A, Scaffaro R, D'Arrigo M, Botta L, Filocamo A, Marino A and Bisiqnano G 2012 Study on carvacrol and cinnamaldehyde polymeric films: mechanical properties, release kinetics and antibacterial and antibiofilm activities *Appl Microbiol Biotechnol.* **96 (4)** 1029
[12] Chinnan M S and Park H J 1995 Effect of plascticizer level and temperature on water vapor transmission of cellulose-based edible films *Journal of Food Process Engineering* **18** 417-29
[13] Chiou B S, Avena-Bustillos R J, Bechtel P J, Imam S H, Glenn G M and Orts W J 2009 Effects of drying temperature on barrier and mechanical properties of cold-water fish gelatin films *Journal of Food Engineering* **95** 327–31
[14] Gemili S, Yemenicioglu A and Altınkaya S A 2009 Development of cellulose acetate based antimicrobial food packaging materials for controlled release of lysozyme *Journal of Food Engineering* **90** 453–62
[15] Pork H J 2000 *Rate of paper degradation: the predictive value of artificial aging test* (Amsterdam: European Commission on Preservation and Access) p 13-9
[16] Karlsson H 2007 *Some aspects on strength properties in paper composes of different pulps* (Karlstad: Karlstad University Studies) p 17 - 32
[17] Considine J, Skye W, Chen W, Matthys D, Vahey D, Turner K, Rowlan R June 1-4 2009 Enhancing paper strength by optimizing defect configuration *Proceedings of the SEM Annual Conference* Albuquerque New Mexico USA Society for Experimental Mechanics Inc
[18] Rantanen J and Maloney T C 2013 Press dewatering and nip rewetting of paper containing nano- and microfibril cellulose *Paper Physics, Nordic Pulp & Paper Research Journal* **28 (4)** 582 – 7
[19] Stading M, Rinlav-Westling A and Gatenholm P 2001 Humidity-induced structural transitions in amylose and amylopectin *films Carbohydrate Polymers* **45** 209-17
[20] Keddie J L, Braun P V and Giannelis E P 1993 Relationship between water desorption and low-temperature densification of colloidal anatase thin films *Journal of The American Ceramic Society* **76 (10)** 2529 – 33
[21] Tapia-Blácido D, Sobral P J and Menegalli 2005 Effects of drying temperature and relative humidity on the mechanical properties of amaranth flour films plasticized with glycerol *Brazilian Journal of Chemical Engineering* **22 (02)** 249 – 56
[22] Karlovits M and Gregor-Stevec D 2012 Durability of cellulose and synthetic papers exposed to various methods of accelerated aging *Acta Polytechnica Hungarica* **9 (6)** 81-100
[23] Zervos S 2007 Accelerated aging kinetics of pure cellulose paper after washing, alkalization and impregnation with methylcellulose *Restaurator* **28 (1)** 55–69

[24] Havlínová B, Kauscak S, Petrovicova M, Makova A and Brezova V 2009 A study of mechanical properties of papers exposed to various methods of accelerated aging - part I. The effect of heat and humidity on original wood-pulp papers *Journal of Cultural Heritage* **10** 222–31
[25] Petrou A L, Roulia M and Tampouris K 2002 The use of the Arrhenius equation in the study of deterioration and of cooking of foods. Some scientific and pedagogic aspects *Chemistry education: research and practice* **3 (1)** 87-97
[26] Pogaku R, Raman J K and Ravikumar G 2012 Evaluation of activation energy and thermodynamic properties of enzyme catalyzed transesterification reactions *Advances in Chemical Engineering and Science* **2** 150

Design of reflective optical fiber sensor for determining refractive index and sugar concentration of aqueous solutions

Ahmad Marzuki*, Nila Wulan Sari, Riatun
Department of Physics, Sebelas Maret University
Jl. Ir. Sutami 36A, Surakarta 57126 Indonesia

*E-mail: amarzuki@mipa.uns.ac.id

Abstract. A reflective optical fiber sensor designed for measuring refractive index and sugar concentration of aqueous solutions is described. Two strains of parallel polymer optical fibers (POF) were wrapped in a bundle such that one of their fiber's end cross-sections had the same distance to the mirror surface. The light coming out from one strain of the fiber was reflected by the mirror to the second fiber. Sugar concentration of the aqueous solution filling the space between the fiber ends and the mirror was varied (1.0 M, 1.5 M, 2.0 M, 2.5 M, 3.0 M, 4.0 M, and 5.0 M). It was shown from the experiment that light intensity detected by photo-detector is linearly related to the percentage of the dissolved sugar in the solution as well as the variation of the sugar solution refractive index ($R^2 = 0.987$).

1. Introduction

In the last few decades, fiber optic sensors have emerged as the one of the main class of modern sensors. Now, these sensors are available for measuring almost all physical properties at which the already established sensors such as the electronic based sensors have long been used, i.e. mechanical sensor [1,2], temperature sensor [3], biochemical sensors [4]. The fast growing interest in this relatively new class of sensor is mainly driven by the fact that fiber optic has some unique physical properties related to their small size, high flexibility, immunity to electromagnetic and radio interference, and high resistance level to harsh environments. Having these properties, fiber sensors have found widespread applications which comprise the use of the sensors in physical, chemical and biological sensors.

Fiber sensors used for chemical substance monitoring have been developed by employing varies types of light-material interactions, i.e., reflection, refraction, absorption, emission, and polarization. Among them, fiber sensors work based on absorption and emission are mostly used. Examples of fiber sensors developed under these working principles are fiber sensors used for detecting CO_2 [5, 6] and glucose [7, 8]. Due to the nature of the absorption and emission processes, these types of the fiber sensors work only at a particular wavelength corresponding to the energy difference between the ground electronic state and an excited electronics state of a substance under investigation. Using these sensors, a chemical substance contained in any mixtures can be theoretically determined. In a process involving different flows of different single solution, employing these types of sensors mean that many different chemical sensors must be available at any time.

In this paper, a reflective optical fiber sensor which has the ability to determine the refractive index and sugar concentration in aqueous solution is reported. This type of sensor works based on the fact that solution with different sugar concentration will result in the different refractive index.

IOP Conf. Series: Materials Science and Engineering **107** (2016) 012032 doi:10.1088/1757-899X/107/1/012032

2. Experiment

Figure 1 is the experimental setup used to determine the optimum configuration of the fiber sensor for measuring refractive index and concentration of sugar solution. Light from the light source is split into two paths: modulation path and reference path. Light in modulation path is directed to a reflector through fiber transmitter such that reflection occurs. Part of the reflected light will then go to the fiber receiver adhered to the fiber transmitter. Light from these two paths are captured by photodetector detectors (PD1 and PD2) for further processing in the oscilloscope.

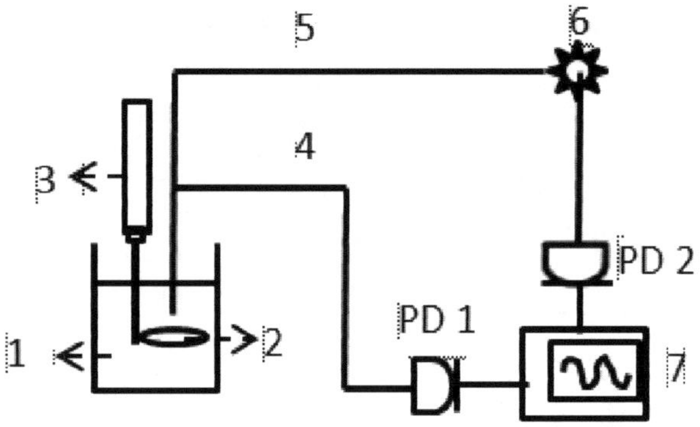

Figure 1. Schematic arrangement for refractive index fiber sensor (1. Solution, 2. Mirror, 3. micrometer, 4. Modulated path optical fiber, 5. Reference path optical fiber, 6. The light source, 7. Oscilloscope)

Light transmittance (T) is then defined based the ratio between the intensity of light coming from modulation path (I_m) and reference path (I_r) as given by equation (1).

$$T = \frac{I_m}{I_r} x\ 100\% \qquad (1)$$

T is set to 100% if the fiber sensor is in the air. As the fiber sensor is dipped into sugar solution the range of angles over which the fiber transmitter emits the light (numerical aperture) is decreasing. The numerical aperture (NA) of a fiber whose core and cladding refractive indices are n_{co} and n_{cl} is given by equation (2).

$$\mathrm{NA} = \sin\alpha = \frac{1}{n_m}\sqrt{{n_{co}}^2 - {n_{cl}}^2} \qquad (2)$$

Where α is the acceptance angle and n_m is the refractive index of sugar solution. Variation of refractive were made by making a solution with a different concentration of sugar (1.0 M, 1.5 M, 2.0 M, 2.5 M, 3.0 M, 4.0 M, and 5.0 M). LED of 632 nm is used.

3. Results and Discussion

As shown in figure 1, the intensity of light transmitted through receiver fiber as well as reference fiber is measured by oscilloscope as voltage. Figure 2 shows how light transmittance change as the distance of fiber ends facing to the mirror changes. Due to the simplicity of the experiment, measurement repetition produced the type of curve but different in their path. Interestingly, the peaks of all curves for each sugar concentration are located at the same mirror-fiber ends distance.

Looking back to Figure 2, it can be seen that the peak of the curve obtained for 5.0M of sugar solution is located at higher value of displacement (distance between fiber ends and the mirror) compared to that of 1.0 M. The way the peak positions change as function of sugar concentration is shown in figure 3. It is seen that peak position shift is linearly related to the sugar concentration (R^2 = 0.987).

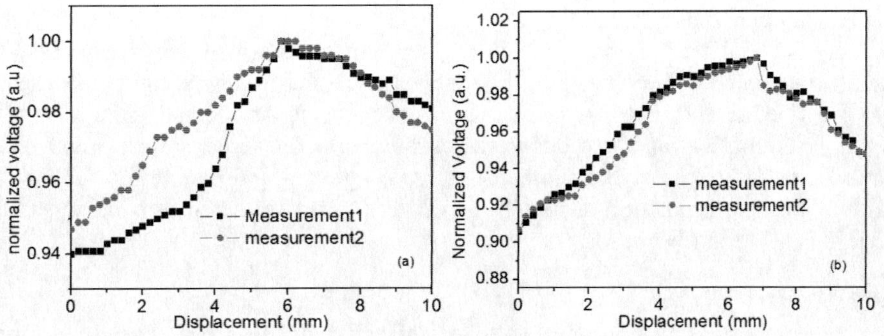

Figure 2. Normalized voltage change measured as a function of the distance between the mirror and the fiber ends for (a) 1 molar solution, (b) 5 molar solution.

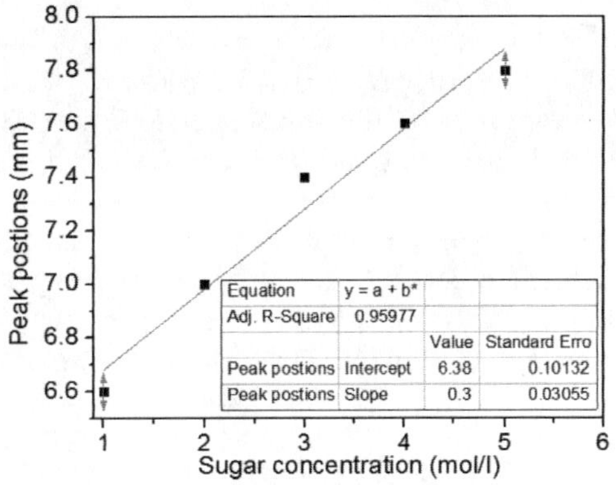

Figure 3. The peak position of the curve as a function of sugar concentration.

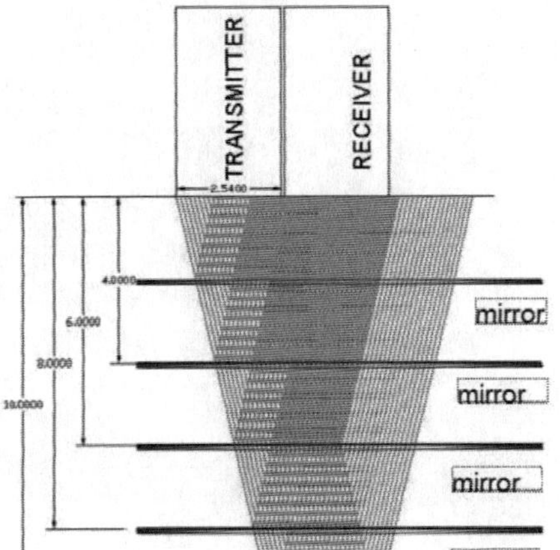

Figure 4. Reflection model explaining the position of mirror relative to the fiber ends giving maximum transmittance

IOP Conf. Series: Materials Science and Engineering **107** (2016) 012032 doi:10.1088/1757-899X/107/1/012032

The above results can be understood by applying equation (2) and figure 4. For a given sugar concentration and thus at the solution refractive index is fixed, a light path for a given light guide mode will leave the fiber transmitter at the same acceptance angle (α). All the reflected light will enter into the receiver fiber only at a certain distance. As the mirror is positioned very close to the fiber ends, only small fraction of light comes into the receiver. This fraction gets higher as the distance from the mirror to the fiber ends bigger. After a maximum value is reached, this faction will decrease. Furthermore, fiber Numerical aperture decreases as the sugar refractive index (sugar concentration) increases. This leads to decreasing the same acceptance angle (α). Consequently, the mirror has to be displaced further down to a position at which the far left of the reflected light as shown in Figure 4 falls into the left wall of the fiber receiver.

4. Conclusions

An experimental setup used to configure a sugar refractive index fiber sensor working based on reflectively modulated fiber sensor have been shown. Analysis has been given to see the unique character of the reflected light by the mirror dipped into the sugar solution. For a given solution, it is found that the maximum transmission is obtained at a particular distance. This distance is displaced farther as the sugar concentration and thus the refractive index of solution increases.

References

[1] Gage J R, Wang H F, Fratta D and Turner A L 2014 *Int. J Rock. Mech. Min. Sci* **71** 350-361
[2] Marzuki A, Heriyanto M, Setiyadi I D and Koesuma S 2015 *JPCS* **622** 012059
[3] Ma L, Kang Z, Qi Y and Jian S 2015 *International Journal of Light and Electron Optics* **126** (9–10) 1044-1046
[4] He Y J 2015 *Sensor Actuator* B 47 530-536
[5] Starecki F, Charpentier F, Doualan J L, Quetel L, Michel K, Chahal R, Troles J, Bureau B, Braud A, Camy P, Moizan V and Nazabal V 2015 *Sensor Actuat B-Chem* **207** 518–525
[6] Mulrooney J, Clifford J, Fitzpatrick C, Lewis E 2007 *Sensor Actuat A-Phys* **136** (1) 104–110
[7] Li D, Sun Y, Yu S, Sun C, Yu H, and Xu K 2015 *Sensor Actuat B-Chem* **220** 1033–1042
[8] Khan F and Pickup J C 2013 *Biochem. Biophys. Re.s Commun.* **438** (2013) 488–492.

Synthesis of $LiNi_xMn_{2-x}O_4$ by low-temperature solid-state reaction and its microstructure

D Purwaningsih[1,2*], R Roto[1] and H Sutrisno[2]

[1]Department of Chemistry, Faculty of Mathematics and Natural Sciences, Gadjah Mada University, Sekip Utara BLS 21, Yogyakarta 55584 Indonesia
[2]Department of Chemistry Education, Faculty of Mathematics and Natural Sciences, Yogyakarta State University, Jl. Colombo 1, Yogyakarta 55281 Indonesia

[*]E-mail: dyahuny@yahoo.com

Abstract. This study aims to synthesis the Ni-doped $LiNi_xMn_{2-x}O_4$ (x = 0; 0.02; 0.04; 0.06; 0.08; 0.1) by low temperature solid-state reaction. The microstructure of the product was evaluated based on the mole ratio of Ni/Mn of the precursors. The structural analysis of $LiNi_xMn_{2-x}O_4$ was analyzed by x-ray diffraction equipped with the Direct Method of win PLOTR package program and Diamond. It was found that doping with Ni could change the size, crystallinity and microstructure of $LiNi_xMn_{2-x}O_4$. The $LiNi_xMn_{2-x}O_4$ solids have a cubic structure with a space group of *Fd3m*. The increase in the doping content does not affect the structure. The particle size of the products is about 150-500 nm. The crystallinity of the solids tends to increase with the increase in Ni content. However, the increase of Ni content in the product causes the lattice parameters of the unit cell to decrease.

1. Introduction

Lithium-ion batteries are widely used for energy storage, portable devices, and electric vehicles. Although many new cathode materials are developed, $LiMn_2O_4$ is still well recognized as a promising cathode material for these batteries [1]. The spinel-structured $LiMn_2O_4$ offers attractive alternative material over commercially available $LiCoO_2$. $LiMn_2O_4$ is cheap, non-toxic, and safe [2]. However, it has severe capacity fading in the application [3]. The charge-discharge cycling capability of $LiMn_2O_4$ at high temperatures can be improved by the substitution of Mn by other transition metals to give the corresponding $LiM_xMn_{2-x}O_4$ where M = Co, Mg, Al, Cr, Ni, Fe, Ti and Zn [4]. $LiNi_xMn_{2-x}O_4$has a discharge capacity of 140 $mAhg^{-1}$ and open circuit potential of 4.7 V. This energy density is tunable by doping [5].

It is documented well that electrochemical performance of the battery cathode is affected by the powder properties such as structure, particle size, grain morphology, specific surface area and crystallinity [6]. In addition to its large surface area, fast transport of mass and charge, there are many properties to explore. The large surface area and small particle size can improve solubility in the electrolyte. The alteration in chemical properties may make the electrode materials sensitive to the presence of impurities[7].

The electrochemical property of $LiMn_2O_4$–based spinel highly depends on its synthetic routes, such as the Pechini process, sol-gel, emulsion method, the citric method [8], etc. However, most of these methods involve many treatment processes or expensive reagent, which is time-consuming and high cost for commercial applications. We report here on the synthesis of series of $LiNi_xMn_{2-x}O_4$ using a low-temperature solid-state reaction at various Ni/Mn mole ratios. The physical properties and microstructures of the synthesized materials were investigated.

2. Experimental Section

2.1. Synthesis of MnO_2 and $LiNi_xMn_{2-x}O_4$.

An analytical grade of $Mn(CH_3COO)_2$ and $Na_2S_2O_8$ (Aldrich) were used. All other chemicals were used without further purification. In a typical synthesis, $Mn(CH_3COO)_2$ and $Na_2S_2O_8$ with a molar ratio of 1:1 were dissolved at room temperature in 80 mL deionized distilled water. The mixture was stirred to form a clear homogeneous solution. The solution was transferred to the flask and heated at 120 °C for 12 h. The obtained powder was subsequently dried at 300 °C for 1 h in the oven.

Synthesis of $LiNi_xMn_{2-x}O_4$ is the following; LiOH of 0.00143 moles, 0.0000286 moles of $Ni(CH_3COO)_2$ and 0.0028314 moles were dispersed into high purity ethanol to form a thick slurry and stirred. The product was separated and dried at room temperature. The above process was repeated three times. The $LiNi_{0.02}Mn_{1.98}O_4$ powder was ignited at 750 °C for 10 hours. The same procedure was followed for synthesizing the materials with different mole ratio.

2.2. Characterization of the $LiNi_xMn_{2-x}O_4$ microstructure.

The obtained MnO_2 and $LiNi_xMn_{2-x}O_4$powders were analyzed using X-ray powder diffractometer (XRD). The XRD patterns were obtained on Rigaku Miniflex 600-Benchtop XRD instrument using Cu Kα radiation (λ = 1.5406 Å) at ambient temperature. The XRD instrument was set to operate at 40 kV and 15 mA. The XRD data was obtained with a 2Θ interval ranging from 20° to 90°. The Rietveld analysis was conducted with the Fullprof software package [9] to refine the X-ray diffraction data. The refined parameters are a unit cell, scale factor and full width at half-maximum (FWHM). The SEM images were obtained using JEOL JSM-6510LASEM. The effect of the nickel content on the structure of $LiMn_2O_4$ was studied using energy dispersive X-ray spectroscopy (EDX). The EDX analysis was also used to analyze the presence of Mn, O and Ni elements in the prepared materials. The Brunauer-Emmett-Teller (BET) data was collected based on adsorption data in relative pressure (P/Po) of about 0.30 [10]. The BET calculation was done in ASAP 2020 V4.01 software from Micrometrics.

3. Results and discussion

3.1. XRD Patterns

Figure1 shows XRD patterns of produced $LiNi_xMn_{2-x}O_4$. High peak intensity suggests that the $LiNi_xMn_{2-x}O_4$ has good crystallinity. Figure 2 shows the variation of the lattice constant of the materials with different compositions. The Rietveld refinement data obtained using Full Prof indicates that lattice parameter decreases with the increase in doping content. The substitution of Mn^{3+} by Ni^{2+} in the octahedral sites leads to the decrease in the lattice parameters. Note that ionic radius of Ni^{2+} is 0.560 Å, smaller than that of Mn^{3+}, which is 0.645 Å. The difference in the ionic radius between Ni^{2+} and Mn^{3+} is small, therefore only a few of change the lattice parameter.

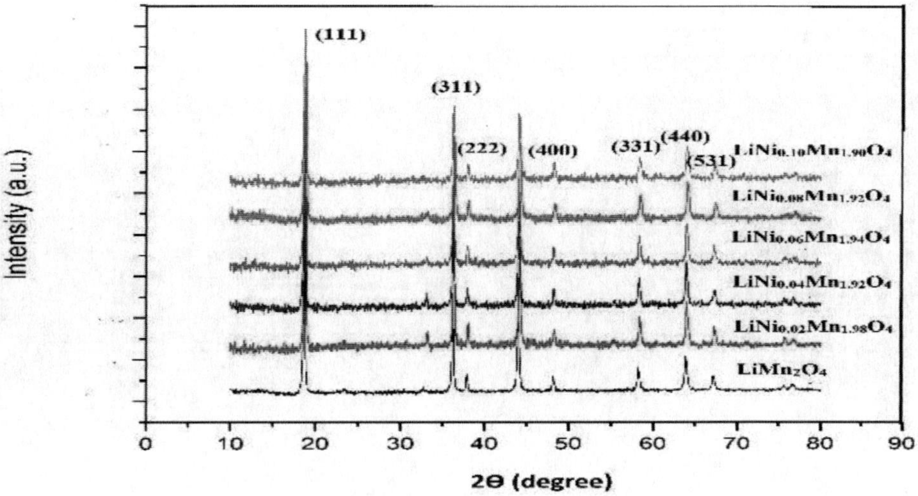

Figure 1. XRD Patterns of $LiNi_xMn_{2-x}O_4$ Ni/Mn mole ratios

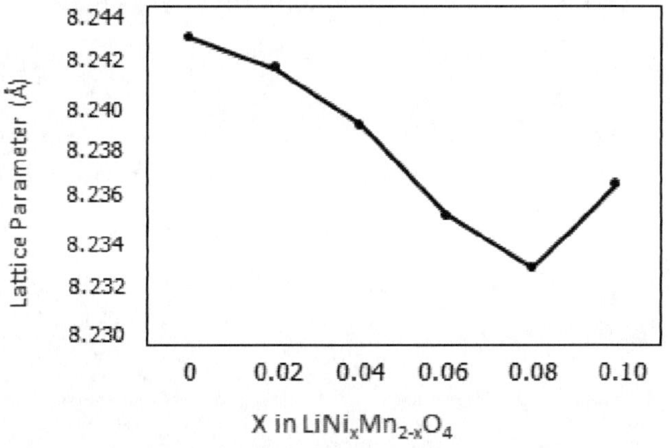

Figure 2. Variation of lattice parameter with x in $LiNi_xMn_{2-x}O_4$

Figure 3 shows the XRD patterns of $LiNi_xMn_{2-x}O_4$. The products have a cubic phase with space group of *Fd3m* as indicated by results of analysis using winPLOTR package program and Diamond. The results of Rietveld are shown in Fig. 3. The experimental points are given as dot (.) and theoretical data are shown as solid line. The difference between theoretical and experimental data is shown as a bottom line. The vertical lines represent the Bragg's allowed peaks. The X-ray diffraction data show at 31-33°, which can be assigned to a Mn_2O_3 phase. The presence of Ni in the structure causes the appearance of $NiMnO_3$, which is indicated by the presence of the peaks at 31-33° [11]. It seems that lithium ions occupy the tetrahedral (8a) site. After the modification, the lattice shrinkage is due to the smaller ionic radii of Ni^{2+}, which replaced the Mn^{3+} at 16d sites. On the other hand, Ni^{2+} substituting part Mn^{3+} can enhance the content of Mn^{4+} to keep the charge balance, and Mn^{4+} is with smaller ionic radii than that of Mn^{3+} [12]. Oxygen atoms are arranged in the cubic-closed packing 32e.

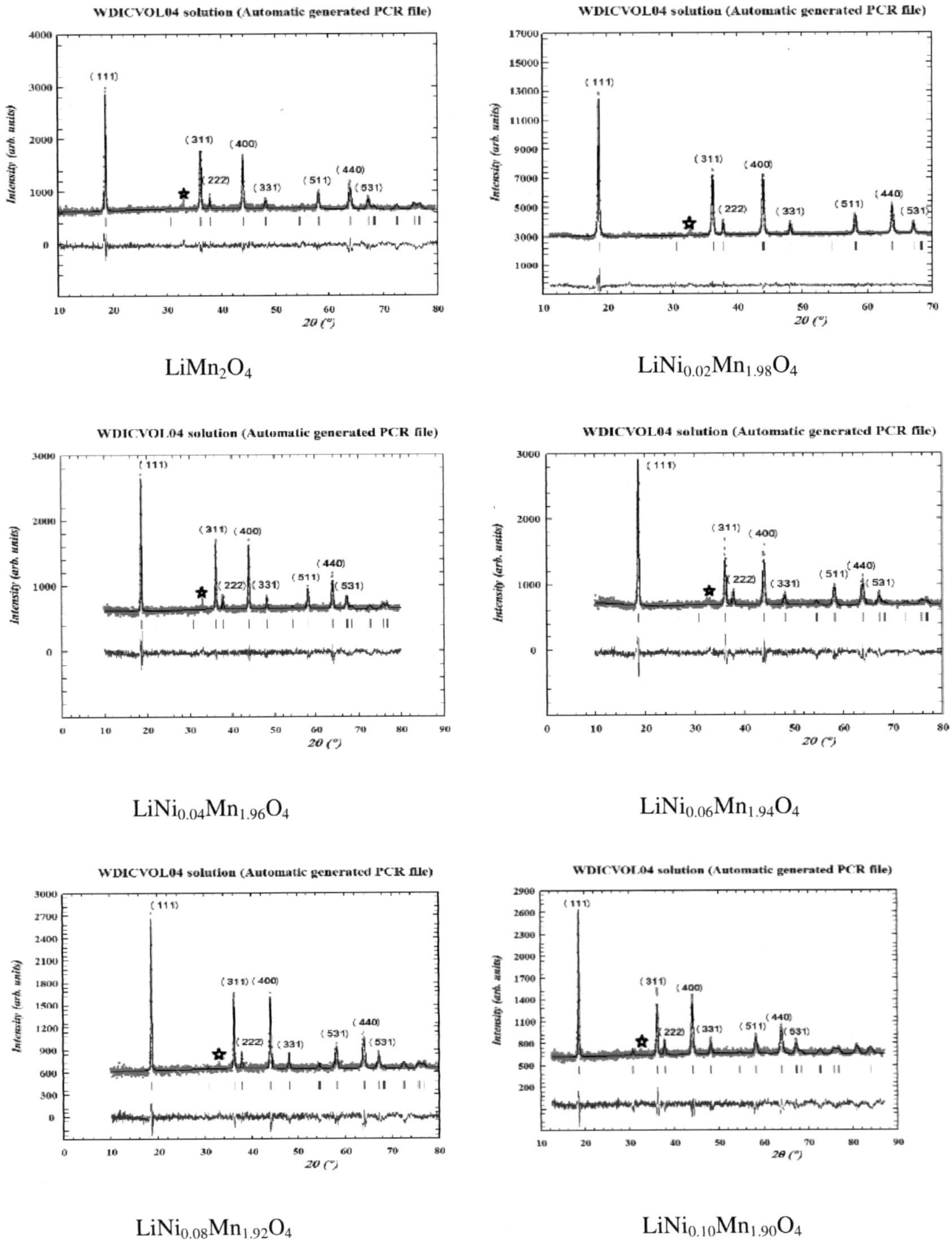

Figure 3 XRD refinement data of LiNi$_x$Mn$_{2-x}$O$_4$ with various Ni/Mn mole ratios

IOP Conf. Series: Materials Science and Engineering **107** (2016) 012033 doi:10.1088/1757-899X/107/1/012033

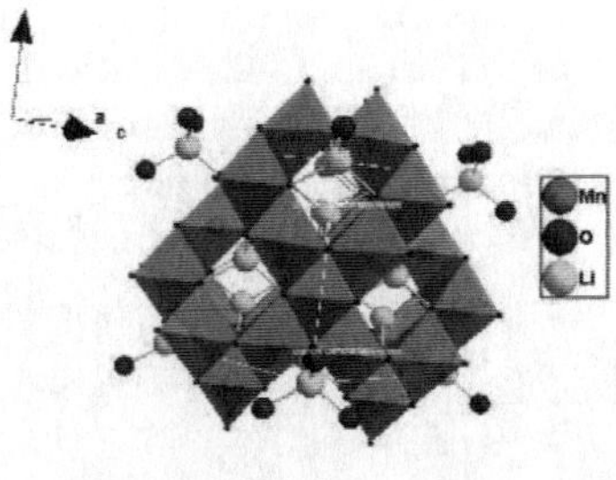

Figure 4. Microstructure of $LiNi_{0.1}Mn_{1.9}O_4$

3.2. SEM Images

Figure 5. SEM photographs of $LiNi_xMn_{2-x}O_4$

The SEM images of the synthesized powders are shown in Figure 5. The average particle size of the nickel-doped samples is smaller than that of undoped $LiMn_2O_4$. The doping of $LiNi_xMn_{2-x}O_4$ with Ni cause the tendency of nucleation process overcomes the tendency of crystal growth. The Ni-doped $LiMn_2O_4$ powders have irregular particle size. The particle sizes of $LiNi_xMn_{2-x}O_4$ are in the range of 150 to 500 nm. The increase in the nickel doping leads to increase in $LiMn_2O_4$ particle size. The morphology is almost similar.

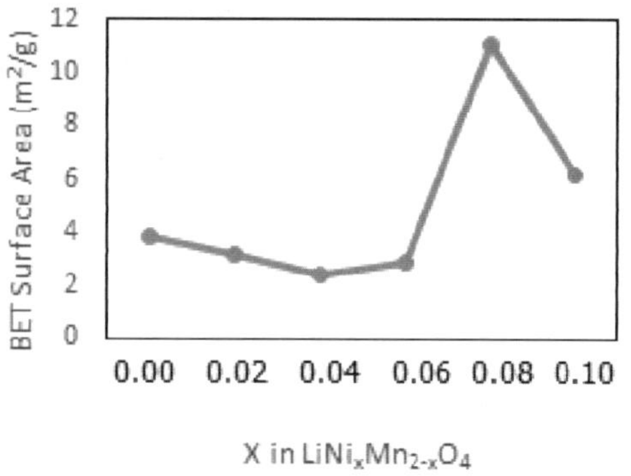

Figure 6. BET surface area of $LiNi_xMn_{2-x}O_4$

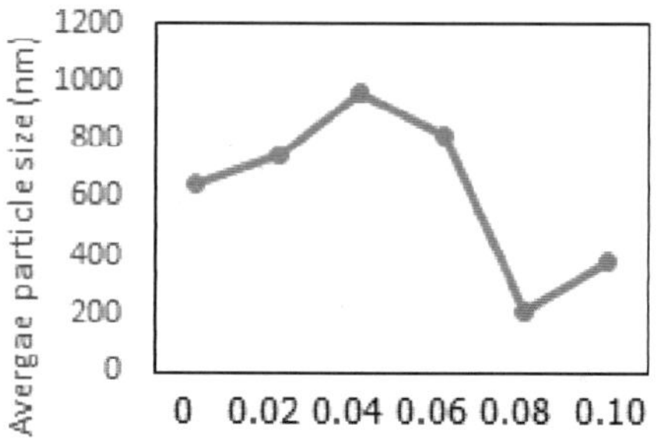

Figure 7. Variation of the average particle size with x in $LiNi_xMn_{2-x}O_4$

Figure 6 shows the BET surface area of $LiNi_xMn_{2-x}O_4$ at various mole ratios. The addition of Ni does not change the BET surface area. However, at a mole ratio of 0.08, there is a noticeable change in the BET surface area from 6 m^2/g to 11 m^2/g. Figure 7 shows the average particle size of the material. The particle size is in the range of 200-900 nm (Table 1). At low doping content, the particle size is relatively constant until it reaches x = 0.04. For x>0.04, the particle size decreases along with the increase of doping agent.The particle size is about 300 nm for x = 0.08. Good battery performance requires small particle size of cathodes [13]. In general, the particles have a rod shape. The particles are also agglomerated. The agglomerated particles might be beneficial for the production of lithium-ion batteries with high energy capacity [14]. A further test of the energy capacity of $LiNi_{0.08}Mn_{1.92}O_4$ is required. Calculation using Debye-Scherer equation gives crystallite size of $LiMn_{2-x}O_4$ in the range of 30-40 nm. The particle size obtained from the calculation (**) is around 200-900 nm.

Table 1. The crystallite size of the prepared materials as calculated by Debye-Scherer method, D, and the particle size as calculated from BET surface area data.

x	XRD	BET	
	D (nm)*	L (nm)**	$S_{BET}(m^2/g)$
0	39.42802	651.2078	3.6454
0.02	41.39063	746.8011	3.1756
0.04	43.1764	965.8867	2.4553
0.06	33.45325	817.3502	2.9015
0.08	36.38719	213.5754	11.104
0.1	36.21143	381.4117	6.2178

*D= (K.λ)/B.cos Θ

** L= 6000/(ρ. S_{BET})

4. Conclusion

$LiNi_xMn_{2-x}O_4$ has been synthesized by the low-temperature solid-state reaction. The $LiNi_xMn_{2-x}O_4$ powders have a cubic crystal structure with an *Fd3m* space group. The particle size of the prepared materials is not homogeneous. Doping with Ni leads to changes in size, crystallinity, and microstructure of the products. The average particle size of $LiNi_xMn_{2-x}O_4$ is about 150-500 nm. The crystallinity of the materials tends to increase with the increase in the Ni doping content. When Ni content in the compound increases, the lattice parameters decrease.

References

[1] Fergus J 2010 *J Power Sources* **195** 939-954
[2] Julien C M, Mauger A, Zaghib K and Groult H 2014 *Inorganics* **2** 132-154
[3] Hu M, Pang X and Zhou Z 2013 *J. Power Sources* **237** 229-242
[4] Peng C, Bai H, Xiang M, Su C, Liu G and Guo J 2014 *Int. J. Electrochem. Sci.* **9** 1791-1798
[5] Liu Q, Wang S, Tan H, Yang Z and Zeng J 2013 *Energies* **6** 1718-1730
[6] Guoqiang L 2010 $LiNi_{0.5}Mn_{1.5}O_4$ spinel and its derivatives as cathodes for Li-Ion batteries http://cdn.intechopen.com/pdfs-wm/29289.pdf
[7] Ding Y, Xie J, Cao G S, Zhu T J, Yu H M and Zhao X B 2011 *Adv. Func. Mat.* **21** 348-355
[8] Gu X, Li X, Xu L, Xu H, Yan J and Qian Y 2012 *Int. J. Electrochem. Sci* **7** 2504-2512
[9] Roisnel T and Ridriguez-Carvajal L 2001 in *WinPLOTR a graphic tool for powder diffraction*, CNRS-Lab. De Chimie du solide et inorganique moleculaire Universite de Rennes
[10] Brunauer S, Emmett P H and Teller E 1938 *J. of the American Chemical Society* **3** 309-319
[11] Purwaningsih D, Roto R, Narsito and Sutrisno H 2015 *Advanced Materials Re*search **1101** 134-137
[12] Li X 2009 *Journal of Alloys and Compounds* **479** 310-313
[13] Wang F X, Xiao S Y, Shi Y, Liu L L, Zhu Y S, Wu Y P, Wang J Z and Holze R 2013 Spinel $LiNi_xMn_{2-x}O_4$ as cathode material for aqueous rechargeable lithium batteries http://ro.uow.edu.au/aiimpapers/608
[14] Yi T, Xie Y, Ye M, Jiang L, Zhu R and Zhu Y 2011 *Ionics* **17** 383-389

Synthesis and characterization of *bis*-(2-cyano-1-methyl-3-{2-{{(5-methylimidazol-4-yl)methyl}thio}ethyl)guanidine copper(II) sulfate tetrahydrate

Sentot B Rahardjo*, Teguh Endah Saraswati, Edy Pramono, Nur Fitriana

Department of Chemistry, Faculty of Mathematics and Natural Sciences, Sebelas Maret University Jl. Ir. Sutami 36 A Surakarta, 57126, Indonesia

*E-mail: sentotbr@yahoo.com

Abstract. Complex of copper(II) with 2-cyano-1-methyl-3-{2-{{(5-methylimidazol-4-yl)methyl}thio}ethyl)guanidin(xepamet) had been synthesized in 1 : 4 mole ratio of metal to the ligand in methanol. The complex was characterized by metal analysis, thermal gravimetry/differential thermal analyzer (TG/DTA), molar conductivity meter, (Fourier transform infrared spectroscopy) FT-IR, UV-Vis spectroscopy, and magnetic susceptibility balance. The molar conductivity measurement shows that the complex was 2: 1 for electrolyte and SO_4^{2-} which was acting as a counter ion. The thermal analysis by Thermogravimetric (TG) indicates that the complex contained four molecules of H_2O. The Infrared spectral data indicates that functional groups of (C=N) imidazole and (C-S) are coordinated to the center ion Cu^{2+}. Magnetic moment measurement shows that the complex is paramagnetic with μ_{eff} = 1.78 ± 0.01 BM. Electronic spectra of the complex show a broad band at 608 nm (16447.23 cm^{-1}) are due to $Eg \rightarrow T_{2g}$ transition. Based on those of characteristics, The complex formula was estimated as $[Cu(xepamet)_2]SO_4.4H_2O$. The structure of $[Cu(xepamet)_2]SO_4.4H_2O$ complex is probably square planar.

1. Introduction

Recently, the research field in solar energy has been studied and widely developed. Some studies are focusing on a photocatalytic system using TiO_2 [1]. However, TiO_2 possesses ultraviolet absorbance, while the solar light energy possesses wider range absorbance. To enhance the sensitivity of TiO_2 in the wider range, the addition of other substances that have visible area absorbance is required. Dyes are one of the substances that can be added to modify the UV absorbance-possessed material due to their optical properties in visible area absorbance. Organic dyes are usually made of chlorophyll and coumarin. However, these organic substances are vulnerable to high temperature [2]. Other substance such complex compounds might be able to be used as a sensitizer that not vulnerable to high temperature and able to absorb visible light.

Complex compounds that absorb visible light area are mainly colored transitional metal ion complexes, for example, complexes of copper. Copper (II), has visible light area absorbance ranging from 600 nm – 820 nm. The maximum of absorbed wavelength depends on the ligands bonded to the complexes [3]. The ligands also determine the geometry complex [4]. The complexes of copper (II) are generally high in stability compared to that of the first periodic transitional metal.

The Cu (II) complex can create many geometries depending on the bonded ligands. For example complex Cu[[*N,N'-bis(3,5-di-t-pentylsalicyliden)-1,2-etilendiamin*]-*N,N',O,O'*] has a flat rectangular geometry, of which two nitrogen atoms (C=N) and two O atoms ligand *N,N'-bis(3,5-di-t-pentylsalicyliden)* are coordinated to the central ion Cu(II) tetra dentately [5]. Complex Cu[(3-O_2Nbz)$_2$(nia)(H_2O)$_2$] (nia=*nicotinamid, 3-O₂Nbz = 3-nitrobenzoate*) has a square pyramidal geometry, of which two molecules *3-nitrobenzoat* are coordinated by two O atoms, while nicotinamide by cyclic atom N is coordinated to the central ion Cu(II) mono dentately [6]. Complex [M(L^1)SO_4], M = Co(II), Ni(II) and Cu(II), L^1 = *3,3'-thiodipropionic-acid-bis(4-amino-5-ethylimino-2,3-dimethyl-1-phenyl-3-pyrazoline),* has a octahedral geometry. L^1 is coordinated to the metal ion Penta dentately by two azomethine nitrogen atoms, two groups of NH and one atom of sulfur (C-S) [7]. Complex [(BPCA)Cu(MA)] (*MA = racemic mandelate, BPCA=bis(2-pyridylcarbonyl)amidate*) has distorted square pyramid geometry; ligand MA is coordinated by atom O while BPCA is coordinated by two groups of (C=N) and one (N-H) trientately [4]. Complex [M(CA)$_2$(Ad)X_2], M = Co(II), Ni(II), Zn(II), Cd(II), and Cu(II), CA= caffein, Ad= adenine and X = SCN^- show how caffeine is coordinated to metal ions mono dentately by atom nitrogen group (C=N), while adenine is coordinated bidentately by atom nitrogen group (C=N). Two thiocyanate ions are coordinated to the metal ions by sulfur atoms. This coordination results in octahedral geometry complexes [8].

The examples of complexes explained above reveals how various groups (C=N), (C-S), (N-H) that are coordinated to the central ion Cu(II) result in complex Cu(II) with various flat rectangular, tetragonal pyramidal or octahedral geometry, coordinated mono dentately or chelatedly. One of the ligand compounds containing groups (C=N), (C-S), (N-H) and (C≡N) coincidently is xepamet, of which the structure is shown in figure 1. Hence, many possibilities of complex structure Cu(II)-xepamet may be formed. Therefore, here we report the synthesis and characterization of a complex of Cu(II)-xepamet which was successfully synthesized and investigated to its potential for solar cell application.

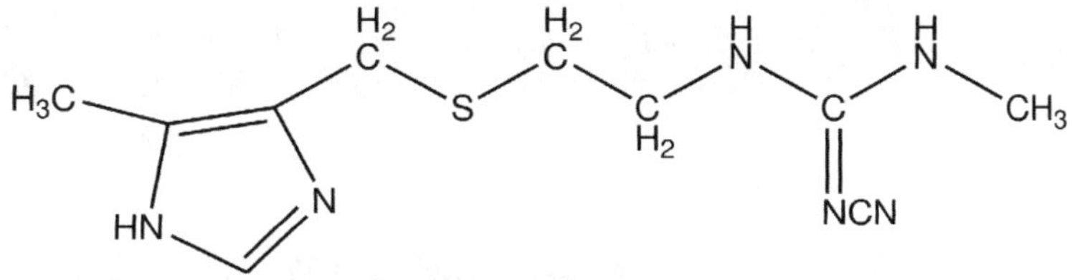

Figure 1. Structure of Xepamet

2. Experimental

All chemicals were purchased from E.Merck and used without further purification.The IR spectra were recorded on a Spectrophotometers Prestige-21 Shimadzu. The copper content was determined by Atomic Absorption Spectrometer (AAS) Shimadzu AA-6650. The thermogravimetric analysis of the metal complex was performed by a TG/DTA Diamond Perkin Elmer analyzer. Spectra UV-VIS was performed with a Shimadzu UV-3601 spectrophotometer. Molar conductivity (Λ^*_m) of 1 mM solution in water was measured at 25 °C with a Jenway CE 4071 Conductivity meter. The magnetic moment was measured with Auto Sherwood Scientific 10169 Magnetic Susceptibility Balance.

2.1. Determination of Coordinated Numbers of Cu(II)

Coordinated numbers for Cu(II) complex with xepamet were determined by the comparison method of metal moles and ligand moles. This methodology was exercised by creating a solution of $CuSO_4.5H_2O$ and xepamet. The comparisons between mole of $CuSO_4.5H_2O$ and mole of xepamet are 1:0, 1:1, 1:2, 1:3, 1:4, 1:5, 1:6. Each solution was measured using UV-Vis spectrophotometers to obtain maximum wavelength (λ_{maks}). A graphic was then made comparing maximum wavelength as an ordinate and the comparison between ligand mole and metal mole as the abscissa. By using the graphic, it was then determined the comparison between the amount of moles of the metal ions and the ligands in the complexes on the equal basis. The mixture composition of $CuSO_4.5H_2O$ and xepamet in methanol are given in table 1.

Table 1. The mixture composition of $CuSO_4.5H_2O$ and xepamet in methanol

No.	Comparison of mmol $CuSO_4.5H_2O$ and xepamet	Cu^{2+} (mmol)	Cu^{2+} (mg)	xepamet (mmol)	xepamet (mg)
1.	1:0	0.1	25	0	0
2.	1:1	0.1	25	0.1	25.2
3.	1:2	0.1	25	0.2	50.4
4.	1:3	0.1	25	0.3	75.6
5.	1:4	0.1	25	0.4	100.8
6.	1:5	0.1	25	0.5	126.0
7.	1:6	0.1	25	0.6	151.2

2.2. Complex Synthesis

A solution of copper(II) sulfate pentahydrate (0.249 g, 1mmol) in methanol (15 mL) was added to a hot solution of the ligand (1.009 g, 4 mmol) in methanol (30 mL) and the resulting solution was refluxed for 1 h. The volume was reduced to half and left for about 24 hours. The green precipitate, which formed was separated, washed with methanol and dried under vacuum.

3. Results and Discussion

3.1. Determination of Coordinated Numbers of Cu (II)

The result of spectra of the determined coordinated numbers of Cu(II) and xepamet is shown in figure 2 while the graphic λ_{max} versus comparison of mol $CuSO_4.5H_2O$ and xepamet is shown in figure 3.

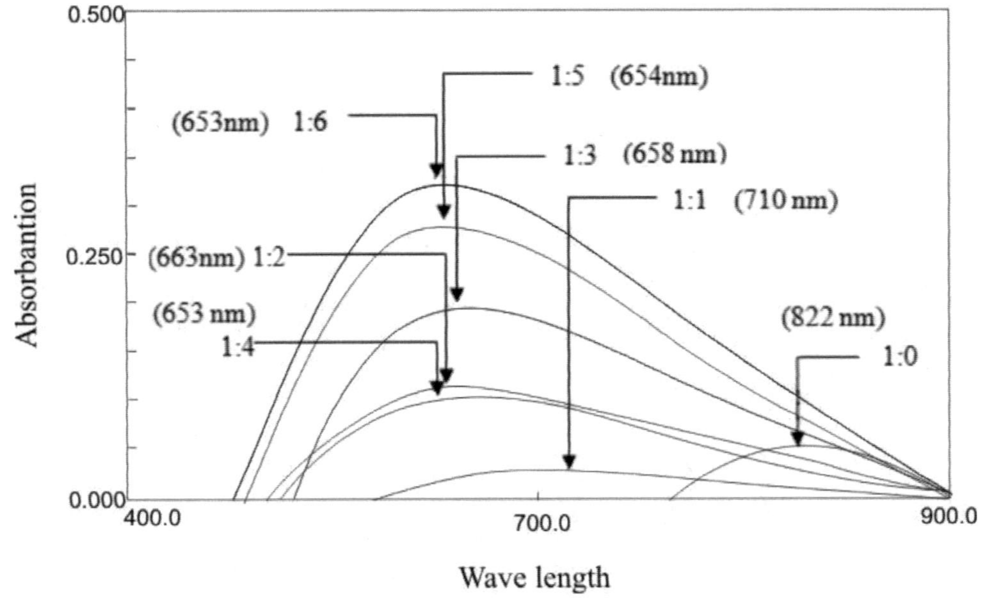

Figure 2. The spectra of complexes of Cu(II)-xepamet with various mole ratios.

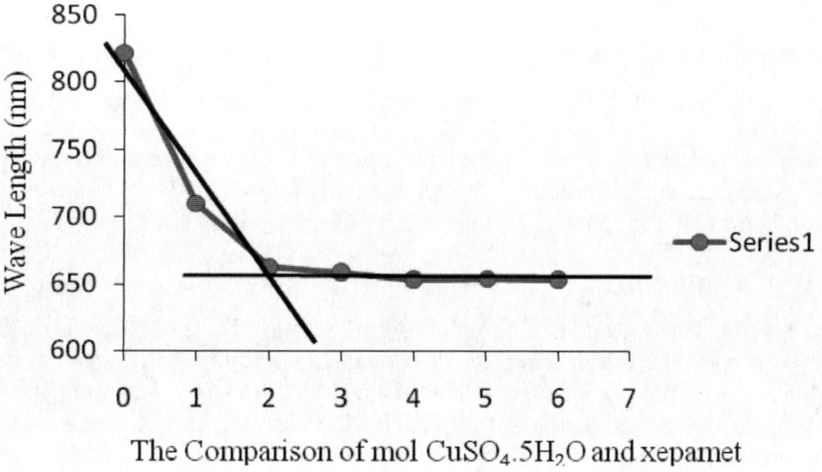

Figure 3. Graphic λ_{max} versus the comparison of moles of $CuSO_4.5H_2O$ and xepamet

Figure 3 shows a friction of λ_{max} absorbance of solution mixture Cu(II) and xepamet. The graphic depicts how the value of λ_{max} comparison of mol Cu (II) and xepamet by 1:2 is in the constant value, suggesting the complex was successfully formed by the comparison of mol Cu(II) and xepamet by 1:2.

3.2. Complex Synthesis

The mixture of $CuSO_4.5H_2O$ (0.249 g) solution in methanol (15 mL) and solution of xepamet (1.009) in hot methanol (30 mL) creates a green crystal. The electronic spectrum can be seen in figure 4.

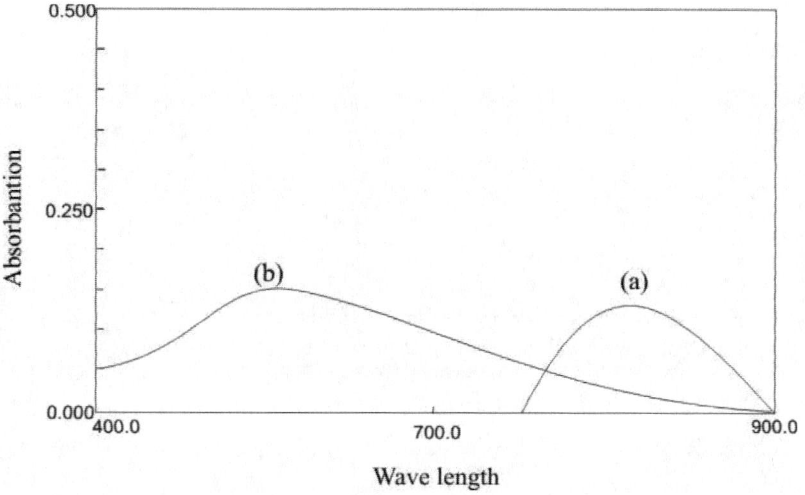

Figure 4. Electronic spectrum (a) solution of $CuSO_4.5H_2O$ and (b) solution of complex Cu(II)-xepamet within the water.

IOP Conf. Series: Materials Science and Engineering **107** (2016) 012034 doi:10.1088/1757-899X/107/1/012034

Figure 4 shows a shifted of maximum wavelength absorbance of $CuSO_4.5H_2O$ (806 nm) to a smaller wavelength (608 nm) in the complex Cu(II)-xepamet. The shifted of λ_{max} reveals the formation of complexes between Cu(II) and xepamet and also reveals that xepamet is a stronger ligand than H_2O.

3.3. Determination of Formula and Characteristics of Complexes

The measurements of the amount of Cu in the complexes of Cu(II)-xepamet are of 8.71±0.24%. If the result of the measurement is compared to the theoretical amount of Cu (8.62%), the estimated complex compound formula for Cu(II)-xepamet is possibly $Cu(xepamet)_2.SO_4.(H_2O)_4$.

3.4. Thermal Analysis with TG/DTA

TG/DTA analysis for Cu(II)-xepamet complex is shown in figure 5. Figure 5 shows that the complexes are decomposed through several levels. The TG curve of Cu(II)-xepamet complex depicts the reduction of 9.5% of the mass in the first level at 79.9-146.4 °C, indicating the release of four molecules of H_2O (of which the theory is 9.78%). This is supported by the appear of the endothermal peak at 105 °C on the DTA curve. The next level shows two peaks approximately at 180 °C and 260 °C. In this level, mass reduction is hardly seen. The TG/DTA analysis supports $Cu(xepamet)_2.SO_4.4H_2O$ complex formula.

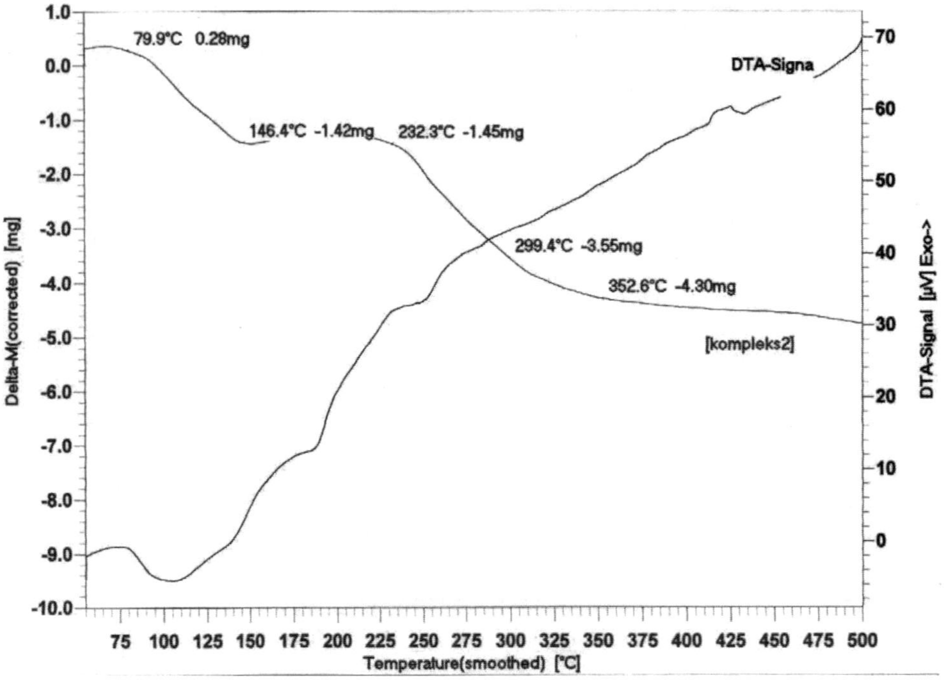

Figure 5. Analysis of DTA (a) and TG (b) of Cu(II)-xepamet complex.

3.5. Electrolytic Conductivity Measurement

The Electrolytic Conductivity Measurement is shown in table 2. The electrolytic conductivity of complex Cu(xepamet)$_2$.SO_4.4H_2O is similar to copper(II) sulfate and nickel(II) sulfate indicates that the complex is an electrolyte corresponds to 1 : 1 electrolyte. Sulfate is a counter ion and not coordinate to Cu(II) as in [Cu(INHCBA)$_2$(H_2O)$_2$](SO_4).2H_2O, INHCBA = *isonicotinamido-4-chlorobenzalaldimine* complex [9]. Therefore, these complexes may be formulated as [Cu(xepamet)$_2$]SO_4.4H_2O.

Tabel 2. The Electrolytic conductivity of standar solution and complex Cu(xepamet)$_2$.SO_4.4H_2O±1.10^{-3} M in water

Solution	Λ^*_m ($S.cm^2.mol^{-1}$)	Number of ions
$CuSO_4.5H_2O$	144.2 ± 0.2	2
$NiSO_4.6H_2O$	162.4 ± 0.2	2
$CuCl_2.2H_2O$	235.7 ± 1.0	3
$NiCl_2.6H_2O$	257.7 ± 1.0	3
$AlCl_3.6H_2O$	349.7 ± 1.0	4
Cu(xepamet)$_2$.SO_4.4H_2O	133.2 ± 0.3	2

3.6. Infra-Red Spectra

The infrared spectra of xepamet and complex [Cu(xepamet)$_2$]SO_4.4H_2O are shown in figures 6 and 7, while IR absorption bands are shown in table 3.

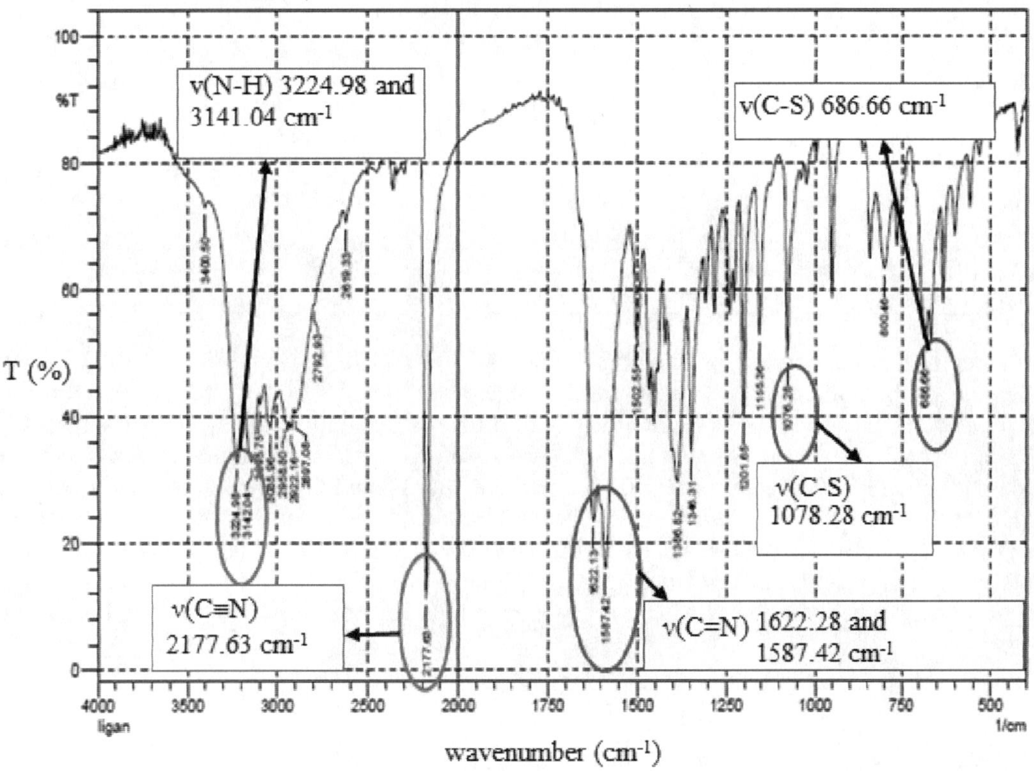

Figure 6. The infrared spectra of xepamet.

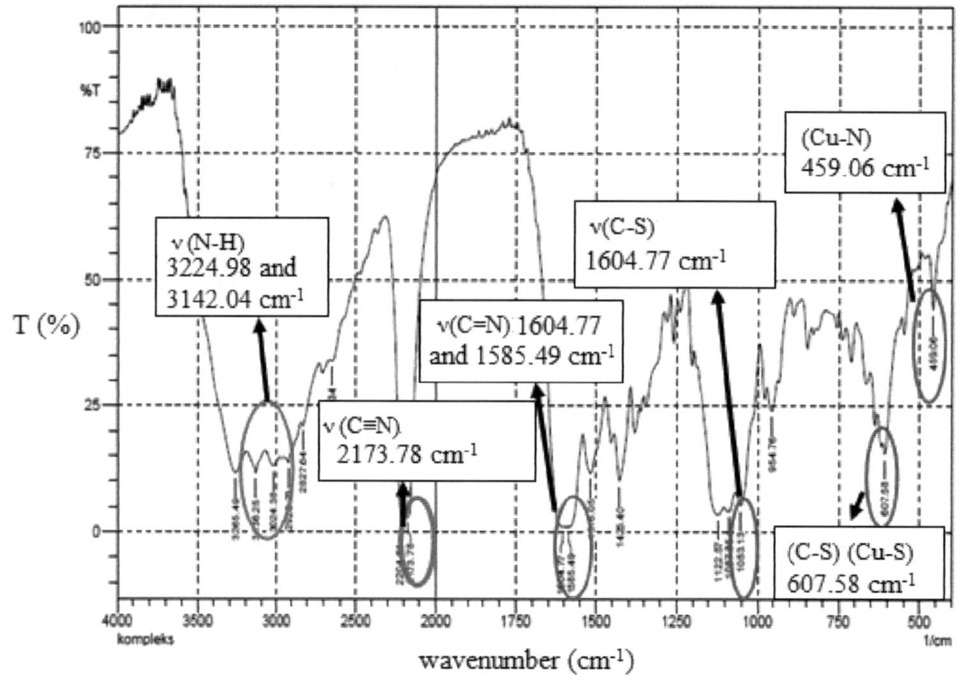

Figure 7. The infrared spectra of complex[$Cu(xepamet)_2]SO_4.4H_2O$.

Tabel 3. IR absorption bands (cm^{-1}) of xepamet and complex [$Cu(xepamet)_2]SO_4.4H_2O$

Compound	IR spectra (cm^{-1})					
	(C=N)	(C-S)	(N-H)	(C≡N)	(Cu-N)	(Cu-S)
Xepamet	1622.13	1078.28	3224.98	2177.63	-	-
	1587.42	686.66	3142.04			
[$Cu(xepamet)_2]SO_4.4H_2O$	1604.77	1053.58	3265,49	2173.78	459.06	607.58
	1585.49	607.58	3136.25			

The IR spectrum of the complex [$Cu(xepamet)_2]SO_4.4H_2O$ in the 1604.77; 1585.49; 1053.58; 607.58 cm^{-1} region exhibit shifted to lower than free xepamet (1622.13;1587.42; 1078.28; 686.66 cm^{-1}). In the two band 1604.77 and 1585.49 cm^{-1} attributed to C=N stretching while in the band 1053.58 and 607.58 cm^{-1} attributed to C-S stretching. The bands shift to the lower frequency than free xepamet indicates that the functional group of C=N and C-S are coordinated to the metal ion Cu^{2+}. New bands appeared in the low-frequency region for Cu-N at 459.06 cm^{-1} and Cu-S at 607.58 cm^{-1} which give good evidence of Cu-N and Cu-S bonds in the studied complex as in [$Cu(L^2)](ClO_4)_2$, L^2= *2-(1,3-benzothiazole-2-yl-methyl)-1,3-benzothia-zole* Cu-N at 450 $cm^{-1}$and Cu-S at 610 cm^{-1} [10]. IR absorption bands other complexes which are similar to [$Cu(xepamet)_2]SO_4.4H_2O$ complex are shown in table 4.

Tabel 4. IR absorption bands other complexes which is similar to $[Cu(xepamet)_2]SO_4.4H_2O$ complex

No.	Complex	IR absorption band of
1.	[Cu(*bemims*)X_2], bemims = *bis[(2-ethyl-5-methyl-imidazo-4-yl)methyl]sulfide* 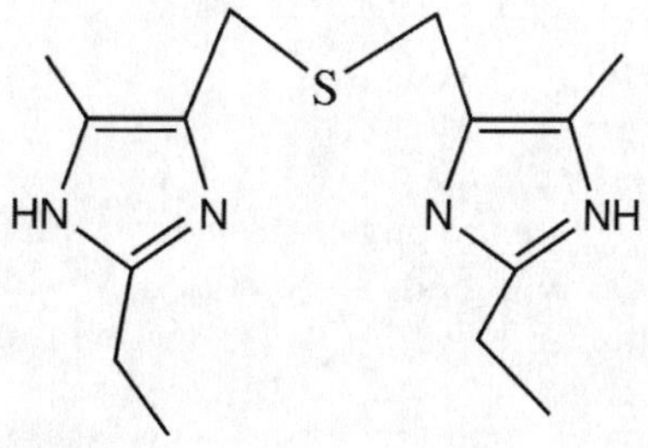	(C=N) in 1637 and 1544 cm^{-1}, (C-S) in 1032 cm^{-1} indicates coordination of nitrogen atom imidazol and sulfur atom to Cu^{2+} ion [11].
2.	$[Cu(DIPO)Cl_2]$, DIPO = *1,3-di(1H-imidazol-1-yl)-2-propanol* 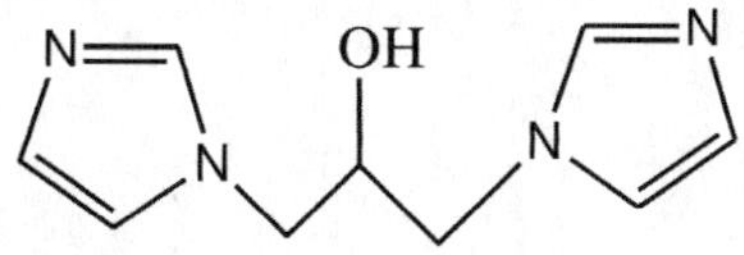	(C=N) in 1521 cm^{-1}(sh) and 1506 cm^{-1}, (C-S) in 1032 cm^{-1} indicates coordination of nitrogen atom imidazol to Cu^{2+} ion [12].
3.	$[CuL](ClO_4)_2$, L=*2-(1,3-benzothiazole-2-yl-methyl)-1,3-benzothiazol* 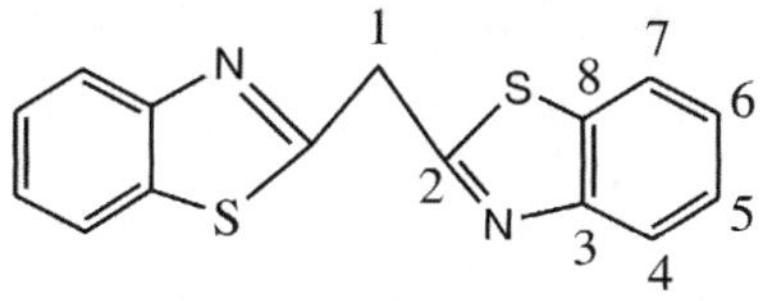	(C=N) in 1621 cm^{-1}, (Cu-N) in 450 $cm^{-1}$and Cu-S in and in 610 cm^{-1} [10].

3.7. Magnetic properties

Magnetic susceptibility measurements of the solid complex gave magnetic moments of 1.78, B.M., showing that S = ½ for Cu(II). This value indicates that the complex is mononuclear and lack of Cu-Cu bond. This value is similar to other complex such as$[Cu(L^3)_2]$, L^3 = *2-(4-nitrophenylaminocarbonyl) benzoic acid* which has magnetic moments of 1.79 BM and the complex is a square planar structure [13].

3.8. Electronic spectra

The electronic spectra of the complex recorded in water and the observed valuesare shown in Table 5. The electronic spectra of the complex $[Cu(xepamet)_2]SO_4.4H_2O$ shows one broad peak atλ_{max} 608 nm (16447.39 cm^{-1}). The absorption band shows *d-d* transition of $^2B_{1g} \rightarrow {}^2A_{1g}$as in $[Cu(L^4)_2]$, L^4 = *1-(2-hydroxyphenyl)-3-phenyl-2-propen-1-one,N^2-[(3,5-dimethyl-1H-pyrazol-1-yl)methyl]-hydrazone* at 16120 cm^{-1} [14] and in $[Cu(L^5)]Cl_2$, L^5 = *(2E,3E)-N1,N2-bis(1-(2-oxo-2H-chromen-3-yl)ethylidene)ethane-1,2-diammine* at 16120-16570 cm^{-1} which both complexes indicate square planar geometry [15].

Table 5. The maximum absorption ((λ_{maks}), Absorbance (A), Wave number (ν) and Absorbtivity Molar (ε) of $CuSO_4$ and complex[Cu(xepamet)$_2$]SO_4.$4H_2O$

No.	Compound	λ_{maks} (nm)	A	ν(cm^{-1})	ε (Lmol^{-1}cm^{-1})
1.	$CuSO_4.5H_2O$	806	0.0162	12406.95	20.23
2.	[Cu(xepamet)$_2$]SO_4.$4H_2O$	608	0.0249	16447.39	70.49

3.9. Complex Structure

The suggested structure of [Cu(xepamet)$_2$]SO_4.$4H_2O$ complex is a square planar geometry with (C=N) imidazole and (C-S) coordinated to copper(II) ion as shown in figure 8. According to IUPAC nomenclature, the suggested complex is named by bis-(2-cyano-1-methyl-3-{2-{{(5-methylimidazol-4-yl)methyl}thio}ethyl)guanidine copper(II) sulfate tetrahydrate.

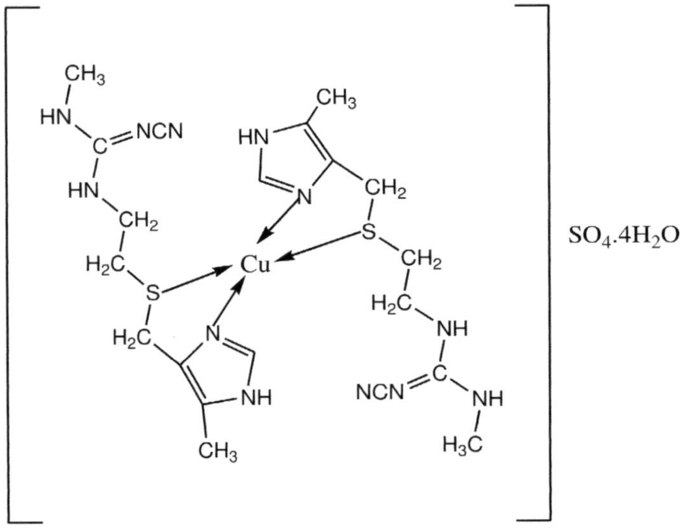

Figure 8. Suggested structure of [Cu(xepamet)$_2$]SO_4.$4H_2O$ complex

4. Conclusion

Complex [Cu(xepamet)$_2$]SO_4.$4H_2O$ can be synthesized by refluxing solution of copper(II) sulfatepentahydrate and xepamet in methanol for one hour. The complex is paramagnetic, probably a square planar structure with the functional group of C=N and C-S are coordinated to the metal ion Cu^{2+}.

Acknowledgement

The author is gratefully acknowledged to Indonesian Directorate General of Higher Education (DIKTI) for financial support.

References

[1] Wongcharee, Khwanchit, Meeyoo V, Chavadej S 2007 Dye-sensitized Solar Cell Using Natural Dyes Extracted from Rosella and Blue Pea Flowers *Sol Energ Mat Sol C.* **91** 566-571

[2] Zhou H, Wu L, Gao Y and Ma T 2011 Dye-Sensitized Solar Cells Using 20 Natural Dyes as Sensitizers *PhotochemPhotobiol A-Chem* **219** 188-194

[3] Akbar A M, Hjh Junaidah, Hj Bakar A, Mirza A H, Smith S J, Gahan L R and Bernhardt P V 2007 Preparation, spectroscopic characterization and X-ray crystaland molecular structures of nickel(II), copper(II) and zinc(II)complexes of the Schiff base formed from isatinand S-methyldithiocarbazate (Hisa-sme) *Polyhedron* **27** 71–79

[4] Halder P, Zangrando E and Paine T K 2010 Copper(II) a-hydroxycarboxylate Complexes of bis(2-pyridylcarbonyl)amine ligand: From Mononuclear Complex to One-dimensional Coordination Polymer *Polyhedron.* **29** 434-440

[5] Dyers L, Que S Y, Derveer D V and Bu X R 2006 Synthesis and Structures of New Salen Complexes with Bulky Groups *Inorganica Chimica Acta* **359** 197-203

[6] Vaskova Z, Moncola J, Korabik M., Valigura D, Švorec J, Lis T, Valko M and Melnik M 2010 Supramolecular Dimer Formation through Hydrogen Bond Extensions of Carboxylate Ligands – Path for Magnetic Exchange *Polyhedron* **29** 154–163

[7] Chandra S, Jain D, Sharma A K and Sharma P 2009 Coordination Modes of a Schiff Base Pentadentate Derivative of 4-Aminoantipyrine with Cobalt(II), Nickel(II) and Copper(II) Metal Ions: Synthesis, Spectroscopic and Antimicrobial Studies *Molecules* **14** 174-190

[8] Shaker S A, Farina Y, Mahmmod S and Eskender M 2010 Synthesis and Characterization of Mixed Ligand Complexes of Caffeine, Adenine and Thiocyanate with Some Transition Metal Ions *SainsMalaysiana* **39**(6) 957–962

[9] Ababei L V, Kriza A, Andronescu C and Musuc A M 2011 Synthesis and Characterization of New Complexes of Some Divalent Transitional Metals with Isonicotinamido-4-Chlorobenzalaldimine *J. Serb. Chem. Soc.* **2** 1-24

[10] Karabocek N, Karabocek S, Mazlum H, Degirmenciogiu I and Serbest K 2004 Synthesis and Characterization of Cu(II) Complexes of Two Ligands Derived from Malonyl Dichloride *Turk J Chem.* **28** 87-94

[11] Garc´es J D B, D´ıaz G M, Aguado F V and Bern`es S 2009 Synthesis and Characterization of bis[(2-ethyl-5-methyl-imidazo-4-yl)methyl]Sulfide and Its Coordination Behavior toward Cu(II) as a Possible Approach of a Copper Site Type I. *Hindawi Publishing Corporation Bioinorganic Chemistry and Applications* **2009** 1-5

[12] Rehman S, Ikram M, Rehman S, Faiz A and Shahnawaz 2010 Synthesis, Characterization and Antimicrobial Studies of Transition Metal Complexes of Imidazole Derivative *Bull. Chem. Soc. Ethiop.* **24**(2) 201-207

[13] Ashraf M A, Maah M J and Yusoff I 2011 Synthesis, Spectral Studies and Biological Applications of Benzoic Acid and Its Derivatives Thermal Decomposition *International Conference on Biology, Environment and Chemistry.* **1** 298-301

[14] Tharmaraj P, Kodimunthiri D, Sheela C D and Priya C S S 2009 Synthesis, Spectral Studies and Antibacterial Activity Of Cu(II), Co(II) and Ni(II) Complexes of 1-(2-hydroxyphenyl)-3-phenyl-2-propen-1-one,N^2-[(3,5-dimethyl-1*H*-pyrazol-1-yl)methyl]hydrazone *J. Serb. Chem. Soc.***74**(8–9) 927–938

[15] Akkasali R, Patil N and Angadi S D 2009 Synthesis, Characterization and Microbial Activities of Metal Complexes with Coumarine Derivatives *Rasayan J Chem.* **2** 81-86

IOP Conf. Series: Materials Science and Engineering **107** (2016) 012035 doi:10.1088/1757-899X/107/1/012035

The single cell of low temperature solid oxide fuel cell with sodium carbonate-SDC (samarium-doped ceria) as electrolyte and biodiesel as fuel

F Rahmawati*, A Nuryanto, K D Nugrahaningtyas
Research Group of Solid State Chemistry & Catalysis, Chemistry Department, Sebelas Maret University, Jl. Ir. Sutami 36 A Kentingan Surakarta 57126

*E-mail: fitria@mipa.uns.ac.id

Abstract. In this research NSDC (composite of Na_2CO_3-SDC) was prepared by the sol-gel method to produce NSDC1 and also by the ceramic method to produce NSDC2. The prepared NSDC then were analyzed by XRD embedded with Le Bail refinement to study the change of characteristic peaks, their crystal structure, and their cell parameters. Meanwhile, the measurement of impedance was conducted to study the electrical conductivity of the prepared materials. A single cell was prepared by coating NSDC-L (a composite of NSDC with $Li_{0.2}Ni_{0.7}Cu_{0.1}O_2$) on both surfaces of NSDC. The NSDC-L was used as anode and cathode. The ionic conductivity of NSDC1 and NSDC2 at 400 °C are 4.1109 x 10^{-2} S.cm^{-1} and 1.6231 x 10^{-2} S.cm^{-1}, respectively. Both electrolytes have ionic conductivity higher than 1 x 10^{-4} S.cm^{-1}, therefore, can be categorized as good electrolyte [1]. However, the NSDC1 shows electrode-electrolyte conduction. It indicates the existence of electronic migration from electrolyte-electrode or vice versa. Those may cause a short circuit during fuel cell operation and will reduce the fuel cell performance fastly. The single cell tests were conducted at 300, 400, 500 and 600 °C. The single fuel cell with NSDC1 and NSDC2 as electrolyte show maximum power density at 400 °C with the power density of 3.736 x 10^{-2} mW.cm^{-2} and 2.245 x 10^{-2} mW.cm^{-2}, respectively.

1. Introduction

Over decades, the fossil fuel combustion occurs to produce energy for life. Meanwhile, the combustion process releases carbon as pollution to the environmental and allows the global warming effect. Therefore, a clean technology in energy conversion is required. A fuel cell is a promising device for energy conversion due to its flexibility of fuel options and low carbon emission. The fuel cell convert fuel into electrical energy through reduction and oxidation reaction [2]. Solid oxide fuel cell, SOFC, has high efficiency of converting energy, i.e. 65%. The SOFC also can use many fuels whether fossil fuels or new and renewable fuels. The SOFC design also simple. However, it operates at a high temperature of 800 – 1000 °C [3]. In order to reduce the operational temperature of SOFC, some efforts have been conducted to find the electrolyte material that has high ionic conductivity at low temperature. Samarium-doped ceria, SDC, has been known as a promising one. SDC has ionic conductivity at around 0.1 S.cm^{-1} at low temperature (< 600 °C) [4]. Unfortunately, doped-ceria features mixed ionic-electronic conductivity that leads to reducing fuel cell performance due to a short circuit of anode-cathode that cause electronic conduction [5].

Recently, dual phase- materials are known to have a higher efficiency for Intermediate-Low Temperature Solid Oxide Fuel Cell, ILT-SOFC [6]. The composite of ceria-carbonate, NSDC, which was used as electrolyte provide ionic conductivity at around 10^{-1} $S.cm^{-1}$ at 600 °C, and it was developed for an intermediate-SOFC [7]. A nano core-shell samarium-doped ceria with amorphous Na_2CO_3 has an ionic conductivity of 0.1 $S.cm^{-1}$ at above 300 °C [8]. Meanwhile, the SDC-Li_2CO_3-Na_2CO_3 shows ionic conductivity of 0.02 $S.cm^{-1}$ to 0.9 $S.cm^{-1}$ at 400 – 600 °C [9]. The carbonate can inhibit the electronic conduction of doped- ceria. The electronic conduction occurs due to a reduction of Ce^{4+} into Ce^{3+} during fuel cell running [10]. The sodium carbonate addition also does not change the crystal structure of SDC due to the amorphous properties of carbonate [8].

The composite of sodium carbonate-SDC (NSDC) with $Li_{0.2}Ni_{0.7}Cu_{0.1}O$ (LNC) may become a potential material for electrode due to it's mixed ionic-electronic conduction and also the catalytic properties of nickel and copper [11]. The composite of NSDC with LNC (NSDC-L) has been used as an electrode for low temperature-SOFC with glycerol and bioethanol as fuels. The maximum power density with glycerol fuel is 215 $mW.cm^{-2}$ at 580 °C. It may have a great impact on sustainable energy and environmentally friendly energy [11]. However, the hygroscopic property of the sodium carbonate, Na_2CO_3, might become a problem. The sodium carbonate is a hygroscopic molecule, in which it will spontaneously absorb water molecules when it is exposed to air [12]. In addition, its diffraction pattern does not confirm the presence of carbonate phase [13, 14]. Even, the presence of the sodium carbonate phase only interpreted from the presence of Na peak in the EDX spectrum [14]. Meanwhile, the Na peak might come from sodium oxide.

In this research, the NSDC was prepared through two methods i.e. sol-gel preparation with pre-heating treatment on Na_2CO_3 producing NSDC2 and without pre-heating treatment on Na_2CO_3 producing NSDC1. The different method of preparation objects to find the less hygroscopic NSDC. This paper also discusses the crystal structure change after composite formation. The Le Bail refinement was conducted to analyze the phase content in the composite. This paper also analyzes the change of ionic conductivity. A single cell was constructed with the prepared NSDC as electrolyte and NSDC-L as an electrode. The single cell performance was tested with biodiesel as fuel.

2. Experiment

The samarium-doped ceria, $Ce_{0.8}Sm_{0.2}O_{1.9}$, was synthesized through sol-gel method. The precursors were cerium nitrate hexahydrate ($Ce(NO_3)_3.6H_2O$, 99%, Sigma-Aldrich) and samarium nitrate hexahydrate ($Sm(NO_3)_3.6H_2O$, 99.9%, Sigma-Aldrich). The precursors were mixed stoichiometrically in deionized- water. The citric acid and Polyethylene Glycol, PEG were added as a chelating agent. The mixing process was conducted under continuous stirring at 50 °C until forming a gel. Then followed by heating treatment at 400 °C that produced a yellow powder. The yellow powder then to be heated at 800 °C for 10 hours.

The composite of SDC with sodium carbonate (Na_2CO_3) was prepared by dissolving $Ce(NO_3)_3.6H_2O$ (99.99%) and $Sm(NO_3)_3.6H_2O$ (99.9%) in deionized water with a molar ratio of Ce^{3+}: Sm^{3+} = 4:1 to produce a 100 mL of synthesis solution. The 200 mL of carbonate solution (1.0 M) was added gradually with addition rate of 10 $mL.min^{-1}$ under continuous stirring for 2 hours at 100 °C. The mixture then was heated at 800 °C for 2 hours to produce NSDC1. The second type of NSDC, i.e., NSDC2 was prepared with similar method. However, the Na_2CO_3 was heated at 800 °C before used.

Synthesis of $Li_{0.2}Ni_{0.7}Cu_{0.1}O$ was conducted by sol-gel method. A stoichiometric weight of Li_2CO_3 (99.9 %), Cu(OH)2 (95 %) and $NiCO_3.2Ni(OH)_2.4H_2O$ (46%, Ni-based) were dissolved in deionized-water to produce 1.0 M of the synthetic solution. Citric acid solution (99.5 %) was added at the ratio of total ions to the cations of citric acid at 1:2.The addition was conducted under high rate stirring to produce a homogeneous solution, and the continued with heating at 100 °C for 2 hours under stirring condition and followed with heating at 800 °C for two hours.

The anode and cathode were prepared by mixing NSDC with LNC-271 at a ratio of 1:1. The mixture was then mixed with α-terpineol at ratio of 1:1 to form a slurry. A pellet of the electrolyte with a diameter of 1 cm and thickness of 1.5 cm was prepared by pressing NSDC with a hydraulic press and then sintered at 800 °C. The NSDC pellet was coated with NSDC-L at both surfaces by screen printing method and then heated at 600 °C for 2 hours. Silver paste and Ag mesh were applied on both surfaces of NSDC-L |NSDC|NSDC-L. Meanwhile, silver wire was attached to the single cell and connected to the measurement device. The scheme of single cell measurement is described in Figure 1(b). A micrometer (mA) was connected in series with the single cell. Meanwhile, a voltmeter was connected in parallel to the anode-cathode. A potentiometer was used to discard current by switching the resistance. Biodiesel vapor flowed into the single cell as fuel, and oxygen in the air was used oxidator. The single cell test was conducted at 300, 400, 500 and 600 °C.

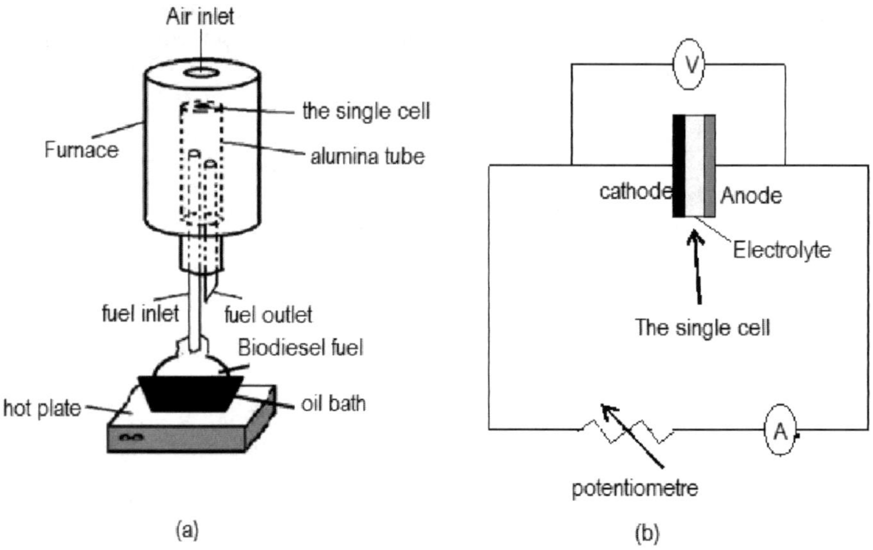

Figure 1. The scheme of single cell testing (a) and the electronic network on the power density measurement (b)

3. Result and Discussion

The samarium-doped ceria, SDC that was synthesized in this research has a diffraction pattern that is in agreement with the standard diffraction of SDC #28791 from ICSD (Figure 1). The result of Le Bail refinement shows that SDC was crystallized in cubic structure with a space group of *FM3M* and the cell parameter 5.437909(2) Å. The refinement result was depicted in Table 1.

The two types NSDCs have been synthesized which were defined as NSDC1 and NSDC2. The NSDC1 was synthesized without heating treatment on sodium carbonate. Meanwhile, in NSDC2 synthesis, the sodium carbonate was pre-heated at 800 °C. The diffraction patterns of each in comparison to the SDC pattern are depicted in Figure 2. The diffraction pattern does not show any peaks of Na_2CO_3. The amorphous phase might allow the sodium carbonate peaks did not presence in the patterns [9]. The amorphous structure might occur due to thermal treatment would fuse the sodium carbonate and then layered the SDC particles [10]. However, Le Bail refinement only proceed well when the Na_2CO_3 phase was included in the calculation. The refinement also detects the presence of sodium oxide, Na_2O. The results of refinement are listed in Table 1 in which the prepared NSDCs consist of three phases of SDC, Na_2CO_3, and Na_2O.

IOP Conf. Series: Materials Science and Engineering **107** (2016) 012035 doi:10.1088/1757-899X/107/1/012035

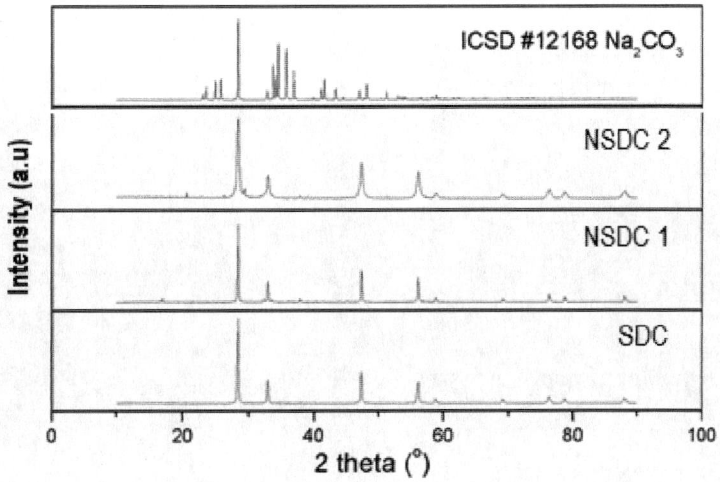

Figure 2. The diffraction patterns of NSDC1 and NSDC2 in comparison with SDC pattern dan standard diffraction of Na_2CO_3 #12168

Table 1. The cell parameters of SDC, NSDC1 and NSDC2 as resulted from Le Bail refinement

Cell parameters	SDC	NSDC1			NSDC2		
	Cubic FM3M	SDC	Na_2CO_3	Na_2O	SDC	Na_2CO_3	Na_2O
		Cubic FM3M	*Monoclinic C12/M1*	*Cubic FM3M*	*Cubic, FM3M*	*Monoclinic C12/M1*	*Cubic FM3M*
a (Å)					5.433000	9.013000(0)	5.550000(0)
	5.433000(0)	5.433000(0)	9.013000(0)	5.550000(0)	(0)	5.237000(0)	
b (Å)			5.237000(0)			6.312000(0)	
c (Å)			6.312000(0)				
V ($Å^3$)	160.36853(0)	160.36853(0)	295.818878(0)	166.374954(0)	160.36853(0)	295.818878(0)	166.374954(0)
Rp (%)			6.92			6.92	
Rwp (%)			6.85			6.85	

The impedance analysis on the prepared NSDCs found that the materials show different Nyquist plot as described in Figure 3 and Figure 4. Fitting on the impedance data found that the SDC consist of only ionic conductivity at 300 – 500 °C and the electrode-electrolyte conductivity present at 600 °C (Table 2). Meanwhile, the ionic conductivity of NSDC1 is on the order of 10^{-6}. The ionic conductivity values are not sufficient for the electrolyte. Also, the NSDC1 is very hygroscopic then it was not proper for single cell test. The NSDC1 was prepared by mixing the sodium carbonate solution with the solution of SDC precursors. The aqueous phase of sodium carbonate might become the reason of the hygroscopic property of the NSDC1. Meanwhile; the NSDC2 was prepared by mixing the pre- heated sodium carbonate powder with the solution of SDC precursors. Then, the NSDC2 is not hygroscopic like the NSDC1. Also, the NSDC2 shows higher ionic conductivity than SDC at various temperature. However NSDC2 also shows the electrode-electrolyte conduction at 300 – 600 °C. The electrode-electrolyte conduction may indicate the presence of electronic migration that even started since 300 °C. The conduction might cause cell performance degradation due to short circuit.

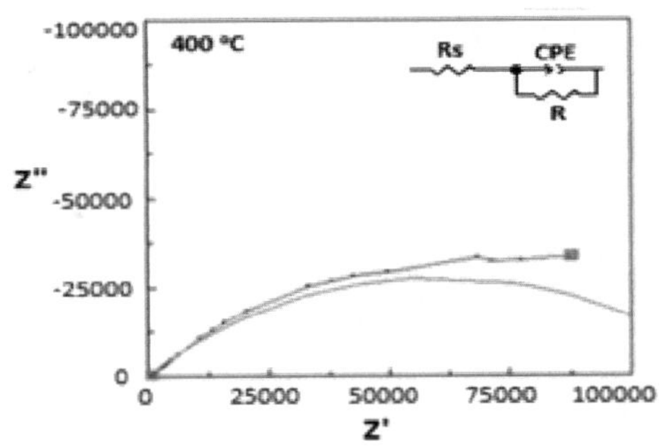

Figure 3. The Bode plot of impedance data of NSDC1 at 400 °C

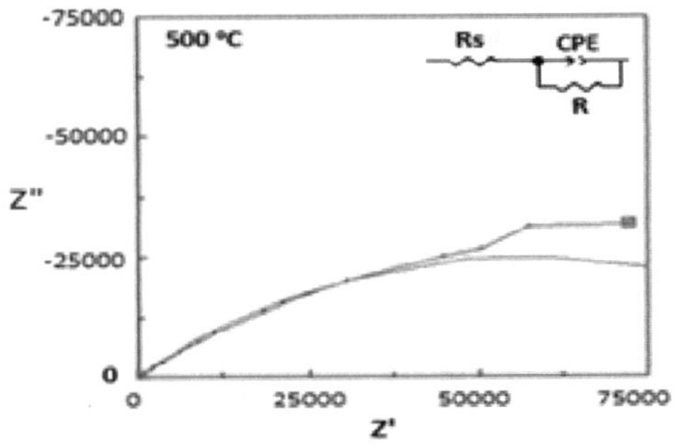

Figure 4. The Bode plot of impedance data of NSDC1 at 500 °C

Table 2. The electrical conductivity of SDC, NSDC1, and NSDC2.

Materials	Temperature (°C)	Ionic conductivity (S.cm^{-1})	Electrode-electrolyte conductivity (S.cm^{-1})	Capacitance (Farad)	
				CPE*_1	CPE*_2
SDC	300	3.0233x10^{-6}	-	2.012x10^{-10}	-
	400	3.9237x10^{-5}	-	6.701x10^{-10}	-
	500	6.3901x10^{-4}	-	8.066x10^{-8}	-
	600	3.6439x10^{-3}	2.15x10^{-2}	5.556x10^{-8}	3.931x10^{-5}
NSDC1	400	1.8743x10^{-6}	-	1.515x10^{-7}	-
	500	1.9448x10^{-6}	-	3.928x10^{-7}	-
	600	4.5302x10^{-6}	-	3.464x10^{-7}	-
NSDC2	300	9.0440x10^{-3}	2.0667x10^{-2}	2.371x10^{-11}	5.621x10^{-4}
	400	4.1109x10^{-2}	2.6318x10^{-3}	3.102x10^{-15}	1.544x10^{-3}
	500	6.4600x10^{-2}	6.3280x10^{-3}	9.511x10^{-15}	1.381x10^{-3}
	600	1.5073x10^{-1}	1.4184x10^{-2}	3.062x10^{-12}	2.168x10^{-3}

The single cell testing on NSDC2 as electrolyte found that the optimum power density occurred at 400 °C with a value of 3.736 × 10^{-1} mW.cm^{-2}, as described in Figure 5. The open circuit voltage is 0.760 Volt. Meanwhile, the power density at 300 °C is 3.138 ×10^{-1} mW.cm$^{-1,}$ and the open circuit voltage is 0.97 Volt. The power density is still too low compare to the single cell with glycerol as fuel i.e. 215 mW.cm^{-2} at 580 °C [11]. The high molecular weight of biodiesel might cause low gas phase diffusion in the functional materials. The low grade of biodiesel may also cause less conversion of liquid phase into the gas phase. The power density dropped at a high operational temperature of 500 °C and 600 °C. Those might occur due to at high temperature the reduction reaction of Ce^{4+} into Ce^{3+} occurred and allowed a short circuit that reduce the fuel cell performance. The conductivity measurement of NSDC2, as it was calculated from impedance data, confirms that NSDC2 provide the electronic-ionic conductivity even at 300 °C. It might occur due to the transportation of electrons from the electrode cross to the electrolyte. It also confirms that the composite electrolyte can be provided as a medium for electronic migration.

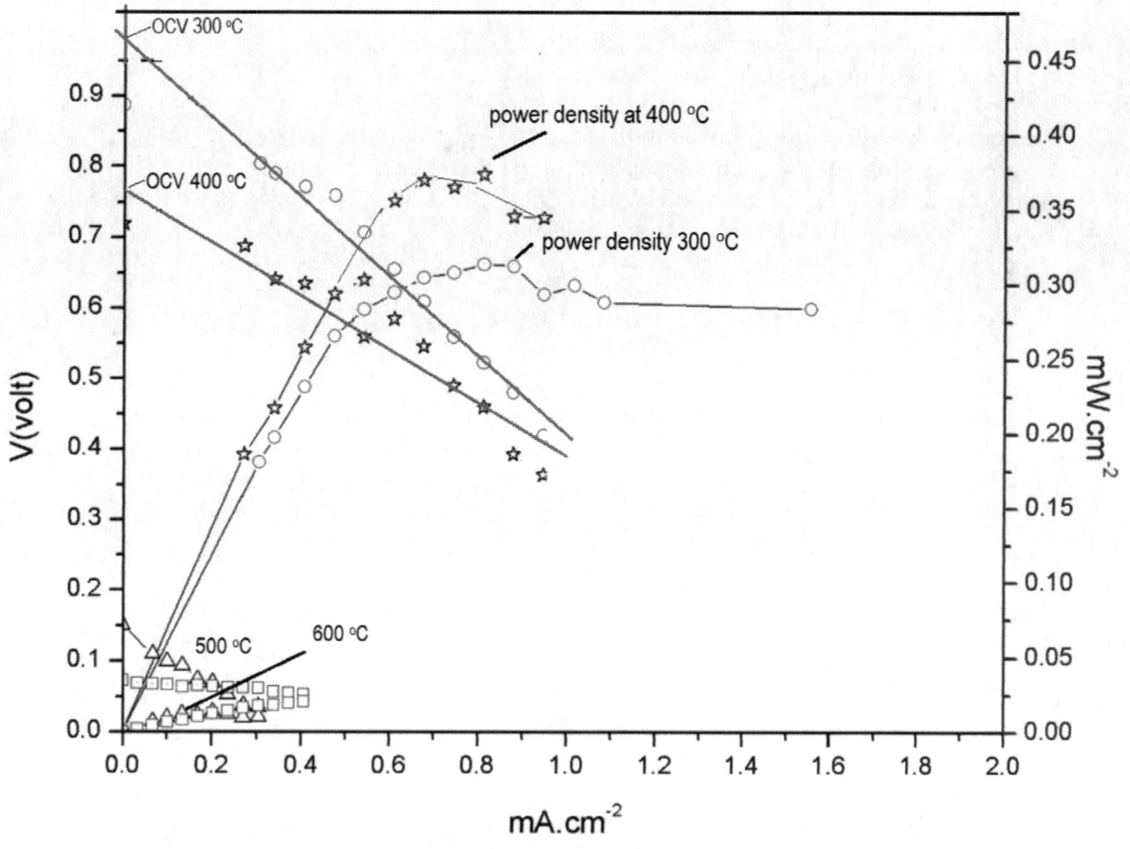

Figure 4. The power density (mW.cm^{-1}) and *open circuit voltage* (volt) NSDC2 at various temperature.

4. Conclusion

The composite of sodium carbonate with samarium-doped ceria was successfully synthesized by sol-gel method with pre-heating treatment or without pre-heating treatment on sodium carbonate. The composites consist of three phases of SDC, Na_2CO_3 and Na_2O. The presence of sodium carbonate and sodium oxide did not change the crystal structure of SDC. The ionic conductivity of composite with the pre-heating treatment of sodium carbonate i.e. NSDC shows higher ionic conductivity than the composite prepared without pre-heating treatment, i.e. NSDC1. The ionic conductivity is also higher than SDC at various temperatures. NSDC1 is very hygroscopic, and it is not proper for single cell fabrication. The power density of single cell with NSDC2 as electrolyte and NSDC-L as electrodes is 3.736×10^{-1} mW.cm^{-2} at 400 °C with biodiesel as fuel.

References

[1] Agrawal R C and Gupta R K 1999 *J. Mater. Sci.* **34** 1131
[2] Steele B C H and Heinzel A 2001 *Nature* **414** 345
[3] Raharjo J, Dediknari R and Wan 2008 *Indonesian Journal of Material Science* **10** 28
[4] Jeffrey F, Rob H, Li X, Wilkinson D P and Zhang J 2009 *Solid oxide fuel cells: materials properties and performance* Boca Raton: CRC Press

[5] Rahmawati F, Syahroni A and Syarif D G 2015 *Advanced Materials Research* **1101** 15
[6] Wei Z, Xia C, Ding D, Shi X and Guangyao M 2006 *Mater Res Bull* **41** 2057
[7] Zhu B 2006 *Int. J. Energy* **30** 895
[8] Wang X, Ma Y, Raza R, Muhammed M and Zhu B 2008 *Electrochem Commun* **10** 1617
[9] Zhu B 2013 *J. Power Source* **114** 1
[10] Zhu B, Yang X T, Xu J, Zhu Z G, Ji S-J and Sun M T 2003 *J. Power Source* **118** 47
[11] Qin H, Zhu Z, Liu Q, Jing Y, Raza R, Imran S, Singh M, Abbas G and Zhu B 2011 *Energy & Env Sci* **4** 1273
[12] Sodium carbonate- properties of sodium carbonate, http://science.jrank.org/pages/6232/Sodium-Carbonate-Properties-sodium-carbonate.html, [access date is November 14, 2015]
[13] Ma Y, Wang X, Raza R, Muhammed M and Zhu B 2010 *Int. J. Hydrogen Energy* **35** 2580
[14] Zhu B and Mat M D 2006 *Int. J. Electrochem Sci* **1** 383

IOP Conf. Series: Materials Science and Engineering **107** (2016) 012036 doi:10.1088/1757-899X/107/1/012036

Physical and optical properties of lead doped tellurite glasses

Riyatun[1], Lita Rahmasari[2] and Ahmad Marzuki[1*]

[1]Department of Physics, Sebelas Maret University, Jl. Ir. Sutami 36 A Surakarta 57126 Indonesia
[2]Department of Physics Education, Sebelas Maret University, Jl. Ir. Sutami 36 A Surakarta 57126 Indonesia

[*]E-mail: amarzuki@mipa.uns.ac.id

Abstract. Physical and optical properties of lead telluride (Pb:TZBN) glasses with composition $55TeO_2$-(41-x)ZnO-$2Bi_2O_3$-$2Na_2O$-xPbO where x = 1.0, 1.5, 2.0, 2.5% mol are presented. UV-VIS-NIR spectra of the glasses in the range of 300 – 800 nm along with their densities and refractive indices at 746 nm were recorded at room temperature. The optical bandgap energy (E_g) has been calculated from the fitting of Tauc plot. On the basis of these results we found that with the increase of Pb^{2+} content, their refractive indices are increased while their optical bandgaps are decreased. From this experiment, no distinct relationship between the Pb^{2+} content variation and the electronic polarizability (α_{O2-}) as well as their optical basicity values (Λ) were observed.

1. Introduction

In the last view decades, tellurite glasses have received considerable attention due to their unique properties, i.e., they have high glass transition temperature, high refractive index, wide infrared transmissions, relatively low melting point (725 °C), and high rare earth ion solubility [1,2,3]. Considering these properties, tellurite glasses have been widely researched and applied to the laser, photonic, communications and nonlinear applications [4,5].

Possessing high refractive index is one of the main reason tellurite glasses are considered as the important material for non-linear optical (NLO) applications. It has been shown that incorporating heavy metals oxides of Bi^{2+} or Pb^{2+} into TeO_2 glass enhances NLO properties [6].

This paper presents physical and optical properties of Pb:TZBN glasses with composition $55TeO_2$ - (41- x) ZnO-$2Bi_2O_3$-$2Na_2O$-xPbO where x = 1.0, 1.5, 2.0, 2.5% mol to predict the NLO properties.

2. Experiment

Tellurite glasses that we call Pb:TZBN with composition $55TeO_2$-(41-x)ZnO-$2Bi_2O_3$-$2Na_2O$-xPbO) here x = 1.0, 1.5, 2.0, 2.5% mol were obtained by melting 10 g batches in platinum crucible at temperature of 900 °C for 30 minutes in air atmosphere. The melt was poured onto a pre-heated stainless steel parallel plates and then annealed at a temperature of 350 °C for 6 hours allowing the glass to release its thermal and mechanical stress followed by cooling to room temperature at the cooling rate of 1 °C/min. The densities of all the glass samples (ρ) were measured based on Archimedes method carried out at room temperature using pycnometer. Glasses refractive indices (n) were calculated from Brewster Angle θ_p using the equation (1).

$$n = tan\theta_p \quad (1)$$

The optical absorption of the samples was measured at wavelength range 300 - 800 nm at room temperature in order to calculate the optical bandgap energies (E_g). The results were then compared to those predicted by following Lorentz-Lorenz equation (2).

$$E_g = 20(1 - R_m/V_m)^2 \quad (2)$$

The value of the electronic glass polarizability (α_{O2-}) determined based on the measured density and the refractive index was calculated using equation (3).

$$\alpha_{o2-}(n_o) = \left[\left(\frac{V_m}{2{,}52}\right)\frac{(n_o^2-1)}{(n_o^2+2)} - p\alpha_i\right]q^{-1} \quad (3)$$

$$\alpha_{o2-}(E_g) = \left[\left(\frac{V_m}{2{,}52}\right)\left(1 - \sqrt{\frac{E_g}{20}}\right) - p\alpha_i\right]q^{-1} \quad (4)$$

Similarly, their corresponding value of optical basicity (Λ) was calculated using equation (5).

$$\Lambda(no) = 1{,}67\left(1 - \frac{1}{[(V_m/2{,}52)(n_o^2-1)/(n_o^2+2) - p\alpha_i]q^{-1}}\right) \quad (5)$$

$$\Lambda(E_g) = 1{,}67\left(1 - \frac{1}{\left[\left(\frac{V_m}{2{,}52}\right)(1-\sqrt{E_g/20} - p\alpha_i\right]q^{-1}}\right) \quad (6)$$

Where p is the cation number, q is the number of oxide ions, and α_i is the cationic polarizability.

3. Results and Discussion

3.1. Density and Molar Volume

The experimental values of the density and molar volume of all the glasses are given in Figure 1. Density and molar volume of the glasses increase with an increase in Pb^{2+} content. These results may be attributed to the change of the atomic mass content and interatomic interspacing between atoms in the glass. The increase of the molar volume and density with the addition of the heavy and large ionic radius Pb^{2+} into the glass network indicates that the glass network is shrinkage and creation of non-bridging oxygen is resisted.

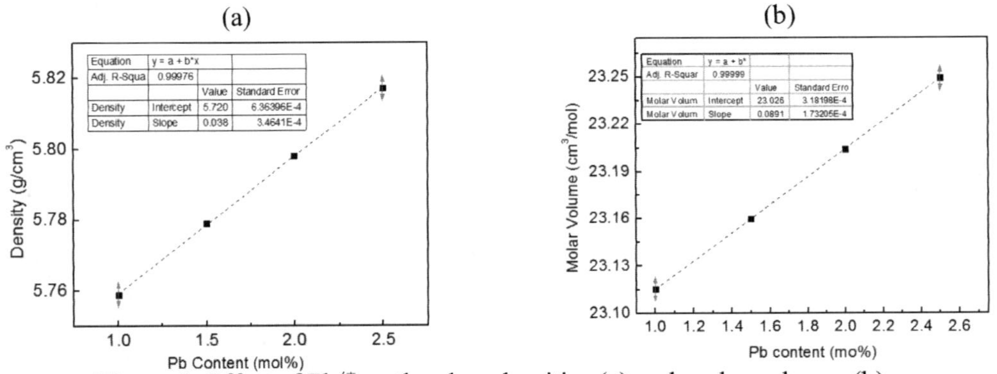

Figure 1. Effect of Pb^{2+} on the glass densities (a) and molar volumes (b).

IOP Conf. Series: Materials Science and Engineering **107** (2016) 012036 doi:10.1088/1757-899X/107/1/012036

3.2. Refractive index

The values of refractive index as shown in Figure 2 indicate that addition of Pb^{2+} to TZBN glasses increases their refractive index. Similar results were also obtained if Pb^{2+} ions were added to other glasses. These results can be understood by applying Lorentz-Lorentz equation that can be expressed as equation (7).

$$n = \sqrt{\frac{V_m + 2R_m}{V_m - R_m}} \quad (7)$$

where R_m is the molar refractivity and V_m is the molar volume. Molar refractivity can be expressed as a function of polarizability of a molecule (α_m) as given by equation (8).

$$R_m = \frac{4\pi\alpha_m N_A}{3} \quad (8)$$

where N_A is the Avogadro number. For a multi-component glass, total molar refractivity (R_{mt}) is an additive function of its individual component molar refractivity (R_{mi})

$$R_{mt} = \Sigma f_i R_{mi} \quad (9)$$

where f_i is a molar fraction of a component composing the glass. Tellurite glass is an oxide glass composed of many individual oxides A_pO_q each has molar refractivity

$$R_m = p.R_{m-A^{q+}} + qR_{O^{2-}} = 2.52(p.\alpha_{m-A^{q+}} + q\alpha_{O^{2-}}) \quad (10)$$

Where $\alpha_{m-O}{}^{2-}$ and $\alpha_{m-A}{}^{q+}$ are polarizability of the cation and the anion, respectively. Among the cationic component in the glass, the cationic polarizability of Pb^{2+} is the highest. Their values are $Te^{4+} = 0.242$ Å^3, $Zn^{2+} = 0.283$ Å^3, $Bi^{3+} = 1.508$ Å^3, $Na^{+} = 0.175$ Å^3, and $Pb^{2+} = 3.623$ Å^3 [7]. Looking back to equation (10) and (7), it is easily understood that incorporating Pb^{2+} into the glass can increase the glass refractive index significantly.

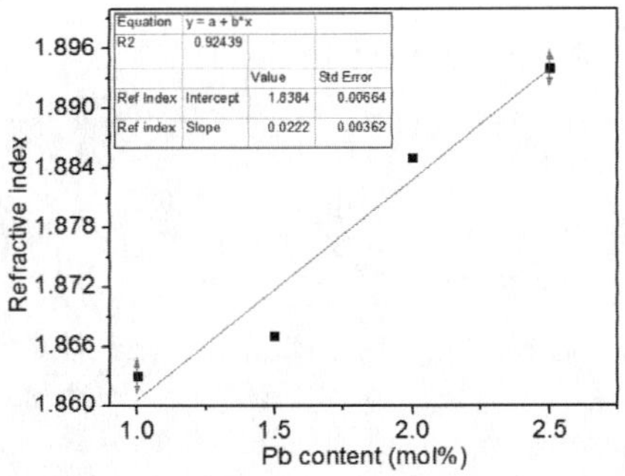

Figure 2. Refractive index change in TZBN glass as a function of Pb^{2+} content.

3.3. Optical energy bandgap

Figure 3 is absorption spectra of Pb-incorporated TZBN glasses at 200 – 400 nm measured at room temperature. The Higher absorption coefficient was observed for a glass with higher Pb^{2+} concentration. Red shifted UV edge was also observed as Pb^{2+} ions were added to the glass. Converting the graph shown in Figure 3 into $(\alpha h\nu)^{1/2}$ versus hν where α, h, and ν are absorption coefficient, Plank constant, and light frequency, respectively; optical energy bandgap can be calculated. The typical graph obtained for this purpose is shown in Figure 4. Optical energy bandgaps obtained in this way along with those obtained from equation (2) are given in table 1. It can be seen that the optical energy bandgap decreases with respect to increasing the Pb^{2+} content.

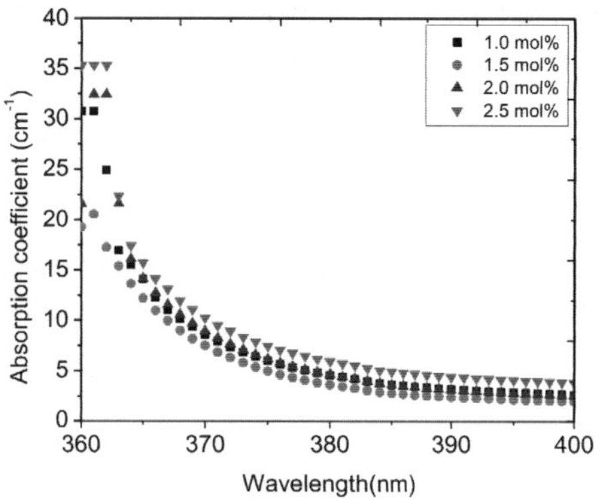

Figure 3. Absorption spectra of Pb-incorporated FTZBN glass at 200 – 400 nm.

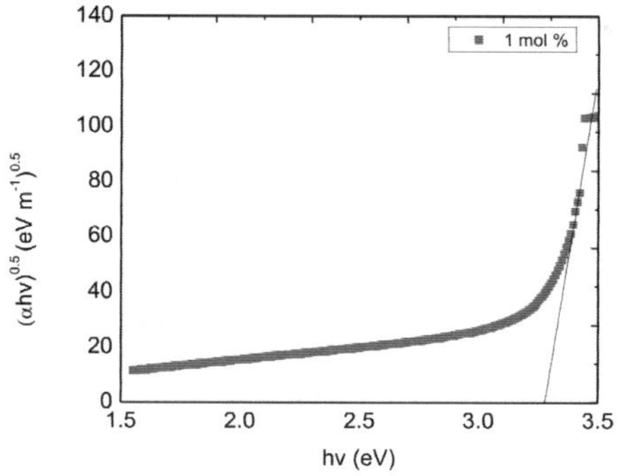

Figure 4. Calculating optical energy bandgap using Tauc Plot.

Tabel 1. Calculated and measured optical energy bandgaps

Samples	E_{gap}		Relative difference (%)
	Calculation (eV)	Measurement (eV)	
1.0 mol%	3.759	3.277	12.82
1.5 mol%	3.741	3.269	12.62
2.0 mol%	3.723	3.248	12.76
2.5 mol%	3.707	3.227	12.95

Table 2. Electronic polarizability and basicity of TZBN glass.

	Pb^{2+} content (mol%)			
	1.0	1.5	2.0	2.5
Electronics Polarizability (α_{O2-}) (Å^3)	2.486	2.485	2.491	2.461
Optical basicity (Λ)	0.978	0.9775	0.981	0.9725

From the value of the refractive index and optical energy bandgap and by applying equation (3) – (6) we can predict the glass electronic polarizability optical and optical basicity (Table 2) [8]. Typical oxide electronic polarizabilities of common oxide glasses are within the range of 1.4-1.6 Å^3) while their corresponding optical basicity values are in the range of 0.4-0.6. Since electronic polarizability represents an ability to donate electrons to the surrounding cations, an increase in electronic oxide polarizability means that an oxide ion has a stronger electron donor ability and thus easily be polarized. High values of oxide electronic (ionic) polarizability, as well as optical basicity as shown in Table 2, indicate that Pb-incorporated TZBN glasses are potentially used as the important material for non-linear optical (NLO) applications. High optical basicity of the glasses means that the glasses have the small single bond strength and thus easily polarized.

4. Conclusions

We have investigated the physical and optical properties of Pb:TZBN glasses with composition $55TeO_2$-(41-x)ZnO-$2Bi_2O_3$-$2Na_2O$-xPbO where x = 1.0, 1.5, 2.0, 2.5% mol. We found that refractive index and molar volume of the glass increases with the increase of Pb^{2+} content. We also calculated the oxide ion polarizability (α_{O2-}) as well as their optical basicity value (Λ). The way their values change as a function of Pb^{2+} addition in the glass is not obvious. High values of α_{O2-} and Λ obtained in the experiment here in shows that these glasses have potential use in non-linear optical (NLO) applications.

References

[1] Mohamed N B, Yahya A K, Deni, M S M, Mohamed S H, Halimah M K and Sidek H A 2010 *J. Non-Cryst. Solid* **356** pp 1626-1630
[2] El-Mallawany R *Tellurite glasses handbook* Physical Properties and Data CRC Press LCC 2002
[3] Yamane Masayuki and Asahara Yoshiyuki 2000 *Glasses for photonics* United Kingdom at the University Press Cambridge
[4] Inoue S and Nokui 1991 A *Proceedings of the int, conf,on science and technology of new glasses* ed. Sakka S and Soga N Tokyo p 77
[5] Nasu H, Ibara Y and Kubodera K 1989 *J. Non-Cryst.Solids*, **110** 229
[6] Suhara, Hayataka and Nogami 2011 *Journal of the University of Chemical Technology and Metallurgy* **47** 4 pp 369-373
[7] Demitrov V and Komatsu T 2010 Journal of the University of Chemical Technology and Metallurgy **45** 3
[8] Demitrov V and Komatsu T 1999 *J. Non-Crystalline Solids* **249** 160 ± 179.

The influence of the composition of P3HT:TiO_2 on the characteristics of hybrid polymer solar cell

Erlyta Septa Rosa[1*], Shobih[1], Hanik Munawaroh[2] and Pranoto[2]

[1]Research Centre for Electronics and Telecommunication-Indonesian Institute of Sciences (PPET–LIPI) Kampus LIPI Gd. 20 Jl. Sangkuriang Bandung 40135
[2]Department of Chemistry, Sebelas Maret UniversityJl. Ir. Sutami 36A Kentingan Jebres Surakarta 57126

[*]E-mail:erlytasr@gmail.com

Abstract. In this paper, a study on the fabrication of hybrid polymer solar cells based on organic semiconductor materials P3HT (poly-3-hexylthiophene) and inorganic semiconductor materials TiO_2 (titanium dioxide) has been carried out. The study is focused on the influence of the composition of P3HT and TiO_2 on the optical and electrical characteristics of hybrid polymer solar cells. The composition of P3HT and TiO_2 are varied with ratios of (1:1), (2:1), and (1:2), respectively, in a concentration of 10 mg/ml. The optical characterization using a UV-Vis spectrometer shows that the higher absorption of the active layer results from the (1:1) ratio of P3HT:TiO_2. Based on the electrical characterization, using solar simulator on hybrid polymer solar cells, can be concluded that a mass ratio of P3HT:TiO_2 (1:1) gives the best performance, with an open-circuit voltage of 0.2437 volts, a short-circuit current of 0.0029 milliamperes, a maximum power of 0.0002 milliwatts, and a power conversion efficiency of 0.00006%, at the light intensity of 500 W/m^2.

1. Introduction

Polymer-based solar cells could be good candidates for the next generation of the photovoltaic device due to their various advantages particularly such as low cost and excellent mechanical property. Polymeric materials could be deposited by the solution-based technique or electrochemical polymerization onto printable, flexible and large-scale film, promising for various applications and commercialization of solar cells [1]. However, the disadvantages of polymer solar cells are seriously hindering their performance as follows (i) the poor spectral match between the absorption of the polymer layer and the solar emission results in relatively low current density and energy conversion efficiency; (ii) the extremely large binding energy of photogenerated quasi-one-dimensional excitons (0.4 eV) in conjugated polymers makes direct dissociation of excitons into free carriers very inefficient; (iii) the short diffusion length of 10 nm leads to high recombination, low photocurrents and energy conversion efficiency; (iv) the conversion efficiency is also limited in single layer diodes by the surface of the electrode/polymer contact where the local electric field induces charge separation; (v) allow resistance towards oxygen, high temperature and UV light, etc. causes poor stability and degradation [1].

One promising alternative approach to overcome the problems is to use a nanostructured inorganic semiconductor as the electron-transport component. Electron-transporting metal oxide such as

titanium dioxide (TiO_2), tin oxide (SnO_2), and zinc oxide (ZnO) is attractive materials in this content on account of their low cost, good stability, ease of fabrication, and their potential that exists for controlling their film morphology on the nanoscale [2].
Hybrid polymer-inorganic solar cells offer the opportunity to combine the beneficial properties of the two materials in charge generation and transport with easy and cheap processing in large areas [3]. The use of two materials with complementary donor and acceptor electronic properties is crucial to the operation of the hybrid polymer solar cell, because photoexcitation of a conjugated polymer provides a bound electron-hole pair or exciton rather than free charges. This exciton can be dissociated efficiently at the interface with a second material via a hole or electron transfer to produce the free charges that generate the photovoltaic effect [4-5]. Compared to organic-fullerene cells, hybrid polymer solar cells may have the advantages of being morphologically more stable and being able to utilize the high charge carrier mobility of the inorganic material. Of course, the challenges for hybrid polymer solar cells remain to create a morphology in which the inorganic and polymeric materials are intimately mixed on a nanometer scale but with continuous phases [4]. In recent years, various solution processed bulk heterojunction hybrid polymer solar cells have been explored [6]. Many of these methods aim at creating an intimately mixed bulk-heterojunction microstructure of the two materials on a nanometer scale, ensuring efficient charge generation at the interface and effective percolating pathways for photogenerated electrons and holes to the electrodes [3].

Poly(3-hexylthiophene) or P3HT is a conjugated polymer with a band-gap of approximately 1.9 eV, and now often used as an active layer for polymer solar cells [7]. The highest efficiency of bulk-heterojunction structure of polymer solar cells was made with P3HT as the donor electron [8]. TiO_2 is an acceptor because it has a high absorption coefficient and size-tunable [9]. It was also reported that addition of titanium dioxide (TiO_2) could not only significantly improve the stability of devices composed of P3HT but also enhanced the charge transfer and thus the efficiency of hybrid solar cells [10]. In this research, hybrid polymer solar cells based on P3HT and TiO_2 has been fabricated. In order to obtain the optimum separation of exciton and charges transfer, it is needed to adjust donor/acceptor ratio. Therefore, the research was focused on the variation of P3HT and TiO_2 ratios. The correlation between P3HT:TiO_2 ratios on the optical and electrical characteristics of the cells under air mass 1.5 illuminations at a light intensity of 500 W/m^2 were investigated.

2. Method
Hybrid polymer solar cells were fabricated with the structure of PET/ITO/PEDOT:PSS/P3HT:TiO_2/Al/PET. The fabrication process consists of several steps, i.e. ITO etching, PEDOT:PSS spin coating, P3HT:TiO_2 spin coating, aluminum evaporating, cells lamination, and characterization.

2.1. ITO etching
The substrate used is an ITO coated PET with a sheet resistance of 60 Ω/square. ITO is a transparent conductive oxide which serves to drain the generated hole. In order to avoid a short circuit of the cell, the ITO pattern was etched partially. A black paint is sprayed on the ITO surface which has been patterned by adhesive tape mask previously. The substrate is then dried in room temperature overnight, and when it has dried the mask is removed. The next step is an etching process which aims to eliminate undesired ITO layer. The etching process is performed by dipping the substrate into 50% HCl solution for 5 minutes. After the undesired ITO layer has disappeared, the substrate is washed in DI water and dried. The remaining paint then removed by immersing the substrate into thinner solution, isopropyl alcohol, and DI water, respectively, for 10 minutes in the ultrasonic bath.

2.2. PEDOT:PSS spin coating
PEDOT:PSS is used as hole transporter and to prevent the diffusion of ITO into the active layer of the solar cell. The spin coating process of PEDOT:PSS is performed by dripping a PEDOT:PSS solution on ITO surface which has been covered with adhesive tape previously as PEDOT:PSS mask pattern

and then rotated at 1000 rpm for 60 seconds. The PEDOT:PSS layer is then dried in a vacuum oven at 120°C for 60 minutes.

2.3. P3HT:TiO2 spin coating

The mixture of P3HT:TiO_2 serves as an active layer of the solar cell, so the fusion of these two materials greatly influences the optical and electrical characteristics of solar cells. The mixtures were varied in (1:1), (2:1), and (1:2) ratios, respectively, in a constant concentration of 10 mg/ml. The mixtures were spin coated on PEDOT: PSS layer that have been covered with the masking tape. The spin coating process was done on 1000 rpm for 1 minute. The dried layer is then annealed in a vacuum oven at 120 °C for 10 minutes.

2.4. Aluminum evaporating

Aluminum (Al) layer serves as a cathode in hybrid polymer solar cell devices. The aluminum layer is grown on the surface of the P3HT:TiO_2 using shadow mask patterns. The evaporation process is done until the chamber pressure of 5 x 10^{-5} mBar and takes place for 5 minutes to produce an aluminum layer with a thickness of 60 nm.

2.5. Lamination

Lamination process aims to protect the cell from the outside air and moisture. Before laminated, two tabbing wires with 15 mm long are attached to ITO and Al electrodes respectively using Ag paste glue as the external contacts for I-V characterization. The lamination process is done by attaching thermoplastic sealant between device and PET and then clamped using two glasses on both sides. Finally, it is heated in a vacuum oven at 100 °C for 10 minutes.

2.6. Characterization

Measurement of the electrical characteristics of the cells was performed by illuminating the cells under AM 1.5 solar spectrum of Oriel solar simulator using a xenon lamp, which is equipped with National Instrument I-V measurement systems at the light intensity of 500 W/m^2. Characterization of optical properties is carried out using UV-Vis spectrometer, whereas the active layer surface observation is performed using scanning electron microscope (SEM).

3. Result and Discussion

Figure 1 shows 20,000X magnification SEM photographs of surface morphologies of P3HT and TiO_2 in ratios of (1:1), (2:1), and (1:2), respectively.

(a) (b) (c)

Figure 1. SEM photographs of surface morphologies of P3HT and TiO_2 in ratios of 2:1 (a); 1:1 (b); and 1:2 (c).

Based on the results of SEM, the active layer of each sample showed a different morphology. On the entire surface of the layers shows agglomerate structures. The agglomerates may be formed from TiO_2 nanoparticles which covered by the P3HT film. Although TiO_2 used in this research has a particle size of 25 nm (Aldrich P25), but the distribution of the TiO_2 nanoparticles in the P3HT film is not fully

IOP Conf. Series: Materials Science and Engineering **107** (2016) 012037 doi:10.1088/1757-899X/107/1/012037

spread. Thus, agglomeration forming cannot be avoided. It is also reported by Yu et al, that the size distribution of the inorganic nanoparticles in the active layer has not been precisely controlled [10].
In the ratio of P3HT:TiO_2 (2:1), a smaller agglomerates structure obtained, indicates the most porous surface. The porous surface will cause high absorption of light. In the surface of (1:1) and (1:2) ratios, the structures obtained are relatively similar, The surfaces were formed a larger agglomerates.

Characterization results of an optical property of hybrid polymer solar cells based on P3HT:TiO_2 using UV-Vis spectroscopy are shown in figure 2.

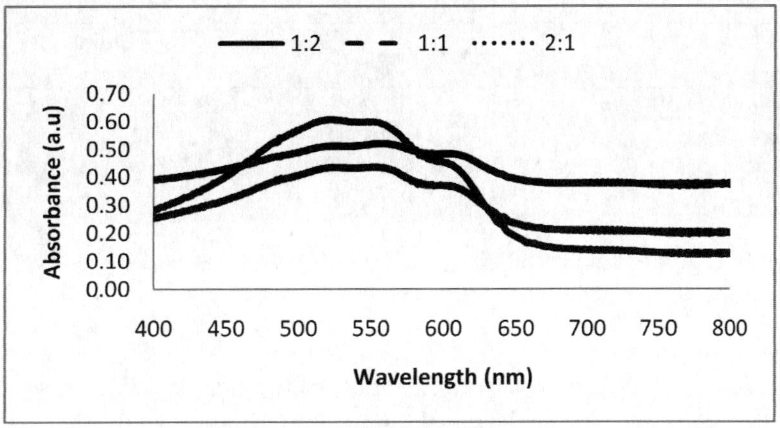

Figure 2. The absorption spectrum of the active layer varying ratios of P3HT:TiO_2.

In figure 2, note that P3HT:TiO_2 ratios of (2:1), (1:1), and (1:2) are resulting the similar spectrum in the visible. The maximum peak of the (2:1) ratio is higher compared to (1:1) and (1:2) ratios, in the range of 465-590 nm. However, in the range of 400-465 nm and 590-800 nm, (1:1) ratio has higher absorption, and the average absorption of (1:1) ratio is equal to the (2:1) ratio, which is around 0.4 au. According to the standard spectrum, pure P3HT show strong absorption in the visible wavelength range of 450-650 nm, while pure anatase TiO_2 absorbs light in the ultraviolet wavelength range of 200-300 nm [11]. The addition of TiO_2 is decreasing the absorption of P3HT and shifts the maximum peak to the shorter wavelength. Kim et al reported that the addition of excess TiO_2 into P3HT provides the results for phase separation of larger scale and damages the structure of P3HT chains [12]. The same results were also confirmed by Yun and Sulaiman which showed that the addition of TiO_2 in the active layer interferes intermolecular structure of P3HT chains which can reduce the absorption of light contributed by P3HT in the visible wavelength range [11]. This is caused by a reduction in the volume occupied by the benzene ring of the P3HT polymer in the mixed film.

Regarding light absorption by P3HT and charge transport by TiO_2, if P3HT is higher than TiO_2, more light will be absorbed, and more exciton will be generated. However, at same the time, the electron charge transport towards the cathode will be decreased because of the number of electrons produced is not proportional to the amount of TiO_2 particles. The generated electrons will recombine partially and only small portion of electrons will reach the cathode. Conversely, if TiO_2 is higher than P3HT, then the exciton will be less generated. As the consequence, free charges will be less generated, although there are many numbers of TiO_2 particles, only a few charges are transported toward the cathode. By using a balanced ratio of (1:1) may be the problem above can be improved, although there is a risk of reduction on the light absorbed by P3HT. This can be seen from the absorption spectrum of the (1:1) ratio, although the light absorption in the range of 465-590 nm is lower than the (2:1) ratio, but the absorption in the range of 400-465 nm and 590-800 nm is higher than that of the (2:1) ratio. Although the average absorption of the (1:1) ratio is equal to of the (2:1) ratio, the spectrum of the

(1:1) ratio is broader than of the (2:1) ratio. Therefore, the (1:1) ratio will result in the generation and separation of exciton more efficiently than the others into electrons and holes free-charges.

Table1 shows electrical characteristics of hybrid polymer solar cell based on P3HT:TiO_2. These data are the results from the calculation of the I-V curve using Labview software facilitated by National Instrument.

Table1. The electrical characteristics of hybrid polymer solar cells based on P3HT:TiO_2 by measuring under light intensity conditions of 500W/m^2 at room temperature.

Electrical Characteristics	P3HT:TiO_2 Ratios		
	(2:1)	**(1:1)**	**(1: 2)**
Open-circuit voltage (V_{oc})	0.1615 V	0.2437 V	0,1211 V
Short-circuit current (I_{sc})	0.0020 mA	0.0029 mA	0.0019 mA
Maximum power (P_m)	0.0001 mW	0.0002 mW	0.00004 mW
Power conversion efficiency (PCE)	0.00002%	0.00006%	0.00001%

From the table, hybrid polymer solar cells using P3HT: the TiO_2 ratio of (1:1) gives better electrical characteristics than the others. The (1:1) ratio has an open-circuit voltage of 0.2437 volts, a short-circuit current of 0.0029 milliamperes, a maximum power of 0.0002 milliwatts, and a power conversion efficiency of 0.00006%.

In general, the fabricated cells have low electrical characteristics especially in short-circuit current (I_{sc}) and open-circuit voltage (V_{oc}). The short-circuit current generated only reach the order of microampere. As a result, the maximum power generated is also reaching the order of microwatt, and the efficiency as well. The photocurrent that is extracted from the solar cell at under short-circuit condition can be affected by several factors [13]. Foremost of them is the light absorption of the active layer and hence the thickness of the film. The absorption of the entire film determines the short-circuit current. However, an optimization of the active layer thickness is also needed. A thicker film result in the lower electric field and charge collection, hence lower short circuit current [13].

The next factor is morphology, the interpenetrating network of polymer donor and inorganic acceptor which determines how easily the exciton gets separated before geminate recombination takes place [12]. In this study, the existence of agglomerates (as seen in the SEM picture) may result in recombination and also, cause external light scattering. Processing parameters such as the choice of solvent, spin speed, and annealing treatment also play key roles in determining the film morphology [14].

Polymer solar cells are highly susceptible to degradation. Degradation processes involving oxygen which is then bonded to the carbon most tip of the alkyl chain P3HT. Then the bond forming reactions will indirectly interfere with the conjugation of P3HT backbone [15]. Environmental condition around the moist sample also causes the penetration of the free molecules to the devices. In addition, the degradation of the devices could be caused by the diffusion of aluminum as electrodes into the active layer, so that the possibility of leakage currents and high resistance [16].

4. Conclusions

Hybrid polymer solar cells based on P3HT and TiO_2 with the configuration of PET/ITO/ PEDOT:PSS/P3HT:TiO_2/Al/PET have been fabricated. The ratio of the P3HT:TiO_2 active layer affects the optical and electrical characteristics of polymer solar cells. A (1:1) ratio of P3HT:TiO_2 gives the best performance, with an open-circuit voltage of 0.2437 volts, a short-circuit current of 0.0029 milliamperes, a maximum power of 0.0002 milliwatts, and a power conversion efficiency of 0.00006%.

Acknowledgement

This work was supported by Research Center for Electronics and Telecommunication Indonesian Institute of Sciences (PPET LIPI) through DIPA Tematik Program 2015.

IOP Conf. Series: Materials Science and Engineering **107** (2016) 012037 doi:10.1088/1757-899X/107/1/012037

References

[1] Wang Y, Wei W, Liu X and Gu Y 2012 *Sol. Energy. Mater. & Sol. Cells.* **98** 129–145

[2] Ravirajan P, Haque S A, Durran J R, Bradley D D C and Nelson J 2005 *Adv. Funct. Mater.* **15** (4) 609-618

[3] Beek W J E and Janssen R A J 2009 in *Hybrid Nanocomposites for Nanotechnology* Springer Science + Business Media pp 324

[4] W J E Beek, Wienk M M and Janssen R A J 2005 *J. Mater. Chem.* **15** 2985-2988

[5] Saunders B R 2012 *J. Colloid and Interface Sci.* **369** 1-15

[6] Ji L W, Shih W S, Fang T H, Wu C Z, Peng S M and Meen T H 2010 *J. Mater. Sci.* **45** 3266-3269

[7] Krebs C F 2009 *Sol. Energy Mater. & Sol. Cells* **93**, 465-475

[8] Kim H, Soo W and Moon S 2006 *J. Korean Phys. Soc.* **48**(3) 441-445

[9] Huynh W U, Peng X G and Alivisatos A P 1999 *Adv Mater* **11** 923-927

[10] Yu Y Y, Chien W C, Ko Y H, Chan Y C, Liao S C 2012 *Current Appl. Phys.* **12**(S2) S7-S13

[11] Yun T W and Sulaiman K 2011 *Sains Malaysiana* **40**(1) 43–47

[12] Kim H, Soo W and Moon S 2006 *J. of the Korean Phys. Soc.* **48**(3) 441-445

[13] Bhattacharya J 2013 Understanding the physics of degradation of polymer solar cells, Graduate Theses and Dissertations *Iowa State University,* Paper 13475 pp 33

[14] Chen H J, Wang L and Chu W Y 2007 *European Polymer* J. **43** 4750–4761

[15] Grossiord N, Kroon M J, Androessen R and Blom P M W 2012 *Org. Elec.* **13**, 432-456

[16] Xiaoyin X, Ju H and Lee E 2010 *J of the Korean Phys Soc* **57**(1) 144-148

Effect of cyclical thermal to mechanical properties of Hybrid Laminate Composites (HLC) with skin recycle polypropylene/natural fiber/halloysite and core PP/KF composites

Neng Sri Suharty[1*], Hanafi Ismail[2], Desi Suci Handayani[1], Kuncoro Diharjo[3], Fajar Rachman Wibowo[1], Margaretha Arnita Wuri[1]

[1]Chemistry Department, Faculty of Mathematics and Natural Sciences, Sebelas Maret University, Jl. Ir Sutami 36A Surakarta, Indonesia
[2] School of Materials and Mineral Resources Engineering, Engineering Campus, University Sains Malaysia, 14300 Nibong Tebal, Pulau Pinang, Malaysia
[3] Faculty of Engineering, Sebelas Maret University, Surakarta, Indonesia

[*]E-mail: nsuharty@yahoo.com

Abstract. This research has successfully synthesized six hybrid laminate composites (HLC). These HLC consist of two layers skin composites and one layer of core PP/KF composites. There are sticked with epoxy adhesive by using cold press method. In this research 6 types of skin are used, namely the rPP (recycled polypropylene, HC1); rPP/DVB/PP-g-AA/KF (HC2); rPP/DVB/PP-g-AA/Hall (HC3); rPP/DVB/PP-g-AA/Hall+ZB (HC4); rPP/DVB/PP-g-AA/KF/Hall (HC5) and rPP/DVB/PP-g-AA/KF/Hall+ZB (HC6) composites. The mechanical properties assessment tensile strength (TS) of various HLC before and after cyclical thermal (CT) was done by ASTM D638. While testing the flame retardant: such as time to ignition (TTI) and burning rate (BR) was done by ASTM D635. Heat stability of HLC can be recognized by conducting the CT treatment. It is to determine the effect of fluctuating heat loads on mechanical properties of HLC materials. The TS result of five HLCs (HC2, HC3, HC4, HC5 and HC6) before CT treatment were higher than HC1 (blank HLC). Those five HLC are also able to increase the TTI and reduce the BR compared to HC1. The CT treatment conditions performed at 45 ^{o}C as much as 125 times. After CT treatment, the TS values only slightly decline compared to before CT treatment.

1. Introduction

By the development of science and technology, composites design leads to the development of hybrid composites. Hybrid composite is a composite which uses two or more kinds of fillers in a single matrix [1]. One of the means to overcome the weaknesses of the mechanical properties of the composite is to develop a new type of composites, namely Hybrid Laminate Composites (HLC). HLC is a combination of at least two layers of the composite laminate skin and core materials which glued together with an adhesive. The skin is a layer on the HLC surface which coats the core of HLC, Figure 1 [2-4]. Currently, the gluing method used to combine layers of the laminate in the HLC is the most suitable method, because the adhesion is lightweight, it does not require high technology and the stresses distributed more evenly, so it can produce a highly strong HLC [3,5]. One type of adhesive used is an epoxy which has a high adhesion capabilities, robust and heat resistant [4,6]. Generally, the main properties of the core material are able to increase the thickness of HLC, so that it can improve the mechanical properties without increasing significant weight [2].

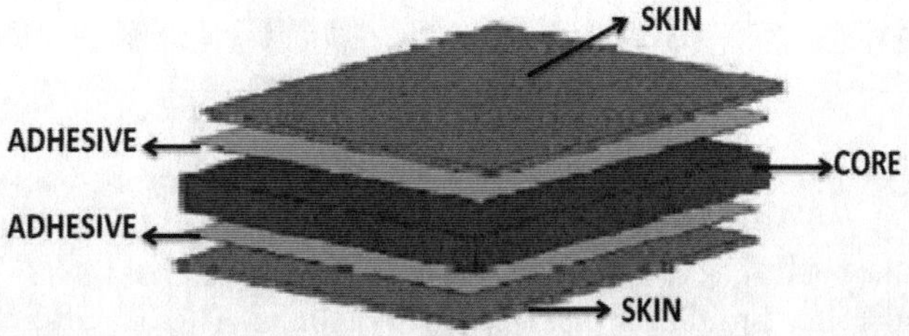

Figure 1. HLC Three Layer Skin-Core-Skin Construction

The composition of the skin composites to produce HLC must fulfill the criteria, such as has high mechanical properties and high flame resistance [2]. In the automotive industry, composite materials used must have strong mechanical properties, flame retardant (heat stability), and lightweight. The fuel efficiency of the automotive construction must be as light as possible [1]. According to some researchers, improved mechanical properties can be done with the addition of natural fibers such as kenaf fiber (KF) [7-10]. Suharty *et al.*[11] had synthesized some composites: [rPP/DVB/PP-g-AA/Hall], [rPP/DVB/PP-g-AA/Hall+ZB], [rPP/DVB/PP-g-AA/KF], [rPP/DVB/PP-g-AA/KF/Hall], [rPP/DVB/PP-g-AA/KF/Hall+ZB], which is able to significantly improve the mechanical properties. It is found that cellulose of KF and halloysite (Hall) can interact with PP matrix through coupling compound PP-g-AA and compatibilizer DVB. Other researchers had also reported that natural flame retardant compound such as kaolin, bentonite, montmorillonite and Hall is friendly environment [12-16]. Hand et al.[17] and Rawtani et al. [18] reported that the Hall is a potential inhibitory compound because it has high thermal stability and acts as an oxygen barrier to the formation of flame. Suharty et al. [19] also reported rPP/pineapple-leave fiber/bentonite+ZB composites can improve flame resistance due to the cellulose, bentonite, and zinc borate (ZB) compounds can inhibit the combustion process.

Nowadays, the need of automotive increases every year [20]. However, the supporting material such as metal in nature recently was limited. On the other hand, the availability of fuel is also depleting [21, 22]. To overcome those problems alternative material which has good mechanicalproperties, high heat stability and light in mass for metal replacement, such as HLC material is needed [9, 22]. Some researchers reported that repeated heating on glass-fiber reinforced concrete (GFRC) and bismaliemede (BMI) carbon-fiber composites below the distortion temperature of the primary material will degrade physical and mechanical properties [23, 24]. Susanto [25] also reported that 10 times CT treatment on GFRC and ramie fiber reinforced concrete (RFRC) below distortion temperature of the main matrix causes thermal fatigue and damage, thereby reducing the mechanical properties of composites. To be able to identify heat stability of materials, the CT treatment under distortion temperature of the main polymer matrix must be performed.

In this study, (1) HLC is made by using skin of rPP/KF/Hall+ZB composites which has been synthesized in our previous work [11] and core PP/KF composites:

a)[rPP]/[PP/KF]/[rPP];**b)**[rPP/DVB/PP-g-AA/Hall]/[PP/KF]/[rPP/DVB/PP-g-AA/Hall];**c)**[rPP/DVB/PP-g-AA/Hall+ZB]/[PP/KF]/[rPP/DVB/PP-g-AA/Hall+ZB];**d)**[rPP/DVB/PP-g-AA/KF]/[PP/KF]/[rPP/DVB/PP-g-AA/KF];**e)**[rPP/DVB/PP-g-AA/KF/Hall]/[PP/KF]/[rPP/DVB/PP-g-AA/KF/Hall];**f)**[rPP/DVB/PP-g-AA/KF/Hall+ZB]/[PP/KF]/[rPP/DVB/PP-g-AA/KF/Hall+ZB], (2)

The mechanical and inflammability properties of HLC is tested before CT treatment.(3) The CT treatment of HLC materials is conducted under distortion temperature of PPto understand their heat stability(4) The mechanical properties of HLC after CT treatment is tested, (5) This research will find the effect of CT treatment on mechanical properties of HLC.

2. Methods

2.1. Materials

The starting compounds which belong to pro-analysis grade except mentioned is used without any refining. In this work recycled PP (rPP) used is the cup waste of mineral water packaging. The rPP waste after cleaned up then chopped in 2 x 2 mm. The kenaf fiber (KF) used was bark fiber. It was supplied with Centre of Kenaf Fiber Product in Lamongan Indonesia. The KF after been cut to 5 mm size, then being immersed for 24 hours in NaOH 4%. The halloysite powder was supplied by Applied Minerals Inc., USA. After sieving through 200 mesh, the halloysite powder then calcinated at 800 ^{o}C for 1 hour. The core materials PP/KF composites were purchased from Toyota Boshoku, Malang Indonesia. The epoxy A and B pro-technical adhesive were purchased from PT. Jutus Kimia Raya Semarang Indonesia. Other chemical compounds, such as acrylic acid (AA) and divinyl benzene (DVB), xylene derived from Aldrich; while benzoyl peroxide (BPO) and zinc borate (ZB) purchased from E.Merck.

2.2 Synthesize of the HLC

Preparation of six types skin composites (see Table 1) refers to Suharty et al. [11]. Synthesizing of six type of HLC are as follow: (1) Preparation of a specimens skin composite with an area 12 x 12 cm^2 and the surface made rough, (2) Preparation of specimen core composite with an area 12 x 12 cm^2 and the surface made rough, (3) Preparation of adhesive by mixing epoxy A and B in ratio 1:1, by sudden, (4) Layer up the surface specimens either skin or core each one with 2 grams epoxy (A+B) by hand lay-up. These layers are allowed to stand for 7 minutes (time starting to pressure), (5) Each specimen glued following skin-core-skin construction and then cold pressed with a load of 3 kN/m^2 for 24 hours.

Table 1. Formulation of Hybrid Laminate Composite (HLC)

Code	Skin	Core
HC1	rPP	PP/KF
HC2	rPP/DVB/PP-g-AA/Hall	PP/KF
HC3	rPP/DVB/PP-g-AA/Hall+ZB	PP/KF
HC4	rPP/DVB/PP-g-AA/KF	PP/KF
HC5	rPP/DVB/PP-g-AA/KF/Hall	PP/KF
HC6	rPP/DVB/PP-g-AA/KF/Hall+ZB	PP/KF

2.3 Cyclical Thermal (CT) Treatment, Mechanical, and Inflammability Testing of HLC

Criteria of one time CT with duration 30 minutes are: (1) Heating the sample from the temperature 25 – 45 °C for 4 minutes, (2) then held for 12 minutes at 45 °C, (3) After that, cooled down to 25 °C for 12 minutes, (4) Finally, held at 25 °C for 2 minutes. Cyclic thermal (CT) testing carried out by an automatic cyclic thermal equipment. The CT conditions performed at 45 ^{o}C as much as 125 times. Mechanical properties assessment was done by using ASTM D638 and inflammability testing following ASTM D635.

3. Result and Discussion

3.1 Mechanical Properties Test BeforeCT Treatment

The mechanical properties testing of TS and % elongation can be seen in Figure 2. The TS comparison and comparative % elongation against HC1 before CT treatment can be seen in Figure 3. The TS value of HC1 (which is made from rPP skin, referred as HLC blank) was 41.34 MPa. TS of all

HLCs such as: HC2, HC3, HC4, HC5 and HC6 increased to 5%, 12%, 9%, 22% and 30%, respectively. Percentages elongation of HC1 was 3.6; the %elongation of HC2 and HC3 composites 6%, and 11% decreased, respectively; in the other side, the % elongation of HC4, HC5 and HC6 composites 14%, 8% and 3% increased, respectively.

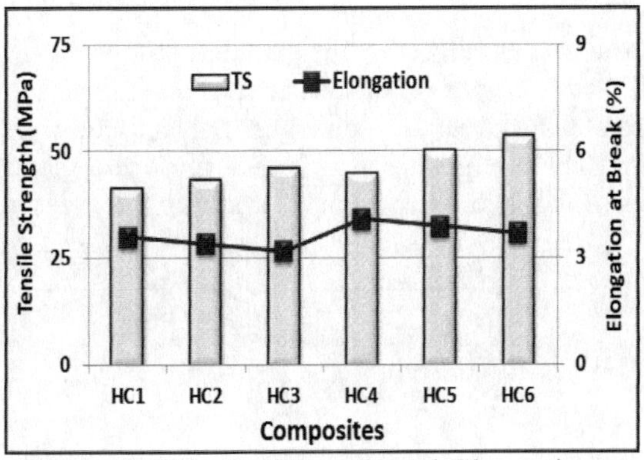

Figure 2. Tensile Strength and Elongation at Break of Various HLC Before CT Treatment

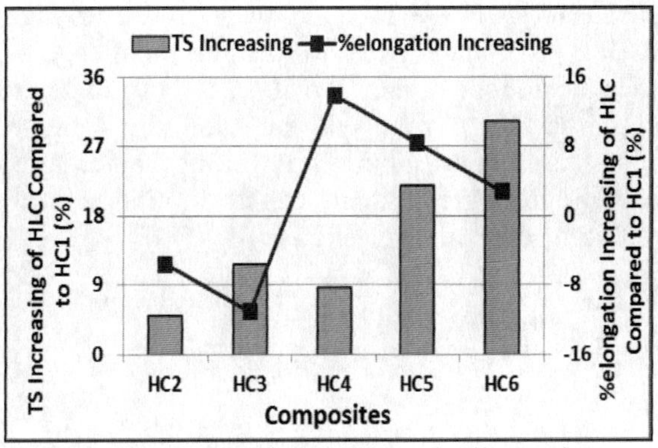

Figure 3. Comparison TS and Elongation at Break of HLC to that of blank HLC Before CT Treatment

3.2 The Inflammability of HLC Before CT Treatment

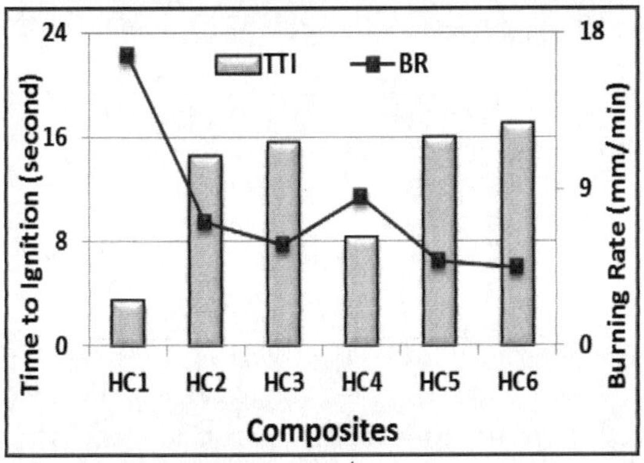

Figure 4. TTI and BR Values of Various HLC Before CT Treatment

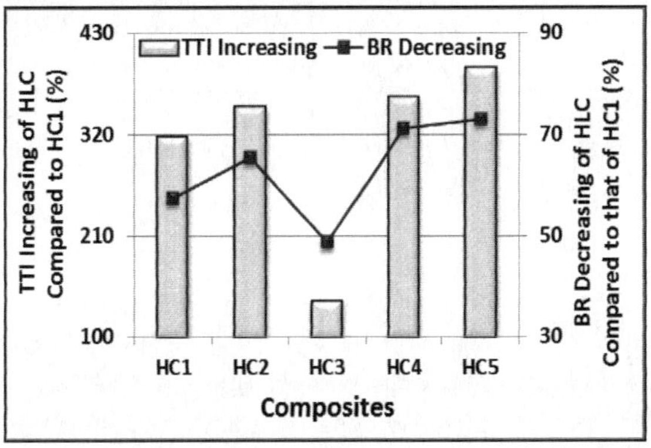

Figure 5. Comparison BR of HLC to that of blank HLC Before CT Treatment

The inflammability test result TTI and BR of HC1-HC6 and their comparison before CT treatment to that of HLC blank can be seen in Figure 4 and 5. The TTI values of HC1 is 3.5 seconds; TTI value of HC2, HC3, HC4, HC5 and HC6 composites317%, 350%,139%, 361% and 392% increased, respectively. The BR of HC1 is 16.7 mm/min; BR of HC2, HC3, HC4, HC5 and HC6 composites 57%, 65%, 49%, 71% and 73% decreased, respectively. The cause of the increasing TTI value and the declining BR value is due to the flame retardants on each skin composites HC2 to HC6.The KF content of the skin and core of HC4 will produce charcoal when it burned. The charcoal acts as a barrier to the entry of oxygen, so the flame formation is inhibited. The flame inhibition also has an effect on the combustion process, to decrease the BR. In the HC2 and HC3, which their skin contains Hall and ZB, in the combustion process, Hall can undergo endothermal reaction and also can produce charcoal as a barrier oxygen [12, 14, 17]. In burning process, ZB produced B_2O_3 and water vapor to inhibit combustion [19].

3.3 Mechanical Properties of HLC After Cyclic Thermal (CT)

The mechanical properties of TS and % elongation values after CT treatment can be seen in Fig. 6. The TS comparison and % elongation comparative results against blank HLC after CT treatment can be seen in Fig. 7. The effect of 125x CT treatment at temperature 45 °C, for each TS value of HC1, HC2, HC3, HC4, HC5, and HC6 decreased slightly. The decreasing values are 6.5%, 3.2%, 2.8%, 2.5%, 3.3% and 3.9 %, respectively. After CT treatment, the result of mechanical properties as follows: almost all the TS of HC1 to HC6 undergo a slight decline. It is because when there is heating in composites, there will be molecule's movement. This movement changes the structure of composites and resulting in lower mechanical properties. The composites which undergo heating and cooling process repeatedly can lead to thermal fatigue and damage its structures. It slightly decreases the TS of the composites [25].

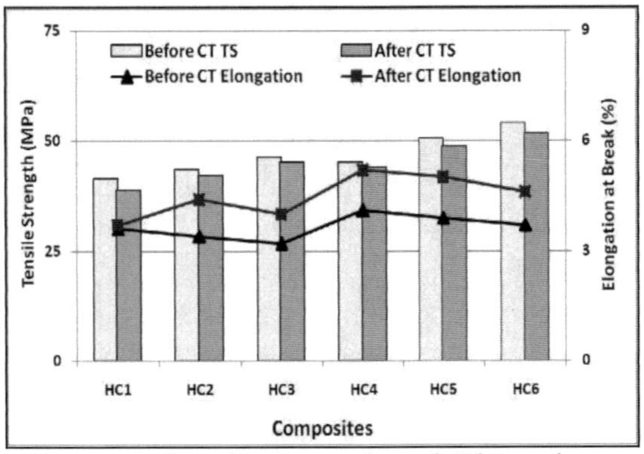

Figure 6. Tensile Strength and Elongation at Break of Various HLC Before and After CT Treatment 125x at 45 °C

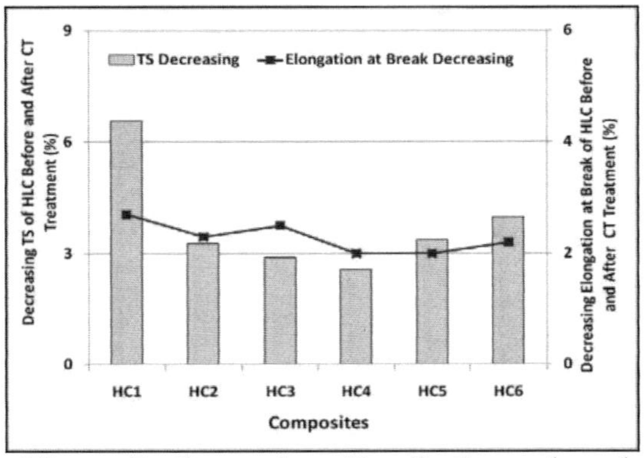

Figure 7. Comparison Tensile Strength and Elongation at Break of HLC Before and After CT Treatment 125x at 45 °C

The effect of 125x CT treatment at temperature 45 °C, for each % elongation value of HC1, HC2, HC3, HC4, HC5, and HC6 increased slightly. The increasing values are 2.7%, 2.3%, 2.5%, 2.0%, 2.0% and 2.2%, respectively. After CT treatment, the result of % elongation as follows: almost all the TS of HC1 to HC6 undergo a slightly ascending. In the CT treatment, HLC which is heated and cooled repeatedly will cause expansion and contraction. HLC, which made from various raw materials having different expansion and shrinkage coefficients, so that it will degrade the initial structure of HLC, including the elongation, was also changed slightly bigger, but this change is not so significant. Chemically, when the compound is heated the molecules will absorb energy, which lead to the movement of molecules irregularly. The irregularity of molecule will affect the strength of the bonds between the polymer molecules either adhesion or cohesion that it can change the structure of the polymer and cause degradation [26]. Thus, the repeating process of heating and cooling to the composites will lead to different expansion and contraction, therefore the bond strength decreased. This resulted in the bond extension so that the elongation of HLC can slightly increase [27]. The slightly changes in the TS as an effect of CT is due to the raw material of HLC each skin and core contain flame retardant compounds significantly. This can be seen in the high TTI value and low BR values of the various HLC. Thus, the absorption of energy on heating is relatively small and is used in an endothermic reaction of the Hall and the charcoal formation of KF and formation B_2O_3 of ZB, where these compounds inhibit the combustion process.

4. Conclusion

This research has successfully synthesized six types of HLC namely HC1, HC2, HC3, HC4, HC5, and HC6. Their mechanical properties of TS and inflammability of TTI and BR before and after CT was done by using ASTM D638 and D635. Before CT the five of HLC (HC2, HC3, HC4, HC5, and HC6) produced materials which able to increase the TS compared to blank HLC. The TTI and BR of HLC before CT treatment were increasing and decreasing respectively, compared to HLC blank. After CTtreatment, the TS values of HLC were slightly decline compared to before treatment.

Acknowledgment

The author would like to thank for financial support in this project through International Research Collaboration and Scientific Publication Grants to Directorate General of Higher Education Indonesia.

References

[1] Zhang J, Chaisombat K, He S and Wang C H 2012 *Mater. Design.* **36** 75
[2] Lukkassen D and Meidell A 2007 *Advanced Materials and Structures and their Fabrication Processes: Part II Sandwich Constructionsed* 3 (Narvik: Narvik University College HiN)
[3] Banea M D and da Silva L F M 2009 *J. Mater. Design. App.* **223** 1
[4] Verma C S and Charian V M 2012 *Compos. : Part B* **43** 1063
[5] Cherian Z and Lehman R 2005 *Int. J. Adhes. Adhes.* **25** 502
[6] Vick C B 1995 *Proceeding of the Symposium sponsored by U.S. Department of Agriculture Forest Service* (USA: Forest Product Laboratory and the Forest Product Society)
[7] Wambua P, Ivens J, and Verpoest I 2003 *Compos. Sci. Technol.* **63** 1259
[8] Bledzki A K, Mamun A,Faruk O 2007 *eXPRESSPolym. Lett.* **1** 755
[9] Suharty N S, Wirjosentono B, Maulidan F, Handayani D S, Sholikhah J and Maharani Y A 2008 *J. Phys. Sci.* **19** 105
[10] Suharty N S, Almanar I P, Sudirman, Diharjo K and Astasari N 2012 *Proceed. Chem.* **4** 282
[11] Suharty N S, Ismail H, Diharjo K, Handayani D S and Firdaus M 2016 *Proceed. Chem.* In Progress
[12] Gilman J W, Jackson C L, Morgan A B and Harris Jr R 2000 *Chem. Mater.* **12** 1866
[13] Du M, GuoB and Jia D 2006 *Eur. Polym. J.* **42** 1362
[14] Sarkar M, Dana K, Ghatak S and Banerjee A 2008 *Bull. Mater. Sci.* **31** 23
[15] Al Hassany Z, Genovese A and Shanks R A 2010 *eXPRESSPolym. Lett.* **4** 79
[16] Marney D C O, Russel L J, Wu D Y, Nguyen T, Cramm D, Rigopoulos N, Wright N and Greaves M 2008 *Polym. DegradStabil.* **93** 1971
[17] Handge U A, Hedicke-Hochstotter K and Altsadt V 2010 *Polym.* **51** 2690
[18] Rawtani D and Agrawal Y K 2012 *Rev. Adv. Mater. Sci.* **30** 282
[19] Suharty N S, Ismail H, Wibowo F R, Dihardjo K, Handayani D S, Firdaus M and Lathifah L 2015 *App. Mech. Mater* **736** 13
[20] McAule J W 2003 *Environ. Sci. Technol.* **37** 5414
[21] Hestiawan H, Asrul F 2014 *Mechanical Journal* **5** 1
[22] Zampalony M, Pourboghrat F,Yankovick S A, Rodgers B N, Moore J, Drzal L T, Mohanty A K and Misra M 2007 *Compos.: Part A* **38** 1569
[23] Widyanto S A, Sulardjaka and Arijanto 2004 *Report of Basic SciencesResearch DirectorateGeneral of Higher Education*(Semarang: Engineering Faculty UNDIP)
[24] Ju J and Morgan R J 2004 *J. Compos. Mater.* **38** 2007
[25] Susanto I 2004 *Thermal Cyclic Influence on Mechanical Strength Composite Matrix (Al2O3 + 20 wt% SiO2) which is Reinforced with Particles Cr* (Yogyakarta: Gadjah Mada University)

[26] Liza 2000*Damage Analysis Techniques (Failure Analysis) on Plastic Components*(Bandung: Centre of Polymer Technology)
[27] Ray B C 2005 *J. Reinf. Plas. Compos.***24** 111

IOP Conf. Series: Materials Science and Engineering **107** (2016) 012039 doi:10.1088/1757-899X/107/1/012039

Physical characteristics of chitosan-silica composite of rice husk ash

Woro Sumarni[1*], Retno Sri Iswari[2], Putut Marwoto[3], Endah F Rahayu[1]

[1]Department of Chemistry, Faculty of Mathematics and Natural Sciences, Semarang State University, Semarang, 50229, Indonesia
[2]Department of Biology, Faculty of Mathematics and Natural Sciences, Semarang State University, Semarang, 50229, Indonesia
[3]Department of Physics, Faculty of Mathematics and Natural Sciences, Semarang State University, Semarang, 50229, Indonesia

[*]E-mail: woro.kimia.unnes@gmail.com

Abstract. Some previous studies showed that the characteristics of chitosan membranes have a very rigid and non-porous structure so that its utilization is not maximized, particularly in the filtration process. Hence, it needs modification to improve the quality of the chitosan membranes. Adding the silica into the chitosan membranes is one of the offered solutions to overcome the problems of physical and mechanical properties of chitosan. This study aims to investigate the effect of variations in the silica composition to the physical characteristics of the chitosan-silica membranes of rice husk ash that were synthesized. The chitosan used is derived from the chitin of Vannamei shrimps' shell with 82% degree of de-acetylation, while the silica was synthesized from rice husk ash with rendering of silica (SiO_2) by 51% and the results of XRD analysis showed an amorphous phase. Membrane synthesis was performed using the phase inversion method with chitosan-silica mass ratios of rice husk ash, which were 1:0.0; 1:0.5; 1:1.0; 1:1.5 and 1:2.0. The results showed that the addition of silica increases the swelling index and the membrane permeability. The results of the analysis, FTIR spectra, obtained a new functional group after the addition of silica, they are Si-OH, Si-O-Si, and CO-NH2. The morphology test using CCD Microscope MS-804 results in the very tight chitosan membranes without the silica surface, it has no pores, smooth and homogeneous, while the chitosan-silica composite membrane of rice husk ash obviously has cracks and small cavities that seemed to spread out.

1. Introduction

As a maritime country, Indonesia produces massive fishery products such as crabs, shrimps, and scallops. The waste of crab, shrimp, and scallop shells poorly optimized; therefore, it would cause environmental pollution. The general society does not know that in the crab, shrimp and scallop shells waste contains chitin which could be processed into chitosan that is able to be used for various purposes.

Chitosan is a derivative component from chitin that is called (2-amino-2-deoxy-β (1.4)-D-glucopyranose) in the poly structure. Chitosan is a polysaccharide polymer amine unit composed of glucosamine and N-acetyl glucosamine which is a non-toxic, antibacterial, biocompatible and biodegradable hydrophilic polymer [16]. Chitosan has good chemical reactivity as it has a number of hydroxyls (-OH) groups and amine ($-NH_2$) groups on its chain. Both of these groups allow modifications to the chitosan to produce physical and chemical properties desired [31].

At the moment, chitosan has a lot of uses, among others are in the area of health, water treatments, membranes, hydrogels, adhesives, antioxidants, and food packaging. Chitosan is insoluble in water but soluble in organic acid solvents under pH 6 include formic acid, acetic acid, and lactic acid. As one of the organic polymer materials which can be used as a membrane, chitosan serves as a thin barrier to separate the two phases, which can release certain components and hold the other components of a fluid flow. Some researchers suggest that chitosan membrane can be used as an adsorbent of heavy metal ions [2, 18], the adsorbent of the urea [19, 11] edible film [21] and can be used as supporting material to produce alternative energy called fuel cell [31, 28].

Chitosan membrane will have pores when added additives or pore-formers (porogen). The pore size of the chitosan membrane can be formed in accordance with the needs of the application. Modifications are done towards the hydroxyl (-OH) group and amine ($-NH_2$) groups on the glucosamine unit of chitosan.

Some researchers have carried out some modifications to the chitosan membrane. Some of the modifications is done by adding materials or polymers that can form crosslinks with chitosan molecules, such as the addition of glutaraldehyde or genipin [13], the establishment of tissue between the chitosan molecules with other polymer molecules such as (ethylene oxide) (PEO) poly [13], and coating the surface of the chitosan membrane with alkane steam plasma (petroleum ether). Chitosan membrane modification is expected to produce membranes with better characters such as increased stability of the membrane [13], reduced size of the membrane pores to make the separation of molecules or macromolecules rejection of a solution by the membrane more effective. The cross-linked of chitosan is a proposed solution to overcome the problems of the mechanical properties of chitosan to increase the adsorption of metal [8, 2, 25] conducted a synthesis of the chitosan-zeolite composite membrane. Zeolite composite chitosan is based on morphological analysis including non-pore membrane. FTIR analysis result does not show the formation of new functional groups which means that the composite is the result of physically mixing [34] conducted a synthesis of chitosan-silica composite membrane of tetra ethoxy ortosilan (TEOS) to determine the effect of chitosan concentration on the properties of chitosan-silica composite membranes for fuel cells.

Silica can be synthesized from a wide variety of natural resource-rich silica. One source of silica which has so far not fully utilized is rice husk. As an agricultural country, Indonesia is a rice producing country. Rice milling process produces rice husks which are usually discarded and become waste that cause environmental pollution or burned away in the rice milling area. In the process of burning rice husks, some compounds such as hemicellulose {xylose-β (1,4) - mannose-β (1,4) - Glucose-α (1,3) -Galactose}, cellulose (β-1,4-poly glucose) and others will be converted to CO_2 and H_2O. The ash obtained from the combustion of rice husk is about 13.1% - 29.04% in dry weight [14]. This rice husk ash can be used as a source of silica (SiO_2). The research conducted by [10] shows the content of silica (SiO_2) in rice husk ash is 94-96%. Silica (SiO_2) that is inert, hydrophobic, and transparent has a mechanical strength and high thermal stability. It does not swell in organic solvents.

Silica (SiO_2) is as one of the ingredients to form a solid membrane which can be used for metal ion absorption of manganese (Mn^{2+}), cadmium (Cd^{2+}) and nickel (Ni^{2+}) in waste water. The use of rice husk ash silica alone in membrane synthesis will produce fragile membranes because inorganic membranes have the disadvantage of limited application and are fragile [33] as well as have low selectivity level [4]. The addition of silica into the polymer matrix can increase the mechanical strength such as water resistance, strong stretching and tensile strength [30]. Modification of chitosan and silica can be used as raw materials for the synthesis of silica-based material for decolorization dye process and the reduction of heavy metal ions by filtration [31, 22].

Referring to the characteristics of chitosan and silica, and then both of the materials have the potential to be combined in an effort to get a filtration membrane with several improved characteristics. As has been done by [35] in his study, the chitosan-silica membrane is prepared by phase inversion technique.

Based on the explanation above, it has carried out research aimed 1) to synthesize silica from rice husk ash, 2) to synthesize chitosan fromVannamei shrimp shells, 3) to synthesize composite silica membrane of chitosan-rice husk ash, and 4) to characterize the result of the synthesized chitosan-silica membrane. As for the physical characterization of chitosan-silica, composite membranes of rice husk ash include swelling index, permeability, group function and morphology of the membrane. The expected benefits of this research are 1) Improving the economic value of rice husks and shells waste; 2) Providing information about the physical properties of chitosan-silica composite membranes of rice husk ash; 3) Producing a qualified composite membrane product which is environmentally friendly.

2. Method

2.1. Rice Husk Silica Manufacturing

Silica (SiO_2) of rice husk obtained through the charcoal using process and incineration with high temperature. The ashing process is carried out in the furnace at a temperature of 600 degree Celsius for an hour. The result of this process is white rice husk ash [23]. To determine the composition of the rice husk ash, X-Ray fluorescence (XRF) is used [26]. The ash obtained from the ashing process is then purified with acid. This aims to eliminate the metal and non-metal content from the rice husk ash. The result of the purification is the brown rice husk ash. The ash that has been through acidification is reheated by 600 degree Celsius for 1 hour [8]. Synthesized silica is then characterized by using a Shimadzu XRD-7000 X-ray Diffractometer to obtain its crystalline results.

2.2. The manufacturing of sodium silicate solution as the silica source

To make a solution of sodium silicate, rice husk ash is reacted with NaOH 4M solution. By heating and stirring the solution, the brownish viscous mixture is obtained. After the half-dried product, it is put in the furnace with 600 degree Celsius for 30 minutes to obtain solid sodium silicate with greenish white color. Solid sodium silicate is then dissolved in demineralized water. This process takes overnight until its color becomes brownish-yellow.

2.3. The manufacturing of Vannamei crusts shrimp Chitosan

The preparation of van named shrimp shell sample was taken from the shrimp shell waste in Tambakrejo. The amount of the shrimp shell sample used was 1 kg. Then, this shrimp shell sample was washed, boiled, and dried in an oven at 120 degree Celsius for approximately one hour. After dried, it was then milled, and it was also sieved by using a 50 mesh sieve to obtain dried shrimp shell powder. After that, the powder obtained was processed through deproteination, demineralization, depigmentation, and deacetylation. The powder obtained was then washed by using arcades until neutral. It was also put at 60 degree Celsius until completely dry (± 8 hours). Furthermore, chitosan was characterized using FTIR to provide information about DD value, water level, and ash level. To make a chitosan solution, chitosan was dissolved in CH_3COOH 2% v/v, stirred using a magnetic stirrer for 1 hour until it formed a homogeneous solution [5, 19].

2.4. The manufacturing of Chitosan-silica composites membranes

Chitosan-silica composite membrane was made by mixing chitosan solution with sodium silicate solution which is the source of silica [5]. To obtain the homogeneity of the chitosan-silica solution, stirring this mixture for 1 hour will obtain a yellowish colored clear solution which has a thick texture. 100 ml of a solution of chitosan-silica mixture is poured into a mold, and make sure that the surface is even. The synthesis of membranes with the variation in composition of chitosan solution: silica 1:0; 1: 0.5; 1:1; 1:1.5; 1:2. Once it is formed, the chitosan-silica solution mixture is dried at normal temperature [21]. Dried Chitosan-silica membrane is still acidic and readily soluble in water, so it needs to be neutralized with a base. According to [18], soaking the membrane in alkali solution aims to transform the amine group of cation ($-NH_3^+$) into free amine group ($-NH_2$). The membrane obtained is then washed with aquades to make it neutral. This process is done in order to facilitate the next stage. To determine the physical characteristics of the membrane, it will undergo swelling index test, flux membrane test, and the identification of functional groups using FTIR (Fourier Transform Infra-Red) and Morphology Test of chitosan-silica composite membrane section using a CCD Microscope MS-804.

3. Results and Discussion

3.1. The results of silica manufacturing from rice husk ash

Silica (SiO_2) rendering of rice husk from the synthesis results are as shown in table 1.

Table 1. Data Results of rice husk silica ash yield

Rice Husk Charcoal Mass (gram)	Rice Husk ash Silica Mass (gram)	Rice Husk Ash Silica Level (%)
180	90.6482	49.6
250	122.0160	51.2
250	119.3080	52.3
The average levels of rice husk ash silica are **51.0**		

Based on table 1, this process produced rice husk silica (SiO_2) up to 51.0% from 680 grams of rice husk charcoal. By using X-Ray fluorescence (XRF), the characterization result of the chemical content of the rice husk ash silica obtained SiO_2= 97.24%.

Silica crystallinity of rice husk ash was characterized using XRD. This qualitative analysis was conducted with 1.54060 Å wavelength. The operating condition involved Cu radiation, with a step size of 0.0200 degrees, voltage of 40.0 kV and current of 30.0 mA. The samples were scanned from the observation area between 5.0000 to 90.0000 degrees. The data obtained were in the form of inter-field distance, intensity, and angle magnitude (2θ) which is then matched with the pattern of X-ray diffraction on JCPDS (Joint Committee on Powder Diffraction Standard) number 39-1425. Diffractogram of rice husk ash silica is presented in figure 1.

The diffractogram result of silica (SiO_2) obtained in Figure 1, appears a sharp peak at $2\theta = 21.7200$; $2\theta = 20.8800$; and $2\theta = 22.5200$, it shows the silica diffraction pattern of hydrated amorphous phase [7]. The sharp peak at $2\theta = 20.8800$ indicates that the silica contains the SiO_2 compound in the form of cristobalite mineral. According to [15], the broad peak shape with a central peak around $2\theta = 22$ indicates that the silica (SiO_2) are amorphous. Amorphous structure of the silica surface is described as random network with a low regularity degree [8, 20]. This kind of amorphous phased silica is more preferable because it is more soluble. The more soluble the silica, the better of the silica-based material making, in this case, is sodium silicate.

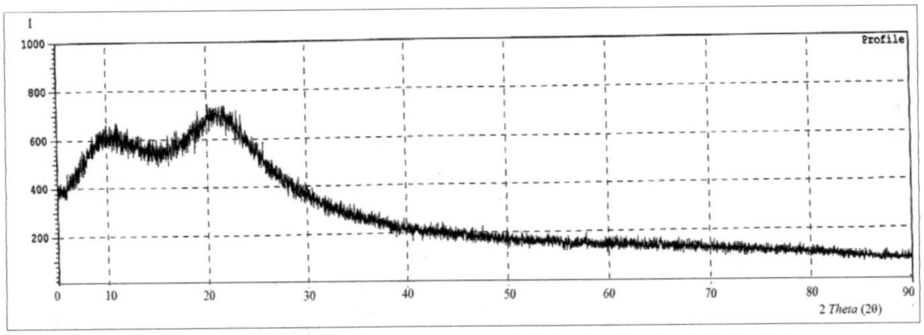

Figure 1. The Result of X-ray diffractogram Rice Husk Silica Synthesis.

3.2. The result of chitosan characterization from Vannamei shrimps crust waste
Table 2 presents the results of the synthesized chitosan characterization from Vannamei shrimp shell.

Table 2. The characteristics of synthesized chitosan

Parameter	**Chitosan**
Base material	Vannamei shrimp crust
Partikel size	Powder (50 mesh)
Water level	5.8%
Ash level (%)	0.12%
Solubility in acetic acid 1 %	> 99%
Degree of deacetylation	82%

The result shown in table 2 indicates that chitin from Vannamei shrimp shell has been converted into chitosan.

3.3. The result of rice husk ashes chitosan-silica characterization
The results of the synthesis of rice husk ash chitosan-silica composite in the form of membranes of which the shape is like a thin sheet of plastic, transparent, mushy when exposed to water, and rigid when dried. In general, the synthesized rice husk ash chitosan-silica composite membrane product has good water resistance. This is indicated by the unbroken, strong membrane. The result of swelling index test Swelling index results is presented in table 3.

Table 3. Data Results of Chitosan-Silica Membrane Swelling Index

Variation chitosan: silica	***Swelling index*** (%)
1:0	45.538
1:0.5	51.333
1:1	53.486
1:1.5	54.518
1:2	54.936

Based on table 3, it can be seen that the chitosan-silica composite membrane will cause higher water absorption capacity than the membrane with chitosan composition only. This is based on [12]. This is possible because silica (SiO_2) is a compound which has a strong affinity towards water molecules. It is hygroscopic and has a broad surface area. These properties make the membrane easily absorb water on the surface, and increase the water resistance properties [9].

Another thing that also affects the swelling degree of chitosan membrane is its hydrophilic characteristic [24]. In general, the membranes with chitosan as the basis have higher hydrophilicity properties. This is possible because of the active functional group in chitosan such as a group of -OH and $-NH_3^+$ which can bond the H_2O molecules. Hydrophilicity properties of chitosan-silica membrane allow it to be used as the filtration media in decolorization. The membrane of which water absorption degree is more than 50% can be used for absorption or filtration [8].

The water content in the membrane is also one important part of the characterization of polymer electrolyte membrane for fuel cell application as it relates to its conductivity ability [6]. According to [8], the good membrane used for fuel cell application is ones who have a water absorption of less than 50%. If the water absorption is too high (more than 50%), the membrane will become really soft. If so, the membrane's lifetime will be shorter. The soft membrane cannot be used in fuel cells because it cannot function as an insulator between the two electrodes. According to [29] the higher the water uptake (percentage of water swelling), the higher the proton conductivity will be like the more water molecules in the membrane which can be the proton transfer media.

3.4. Water Vapor Permeability (WVP) of chitosan-silica membrane

In this permeability test, 0.0025 L of permeate volume with diameter membrane of 0.04 m was used, the area of the membrane was 1.256×10^{-3} m^2.Table 4 shows the result of permeability test of chitosan-silica membrane in a synthesized rice husk ash.

Table 4. Data of Membrane Permeability Test

Membrane Variation	Flukes (L/m^2.Hour)
1:0	13.2733
1:0.5	16.9138
1:1	19.0346
1:1.5	20.9051
1:2	23.8127

Table 4 shows that there is an improvement of membrane permeability/ fluke value along the material addition in the chitosan membrane.

The added silica gave effect to the structure of chitosan. The silica will form some pores with intermolecular silica-oxygen bonds, so it forms a long chain of SiO_2 and bonds between silica and oxygen in –OH group [29]. The formed pores influences permeate to pass quickly through the membrane [3]. The result shows that chitosan-silica membrane was the appropriate membrane that was used in the filtration process because it had pores, so it was a good fluke value. However, permeability value should not more than 75 L/m^2.hours. That membrane is not good to use as filtration media [8]. This is because membrane with big permeability value could not resist unwanted species.

3.5. The function group test of Chitosan-silica membrane

Based on result analysis of group function in FTIR spectra in a combination of the membrane in figure 2, it is specifically presented in Table 5.

Table 5. The resulting analysis of group function of chitosan membrane and chitosan-silica membrane

Types of absorption	**The number of the wavelength (cm^{-1}) of membrane**				
	1:0	**1:0.5**	**1:1**	**1:1.5**	**1:2**
Si-OH	-	902.69	902.69	902.69	902.69
Si-O-Si stretch	-	1072.42	1072.42	1072.42	1072.42
CH_2 bending	1427.42	1411.89	1427.42	1419.61	1381.03
$CO-NH_2^-$	-	1658.78	1658.78	1635.64	1635.64
CH stretch	2885.51	2885.51	2877.79	2885.51	2924.09
OH stretch	3425.58	3448.72	344872	3417.86	3441.01
NH_3^+	1604.77	-	-	-	-

From Table 5, it can be seen that there is some new absorption in Chitosan-silica membrane: Si—O—Si group, Si—OH, Si—O—Si stretch. The existence of new function group after adding the silica shows that the added silica has interacted with chitosan chemically. The existence of the Si—OH group shows hydrogen silanol groups of the silica interacts with the amide group or oxy group in chitosan [25, 27] while a combination of functional groups between the constituent components of composite indicatesphysical mixing. Ionic bond happens between amino group from chitosan of silanol group, and covalent group happens because of the reaction between hydrosol groups from chitosan of the silica network [1]. The mechanism of cross binder of chitosan silica can be showed as follow in figure 3.

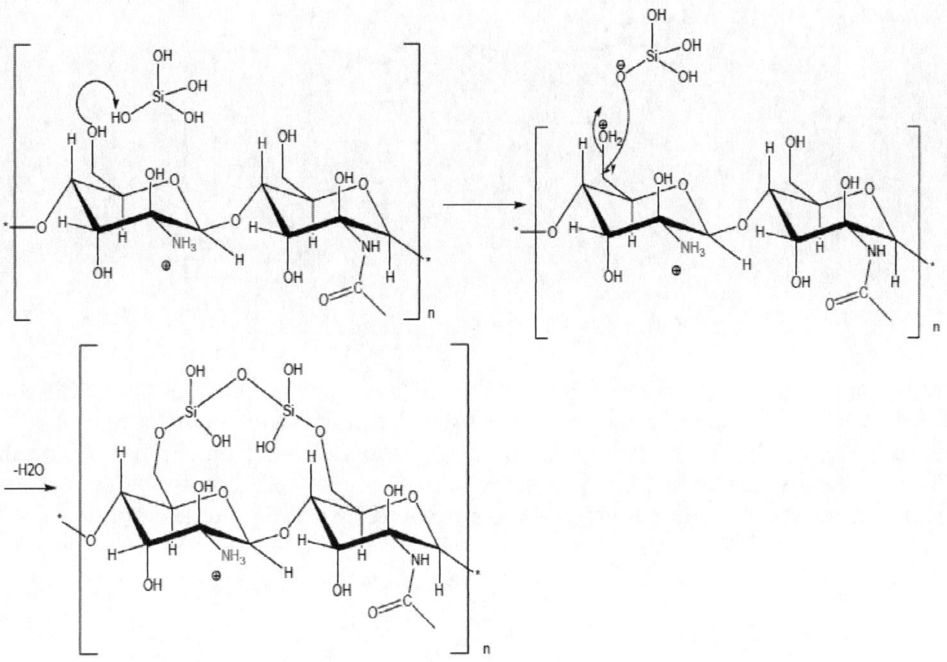

Figure 3. Mechanism of the cross binder of Chitosan-silica[3].

Besides covalent bonding occurs, most likely the silica enter to matrix chitosan through ionic interaction. However, this interaction could not be seen in the infrared spectrum. Another interaction probably happened was showed in figure 4 [3].

In figure 4 also shows there is Vander Waals interaction between amino group and silanol group (Si-OH), and hydrogen bonding in acetamide group with silanol group that contribute power to chitosan-silica membrane

Silica interaction to Chitosan also can be seen from the decreasing of the stretch intention of N-H in the 3270-3290 cm-1 area that overlaps with the absorption of –OH. It is because of the group N-H interacted with the added silica. In Chitosan-Silica 1:0 contained NH_3^+ absorption in 1597.06 cm-1 area, and it was not found $-NH_2$ absorption. It was because, in chitosan-silica membrane, there is no coagulation that change NH_3^+ to be a $-NH_2$ group, while in the membrane with the adding silica has passed the coagulation process with NaOH solution in making natrium silicate as Silica source.

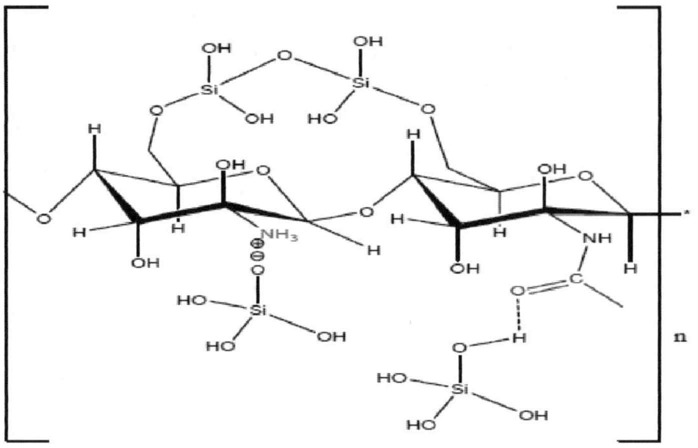

Figure 4. Chitosan Silica Interactions[3].

The absorption in the form of bending in 900-1000 cm-1 indicates symmetry stretch vibration of –Si-OH [17]. From the data obtained comes another absorption long wavelength 594.08; 594.08; 563.21 cm^{-1}. This absorption identifies the symmetry stretch vibration to Si-O from–Si-O-C. Wide absorption in 500-700 cm^{-1} area with special characteristic on it's not too sharp bending was the absorption that indicated symmetry stretch vibration of –Si-O from–Si-O-C [8]. From the absorption of –Si-O-C, it can be identified that the added silica has interacted with chitosan. The interaction occurs between Silica and Chitosan affected the opening of the cavities or cracks that make the synthesized membrane with pores.

4. Conclusion

Based on the test results of the physical characteristics of the composite membrane of chitosan-silica of rice husk ash which is synthesized, it can be concluded that the silica has properties as porogen to provide the pores on chitosan membrane so as to increase the capacity of water absorption, permeability, and increase the active sites of chitosan membrane. Thus, the chitosan-silica composite membranes of rice husk ash that are synthesized are very high potential to be used as a filter to reduce metal ions and dye textiles.

5. Acknowledgement

I thank Mrs. F. Widhi Mahatmanti, M. Si, Fransiska Hesti, Dyah Listyaningsih, and Lysa Setyaningrum, who are helping collect the data and writing this article.

References

[1] Al-Sagheer F and Muslim S 2009 *J.Nanomaterials* **1** 7
[2] Benavente M Luis M and Joaquin M 2011 *J.Taiwan.Institute.Chemical.Engineers* **339** 1-13
[3] Berghuis N T 2008 Sintesis membrane chitosan-tetraortosilikat (TEOS) sebagai membrane fuel cell pada such Tinggi (Bandung: ITB)
[4] Buhari Narsito and Eko S K 2009 Indo. *J. Chem.* **9** 2 170
[5] Cregg J, Sherri L, Wiseman B, Nicole M, Pietrzak-Goetze B, Martyn R, David B, Jaroch M, Daniel C, Ryan J and Gilbert 2009 *Tissue.Engineering* **167**
[6] Dhuhita A and Arti D K 2010 Karakterisasi dan Uji Kinerja SPEEK, cSMM dan Nafion untuk Aplikasi Direct Methanol Fuel Cell (DMFC) (Final Project Semarang: Chemistry Technic Department of Diponegoro University)
[7] Hanafi A and Nandang R 2010 *J. Kimia Indonesia* **51** 35
[8] Handayani E 2009 Sintesa Membran Nanokomposit Berbasis Nanopartikel Biosilika Dari

Sekam Padi dan Kitosan sebagai Matriks Biopolimer (Thesis. Institut Pertanian Bogor: Bogor)

[9] Handayani S, Dewi E L, Purwanto W W dan Soemantojo R W 2007 *J. Teknik Kimia Indonesia* **61** 563

[10] Harsono H 2002 *J. Ilmu Dasar 3* **2** 98

[11] Heryanto S A 2012 Modifikasi membran kitosan-silika Cu sebagai filter dan adsorben urea (Bandung : Universitas Pendidikan Indonesia)

[12] Huang K S, Jeng-Shian, Cheng Fun-Eu Lin and Shyh-Jer Lin 2009 Preparation and properties of PAA-chitosan/SiO2 hybrid materials Kun Shan University Taiwan 120

[13] Jin J, Song Mand and Hourston D J 2004 *Biomacromol* **51** 65

[14] Joddy A L 2002 *Prosiding Seminar Tantangan Penelitian Kimia* **298**

[15] Kalapathy U, Proctor A and Shultz J 2000 Biores.Technol **72** 99

[16] Kurita K 2006 Marine *Biotechnology* **8** 203

[17] Lee J, Shin, Hyoun, Hae, Young and Jun 2009 *J. Biomaterials* **30** 743..

[18] Lin E 2007 Study of The Mobility of Silver Ions in Chitosan Membrans (Thesis University of Waterloo: Canada)

[19] Liu J, Xin C, Zhengzhong S, Ping Z 2003 Preparation and characterization of chitosan/Cu (II) afinity membrane for urea adsorption Inc *j Appl Polym Sci* **90** 1108-1112

[20] Mujiyanti D R, Nuryono and Eko S R 2010 *Sains dan Terapan Kimia* **42** 150

[21] Nadarajah K 2005 Development and Characterization of Antimicrobial Edible Films from Crawfish Chitosan(Disertasi Louisiana State University)

[22] Noralia E and Dina K M 2013 UNESA *J. Chemistry* **21**

[23] Nuryono 2004 Effect of NaOH concentration on destruction of rice husk ash with wet technique (Proceeding Seminar Nasional Hasil Penelitian MIPA 2004, FMIPA Undip, Semarang)

[24] Pierog M, Magdalena and Jadwika 2009 Effect of ionic cross linking agent on swelling behaviour of chitosan hydrogel membrans Progress on Chemistry and Application of Chitin and its 14 14 75

[25] Putro A S 2013 Membran Komposit Kitosan-Zeolit untuk Aplikasi Direct Methanol Fuel Cell (Skripsi. Bogor: Departemen Kimia Fakultas Matematika dan Ilmu Pengetahuan Alam Institut Pertanian Bogor)

[26] Rafiee E, Shahebrahimi S, Feyzi M and Shaterzadeh M 2013 *Int.Nano. Letters* **229** 1

[27] Sudiarta I W, Diantariani N P and Suarya P 2013 Modifikasi silika gel dari abu sekam padi dengan ligan difenil karbaz on *Jurnal Kimia* **71** 57

[28] Smitha B, Sridhar S and Khan A A 2004 *Macromolecules* **376** 2233

[29] Suka I G, Simanjuntak W and Dewi E L 2010 *J. Natur Indonesia* **131** 3

[30] Teoh K H, S Ramesh and A K Arof *J.Solid.State.Electrochem* **16** 3165

[31] Vaghari H, Javarizadeh-Malmiri H, Berenjian A and Anarjan A 2013 *Sustainable Chemical Processes* **116** 1

[32] Yunianti S and Maharani D K 2012 *J.Chem* **51**

[33] Zulfikar M A and Mohammad A B 2006 *J. Matematika Sains* Bandung ITB

[34] Zulfikar M A, Wahyunigrum D and Berghuis N T 2009 Pengaruh Kitosan terhadap Sifat Membran Komposit Kitosan-Silika untuk Sel Bahan Bakar.(Prosiding Seminar Kimia Bersama UKM-ITB VIII. Bandung: Institut Teknologi Bandung)

[35] Zulti F, Kiagus D and Purwatiningsih S 2009 Sintesis dan Karakterisasi Membran untuk Filtrasi Limbah Prosiding Seminar Nasional Sains II(Bogor Hal 286)

IOP Conf. Series: Materials Science and Engineering **107** (2016) 012040 doi:10.1088/1757-899X/107/1/012040

Study of the acidic properties of ZrO_2-pillared bentonite

Ahmad Suseno[1*], Priyono[2], Karna Wijaya[1], and Wega Trisunaryanti[1]

[1] Department of Chemistry, Faculty of Mathematics and Natural Sciences, Gadjah Mada University, Sekip Utara BLS 21, Bulaksumur, Jogjakarta, Indonesia

[2] Department of Physic, Faculty of Sciences and Mathematics, Diponegoro University, Jl. Prof Soedarto, SH Tembalang, Semarang 50275, Indonesia

*Email: susen_undip@yahoo.com

Abstract. Research on pillared clays prepared from purified bentonite of Boyolali Central Jawa, Indonesia, and polycation Zr at various concentration and calcination temperature had been done. Effect of acidity characteristic and structure of resulting materials were studied. The nature of acidic site of the material was identified on the basis of FTIRspectra of pyridine adsorbed on ZrO_2-pillared bentonite catalysts. Analysis showed that increasing calcination temperature was followed by decreasing acidity and increasing ZrO_2 content in the pillared bentonite accompanied by the increase of its acidity. FTIR spectra showed there was an intensity increase of the characteristic band of 1635 cm^{-1}that indicates a Brønsted acid.

1. Background

Zirconium pillared clay is a microporous solid with a large surface area. The nature of zirconium itself is good as acid sites or as co-catalysts. Zirconium chloride octahydrate solution ($ZrOCl_2.8H_2O$) has been used to form stable oxide pillars between layers in bentonite clays. Pillarization impact is able to increase the surface area, acidity, and high selectivity. A $ZrOCl_2.8H_2O$ solution is normally used in ion exchange. However, the hydroxy zirconium structure on intercalation process and its transformation into ZrO_2 are not so easy to understand until at present [1-4]. In pillared clays, it was shown that the number of acid-Lewis centers that are present in the pillars predominates over that of Brønsted type, which mainly arises from the structural hydroxyl groups of the layers [5-7]. Acidity properties of catalyst surface play an important role in bringing various organic transformations. The nature of the acidic site in clays is different from that of metal oxides and zeolites. The study of acid centers on clays is still a subject of investigation primarily due to the presence of different kinds of water molecules and their indefinite contribution to a different type of acidities on the surface. Desorption analysis of pyridine can also be used for rapid evaluation of the amount and type of acid sites in clay minerals [8]. Brønsted and Lewis acid sites can be distinguished from FTIR spectrum of pyridine adsorbed on the clay surface. The spectrum shows well resolved and characteristic peak for pyridine adsorbed on both the types of acid sites [9].

The acidity characterization of pillared clays is frequently made by commonly used methods of acidic characterization of solids, namely, adsorption of basic molecules followed by FTIR spectroscopy analysis [10-11]. The present study focuses characterization of the surface acidity adsorbed pyridine, *basal spacing* (d_{001}) and morphology after pillarization process of ZrO_2/bentonite.

2. Experiment

2.1. Clay source and sampling
A sample of natural clay was collected from an open clay deposit in Boyolali, Central Java, Indonesia. The clay sample was dissolved using deionized water for 24 hours. Furthermore colloidal solution filtrate was centrifuged to obtain its precipitate that was separated from its filtrate by evaporation to get the bentonite. The solid bentonite was dried at 110 °C and sieved to 150 mesh size.

2.2. ZrO_2-pillared bentonite synthesis
Transformation of natural bentonite into pillared clay required the use of pillaring solution prepared at a variation in concentrations of 0.05; 0.10; and 0.20 M $ZrOCl_2.8H_2O$ for 2 hours as well as refluxed it under constant stirring at 70 °C. Thereafter, the pillaring solution was added dropwise to *ca.* 4 g of clay in 200 mL suspensions with constant stirring and aged overnight at room temperature. Afterward, the pillared clay suspension was filtered and washed with distilled water and neutralized until Cl^-free. The pillared clay was air-dried at ambient temperature and 80 °C for 8 hours, then calcinated at 400 °C for 4 hours. The resulting product were sequentially named as BZR05; BZR10; BZR20. Furthermore, calcination was performed at temperature variations of 400 °C, 500 °C, 600 °C. The product were sequentially named as BZR10-400; BZR10-500; BZR10-600.

2.3. Pyridine adsorption
Bentonite and ZrO_2-pillared bentonite samples prepared by the above process were exposed to pyridine vapor. In vacuum desiccator, 25 mL of dry pyridine and 0.3 g of bentonite were kept under vacuum overnight. Subsequently, the samples were evacuated for 3 hours at room temperature to desorb physisorbed pyridine.

2.4. Characterization ofZrO_2- pillared bentonite
The vibrational framework of the clay samples was studied using FTIR Spectroscopy (Shimadzu 8201 PC). Determination of basal spacing of the product were characterized by X-ray diffraction (XRD Shimadzu model X-RD 6000 using Cu Kα radiation). The Transmission Electron Microscopy (TEM) micrographs were obtained using JEOL Hitachi H-600 at 120 kV to observe the surface morphology of the pillared clays.

3. Results and discussion
Results of infrared spectra of bentonite are presented in figure 1 and 2. Considering the figures it can be seen that the peak of bentonite spectra which occur in the wave numbers of 3626.17 cm^{-1} and 3448.72 cm^{-1} that are identified as a ^-OH band stretching vibrations of octahedral and ^-OH band stretching vibration of water molecules, respectively. Pyridine vibration bands appear in the IR region of 1400–1700 cm^{-1}. The IR bands at 1637 and 1545 cm^{-1} are assigned to Brønsted acid. The IR bands at 1450–1455cm^{-1} are characteristic for Lewis acid sites and band at 1490 cm^{-1} indicates the combined presence of all acid sites. The absence of a band at 1450–1455 cm^{-1} which is characteristic for Lewis acid showing that at 1640 and1540 cm^{-1} correspond to Brønsted acid sites. The band at 1540 cm^{-1} is attributed to –NH bending of pyridinium ions is characteristic of Brønsted acid sites [12-13].

In this method, the number of Brønsted or Lewis acid sites was determined and depended on pyridine bonded on the sites in mmol/g catalyst. Acid site or acidity of zirconium-pillared bentonite are related to three sources, namely cations M^{n+} that equilibrate the electrical charges of the material, specific sites at the edge of the crystal due to bond breaking or crystal defect, and pillaring metal oxide. Dehydration process may cause Brønsted sites of Si-OH-Al change to Lewis acid. Besides, those processes remove water molecules from its coordination bond to metal M^{n+} leaving the ion as Lewis acid [14]. Therefore, the change in the acid properties is related to dehydration process of Si-OH-Al group. Considering gravimetric acid analysis as depicted in table 1 it comes to the conclusion that the more pillaring metal oxide, the more acidic the material is, the increase was from 1.1490 to 1.5488 mmol/g (BZR20).

Moreover, increasing in calcination temperature was followed by decreasing its acidity, from 1.2772 to 0.9072 mmol/g.

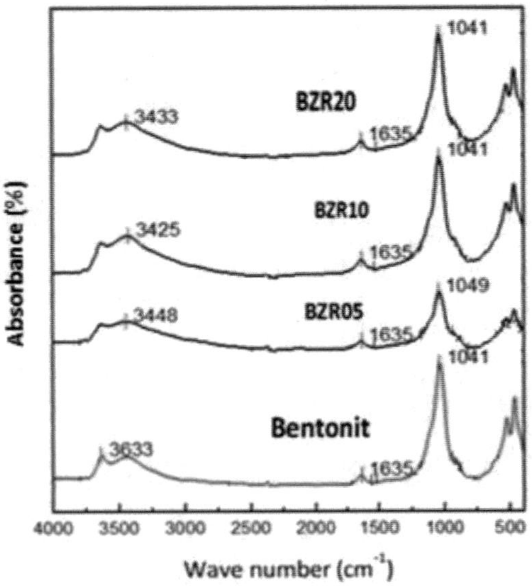

Figure 1. IR spectra of pyridine adsorbed on Bentonite and ZrO_2-Bentonite.

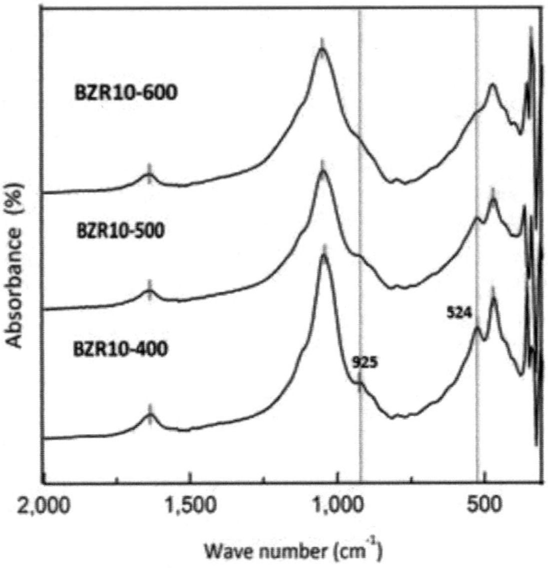

Figure 2. IR spectra of pyridine adsorbed om ZrO_2-Bentonite with different calcination temperature.

Tabel 1. Determination of acidity with pyridine adsorption

	Acidity (mmol/gr)
Bentonite	1.1490
BZR05	1.5558
BZR10	1.2771
BZR29	1.5488
BZR10-400	1.2771
BZR10-500	0.9130
BZR10-600	0.9072

Tabel 2. Basal Spacing d_{001} of Bentonite and ZrO_2-pillared Bentonite

	2θ	d_{001} (Å)
Bentonite	7.74	11.41
BZR05	4.32	20.41
BZR10	4.01	22.01
BZR29	4.42	19.96
BZR10-400	4.01	22.01
BZR10-500	4.19	21.03
BZR10-600	4.14	21.29

The identification results of bentonite and ZrO_2–pillared bentonite with X-ray diffraction methods can be seen in figure 3 and 4. From the diffractogram can be seen abroad reflection on the regions of 2θ= 7.74° and19.43° with a values of d= 11.41 Åand 4.56 Å which area typical reflection of montmorillonite sample. From these results can be concluded that the bentonite contains montmorillonite minerals. In addition, bentonite also composes of feldspar and quartz minerals as impurities. The X-ray diffraction analysis shows that there has been peak shifting of bentonite on angle 2θ of 7.74° to 2θ = 4.01° in pillarization of BZR10 stage which means the pillarization process has been done. Characterization of zirconium pillared product into silicate bentonite interlayer using X-ray diffraction are shown in figure 3 and 4. It is characterized by increasing in *basal spacing* of bentonite d_{001} = 11.41 Å to d_{001} = 22.01 Å for pillared bentonite. Pillarization result at different concentration of zirconium shows similar effects, an increase in basal spacing.

On the other hand, the changes in temperature of calcination at ZrO_2- pillared bentonite (BZR10-400; BZR-500; BZR10-600) does not change the structure and the basal spacing, it indicates that the product pillarization has a good thermal stability as shown in table 2.The b Basal spacing change is determined by calculating the difference between *basal spacing*$d_{001}$of point of zirconia-bentonite intercalation product with the thickness of the silicate–bentonite interlayer hydrated or usually written as Δd=(d_{001}– 9.6) Å[15]. The results of the analysis for pillarization stage as shown in table 2 also showed that there has been shifting the peak of bentonite before pillarization at an angle of 2θ = 7.74° shifted towards conditions ZrO_2 - bentonite of 2θ = 4.01° (BZR10). The shifting peak was followed by a shifting in *basal spacing* bentonite d_{001}= 11.41 Å or Δd_{001}= (11.41 – 9.6) Å = 1.81 Å *basal spacing* ZrO_2-bentonite into d_{001}=22.01Å or Δd_{001}= (22.01 – 9.6) Å = 12.41 Å. The changes in basal spacing showed that pillarization zirconia oxide in silicate interlayer of bentonite had been successfully carried out with an increase in the silicate interlayer distance of (12.41–1.81) Å = 10.60 Å.

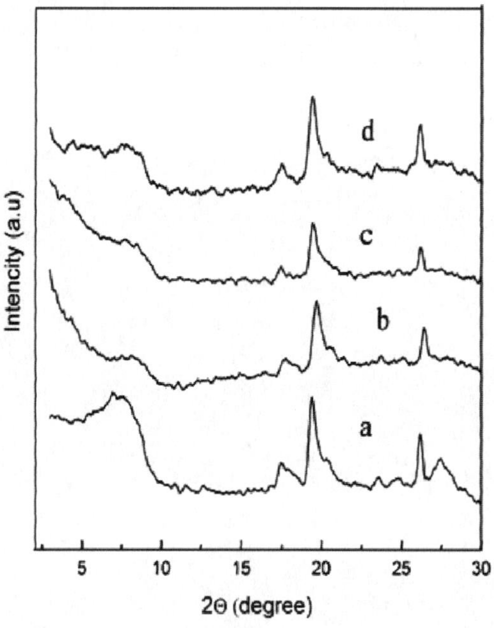

Figure 3. XRD pattern of Bentonite (a); BZR05 (b); BZR10 (c); BZR20 (d).

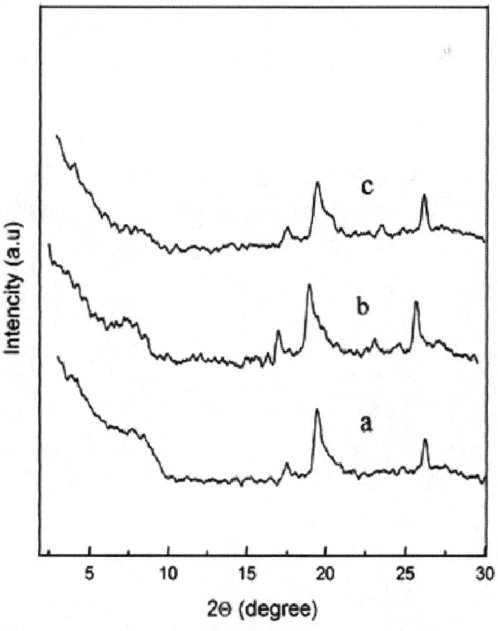

Figure 4. XRD pattern of ZrO_2- Pillared Bentonite (a); BZR10-400 (b); BZR10-500 (c); BZR10-600.

4. Conclusion

Zirconium pillared bentonite has been synthesized successfully as proved by characterization result performed using XRD and acidity of pyridine adsorption. The analysis showed that the presence of metal oxide $ZrO_2$was able to increase the basal spacing of 11.41 **Å** to 22.01 **Å** and related to the raising damage of the bentonite structure. Moreover, increasing in calcination temperature is followed by decreasing in its acidity, from 1.2772 to 0.9072 mmol/g.

Acknowledgment

This research as well as the publication were partly funded by a grant PENELITIAN FUNDAMENTAL, DITLITABMAS KEMENRISTEK DIKTI 2015, Republic of Indonesia.

References

[1] Burch R and Warburton C I 1986 *J.Catal.* ***97*** 503-510
[2] Kloprogge J T 1998 *J. Mater Porous* ***5*** 5-10
[3] Miehe-Brendle J, Khouchaf L, Baron J and Tuilier M H 1997 *J. Phys IV France* **7**
[4] Gil A, Vicente M A and Gandia L M 2000 *Micropor. Mesopor. Mater.* **34** 115-125
[5] Jerónimo D, Guil J M, Corbella B M, Vesques H, Miranda A, Silva J M, Lobato A, Pires J and Carvalho A P 2007 *Appl. Catal. A: General* **330** (2007) 89–95
[6] Figueras F 1988 *Catal. Rev. Sci. Eng.* **30** 457
[7] González F, Pesquera C, Blanco C, Benito I and Mendioroz S 1992 *Inorg. Chem.* **31** 727
[8] Yue-Xiu J, Xiao-Mei C, Yun-Fen M and Zhang-Fa T 2004 *J. Mol. Catal. A: Chem.* **213** 231
[9] Guil J M, Perdigon-Melon J A, Carvalho M B, Carvalho A P and Pires J 2002 *Micropor. Mesopor. Mat.* **51** 145
[10] Carvalho A P, Martins A, Silva J M, Pires J, Vasques H and Carvalho M B 2003 *Clays Clay Min.* **51** 340
[11] Gil A, G Guiu, P Grange and M Montes 1995 *J. Phys. Chem.* **99** 301
[12] Caglar B, Cubuk O, Demir E, Coldur F, Catir M, Topcu C and Tabak A 2015 *J. Mol. Struct.* 1089 59–65
[13] Djomgoue P and Njopwouo D 2013 *J. Surf. Eng. Mater. Adv. Technol.* **3** 275–282
[14] Bergaya F, Theng B K G and Lagaly G 2006 *Handbook of Clay Science* 1sted. Elsevier Amsterdam
[15] Ruslan Wijaya K and Triyono 2013 *Int. J. App. Chem*, **9** (1) 15-35

Effect of glycerol on mechanical and physical properties of silver-chitosan nanocomposite films

E Susilowati[1,2*], I Kartini[2], S J Santosa[2] and Triyono[2]
[1] Chemistry Education Department FKIP Universitas Sebelas Maret, Jl. Ir Sutami 36 A Surakarta Indonesia
[2] Department of Chemistry FMIPA Universitas Gadjah Mada, Sekip Utara BLS 21 Yogyakarta

*Email: endwati@yahoo.co.id

Abstract. The effect of using glycerol as plasticizer on mechanical and physical properties of silver-chitosan nanocomposite films have been studied. The nanocomposite films were prepared via three steps consisting of silver-chitosan colloidal nanocomposites preparation, adding of glycerol to colloids and silver-chitosan nanocomposites films formation. During the first step, silver ions were reduced by glucose and accelerated by sodium hydroxide (NaOH). Chitosan of 1 % (v/v) act as stabilizing agent. Glycerol with volume variation of 0.2, 0.4, 0.6, 0.8 and 1.0 mL was added colloidal nanocomposites of 60 mL on the second step. On the third step, colloidal nanocomposites were cast on the polypropylene plate and dried at room temperature. The as-prepared films were then neutralized by NaOH and rinsed with distilled water until the filtrate reached the pH of 7. The colloidal nanocomposites were characterized by UV-Vis spectroscopy and transmission electron microscopy (TEM). The film were characterized by X-ray diffraction (XRD) and thermogravimetric analysis (TGA). The mechanical properties, swelling capacity, water vapor permeability (WVP) of the films were also studied. The results indicated that the addition of different amounts of glycerol on colloidal nanocomposites effects on mechanical and physical properties of the resulted nanocomposite films. The elongation and tensile strength were gradually increased as the glycerol amount. Meanwhile, the swelling capacity, WVP, and crystallinity of the film also showed enhancement at increasing glycerol amount. However, the thermal stability decreased.

1. Introduction

Chitosan is available natural polymer derived by deacetylation of chitin, which is the second most abundant polysaccharide found in nature [1]. Compared to other polysaccharides, chitosan has several important advantages, including biocompatibility, biodegradability, and no toxicity[2]. Moreover, several studies have indicated that chitosan possesses antibacterial [3,4,5] and antifungal[6]activities. This resulted in widespread use of chitosan in food, pharmaceutical, and environmental industries [7].

Like the film, chitosan finds specific applications such as wound dressings and food packaging. Chitosan is insoluble in most common solvents (including water), but it can be dissolved in dilute aqueous acidic solutions [7]. It has been established that aqueous acetic acid is one of the most suitable solvents for film preparation resulting in good film properties such as tensile strength, strain-at-break (extensibility), resistance to water and water vapor permeability [8]. Several researchers have incorporated chitosan with nanoparticles metal to form nanocomposite films to improve its antibacterial [9,10,11,12]. One of the widely studied nanoparticles is silver (Ag).

Recent work has shown that combining chitosan or modified chitosan with Ag resulted in films and hydrogels materials with enhanced antimicrobial activity, increased tensile strength but decreased water vapor permeability [9,13,14]. It was also found, that combining low molecular weight chitosan with Ag nanoparticles yielded materials with high antibacterial activity against Gram-positive bacterium of *Staphylococcus aureus* and Gram-negative bacterium of *Escherichia Coli* as well as displayed promising wound healing characteristics [5].

Unfortunately, chitosan films are rigid and need plasticizers to reduce frictional forces among polymer chains, such as hydrogen bonds or ionic forces, in order to improve mechanical properties. Several plasticizers have been used to improve the mechanical properties film, including sorbitol, PEG, and glycerol [14]. Glycerol is a polyol compound which is widely used in a diverse range of industries, for example, in food and cosmetics industries. It is added as a solvent, plasticizer, humectant, emollient, and lubricant. The addition of glycerol can improve film biopolymer film properties. The addition of glycerol reduced the tensile strength of the chitosan film [15]. The presence of glycerol resulted in less resistant, more elastic and more permeable films. It is also observed that WVTR decreased, but the elasticity improved [16].

In this study, the nanocomposite silver-chitosan films were prepared by adding glycerol as plasticizer. The effect of the presence of various amount of glycerol as plasticizer on mechanical and physical properties such as tensile strength, elongation at break, crystallinity, swelling, water vapor permeability (WVP) and thermal properties of the film is described.

2. Experimental

Chitosan with a molecular weight (MW) of 1,077,919.28 Da and degree of deacetylation (DD) of 73.44% was purchased from Biotech Surindo Cirebon Indonesia. Silver nitrate ($AgNO_3$), acetic acid (CH_3COOH), glucose ($C_6H_{12}O_6$) sodium hydroxide (NaOH) and glycerol were purchased from Merck.

A solution of chitosan (1% w/v) in acetic acid solution (1% v/v) was firstly prepared by stirring and then kept overnight at room temperature. The solution was filtered to remove any impurity before use. 1.0 mL of aqueous solution of $AgNO_3$ (0.012 g/mL) was added to 25 mL of Chitosan (1% w/v) in a flask. The solution was stirred at room temperature for 10 min, and then an aqueous solution of glucose was added so that the molar ratio of $AgNO_3$/glucose is 1:4. The solution was stirred for 10 min. After that, 2.5 mL of aqueous solution of NaOH (2M) were added. The gel was immediately formed in the solution, and then the color was turned brown. The reaction was continued for 25 min under stirring. Then, the resulted gel was dissolved in 35 mL chitosan (1% v/v) and stirred to achieve complete dissolution forming a solution of silver-chitosan colloidal nanocomposites. Glycerol with volume variation of 0.2; 0.4; 0.6; 0.8 and 1.0 mL were added to the colloidal solution and stirred for 30 min. Then, the colloidal solutions were cast onto polypropylene plates and dried at room temperature for 6 days to obtain the composite film. To get a neutral film, the films were neutralized using NaOH then rinsed with distilled water until the filtrate reached pH of 7. The films dried on glass board until dry. The films were denoted as Ag/CG1, Ag/CG2, Ag/CG3, Ag/CG4, Ag/CG5.

Localized surface plasmon resonance (LSPR) colloidal nanocomposite was determined by Shimadzu UV3150 UV-Vis spectrophotometer, operating in the absorbance mode. UV-Vis absorption spectra of the samples were recorded in the wavelength range of 300 to 600 nm. For colloidal nanocomposite, the sample was diluted 10 times before analysis. The nanocomposite film, the sample was placed in the cuvette holder.

IOP Conf. Series: Materials Science and Engineering **107** (2016) 012041 doi:10.1088/1757-899X/107/1/012041

Size and distribution of particles were characterized by Transmission electron microscopy JEM-2000EX at an accelerating voltage of 120 kV. Particle size analyzes were performed using ImageJ 1.43u software.

The tensile strength (TS) and elongation at break (E) of the films were measured with Zwibk Universal Testing Machine type of Z 0.5. Tensile strength values were reported as measured maximum load (N) divided by film cross-sectional area (mm^2) with a unit of MPa. Elongation values were obtained by recording elongation at break divided by initial length of the specimen and multiplied by 100 [17].

The X-ray diffraction method was used to identify the formation of nanoparticles in the films and their crystallinity. These measurements were carried out for dried and finely grounded samples on XRD-6000 Shimadzu. Measurement of X-ray diffraction method is done by using Cu Kα radiation source (= 1.54060Å) at a voltage of 40 kV and current of 30 mA, nickel metal filter, the scanning rate of 5/min at 3-80° 2θ region.

The swelling ratio (Q) was determined by gravimetric method [18]. Dried films were swollen in (100ml) phosphate buffer (pH 7.4) solution at 25 °C. The weight of swollen films was measured at equilibrium swelling after removing the surface solution with filter paper. The swelling ratio (Q) was calculated using Equation (1):

$$Q = We/Wd \tag{1}$$

where We are the weight of the swollen film at equilibrium and Wd is the dry weight of the film.

Thermal degradation processes were investigated using TGA Parkin Elmer at 25.00°C to 700.00°C with the heating rate of 10 °C/min under nitrogen atmosphere.

Water vapor permeability (WVP) was determined gravimetrically according to the standard method ASTM E 96-95 with some modifications [19]. The films were fixed on top of test cells containing a desiccant (silica gel). Test cells then were placed in a relative humidity chamber with controlled temperature and relative humidity (28 °C and 75% RH). After steady-state conditions had been reached, the weight of test cells was measured every 1 h over 8 hours. The water vapor transmission rate (WVTR) was determined using Equation (2).

$$WVTR = \frac{\left(\frac{G}{t}\right)}{A} g/h\,m^2 \tag{2}$$

where G, weight change (g); t, time (h) and A, test area (m^2). WVP was then calculated using Equation (3).

$$WVP = \frac{WVTRxd}{\Delta p} g\ mm/kPa\ h\ m^2 \tag{3}$$

where d, film thickness (mm) and Δp, the the partial pressure difference across the films (kPa).

3. Result and discussion

Silver-chitosan nanocomposite films were prepared by casting silver-chitosan colloidal nanocomposites at room temperature. The silver-chitosan colloidal nanocomposites have been performed at room temperature by chemical reduction using NaOH as an accelerator, glucose as the reducing and chitosan as stabilizer agent. The color of the colloidal solution was from colorless to brown indicating the formation of Ag nanoparticles.

The brown color was due to the localized surface plasmon resonance (LSPR) absorption of silver nanoparticles [20]. The surface of a metal is like a plasma, having free electrons in the conduction band and positively charged nuclei. LSPR is a collective excitation of the electrons in the conduction band near the surface of the nanoparticles [20]. The electrons are limited to specific vibrations modes by the particle's size and shape. Metallic nanoparticles have the characteristic of optical absorption spectrums in the UV-Vis region [21].

The UV–Vis absorption spectra of silver-chitosan colloidal nanocomposite solution and film are shown in Figure 1with the related value of λ_{max} (nm) and absorbance. The result silver nanoparticles displayed absorption peaks, centered at 404.6 and 424.02 nm for colloidal solution and film respectively.The observed single LSPR band indicates that the silver nanoparticles are spherical in shape [20]. It is confirmed by the TEM images (Figure 2).

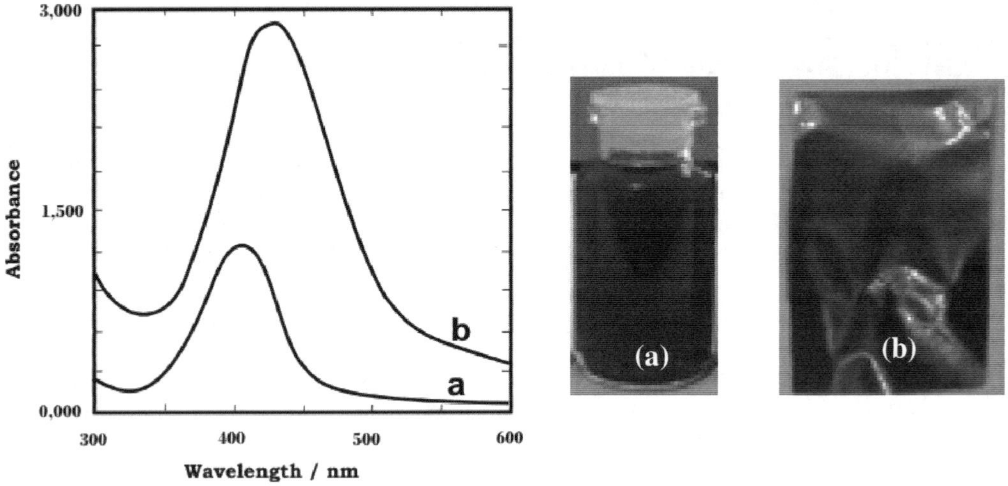

Figure 1. The UV-Vis spectra and image of silver-chitosan nanocomposite (a) colloidal solution and (b) film (Note: the colloidal nanocomposite was diluted 10 times before analysis)

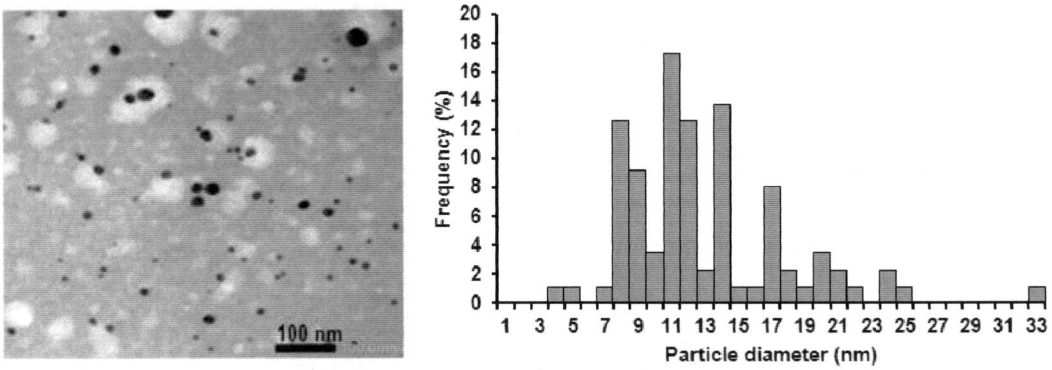

(a) (b)

Figure 2.(a) TEM images of silver nanoparticles and (b) the corresponding particle size distribution

Figure 1shows UV–Vis absorption spectra of silver-chitosan nanocomposite films prepared from the colloidal nanocomposites sample. The film also showed a strong band in the visible region due to the characteristic surface plasmon resonance of silver nanoparticles. Compared to the colloidal nanocomposites, the film shows red-shifted significantly that might be caused by increase particle size of silver due to aggregation during the evaporation process.

The mechanical properties of silver-chitosan nanocomposite films were improved by adding glycerol as plasticizer. Figure 3 shows the tensile strength and elongation at break of silver-chitosan nanocomposite films at various glycerol amounts. The results show a slight increased at the initial addition of glycerol (for sample Ag/C to Ag/CG3) result in a slight increase in tensile strength and elongation. But, at higher plasticizer amount of plasticizer (for sample Ag/CG4 and Ag/CG5)tensile strength and elongation of films significantly increase. The increasing both tensile strength and elongation indicated that plasticizer of glycerol strengthens the film structure and integrity. Glycerol molecule in the film may construct the crystalline formation during the process and strengthen intermolecular hydrogen bonding among chitosan molecule.

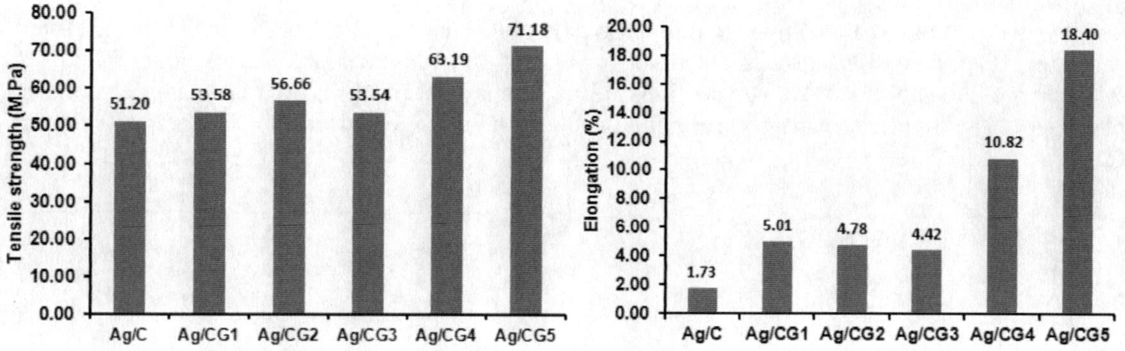

(a) (b)

Figure 3. The effect of plasticizer concentration on mechanical properties of silver-chitosan nanocomposite films ((a) tensile strength and (b) elongation)

The swelling capacity of the films is one of the important criteria for maintaining a moist environment. Figure 4 shows the effect of glycerol content on nanocomposite films. It is observed that swelling capacity of the films increases increasing glycerol content in the films. This could be due to the formation of the hydrophilic structure of nanocomposite films with the addition of glycerol. Three hydroxyl groups in glycerol structure may contribute to enhanced hydrophilicity.The presence of more hydrophilic groups in the film network will assist in improving the swelling characteristic of the film [22].

However, the water vapor permeability (WVP) of the films do not significantly increase as the glycerol amount increase (Figure 5). WVP is a value for water permeation at a given temperature. The permeability of a film depends on its chemical structure and morphology, the nature of permanent and temperature of the environment [23]. The significant increase of WVP is observed for the silver-chitosan-glycerol film in comparison to the film without glycerol. This could be due to an additional hydroxyl group in polymeric chains. Hydrophilicity induces an increase in water vapor permeability

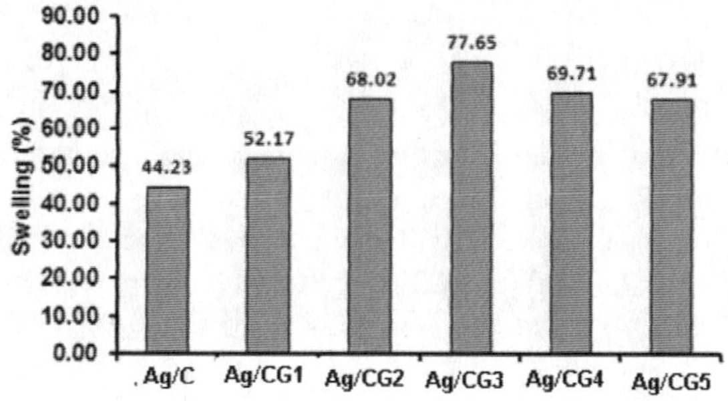

Figure 4. Swelling capacity of nanocomposite films at various plasticizer amounts

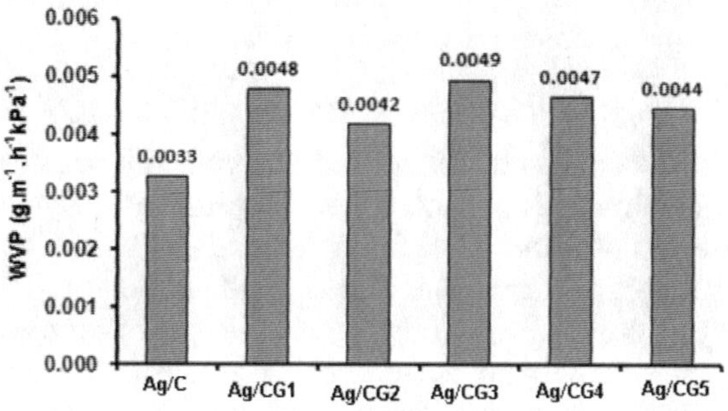

Figure 5.WVP of nanocomposite films at various plasticizer amounts.

The X-ray diffraction (XRD) is used to confirm the formation of silver nanoparticles. The XRD pattern of chitosan film, silver-chitosan nanocomposite film and silver-chitosan nanocomposite film plasticized by glycerol are shown in Figure 6. The amorphous structure of chitosan film is confirmed by broad peaks appeared at a diffraction angle of 9.88°; 19.86°; 20.96°; 22.06°. While, the XRD pattern of silver-chitosan nanocomposite films also exhibits a sharp peak at 38.08°, which is consistent with the (111) plane diffraction of the silver crystal [24,25]. Narrow and sharp peaks of chitosan structure are indicating improved crystallinity of the film are observed for the films using glycerol as plasticizer. This indicates that glycerol makes more good arrangement in the structure of chitosan film.

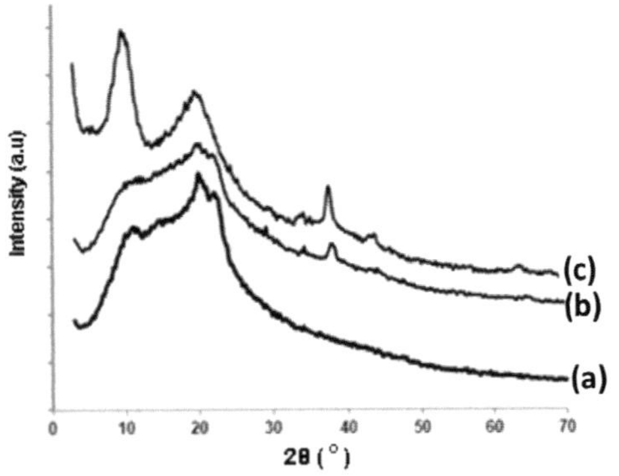

Figure 6. XRD pattern of chitosan film (a) silver-chitosan nanocomposite film (b), silver-chitosan-glycerol nanocomposite film (c).

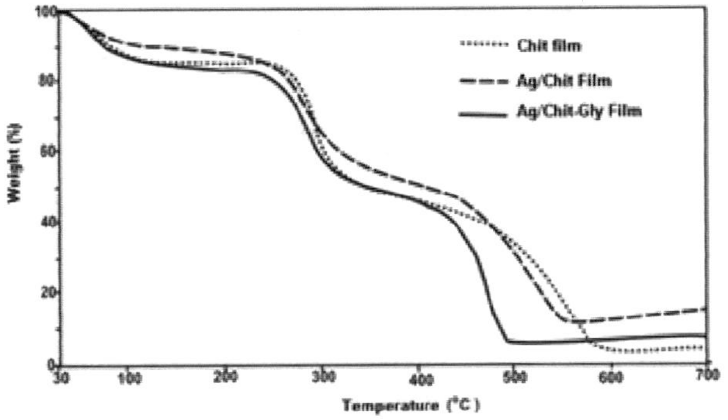

Figure 7. TGA thermogram of chitosan film, silver-chitosan nanocomposite film and silver-chitosan-glycerol nanocomposite film.

Thermal degradation of the chitosan film, silver-chitosan nanocomposite film and silver-chitosan nanocomposite film plasticized by glycerol are shown in Figure 7. It can be seen that the films have similar thermal properties. Three weight losses areas are observed in the films. The initial weight loss observed at 30 – 100 °C was attributed to the water molecules. The second, weight loss at 200 – 400 °C may be related to the decomposition of amine and $-CH_2OH$ group [26]. The third stage, at 500 – 600 °C for chitosan and silver-chitosan may be due to total degradation of the glucopyranose of chitosan. However, silver-chitosan nanocomposite film plasticized by glycerol degrades at a lower temperature of 420 – 500 °C. This indicates that the presence of glycerol in the chitosan structure causes the chitosan polymer is easier to break than that without glycerol. This is reasonable since the glycerol reduces the intermolecular hydrogen bonding of chitosan backbone, thus weakening the bond between the polymer chains.

4. Conclusion

It is concluded that addition of glycerol effects on structural properties of chitosan and their corresponding silver nanocomposite films. The glycerol is significant increasing on the mechanical properties of films. Elongation at break and tensile strength of the films were enhanced at high concentration of glycerol of 0.8 to 1.0 mL in 60 mL 1% chitosan solution or 160 to 200% (w/w) (glycerol/chitosan). Similarly, the swelling capacity, WVP, and crystallinity of the films also increase by increasing the concentration of glycerol. However, as the consequence, the thermal resistance of the plasticized film is lower than that of unplasticized films due to the weak intermolecular forces among chitosan structures.

Acknowledgement
The authors would like to acknowledge Directorate General of Higher Education, Ministry of National Education Indonesia that has given scholarship for the doctoral program.

References

[1] Park J K, Chung M J, Choi H N and Park Y I 2011 *Int.J.Mol.Sci.* 266-277
[2] Kim M K, Son, J H and Kim S K 2006 *J. Food Sci. E: Food Eng. Phys. Prop.* **71(3)** 119–124
[3] Fernandez-Saiz P, Lagaron J M, Hernandez-Munoz P and Ocho M J 2008 *Int. J. Food Microbiol.* **124** 13-20
[4] Islam M D, Masum S M D and Mahbub K R 2011 *J. Bangladesh Chem. Soc.* **24(2)**185-190
[5] Susilowati E, Maryani and Ashadi 2015 *Indonesian J. Pharm.* **26(1)** 37 – 44
[6] Rodriguez 2003 *Cultivos Tropicales* **24(2)** 85-88
[7] Rinaudo M 2006 *Prog.Polym.Sci.* **31** 603-632
[8] Rhim J W, Weller C L and Ham K S 1998 *Food Sci. Biotechnol.* **7** 263–268.
[9] Rhim J W, Hong S-I, Park H M and Ng P K W 2006 *J.Agric. Food Chem.* **54** 5814- 5822
[10] Ragiel A, Irusta S, Kyziol A and Arruebo M 2013 *Nanotechnology* **24**1-13
[11] Manikandan, A and Muthukrishnan S M 2015 *J.Nanomed. Nanotechnol.* **6**1
[12] Kavitha A L, Prabu G H and Babu S A 2013 *Int.J.Polym.Mater* **62(1)** 45-49
[13] Liu B S and Huang T B 2008 *Macromol. Biosci.* **8** 932–941.
[14] Bourtoom T 2008 *Songklanakarin J. Sci. Technol.* **30 (1)** 149-165
[15] Ziani K, Oses J, Coma V and Mate J I 2008 *Food Sci. Technol* .**41(10)** 2159-2165.
[16] Adila S N, Suyatma N E, FirlieyantiA S and Bujang A 2013 *Adv.Mat.Res* **748** 155-159
[17] Park S I, Daeschel M A and Hao Y Z 2004 *J. Food Sci.* **69(8)** 215-221
[18] Varaprasad K, Vimala K, Ravindra S, Reddy, N, Reddy G V S and Raju K M 2011 *J Mater Sci: Mater. Med.* **22** 1863–1872
[19] Bertuzzi M A, Vidaurre E F C, Armada M and Gottifredi J C 2007 *J. Food Eng.* **80** 972-978
[20] Guzmán M G, Dille J and Godet S 2009 *Int. J. Chem. Biol. Eng.* **2(3)** 104-111
[21] Šileikaitė A, Puišo J, Prosyčevas I and Tamulevičius S 2009 *Mater. Sci., (Medžiagotyra)* **15(1)** 21-27
[22] Vimala K, Sivudu K S, Mohan Y M, Sreedhar B and Raju K M 2009 *Carbohyd Polym* **75** 47–463.
[23] Siripatrawan U and Harte B R 2010 *Food Hydrocolloids* **(24)** 770-775
[24] Sharma P, Sanpui P, Chattopadhyay A and Ghosh S 2012 *RSC Adv* **2** 5837–5843
[25] Sathishkumar M, Sneha K and Yun Y S 2010 *Bioresource Technology* **101** 7958–7965
[26] Martinez-Camacho A P, Rocha M O C, Brauer J M E, Verdugo A Z G, Felix F R, Ortega M C, Gomez M S Y and Jatomea M P 2010 *Carbohydr. Polym.* **82** 305-515

Functionalization of C-4-methoxyphenylcalix[4]resorcinarene with several ammonium compounds

Suryadi B Utomo[1*], Agung N C Saputro[1], and Yudi Rinanto[2]

[1]Chemical Education Study Program, FKIP, Sebelas Maret University, Ir. Sutami 36 A, Surakarta, 57126, Indonesia
[2]Biology Education Study Program, FKIP, Sebelas Maret University, Ir. Sutami 36 A, Surakarta, 57126, Indonesia

[*]Email: sbukim98@yahoo.com

Abstract. The synthesis of new quaternary ammonium modified calix[4]resorcinarenes were accomplished by the reaction of C-4-methoxyphenylcalix[4]resorcinarene (*I*) with various ammonium compounds i.e. (1) (3-chloro-2-hydroxypropyl)trimethylammonium chloride, (2) glycydiltrimethylammonium chloride, and (3) N-hexadecyl-N,N,N-trimethylammonium bromide. The modification processes were carried out through just over one-step reaction by using minimal solvent and suitable reaction conditions to yield resorcinarene *II, III*, and *IV*, respectively. All products were characterized by FT-IR, ^{1}H-NMR, and ^{13}C-NMR data. Solid morphology of some resorcinarenes was determined using SEM equipped with EDX detector. Based on the spectral properties of ^{1}H-NMR, all calixresorcinarene compounds were isolated in the molecular shape of the chair (C_{2h}) conformation.

1. Introduction

Modifications on calixarene compounds have been carried out primarily for phenol and substituted phenol derivatives calixarenes [1-6]. However, it is different with resorcinol derivatives calixarenes (resorcinarenes) due to their low solubility in various organic solvents [7,8]. C-arylcalix[4]resorcinarenes, for examples, only dissolve in warm DMF and DMSO. In fact, there are a lot of resorcinarenes which can be synthesized from resorcinol and its derivatives with various of both aliphatic [9-11] and aromatic aldehydes [7, 10, 12].

In the field of supramolecular chemistry, it is very important to make the calixarenes have multifunction as hosts for cations [13], anions, and neutral organic molecules via further functionalization or modification. Calix[4]resorcinarenes provide a cavity, which in combination with several active sites, can be modified by suitable substitution at their upper rim of methylene bridges [7, 12] and extra annular –OH groups [14, 15]. Functionalization on these sites has been intensively studied through aminomethylation by refluxing the C-methylcalix[4]resorcarene and C-phenylcalix[4]resorcinarene with a primary amine and formaldehyde in mixtures of benzene with the minimum amount of ethanol to achieve a homogeneous reaction mixture [15].

In our previous work, we then introduce this method in order to obtain quaternary ammonium modified resorcinarenes by reacting C-3,4-dimethoxyphenylcalix[4]resorcinarene with formaldehyde and dimethylamine in the presence of HCl according to the Mannich procedure followed by alkylation with MeI. This resorcinarene, tetrakis-N,N,N-trimethyl-ammoniummethyl-C-3,4-dimethoxyphenyl-calix[4]resorcinarene iodide, has been proven to chelate and reduce the toxic chromium (VI) content in serum, liver, and kidney of mice [12]. Unfortunately, this procedure synthesis needs long steps wise of reactions and consume more solvents and reagents. For this reason, this research study the synthesize of similar compounds were conducted through just over one-step reaction. The modification processes are expected to take place directly to the C-4-methoxyphenylcalix[4]resorcinarene using various quaternary ammonium compounds with minimal solvent and suitable reaction conditions.

The existence of two types of active groups located far apart in the targeted resorcinarenes (Figure 1 and 2), hydroxyl and the quaternary ammonium groups, make the compounds can act as host molecules for both of anionic and cationic heavy metals. Quaternary ammonium active sides will be able to bind anionic species while the hydroxyl groups will be able to bind cationic species.

2. Experimental

2.1. Materials

Resorcinol, 4-methoxybenzaldehide, ethanol, fuming hydrochloric acid, dichloromethane, sodium hydroxide, ethyl acetic, acetone, dimethylformamide, dimethyl sulfoxide, glycidyltrimethylammonium chloride, hexadecyltrimethylammonium bromide (HDTMA-Br), and (3-Chloro-2-hydroxypropyl)trimethylammonium chloride solution. All reagents of analytical grade were obtained from E Merck Co Inc. (Germany), and Sigma-Aldrich used without further purification.

2.2. Instruments

Melting points were determined with an electrothermal IA9100I and were uncorrected. The homogeneity of all the newly synthesized compounds was routinely checked by TLC on silica gel 7730 $60GF_{254}$ plates and spots were observed by short and long wavelength ultraviolet light. IR spectra were recorded on a Fourier-transform infrared spectrometer (Shimadzu FTIR-8201 PC) in KBr pellets. ^{1}H-NMR and ^{13}C-NMR spectra were performed on a Bruker AC300F and an Aligent400 instruments using DMSO-d_6 as solvent and Tetramethylsilane (TMS) as an internal reference. Solids morphology were determined using SEM supplied with EDX detector (JSM-6360).

2.3. Procedures

2.3.1. Synthesis of C-4-methoxyphenylcalix[4]resorcinarene (I). Compound *I* was synthesized from 4-methoxybenzaldehide according to the previous procedure [7] with slight modification. The mixture of 4-methoxybenzaldehide (1.66 g; 10 mmol), resorcinol (1.1 g; 10 mmol), ethanol absolute (100 mL), and fuming hydrochloric acid (0.5 mL) was refluxed under inert condition for 18 hours until the spot of 4-methoxybenzaldehide totally used up at the reaction (monitoring by TLC). The mixture was cooled, and the product was filtered followed by washing with ethanol-water (1:1) and acetone, respectively. Finally, the product was dried to give the desired compound as a light purple crystal in 94.26%, mp 339-341 ^{o}C (dec); FTIR (KBr) ν (cm^{-1}): 3397 (OH group), 3007 (C_{sp2}-H), 1605 and 1505 (C=C aromatic), 2910-2834 (C_{sp3}-H aliphatic), 1432 (-CH methine bridge), 1368 (-CH_3); ^{1}H-NMR (500 MHz, DMSO-d_6) δ (ppm): 7.92 and 8.03 (8H,s,OH), 6.16-6.56 (24H,m,ArH), 5.53 and 5.68 (4H,s,Ar_2CHAr), 3.66 (12H,s,OCH_3).

2.3.2. Synthesis of 5,17-di(2-hydroxypropyl trimethyl ammonium chloride)-C-4-methoxyphenylcalix[4]resorcinarene (II). Resorcinarene *I* (0912 g, 1.0 mmol) was dissolved in 30 mL of warm DMF accompanied by stirring. Into a solution of resorcinarene *I* was then added (3-Chloro-2-hydroxypropyl)trimethylammonium chloride and 0.5 mL of fuming hydrochloric acid respectively. The mixture was stirred and heated at 100 ^{0}C under inert condition over the night (18 hours) until the stain material on the basis of TLC spots has completely reacted. The solvent was then removed on the proper evaporator and the residue was triturated with distilled water. The mixture was cooled, and the precipitate was filtered off. The precipitate that formed was washed with ethanol: water (1: 1), recrystallized from acetone-water and dried to yield 0.705 g (58.02%) of resorcinarene *II*; mp. 362 °C; FTIR (KBr) ν (cm^{-1}): 3245 (OH), 3026 (C_{sp2}-H), 1607 and 1511 (C=C aromatic), 3000-2756 (C_{sp3}-H aliphatic), 1474 ($-CH_2-$), 1425 (-CH methine bridge), 1375 ($-CH_3$), 1249 and 1201 (C-N), 1077 ($-O-CH_3$); ^{1}H-NMR (300 MHz, DMSO-d_6) δ (ppm): 8.47 (8H,s,OH resorcin), 8.00 (2H,s,OH hydroxypropyl), 6.13-6.54 (22H,m,ArH), 5.50 (4H,d of s,Ar_2CHAr), 4.15 (2H,m,-C<u>H</u>(OH)-), 3.65 (12H,s,OCH_3), 3.18 (18H,s,N-CH_3), 2.91 (4H,d,-CH_2-N), 2.75 (4H,s,ArCH_2-); ^{13}C-NMR (125 MHz, DMSO-d_6) δ (ppm): 156.8 (-C-OMe), 152.9, 152.8 (Ar-C-OH), 136.6, 130.1, 121.7, 121.3, 112.7 (Ar), 54.9 (-C-N), 41.6 (-C-OH), 40.7 (O-CH_3), 36.2 (N-CH_3), 31.0 (Ar_2CHAr), 21.5 (Ar-CH_2-).

2.3.3. Synthesis of 6,18-di(2-hydroxypropyl trimethyl ammonium chloride)-C-4-methoxyphenylcalix[4]resorcinarene (III) (solvent free). Into 30 mL of NaOH solution (0.01 M) was added resorcinarene *I* (0.912 g, 1.0 mmol) then stirred and heated at temperatures of 35 ° C until the sample dissolved completely. To this homogeneous mixture, glycidyltrimethylammonium chloride (0.671 mL; 2.5 mmol) was inserted carefully and refluxed under the mild condition for 15 hours (by TLC observation) with constant stirring. The reaction mixture was cooled to room temperature, and the resulting pink precipitates were filtered instantly over a Buchner filtration with the help of vacuum and washed thoroughly with acetone. The product was dried under reduced pressure to give 0.764 g (62.88%) of ammonium modified resorcinarene *III*; mp. 369 °C; FTIR (KBr) ν (cm^{-1}): 3373 (OH), 3231 (C_{sp2}-H), 1608 dan 1511 (C=C aromatic), 2993-2838 (C_{sp3}-H aliphatic), 1429 (-CH methine bridge), 1383 ($-CH_3$), 1257-1183 (C-N), 1077 ($-O-CH_3$), 1030 (Ar-O-C); ^{1}H-NMR (400 MHz, DMSO-d_6) δ (ppm): 8.60 and 8.55 (6H,s,OH), 8.51 (2H,s,OH hydroxypropyl), 6.05-6.60 (24H,m,ArH), 5.45 (4H,d of d,Ar_2CHAr), 4.57 (4H,d,-OCH_2-), 4.02 (2H,m,-C<u>H</u>(OH)-), 3.64 (12H,s,OCH_3), 3.57 (4H,d,-CH_2-N), 3.39 (18H,s,N-CH_3); ^{13}C-NMR (125 MHz, DMSO-d_6) δ (ppm): 162.7 (-C-OMe), 157.6 (Ar-C-OH), 152.6 (-C-O-), 133.9, 133.4, 130.3, 129.8, 124.4, 123.5, 113.3, 107.7, 107.2 (Ar), 55.1 (Ar-O-C-), 51.3 (-C-N), 43.7 (-C-OH), 42.6 (-O-CH_3), 36.2 (-N-CH_3), 31.1 (Ar_2CHAr).

2.3.4. Functionalization of C-4-methoxyphenylcalix[4]resorcinarene with HDTMA-Br (IV) (solvent free). Resorcinarene *I* (0.912 g; 1.0 mmol) was added to a solution of 35 mL HDTMA-Br (1 mol/L) in a round bottom flask. The critical micelle concentration (CMC) of HDTMA-Br at 30 ^{0}C was 0.94 mmol/L as estimated by conductivity measurements of aqueous HDTMA-Br solutions at different concentrations [17]. The mixture was stirred at room temperature and then added 0.1 M NaOH dropwise until completely dissolved. The mixture was further stirred at a temperature of 70 °C for 18 h, and the reaction was monitored by TLC. Trituration with 0.01M HCl until neutral and the precipitate was filtered, recrystallized using ethanol: water (1: 1), and dried to obtain a fixed weight of the desired product (0.814 g; 60.85%); mp. 358 °C; FTIR (KBr) ν (cm^{-1}): (3396 (OH), 3064 (C_{sp2}-H), 1608 dan 1510 (C=C aromatic), 3000-2841 (C_{sp3}-H aliphatic), 1427 (-CH methine bridge), 1467 ($-CH_2-$), 1379 ($-CH_3$), 1326-1147 (C-N); ^{1}H-NMR (400 MHz, DMSO-d_6) δ (ppm): 8.51 (2H,m,OH resorcin), 6.05-6.44 (24H,m,ArH), 5.40 (4H,d of d,Ar_2CHAr), 3.64 (12H,s,OCH_3), 3.40 (12H,d,-CH_2-N), 3.55 (40H,s,N-CH_3), 1.10 (174H,m,alkane); ^{13}C-NMR (125 MHz, DMSO-d_6) δ (ppm): 152.9 (-C-OMe), 152.7 (-C-OH), 143.7, 137.0, 120.9, 120.7, 114.1, 112.7, 102.1 (Ar), 65.9 (-C-N), 55.9 (OCH_3), 37.6 (N-CH_3), 31.6 (Ar_2CHAr), 29.5, 20.5 (alkane).

3. Results and discussion

Synthesis of resorcinarene *I* in this study has a slightly different technique with the method that we have previously reported [7,16]. At this time, the synthesis reaction is made in an inert atmosphere so that the catalyst consumption can be minimized. The effectiveness of HCl as the catalyst was strongly influenced by the presence of water vapor and oxygen. In inert conditions, the two inference components do not exist so that the reaction can be carried out effectively, minimum catalyst consumption, faster time, and more yield. The product of resorcinarene *I* was obtained as a light purple crystal in 94.26% yield with a melting point of 339-341 °C. Based on the proton NMR investigation, resorcinarene *I* was observed in chair conformation proven by the exhibit of two singlet signals on four methine bridges.

Based on the structure and functional groups that are owned by resorcinarene *I*, the compound allows to be modified through some reaction on the accordingly active sides. Not only at the ortho position of the hydroxyl group alone (aromatic ring), as has been reported previously [18], but the eight hydroxyl groups can also be further functionalized through the base catalyzed the reaction. However, in contrast, to the methoxy groups on the resorcinarene compound are difficult to be modified due to the low reactivity. In this publication, we report several possible functionalization on compound *I*.

3.1. Modification on C atom number 5 and 17 (aromatic ring)

3.1.1. Synthesis of 5,17-di(2-hydroxypropyl trimethyl ammonium chloride)-C-4-methoxyphenylcalix[4]resorcinarene (II). Due to the existence of two reactive sites on resorcinarene *I*, the synthesis of ammonium modified resorcinarene *II* can only accomplish under certain conditions, i.e. under acid catalyzed reaction. In acidic condition, the eight phenolic hydroxyl groups, that are also acidic, are inactive sides. Therefore, the central of reaction occurs only on the aromatic rings. In this case, the most reactive position held by carbon atoms of 5, 11, 17, and 23 in resource residues where the atoms are located at the ortho position of the hydroxyl groups.

At this stage of the research, modification process was done by using 3-Chloro-2-hydroxypropyl) trimethylammonium chloride solution in the presence of fuming hydrochloric acid. Reactions involved in this process is the electrophilic aromatic substitution reaction as shown in Figure 1. Functionalization was held only on two position i.e. carbon atoms of number 5 and 17 by using a molar ratio of 1:2.5 of resorcinarene *I*: 3-Chloro-2-hydroxypropyl) trimethylammonium chloride.

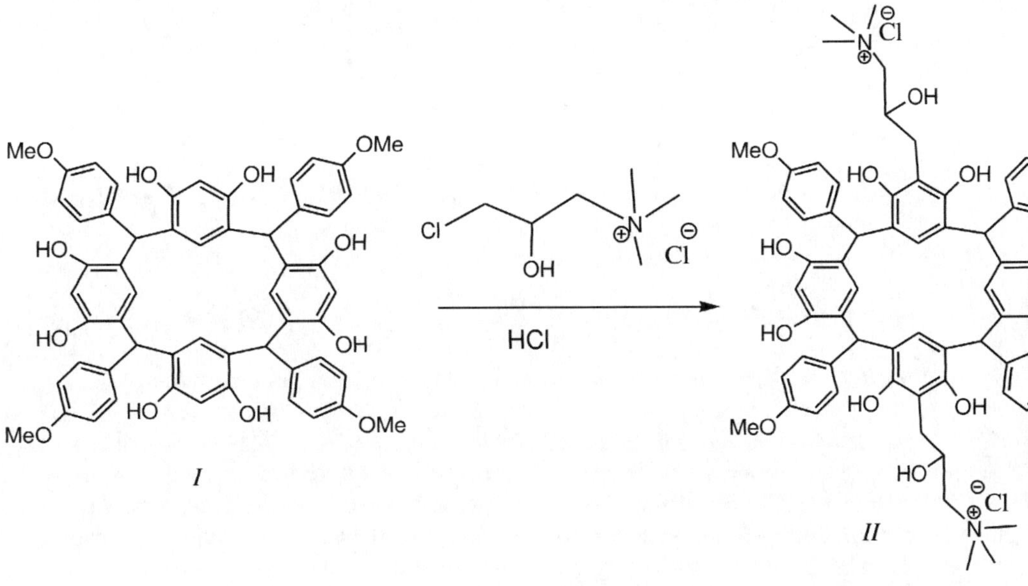

Figure 1. Acid-catalyzed reaction in the synthesis of resorcinarene *II*.

From the research that has been conducted, the product compound was obtained as brown crystalline solids in 58.02% yield with the melting point of 362 °C. Based on the solubility assessment, resorcinarene *II* was not soluble in water, chloroform, and dichloromethane, slightly soluble in dimethyl sulfoxide and acetonitrile, but soluble in NaOH solution and hot dimethylformamide.

3.2. Modification of hydroxyl groups

3.2.1. Synthesis of 6,18-di(2-hydroxypropyl trimethyl ammonium chloride)-C-4-methoxyphenylcalix[4]resorcinarene (III) (solvent free). In alkaline conditions, from three active sites on resorcinarene I only hydroxyl groups that became the center of the reaction. Eighth phenolic OH groups were acidic, with the presence of a base causes the protons (H^+) will be released so that the resulting of phenoxide ion capable of acting as a nucleophile towards reagent, in this case, was glycidyl trimethyl ammonium chloride.

Reactions involved in the synthesis of compound *III* from *I* was the epoxide ring opening reaction of glycidyl trimethyl ammonium chloride (Figure 2). The carbons in an epoxide group are very reactive electrophiles, due to in large part to the fact that substantial ring strain is relieved when the ring opens upon nucleophilic attack (phenoxide ion of the resorcinarene *I*). Ring-opening reactions can proceed by either S_N1 or S_N2 mechanisms, depending on the properties of the epoxide compound as well as on the reaction conditions. When an asymmetric epoxide (in this case: glycidyl trimethyl ammonium chloride) undergoes solvolysis in basic solution, the ring-opening occurs by an S_N2 mechanism, and the less substituted carbon is the site of nucleophilic attack, leading to what we will refer to as the product of resorcinarene *III*.

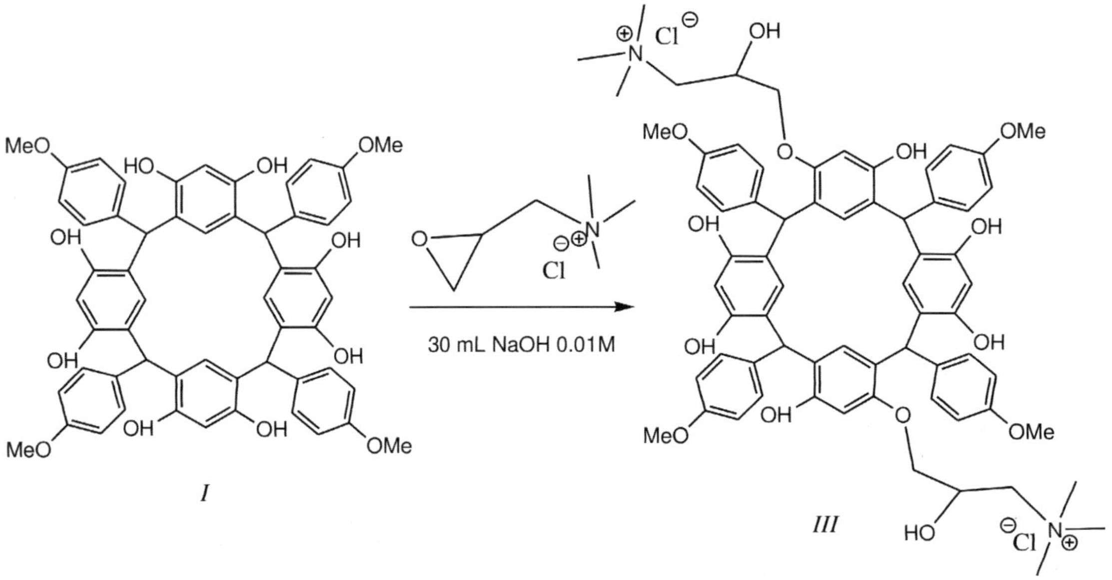

Figure 2. Based-catalyzed reaction in the synthesis of resorcinarene *III*.

3.2.2. Functionalization of C-4-methoxyphenylcalix[4]resorcinarene with HDTMA-Br Surfactant (IV) (solvent free). As Mentioned before, resorcinarene *I* have multiple active sites, namely hydroxyl (-OH), methoxy (-OMe), and an aromatic ring. All of these groups have a negative character, its mean that the surface of resorcinarene *I* have a negative surface as can be described in figure 3. Meanwhile, the surfactant of HDTMA-Br has two distinct parts of characters. The tail of the alkyl chain is nonpolar (uncharged) while the head is polar side (+). Thus, when the HDTMA-Br were added into a solution of resorcinarene *I*, the negative sides on the resorcinarene will interact each other and form bonds with the polar end (+) of HDTMA-Br.

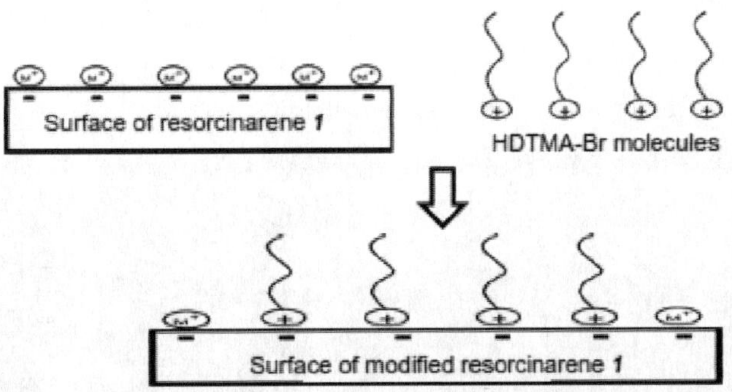

Figure 3. Mechanism of interaction between $HDTMA^+$ with the surface of resorcinarene *I*

Attachment of $HDTMA^+$ on the resorcinarene surface is strongly influenced by the degree of surfactant adsorption on a solid surface [17]. At very low concentrations, the $HDTMA^+$ molecules move freely and can be lined up parallel to the surface. By increasing the concentration of $HDTMA^+$, then the number of surfactant molecules on the surface also increases, so that there is no longer space for the surfactant for parallel lining, so it starts moving in one direction, in which direction depends on the nature of the hydrophilic group and the surface. At higher concentrations, in this study, a bilayer was formed due to the amount of $HDTMA^+$ was higher (1 mol/L) than its Critical Micelle Concentration (CMC, 9.4 x 10^{-4} mol/L). In this condition, the hydrophobic tails of the surfactant molecules will associate to form a bilayer. Surfactant molecules form a monolayer at the solid-aqueous interface via strong Columbic interaction below its CMC. The formation of monolayer and bilayer is represented in figure 4.

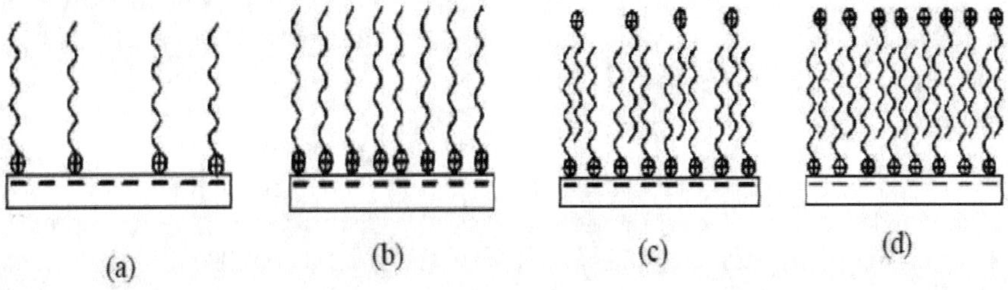

Figure 4. Attachment of $HDTMA^+$ on the resorcinarene surface. (a), (b) $HDTMA^+ < CMC$. (c), (d) $HDTMA^+ > CMC$.

3.3. The Molecular Properties of The Resorcinarenes

The success of the modification processes above was confirmed by the results of the data analysis by FTIR and NMR apparatus. The IR spectrum of resorcinarene *II* and *III* were displayed in Figure 5. The most notable difference between resorcinarene *II* and *III* is located in the absorption area on 3300 cm^{-1} correspond to vibrational of hydroxyl groups (-OH), in which its intensity on resorcinarene *II* higher than that of resorcinarene *III*. This fact strengthens the evidence that the acid-catalyzed functionalization only occurs on an aromatic ring such that the abundance of the hydroxyl groups was not reduced even increased with the inclusion of (3-Chloro-2-hydroxypropyl) trimethylammonium chloride. In contrast, base-catalyzed modification only activates the hydroxyl groups on resorcinarene *I*. This resulted in the replacement of the hydroxyl groups by quaternary ammonium groups so that the absorption intensity of hydroxyl groups was reduced (Figure 5).

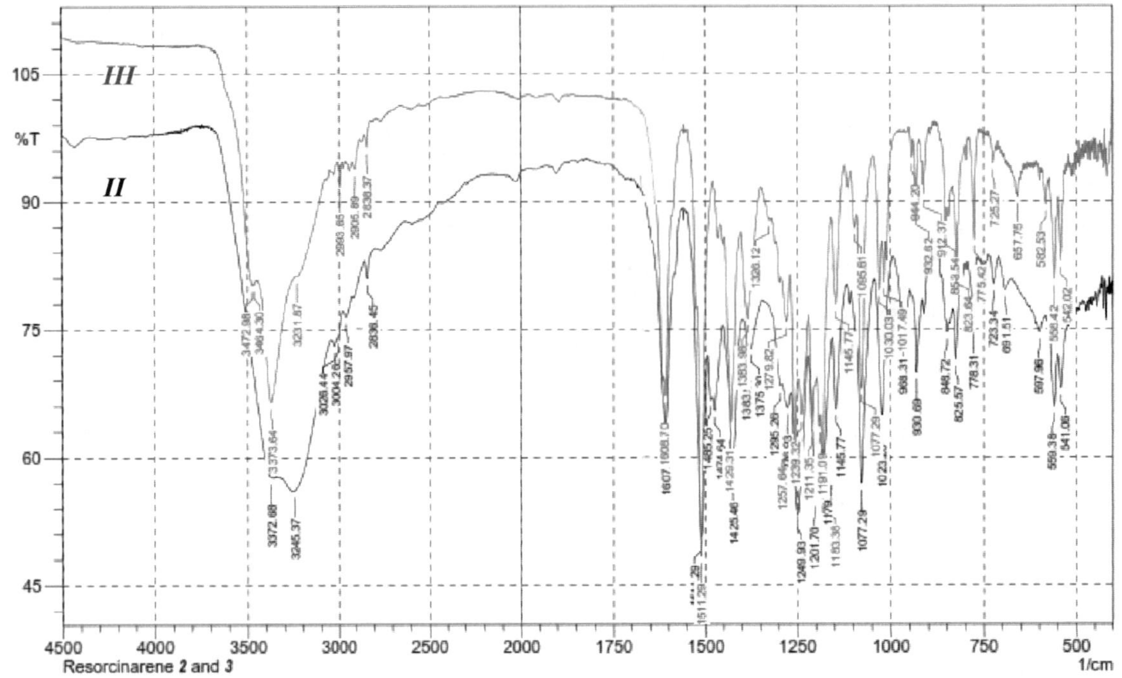

Figure 5. IR spectrum of resorcinarene *II* and *III*.

Clear evidence of the formation of resorcinarene *II* and *III* were further confirmed by their NMR data (figure 6). The existence of several new absorption at chemical shifts on 4.15 (2H,*m*,-C<u>H</u>(OH)-), 3.18 (18H,*s*,N-CH_3), and 2.91 (4H,*d*,-CH_2-N) indicates that the functionalization with quaternary ammonium compound has been reached. Similar to the IR spectrum, resorcinarene *II,* and *III* also has differences in their H-NMR spectrum, mainly on the hydroxyl proton signals and the appearance of the aryl protons. In resorcinarene *II*, the protons of hydroxyl group are identified by absorption in two different areas (i.e., 8.47 for OH resorcin and 8.00 for OH hydroxypropyl). While on resorcinarene *III* appears only in one area that is at δ 8.51-8.60 ppm with the smaller integration of protons than that of resorcinarene *II*. Furthermore, aryl proton signal at resorcinarene *II* more simple than resorcinarene *III*. This means that the type of proton on resorcinarene *II* less than that on resorcinarene *III*. This indicates that the functionalization on resorcinarene *II* occurs in aromatic rings, while the resorcinarene *III* does not.

Based on the signal at δ 5.5 attributed to methine bridge protons at both resorcinarene *I*, *II*, and *III* have the same appearance i.e., two singlets This is consistent with a chair conformation, adopted by the calix[4]resorcinarene as reported previous [11], where two protons at methine bridges are placed at opposite sides in relation to the others.

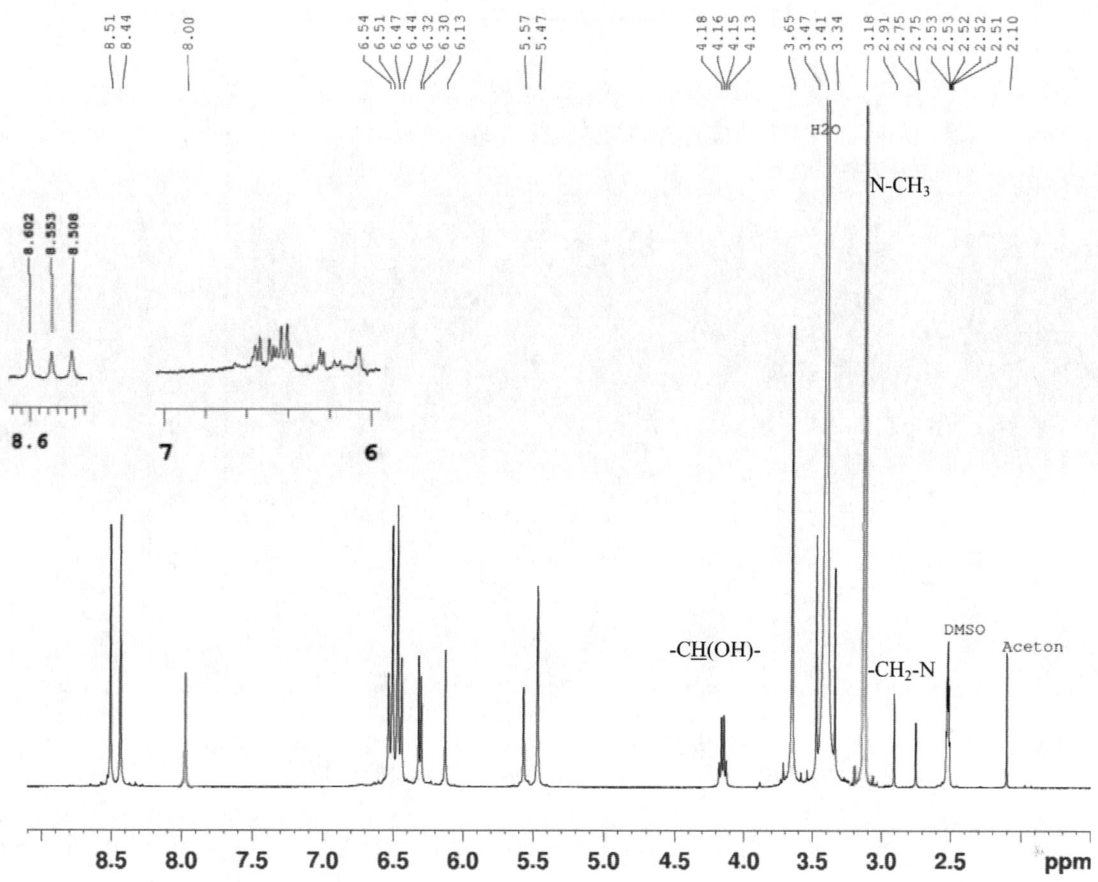

Figure 6. H-NMR spectrum of resorcinarene *II* and *III*.

The IR spectra of resorcinarene *I* and its modified forms (resorcinarene-$HDTMA^+$, *IV*) are presented in figure 7. It can be noticed that there is net difference between the original sample and the resorcinarene *IV* characterized by the appearance of new bands at 2841-2974 cm^{-1} related to the symmetric and asymmetric CH_2 and CH_3 stretching vibration. Their two bending vibrations located at 1467 and 1379 cm^{-1}, due to the methylene and methyl scissoring modes, could be observed in IR spectra of resorcinarene *IV*. These new bands indicating that HDTMA surfactant has been substituted on the surface of resorcinarene *I* to yield resorcinarene *IV*.

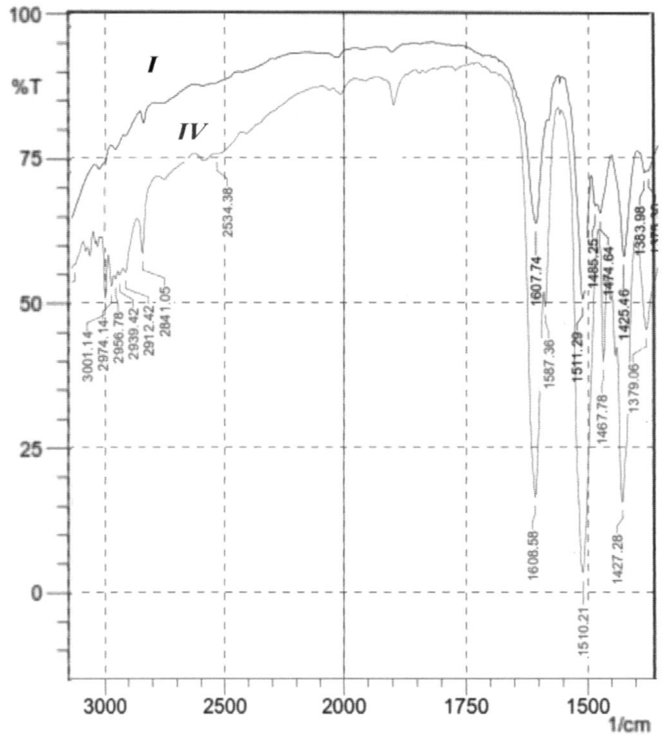

Figure 7. IR spectra of resorcinarene *I* and *IV*

Other evidence for the modification of resorcinarene *I* by $HDTMA^+$ was obtained by Scanning Electron Microscopy- Energy Dispersive X-ray Spectroscopy (SEM/EDX). Based on SEM images (Figure 8), there are notable morphological differences between resorcinarene *I* and the prepared resorcinarene *IV*. For resorcinarene *I* (figure 8a) considerable pores are clearly observed. These characteristics are not visible in the SEM micrographs of the resorcinarene *IV* due to surfactant coverage on the external crystal surface (Figure 8b) confirming, thus, a good homogeneity within the resorcinarene-$HDTMA^+$.

(a) (b)

Figure 8. SEM images of resorcinarene *I* (a) and *IV* (b).

4. Conclusion

Functionalization of C-4-methoxyphenylcalix[4]resorcinarene (*I*) has been conducted with several ammonium compounds with differences reaction conditions. Modification of resorcinarene *I* with (3-Chloro-2-hydroxypropyl)trimethylammonium chloride has been done in the presence of fuming hydrochloric acid via electrophilic aromatic substitution to yield resorcinarene *II*. Meanwhile, base-catalyzed functionalization of resorcinarene *I* by using glycydiltrimethylammonium chloride results resorcinarene *III*. The modification process involves the epoxide ring opening reaction of the ammonium compound. Attachment of $HDTMA^+$ on the resorcinarene *I* surface has been accomplished in the alkaline condition in the presence of an excess HDTMA-Br surfactant. Based on the spectral properties of ^{1}H-NMR, all calixresorcinarene compounds were isolated in the molecular shape of the chair (C_{2h}) or flattened partial cone conformation. FT-IR, NMR, and SEM analyzes indicated that the N-hexadecyl-N,N,N-trimethylammonium bromide molecules were substituted on the surface of initially C-4-methoxyphenylcalix[4]resorcinarene.

Acknowledgements
The authors gratefully acknowledge the Directorate General of Higher Education, Republic of Indonesia, for financial support through the research grant program on 2014/2015.

References
[1] Almi M, Arduini A, Casnati A, Pochini A and Ungaro R 1989 *Tetrahedron* **45** 2177-2182
[2] Shu C M, Lin W L, Lee G H, Peng S M and Chung W S 2000 *J Chin Chem Soc* **47** 173-182
[3] Dondoni A, Kleban M, Hu X, Marra A and Banks H D 2002 *J Org Chem* **67** 4722-4733
[4] Utomo S B, Jumina and Wahyuningsih T D 2009 *Indo J Chem* **9** 437-444
[5] Edwards N Y and Possanza A L 2013 *Supromol Chem* **25** 446-463
[6] Poms D, Itzhak N, Kuno L and Biali S E 2014 *J Org Chem* **79** 538-545
[7] Utomo S B, Jumina, Siswanta D, Mustofa and Kumar N 2011 *Indo J Chem* **11** 1-8
[8] Utomo S B, Jumina, Siswanta D and Mustofa 2012 *Indo J Chem* **12** 49 -56
[9] Aoyoma Y, Tanaka Y and Sugahara S 1989 *J Am Chem Soc* **111** 5397-5404
[10] Tunstand L M, Tucker J A, Dalcanale E, Weiser J, Bryant J A, Sherman J C, Helgeson R C, Knobler C B and Cram D J 1989 *J Org Chem* **54** 1305-1312
[11] Cram D J, Karbach S, Kim H E, Knobler C B, Maverick E F, Ericson J L and Helgeson R C 1988 *J Am Chem Soc* **110** 2229-2237
[12] Utomo S B, Jumina, Siswanta D and Mustofa 2013 *Indo J Chem* **13** 158-165
[13] Yin R H, Wu Q S and Chen Y 2007 *Chem Pap* **61** 224-227
[14] Ngurah B I G M, Jumina, Anwar C, Mustofa and Sahadewa 2014 *Indo J Chem* **14** 160-167
[15] Weinelt F and Schneider H J 1991 *J Org Chem* **56** 5527-5535
[16] Jumina, Sarjono R E, Paramita B W, Siswanta D, Santosa S J, Anwar C, Sastrohamidjojo H, Ohto K and Oshima T 2007 *J Chin Chem Soc* **54** 1167-1178
[17] Rožic M, Sipuši D I, Sekovanic F, Miljanic S, Curkovic L and Hrenovic J 2009 *J Colloid Interf Sci* **331** 295-301
[18] Kusumaningsih T, Jumina, Siswanta D, Mustofa, Ohto and Kawakita H 2011 *Indo J Chem* **11** 186-190

IOP Conf. Series: Materials Science and Engineering **107** (2016) 012043 doi:10.1088/1757-899X/107/1/012043

The Effects of plasticizers and palmitic acid toward the properties of the carrageenan Film

Atmanto Heru Wibowo*, Oktaviana Listiyawati and Candra Purnawan
*Department of Chemistry, Faculty of Mathematics and Natural Sciences, SebelasMaret University, Jl. Ir. Sutami 36A Surakarta 57126, Indonesia

*Email: aheruwibowo@staff.uns.ac.id

Abstract. Varied plasticizers and palmitic acid additive have been added in the carrageenan film. The film was made by mixing of the carrageenan and plasticizers (glycerol, polyethylene glycol, polyvinyl alcohol) with composition of 92:3, 90:6, 87:9, 84:12, 81:15(%w/w) and in the presence of palmitic acid as additive with 1%, 2%, 3%, 4%, 5% of total weight. Casting method was used for the film molding and drying at 60°C with the oven for 12 hours. To investigate the effects of plasticizers and additive, some mechanical tests on film were performed. The test result concludes that plasticizers in the film decreased the tensile strength and increased the elongation break of the carrageenan film. The additive of palmitic acid decreased the tensile strength of the carrageenan film and also decreased the-the water absorbance of the film. The highest tensile strength of films made was with the formulation of carrageenan: PEG with composition of 92:3 (% w/w). The highest elongation break of the film was for carrageenan:PVA with the composition of 81: 15 (%w/w) and carrageenan:palmitic acid:PEG with the composition of 92: 3: 1 (%w/w). The lowest water absorption of the film was achieved for carrageenan:PVA:palmitic acid with the composition of 87: 3: 5 (%w/w).

1. Introduction

Carrageenan is the generic name for a family of high molecular weight sulfated polysaccharides obtained by extraction of certain species of red seaweeds [1]. This extracted carrageenan is a hydrocolloid composed from esther, sodium, potassium, magnesium, and calcium sulfate with a copolymer of 3,6 anhydrogalactosa. K-Carrageenan is a sulfated galactan polymer that is often used in the food industry and is expected to exhibit good adhesion on the surface of membrane filters via hydrogen bonding as well as electrostatic interactions [2]. The combination of gelatin and carrageenan also results in a hydrogel. Hydrogels are nothing but hydrophilic three-dimensional polymer networks capable of absorbing large amounts of biological fluids. Because of their relatively high water contents and their soft, rubbery consistency, hydrogel materials more closely resemble their physical properties to living tissue [3]. Some papers reported that carrageenan is suitable to be used for edible film or coating. Carrageenan-based coatings have been proven efficient to extend the shelf life of fresh-cut and fresh whole fruits, observed that carrageenan films present lower oxygen permeability than starch films. Moreover, carrageenan-coated strawberries presented lower weight loss and firmness loss when compared to starch-coated strawberries, probably reflecting a better moisture barrier of carrageenan coatings [4].

The plasticizer is added into the carrageenan film in order to increase the flexibility of the biopolymer film by reducing the intermolecular and intramolecular interaction within the polymer in the film, decreasing brittleness and increasing the elasticity of the film. Some plasticizers have been added in the biopolymer film such as glycerol, polyethylene glycol, polyvinyl alcohol, and sucrose [5]. However, the addition of plasticizer with high concentration is able to reduce the functionality of the edible film such as the water vapor resistance and the mechanical properties [6]. The additive agent added into the film is common with highly lipophilicity or lipid. The addition of the lipid is able to increase the hydrophobicity of the film system. Lipid plasticizer that is commonly used for the edible film is glyceride, wax and a fatty acid such as stearic and palmitic acid due to the hydrophobicity and high melting point [7]. This study is aimed to investigate the influence of the plasticizer concentration and the fatty acid toward mechanical properties of edible film from carrageenan film.Besides, the fatty acid added into the carrageenan-based film is to increase the lipophilicity of the film to resist the water pass through where plasticizers used might be a lack of their lipophilicities. The addition of the palmitic acid into the carrageenan film together with the palmitic acid as long as the knowledge of the author has been not reported yet.

2. Methods

2.1. Materials

Carrageenan used for the study was kappa (κ) form obtained from PT. Bratachem (Solo Branch) for pharmaceutical, cosmetic, food, and other industries Solo, Indonesia). Plasticizers of glycerol, polyethylene glycol (PEG), polyvinyl alcohol (PVA) 98% were purchased from Bratachem (Indonesia), Merck (Germany), and Merck (Germany), respectively. The palmitic acid additive was obtained from PT.Bratachem (Solo Branch) for pharmaceutical, cosmetic, food, and other industries. Potassium chloride was obtained from Merck (Germany).

2.2. Instruments

The instrument used for the study is infra-red spectrophotometer from FT-IR IR prestige Shimadzu 8201 PC. Tensile strength measurement device is from Pearson Panke Equipment LTD and oven from Barnstead International 3513-1

2.3. Edible film casting with various plasticizers

Edible film formulation was divided based on the plasticizer used. Three kinds of carrageenan film with formulation glycerol plasticizer, PEG plasticizer, and PVA plasticizer.A film with glycerol was formulated by mixing 3 g κ-Carrageenan, 0.144 g potassium chloride and glycerol plasticizers with various concentrations (w/w) of 3%, 6%, 9%, 12%, 15%. A film with PEGwas formulated by mixing 3 g κ-Carrageenan, 0.144 g potassium chloride and PEG plasticizers with various concentrations (w/w) of 3%, 6%, 9%, 12%, 15%. A film with PVAwas formulated by mixing 3 g κ-Carrageenan, 0.144 g potassium chloride and PEG plasticizers with various concentrations (w/w) of 3%, 6%, 9%, 12%, 15%.

K-Carrageenan, potassium chloride, and plasticizers were solved with 100 mL of distilled water and stirred until homogeny solvent of suspension-like achieved. The solvent was heated at about 80°C and exposed with homogeny thickness on the 21.5 cm x 16 cm glass plate until room temperature. Heating the film on the plate was done for 24 hours at 60 °C. The casted film was than stick off from the glass for the further test.

2.4. Edible film with palmitic acid additive

The influence of palmitic acid additive toward the character of the film was tried to select films with a plasticizer of the highest tensile strength value of tested film. The casting method of the edible film with the addition of palmitic acid is the same with above method. The addition of palmitic acid to the films is with various compositions of 1%, 2%, 3%,4%,and 5% (w/w)palmitic acid.

IOP Conf. Series: Materials Science and Engineering **107** (2016) 012043 doi:10.1088/1757-899X/107/1/012043

2.5. Characterization of edible film

Films cast with various plasticizers and with the addition of palmitic acid were tested for some mechanicals parameters such as tensile strength (TS), Young's modulus with Pearson Panke Equipment LTD and elongation to break with Ruller. The swelling test was also done to investigate the resistance of film towards the water. Fourier Transform Infra-Red (FTIR) was also done to investigate the kind of mixture binding of the films.

2.5.1. Test of tensile strength (TS), % elongation to break and Young's modulus

Tensile strength and elongation test is carried out based on ASTM D-638. The equations used for the tensile strength determinationare listed as:

a. Tensile Strength value of thefilm is determined by formula (equation (1)).

$$\sigma = N / (T \times L) \tag{1}$$

Where:
σ =tensile strength of films (MPa)
N = force applied (N)
T = thickness of film (mm)
L = length of the film (mm)

b. Elongation value of thefilm is determined by equation (2).

$$\{(Li - Lo) / Lo\} \times 100\% \tag{2}$$

Where:
Li = length of the film after test (cm)
Lo = length of the film before test (cm)

c. Young's Modulus is determined by equation (3).

$$E = \frac{\sigma}{\varepsilon} \tag{3}$$

Where:
E = Rigidity of the film (MPa)
σ = Strength of drag force (MPa)
ε = Elongated film (%)

2.5.2. Swelling test of water

Swelling test is determined by equation (4).

$$\text{Swelling Index (\%)} = \frac{W - Wo}{Wo} \times 100\% \tag{4}$$

W = weight of edible film after water soaking (gram)
Wo = weight of edible film before water soaking (gram)

3. Result and Discussion

3.1. Carrageenan Edible Film with Variation of PlasticizerConcentration

The edible film of Carrageenan is with light brown color, rather a rough, a little bit oily with the thickness of 0.8 – 1.5 mm. Products of the edible film with the variation of plasticizers are showed in Figure 10.

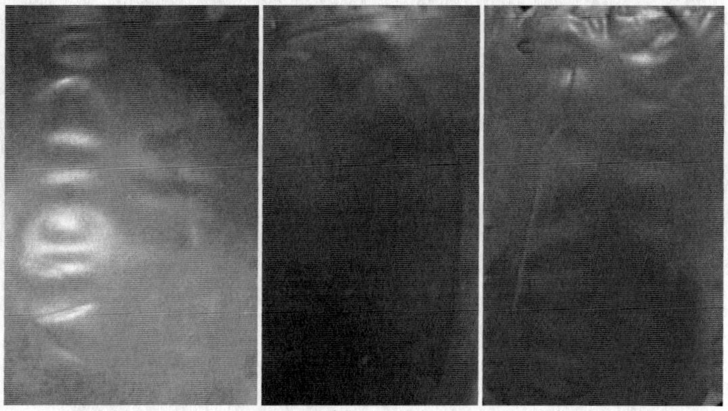

(a) (b) (c)

Figure 1. Edible film of carrageenan with glycerol (a) with PEG (b) with PVA (c)

3.1.1. Tensile Strength Test

Figure 2 showed tensile strength value of film which was tested with ASTM D-638 standard method.

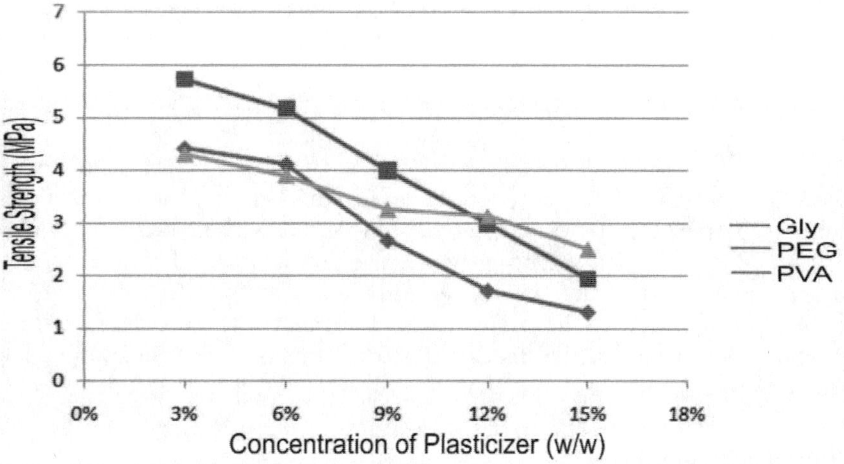

Figure 2. The effect of plasticizer toward tensile strength in the carrageenan film

Figure 2 showed that all kind plasticizers with increasing concentration of 3% to 15% (w/w) decreased the tensile strength of the film. This result indicated that the plasticizer as the common basic property of plasticizer just afford the elasticity but not simultaneously increased the tensile strength. The tensile strength obtained from the film was around 4.418- 1.308 Mpa with glycerol, 5.725 – 1.942 Mpa with PEG and 4.297 – 2.5 Mpa with PVA. Plasticizers behaved on the film as an agent which changes the mechanical property of the film and decreased the cohesion and mechanical resistant of the chain polymer [8]. Plasticizers also caused internal hydrogen bonding strength and reduced the polymer intermolecular bonding strength which reduced the strength of the strain-brake strength of the film. This bonding change within the film was showed on the infrared spectrum, in which intensity of OH group become wide after components blending to form a film. Plasticizer dispersed in the carrageenan was depended on the concentration added. More plasticizers added, the strength of intermolecular bonding decreased as more plasticizers filled between the carrageenan-film layers.

3.1.2. Elongation to break test

Elongation to break test was done following ASTM D-638 standard. Figure 3 showed the percentage of elongation of the Carrageenan with varied concentration.

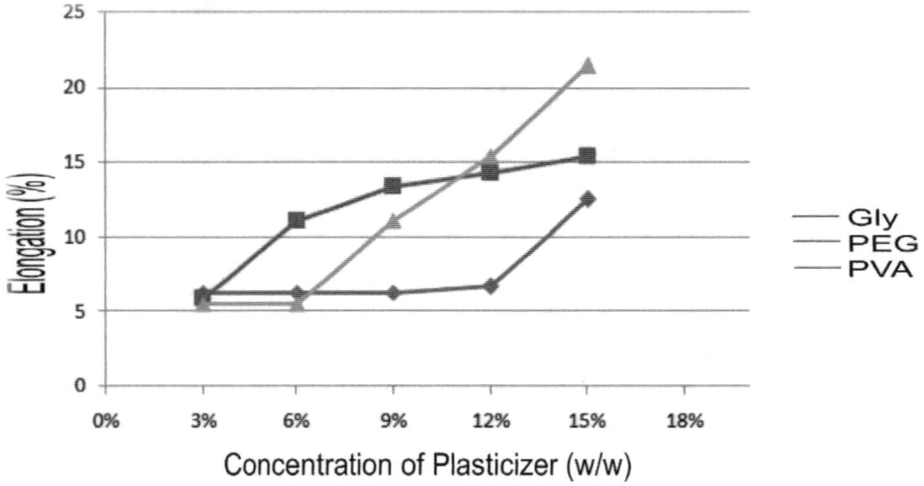

Figure 3.Effect of plasticizer and concentration toward % elongation of carrageenan film

Figure 3 showed that increasing concentration of plasticizer 3% to 15% (w/w) increase also the elongation value. Elongation value obtained was 6.25% – 12.5% with glycerol, 5.88% - 15.38% with PEG and 5.55% - 21.43% with PVA. The phenomena of the change of tensile strength were also similar to the elongation value, in which reducing the strength of hydrogen bonding and intermolecular force was the cause of the strain-brake force and the increase the flexibility of the film. The present of the plasticizer randomized the structure of film molecule which affected to the reduce of the film matrix rigidity. The increase of plasticizer will also increase the elongation value.

3.1.3. Young Modulus Test

Modulus Young Test was done following the ASTM D-638 standard. Young modulus value of the film was obtained from the calculation of tensile strength divided by elongation to break. Figure 4 showed the modulus young of the film with the concentration of the plasticizer.

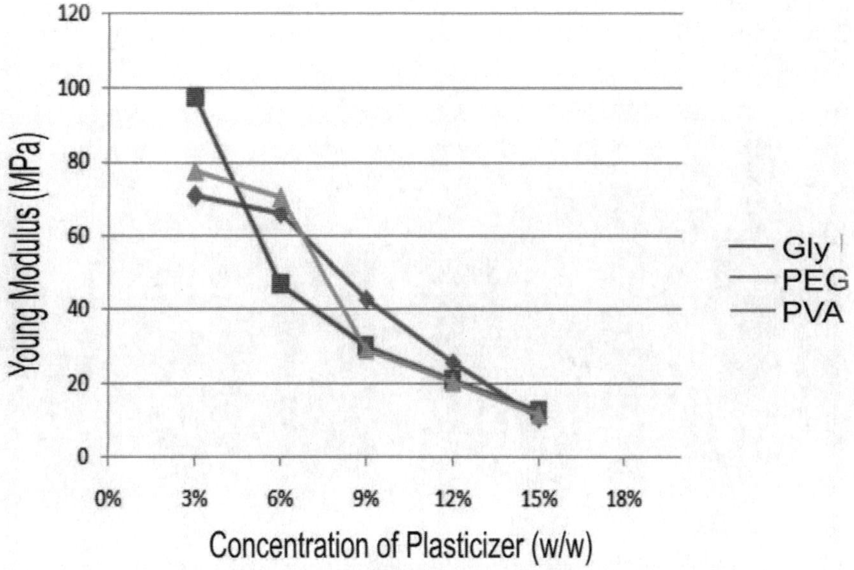

Figure 4. Youngs Modulus of the film vs. concentration of the plasticizer.

Figure 4 showed that Youngs modulus value of film tended to decrease with the increase of plasticizer concentration added. Young Modulus obtained of test on the film was 70.688 – 10.464 Mpa with glycerol, 97.364 – 12.627 Mpa with PEG and 77.423 – 11.666 Mpa with PVA. The increase of plasticizer concentration added (3% - 15% (w/w), the rigidity of the material tend to decrease. This was due to the role of the plasticizer to increase the elasticity of film.

3.2. The effect of the additive concentration of palmitic acid

The effect of palmitic acid as an additive agent was tested toward the film which has the highest tensile strength of each plasticizer. The aim of the test on the film with high elongation is to investigate the ability of the film to protect the product from the mechanical accident. From the physical figure of figure 5, white granules with thickness 0.8 – 1.5 mm appeared on the surface. This seemed that the lipophilicity difference was still a problem. The carrageenan film products with the variation of palmitic are shown in figure 5.

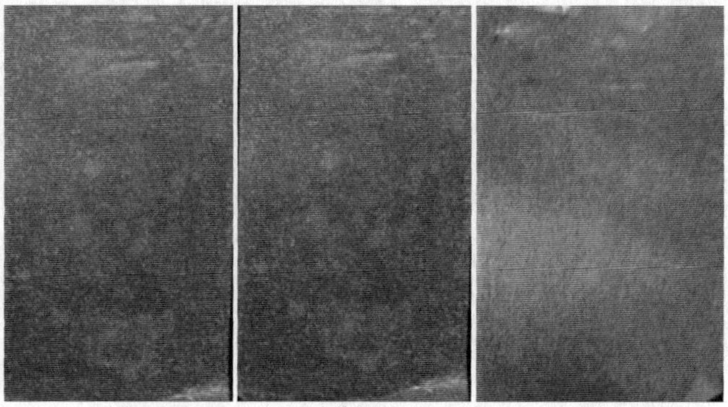

(a) (b) (c)

Figure 5. Edible film of carrageenan with palmitic acid and glycerol (a*)* with palmitic acid, PEG and glycerol (b) with palmitic acid, PVA, and glycerol (c)

3.2.1.Tensile Strength Test

Figure 6 showed the effect of the palmitic acid additive on the tensile strength of the film.

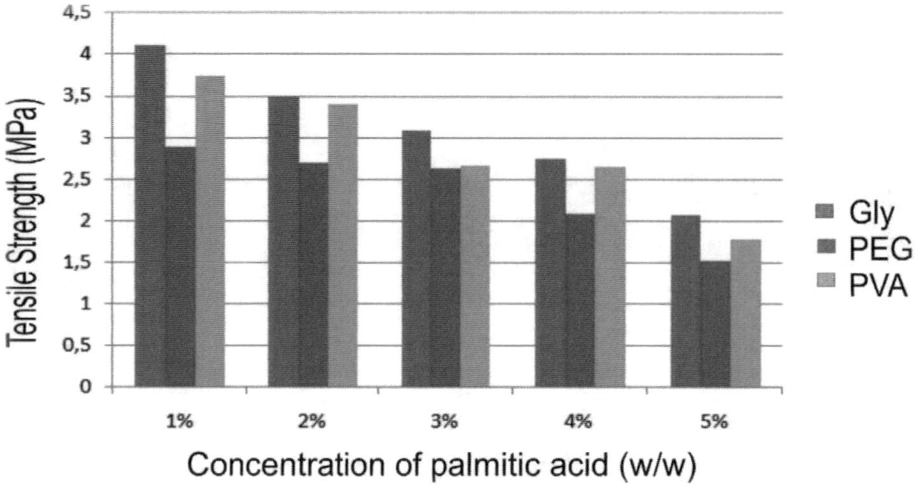

Figure 6. Effect of the tensile strength of the carrageenan film at various palmitic acid concentration

Figure 6 showed that the tensile strength of the film tended to decrease as palmitic concentration from 1% to 5% (w/w)increased. The tensile strength obtained with the palmitic acid additive was 4.111 – 2.074 Mpa for glycerol plasticizer, 2.9 – 1.533 Mpa for PEG plasticizer, and 3.745 – 1.785 Mpa for PVA plasticizer. According to Yang and Paulson (2000), the addition of fatty acid causes the decrease of tensile strength value of the film. The decrease in the tensile strength causes the film more fragile or broken

3.2.2. % Elongation test

Figure 7 showed the result of elongation test after addition of palmitic acid toward Carrageenan film.

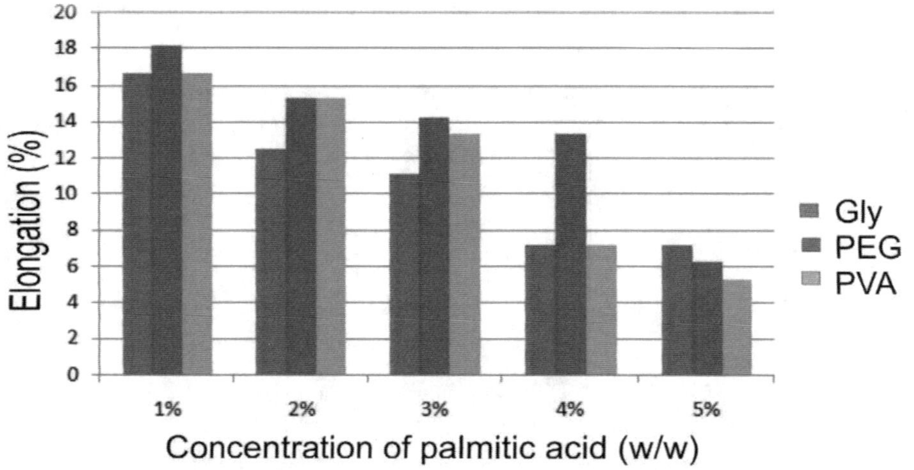

Figure 7. Elongation of carrageenan film after addition of palmitic acid

The figure 7 showed that the increase of the palmitic acid concentration from 1% to 5% (w/w) tended to decrease the elongation value. The elongation value of film with palmitic acid was 16.67%–7.143% with glycerol plasticizer, 18.18% - 6.25% with PEG plasticizer and 16.67% - 5.26% with PVA plasticizer. Palmitic acid seemed to hinder the interaction of carrageenan molecule in the film.

3.2.3. Youngs Modulus Test

Figure 8 showed the correlation of Young Modulus Test of the film after addition of palmitic acid.

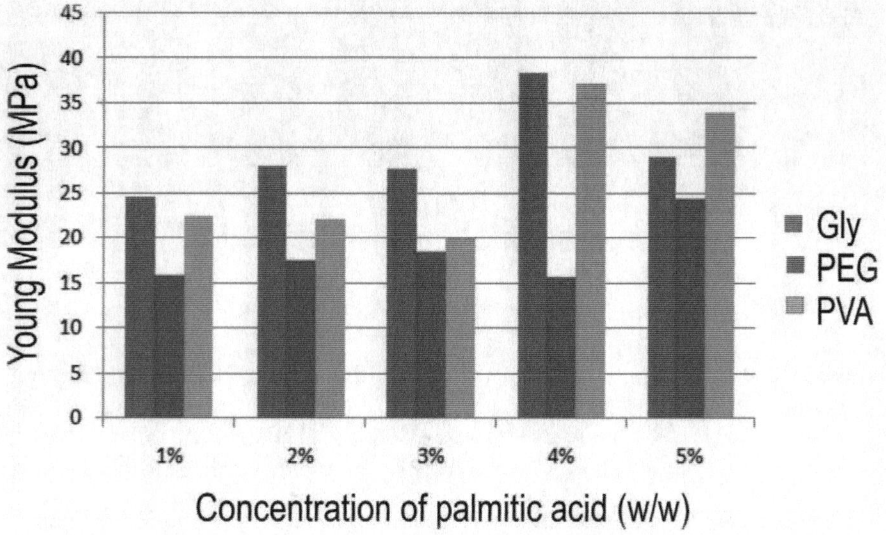

Figure 8.The Young modulus value of the film after the addition of palmitic acid.

Figure 8 showed that the lowest Youngs modulus value of the film after the palmitic acid addition was with the composition of film 3% PEG and 4% palmitic acid. The Youngs modulus value of the film was 24.661 – 29.035 Mpa with glycerol plasticizer, 15.952 – 24.524 Mpa with PEG plasticizer and 22.466 – 33.935 Mpa with PVA plasticizer. The addition of palmitic acid from 1% to 5% (w/w) is generally able to increase film elasticity.

3.2.4. Swelling test

The swelling test was done to investigate the interaction between polymer and uniformity of bonding within polymer which is determined by percentage of weight increase after swelling. Diffusion of the solvent molecule into polymer causes the swelling of the polymer. Figure 9 showed that the swelling test (swelling index) of the film in the present of varied palmitic acid.

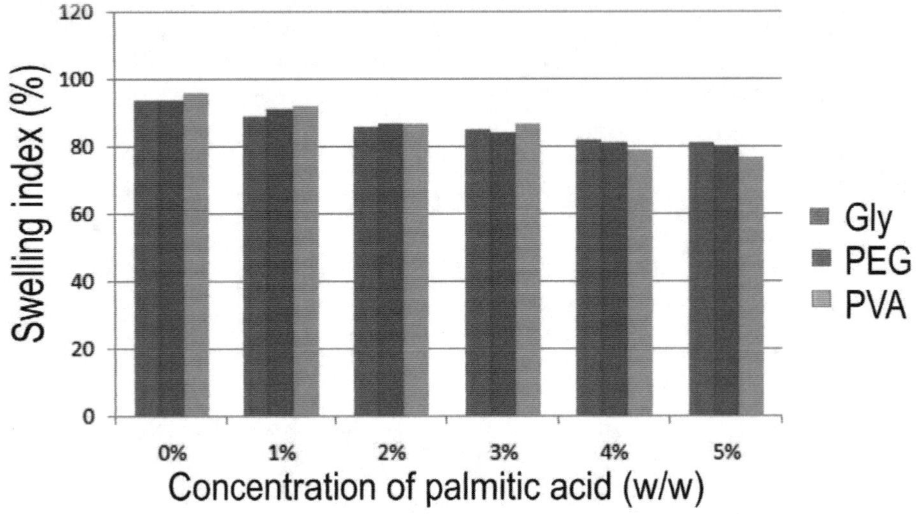

Figure 9. The effect of palmitic acid concentration toward swelling index (%) of the film.

Figure 9 showed that 5% palmitic acid in the film affected the best water resistant toward other additions. Swelling index (%) of film with various concentration of palmitic acid was 94% - 81% for glycerol plasticizer, 94% - 80% for PEG plasticizer, 96% - 77% for PVA plasticizer. Palmitic acid in the film decrease the swelling index as hydrophilicity of the palmitic acid hindered water diffusion into the film.

3.3. Analysis of Film Structure
Structure analysis of the film is aimed to investigate the content and interaction of functional group in the film. IR spectra of the film are showed in figure 9.

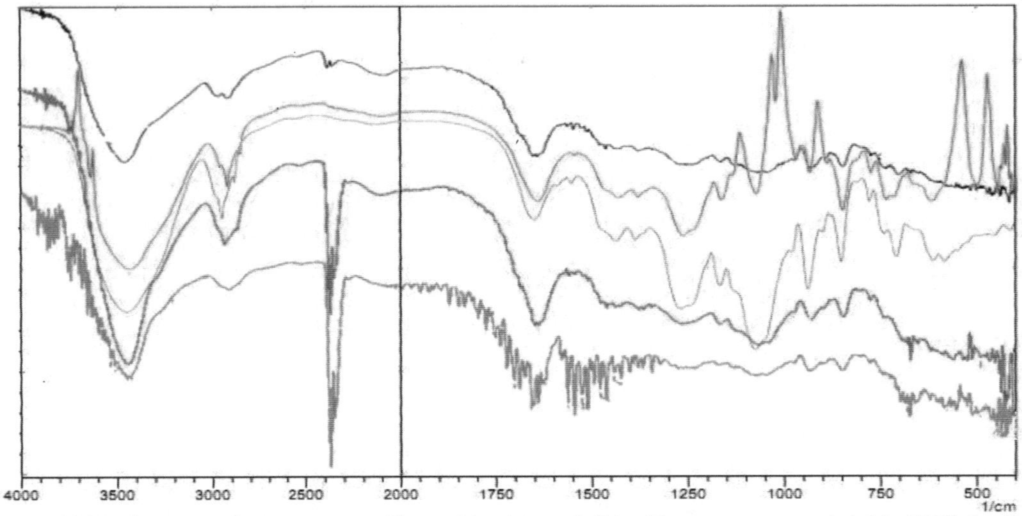

Figure 10.IR Spectra of carrageenan film with glycerol (black), carrageenan (pink), PEG (red), palmitic acid (green)

Figure 10 showed that hydroxyl group of the film at about 3400- 3450 $cm^{-1}$was seen from carrageenan, plasticizer and additive (plasticizer glycerol, PEG, and PVA). C-H stretch of glycerol at 2945.30 cm^{-1}, PEG at 2920.23 cm^{-1}, PVA at 2902.87 cm^{-1}, sulphate ester at 1219 – 1259 cm^{-1}, glycosidic at 1035 - 1072 cm^{-1} (for all carrageenan). 3,6-Anhydro D-galactose bond at 910–925 cm^{-1}. D-Galactosa-4 sulfate at 844 – 846 cm^{-1}. C=O Carbonyl on glycerol at 1627.92 cm^{-1}, PVA at 1635.64 cm^{-1}. S=O group at 1219.01 cm^{-1} for glycerol, PEG at 1163.08 cm^{-1}, PVA at 1035.77 cm^{-1}. O-H stretch of glycerol at 3419.79 cm^{-1}, PEG at 3411.01 cm^{-1}, PVA at 3444.87 cm^{-1}. C-O ester stretch of glycerol at 910.40 cm^{-1}, PEG at 1062.78 cm^{-1}, PVA at 923.90 cm^{-1}. The additional of glycerol, PEG, PVA and palmitic acid are aimed for the development of carrageenan film on mechanical properties and water resistant where the IR spectra showed just the physical interaction between additive and matrix in the film occurred.

4. Conclusion

Almost all plasticizers in the carrageenan behave to reduce tensile strength and Young modulus properties and increase the elongation value of the film (3% and PVA 15%). Increased concentration of palmitic acid in the film decrease the tensile strength, % elongation and Young modulus of the carrageenan film. The best composition of film that affords the value of swelling index, tensile strength, elongation and Young modulus is obtained for the film (3% glycerol, 1% palmitic acid and 3% PEG, 1% and 4% palmitic acid) and the best value of swelling index of film is obtained for the film (5%palmitic acid and PVA plasticizer). Palmitic acid is able to be used together with a plasticizer to increase the water resistant of the carrageenan film.

References

[1] Li Z-Q, Yang J-C, Sui K-W and Yin N 2015 Facile synthesis of metal-organic framework MOF-808 for arsenic removal *Mater. Lett.* **160** 412-414

[2] Yamazaki S, Mori T, Ogino I and Mukai S R 2015 Flexible film-type catalysts encapsulating urease within κ-carrageenan hydrogel network *Tailoring Sustain. Chem. React. Eng.* **278** 122–128

[3] Varghese J S, Chellappa N, and Fathima N 2014 Gelatin–carrageenan hydrogels: Role of pore size distribution on drug delivery process *Colloids Surf. B Biointerfaces* **113** 346–351

[4] Paula G A, Benevides N M B, Cunha A P, de Oliveira A V, Pinto A M B, Morais J P S and Azeredo H M C 2015 Development and characterization of edible films from mixtures of κ-carrageenan, ι-carrageenan, and alginate *Food Hydrocoll.* **47** 140–145

[5] Krochta J M, Baldwin E A, dan Nisperos-Carriedo M O 1994 Edible Coating and Film to Improve Food Quality (NewYork: Technomic Publishing Company)

[6] Glicksman M 1984 Food Hydrocolloids **3** 1983 Boca Raton Fl: CRC Press as Plasticizers Affect Mechanical and Water Vapour Barrier Properties of Edible Wheat Gluten Film *J. Food Sci.* **1** 206-210

[7] Hagenmaier R D and Shaw P E 1990 Moisture permeability of edible film made with fatty acids and hydroxypropyl methylcellulose *Journal of Agricultural and Food Chemistry* **38** 1799-1803

[8] Lieberman E R and Gilbert S D 1973 Gas Permeation of Collagen Films as Affected bt Cross Linkage Moisture and Plasticizer Content *Journal of Polymer Science* **41** 33–43

[9] Yang L and Paulson AT 2000 Effects of lipids on mechanical and moisture barrier properties of edible gellan film *J. Food Research International* **33** 571-578

IOP Conf. Series: Materials Science and Engineering **107** (2016) 012044 doi:10.1088/1757-899X/107/1/012044

Synthesis H-Zeolite catalyst by impregnation KI/KIO_3 and performance test catalyst for biodiesel production

W. Widayat[1,2,*], Adit Rizky Wicaksono[1], Lukman Hakim Firdaus[1] and Ndaru Okvitarini[1]

[1]Department of Chemical Engineering, Faculty of Engineering, Diponegoro University, Jl Prof H. Soedharto SH Tembalang, Semarang 50239, Indonesia

[2]Material Technology Laboratory Center of Research and Service, Diponegoro University, Jl Prof H. Soedharto SH Tembalang, Semarang 50239, Indonesia

*E-mail: yayat_99@yahoo.com

Abstract. The objective of this research is to produce H-catalyst catalyst that was impregnated with KI/KIO_3. The catalyst was analyzed about surface area, X-Ray Diffraction (XRD) and performance test of catalyst for biodiesel production. An H-Zeolite catalyst was synthesized from natural zeolite with chemical treatment processing, impregnation KI/KIO_3 and physical treatment. The results shows that the surface area of the catalyst by 27.236 m^2/g at a concentration of 5% KI. XRD analysis shows peak 2-θ at 23.627° indicating that KI was impregnated on H-zeolite catalyst. The catalyst was tested in production of biodiesel using palm oil with conventional methods for 3 hour at temperature of 70-80 °C. The result for conversion Fatty Acid Methyl Ester (FAME) reached maximum value on 87.91% under production process using catalyst 5% KIO_3-H zeolite.

1. Introduction

Fatty Acid Methyl Ester (FAME) is an alternative energy used replacing diesel oil. Commonly, this compound calling as biodiesel. FAME can used as a mixture or pure. In application, biodiesel has composition 10-100% [1]. Some of the advantages of biodiesel is able to lower the greenhouse effect, biodegradable, non-toxic, and specifically reduce levels of CO 65%, CO_2 78%, SO_2 90% and unburned carbon 50% when compared to fossil fuels [2]. Biodiesel is a renewable fuel substitution are biodegradable, environmentally friendly, energy efficient, and can meet the needs of energy security without compromising operational engine performance [3].

Commonly, biodiesel produced by using homogeneous catalyst like as sulfuric acid, sodium hydroxide (NaOH) and potassium hydroxide (KOH). Biodiesel production process can be done by using a homogeneous acid catalyst process [4, 5], an acid and base catalyst process [6], supercritical process [7], enzymatic process [8], heterogeneous acid catalyst [9] and sonochemical [1,10]. The combination heterogeneous catalyst and enzymatic process was also done in biodiesel production. It was used *Jatropha Curcas* seed with lipase [11]. Biodiesel also can be produced by using in situ processing with acid /base catalyst [12, 13]. Biodiesel production is done with the mechanism of the chemical reaction of vegetable oils with methanol to produce Fatty Acid Methyl Ester (FAME) and glycerol as a byproduct [1, 14]. The most commonly used to produce biodiesel is by transesterification of triglycerides from oil-containing material and reacted using the alcohol with the addition of the catalyst. Here is a transesterification reaction mechanism.

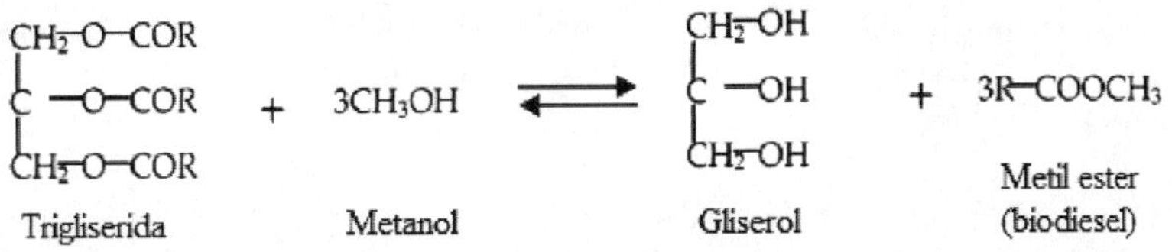

The biodiesel production using homogeneous catalyst has disadvantages, such as formation of soaps, biodiesel separation from catalyst very complexly [15, 16]. To overcome the disadvantages, utilization a heterogeneous catalyst in biodiesel production was developed. Heterogeneous catalysts are very easily separated from biodiesel and can be re-used. In addition, biodiesel production by using heterogeneous catalysts did not found soap formation [17].

Heterogeneous catalyst that used in biodiesel production must has acid or base catalyst like as calcium oxide, magnesium oxide, aluminum oxide and zeolite. Some researchers studied biodiesel production by using heterogeneous catalyst. Boz et.al.(2009) used γ-alumina catalyst that added with potassium fluoride. The results obtained yield of biodiesel 97% [18]. Furuta et.al. (2004) [19] used W/Zr-γ-Al_2O_3 and Yan et.al.(2009) [20] used lanthanum impregnated in calcium oxide, CaO. Commonly, the catalyst is manufacturing in industry and expensive. To achieve maximum yield yet, the development of catalyst technology are still required. Indonesia has potential material for developing an catalyst. This material is natural zeolite that is naturally deposited in large enough amount [21]. Natural zeolite itself has a content of Si and Al. Si and Al components can act as a catalyst.

Natural zeolite from Indonesia has silica content at minimum 60%. Commonly, it use as an adsorbent in drinking water and wastewater treatment and fertilizer [21]. Zeolite pore surface varies in structure and electric field, it is that causes the zeolite catalyst has different properties [22]. Zeolites have been tested as catalysts for the conversion of oil-rich Free Fatty Acid (FFA) to be converted into (FAME). Marchetti and Erazzu (2008) [23] developed natural zeolites in esterification to minimize the presence of water. Water content can affect in biodiesel production i.e. decreasing conversion of esterification reaction. Therefore, preparation steps includes chemical and physical treatment are needed before using natural zeolite in catalyst application [21].

Chemical treatment is intended to control the activity and selectivity of the zeolite acidity. Comparison between the silica and alumina will leads to property of adsorption activity. Number of brönsted acid sites of zeolite will grow with increasing aluminum content. Chemical treatment on the zeolite catalyst activation can be done either by acid treatment using HNO_3, HCl or H_2SO_4. Acid treatment is basically to increase the Si/Al ratio. The activity and selectivity of the catalyst is influenced by the characteristics of the catalyst where these characteristics depend on the method of preparation. One method that is often used in the preparation of the catalyst is impregnation method, because in terms of the manufacturing process, it is more practical than other methods of preparation (sol-gel and co-precipitation) [24].

Calcination is the process of heating an object until the temperature is high, but still below the melting point to eliminate compounds that can evaporate. Zeolite activation process through calcination will release water and increase crystallinity. Therefore, calcination of natural zeolite can improve the specific properties of zeolite.

The objective of this research is to study the preparation impregnated H-zeolite catalyst and the characterization and performance of catalysts in biodiesel production.

2. Research Methods

2.1. Materials

The material used here is a natural zeolite from district Wonosari Gunung Kidul. The material size were first reduced and followed by composition, water, and crystallography analysis. Solvent for chemical treatment used here is 4 M hydrogen chloride acid. $AgNO_3$ 0.1 M solution was used as an indicator for washing of catalyst product. Aquadest was used for washing the synthesized material during catalyst preparation. KI and KIO_3 used are in analytical grade from MERCK that used as impregnation compounds. Palm oil used as raw material in performance testing of catalyst. Contained free fatty acid < 0.03%. This schematic view of installed equipment is presented in Figure 1. This equipment also used in biodiesel production. Oven and muffle furnace were used in physical treatment.

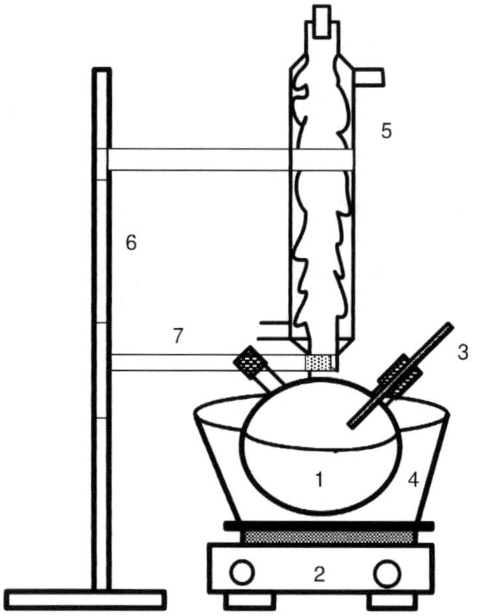

Figure 1. Equipment for chemical treatment and impregnation process.

2.2. Preparation of H-zeolite Catalyst

The method in preparation of H-zeolite was reffered to methods published in our previous papers [21, 25]. The method applied are chemical and physical treatment. Chemical processing was used dealumination process. The dealumination process of acid treatment was carried out with a natural zeolite using 20 ml of hydrogen chloride solution 4 M. The dealumination process was done by reflux at temperature (90 °C). After that, solid was filtered and washed with distilled water repeatedly until chloride was totally removed. The samples were then calcined in inert condition on isothermal temperature at 600 °C until four hours.

2.3. Impregnation of KI and KIO_3

The impregnation method was done by attaching the active compound component on a porous support material. In this research KI and KIO_3 were used as active compounds. The initial step impregnation process carried out by dissolving KI/KIO_3 in 50 mL of distilled water at a concentration of 1-5%. Furthermore, H-zeolite was mixed with a solution of 50 grams of KI/KIO_3 on three neck flask for stirred and heated for 2 hours at 90 °C. The mixture is filtered with suction strainer for further dried sludge screening results. Finally, KI/KIO_3-H-zeolite was obtained after catalyst product was dried in oven at temperature 110 °C.

2.4. Catalyst Characterization

The crystallography of KI/KIO_3-H-zeolite catalyst product were analyzed using XRD photographs a Philips 57.3 mm diameter, camera, with Cu Kα radiation. The analysis process was conducted at Research Center Institute Technology Sepuluh Nopember Surabaya. The surface area analysis was performed by the Brunauer Emmett Teller (BET) analysis using *Quantachrome NOVA 1000 High-Speed Gas Sorption Analyser* with Po = 711.65 mmHg and nitrogen as carrier gas. The analysis was conducted at Instrumentation Laboratory Department of Chemical Engineering, Institute Technology Sepuluh Nopember Surabaya.

2.5. Catalytic Activity Test.

KI/KIO_3-H-zeolite catalyst was tested for biodiesel production. Biodiesel production process used transesterification reaction by conventional methods. The palm oil must be checked free fatty acid contents < 0.5% [15, 16]. The initial step for the production of biodiesel, all material such as cooking oil, methanol and catalyst are mixed in a three-neck flask. Mole ratio between oil with methanol is 1: 6. The mixture was stirred and heated at temperature of 70-80 °C for 3 hours. The product of the transesterification reaction was separated using a decanter to separate biodiesel and glycerol.

Biodiesel product was analyzed by using Gas Chromatography (GC). The analysis process was conducted at Chemical Analysis Laboratory Polytechnic Malang using HP 5890, with OV-07 column and FID (Flame Ionization Detector). The operation used nitrogen as gas carrier with flow rate 28 ml/minute, temperature 125-275 °C and detector temperature 275 °C. The initial temperature was kept constantly for 3 minute, then increased at a rate of 15°/minute until final temperature reached to 275 °C. The compounds will be analyzed free fatty acid ester with internal standard methods [1, 13]. The data was used for calculation biodiesel conversion with this equation (1)

$$X_A = \frac{Triglyceride\ reacted}{Triglyceride\ initial} x100\% \qquad (1)$$

3. Results and Discussion

3.1. Characterization Catalyst

The results of BET analysis using H-zeolite catalyst with KI concentration of 5 wt% presented in Figure 2. Figure 2 shows the relationship relative pressure with volume at standard conditions (STP). Volume of nitrogen gas adsorbed by H-zeolite catalyst was measured as 28.013 cc. From figure 2, the specific surface area of KI/ H-zeolite catalyst was obtained 27.236 m^2/g. Total surface area is more smaller than H-zeolite that produced Widayat et.al, 2010 [26]. The data is then compared in several studies conducted by using other types of catalyst as shown in Table 1.

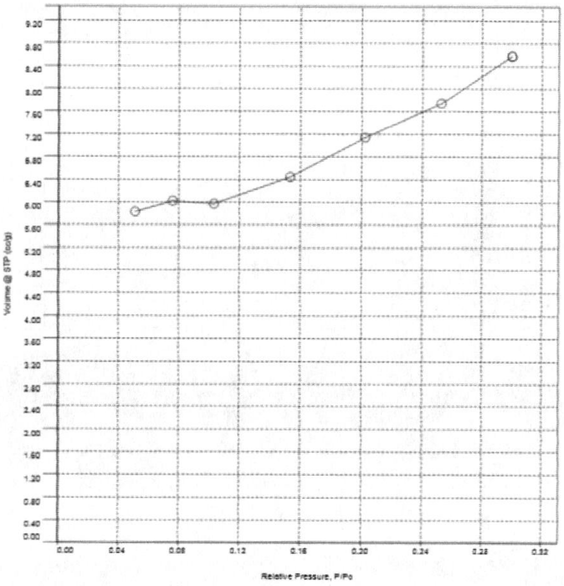

Figure 2. Volume gas versus relative pressure (P/Po).

In the previous studies, natural zeolite catalyst has a surface area of 172.92 m^2/gram [25]. H-zeolite that syntheses from natural zeolite from Gunung Kidul district obtained surface area of 199.76 m^2/gram. KI compound that impregnated in H-zeolite caused the increasing the basicity of H-zeolite catalyst and affected to the activity and active surface area of the catalyst [9]. In this research, KI compound caused decreasing on specific surface area of H-zeolite because after impregnation process H-zeolite was dried at temperature 115 °C without further calcination. Commonly, natural zeolite has a surface area of about 20 m^2/gram. The dealumination and calcination processing affected to change the surface area and pore volume. Dealumination process caused an increasing in ratio of Si/Al and dissolution of CaO impurity. The increasing ratio SiO_2 to Al_2O_3 leads to change the framework of the zeolite as a result of dissolution in acid solution or reaction and calcination process. The changes in the zeolite framework will lead to change the pores of the zeolite resulting the changes on its surface area and pore volume. Acid treatment will increase the surface area [21, 25].

Table 1. Surface Area on the type Catalysts

Catalyst	Surface Area (m^2/g)	Reference
Natural zeolite	172.929	[25]
Sodium Silicate ($Na_2O.nSiO_2$)	5.91	[26]
Calcium Aluminate	21.29	[27]
$Ca/Al/Fe_2O_3$	25.89	[28]
KI-H-zeolite	27.236	This research
H-zeolite	199.760	This research

H-zeolite and KI-H-zeolite catalyst also was analyzed by using XRD. The diffractogram was presented in Figure 3. This figure shows H-zeolite and KI-H zeolite catalyst has the same XRD patterns. Impregnation process affected in some 2Θ, indicating that KI compound impregnated in H-zeolite. After impregnation process, KI-H-zeolite was dried in oven and was not followed by calcination. The influence of impregnation of KI in the process of a solid catalyst detected on 2-θ with a peak value of 22-24 [29].

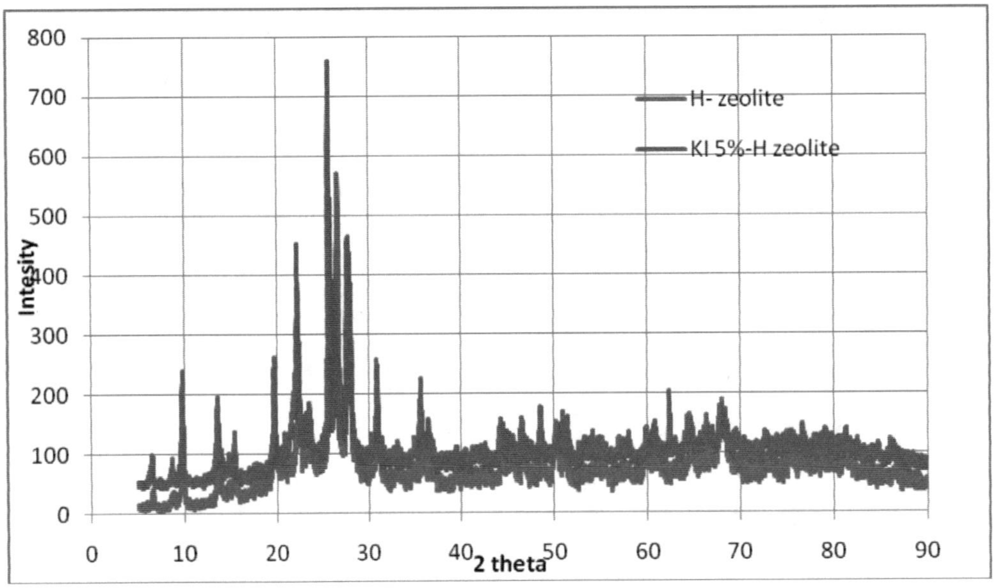

Figure 3. XRD Diffractogram of KI-H-Zeolite and H-zeolite Catalysts.

The others research conducted by Xie and Li (2006) [9] about heterogeneous catalyst impregnated by KI for biodiesel production from soybean oil reported that KI compounds was shown in 2θ between 22-24. Peak at 2θ 23.62 of KI-H-zeolite has higher intensity than H-zeolite. KI-H-zeolite has dominant peak and the highest intensity at 2θ is 25.62. This 2θ is a SiO_2 compound as the main component of natural zeolite from Gunung Kidul. Therefore, in peak at 23.62 referred to KI that impregnated in H-zeolite catalyst.

3.2. Qualitative Analysis

The results of biodiesel products presented in Figure 4. Figure 4(a) is a chromatograph biodiesel standard on retention time at 7.76 to 15.73. Figure 4(b) is a chromatograph for biodiesel that produced by using conventional method for 2 hours at temperature 50 °C. In this research, FAME had a retention time of 7.90 to 12. Intensity in Figure 4(b) is less intense than intensity of figure 4(a). When compared to standard in Widayat et.al., (2013b), the expected results of biodiesel analysis with GC: 8.77 minutes was methyl stearic ($C_{18:0}$); 9.23 minutes was methyl oleic ($C_{18:1}$); 10,28 minutes was methyl linoleic ($C_{18:2}$); and 11.99 minutes was methyl arachidic ($C_{20:0}$). Figure 4(b) has retention time for FAME in 7.90; 8.69; 10.12; 11.36 and 12.20. For the same retention time in Figure 4(b) with Widayat et al (2013b), shown the biodiesel product contain methyl stearate (RT 8.69), 10.12 minutes was methyl linolenic($C_{18:3}$), 11.36 minutes was methyl nonadecanoic ($C_{19:0}$) and 12.20 minutes was methyl eicosanoic ($C_{20:1}$). Although the entire peak was not obvious, but this shows that the FAME was indeed formed in biodiesel sample and triglyceride was not detected again.

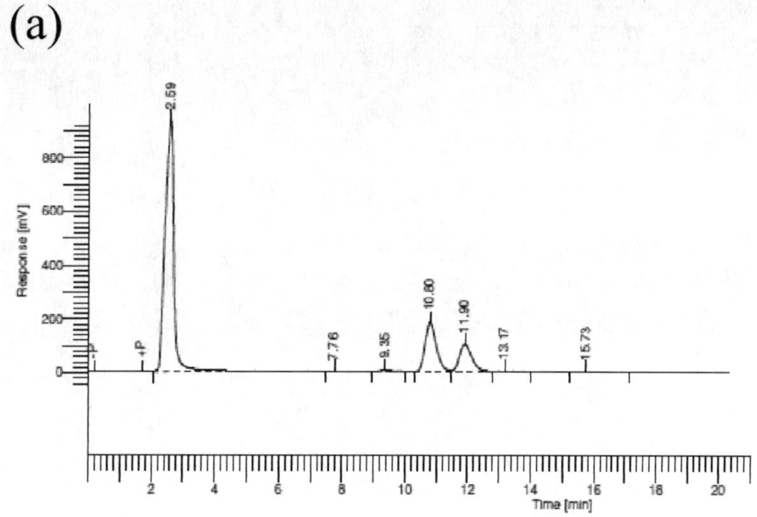

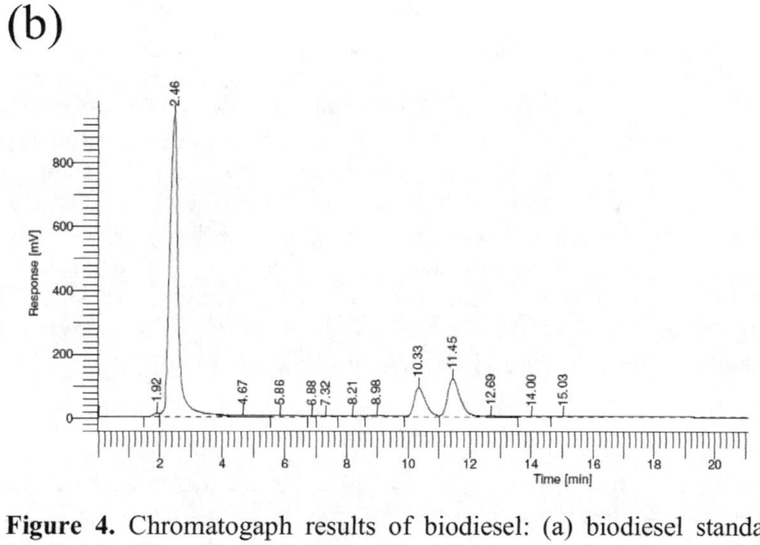

Figure 4. Chromatogaph results of biodiesel: (a) biodiesel standard and (b) biodiesel product by using KIO_3 –H zeolite

3.3. Catalytic Testing for Biodiesel Production

The results of catalyst testing for biodiesel production was presented in figure 5. Figure 5 shown a comparison of biodiesel production with potassium hydroxide (KOH), natural zeolite, KI-H-zeolite and KIO_3-H-zeolite. Biodiesel production with KOH catalyst obtained conversion of 84.16% and H-zeolite as catalyst obtained 56.08%. Biodiesel production with KIO_3 –H zeolite as catalyst obtain the higher 87.91% at catalyst concentration 5%. The results of biodiesel production with KOH catalyst is greater than H-zeolite catalyst with impregnation KI and KIO_3. However, in the case of catalyst separation is easier to use heterogeneous catalyst such H-zeolite. H-zeolite catalysts without impregnation has lower biodiesel conversion than the KI-H-zeolite and KIO_3–H-zeolite. This is because the K^+ ion in KI fill vacant sites in the zeolite pores and further can accelerate dispersion and decomposition of KI from the basic sites on the activation process [9].

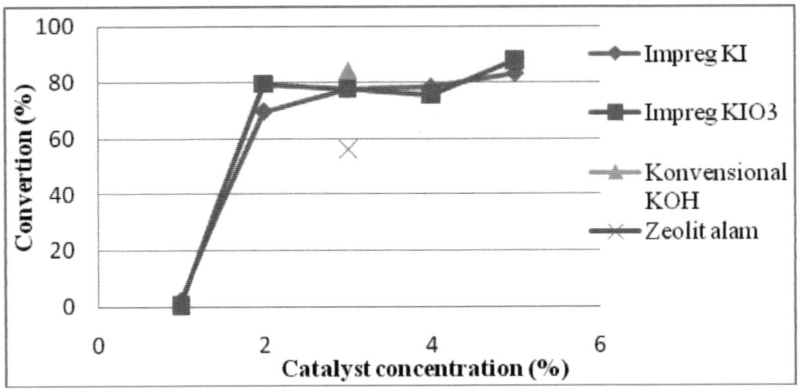

Figure 5. Effect of catalyst concentration on biodiesel conversion.

Table 2. Biodiesel Characterization

Parameter	Product		Overall	SNI 04-7182-2006
	KI	KIO_3		
Density	0.878	0.879	0.866 – 0.9 g/ml	0.85 – 0.89 g/ml
Viscosity	2.79	2.72	2.32 – 10.92 Cst	2.3 – 6 Cst
Acid Number	0.25	0.36	0.19 – 0.42mg-KOH/gr	Max 0.8 mg-KOH/gr

The biodiesel products have density between 0.866-0.9 g/ml. Density of biodiesel products parameter included in the standard density of biodiesel according to SNI04-7182-2006 ie0.85-0.89g/ml. The concentration of KI/KIO_3 in H-zeolite affected the activity in methyl ester reaction. So, KI/KIO_3 concentration affected in quality of biodiesel on parameter includes density, viscosity and acid number. These results are in good agreement with the results reported by Cao et al. (2008) [30].

The result of biodiesel viscosity is out of the range of biodiesel viscosity based on SNI 04-7182-2006. Biodiesel production using KI/KIO_3 and H-zeolite catalyst in small concentration caused lower methyl ester production. This caused viscosity biodiesel higher than biodiesel standard. The acid number of biodiesel is important parameter. Acid number can show possibility of oxidation process to form short-chain acids as a result of the decomposition of peroxide and hydroperoxide compounds. In real application, the accumulation of acid in biodiesel can cause a condition that can damage the engine. Overall, characters such as acid number, density and viscosity of the biodiesel produced using synthesized catalyst of KI/KIO_3-H-zeolite meet the requirements of parameter value of SNI 04-7182-2006.

4. Conclusions

KI/KIO_3-H-zeolite catalyst was synthesized by chemical, physical treatment and impregnation processing. KI–H-zeolite catalyst has surface area of 27.236 m^2/gram. XRD analysis reveals the evidence of the successful KI impregnation in H-zeolite. The obtained parameters of the produced biodiesel meet the parameter of biodiesel standard SNI 04-7182-2006. The chromatograph shows Fatty Acid Methyl Ester (FAME) production in low concentration (below 10%). The biodiesel conversion obtained by KI/ KIO_3-H-zeolite as catalyst reached 87.91%.

5. Acknowledgement
This research was supported by research grant of Kompetensi (HIKOM) from Directorate General of Higher Education, Ministry of National Education Indonesia. Authors would also like to thank Department of Chemical Engineering Polytechnic Malang for Gas Chromatography analysis.

Refference

[1] Widayat W, H Satriadi, O Yuariski and D Murwono 2013 Biodiesel Production From Bulk Frying Oil With *Ultrasound Assisted Research Journal of Applied Sciences Engineering and Technology* **6**(10) pp 1732-1739
[2] Peterson C L and Gregory Möller 2005 Biodiesel Fuels: Biodegradability Biological and Chemical Oxygen Demand and Toxicity on the Biodiesel *Handbook edited by G Knothe Jon Van Gerpen and Jürgen Krahl Champaign Illinois* USA: AOCS Press
[3] Van Gerpen J H 2005 Biodiesel processing and production *Fuel Processing Technology* **86** (10) 1097-1107
[4] Furukawa S, Uehara Y and Yamasaki H 2010 Variables affecting the reactivity of acid-catalyzed transesterification of vegetable oil with methanol *Bioresour.Technol* **101** 3325–3332
[5] Widayat and Suherman 2012 Biodiesel Production from Rubber Seed Oil via Esterification Process *IJRED* **1** (2)
[6] Ramadhas A S, Jayaraj S and Muraleedharan C 2005 *Biodiesel production from high FFA rubber seed oil* Elsevier Ltd 0016-2361/$, 335-339
[7] Deshpande A, Anitescu G, Rice P A and Tavlarides L L 2010 Supercritical biodiesel production and power cogeneration: Technical and economic feasibilities *Bioresour. Techno.* **101** 1834–1843
[8] Sotoft L F, Rong B, Christensen K V and Norddahl B 2010 Process simulation and economical evaluation of enzymatic biodiesel production plant *Bioresour.Technol* 101 5266–5274
[9] Xie W and Li Haitao 2006 Alumina-supported potassium iodide as heterogeneous catalyst for biodiesel production from soybean oil *Journal of Molecular Catalysis A: Chemical* **255** 1–9
[10] Ragavan, S Nivetha and R D Vetha 2011 Transesterification of rubber seed oil by sonication technique for the production of methyl esters *Springer-Verlag* DOI 10.1007/s13399-011-0012-4
[11] Gizine I E D, Shadia A E, Salwa I H and Sohair H 2010 Biodiesel Preparation from Jatropha Seeds by Enzymatic Reactive Reaction and Dimethyl Carbonate *International Review of Chemical Engineering (IRECHE)* **2** n 6 pp 670-682
[12] Widayat, Wibowo A D K, Hadiyanto and Jos B 2012 Study On Production Process Of Biodiesel From Rubber Seed (*HeveaBrasiliensis*) By In Situ (Trans)esterification Method, *IREME* Vol **6**
[13] Widayat, Wibowo A D K and Hadiyanto 2013b Study On Production Process Of Biodiesel From Rubber Seed (Hevea Brasiliensis) By In Situ (Trans)esterification Method With Acid Catalyzed *Energy Procedia*
[14] Intarapong, P K Noiroj, A Luengnaruemitchai and S Jai-In 2009 A comparative study of KOH/Al_2O_3 and KOH/NaY catalysts for biodiesel production via transesterification from palm oil *Renew. Energy* vol **34** pp 1145–1150
[15] Krawczyk T 1996 Biodiesel-Alternative fuel makes in road but hurdles remain *INFORM* **7** pp 801-829
[16] Kalam M A and Masjuki H H 2002 Biodiesel from palmoil-an analysis of itsproperties and potential *Biomass and Bioenergy* **23** p 471-479
[17] Dennis Y C, Xuan W and M K H Leung 2009 A Review on Biodiesel Production Using Catalyzed Transesterification *Applied Energy* **87** 2010 1083–1095
[18] Boz, N N D dan D M Kalyon 2009 Conversion of biomass to fuel: Transesterification of vegetable oil to biodiesel using KF loaded nano-g-Al2O3 as catalyst, *Applied Catalysis B Environmental* **89** pp 590–596
[19] Furuta S H and Matsuhashidan K A 2004 Biodiesel fuel production with solid superacid catalysis in fixed bed reactor under atmospheric pressure *Catalysis Communications* **5** pp 721–723

IOP Conf. Series: Materials Science and Engineering **107** (2016) 012044 doi:10.1088/1757-899X/107/1/012044

[20] Yan S, M Kim, S O Salley and K Y Simon Ng 2009 Oil transesterification over calcium oxides modified with lanthanum *Applied Catalysis A: General* **360** 163–170

[21] Widayat R A and Rachimoellah M 2009 The Effect of Time dealumination and Solvent Concentration in Synthesis of Zeolite Catalyst andCatalytic Test for Diethyl Ether Production Process *AIP Proceeding Internation Workshop on Advanced Material for New and Renewable Energy* editor L T handoko and Masbah R T S **1169** 106-112

[22] Yizhou H, Rongxian B and Yisheng T 2004 Study on the carbon dioxide hydrogenation to iso-alkanes over Fe–Zn–M/zeolite composite catalysts *Fuel Process Technol* **86** pp 293–301

[23] Marchetti J M and Errazu A F 2008 Comparison of different heterogeneous catalysts and different alcohols for the esterification reaction of oleic acid. *Fuel* **87** pp 3477–80

[24] Krishnan B and Richard D G 1993 *Journal of Catalyst* **144** 395

[25] Widayat R A and Rachimoellah M 2010 Pengaruh Waktu Dealuminasi dan Jenis Sumber Zeolit Alam terhadap Kinerja H-Zeolit untuk Proses Dehidrasi Etanol *Reaktor* Vol **13** No 1 pp 51-57

[26] Guo F, Peng Z G, Dai J Y and Xiu Z L 2010 Calcined sodium silicate as solid base catalyst for biodiesel production *Journal of Fuel Processing Technology* **91** p 322-328

[27] Tang S, Wang L, Zhang Y, Li S, Tian S and Wang B 2012 Study on preparation of Ca/Al/Fe_3O_4 magnetic composite solid catalyst and its application in biodiesel transesterification *Journal of Fuel Processing Technology* **95** p 84-89

[28] Shengyang H, Yanping G, Wang Y and Heyou H 2011 Nano-magnetic catalyst KF/CaO-Fe3O4 for biodiesel production *Applied Energy* **88** p 2585-2690

[29] Jairam S, Kolar P, Sharma S R, Osborne J A and Davis J P 2012 KI-impregnated oyster shell as a solid catalyst for soybean oil transesterification *Bioresource Technology* **104** 329–335

[30] Cao F Y, Chen F, Zhai J, Li J, Wang X, Wang S,Wang W and Zhu 2008 Biodiesel production from high acid value waste frying oil catalyzed by superacid heteropolyacid *Biotechnol* **101** 93–100

Nanocellulose prepared by acid hydrolysis of isolated cellulose from sugarcane bagasse

W T Wulandari*, A Rochliadi and I M Arcana
Inorganic and Physical Chemistry Research Division, Institut Teknologi Bandung, Jl. Ganesha 10, Bandung 40132-Indonesia

*E-mail: windatrisnawulandari@yahoo.com

Abstract. Cellulose in nanometer range or called by nano-cellulose has attracted much attention from researchers because of its unique properties. Nanocellulose can be obtained by acid hydrolysis of cellulose. The cellulose used in this study was isolated from sugarcane bagasse, and then it was hydrolyzed by 50% sulfuric acid at 40 °C for 10 minutes. Nanocellulose has been characterized by Transmission Electron Microscope (TEM), Particle Size Analyzer (PSA), Fourier Transform Infrared Spectroscopy (FTIR) and X-Ray Diffraction (XRD). Analysis of FTIR showed that there were not a new bond which formed during the hydrolysis process. Based on the TEM analysis, nano-cellulose has a spherical morphology with an average diameter of 111 nm and a maximum distribution of 95.9 nm determined by PSA. The XRD analysis showed that the crystallinity degree of nano-cellulose was higher than cellulose in the amount of 76.01%.

1. Introduction

Cellulose is one of the most abundant, natural, renewable and biodegradable polymers [1, 2]. Cellulose can be found in wood, cotton, rice straws, and sugarcane bagasse [2-5]. Sugarcane bagasse is a residue from the sugar and alcohol industry, and it was produced in large quantities every year. The sugar production process utilizes sugarcane bagasse as a fuel for the sugar mill. Some industry uses sugarcane bagasse as a raw material for electric generation, paper production, fermented products, and being developed for biomass [3].

Sugarcane bagasse contains 40-50% cellulose, much of which is in the crystalline structure. Another component in sugarcane bagasse is hemicellulose as much as 25-35% which is an amorphous polymer and mainly composed of xylose, arabinose, galactose, and mannose. The rest is mostly lignin with about 18-24% [3,6].

Cellulose is a polysaccharide composed of D-glucopyranose linked by 1,4-β glycoside bond. There are three hydroxyl groups in cellulose; secondary -OH at the C-2 and at the C-3, primary -OH at the C-6 position. [3,7,8].These hydroxyl groups in cellulose have an important role in the compactness of the crystalline structure and determines physical properties of cellulose. At the plant fiber, cellulose can be found in an amorphous phase which are connected with crystalline phase through intermolecular and intramolecular hydrogen bonds. This is causing cellulose not to melt the thermal degradation temperature [3,9,10]. Cellulose has a good mechanical properties, low density, and biodegradable. There are several types of cellulose (I, II, III, IV and V), cellulose I and II are commonly found in nature. Cellulose I has a best mechanical properties and a parallel chain orientation, while cellulose II has anti-parallel chain [3,11,10].

Cellulose is a natural polymer which insoluble in water because it has a long chain and high molecular weight (more than 500.000 Da). Cellulose in nanometers or nanocellulose has a size range from 10 nm to 350 nm. Nanocellulose has a higher surface area than cellulose, so nano-cellulose can be one of the alternative solutions to make cellulose more easily to modified and dissolved in water [2,3,5].

Various methods can be used to obtained nano-cellulose, such as acid hydrolysis, ultrasonic technique, and enzymatic hydrolysis [1,4,13,14]. The method that most widely used is acid hydrolysis [2]. This method is easy and fast to produce nanocellulose that has a better properties. Some researchers have reported that crystallinity index of nano-cellulose produced by acid hydrolysis was higher than other methods. The nano-cellulose obtained from the acid hydrolysis also has a smaller size. These are the reasons that acid hydrolysis method is selected to obtained nano-cellulose.

A Strong acid such as H_2SO_4 and HCl is commonly used to break the glycoside bonds in cellulose. There are several steps in acid hydrolysis: (1) strong acid hydrolysis of cellulose under controlled conditions such as acid concentration, time, temperature, and ratio of the acid to cellulose; (2) dilution with some water to stop hydrolysis process and repeated washing with successive centrifugation; (3) dialysis which aims to fully remove free acid molecules; (4) sonication in order to form a stable suspension of nano-cellulose; (5) drying of the suspension to yield solid nano-cellulose [13,15].

Nanocellulose has unique properties such as low density, biodegradable, and good mechanical properties. In addition, nano-cellulose is also easily modified and has high surface area and typical morphology [1,2]. Nanocellulose with a small particle size and high crystallinity has a better properties and usually used as nanofiller for some polymer materials.

Nanocellulose is used in the various field such as a barrier in the separation process of hazardous waste, food wrappers which replace the non-biodegradable plastics, and as nanocomposite to improve properties such as mechanical, thermal, ionic conductivity properties of the polymer [2,16].

2. Experimental

2.1. Materials
Sugarcane bagasse was collected from plantations located at Tasikmalaya, West Java, Indonesia. Other reagents used were: sodium hypochlorite; sodium hydroxide; sulfuric acid.

2.2. Methods

2.2.1. Isolation of cellulose. Sugarcane bagasse was dried in sunlight and then cut into small pieces. The cut bagasse was milled to become powder. The powder of bagasse was bleached with 250 ml of 0.735% (w/v) sodium hypochlorite for 6 hours with constant stirring at 45 °C to remove the lignin. The residue was washed with distilled water until a neutral pH. The neutral residue was refluxed with 150 ml of 17.5% sodium hydroxide for 3 hours with constant stirring at 45 °C to remove hemicellulose. The residue of this process was also washed until reach a neutral pH, and it was dried at room temperature for 2-3 days.

2.2.2. Preparation of nano-cellulose. Isolated cellulose from sugarcane bagasse was hydrolyzed with sulfuric acid with a ratio of cellulose to sulfuric acid 1:25. The hydrolysis of cellulose was carried out in several variations of sulfuric acid concentration, temperature, and time. The hydrolysis process was quenched by adding 10-fold excess distilled water (250 ml) to the reaction mixture. A colloidal suspension which produced was centrifuged at 6500 rpm for 30 minutes. Then, it was dialyzed for 5 days to neutralize and eliminate the sulfate ions. The neutral colloidal suspension was sonicated for 10 minutes to homogenize the generated nano-cellulose.

2.3. Characterization

2.3.1. Particle size analyzer. The particle size of nano-cellulose were determined by Particle Size Analyzer (PSA) with a Beckman Coulter Delsa™ Nano. The nano-cellulose was diluted with distilled water and inserted into the cuvette. The measurements were done at room temperature with the measurement range from 10 nm until 4000 nm.

2.3.2. Morphological analysis. The morphology measurements of the nano-cellulose were carried out on a Transmission Electron Micrograph (TEM), JEOL JEM 1400. The sample of nano-cellulose was measured in colloidal form with distilled water as a solvent.

2.3.3. Infrared spectroscopy (IR) analysis. The FTIR spectra of cellulose and nanocellulose were recorded on Fourier Transform Infrared Spectrometer Shimadzu Prestige 21. The samples were blended with KBr powder and then pressed into thin pellets. The sample was measured in the wavelength range from 4000 cm^{-1} to 500 cm^{-1}.

2.3.4. X-ray diffraction (XRD) analysis. The X-ray diffraction data were collected using PW 1830 Diffractometer. This measurement was performed using Cu Kα with a step size 0.02, and the crystallinity index (CI) was calculated using the equation (1)[17], by measuring the peak height of the crystalline region (I_{002}) and the amorphous region (I_{am}) [18].

$$C_I(\%) = \frac{I_{002} - I_{am}}{I_{002}} x100\% \quad (1)$$

I_{002} is the intensity value for the crystalline cellulose (2θ = 22.5° for cellulose I and 2θ = 20.1° for cellulose II), and I_{am} is the intensity value for the amorphous cellulose (2θ = 18° for cellulose I and 2θ = 16.3° for cellulose II) [3,19,20].

3. Results and discussion

3.1. Infrared spectroscopy (IR) analysis

FTIR spectroscopy was used to showed that the lignin and hemicellulose have been removed during cellulose isolation process through analysis of its functional groups. The FTIR spectra of sugarcane bagasse and isolated cellulose are shown in figure1.

Based on FTIR spectrum, there are several peaks in sugarcane bagasse which is not found in the spectrum of cellulose. The peaks are around 1252 cm^{-1}, 1511 cm^{-1}, and 1738 cm^{-1}. The absorption peak of 1252 cm^{-1} is derived from C-O stretching vibration of aryl group in lignin. The peak at 1511 cm^{-1} is attributed to the C=C stretching vibration of the aromatic ring in lignin. The C=O stretching vibration of carboxylic groups of hemicellulose and lignin is around 1738 cm^{-1} [3,21,22].

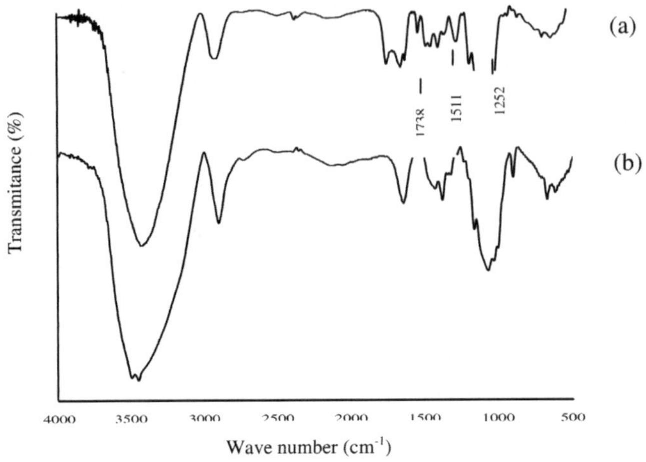

Figure 1. FTIR spectra of (a) sugarcane bagasse (b) cellulose

There are several factors which influence hydrolysis of cellulose such as acid concentration, temperature, and time. This work has tried some variation conditions to produce nanocellulose; the result shows that nano-cellulose was obtained in two conditions, firstly in 60% of sulfuric acid at 40 °C for 5 minutes (nano-cellulose A) and 50% of sulfuric acid at 40 °C for 10 minutes (nano-cellulose B). Both of nano-cellulose were characterized by FTIR, PSA, XRD, and TEM.

The FTIR spectra of nano-cellulose A and B are showed in Fig.2. Both of nano-cellulose FTIR spectra are not different with cellulose. It indicates that there are not new bonds formed during hydrolysis process.

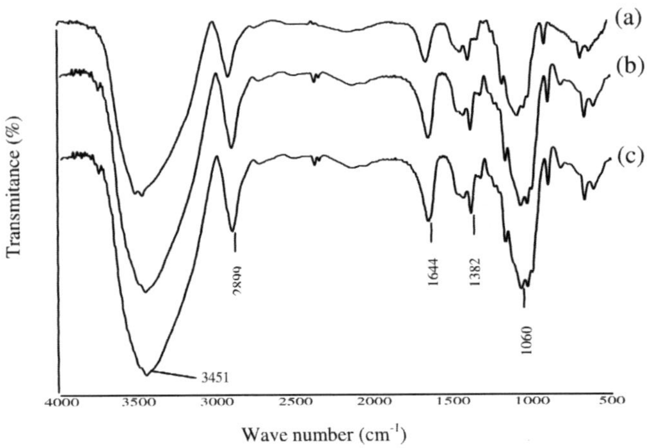

Figure 2. FTIR spectra of (a) cellulose; (b) nano-cellulose A; (c) nano-cellulose B

The absorption peaks at 3451 cm^{-1} and around 2899 cm^{-1} were attributed to the O-H and C-H stretching vibrations, respectively. The peak absorption at 1644 cm^{-1} was reported as the O-H vibration of absorbed water. The peak for C-H and C-O vibrations contained in the polysaccharide rings of cellulose is around 1382 cm^{-1}. The vibration of C-O-C in pyranose ring indicated by the absorption peak at 1060 cm^{-1} [3,5].

IOP Conf. Series: Materials Science and Engineering **107** (2016) 012045 doi:10.1088/1757-899X/107/1/012045

3.2. Particle sizes analysis

The particle size analysis shows that nano-cellulose A has an average diameter of 196.7 nm with the maximum distribution of 148.4 nm while nano-cellulose B has an average diameter of 111 nm and the maximum distribution in 95.9 nm. The results of the particle size analysis are shown in figure 3 and figure 4.

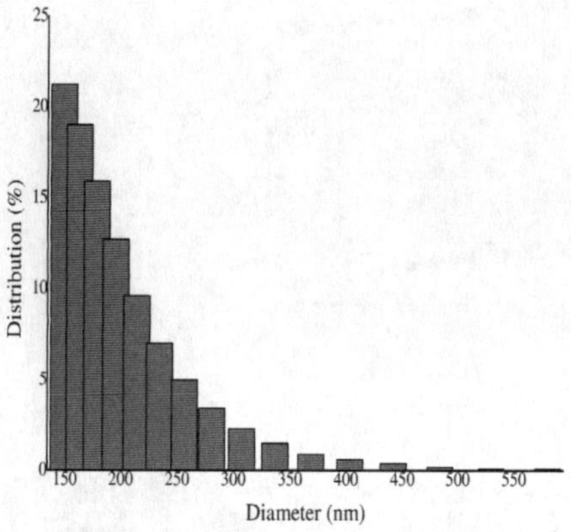

Figure 3. Particle size distribution of nanocellulose A

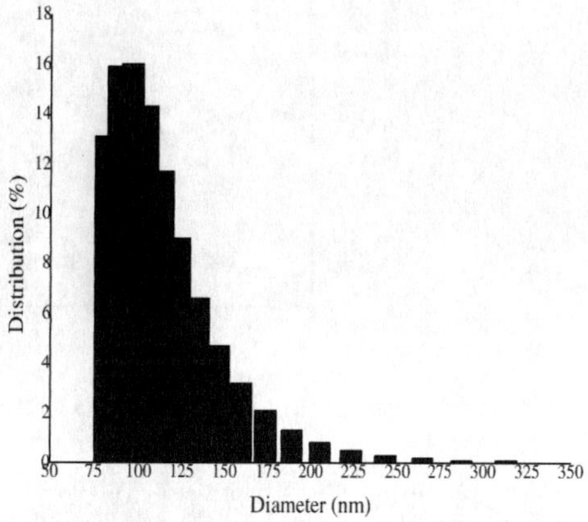

Figure 4. Particle size distribution of nanocellulose B

Nanocellulose A has the bigger size than nano-cellulose B, due to the hydrolysis time to produce nano-cellulose B was longer than nano-cellulose A so that the chain of cellulose B was more hydrolyzed than cellulose A and generates a smaller particles size for nano-cellulose B. The results of PSA indicate that acid concentration does not significantly influence in particle size of nano-cellulose. This is evidenced by the particle size of nanocellulose A is bigger than nano-cellulose B, whereas the acid concentration to produce nano-cellulose A was higher than nano-cellulose B (60% of sulfuric acid for nano-cellulose A and 50% for nano-cellulose B). Some studies have reported that acid concentration influenced on the value of index crystallinity of nano-cellulose, the higher of acid concentration used during hydrolysis process produces the nano-cellulose with the lower crystallinity index [13].

3.3. X-ray diffraction analysis

The crystallinity index of isolated cellulose, nanocellulose A, and B was analyzed by *X-Ray Diffraction (XRD)*. The diffractograms are shown in Fig. 5. Cellulose and both of nano-cellulose have three diffraction peaks; they are $2\theta = 12.2$ °; 20.1 °; 22°. The peaks are typical for cellulose II, which has antiparallel structure. The crystallinity index for cellulose and nanocellulose was calculated by using equation (1) with I_{200} at $2\theta = 20.1°$ and I_{am} at $2\theta = 16.3°$, the results are shown in Table. 1.

The crystallinity index for isolated cellulose from sugarcane bagasse is about 70.62%. Nanocellulose A has the lower index crystallinity in the amount of 67.83% while the crystallinity index for nano-cellulose B is 76.01%. This difference is due to the different of acid concentrations used in hydrolysis process, nano-cellulose A by using 60% of acid concentration while nano-cellulose B by using acid concentration of 50%. The higher acid concentration produces nano-cellulose with lower crystallinity, and it indicates that the addition of acid with higher concentration is not only to break the amorphous region of the cellulose but also to damage parts of the crystalline structure.

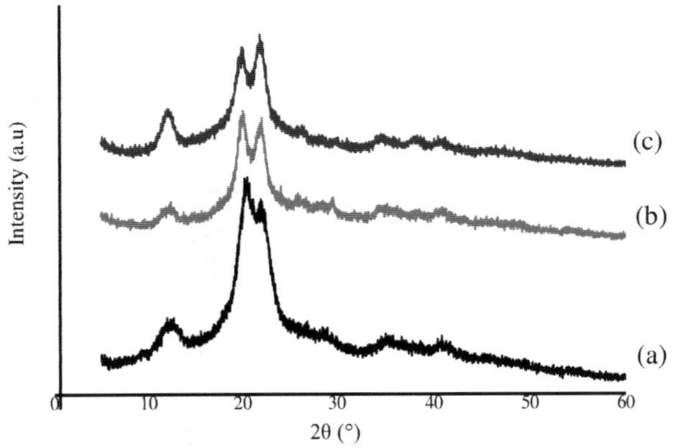

Figure 5. X-ray diffraction pattern of (a) cellulose; (b) nanocellulose A; (c) nanocellulose B

Cellulose has amorphous and crystalline parts; the amorphous will be more vulnerable to hydrolysis process. The hydrolysis by using 50% of acid concentration will break the amorphous region of cellulose to produce nano-cellulose B with higher crystallinity index, however when the acid concentration that used is higher such as 60%, the crystalline part can be damaged during hydrolysis process, and the crystallinity index is decreased.

Table 1. Crystallinity index of cellulose and nanocellulose

	Crystallinity Index (%)
Cellulose	70.62
Nanocellulose A	67.83
Nanocellulose B	76.01

3.4. Morphological analysis

The morphology of nano-cellulose A and B was characterized by Transmission Electron Microscope (TEM). The results can be seen in figure 6 and figure 7.

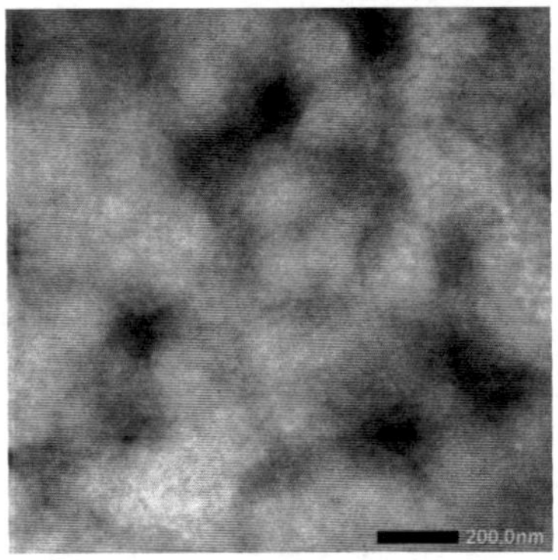

Figure 6. Morphology structure of nanocellulose A

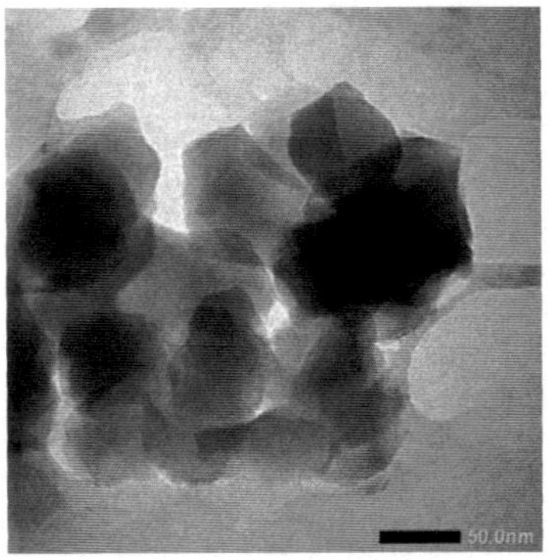

Figure 7. Morphology structure of nanocellulose B

Based on the results of TEM, both of produced nano-cellulose have a spherical morphology, but nano-cellulose A has less clear morphologies than nano-cellulose B. This shows that nano-cellulose A has more amorphous region than nano-cellulose B. It is correlated with the crystallinity index of nano-cellulose that has discussed before.

4. Conclusion

Nanocellulose can be obtained from isolated cellulose from sugarcane bagasse and followed by sulfuric acid hydrolysis. There are two conditions used in this report to produce nano-cellulose, that are sulfuric acid concentration of 60% of at 40 °C for 5 minutes for nano-cellulose A and 50% at 40 °C for 10 minutes for nano-cellulose B. Nanocellulose B has better properties than nano-cellulose A. Nanocellulose B has the average diameter of 111 nm with the maximum distribution of 95.9 nm. The crystallinity index of nano-cellulose B is higher than the isolated cellulose in the amount of 76.01%. The result of TEM analysis shows that nano-cellulose B has a spherical morphology.

References

[1] Salas C, Nypelö T, Abreu C R, Carrillo C and Rojas O J 2014 *Current Opinion in Colloids & Interface Science.* **19** 383-396
[2] Li W, Yue J and Liu S 2012 *Ultrasonics Sonochemistry.* **19** 479-485
[3] Mandal A and Chakrabarty D 2011 *Carbohydrate Polymers.* **86** 1291-1299
[4] Habibi Y, Lucia L A and Rojas O J 2010 *Chemical Reviews.* **110** 3479-3500
[5] Li J, Wei X, Wang Q, Chen J, Chang G, Kong L, Su J and Liu Y 2012 *Carbohydrate Polymers.* **90** 1609-1613Kadla J F, Gilbert R D 2000 *Cellulose Chemistry and Technology.* **34** 197
[6] Jacobsen S E and Wyman C E 2002 *Industrial and Engineering Chemistry Research.* **41** 1454
[7] Kadla J F and Gilbert R D 2000 *Cellulose Chemistry and Technology*. **34** 197
[8] Khalil H P S A, Bhat A H and Yusra I 2012 *Carbohydrate Polymers.* **87** 963-979
[9] Fengel D and Wegner G 1989 *Wood-chemistry, Ultrastructure, Reactions* (Berlin, New York: Walter de Gruyter)
[10] Klemm D, Heublein B, Fink H P and Bohn A 2005 *Angewandte Chemie-International Edition.* **44** 3358-3393
[11] Lenholm H, Iversen T 1955 *Nordic Pulp & Paper Research Journal.* **10** 104
[12] Northolt M G, Boerstoel H, Maatman H, Huisman R, Veurink J and Elzerman H 2001 *Polymer.* **42** 8249-8264
[13] Brinchi L, Cotana F, Fortunati E and Kenny J M 2013 *Carbohydrate Polymer.* **94** 154-169
[14] Wang N D and Cheng R 2007 *Polymer.* **48** 3486-3493
[15] Dong X M, Revol J F, and Gray D G 1998 *Cellulose.* **5** 19-32
[16] Favier V, Chanzy H and Cavaille J Y 1995 *Polymer Advanced and Technology.* **6** 351-355
[17] Segal L, Creely J, Martin Jr A E and Conrad C M 1959 *Textile Research Journal.* **29** 786-794
[18] Park S, Baker J O, Himmel M E, Parilla P A and Johnson D K 2010 *Biotechnology for Biofuels.* **3** 1-10
[19] Cho M J and Park B D 2011 *Journal of Industrial and Engineering Chemistry.* **17** 36-40
[20] Man Z, Muhammad N, Sarwono A, Bustam M A, Kumar M V and Rafiq S 2011 *Journal Polymer Environment* **19** 726-731
[21] Sain M and Panthapulakkal S 2006 *Industrial Crops and Production.* **23** 1-8
[22] Garside P and Wyeth P *Studies in Conservation* **48** 269-275

Laboratory investigation on the role of tubular shaped micro resonators phononic crystal insertion on the absorption coefficient of profiled sound absorber

I Yahya*, J I Kusuma, Harjana, R Kristiani, and R Hanina
The Iwany Acoustics Research Group (iARG), Physics Dept., Sebelas Maret University, Jl. Ir. Sutami 36 A Kentingan Surakarta 57126 Indonesia

*E-mail: iyahya@mipa.uns.ac.id

Abstract. This paper emphasizes the influence of tubular shaped microresonators phononic crystal insertion on the sound absorption coefficient of profiled sound absorber. A simple cubic and two different bodies centered cubic phononic crystal lattice model were analyzed in a laboratory test procedure. The experiment was conducted by using transfer function based two microphone impedance tube method refer to ASTM E-1050-98. The results show that sound absorption coefficient increase significantly at the mid and high-frequency band (600 – 700 Hz) and (1 – 1.6 kHz) when tubular shaped microresonator phononic crystal inserted into the tested sound absorber element. The increment phenomena related to multi-resonance effect that occurs when sound waves propagate through the phononic crystal lattice model that produce multiple reflections and scattering in mid and high-frequency band which increases the sound absorption coefficient accordingly

1. Introduction

Surface modification techniques were commonly used for improving the performance of sound absorbers and sound diffusers. In regards to this purposes, the use of micro-perforated panel (MPP) occurs mostly in many reports and publications. The idea of MPP was initially provided by Maa in 1975 and has been a growing interest among researchers due to its ability on absorbing sound energy in a large frequency band [1,2]. Later many different approach and technique were proposed by other researchers to fulfill engineering requirement, especially in low-frequency range.

The use of multi-layer MPP with air banks between the layers [3], flexible membrane insertion [4], cavity, backed MPP by using Helmholtz resonators and compartmented cavities [5,6] and MPP with extended tubes was applied in various sound absorber design.That mentioned approach gave some advantages not only on shifting the absorption to the lower band but increase MPP performance at the mid and high-frequency range as well. The use of multi-layer MPP with different perforation ratio in its layers, for example, increase the total oscillating mass when sound waves are passing through perforations. It converts the sound waves energy into heat and the same time coupled cavity resonance effect due to the air banks occurs. The slightly different phenomena existed with the cavity backed MPP and extended tubes design where the absorption are related to the changes of resonator cavity and the length of extended tubes respectively. Coupled cavity resonance and extended viscous damping occur on the insertion of the flexible perforated membrane. When the membrane oscillates on its resonant modes, sound wave energy dissipated through viscous damping in a more effective way due to the increase amplitudes of the resonant modes. In the same time, the flexible membrane also works as an additional spring to the resonator cavity. It changes the reactance and produce higher absorption in the mid frequency range.

Later a new special low-frequency absorption technique has been growing rapidly in this decade. Resonant properties a new kind of artificial material known as phononic crystals or locally resonant sonic materials was reported recently including multilayer split tube and Helmholtz resonator based structure [7-9]. Researchers found that in such artificial crystal lattice exist some frequency ranges that classical waves were not transmitted. The Periodic arrangement of the crystal lattice and the properties of scatterers causing local resonance effect. Constructive or destructive interference occurs depending on the frequency of the waves [10]. It gives possibilities for fundamental functional and artificial acoustic materials design for shielding, filtering and guiding.

The insertion of tubular shaped microresonators and its impact on the absorption of profiled sound absorber reported in this paper. Two different profiled sound absorber model with two, five and seven wedge-shaped element on the top surface were investigated through laboratory experiments. The sound absorption performance test was conducted by using impedance tube technique refer to ASTM E1050 standard.

2. Experiment

Profiled surface cube-shaped sound absorber model used in the experiment presented in Figure 1. It made from 2 mm thickness hard paper with the dimension (50 x 50 x 50) mm in its long, wide and height respectively. There is two type of wedges shaped element were attached to the sound absorber, i.e., opened and closed element. Stacking of tubular shaped plastics microresonator with the dimension as presented in Table 1 was inserted vertically into the sound absorber model. The influence of horizontal micro-resonator lattice configuration, simple cubic (SC) and body centered cubic (BCC) also investigated.

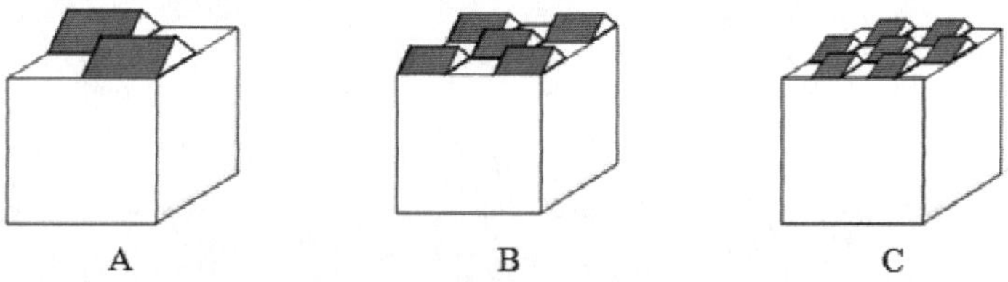

Figure 1. Three different cubes shaped sound absorber model used in the experiment.

Table 1. The tubular microresonator parameters

Sample	Length (mm)	Diameter (mm)
A	20	2
B	20	3
C	30	2
D	30	3
SC	50	5
BCC	50	5

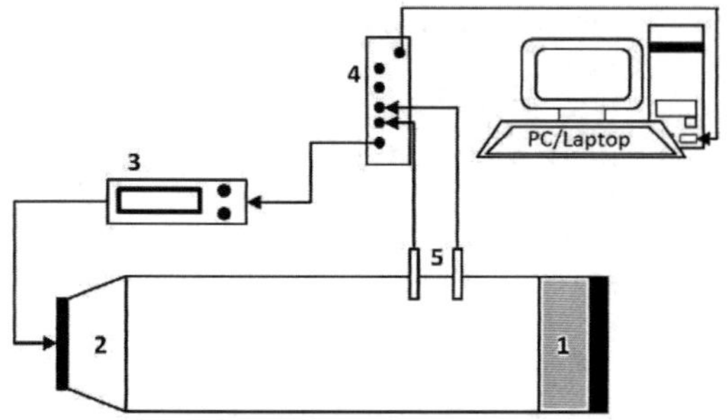

Figure 2. Laboratory set up for sound absorption refer to ASTM E-1050.

The laboratory test set up for of sound absorption measurement by using B&K 4206 impedance tube technique illustrated in Figure 2. Random noise generated from an asignal generator inside B&K Pulse (4) and amplified (3). This amplified sound to be emitted from the loudspeaker (2) inside the tube while test sample (1) placed in the opposite position at the far end of the tube. The two microphones (5) captures both incidents and reflected waves to be analyzed through transfer function based spectral decomposition method. The absorption coefficient than was given by equation (1),

$$\alpha = 1 - |R|^2 \tag{1}$$

where R is the reflection coefficient of the test sample.

3. Result and Discussion

The influence of opened and closed profiled surface to the sound absorption coefficient presented in Figure 3.

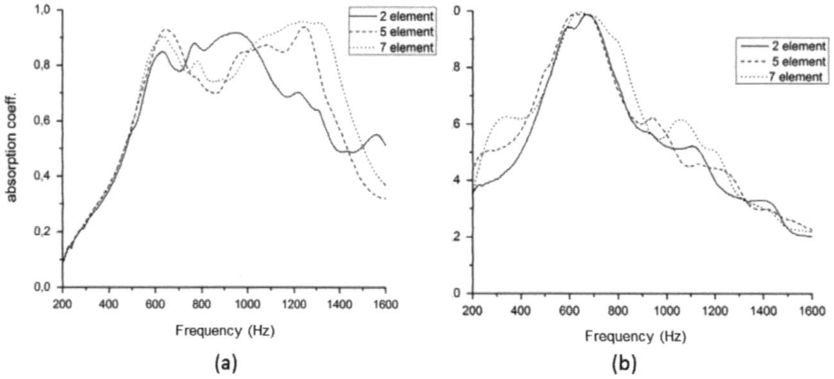

Figure 3. The influence of top sound absorber profile: (a) opened (b) closed.

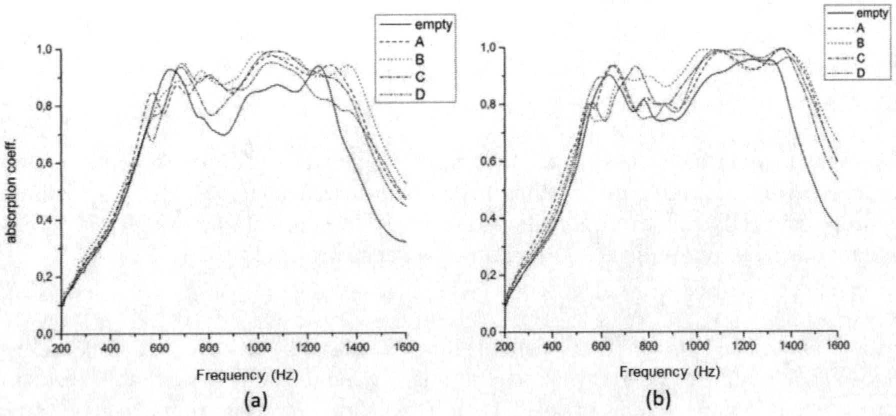

Figure 4. Sound absorption due to microresonator insertion: five (a) and (b) seven element.

The increasing number of profiled element gives a better results to the sound absorption of the sound absorber model. It is clear from Figure 3 the sound absorption increased with the number of profiled wedges. The better performance at higher frequency band for the opened attachment element (a) occurs since the top layer vibrates and oscillate the air resulting resonance and viscous damping effect at the tips of the opened element. Meanwhile, the closed element increased the cavity volume of the sound absorber (b) which takes into account on increasing the sound absorption in the low-frequency range.

Overall sound absorption increment occurs when the tubular shaped microresonators inclusion fitted inside the sound absorber model. One can see from Figure 4 that the inserted tubular shaped resonators with the various dimensions increase the sound absorption at the all frequency range both for the sound absorber model with five (a) and seven (b) profiled elements. The changes in impedance due to microresonator configuration and dimensions resulting multi-local resonance on each of the microresonator elements. According to Zao et al.[7], these local resonance occurs when the incident sound waves scattered from the opened profiled surface and then guided into propagation along the tubular shaped microresonators. Exponential energy losses then occur on this kind of guided waves since energy dissipation increase exponentially with the propagation distance.

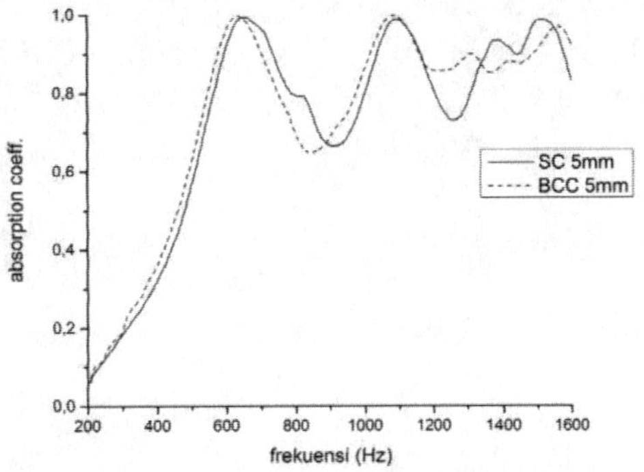

Figure 5. Sound absorption due to SC and BCC lattice model.

The lattice model also investigated in this research. One can see from Figure 5 the lattice structure simple cubic (SC) and body centered cubic (BCC) resulting multiple scattering which increases-the energy dissipation at higher frequency band. These multiple scattering occurs since the lattice normal to the incident wave. But one knows that the incident waves propagates to the lattice structure as oblique incident waves since it is already scattered when wavefront hit the tips of profiled elements on the top of sound absorber model. This result gives the possibility for generating an additional local resonant modes due to the dimension of the microresonators or phononic crystal elements and the lattice constant. It means that functional artificial acoustic materials could be developed based on an array of tubular shaped microresonators to produce a new type of multi-local resonant material that fulfill specific engineering requirement. Further investigation and analysis need to conduct to get better understanding especially on the multiple scattering effect and transmission properties of the multi-local resonant material for application in the low-frequency range.

4. Conclusion

The profile sound absorber with tubular shaped micro resonator inclusion has a better sound absorption performance compared to the empty model without inclusion. Local resonance and multiple scattering occur on the opened profiled surface sound absorber models. The structure of tubular shaped microresonators works as an array of the wave guide that increase energy dissipation at multiple frequencies according to its length and dimensions. In addition, multiple scattering due to the lattice structure model increases energy loss in the high-frequency band. It gives possibilities for development of functional artificial local resonance based sound absorber which is different to the existing MPP based sound absorption techniques.

Acknowledgement
This research funded by Mandatory Research Grant, Research Institute (LPPM) of Sebelas Maret University, 2015.

References

[1] Dah-You M 1975 *Scientia Sinica* **18**(1) 55
[2] Dah-You M 1987 *Noise Control Eng. Journal* **29**(3) 77
[3] Sakagami K, Nakamori T, Morimoto M and Yairi M 2009 *Applied Acoustics* **70** 703
[4] Harjana and Yahya I 2014 *Proc. 7th Int. Conf. on Physics and Its Aplication*vol **1** (Solo, Indonesia: Atlantis Press) p 150
[5] Yahya I and Harjana 2013 *Proc. Int. Cong. on Sound and Vibration* **1** (Bangkok Thailand: Curran Associates Inc.) p 2261
[6] Wang C and Huang L 2011 *J. Acoust. Soc. Am.* **130** (1) 208
[7] Zao H, Wen J, Yu D and Wen X 2010 *J. of App. Phys.* **107** 023519-1
[8] Jing L, Wu J H, Guan D and Gao N 2014 *J. of App Phys.* **116** 103514-1
[9] Li J B, Wang Y S and Zhang C 2013 *J. of Vib and Acoustic* **135** 021015-1
[10] Deymier P A 2013 *Acoustic Metamaterials and Phononic Crystals* (New York: Springer)

IOP Conf. Series: Materials Science and Engineering **107** (2016) 012047 doi:10.1088/1757-899X/107/1/012047

Preparation and properties of arenga starch-chitosan based edible film

Putri Rahmawati, Dwi Ni'maturrohmah, Putri Luthfiana Sari, Putri Puspita Sari, Tetty Nur Indah Sari and Venti Suryanti*

Department of Chemistry, Faculty of Mathematics and Natural Sciences, Sebelas Maret University, Jl. Ir. Sutami 36A Surakarta, Central Java 57126 Indonesia

*E-mail: venty@mipa.uns.ac.id

Abstract. The main objective of this research was to prepare for edible film on the basis of arenga starch-chitosan. Carboxymethyl cellulose (CMC) was added to the blend and glycerol was used as a plasticizer. The tensile strength and water uptake of the film decreased with the increasing of glycerol content in the blend. Contrary, increasing glycerol content in the blend enhanced the elongation and swelling area of the film. The film C1, which has 6% of glycerol in the blend has a tensile strength of 124 KPa, elongation of 6.55%, swelling area of 21% and water uptake of 48%. The film C2, which has 10% of glycerol in the blend has a tensile strength of 58 KPa, elongation of 8.35%, swelling area of 32% and water uptake of 5%.

1. Introduction

An edible film could be defined as a thin layer of edible material which are commonly used for food packaging. Edible films are biodegradable and environmentally friendly. They have other advantageous related with their usage. They are able to extend the shelf life of food since they protect food from deterioration by microbial, oxygen, water and light. Furthermore, edible films also have been applied to control food contamination and enhance sensory perceptions [1-3].

Edible films are produced from renewable materials. The biopolymeric materials commonly applied to produce edible films are polysaccharides, lipids, proteins, or the combination of these components [4-7]. Although edible films have been used widely for coating of fresh fruits, vegetables, and processed foods, they have disadvantages in mechanical or barrier properties and are expensive. Research has shown increased concerning the preparation of edible films from diverse sources to produce films with different characteristics [8-11].

Polysaccharides which are frequently utilized for the preparation of edible films are agar, alginate, carrageenans, chitosan, gum, pectin and starch [3-4]. Amylose is responsible for the film-forming capacity of starch. Amylose starch films have flexible, oxygen impermeable, oil resistant, heat-sealable, and water soluble [12-13]. Chitosan-based edible films exhibit excellent mechanical properties and good oxygen and carbon dioxide permeability. Additionally, chitosan can form cohesive and compact transparent film which has a smooth contour with no pores or cracks on the film surface. Nevertheless, the chitosan based edible film has poor solubility in neutral solutions [14-16].

Blending more than one polysaccharide materials could improve film mechanical functionality. Edible films based on mixtures of many kinds of polysaccharides have been produced, such as chitosan-starch, chitosan-pullulan, CMC-rice starch, CMC-pea starch, starch-alginate and sago-alginate [17-19]. CMC are water-soluble heteropolysaccharides derived from cellulose which are often applied together with starches to enhance overall film quality and stability. The hydroxyl and carboxylic groups in CMC facilitate water binding and moisture sorption properties. Indonesian is consuming arenga or sugar palm (*Arenga pinnata*) starch as their staple food in place of rice. Arenga starch also used as an ingredient in the preparation of cakes, noodles, and other dishes. It has high amylose content which can be used in forming excellent film matrix. Although the film formed was compact and rigid but it is a fairly brittle [20]. The main objective of this research was to prepare an edible film from arenga starch and chitosan. Carboxymethylcellulose (CMC) and glycerol were also added to the blend.

2. Materials and Methods

2.1. Materials
All chemicals were used are analytical grade from e-Merck. The arenga starch was provided by arenga starch home industries in Klaten, Jawa Tengah, Indonesia. Chitosan was supplied from PT Biotech Surindo, Indonesia.

2.2. Film preparation
Arenga starch (1.5 g) and CMC (1 g) were dissolved in 75 mL of distilled water, heated to 90°C and stirred for 24 h. Meanwhile, a mixture of chitosan and acetic acid at different concentration were prepared by stirring at room temperature for 24 h. This two mixture were then blended at different concentration for 1 h. Glycerol at various amount was then added, and the mixtures were stirred for a further 24 h. The sample solution at different volume was poured into casting tray followed by oven drying at 50-60°C for 7 h. The dry films obtained were peeled off and stored for further analysis.

2.3. FTIR Analysis
The spectra of the films were recorded at room temperature as KBr pellets on Shimadzu type FT-IR 8201 PC spectrophotometer. The light source of transmittance was in the range of 4000-500 cm^{-1}. The films were mounted directly in the sample holder.

2.4. Film thickness measurement
Film thickness was determined using a digital micrometer (Krisbow KW06-85) having a precision of 0.001 mm. Measurements were carried out at five different film locations, and the mean thickness value was used to calculate the mechanical properties of the films.

2.5. Mechanical properties
Tensile strength and elongation of the films were determined using a material testing machine (RAY-RAN M500-50CT Testometric Rodhale England type DBBMTCL-5000 Kg Serial No. 36751) according to ASTM standard method D882-91 (ASTM D882-91, 1996). Film specimens, 6 cm x 0.4 cm were cut from each of film samples and were mounted between the grips of the machine. The initial grip separation and cross-head speed were set to 50 mm and 5 mm/min, respectively. The parameter determined were tensile stress (MPa) using equation 1 and elongation (%) using equation 2. The tensile strength was expressed as the force of film rupture, Fmax (N) divided by the cross-sectional area of the film, A (mm^2), and the elongation was expressed as the tensile elongation of the film at rupture, Lmax (mm) divided by the initial length of film, L0 (mm) (equation (1) and (2).

$$\tau = \frac{\text{Fmax}}{A} \tag{1}$$

$$E = \frac{\text{Lmax}}{L0} \tag{2}$$

2.6. Swelling area and water uptake
Pieces of films were cut and were soaked in distilled water for 24 h at room temperature. Swelling area and water uptake of the films were calculated by comparing the surface area or weight before and after soaking. Equation 3 was to calculated the swelling area (*Sa*) and equation 4 was used to

calculated the water uptake (*Wu*), where L1 is the surface area before soaking, L2 is surface area after soaking, W1 is the weight before soaking, and W2 is the weight after soaking (equation (3) and (4)).

$$Sa = \frac{L2 - L1}{L1} x\ 100\% \tag{3}$$

$$Wu = \frac{W2 - W1}{W1} x\ 100\% \tag{4}$$

3. Results and Discussion

3.1. Film preparation

In this studies, the blending of three polysaccharides, which are area starch, chitosan, and CMC, with the addition of glycerol as a plasticizer was carried out. The amount of the each component was varied to find the most suitable composition which could give excellent mechanical properties. The thickness of the film was also modified. Initial attempts were made using a higher amount of a mixture of starch, CMC, and water than that of chitosan. The first attempt was carried out where the composition of the blend was 10.5 g of a mixture of starch, CMC, and water, 1.75 g of chitosan, 9 mL of glycerol and 50 mL of acetic acid. Nevertheless, the film has not formed yet.

The second attempts then performed with the additional amount of glycerol to the mixture. It was though that additional amount of glycerol would produce films since glycerol are widely used as plasticizers which make the brittle films more flexible. The volume of the mixture that were poured into casting tray were varied (30, 40, 60 and 70 mL) to obtain a different thickness of the films (table 1). The results exhibited that films have been formed. However, they are brittle.

Table 1. Variation of mixture composition and the thickness of the produced films.

	Composition					
	A mixture of starch, CMC and water (g)	Chitosan (g)	Glycerol (mL)	Acetic acid (mL)	Mixture volume (mL)	Film thickness (mm)
Film A	10.5	1.75	9	175	50	0.504
Film B1	10.5	1.75	10.5	175	30	0.147
Film B2	10.5	1.75	10.5	175	40	0.493
Film B3	10.5	1.75	10.5	175	60	0.514
Film B4	10.5	1.75	10.5	175	70	0.567

The next attempts were made by changing the concentration of each component. Accordingly, a higher amount of chitosan than that of the concentration of a mixture of starch, CMC and water were applied. The concentration of glycerol was varied while the film thickness was the same (Table 2). The composition of the mixture was 0.175 g of a mixture of starch, CMC, and water, 1.75 g of chitosan, 0.105 (6%) or 0.175 mL (10%) of glycerol and 70 mL of acetic acid. The results revealed that rigid films have been produced. Hence, the FT-IR analysis and properties, such as tensile strength, elongation, swelling degree and swelling area, were then examined for these two films.

Table 2. Different thickness of the produced films.

	Composition				Mixture volume (mL)	Film thickness (mm)
	A mixture of starch, CMC and water (g)	Chitosan (g)	Glycerol (mL)	Acetic acid (mL)		
Film C1	0.175	1.75	0.105	70	40	0.363
Film C2	0.175	1.75	0.175	70	40	0.363

3.2. FTIR analysis

The FT-IR spectra of arenga starch-chitosan based films C1 and C2 are shown in Figure 1. The spectra show similar patterns. An absorption band at 3436 cm^{-1} for film C1 and 3443 cm^{-1} for film C2 appear. An extremely broad peak appeared between 3000-3600 cm^{-1} is the region of hydrogen-bonded hydroxyl groups associated with free, inter- and inter- molecular bound hydroxyl groups.

Both films had a peak for amide I, illustrating C=O stretching/hydrogen bonding coupled with COO, which was a peak at 1638 cm^{-1} and 1641 cm^{-1} for films C1 and C2, respectively. Both films also had a peak at 1529 cm^{-1} for amide II, presenting the bending vibrations of N–H groups and stretching vibrations of C–N groups. Tongnuanchan *et al.* reported that films from fish skin gelatin containing glycerol had peaks at 1630 and 1538 cm^{-1} for amide-I and amide-II, respectively [21]. Hoque *et al.* presented similar spectra for cuttlefish skin gelatin film, where amide-I and II bands were found at the peaks of 1630 and 1539 cm^{-1}, respectively [22].

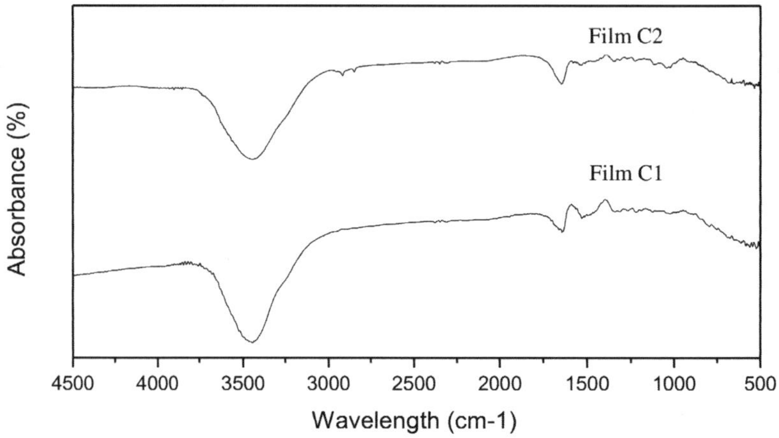

Figure 1. FTIR spectra of arenga starch chitosan-based films C1 and C2 with different concentration of glycerol.

Peaks at 2956, 2918 and 2849 cm^{-1} were shown when the film has more concentration of glycerols in film C2. These peaks represent the methyl, methylene symmetrical and asymmetrical and stretching vibration of the aliphatic C-H in CH_2 and CH_3 groups [23-24]. Both the methylene asymmetrical stretching bands at approximately 2853 cm^{-1} and methylene symmetrical stretching band near 2924 cm^{-1} were present in most lipids [23-24]. Film C2 had higher hydrophobicity properties than that of film C1, as evidenced by the peak appearances at 2849 and 2924 cm^{-1} which associated with hydrophobic interactions.

The band of O-H bending for the alcohols in film C1 that appeared at 1268, 1221 and 1190 cm^{-1} were shifted to 1272, 1228 and 1188 cm^{-1}, respectively, with addition concentration of glycerols in film C2. A peak at 1030 cm^{-1} in film C1 for the C-H aromatic was shifted to 1045 cm^{-1} in film C2. An extra peak at 1021 cm^{-1} was shown in film C2. Other additional peaks were noticed in film C2. An additional peak at 1339 cm^{-1} which was associated with O-H bending of the primary or secondary alcohols appeared in film C2. A peak at 1458 cm^{-1} was shown in film C2, which was ascribed to C-H stretching and bending modes. Furthermore, a peak at 1158 cm^{-1} existed in film C2 which was attributed to C-O-C antisymmetric bridge stretching.

3.3. Film properties

Mechanical properties of films were characterized by assessing the tensile strength and elongation, which are main indicators of films strength and flexibility. The tensile strength can be ascribed to the cohesion between the matrixes of the filmogenic polymer chains. The elongation is a measure of the plasticity of the film or a degree of the film's capacity for stretching. The elongation is film capacity to enlarge to occur before its break [17, 25]. Table 3 presents that the mechanical properties of film C1 and film C2 varying according to the glycerol content. The film C1, which has 0.105 mL of glycerol

(6% of glycerol) in the blend has a tensile strength of 124 KPa, elongation of 6.55%, swelling area of 21% and water uptake of 48%. The film C2, which has 0.175 mL of glycerol (10 % of glycerol) in the blend has a tensile strength of 58 KPa, elongation of 8.35%, swelling area of 32% and water uptake of 5%. The tensile strength results show that tensile strength decreased with the increasing of glycerol content in the blend. In contrast, elongation results show that elongation is increased with increasing glycerol content in the blend. Similar results were reported by other authors [20, 26]. Furthermore, the swelling area increased and water uptake decreased with increasing glycerol content in the blend.

Table 3. Properties of arenga starch chitosan-based films C1 and C2.

Properties	Film C1	Film C2
Tensile strength (Kpa)	124	58
Elongation (%)	6.55	8.35
Swelling area (%)	21	32
Water uptake (%)	48	5

Our results can be explained because glycerol act as a plasticizer. The addition of plasticizers makes the brittle films more flexible, and consequently less rigid and less strong. The glycerol interacts through hydrogen bonding with crystalline amylose and crystalline and amorphous amylopectin when the temperature is increased and also during film storage at room temperature. As a result, matrix mobility increases, viscosity reduces and the material performs like a rubber [27]. Water uptake or water solubility is an important characteristic of edible films. Some applications may require water insolubility to enhance product integrity and water resistance [28]. Our results clearly demonstrated that films with a higher ratio of glycerol exhibited lower water uptake, indicating higher water resistance.

4. Conclusion

The arenga starch-chitosan based edible film have been successfully produced in the presence of CMC. The elongation and swelling area increased with increasing glycerol content in the blend, whereas the tensile strength and water uptake decreased with the increasing of glycerol content. Variation amount of the each component and the thickness of the film still need to be evaluated to produce the film with better mechanical properties.

Acknowledgments

The authors acknowledge their thanks to Sebelas Maret University for the financial support of this research in the form of DIPA UNS Grant 2015.

References

[1] Chandra R and Rustgi R 1998 *Prog. Polym. Sci.* **23** 1273-1335
[2] Han J H 2000 Antimicrobial food packaging *Food Technol.* **54** 56–65
[3] Siracusa V, Rocculi P, Romani S and Rosa M D 2008 *Trends Food Sci. Technol.* **19** 634-64
[4] Janjarasskul T and Krochta J M 2010 *Ann. Rev. Food Sci. Technol.* **1** 415-448
[5] Park S-Y, Lee B I, Jung S T and Park H J 2001 *Mater. Res. Bull.* **36** 511–519
[6] Abdou E S and Sorour M A 2014 *Int. Food Res. J.* **21** 189-193
[7] Kester J J and Fennema O 1986 *Food Technol.* **40** 47-59
[8] Vásconez M B, Flores S K, Campos C A, Alvarado J, and Gerschenson L N 2009 *Food Res. Int.* **42** 762–769
[9] Cutter C N 2006 *Meat Sci.* **74** 131-142
[10] Sallam K I 2007 *Food Cont.* **18** 566–575
[11] Phan T D, Debeaufort F, Luu D and Voilley A 2005 *J. Agric. Food Chem.* **53** 973-981
[12] Dhanapal A, Sasikala P, Rajamani L, Kavitha V, Yazhini G and Banu M S 2012 *Food Sci. Qual. Manage.* **3** 9-17
[13] Rindlav-Westling A, Standing M, Hermansson A M and Gatenholm P 1998 *Carbohydr. Polym.* **36** 217-224
[14] Elsabee M Z and Abdou E S 2013 *Mater. Sci. Eng.* **33** 1819-1841

[15] Srinivasa P C, Ramesh M N and Tharanathan R N 2007 *Food Hydrocoll.* **21** 1113-1122
[16] Xu X-Y, Kim K-M, Hanna MA and Nag D 2005 *Industry. Crops Prod. Int. J.* **21** 185-192
[17] Tongdeesoontorn W, Mauer L J, Wongruong S, Sriburi P and Rachtanapun P 2011 *Chem. Cent. J.* **5** 6
[18] Wu Y, Weller C, Hamouz F, Cuppett S and Schnepf M 2001 *J. Food Sci.* **66** 486-493
[19] Fazilah A, Maizura M, Karim A, Bhupinder K, Bhat R, Uthumporn U and Chew S H 2011 *Int. Food Res. J.* **18** 1027-1033
[20] Sanyang M L, Sapuan S M, Jawaid M, Ishak M R and Sahari J 2015 *Polym.* **7** 1106-1124
[21] Tongnuanchan P, Benjakul S and Prodpran T 2012 *Food Chem.* **134** 1571–1579
[22] Hoque M S, Benjakul S, and Prodpran T 2010 *Food Hydrocoll.* **25** 82-90
[23] Guillén M D and Cabo N 1997 *J. Sci. Food Agric.* **75** 1-11
[24] Guillén M D and Cabo N 2004 *Meat Sci.* **66** 647–657
[25] Seixas F L, Turbiani F R B, Salomão P G, Souzaa R P, and Gimenesa M L 2013 *Chem. Eng. Trans.* **32** 1693-1698
[26] Obasi1 H C, Egeolu F C, and Oparaji O D 2015 *Am. J. Chem. Mater. Sci.* **2** 6-13
[27] Vieira M G A, da Silva M A, dos Santos L O and Beppu M 2011 *Eur. Polym. J.* **47** 254-263
[28] Fazilah A, Maizura M, Abd Karim A, Bhupinder K, Bhat R, Uthumporn U and Chew S H 2011 *Int. Food Res. J.* **18** 1027-1033

Synthesis and characterization of free-isocyanate polyurethane as renewable coating materials

O A Saputra[a], K Apriany, F R Putri, M Firdaus[b]
Department of Chemistry, Faculty of Mathematic and Natural Science, Sebelas Maret University Jl. Ir. Sutami 36A, Surakarta, Indonesia

E-mail: [a]oziadisaputra@student.uns.ac.id; [b]maulidan@mipa.uns.ac.id

Abstract. Green synthesis of diethyl ethane-1,2-diyldicarbamate (M1) as starting dimer of free-isocyanate polyurethane from an excess of diethyl carbonate and ethylene diamine catalyzed by TBD (1,5,7-triazabicyclo-[4,4,0]dec-5-ene) at 80 °C for 4 hours has been carried out. The product has high yield and purity up to 92% and 99,18%, respectively. The product was obtained in the form of rod-shaped white crystals and characterized with ^{1}H-NMR, ^{13}C-NMR and FTIR spectroscopy. Polyurethane (P1) has successfully synthesized via polycondensation methods by reacting of M1 with *cis*-2-butene-1,4-diol. This reaction also catalyzed by TBD organo catalyst. P1 chemical structure has been confirmed using ^{1}H-NMR and FTIR techniques.

1. Introduction

The use of materials which obtained from petroleum resources have to be replaced, due to the availability of petroleum will be depleting [1]. One alternative that should be taken to replace or reduce the use of petroleumas raw material is to re-use natural materials that is renewable and sustainable. In addition, the advantages in using of starting material from natural resources are cheaper, easily available and environmentally friendly [2]. Therefore, some researchers have been using materials from nature to replace material derived from petroleum resources, such as in the synthesis of polyurethane.

Polyurethane is one of the polymeric material which is very important in many applications, such as adhesives, coatings technology, foam, and etc [3-5]. The advantages of this polymer is to have a good anti-corrosion properties that can be used as coating materials in the the transportation industry, petroleum, and household appliances made of metal [4]. The use of polyurethane as a coating material because of its resistance to dissolution, abrasion and give effect in the brightness of the paint [6]. Conventionally synthesis of polyurethane today still use the initial material such as phosgene or isocyanate compound [7-9]. Besides having a toxic properties, both of these compounds are also not environmentally friendly as well as derived from petroleum. Isocyanate compounds have also been reported to cause irritation and long-term asthma [10]. In addition, the isocyanate compounds are also very sensitive to water, so that in its use requires special media and more energy in the synthesis of polyurethanes via the isocyanate [11]. The research also reported that the toluene diisocyanate can release carcinogenic aromatic amines group. Therefore, it is necessary to do the synthesis of non-isocyanate or non-phosgene polyurethane that are environmentally friendly by using green chemistry-based material.

The use of carbonate compounds as raw material in the synthesis of polyurethane to be very interesting, where the compound is classified as green chemistry compound. These compounds can be synthesized from carbon dioxide (CO_2) [12], a derivative of urea [13,14], methanol [15], and others. Some researchers have used the carbonate compound as a replacement for isocyanates or phosgene in the synthesis of polyurethane [16,17]. In addition, other carbonate compounds such as cyclic limonene dicarbonate [18], cyclic carbonate [19], carbonate modified epoxy resin [20] and others, have also been reported as a raw material in the synthesis of polyurethane. Before the polymerization reaction forming a non-isocyanate polyurethane, carbonate compounds such as dimethyl carbonate is usually converted into carbamate compounds [16,21]. Carbamate compounds is an intermediate product and can form polyurethanes when reacted with diol compounds, directly [1]. Firdaus [16] has successfully synthesize carbamates compounds from dimethyl carbonate with the aid of a TBD (1,5,7-triazabicyclo-[4,4,0]dec-5-ene)catalyst. In addition, other studies have also successfully to synthesize carbamates compounds from dimethyl carbonate using acid functional ionic liquid catalyst [21]. Diethyl carbonate compounds have also been reported as a starting material in the synthesis of carbamate compounds in a high yield, which is catalyzed by TBD catalyst [22]. However, the use of di-carbamate compounds from diethyl carbonate as the starting material in the synthesis of polyurethane has not been done.

Therefore, in this study will be carried out of synthesis free-isocyanate polyurethane from diethyl carbonate compounds as raw material and catalyzed by the TBD catalyst.

2. Experimental

2.1. Materials

All reagents were purchased without further purification. The starting materials, diethyl carbonate (DC) and ethylene diamine (ED) were obtained from Merck, as well as for tetrahidrofuran (THF), ethanol, and n-hexane. *Cis*-2-butene-1,4-diols and 1,5,7-triazabicyclo-[4,4,0]dec-5-ene (TBD) were purchased from Sigma Aldrich.

*2.2. Syntesis of Diethyl Ethane-1,2-diyldicarbamate(**M1**)*

All of the starting materials such as diethyl carbonate and ethylenediamine were characterized with FTIR (fourier transform infra red) and TLC (thin layer chromatography) before used. An excess of diethyl carbonate (50 mmol) were reacted with 5 mmol of ethylene diamine. This reaction were catalyzed with TBD (10% mmol eq. to DC) organocatalys. The solution were slowly stirred at controled temperature, 80 ^{o}C. This reaction were monitored with TLC, to stop the reaction until ethylene diamine spot disappeared. The obtained product were characterization with FTIR, GC (Gas Chromatography), and NMR (Nuclear Magnetic Resonance) spectroscopy.

*2.3. Synthesis of Polyurethane (**P1**)*

The polymerization reaction in this study were performed in the vacum reactor. **M1** product from the previous synthesis was used in this reaction. 1 mmol eq of **M1**were mixed with *cis*-2-butene-1,4-diols (1 mmol equiv.). The reaction was hold at 80 ^{O}C with the medium presure (400±5 mbar) until viscous solution succesfully achieved. The temperature was slowly raising to 120 ^{O}C, and holded until polycondensation reaction were completely achieved. In this reaction, TBD organocatlyst also used. The **P1** polymer was purified by precipitation of a concentrated THF solution of **P1** product into cold ethanol.

2.4. Characterization

All of reagents in this study were characterized with FTIR Prestige 21 SHIMADZHU to confirmed the functional groups of each reagent. Also, the M1 and P1 synthesis products were characterized. In this study, we used KBr as background and set the scaning up to 45 times of scan.

^{1}H-NMR and ^{13}C-NMR were used in this study was Agilent with frecuency of 400 MHz. In the caracterized of M1, we used $CDCl_3$ as solvent and for P1 product used warm DMSO solvent. GC-MS analysis were perfomed in Chemistry laboratory of UII. For M1 compound analysis, we used chloroform as solvent and heating up to 350 ^{o}C.

3. Result and Discussion

3.1. Synthesis of dicarbamates

Diethyl ethane-1,2-diyldicarbamate (**M1**) was dicarbamate compound that sucessfully synthesized in this study. **M1** was prepared by reacting diethyl carbonate (DC) with ethylene diamine (ED) in the presence of TBD catalyst (Scheme. 1). TBD catalyst was reported as selective organocatalyst in the forming of carbamates compounds [16,23]. This organocatalyst was able to prepare carbocation as electrophilic, so make it easily to reaction with ethylene diamine or nucleophilic. TLC was used to monitored the reaction. After 4 hours reaction, according to TLC observation showed the disappeared of ethylene diamine spot which indicates the **M1** product has been form.

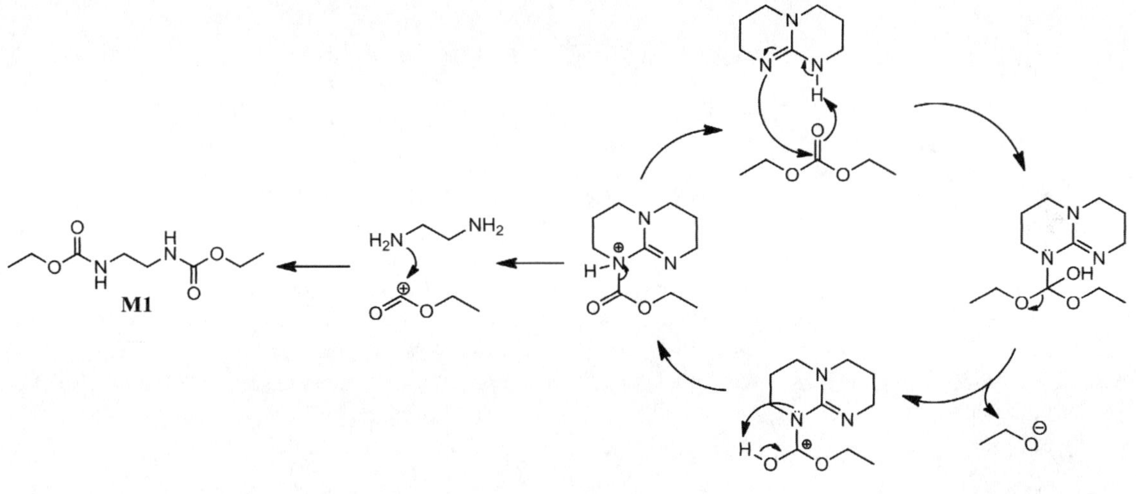

Scheme 1. Possible chemical reaction on synhesis of Diethyl ethane-1,2-diyldicarbamate (**M1**)

M1 product was obtained after settling for 12 hours at room temperature. The **M1** product were solid like crystal and white form. The solid form of **M1** product was very beneficially, due it was easy to separated from the exess of dimethyl carbonate. In the other side, it was also minimizing in energy consumtion. This fact made **M1** product categorized as green compound. In this work, **M1** product has high yield, up to 92% just for 4 hours reaction. This was attributed of TBD organo catalyst, where this catalist was reported as selective catalyst to forming carbamate compound [16,22,23]. This study also investigated of synthesis **M1** product without TBD organo catalyst and these compound was obtained about 4 days. This investigation was evidenced that TBD organo-catalyst plays an important role in the formation rates of **M1** compound.

Gas chromatography technique was performed to determine of **M1** compound purity. These product was recorded using chloroform as solvent. This analysis resulted that **M1** product has high purity, up to 99.18%. It was veru luckly to obtaine the product with high purity. To determine **M1** chemical structure, some analysis has been conducted, such as FTIR technique, ^{1}H-NMR, ^{13}C-NMR and Mass spectroscopy.

Functional group analysis showed a shift of a wavenumber from the precusror material in M1 product indicate the carbamates has been formed. From curve A (spectrum of diethyl carbonat) can be seen that a carbonil group (>C=O) streaching vibration adsorption peak at 1761,08 cm^{-1}. After reaction, the >C=O streaching vibration peak shifted to 1685,86 cm^{-1}. It proved that carbamate group completely formed. Fleischer et al. [12] was confirmed that an urethane bands was obtained in 1680 cm^{-1} and disappeared of carbonate group band at 1700 cm^{-1}verified the complete conversion of all carbonate to carbamates or urethane group. Also, amine group (-NH) bending vibration shifted from 1599 cm^{-1} to 1546 cm^{-1} indicated that carbonil group from diethyl carbonate was interacted with amine group from ethylene diamine to formed carbamates group.

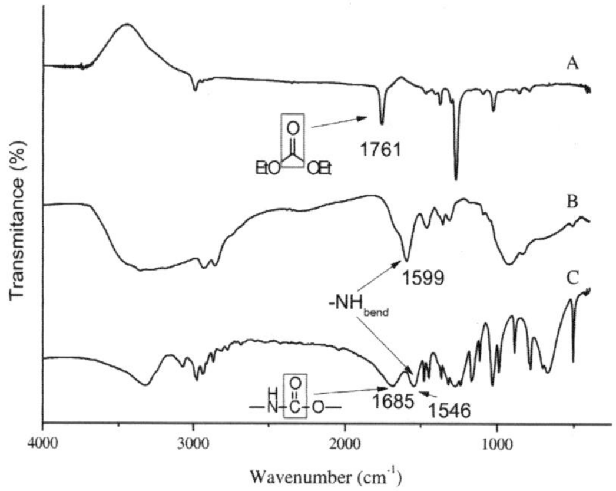

Figure 1. FTIR spectrum of (A) diethyl carbonate, (B) ethylenediamine and (c) M1 product

Figure 2. shows the ^{13}C-NMR spectrum of M1 product. Signal of ^{13}C-NMR was corresponding with chemical structure of M1 product, where a signal arround 14.9 ppm accordance to carbon atom in methyl group (-CH_3). Moreover, the signals at 40.1 ppm and 60.1 ppm were corresponded to carbon atoms in methylene (-CH_2-N-) and methylene (-CH_2-O-) groups, respectively. In Fig. 2 was investigated that only four signal which detected. This cause by a symetry chemical structure of **M1** compound. A peak at 157,06 ppm was assigned to the carbon atom on the carbamates groups. It was also found by Li et al [17].

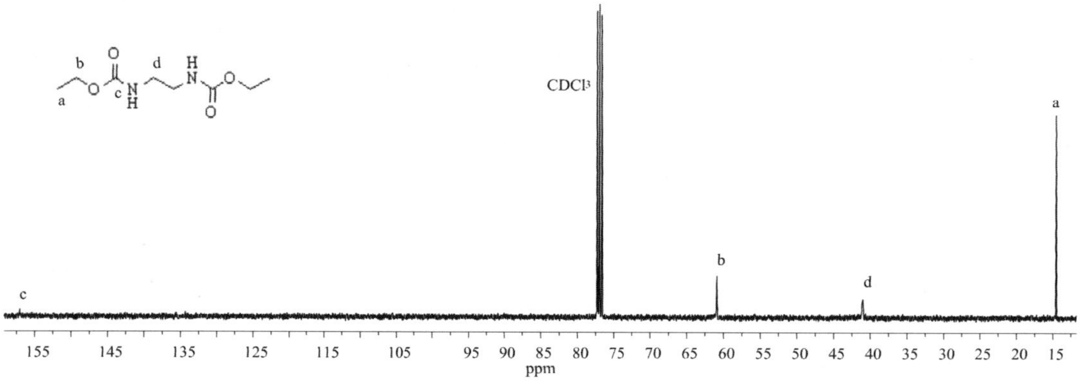

Figure 2. ^{13}C-NMR of M1 product

^{1}H-NMR also performed to strongly prediction of M1 chemical structure (Figure 5). The signal at 5.01 ppm was assigned to hidrogen atom at amide group (-C(O)-NH). The other signals such as 3.36

(t, Ha), 4.03 (q, Hb) and 3.27 (s, Hd) ppm was confirmed as methyl, methylene (CH_2-O-) and methylene (-CH_2-NH-). The existence of this signals was confirmed the M1 chemical structure as seen in Scheme 1. M1 products was high solubility in chloroform, thus, in ^{1}H-NMR analysis $CDCl_3$ used as solvent.

Figure 4 shows a MS signal and possible fragmentation mechanism of M1 product. The MS signals of M1 product was matched with the possible fragmentation which indicated the possible compound was diethyl ethane-1,2-diyldicarbamate (M1). The base peak of M1 can be seen at 115 m/z corresponding to cyclic carbamates. It was found that the cyclic carbamates be a stable structure than the other structures.

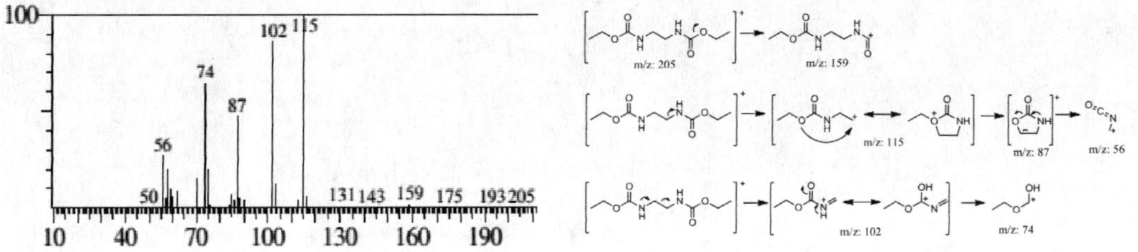

Figure 3. MS and possible fragmented mechanism of M1 product

3.2. *Synthesis of Polyurethane*

Synthesis of polyurethane was performed in the vacum reator to minimized the interaction with other compound in the polycondensation reaction. M1 product was reacted with 2-butene-1,4-diols (DO) with the presence of TBD organocatalyst. Unverferth et al. [23] reported that polycondensation of polyurethane which catalyzed by TBD organocatalyst become an efficient strategy to obtain renewable polyurethane. In this study, temperature of the reaction was raised with slowly from 80 ^{O}C to 120 ^{O}C. In the first step, the polycondenzation reaction was perfomed in 80 ^{O}C about 2 hours and then raised to 120 ^{O}C to completely the polycondenzation reaction. It was found that, at 120 ^{O}C the product was solid form so fastly indicated the polymerization was terminated. The obtained product was yellow and after precipitation process the colours was white to brown, indicated the TBD was dissolved in methanol and the product was pure.

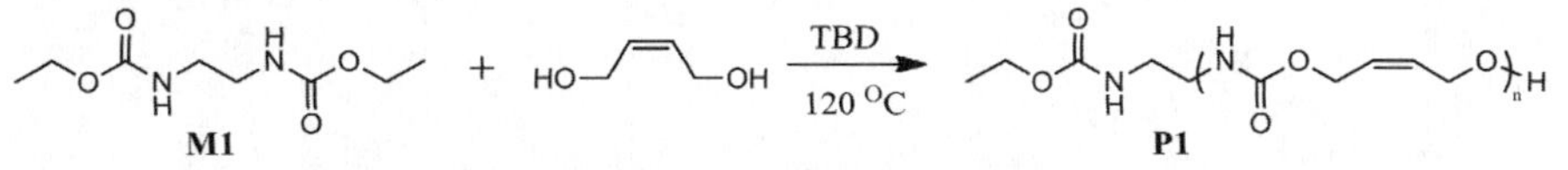

Scheme 2. Polymerization reaction of P1 product

FTIR technique was one of simple methods to characterized of synthesis product by their fuctional group. Several researchers [18, 19, 24] has used this technique as supporting data to analyze and predict of chemical structure. Figure 4. showed FTIR spectrum of M1 and P1 product. The bands at 1645.35 cm^{-1} and 1483.32 cm^{-1} was corresponding to –C=C-H groups. It was indicated that the M1 and DO was reacted to form P1. Also, carbamates group band was shifted to 1722.51 cm^{-1} was clearly the P1 was form.

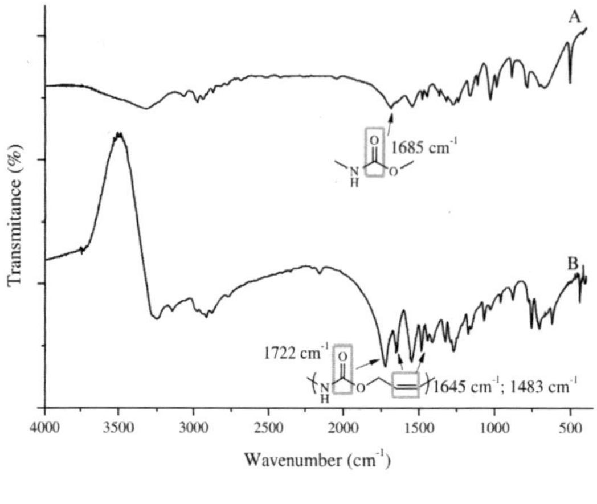

Figure 4. FTIR spectrum of (A) M1 and (B) P1 products

^{1}H-NMR investigated of Polymerizations confirmed the expected structure of P1 products. The ^{1}H-NMR of dicarbamates compound showed a decreasing of signals intensity at 1.11 and 3.95 ppm corresponding to the methyl (Ha) and methylene (Hb), respectively in ^{1}H-NMR spectra of P1 as shown in Figure 5. Not only decrease, the signal also shift, where in M1 products ^{1}H-NMR spectra Ha and Hb has appear in 1.21 and 4.09 ppm, respectively. This was caused by the presence of diol compounds as polymerizing agent which has vinyl group, so the signal has shift but not to much. The P1 products has low solubility in several solvent, such as $CDCl_3$, THF, and DMSO. However, in warm DMSO solution was found that P1 product has good solubility. Thus, before ^{1}H-NMR test, P1 products was dissolved in warm DMSO (40-50 °C, approximately).

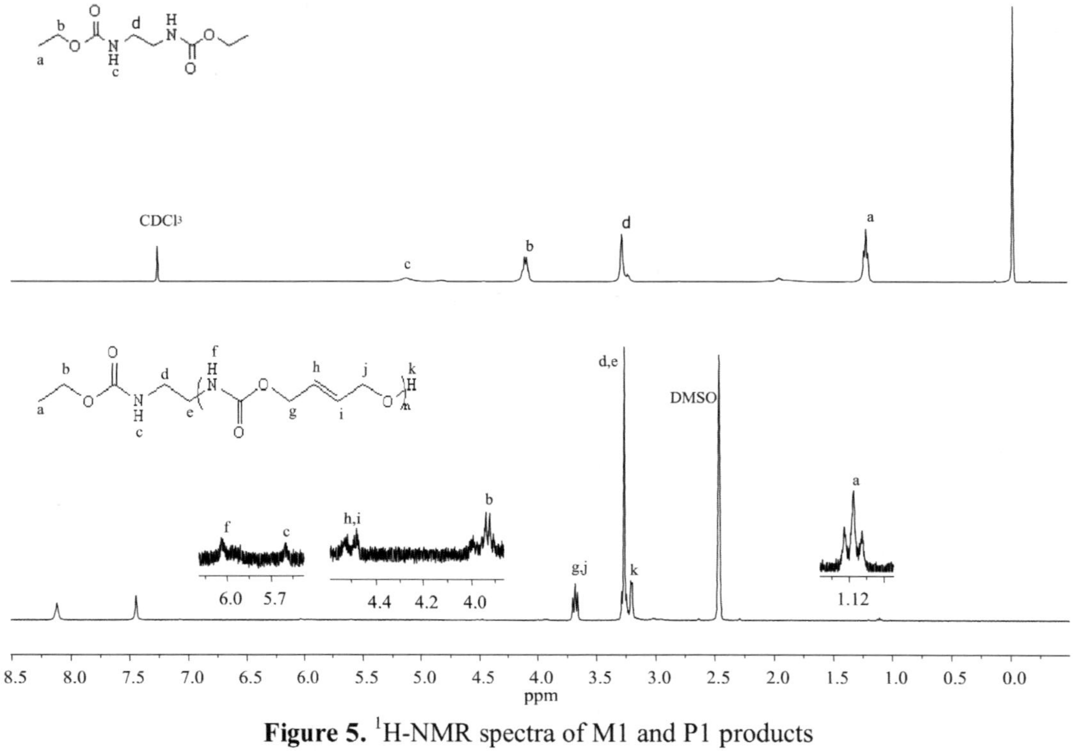

Figure 5. ^{1}H-NMR spectra of M1 and P1 products

4. Conclution

FTIR and NMR technique has confimed that the M1 and P1 product was succesfully form. The purity analysis with GC was confirmed the M1 product has purity up to 99,18%. Also, the MS analysis showed a corresponding signals with the possible fragmentation of M1 product indicated the M1 product surely formed. The polymerization of P1 was confimed using FTIR and ^{1}H-NMR technique. An invistigated showed that both in FTIR and ^{1}H-NMR spectra was prove of P1 structure.

Acknowledgement

Author gratefully acknowledges the Indonesian Directorate General of Higher Education (DIKTI) for financial support of this research by *Program Kreatifitas Mahasiswa* (PKM).

References

[1] Kreye O, Mutlu H and Meier M A R 2013 *Green Chemistry*15 1431
[2] Chaturvedi D 2012 *Tetrahedron* **68** 15
[3] Desai S D, Emanuel A L and Sinha V K 2003 *J. Polym. Res.* **10** 141
[4] Ahmad S, Zafar F, Sharmin E, Garg N and Kashif M 2012*Prog. Org. Coat.* **73**112
[5] Zhang L, Zhang M, Hu L and Zhou Y 2014 *Ind. Crop. Prod.* **52** 380
[6] Countinho F M B, Delpech M C and Alves L S 2001 *J. App. Polym. Sci.* **80** 566
[7] WenX, Mi R, Huang Y, Cheng J, Pi P and Yang Z 2010 *J Coat. Technol.* **7** 373
[8] Hou G X, Chen X G, Liu J J and Sang X M 2011 *Polym-Plast. Technol. Eng.* **50** 1208
[9] Abdel-Hakim A A, Ahmed I S, Moustafa M M, Kamal El-Din A H and Soad A A 2011 *J. App. Sci. Res.* **7** 1424
[10] David A L, Denise E F and Richard J Q 1999 *Prog. Org. Coat.* **35** 109
[11] Blattmann H, Fleischer M, Bahr M and Mulhaupt R 2014 *Macromol. Rapid. Commun.* DOI: 10.1002/marc.201400209
[12] Fleischer M, Blattmann H and Mulhaupt R 2013 *Green Chem.* **15** 934
[13] Wang M, Wang H, Zhao N, Wei W and Sun Y 2006 *Catal. Commun.* **7** 6
[14] Sun J, Yang B, Wang X, Wang D and Lin H 2005 *J.Mol. Chem. A*: *Chemical.* **239** 82
[15] Jia G, Gao Y, Zhang W, Wang H, Cao Z, Li C and Liu J 2013 *Electrochem. Commun.* **34**211
[16] Firdaus M and Meier M A R 2013*Green Chem.* **15** 370
[17] Li C, Li S, Zhao J, Zhang Z, Zhang J and Yang W 2014 *J. Polym. Res.* **21** 498
[18] Bahr M,Bitto A and Mulhaupt R 2012*Green Chem.* **14** 1447
[19] Zhang W Y and Zhang B H 2014*App. Mech. Mater.* **618** 184
[20] Hosgor Z, Apohan N K, Karats S, Menceloglu Y and Cungor A 2010 *Prog. Org. Coat.* **69** 366
[21] Zhao H, Shi F, Tian X, Zhang Q and Deng Y 2007 *J. Mol. Chem. A*: *Chemical.* **271** 89
[22] Carloni S, Vos D E D, Jacobs P A, Maggi R, Sartori G and Sartorio R 2002 *J. Catal.* **205** 199
[23] Unverferth M, Kreye O, Prohammer A and Meier M A R 2013 *Macromol. Rapid. Commun.* **34** 1569
[24] Hwang J Z, Chen C L, Huang C Y, Yeh J T and Chen K N 2013 *J. Polym. Res.* **20** 195

Yeast fuel cell: Application for desalination

Ummy Mardiana[1,2,3*], Christophe Innocent [1], Marc Cretin[1], Buchari Buchari[2] and SuryoGandasasmita[2]

[1]Institut Européen des Membranes, ENSCM-UM2-CNRS, (UMR 5635), Université de Montpellier 2, CC047, place E. Bataillon, 34293 Montpellier Cedex 5, France
[2]Institut Teknologi Bandung, Jalan Ganesha 10 Bandung, West of Java Indonesia 40132
[3]STIKes Bakti Tunas Husada Tasikmalaya, Jalan Cilolohan 36 Tasikmalaya, West of Java Indonesia 46115

[*]E-mail: mardiana.ramdan@gmail.com

Abstract. Yeasts have been implicated in microbial fuel cells as biocatalysts because they are non-pathogenic organisms, easily handled and robust with a good tolerance in different environmental conditions. Here we investigated baker's yeast *Saccharomyces cerevisiae* through the oxidation of glucose. Yeast was used in the anolyte, to transfer electrons to the anode in the presence of methylene blue as mediator whereas $K_3Fe(CN)_6$ was used as an electron acceptor for the reduction reaction in the catholyte. Power production with biofuel cell was coupled with a desalination process. The maximum current density produced by the cell was 88 mA.m^{-2}. In those conditions, it was found that concentration of salt was removed 64% from initial 0.6 M after 1-month operation. This result proves that yeast fuel cells can be used to remove salt through electrically driven membrane processes and demonstrated that could be applied for energy production and desalination. Further developments are in progress to improve power output to make yeast fuel cells applicable for water treatment.

1. Introduction

Current research on energy production has been intensively developed. Microbial fuel cells (MFCs) are a new opportunity for the sustainable production of energy by converting chemical energy to electrical energy by biodegradation using microorganisms as biocatalysts [1–3]. MFCs have advantages as easy to operate without needing a complicated instrument, efficient at ambient or even at low temperature [4] without requiring energy input. In MFCs, microorganisms oxidize organic compounds and produce electrons at the anode, passing through the external circuit to the cathode where the final electron acceptor is reduced. Power density can be produced with various organic compounds such as acetate [3], glucose, sucrose and fructose [5]. Electron mediators such as methylene blue, neutral red are sometimes needed to transfer electrons from microorganism to anode[6–8] although applying mediator-less MFC The have also been studied [9–13].Modified electrodes have been also carried out [10,14–16]. Some microorganisms have been implied in MFC. The use of *Saccharomyces cerevisiae* [9,17,18], *Candida melibiose* [5] and *Arxulaadeninivorans* [19] as microorganisms have been also investigated.

Biocatalyst in the anodic chamber has an important role in cell performance. The mechanisms of electron transfer can take place directly, mediated or endogenous (*i.e.*mediator produced by bacteria themselves). It is known that many prokaryotic microorganisms could utilize oxygen environment for their growth through respiratory pathways in aerobic conditions and fermentation pathways in anaerobic conditions. Yeast can growth in both environments, is easily activated by hydrating or temperature, inexpensive, does not need sterile conditions because good tolerances in different environmental conditions and hence could be stored in the dry state [14].

In our configuration work, a two compartments MFC was constructed with graphite electrode as an anode, and nickel as a cathode. Anode compartment contained the yeast *Saccharomyces cerevisiae,* methylene blue as a mediator as previously designed [7,8]and glucose in phosphate buffer pH 7 as an oxidizer. Methylene blue is first reduced in the outer cell membrane of yeast. The reduced form of mediator captured electrons to the anode, producing electricity and re-oxidizing the mediator. The other half-cell (*i.e.*cathodic compartment) contained potassium ferricyanide, which receives the electrons from the external circuit and becomes reduced. The presence of oxygen in catholyte will oxidize $Fe^{2+}$to Fe^{3+} which causes a cyclic reaction. It is to notice that ferricyanide more favorable as electron acceptor than oxygen because this latter will diffuse through the membrane to the anode compartment to oxidize methylene blue before process transport to the anode [14].Nafion®was used as proton exchange membrane, because of its high ionic conductivity [7].

Recently microbial fuel cell was implicated to be one of promising biotechnology for desalination and wastewater treatment [20–27].The electrical potential gradient is produced by exoelectrogenic bacteria by driving force of ion transport through an ion exchange membrane.

In the first part, we developed here a microbial fuel cell (MFC) to produce electricity from baker's yeast, *Saccharomyces cerevisiae.* The construction of biofuel cell is simple and is available quickly, without conditioning time, unlike microbial biofuel cell based on electroactive biofilm. In a second part, the concept was applied to desalination in an electrodialysis based cell. Simultaneously, we demonstrated that yeast fuel cell can serve as a viable option for integrated energy production and removed salt ion through desalination.

Yeast-based fuel cells have attracted the attention and interest of researchers although they deliver a lower power output than bacterial fuel cell but several studies have reported that by electrodes modification [28,29] or optimization of mediator *i.e.* methylene blue [7], methyl orange and methyl red [30,31], the performances of the biocatalytic systems could improve. As described previously, referring to the nature of yeast as a simple biocatalyst and does not require specific condition then we utilize their activities for energy production and followed for application in desalination.

Nowadays some research based on desalination have been developed [32–34]. The use of microbial desalination cell (MDC) represents a new approach for desalination. Some operational condition and the reactor design have varied widely, but the use of yeast as a biocatalyst is a new application. Yeast fuel cell has a great potential for desalination as a low-cost desalination process with significant environmental benefits. Therefore, it is evidence that some studies should be studied intensively to investigate this novel process and to enhance the performance of yeast fuel cell and desalination process.

MDC have recently drawn attention as a low-energy method of water desalination. The simplest MDC is a microbial fuel cell (MFC) that is modified to contain a middle chamber for desalination, by using two ion exchange membranes between the anode and cathode chamber [35]. In the anode chamber, organic matter is oxidized by a microorganism, with the electrons released by the circuit and protons into solution. Electrons from the anode flow to the cathode where they combine with protons and oxygen to form water [36, 37]. The production of protons at the anode and consumption of protons at the cathode drives desalination of saltwater in the dilute chamber, as salt ions in the saline water in the middle chamber (*i.e* dilute chamber) migrate through the cation and anion exchange membranes to balance charge [38]. As a result, the desalination process in an MDC does not require external energy input for driving in separation like that in conventional electrodialysis, thereby exhibiting significant energy benefits.

We propose here a new method of desalination process through microbial fuel cell with baker's yeast as a biocatalyst. For that, four-compartment desalination cell has been compiled consist anode

IOP Conf. Series: Materials Science and Engineering **107** (2016) 012049 doi:10.1088/1757-899X/107/1/012049

chamber, concentrate, dilute and cathode chamber as illustrated in figure 2. The current is generated at the anode by baker's yeast, and protons are migration through CEM. Therefore, the negatively charged species move from the dilute compartment into concentrate compartment. Positively charged species moved from the dilute compartment through CEM and will react with ferricyanide ion in the cathode compartment. This loss of ionic species from the dilute compartment through concentrate and cathode results in water desalination without external electrical energy supply.

We have been tested several conditions to optimize current and power density of the MFC. The experiments have been focused on the effect of yeast preparation, the behavior of mediator, temperature, and yeast stability. We can achieve two goals as simultaneously; the energy production and the desalination. In our present work, we demonstrated the challenges of yeast as a biocatalyst for practical application of MDC as a sustainable method for water desalination.

2. Material and methods

2.1. Preparation of yeast

A protocol of preparation for *Saccharomyces cerevisiae* has been used as previously designed [39]. It consists of preparing a slurry of 2 g. dried yeast, 1.8 g. peptone (Sigma-Aldrich, France), 1.5 g. dextrose (Sigma-Aldrich, France) and 1 g. malt extract (Sigma-Aldrich, France). The mixture is cultivated for 24h at 30°C in 50 mL phosphate buffer pH 7. Cells were harvested by centrifugation at 5000 rpm for 5 min., washed twice in phosphate buffer pH 7 and then re-suspended in phosphate buffer, and then kept at 4°C as stored cell, and activated at 40°C, 30 min. prior to use in MFC.

2.2. MFC architecture and experimental procedure

The fuel cells were constituted of anode and cathode compartments separated by Nafion® 117 (Dupont, USA) as proton exchange membrane. The Nafion® membrane was pretreated by boiling the film in H_2O_2 (30%), then washing in deionized water and 0.5 M H_2SO_4, each 1h and stored in deionized water until use. A schematic of the MFC system is shown in figure 1.

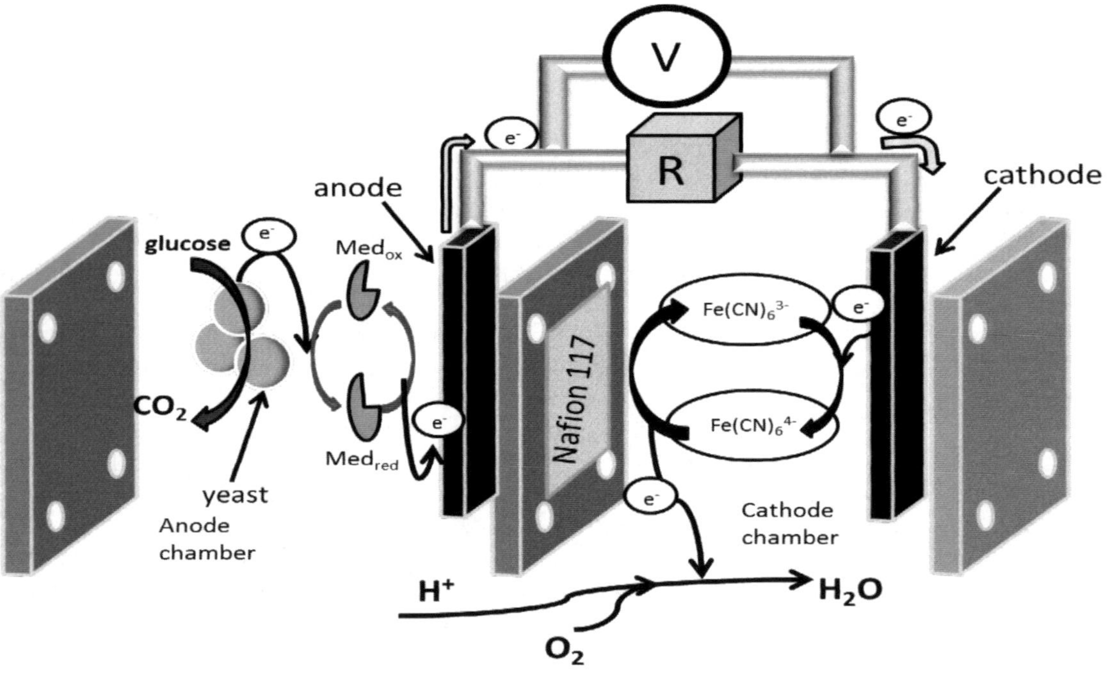

Figure 1. Scheme of yeast fuel cell system.

A two compartment fuel cell similar to one design by Walker [14] was constructed with graphite electrodes separated by a proton conducting membrane. One-half of the cell contained yeast, *Saccharomyces cerevisiae*, glucose and an electron mediator. The oxidation-reduction mediator enters the outer cell membrane and becomes reduced, then leaves in the reduced state. The reduced mediator transfers the captured electrons to an electrode, producing an electric current and re-oxidizing the mediator. The other half of the cell contained an electron acceptor, potassium ferricyanide as observed in a previous study [8, 40] which accepts the electrons from the circuit and becomes reduced. Oxygen in catholyte will oxidize Fe^{2+} back to Fe^{3+}. Ferricyanide was preferred as the electron acceptor over an oxygen-saturated solution because we did not want significant oxygen diffusion through the membrane into the anode compartment where it would oxidize the methylene blue before the methylene blue could transfer electrons to the electrode.

The observations have been made using *Chronoamperometry* and MFC. Chronoamperometry was carried out for 24 hours using graphite as the working electrode, platinum as a counter electrode and saturated calomel electrode (SCE) as a reference electrode. While, the MFC has run for several days and current generated were recorded using a multimeter Voltccraft model VC 850.

The desalination systems displayed in figure 2 was designed a cubic-shaped MFC and similar construction in previous study [41]. The device consists of a four compartments system, with a cation exchange membrane (CEM) next to the anode, an anion exchange membrane (AEM)separating concentrate and dilute compartments and aCEM next to the cathode. The addition of glucose was monitored to prevent the declining value of current. The inside volumes of each compartment were 80 mL. In anode compartment consist of yeast, mediator and glucose 0.1M. meanwhile, the concentrate and dilute compartment contain KNO_3 0.5 M and sea water, respectively. $K_3Fe(CN)_6$ 0.02 M in cathode compartment as electron acceptor.

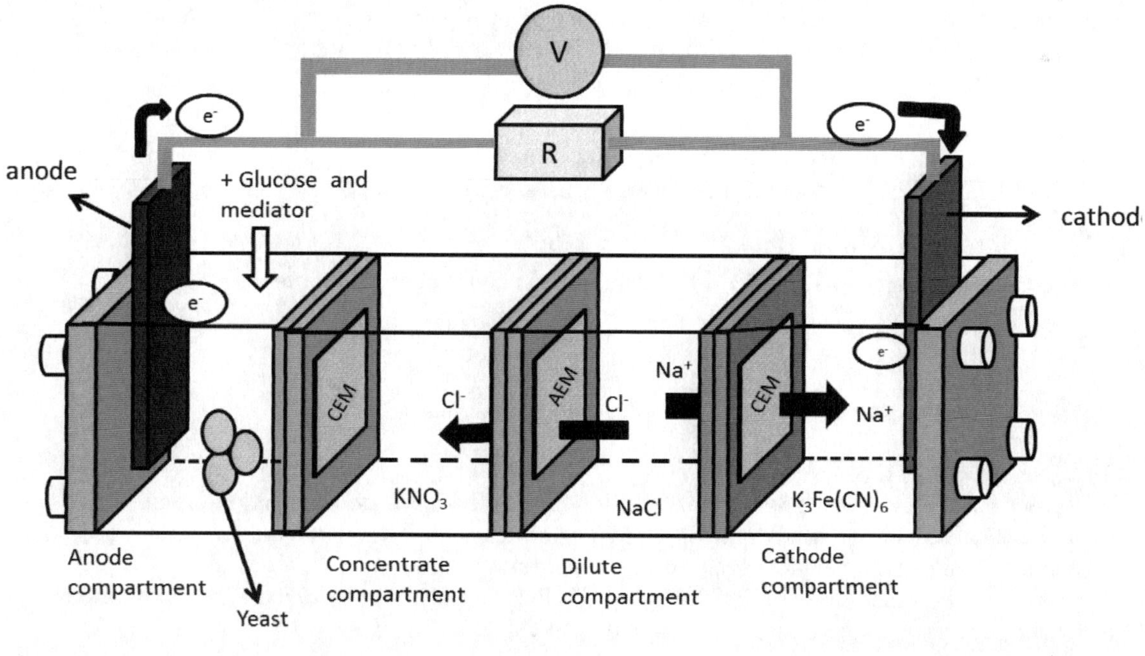

Figure 2. Scheme of yeast fuel cell system for desalination.(CEM: cation exchange membrane, AEM: anion exchange membrane, Mediator used was methylene blue).

The external resistance of the device (figure 2) was fixed at 1kΩ [7]. All desalination process was operated at ambient temperature (25±1°C). The voltage (E) in desalination process was recorded using

IOP Conf. Series: Materials Science and Engineering **107** (2016) 012049 doi:10.1088/1757-899X/107/1/012049

a digital multimeterVoltcraft model VC 850 and the current (I) generated was determined from the equation I = E/R. 2 mL of dilute and concentrate solution have been taken for the measurement of salt concentration by ionic chromatography (DIONEX ICS 900 for cation and DIONEX ICS 1000 for anion).

*2.2.1. Electrodes and membrane preparation.*Carbon graphite (7x1.5x 0.5cm) and nickel (7x1.5x 0,2cm) were used as anode and cathode material respectively. Previously washed in 1M HCl for 48h then rinsed with ultrapure water to clean from trash material. Nafion® membrane was cleaned in 0.5 M $HNO_3$for 24hand kept in ultrapure water prior used.

2.2.2. Anodic and cathodic solutions. The MFC anode chamber was fed with glucose monohydrate (Sigma-Aldrich, France) 0.1 M, yeast 2 g/100 mL, methylene blue (Sigma-Aldrich, France) 0.1mM in phosphate buffer pH 7. The phosphate buffer (PB) was obtained from 4.08 g.Na_2HPO_4 (Sigma-Aldrich, France) and 3.28 g.NaH_2PO_4(Sigma-Aldrich, France) dissolved in 1L ultrapure water. The cathode compartment contained potassium ferricyanide (PF) (Sigma-Aldrich, France) 0.02 M in phosphate buffer pH 7. Dilute and concentrate compartments consist of sea water from Palavas beach Montpellier, France and KNO_3 (Sigma-Aldrich, France)1M.

2.3. Desalination process

The kinetics of transport through the membrane was described as a first-order reaction in saltion concentration as described in equation (1)[42]:

$$ln(C/Co) = -kt \quad (1)$$

where C is the salt ion concentration at some given time in the concentrate compartment, Co the initial concentration of a salt ion in the dilute compartment, *k*is the rate constant (s^{-1}), and *t*is the time of transport (s). values were calculated from the plot of $ln\ (C/Co)$ vs time. The relation of (C/Co) vs time was linear. The permeability coefficient (*P*) was measured as equation (2).

$$P = -\frac{V}{A}k \quad (2)$$

Where V (L) is the volume of the solution in the compartment and A (m^{-2}) is an area of the effective membrane. The initial flux (*Ji*) were defined as equation (3).

$$J_i = PC_o \quad (3)$$

The percentage of salt ion transported was determined according to equation (4).

$$\%\ Transport = \frac{[C](t)}{[Co]} \times 100\% \quad (4)$$

3. Result and discussion

3.1. Preparation of yeast

The aim of the yeast preparation is its activation. The effect of yeast preparation on current response after one day of running is shown in figure 3. The current density generated after yeast preparation is higher than without preparation because of yeast growth. Yeast growth leads then to higher output current because of the increase in the quantity of glucose molecules oxidized. The Observations have also been conducted in MFC after performed during 6 days.

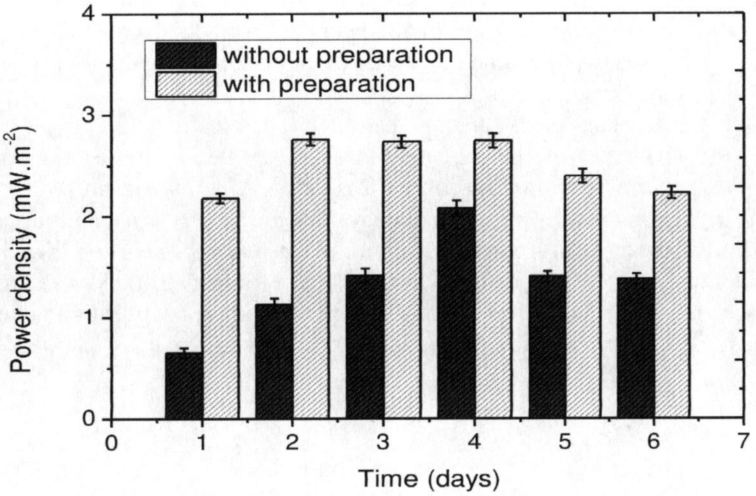

Figure 3. Power density ($mW.m^{-2}$) delivered from MFC as a function of yeast with and without preparation.

In the case of yeast preparation, malt extract and peptone influence the activation of the yeast. The metabolism is also more efficient, and oxidation of glucose starts quickly.

3.2. Determination amount of yeast.

The plate count is one of the most accurate means of enumeration of viable microbes because we get a visual indicator for every cell in the specimen. The technique stems from Robert Koch's insight gained from viewing colonies growing on the surface of a spoiling slice of potato. In practice, a small aliquot of a liquid suspension of microbes is spread on the surface of solidified nutrient medium, which when incubated, leads to each cell 'developing' into a visible colony through repeated fission. By counting the number of colonies that grow on a sterile agar plate, one can determine the number of yeast cells in a packet of yeast. However, there are too many yeast cells to count them so they must first be diluted. In a serial dilution, aliquots of some solution are diluted stepwise such that the first dilution serves as the source from which an aliquot is taken for the second dilution, etc.In this experimental part, the total plate count (TPC) of yeast has been calculated, and it was 3.10^7 /gr-cc. Meanwhile, the number of colonies on the control plate is used to reduce the amount of the dilution plate is a force multiplied by the dilution to obtain the amount of yeast per gram/cc. The number of colonies control plate must not bemorethan5. This method is used only to calculate the amount of live yeast.

3.3. The behavior and toxicity of mediator

In Baker's yeast-based fuel cells, exogenous mediators are a necessity since *S. cerevisiae* is not known to produce such mediators indigenously.To improve the efficiency of the mediated electron transport from the microorganism to the anode and thus to enhance the generated power, methylene blue and neutral red have been selected as an anartificial mediator as described in the previous study[6,43]. In this section, the mediator has been made in suspension solution of PB pH 7 with a concentration of10mM, respectively. It has been reported that power delivered as a function of mediator applied 29.8 mW from methylene blue and 30.4 mW from neutral red respectively.

Results show that both methylene blue and neutral red did not have significant differences. They can be accepted and applied in MFC based *Saccharomyces cerevisiae*. Both methylene blue and neutral red could soluble and chemically stable in the anolyte and have all these requirements and also applicable for yeast species. As studied previously by Babanova [6], methylene blue and neutral red

IOP Conf. Series: Materials Science and Engineering **107** (2016) 012049 doi:10.1088/1757-899X/107/1/012049

have resulted in a good performed when applied in MFC based yeast *Candidamelibiosica.* It is suggested that the redox potential of methylene blue and neutral red has than closer to $NADH/NAD^+$ (-0.320 V vs.NHE) while another artificial mediator such as methylene green, methyl orange, and methyl red is not so. Taking this, we presumed that the compounds with a potential redox closer to the $NADH/NAD^+$ potential could take an electron from NADH.

As a good mediator, the artificial mediator must be electrochemically active, must easily penetrate the cell membrane and the kinetic of the oxidation process at the electrode and non-toxic to the microorganisms. For that purpose, we analyzed cytotoxicity for methylene blue and neutral red as mediators. Sterile filter paper disks were soaked with 1 mM and 10 mM solutions of meditators. Then, they were placed in contact with a constant amount of *Saccharomyces cerevisiae* in a monolayer on agar medium and incubated at 28^{0}C for 24 hours (Kirby-Bauer' disc agar diffusion test)[44]. Results show that the absence of zones of inhibition around the samples suggested biocompatibility of the compound studied (*i.e.* methylene blue and neutral red), and they can be applied in yeast-based fuel cell.

3.4. The influence of temperature
The influence of temperature on MFC is presented in figure 4. The observation has been done after operating during 7 days. The optimal temperature was 40°C. Average power density increased from 1.20 $mW.m^{-2}$ at ambient temperature to about 2.44 $mW.m^{-2}$ at 30°C and reached 4.75 $mW.m^{-2}$ at 40°C. At 50°C, the average power density is only 2.32 $mW.m^{-2}$

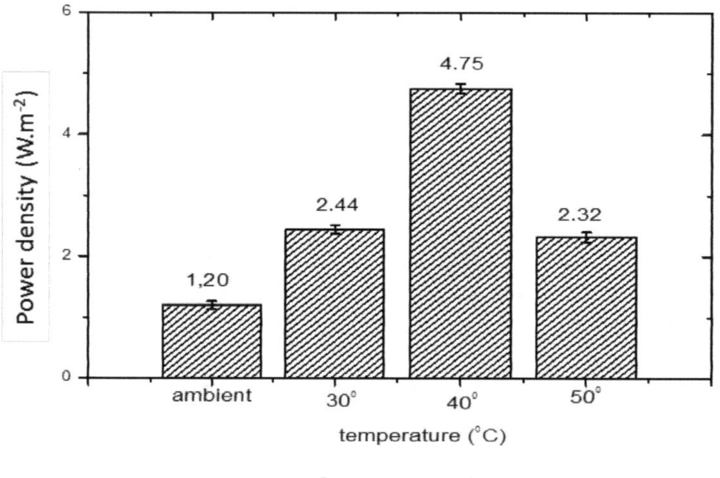

Figure 4. Power density $mW.m^{-2}$ generated by the MFC as a function of temperature.

Yeast is a microorganism which their activity could affect, one of them is by temperature. Results confirm that one of the factors of the microorganism growth is temperature. The optimal value has been achieved at 40°C. If the system applied at the higher temperature, it could cause the damage of microorganism viability and as a consequently the MFC process will be disrupted. This has been evidenced by operating MFC at 50^{0}C; power density has been reduced by 51%.

3.5. Power stability
The stability of the power delivered by the MFC was determined by chronoamperometry at 0.3 V vs. SCE over three weeks and displayed in figure 5.

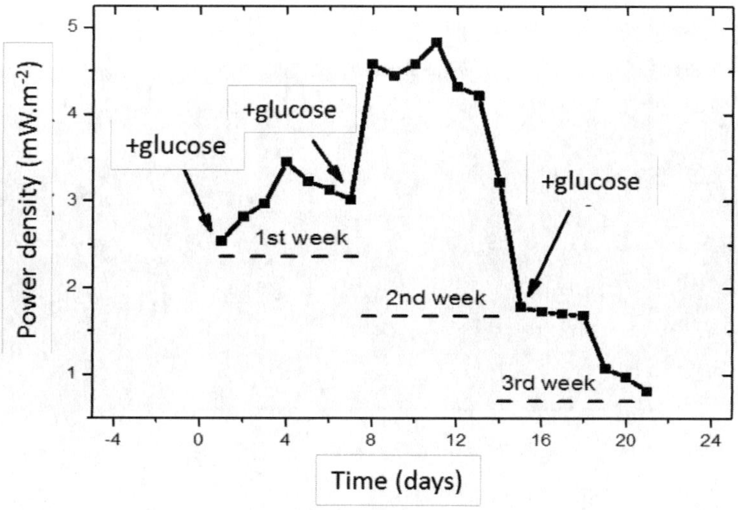

Figure 5. Power density (mW.m^{-2}) generated by the MFC as a function of time.

This experiment has been carried out in the batch model, without refreshment of *Saccharomyces cerevisiae*. The addition of glucose was done at the beginning of each week as a source of fuel. The maximum power density has been achieved in the second week. After the addition of glucose at the beginning of the third week, the results were not an increase in power density. It is assumed that this corresponds to a lifetime of yeast. The coulomb efficiency (CE) defined as the fraction of electrons extracted for conversion into electricity versus that in the starting organic material, calculated by equation (5)[45].

$$CE = = \frac{\int i \, dt}{b F c v} \quad (5)$$

Where I was current over time, b is the number of mol of electrons produced per mole of glucose (b = 24 for glucose), F was faraday 96500 mol^{-1} of the electron, c was consumed substrate concentration (c = 0.1 M), v was the analyte volume (v= 100 mL).The column efficiency for each cycle was shown in Table 1.

Table 1. Coulomb efficiency delivered from each cycle

Time	Coulomb efficiency (CE)
1st week	43.5
2nd week	61.7
3rd week	23.4

From table 1 shows that the second week it has been a maximum of the CE value, but not so with that seen from 3rd week. It was the lowest point. We assumed that there has been a decrease of yeast activities in oxidize substrates.

3.6. Electricity generation from MFC based yeast as biocatalyst

The MFC has been performed for 8 days, and the observation of the current have been recorded every day. During the process, there was no supplemented of substrate and replacement of PF. The result was presented in this figure 6.

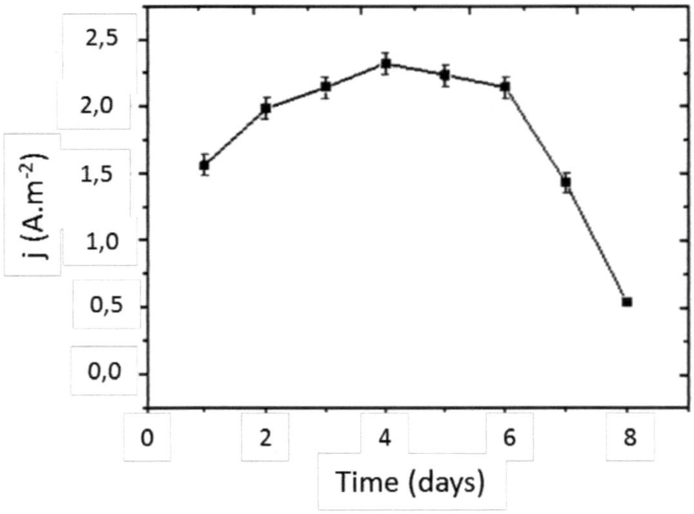

Figure 6. Current density ($A.m^{-2}$) delivered from MFC.

The composition anode chamber consisted of 2 % of yeast suspension solution in 100 mL BP pH 7, methylene blue 10 mM and glucose 0.1 M, respectively. Cathode chamber was consist of PF 0.02 M. The MFC was monitored using a Volt multi meter craft (model VC 850). Voltage was recorded, different resistors were applied, and current have been measured at 1 k Ω as previously studied [7]. The observation of value was recorded for 5-10 min until the voltage stabilized. Electricity was generated with maximum value 2.3 $A.m^{-2}$. The decrease value was assumed the viability of yeast or reducing the concentration of glucose during MFC performed. To further understand the effect of the anolyte recirculation and energy consumption is needed to investigate more.

Figure 6 shows at the beginning, the increase of the current signal in the presence of glucose is due to the activities of *Saccharomyces cerevisiae* allowing the catalytic oxidation of glucose and electron transfer occurs through the facilitation of the artificial redox mediator (*i.e.,* methylene blue) to the surface of the anode. The reactions at the anode and the cathode compartments are depicted in this equation (6).

anode:

$MB_{(oxidized)} + NADH_{(yeast)} \rightarrow NAD^+_{(yeast)} + MB_{(reduced)}$

$MB_{(oxidized)} + NADH_{(yeast)} \rightarrow NAD^+_{(yeast)} + MB_{(reduced)}$

$MB_{(reduced)} \rightarrow MB_{(oxidized)} + 2e^- + H^+$

Cathode:

$4PF_{(oxidized)} + 4e^- \rightarrow 4PF_{(reduced)}$

$$4PF_{(reduced)} + O_2 + 4H^+ \rightarrow 2H_2O + 4PF_{(oxidized)} \quad (6)$$

3.7. Desalination

In our present work, the energy produced from MFC will be applied in desalination process. We have been constructed a desalination cell as displayed in Figure 2, using $K_3Fe(CN)_6$ as an electron acceptor in the cathodic chamber and glucose as an electron donor, methylene blue as mediator and baker's yeast as biocatalyst in the anodic chamber. The electrochemical chain is constituted by the following arrangement: anodic chamber, concentrate compartment, dilute compartment, and cathodic chamber. CEM was placed close to the anodic chamber and AEM among CEM sequentially. Graphite and nickel were used as anode and cathode respectively. Migration of each ion was determined by measuring the concentration of salt by ionic chromatography during the process.

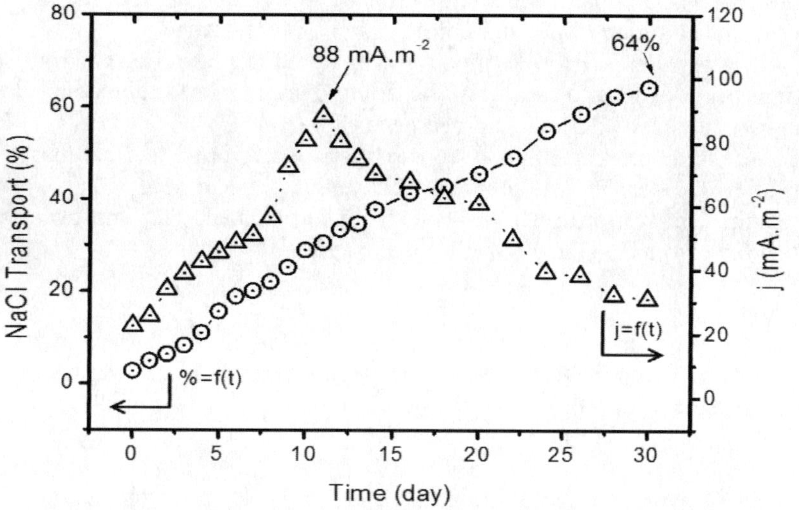

Figure 7. NaCl transport (%) and current density generated from desalination process during 30 days of run.

Results confirm that 64 % of NaCl initially continued in the dilute compartment moved to the concentrate compartment after 1 month under a maximum current density of 88 mA.m^{-2}. There was twice times replacement of PF and the addition of glucose for maintenance performance during desalination. Meanwhile, the batch operation has been applied in anolyte chamber, its mean during the process there is no replacement of yeast. The current generated was recorded every 30 min. The efficiency of the desalination is demonstrated by the flux of salt in the case of seawater demineralization. The flux value was 25 ± 1. A small flux increase is even recorded with sea water probably due to the higher conductivity of sea water. The desalination rate is the one of the important parameters of MDC performance, which greatly dependent on the salt ion sea water. Ideally, the salt concentration in dilute compartment should be higher than the electrolyte concentration (*i.e.,* KNO_3) since a low concentration could deliver in lowering the desalination rate.

The desalination performance can also be increased using multiple ion exchange membrane (IEM) pair which inserted between anode and cathode compartment. The increase in the number of cell pairs reduces the voltage required for each cell allowing a greater net energy obtained. Moreover, the use of thin IEM and desalination compartment, the internal resistance diminishes and the efficient of ion separation and water desalination can be achieved [41].

There are some challenges that must be addressed in the future development of the yeast fuel cell and the application of desalination. First, the improvement of power delivery from this system needs to be further improved; because the efficiency process is necessary to increase. One possible approach to improving performance by modification of anode or cathode. Besides that, a strategy for a system scale up especially in desalination based yeast as biocatalyst should be developed to transform laboratory results into a practical technology.

4. Conclusion

This study has demonstrated a biological fuel cell operating with glucose oxidation by baker's yeast; *Sacharomycescereviceae* was performed. The effect of yeast preparation, mediator, temperature and stability of yeast were evaluated. Application yeast to water desalination was also studied. The ability of yeast to produce electricity was examined. The optimum conditions were obtained at 40°C.methylene blue was accepted and nontoxic for living cell and the lifetime is around 2 weeks and the effect of yeast preparation resulted in improving MFC performance.

Application of yeast fuel cell for desalination was tested to recover NaCl from sea water. After 1 month operation, 64 % of ions moved from the dilute compartment to the concentrate compartment. The maximum current density during the process was 88mA.m^{-2}.

We have presented a simple method of construction of microbial fuel cell based on baker's yeast and the application for desalination process. This autonomous device able to produce power and remove salt simultaneously from real sea water has been studied. The improvement of this process with immobilization of yeast on the electrode is in progress.

Acknowledgements

This work was supported by grant Doctor Scholarships Indonesian Directorate General of Higher Education (DIKTI) in the framework Double Degree Indonesia-France and partially supported by Membranes European Institute (IEM) Montpellier France.

References

[1] Potter AM C, Character B and Sep N 2010 Electrical Effects Accompanying the Decomposition of Organic Compounds Published by *The Royal Society Electrical Effects accompanying the Decomposition of Organic* **84** 260–76

[2] Rabaey K and Verstraete W 2005 Microbial fuel cells: novel biotechnology for energy generation. *Trends Biotechnol.***23** 291–8

[3] Logan B E and Regan J M 2006 Electricity-producing bacterial communities in microbial fuel cells. *Trends Microbiol.***14** 512–8

[4] Cheng S, Xing D and Logan B E 2011 Electricity generation of single-chamber microbial fuel cells at low temperatures. *Biosens. Bioelectron.***26** 1913–7

[5] Hubenova Y and Mitov M 2010 Bioelectrochemistry Potential application of Candida melibiosica in biofuel cells *Bioelectrochemistry***78** 57–61

[6] Babanova S, Hubenova Y and Mitov M 2011 Influence of artificial mediators on yeast-based fuel cell performance*J. Biosci. Bioeng.***112** 379–87

[7] Rahimnejad M, Najafpour G D, Ghoreyshi A, Shakeri M and Zare H 2011 Methylene blue as electron promoters in microbial fuel cell *Int. J. Hydrogen Energy***36** 13335–41

[8] Gunawardena A, Fernando S and To F 2008 Performance of a Yeast-mediated Biological Fuel Cell 1893–907

[9] Sayed E T, Tsujiguchi T and Nakagawa N 2012 Catalytic activity of baker's yeast in a mediatorless microbial fuel cell*Bioelectrochemistry***86** 97–101

[10] Prasad D, Arun S, Murugesan M, Padmanaban S, Satyanarayanan R S, Berchmans S and Yegnaraman V 2007 Direct electron transfer with yeast cells and construction of a mediatorless microbial fuel cell. *Biosens. Bioelectron.***22** 2604–10

[11] Zou Y, Xiang C, Yang L, Sun L, Xu F and Cao Z 2008 A mediatorless microbial fuel cell using polypyrrole coated carbon nanotubes composite as anode material **33** 4856–62

[12] Liu Z and Li H 2007 Effects of bio- and abio-factors on electricity production in a mediatorless microbial fuel cell **36** 209–14

[13] Liu J, Qiao Y, Xian C, Lim S, Song H and Ming C 2012 Bioresource Technology Graphene / carbon cloth anode for high-performance mediatorless microbial fuel cells *Bioresour. Technol.***114** 275–80

[14] Walker A L and Walker C W 2006 Biological fuel cell and application as a reserve power source *J. Power Sources***160** 123–9
[15] Cheng S and Liu H 2006 Power Densities Using Different Cathode Catalysts (Pt and CoTMPP) and Polymer Binders (Nafion and PTFE) in Single Chamber Microbial Fuel Cells **40** 364–9
[16] Cheng S, Liu H and Logan B E 2006 Increased performance of single-chamber microbial fuel cells using an improved cathode structure *Electrochem. commun.***8** 489–94
[17] Raghavulu S V, Goud R K, Sarma P N and Mohan S V 2011 Bioresource Technology Saccharomyces cerevisiae as anodic biocatalyst for power generation in biofuel cell : Influence of redox condition and substrate load *Bioresour. Technol.***102** 2751–7
[18] Nielsen J, Larsson C, Maris A Van and Pronk J 2013 Metabolic engineering of yeast for production of fuels and chemicals *Curr. Opin. Biotechnol.***24** 398–404
[19] Haslett N D, Rawson F J, Barriëre F, Kunze G, Pasco N, Gooneratne R and Baronian K H R 2011 Biosensors and Bioelectronics Characterisation of yeast microbial fuel cell with the yeast Arxula adeninivorans as the biocatalyst *Biosens. Bioelectron.***26** 3742–7
[20] Zhang F, Chen M, Zhang Y and Zeng R J 2012 Microbial desalination cells with ion exchange resin packed to enhance desalination at low salt concentration **418** 28–33
[21] Luo H, Xu P and Ren Z 2012 Bioresource Technology Long-term performance and characterization of microbial desalination cells in treating domestic wastewater *Bioresour. Technol.***120** 187–93
[22] Kim Y and Logan B E 2013 Simultaneous removal of organic matter and salt ions from saline wastewater in bioelectrochemical systems *Desalination***308** 115–21
[23] Kim Y and Logan B E 2013 Microbial desalination cells for energy production and desalination *Desalination***308** 122–30
[24] Du Z, Li H and Gu T 2007 A state of the art review on microbial fuel cells: A promising technology for wastewater treatment and bioenergy. *Biotechnol. Adv.***25** 464–82
[25] Qu Y, Feng Y, Liu J, He W, Shi X, Yang Q, Lv J and Logan B E 2013 Salt removal using multiple microbial desalination cells under continuous flow conditions *Desalination***317** 17–22
[26] Brastad K S and He Z 2013 Water softening using microbial desalination cell technology *DES***309** 32–7
[27] Mehanna M, Saito T, Yan J, Hickner M, Cao X, Huang X and Logan B E 2010 Using microbial desalination cells to reduce water salinity prior to reverse osmosis *Energy Environ. Sci.***3** 1114
[28] Wang K, Liu Y and Chen S 2011 Improved microbial electrocatalysis with neutral red immobilized electrode *J. Power Sources***196** 164–8
[29] Wei J, Liang P, and Huang X 2011 Recent progress in electrodes for microbial fuel cells. *Bioresour. Technol.***102** 9335–44
[30] Hosseini M G and Ahadzadeh I 2013 Electrochemical impedance study on methyl orange and methyl red as power enhancing electron mediators in glucose-fed microbial fuel cell *J. Taiwan Inst. Chem. Eng.***44** 617–21
[31] Park D H and Zeikus J G 2000 Electricity Generation in Microbial Fuel Cells Using Neutral Red as an Electrophore Electricity Generation in Microbial Fuel Cells Using Neutral Red as an Electrophore**66**
[32] Jacobson K S, Drew D M, and He Z 2011 Efficient salt removal in a continuously operated upflow microbial desalination cell with an air cathode. *Bioresour. Technol.***102** 376–80
[33] Luo H, Xu P, Roane T M, Jenkins P E and Ren Z 2012 Microbial desalination cells for improved performance in wastewater treatment, electricity production, and desalination *Bioresour. Technol.***105** 60–6
[34] Qu Y, Feng Y, Wang X, Liu J, Lv J, He W and Logan B E 2012 Simultaneous water desalination and electricity generation in a microbial desalination cell with electrolyte recirculation for pH control. *Bioresour. Technol.***106** 89–94

[35] Miller S, Shemer H and Semiat R 2015 Energy and environmental issues in desalination *Desalination* **366** 2–8

[36] Mehanna M, Kiely P D, Call D F and Logan B E 2010 Microbial Electrodialysis Cell for Simultaneous Water Desalination and Hydrogen Gas Production **44** 9578–83

[37] Liang P, Xiao K, Zhou Y, Zhang X and Logan B E 2009 A New Method for Water Desalination Using Microbial Desalination Cells **43** 7148–52

[38] Zhang F, Chen M, Zhang Y and Zeng R J 2012 Microbial desalination cells with ion exchange resin packed to enhance desalination at low salt concentration *J. Memb. Sci.* **417-418** 28–33

[39] Sayed E T, Tsujiguchi T and Nakagawa N 2012 Catalytic activity of baker's yeast in a mediatorless microbial fuel cell. *Bioelectrochemistry* **86** 97–101

[40] Kong X, Sun Y, Yuan Z, Li D, Li L and Li Y 2010 Effect of cathode electron-receiver on the performance of microbial fuel cells *Int. J. Hydrogen Energy* **35** 7224–7

[41] Kim Y and Logan B E 2011 Series Assembly of Microbial Desalination Cells Containing Stacked Electrodialysis Cells for Partial or Complete Seawater Desalination 5840–5

[42] Kozlowski C a., Walkowiak W and Girek T 2008 Modified cyclodextrin polymers as selective ion carriers for Pb(II) separation across plasticized membranes *J. Memb. Sci.* **310** 312–20

[43] Wilkinson S, Klar J and Applegarth S 2006 Optimizing Biofuel Cell Performance Using a Targeted Mixed Mediator Combination *Electroanalysis* **18** 2001–7

[44] Alagumaruthanayagam A, Pavankumar A R, Vasanthamallika T K and Sankaran K 2009 Evaluation of solid (disc diffusion)- and liquid (turbidity)-phase antibiogram methods for clinical isolates of diarrheagenic E. coli and correlation with efflux. *J. Antibiot. (Tokyo).* **62** 377–84

[45] Yong Y-C, Liao Z-H, Sun J-Z, Zheng T, Jiang R-R and Song H 2013 Enhancement of coulombic efficiency and salt tolerance in microbial fuel cells by graphite/alginate granules immobilization of Shewanella oneidensis MR-1 *Process Biochem.* **48** 1947–51

Fabrication of organic solar cells with design blend P3HT: PCBM variation of mass ratio

Agus Supriyanto[1*], Amrina Mustaqim[1], Maya Agustin[1], Ari H Ramelan[1], Suyitno[2], Erlyta Septa Rosa[3], Yofentina[1], Fahru Nurosyid[1]

[1] Department of Physics, Faculty of Mathematics and Natural Science, Sebelas Maret University, Jl. Ir. Sutami 36A Kentingan, Surakarta, Indonesia

[2] Department of Mechanical Engineering, Faculty of Engineering Sebelas Maret University, Jl. Ir. Sutami 36A Kentingan, Surakarta, Indonesia

[3] Research Center for Electronics and Telecommunication Indonesian Institute of Science (PPET-LIPI), Jl. Sangkuriang - Komplek LIPI Gedung 20, Lt. 4 Bandung 40135, West Java, Indonesia

*E-mail : a_supriyanto@uns.ac.id; agusf22@yahoo.com

Abstract. Organic solar cells of FTO/PEDOT: PSS/P3HT: PCBM/Al has been fabricated, and its performance has been tested in dark and under various illumination of light intensity 1000 W/m^2. The active materials used in this study are poly (3-hexylthiophene) (P3HT) and [6, 6]-phenyl-C_{61}-butyric acid methyl ester (PCBM). P3HT is the donor while PCBM acts as an acceptor. Variation of PCBM and P3HT are 1:1, 1:2, 1:3, 1:4 and 1:5. P3HT: PCBM was mixed by chlorobenzene solvents. The mixing was done by using the ultrasonic cleaner. The absorbance characterization using by UV-Visible Spectrometer Lambda 25 instrument and I-V characterization has been tested using a set of 2602A Keithley instrument. Absorbance characterization shows that two peaks are formed. The first peak in the range of 300 to 350 nm which is a range of PCBM and the second peak range from 450 to 600 nm which is a range of P3HT. As the mass ratio increases, the second peak of P3HT increases while the first peak does not change. The gap energy estimated by the Tauc method is 2.0 eV. I-V characterization of the efficiency was obtained. The efficiency of sample 1, 2, 3, 4, and 5 are 5.80x10^{-2}%; 6.46x10^{-2}%; 7.72x10^{-2}%; 8.25x10^{-2}% and 9.81x10^{-2}%, respectively. The highest value of efficiency was obtained at mass ratio 1:5.

1. Introduction

Important challenges in the field of organic photovoltaic are how to provide a viable alternative for the energy production at large scale including the development processes for up-scaling the fabrication of cells to modules [1]. Another benefit of organic polymeric solar cells is it allows molecular level modification, and the unlimited organic semiconductor materials [2]. The organic materials also can be extracted from some cultivated plants [2]. Another research on organic solar cells (OSC), which is being developed, is a bulk heterojunction (BHJ) device. The BHJ type becomes an option because it has a polymeric active material in a blend, roles as donor and acceptor [3-5]. Currently, the structure of a P3HT (poly(3-hexylthiophene)): PCBM

(6,6-phenyl-C61butyric acid methyl ester) bulk heterojunction (BHJ) cell, which may be considered as one of the most standard organic solar cell (OSC) devices.

An active material can work properly if the donor and acceptor have a proper HOMO (High Occupied Molecular Orbital) and LUMO (Lowest Unoccupied molecular Orbital) [6]. P3HT (*Poly(3-hexylthiophene)*) and PCBM (*Phenyl-C61-Butyric Acid Methyl Ester*) polymer has HOMO 3 eV and 4 eV respectively, facilitating electron transfer from P3HT to PCBM [7]. Poly (3-hexylthiophene) or P3HT polymeric semiconductor has matching criteria as donor material and acts as a p-type. Poly (3-hexylthiophene) has a molecular weight of 65.5 g/mol and a high hole mobility of 3.8~3.9x10^{-4} cm^2/Vs [8]. PCBM or C61-ButyricAcid Phenyl-Methyl Ester is derived from fullerenes. PCBM is typically used as a good acceptor material in organic solar cells. This causes the PCBM be a material uses in solar cell active ingredient compared to other fullerene-derived material [9].

2. Experimental procedure

P3HT and PCBM polymer were blended in the mass ratio of PCBM and P3HT (1:1; 1:2; 1:3; 1:4; and 1:5). P3HT and PCBM were mixed with a 1% concentration of chlorobenzene solvent. A solution of P3HT: PCBM was deposited on glass substrates. This deposition was performed by spin coater instrument at 2500 rpm angular velocity for 30 seconds [10]. The optical absorbance of samples were then analyzed by a UV-Vis Spectrometer (Lambda 25 instrument) with a wavelength ranging from 300-800 nm.

PEDOT: PSS layer was grown on FTO by spin coating method for 20 seconds at 2500 rpm of angular velocity. The method produces FTO / PEDOT: PSS [11]. The next process was growing a P3HT: PCBM active material by spin coater as well producing. FTO / PEDOT: PSS / P3HT: PCBM. The last process was metallization of Al by evaporator instrument. The organic solar cell devices were established with the structure of FTO / PEDOT: PSS/P3HT: PCBM/Al. The absorbance properties of the fabrication layers were then observed.

The current–voltage characteristic in dark at room temperature was analyzed by a 2602A Keithley instrument high voltage source. The measurement was performed under dark and under illumination to study the photosensitizing effect in the device. The device was illuminated by a visible light from xenon light source. The xenon light source was calibrated by adjusting the light intensity until 1000 W/m^2. The intensity was controlled by a solar power meter, meanwhile, the temperature was measured by the thermocouple. The illuminated cell area was 10 mm^2. The photosensitizing parameters of short circuit current density (I_{sc}) and open circuit voltage (V_{oc}) were obtained from the intersection axis of current and voltage from the current-voltage curve under dark and illumination, respectively. Figure 1 shows the structure of organic solar cells of FTO/PEDOT: PSS/P3HT: PCBM/Al.

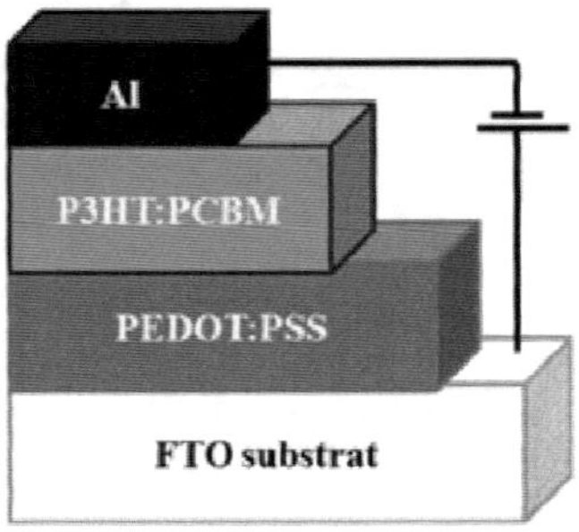

Figure 1. The structure of organic solar cells of FTO/PEDOT: PSS/P3HT: PCBM/Al.

3. Results and discussion

Figure 2 shows the absorption spectra of PCBM and P3HT film. The absorption spectra of PCBM shows that the material has the capability to absorb light at a wavelength of 350 nm,

IOP Conf. Series: Materials Science and Engineering **107** (2016) 012050 doi:10.1088/1757-899X/107/1/012050

which is in a UV range. The absorbance of P3HT is in the range of 450 to 600 nm, which is categorized as a visible light range.

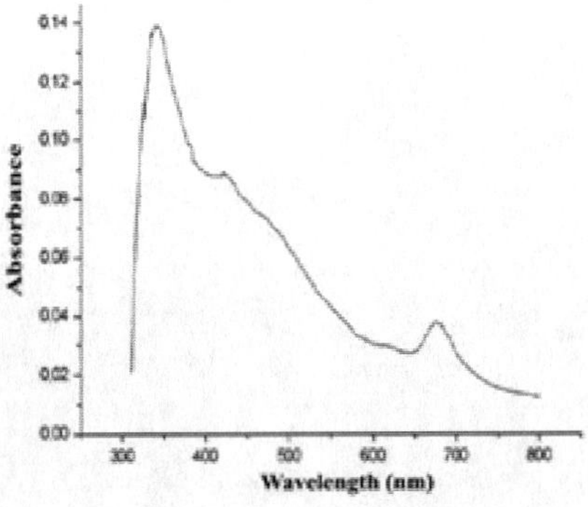

Figure 2. (a) The Absorption spectrum of PCBM

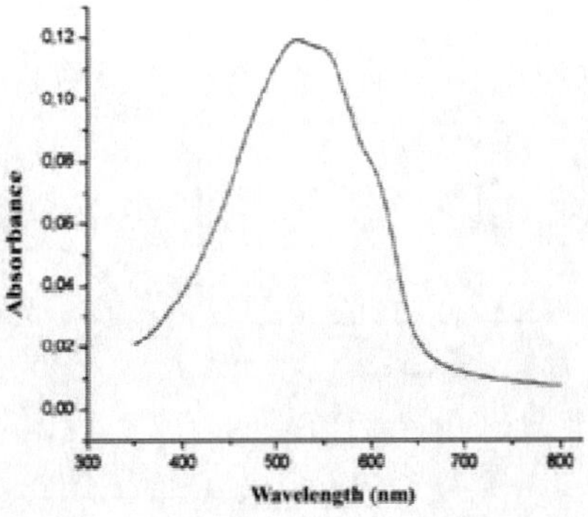

Figure 2. (a) The Absorption spectrum of P3HT

Figure 3 shows the absorption spectra of PCBM and P3HT at various mass ratio (1:1; 1:2; 1:3; 1:4; and 1:5). The first peak is in the absorption region of PCBM and the second peak is in the absorption region of P3HT. The second peak increases as the increasing of the mass ratio of P3HT. Meanwhile, the first peak of the five samples has relatively similar intensities. It means that the addition of P3HT led to increasing the absorbance value of the second peak, which is in the range of visible light.

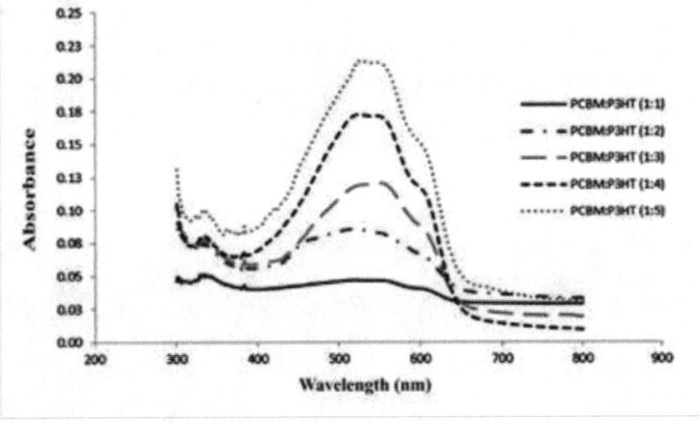

Figure 3. The absorbance spectrum of PCBM: P3HT at various mass ratio

The *I-V* graph characteristic of organic solar cells was measured under dark and under illuminated condition using a 2602A Keithley instrument high voltage source. The *I-V* graph characteristic was used to determine -V_{oc}, I_{sc}, fill factor (FF) and the efficiency of organic solar cells. V_{oc} is the voltage at zero current or in an open loop voltage condition. I_{sc} is the current at zero voltage. The mass ratio of PCBM: P3HT are 1:1, 1:2, 1:3, 1:4 and 1:5, therefore, the *I-V* graph characteristic of organic solar cells have five conditions. The study of *I-V* characteristic of organic solar cells shown in figure 4 used xenon lamp for illumination with the intensity of 1000 W/m^2 and measurement area of 10 mm^2.

The best ratio of PCBM: P3HT is 1:5. The ratio produced high efficiency can be observed in figure 4(e). Figure 4(e) -shows the electrons can be produced more from P3HT due to its properties as an electron donor.

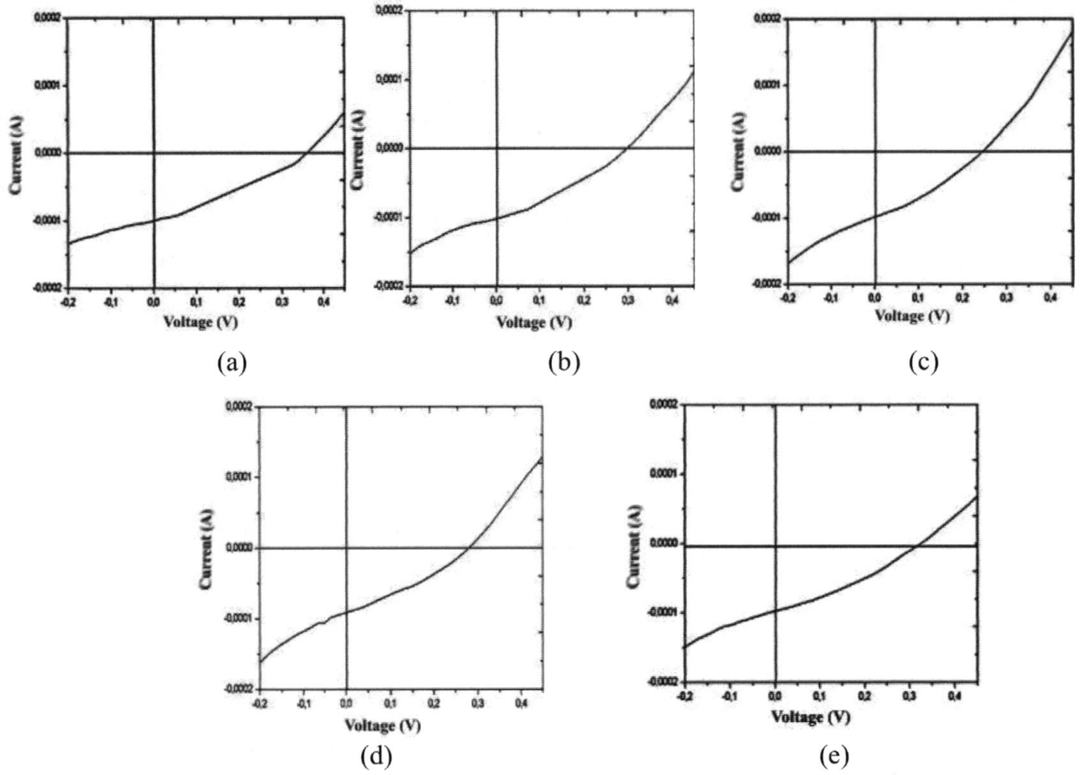

Figure 4. *I-V* graph characteristic of organic solar cells PCBM:P3HT with the light intensity of 1000 W/m^2 (a) ratio 1:1 (b) 1:2 (c) 1:3 (d) 1:4 and (e) 1:5

Meanwhile, the I-V curves can not be observed when the solar cell was operated under dark condition.

The results in figure 4 show that the value of I_{sc} increases, but the value of V_{oc} decreases. V_{max} is the maximum voltage produced by organic solar cells and I_{max} is the maximum current. Table 1 shows the *I-V* properties of organic solar cells characterization. Table 1 shows that the efficiency of organics solar cells were increased as the mass ratio of PCBM: P3HT. It is due to the ability of P3HT to be active in the wavelength range of visible light, and producing so more electrons.

Table 1. I-V characterization of PCBM:P3HT

Ratio PCBM:P3HT	I_{max} (x10^{-2} mA)	V_{max} (V)	I_{sc} (x10^{-2} mA)	V_{oc} (V)	*FF*	*Eff* (x10^{-2}%)
1:1	1.75	0.32	0.98	0.34	0.17	5.8
1:2	2.59	0.25	1.00	0.29	0.21	6.4
1:3	5.94	0.13	1.03	0.25	0.32	7.7
1:4	5.14	0.15	1.06	0.24	0.35	8.2
1:5	6.68	0.14	1.07	0.23	0.36	9.8

Figure 5 shows that the values of I_{sc} increases as the ratio of P3HT polymer increasing. It is due to the absorbance ability of P3HT is at the visible light range. This absorbance influence on the number of photons, which are absorbed by the organic solar cells. The more photons absorbed, the more electrons can be generated. This result is in agreement with the results of the research of Khlyabich (2013) on the donor material.

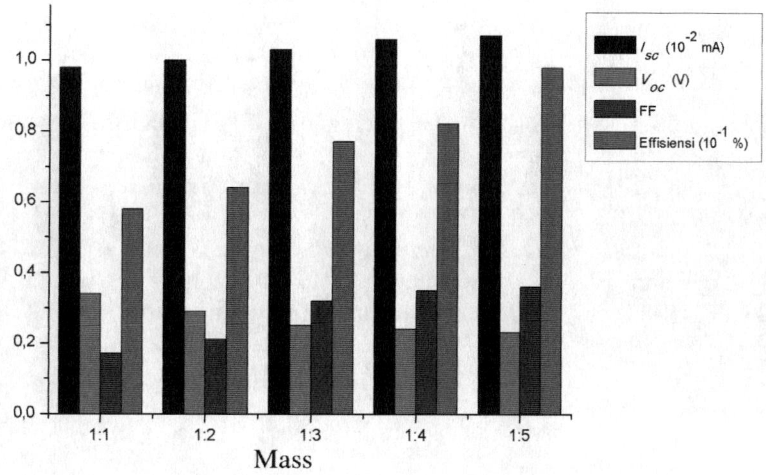

Figure 5. Mass ratio comparison graph of PCBM: P3HT to Isc, Voc, fill factor (FF) and efficiency

Figure 5 shows that the fill factor value increases as the ratio of PCBM: P3HT. Higher fill factor indicates less recombination between holes and electrons because the transportation of free charge is running well. In addition, figure 5 also shows that the efficiency increased as the ratio of PCBM: P3HT increased as well as the value of I_{sc}.

4. Conclusion

The higher ratio of P3HT produces an absorbance peak the range of 450 to 650 nm. Meanwhile, the first peak (peak of PCBM) is in the range of 300 to 350 nm. The estimates gap energy is 2.0 eV. The highest efficiency is on the ratio of PCBM and P3HT 1: 5, i.e. 9.8×10^{-2}%.

Acknowledgement

The authors would like to acknowledge Kemeristek-Dikti Republik Indonesia for financial support in Penelitian Unggulan Perguruan Tinggi (PUPT 2015) of Sebelas Maret University (UNS), Indonesia.

References

[1] Medford, A J Lilliedal, M R Jørgensen, M Aarø, D Pakalski, H Fyenbo and J Krebs, F C 2010 Grid-connected polymer solar panels: initial considerations of cost, lifetime, and practicality, *Optics Express* **18** (S3) A272–A285.

[2] Wanzhu Cai, Xiong Gong and Yong Cao 2010 Polymer solar cells: Recent development and possible routes for improvement in the performance. *Sol Energ Mat Sol C,* **94(2)**: 114-127

[3] Scharber M C and Sariciftci N S 2013 Efficiency of Bulk-Heterojunction Organic Solar Cells *Progress Polymer Science*

[4] Nelson J 2011 Polymer:Fullerene Bulk Heterojunction Solar Cells *Materials* Today, **14,** 462-470

[5] Dennler N, Sariciftci S and Brabec C J 2006 *Conjugated Polymer-Based Organic Solar Cellsin Semiconducting Polymers: Chemistry, Physics and Engineering,* Vol **I** Second Edition edited by G Hadziioannou and G G Malliaras, Weinheim (Wiley-VCH Verlag GmbH & Co Kga) p 511-512

[6] Khlyabich P P, Burkhart B, Rudenko A E and Thompson B C 2013 Optimization and Simplification of Polymere-fullerene Solar Cells Through Polymer and Active Layer Design *Polymer* ***54*****,** 5267-5298

[7] Stubinger T and Brutting W 2001 Exciton Diffusion and Optical Interference in Organic Donor-Acceptor Photovoltaic Cells *Jpn J App Phys* 90 **07** 3632-3641

[8] Kumar G Li, Z Hong dan Y Yang 2009 High efficiency polymer solar cells with vertically modulated nanoscale morphology *Nanotechnology* 5202-5205

[9] Deboisblanc J 2010 Synthesis and Characterization of P3HT:PCBM Organic Solar Cells. *Thesis*: Department of Physics and Astronomy Pomona College

[10] Gupta Anubhav, S Praveen, Kumar Abhishek, Shree Priyanka, Mishra Suchana and Joseph C M 2012 Preparation & Characterization of P3HTPCBM Organic Solar Cells. *International Journal of Electrical an Electronics Engineering,* ***1*****(4),** 13-15

[11] Yue S, YanChun H and JianGang, L 2013 Controlling PCBM Aggregation in P3HT/PCBM Film by a Selective Solvent Vapor Annealing *Chinese Science Bulletin* 2767-2774

IOP Conf. Series: Materials Science and Engineering **107** (2016) 012051 doi:10.1088/1757-899X/107/1/012051

X-ray Absorption Improvement of Single Wall Carbon Nanotube through Gadolinium Encapsulation

Alimin[1,2*], Narsito[2], I Kartini[2], and S J Santosa[2]

[1]Department of Chemistry, Universitas Gadjah Mada, Sekip Utara BLS 21 Yogyakarta 55281, Indonesia

[2]Department of Chemistry, Universitas Halu Oleo, Kampus Hijau Bumi Tridharma Anduonohu, Kendari 93232, Indonesia

[*]E-mail: alkimi_unhalu@yahoo.com

Abstract. X-ray absorption improvement of single-wall carbon nanotube (SWCNT) through gadolinium (Gd) encapsulation has been studied. The liquid phase adsorption using ethanol has been performed for the doping treatment. The Gd-doped SWCNT (Gd@SWCNT) was characterized by nitrogen adsorption isotherms, Raman spectroscopy, Transmission electron microscopy (TEM), and thermal gravimetric analysis (TGA) techniques. A relatively high residual weight of Gd@SWCNT compared to non-doped SWCNT (*n*-SWCNT) indicated that Gd has been doped in the nanotube. Even though Gd nanoparticles could not be observed clearly by TEM image, however, a significant decrease of nitrogen uptakes at low pressure and RBM (Radial Breathing Mode) upshift of Raman spectra of Gd@SWCNT specimen suggest that the metal nanoparticles might be encapsulated in the internal tube spaces of the nanotube.It was found that Gd-doped in the SWCNT increased significantly mass attenuation coefficient of the nanotube.

1. Introduction

As the shortwave electromagnetic, X-rays having wave number ranging from 10^{-11} to 10^{-8} m not only provides benefits to human life but also give potential hazards such as cell mutation, chromosome damage, cancer, etc. Extensive studies have been carried out by many researchers[1-5] for generating materials which are able to project electromagnetic radiation. For instances, sheet lead, concrete, and aluminum-foam have been produced for protecting against electromagnetic radiation. Generally, those materials are applied in such as reactor zone corresponding to electromagnetic radiation to overcome the radiation effect on human body. It is also possible to fabricate electromagnetic radiation protective materials which directly protect the human body from the radiation source such as clothes, gloves, shoes, etc.

Single-wall carbon nanotubes (SWCNTs) are projected to be one of the most promising candidate of the materials due to nanometric dimension, distinctive structure, extraordinary physical properties such as many-fold stronger than steel, harder than diamond, very lightweight, electrical conductivity higher than copper, thermal conductivity higher diamond and also it can easily be formed in bucky paper. Thus, SWCNT can be applied to generate new sustainable technology[6] such as application as transistors and fuel cells[7], as well as for medical sciences[8].

Great efforts have been carried out by many researchers [9-12] for modifying structure of SWCNT in order to be applied in a wide variety of fields. Structural modification of SWCNT can be achieved through two ways,i.e., surface space[9] and internal tube space modifications[10-12].

Regarding the structural modification of carbon nanotubes (CNTs), it has been reported that the modified CNTs have been applied for electromagnetic shielding, particularly at high frequencies ranging in the GHz. In various types of CNT composites have been fabricated by mixing the tubes with metal powders [5]. These studies showed that CNTs could exhibit novel and outstanding electromagnetic effects [13]. It is noted that SWCNT has cylindrical nanotubules that can be nested, and the interaction of radiation with these nanotubules appears to be very different when compared to their interaction with planar or stacked graphene sheets[14]. Interestingly, Fujimori *et al.*[14] found that SWCNT possesses high shielding of X-ray radiation even higher than highly oriented pyrolytic graphite (HOPG). It is believed that the X-ray mass absorption coefficient does not depend on the nanostructure but on the component atoms. However, Fujikawa and Kaneko[15] theoretically showed that the established rule of X-ray absorption is valid within the approximation of the dipole moment transition and the higher order momentum transition depends on nanoscale structure. Hence, we need to continue fundamental studies on well-characterized nanocarbon. In this context, we studied improvement of X-ray absorption of SWCNT through gadolinium encapsulation. So far, improvement of X-ray absorption property of SWCNT through gadolinium encapsulation has never been studied.

In the case of X-ray absorption of gadolinium, experimental data exhibited that the metal has a specific mass absorption coefficientof64.4 cm^2g^{-1} using a monochromatic incident beam of Mo Kα[16]. Therefore, X-ray absorption of SWCNT which has a mass absorption coefficient of 0.515 cm^2g^{-1}[17] is expected to increase significantly through the metal encapsulation method.

2. Method

2.1. Chemicals

$Gd(NO_3)_3.6H_2O$ (99.9 at%), HNO_3 (60 wt%) HCl (36 wt%) and ethanol (99.5 wt%) were purchased from the Wako Chemical Co. The SWCNT synthesized by the arc-discharge method using 1.03 wt% of nickel and 0.26 wt% of yttrium was produced by Meijo-SO. Co.Ltd.

2.2. Opening end cap and purifying of the SWCNT

Methods to open end cap and purify the SWCNTreferred to our previous work [17]. Briefly, the SWCNT samples were thermally oxidized by air in order to obtain opened-end samples. The opened-end SWCNT was then chemically purified using sequential acid treatment technique. The sample was the denoted as *n*-SWCNT.

2.3. Doping of $Gd(NO_3)_3.6H_2O$ solution into the SWCNT

Doping technique of $Gd(NO_3)_3.6H_2O$ solution into the SWCNT has been conducted with the same as a doping technique of gold solution as published in our previous work[17]. Before the *n*-SWCNT was doped by gadolinium solution, the nanotube was evacuated at room temperature under 0.1 Pa for 24 h. The evacuated SWCNT of 5 mg was dipped in $Gd(NO_3)_3.6H_2O$ solution of 10 ppm containing 99.5 wt % ethanol and dispersed ultrasonically by a 28 kHz ultrasound apparatus for 48 h at 298 K. It was then filtered through a Millipore porous filter (0.45 μm), rinsed with ethanol to 200 ml and dried at 383 K for 24 h. The obtained samples are denoted as Gd@SWCNT

2.4. Characterization of non-doped SWCNT and Gd@SWCNT

Thermal gravimetric, as well as differential thermal analysis, were performed on TGA/DTA apparatus, RIGAKU Thermo plus TG 8120 under O_2-N_2 atmosphere. The gas flow and heating rates were 300 $mL\ min^{-1}$ and 3 $K\ min^{-1}$, respectively. The Raman spectrophotometer (RAMAN Renishaw in via Raman microscope) equipped with a red laser (energy 1.58 eV, wavelength 785 nm) was used. In order to obtain good reproducibility, samples were exposed for 3 min with triple accumulation and spectra were taken at different places on the surface of each sample. Nanopore structure change of the samples preheated at 423 K, and 10^{-4} Pa for 2 h was determined by nitrogen adsorption at 77 K using a volumetric apparatus, Quantachrome-Autosorb-iQ2-MP.

2.5. X-ray absorption measurement
The samples were compressed by pelletizer at 10 MPa for 15 min in order to form a disc (tube-like shape) of 5 mm in diameter. Volumes of the samples (tube-like) were calculated by means tube volume formulation so that apparent densities of the compressed samples were obtained. The specimen densities of non-doped SWCNT and Gd@SWCNT were 1.03 and 0.97 gcm^{-3}, respectively. In the present work, measurement technique of X-ray absorption of the samples referred to work of Fujimori *et al.*[14] and our previous work [17].

3. Results and discussion

3.1The effect of gadolinium encapsulation on the pore structure of the nanotube
The pore structure of the nanotube was examined by nitrogen adsorption isotherms. Nitrogen adsorption isotherms of gadolinium encapsulated SWCNT (Gd@SWCNT) and non-encapsulated SWCNT (*n*-SWCNT) at 77 K are presented in figures 1.

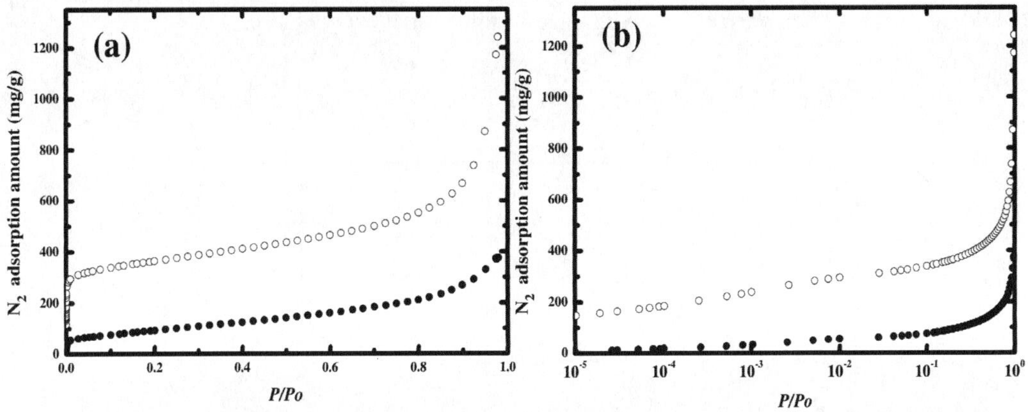

Figure 1. Linear (a) and logarithmic (b) plots of nitrogen adsorption isotherms at 77 K of the *n*-SWCNT (opened circles)[17], and Gd@SWCNT (closed circles).

It can be seen that the metal doping remarkably decreases the nitrogen adsorption amount of about 75% at entire pressure range as shown in figure 1(a). It suggests that the metal was encapsulated in the internal tube space of the nanotube [18,19]. The nitrogen adsorption uptake at low pressure is associated with the internal tube space of the nanotube [18,19]. Thus, the small amounts of nitrogen adsorption uptake of Gd@SWCNT at low pressure as shown in figure1(b) reveals supporting information that the internal tube spaces are filled with the metal particles.

The pore structural parameters of Gd@SWCNT and *n*-SWCNT are summarized in Table 1.Table 1 indicates that the micropore volume of Gd@SWCNT depresses about 14-fold. The surface areas of the Gd@SWCNT is much smaller than that of *n*-SWCNT, being decrease about 5-fold. It could be explained as follows; after doping treatment, the metal particles could fill either the internal tube spaces or doped the external surface or the interstitials of the nanotube. Consequently, the nitrogen molecules uptake of Gd@SWCNTdecrease significantly so that the fewer nitrogen molecule uptake, the smaller surface area and micropore volume of the nanotube. This also supports that the metal has been encapsulated in the internal tube spaces of the nanotube. Therefore, it can be concluded that the small amounts of nitrogen adsorption uptake at low pressure and the drastic depression of micropore

IOP Conf. Series: Materials Science and Engineering **107** (2016) 012051 doi:10.1088/1757-899X/107/1/012051

as well as smaller surface area of the Gd@SWCNT suggest that the metal has successfully been encapsulated in the internal tube space of the nanotube [18,19].

Table 1. The pore structural parameters of Gd@SWCNT and *n*-SWCNT.

Samples	surface area (m^2/g) α_S	Micropore Volume (cm^3/g) α_S
n-SWCNT	19.3±0.1[b]	Natural
Gd@SWCNT	3.8±0.8[c]	99% enriched

The evidence of the metallic particles encapsulated in the internal wall of the nanotubes could also be confirmed by the relatively significant upshifts of the radial breathing mode (RBM) of Raman spectra. The Raman spectroscopy of *n*-SWCNT and Gd@SWCNT, in particular, RBMs are exhibited in figure2. A detailed analysis of the RBM which usually appears at around 100-300 cm^{-1} region reveals the information about encapsulation of metals into SWCNT in particular for determining the SWCNT diameter [20]. It is well-known that the RBM frequency of SWCNT is sensitive to chemical and physical factors as well as the tube diameter of the SWCNT. The relationship between RBM frequency and diameter of the nanotube is given in following equation (1) [20],

$$\omega \text{RBM}\ (cm^{-1}) = A/it\ (nm\ cm^{-1}), \qquad (1)$$

where the A is constant of 248 cm^{-1}.

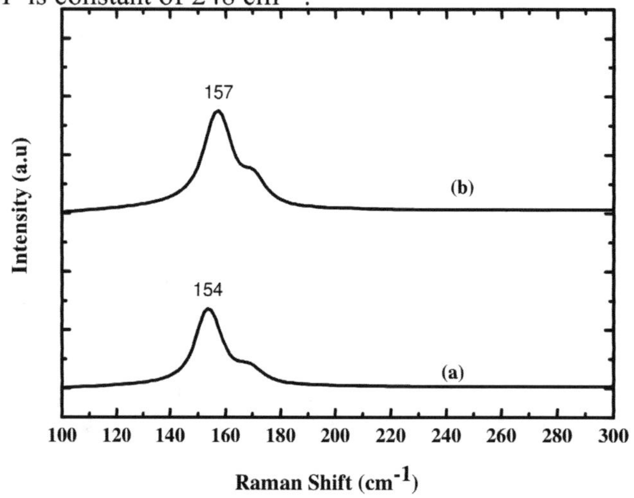

Figure 2. The RBMs of *n*-SWCNT (a)[17], and Gd@SWCNT (b) using laser energy of 1.58 eV.

Figure 2 indicates that the RBM peak-SWCNT appeared at 154 cm^{-1} corresponds to the tubes diameter of 1.61 nm. After encapsulation treatment, the peaks slightly shift to a higher frequency that is observed at 157 cm^{-1}. This peak deals with the tube diameters of 1.58 nm.

The origin of this shifting might also be attributed either to an increase of C–C binding energy due to the charge transfer from the nanotube to the metals [10]. Besides, it is possible that the upshifts originate from van der Waals bond as a result of van der Waals interaction between non-hybridized *p* orbitals of carbon atoms in the nanotube and *d* orbitals of the metal as illustrated in Figure3. If it is assumed that the non-hybridized *p* orbitals are p_z and that the d orbitals are d_z which each those orbitals are occupied by an electron, they are then possible to interact each other through van der Waals attraction. The interaction might be lead to an increase of total binding energy of the nanotube. Consequently, the RBM frequencies of the nanotube went up after doping treatment due to an increase of the binding energy. Owing to the tube diameter is inversely proportional to ωRBM, it is reasonable. Therefore, the upshift of RBM frequency suggests that gadolinium particles have successfully been encapsulated in the internal tube spaces of the nanotube.

In addition, the metal doping shifted RBM of the nanotube to higher frequencies which point out hardening of RBM mode of carbon atoms in semiconducting nanotube[10]. The hardening is caused by the perturbation of radial flexibility of aromatic rings in the nanotube structure that leads to shrinking diameter of the tube[21]. This also implies that the metals have been encapsulated in the internal wall of the nanotube.

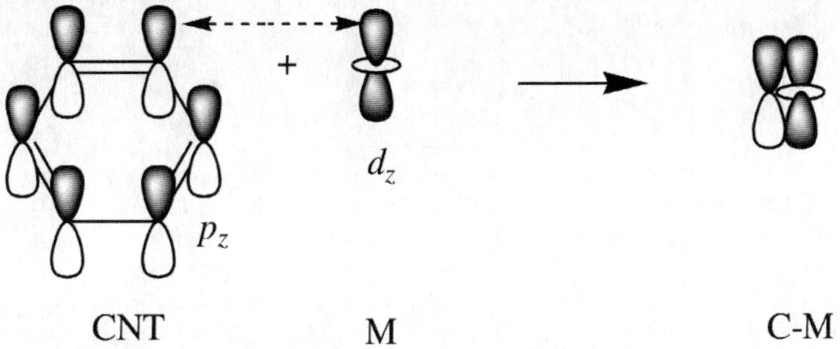

Figure 3. An illustration of interaction between the non-hybridized p_z orbitals of carbon atoms in the nanotube and $d_z$orbital of metals giving van der Waals interaction. Each the orbitals are occupied by an electron. CNT is the carbon nanotube which is represented by a benzene ring and M is the encapsulated metal.

Figure4(a) and (b) show the TEM images of *n*-SWCNT and Gd@SWCNT, respectively. The SWCNT possess a well-ordered and strong bundle structure. Since the SWCNT has strong bundle structure, Gd nanoparticles encapsulated inside SWCNT could be observed clearly.

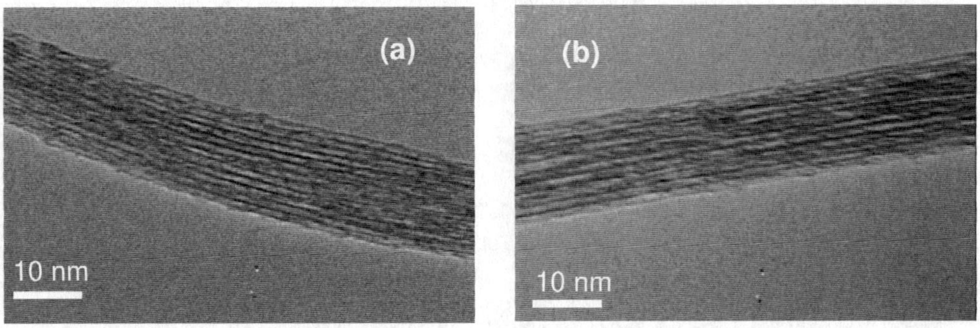

Figure 4. TEM images of *n*-SWCNT (a), and Gd-SWCNT (b)

3.2. Mass attenuation coefficient of Gd-doped SWCNT

The X-ray absorption coefficients (μ) of *n*-SWCNT (a) and Gd@SWCNTsampleswere determined from the Beer-Lambert law using monochromatic X-ray beam MoK_α.

$$I_o / I = e^{(\mu/\rho)\,\rho t} \tag{2}$$

Here ρ and t are the density of the material and the thickness of the specimen for X-ray penetration, respectively. The (μ/ρ) is the X-ray mass attenuation coefficient. The specimen density was used as the material density. The specimen for X-ray absorption experiment was prepared by compression of powdered the nanotube samples under 10 MPa for 15 min forming a disc. The specimen densities of *n*-SWCNT and Gd@SWCNT were 1.03, and 1.05 gcm^{-3}, respectively as shown in table3. The observed (μ/ρ) of *n*-SWCNT and Gd@SWCNT were 0.518 and, 1.548 cm^2g^{-1}, respectively(Table3).

Owing to the Gd@SWCNT contains the metallic dopant and also the catalyst remains, it is necessary to be taken into account those compartments for evaluating mass attenuation coefficient of the nanotube after doping treatment. The (μ/ρ) of the mixed solid consisting of bi-components i and j can be expressed by the linear relation between the X-ray mass attenuation coefficients of the components i and j together with component fraction of Ci and Cj as described in equation (3), when the X-ray absorption property of the mixed solid follow the established law derived from dipole transition approximation.

$$(\mu/\rho) = (\mu/\rho)iCi + (\mu/\rho)jCj \qquad (3)$$

If we have data on Ci and Cj, (μ/ρ) can be calculated using the established (μ/ρ) values of the atoms components. In a similar case, this relation can be extended to a multicomponent-mixed solid by the introduction of the fraction of each component.

The metallic catalyst remained in the SWCNT and the metal particles doped in nanotube were quantitatively evaluated by thermal gravimetric analysis (TGA) while differential thermal analysis (DTA) was qualitatively used for predicting of metal oxides species,as SWCNT was synthesized using 1.03 wt % of nickel and 0.26 wt % of yttrium catalysts [22].The residual weight after burning of the SWCNT sample provides the fraction of the components.

Figure 5 shows TGA as well as DTA profiles of the *n*-SWCNT and Gd@SWCNT under the flow of air of 350 mL/min at 3K/min. The weight losses which appear at around 600 K should correspond to oxidation of carbonaceous. The sharp weight losses of *n*-SWCNT and Gd@SWCNT commencing at approximately 850 K stem from the oxidation of carbon materials. The weight losses of the samples become constant above 1100 K, giving the total weight of the metallic impurities. The residual weights of those specimens are summarized in Table 2. It is reasonable that when metal is oxidized with air (mixing of O_2/N_2), the metal oxides could be produced. Therefore, the residual weight of *n*-SWCNT should correspond to NiO and Y_2O_3 originated from the catalyst remains. Moreover, in the case of the Gd@SWCNT, the residual weight could stem from Gd_2O_3.Thus, the X-ray mass attenuation coefficient of Gd-doped SWCNT can be calculated by using the residual amount of Gd of 1.2 wt %.

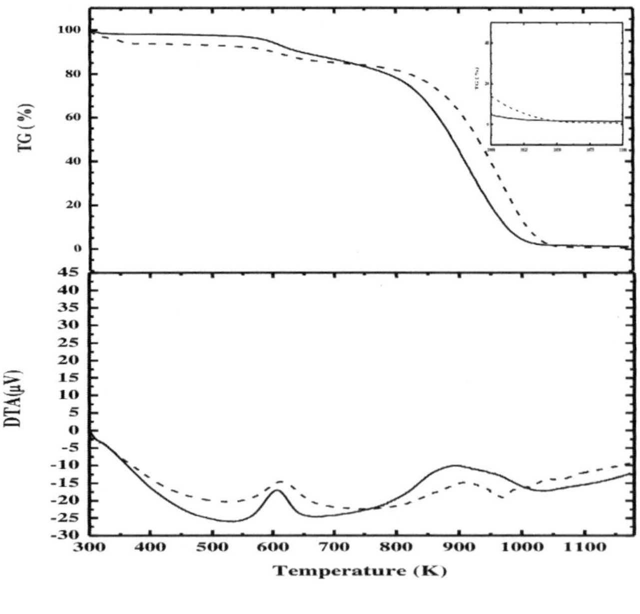

Figure 5. TGA and DTA profiles of *n*-SWCNT (dash line)[17], and Gd@SWCNT (solid line). The inset is TGA profiles above 1000 K

Table 2. The residual weights of SWCNT before and after doping treatment

Specimens	Residual weights (wt %)
n-SWCNT	0.57[17]
Gd@SWCNT	1.20

Table 3. Mass attenuation coefficients (μ/ρ) and densities of the *n*-SWCNTandGd@SWCNT

Samples	Density (gcm^{-3})	Observed (cm^2/g)	Calculated (cm^2/g)
n-SWCNT	1.03	0.518	0.515[17]
Gd@ SWCNT	1.05	1.546	1.200

Table 3 compares the observed (μ/ρ) with the calculated one. The observed (μ/ρ) value of *n*-SWCNT of 0.518 cm^2g^{-1} which are relatively close to (μ/ρ) of highly oriented pyrolytic graphite of 0.576 ± 0.005 cm^2/g[16] is slightly higher than the calculated value. The (μ/ρ) observed value of Gd@SWCNT was approximately 1.3 times larger than the calculated (μ/ρ) derived from the sample dipole-moment transition approximation. Probably, the large observed value of metal-doped SWCNT steams from contributions of both nucleus charges of metal dopants and the remained catalyst in the nanotube. It is likely due to high nucleus charge of the dopants so that (μ/ρ) value of SWCNT increased after doping treatment.

The relatively high X-ray attenuation of the SWCNT is likely originated from the unique cylindrical nanotubules. Carbon atoms embedded in the hexagonal tubular concentric tube of different curvature, are not periodic. The exotic cylindrical of the nanotube induces a significant electron population fluctuation across the carbon tube wall[14]. According to Grüneis*et al.* [23] and Jiang *et al.*[24] that such an electron fluctuation could lead to an enhanced X-ray absorption caused by an increase in the polarization factor. Accordingly, therefore, the presence of metal in the nanotube probably could enrich the electron population fluctuation in the nanotube wall, so that give the increase of X-ray attenuation.

4. Conclusions

We have demonstrated enhanced X-ray absorption property of Gd encapsulated SWCNT. We found that the mass attenuation coefficient of SWCNT increased significantly 2.3–fold might due to Gd nanoparticles encapsulated in the nanotube. Production of Gd@SWCNT nano-composites would be promising and opening a new avenue for fabricating protective electromagnetic radiation materials.

Acknowledgments

The authors thank the Center for Energy and Environmental Science, Shinshu University-Japan for the research project. The authors deeply thank Professor Katsumi Kaneko for giving the great opportunity to study and research under his supervision at the Center for Energy and Environmental Science, Shinshu University-Japan. The authors thank Professor Morinobu Endo of Center for Energy and Environmental Science, Shinshu University-Japan for TEM, SEM and TGA equipment. The authors thank Professor K. Hata for supplying the SG-SWCNT. The authors thank Associate Professor Toshihiko Fujimori for his advice and intensive discussion of Raman spectroscopy and XRD theory. The authors thank DrDaiki Minami for his great support of TEM observation. The authors thank Dr Ryusuke Futamura for his guidance of the XRD experiment. The authors thank Dr Fitri Khoerunnisa for her kindly experimental guidance and intensive discussion. The authors thank Dr. H. Sakamoto, Dr.Suwen Wang, Dr. Fernando Vallejos Burgos, Radovan Kukobat and Li Xiaolin for their supports and advice.

References

[1] Liu Z, Bai G, Huang Y, Ma Y, Du F, Li F, Guo T and Chen Y 2007 *Carbon* **45** 821

[2] Ye Z, Deering WD, Krokhin A and Roberts J A 2006 *J. Phys. Rev B* **74** 075425

[3] Yang Y, Gupta M C, Dudley K L and Lawrence R W 2005 *Nano. Lett.***5** 2131

[4] HiggibothamA L, Moloney P G, Waid M C, Duque J G, Kittrell C, Schmidt H K, Stephenson J, Arepalli S, YowellL L and Tour J M 2008 *Compos. Sci.Tech* **68** 3087

[5] Wen F, Zhang F, and Liu Z 2011 *J. Phys. Chem. C* **115** 14025

[6] Kumar M and Ando Y 2010*J. NanoSci. NanoTech* **10**3739–3758

[7] Rutherglen C, Jain D and Burke P 2009 *Nat. Nanotech* **4** 811-819

[8] Hitoshi K, Katoh M, Suzuki T, Ando Yand Nadia M 2012*Biol. Pharm. Bull* **35**(4) 455-463

[9] Yu J G, Huang K-L, Liu, S-Q and Tang J C 2008 *Physica E* **40** 689-692

[10] EliseevA,Yashina LV, BrzhezinskayaMM, Chernysheva M V, KharlamovaM V, Verbitsky N I, LukashinAV, Kiselev NA, KumskovAS, Zakalyuhin RM, Hutchison J L,FreitagB andVinogradovA S 2010*Carbon* **48** 2708-2721

[11] Kobayashi K, Suenaga K, Saito T, Shinohara H andIijima S 2010 *Advanced Materials***22** 3156-3160

[12] Kharlamova M V and NiuJ 2012 *Appl. Phys. A* **109** 25-29

[13] Imholt T J, Dyke C A,Hasslacher B, Perez J M, Price D W, Roberts J A, Scott A,Wadhawan J B, Ye Z and Tour J M 2003*Chem. Mater* **15** 3969

[14] Fujimori T, Tsuruoka S, Fugetsu B, Maruyama S, Tanioka A, Terrones M, Dresselhaus M S, Endo M and Kaneko K 2011*Mater. Express* **1** (4) 273-277

[15] Fujikawa T and Kaneko K 2012*J. Electr. Spect.& Rel. Phen* **185**509-511

[16] Hubbel J H and Seltzer S M 2004 NIST Standard Data Base **261** NIST Gaithersburg

[17] Alimin, Narsito,Kartini I, and Santosa S J 2015, *Indonesia. J. Chem* (Accepted)

[18] Ohba T, Matsumura T, Hata K, Yumura M, Iijima S, Kanoh H and Kaneko K 2007 *J. Phys. Chem. C* **111** 15660-15663

[19] Hattori Y, Kaneko K and OhbaT 2013*Comprehensive Inorganic Chemistry II* **Vol 5** Oxford Elsevier p. 25-44

[20] Jorio A, Pimenta M A, Souza Filho, A G, Saito R., DresselhausG andDresselhaus M S 2003 *New Journal of Physics* **5** 139.1-139.17

[21] Gotovac S, Honda H, Hattori Y, Takahashi K, Kanoh H, Kaneko K 2007 *Nano. Lett* **7** (3) 583-587

[22] Al-zubaidi A, Inoue T, Matsushita T, Ishii Y, Hashimoto T and Kawasaki S 2012 *J. Phys. Chem.* C **116** 7681-7686

[23] Grüneis A, Saito R, SamsonidzeGG, Kimura T, Pimenta M A,Jorio A, FilhoAGS, Dresselhaus G and Dresselhaus M S 2003 *Phys.Rev.B***67**165402

[24] Jiang J, Saito R, Grüneis A, Dresselhaus G and DresselhausMS 2004 *Chem. Phys. Lett.* **392** 383-389

HLA-DP, HLA-DQ, and HLA-DR-restricted epitopes in GRA5 of *toxoplasma gondii* strains

S Haryati[1,2,3], Y Sari[1,2,3], A APrasetyo[1,2,4*] and R Sariyatun[1,2]

[1]A-IGIC (A-Infection, Genomic, Immunology & Cancer) Research Group, Sebelas Maret University, Jl. Ir. Sutami 36A, Surakarta, 57126, Indonesia
[2]Center of Biotechnology and Biodiversity Research and Development, Sebelas Maret University, Jl. Ir. Sutami 36A, Surakarta, 57126, Indonesia
[3]Department of Parasitology, Faculty of Medicine, Sebelas Maret University, Jl. Ir. Sutami 36A, Surakarta, 57126, Indonesia
[4]Department of Microbiology, Faculty of Medicine, Sebelas Maret University, Jl. Ir. Sutami 36A, Surakarta, 57126, Indonesia

[*]E-mail: afie.agp.la@gmail.com; afieagp@yahoo.com; afie@staff.uns.ac.id

Abstract. The dense granular (GRA) proteins of *Toxoplasma gondii*(*T. gondii*) have been demonstrated as potential sources of *T. gondii* vaccine antigens. However, data of the GRA5 protein are limited. This study analyzed twenty-one complete GRA5 sequences of *T. gondii* GT1, RH, ME49, VEG, MAS, RUB, FOU, p89, VAND, and GAB2-2007-GAL-DOM2 strains to identify potential epitopes restricted by Major Histocompatibility Complex class II (MHC-II) molecules (human leukocyte antigen (HLA)-DP, HLA-DQ, and HLA-DR) in the protein. In all *T. gondii* strains, peptides positioned at amino acid (aa) 15-29, 16-30, 17-31, 18-32, 19-33, 83-97, 84-98, 86-100, 87-101, 89-103, and 90-104 were predicted to pose high affinity and binding with HLA-DRB1*0101, HLA-DRB1*0301 (DR17), HLA-DRB1*0401 (DR4Dw4), HLA-DRB1*0701, HLA-DRB1*1101, HLA-DRB1*1501 (DR2b), and/or HLA-DRB5*0101. Considering the epitope's affinity, ligation strength, and hydrophilicity, LRLLRRRRRRAIQEE sequence (aa 90-104) restricted by HLA-DRB1*0101, HLA-DRB1*0301 (DR17), and HLA-DRB1*0401 (DR4Dw4) was considered as the most potential MHC-II epitope in GRA5 of *T. gondii*. These results would be useful for studies concerning in developing *T. gondii* vaccine and diagnostic method.

1. Introduction

Toxoplasma gondii (*T. gondii*) is the world's most prevalent intracellular protozoan parasite which infects human and many warm-blooded animals [1]. The parasite is distributed in 30-80% of human population globally, commonly as a long-lasting persistent infection [2–4]. The prevalence is higher in tropical areas, as Indonesia [5–8]. *T. gondii* infection appears asymptomatic in healthy people. Nonetheless, it can emerge as a life-threatening infection in immunocompromised individuals [9,10]. The parasite also causes fetal infection and great economic losses in livestock industries due to abortion in farm animals [10–12]. Despite it's considerable medical and economic impacts, no drug can completely cure *T. gondii* infection while a live-attenuated vaccine is commercially available for veterinary use only [10,13,14]. Thus, as one of the bases of the development of an effective vaccine against *T. gondii*, studies identifying prospective antigens of the parasite for vaccine candidate are crucial [15].

The identification of potential epitopes will be useful for the development of vaccines and diagnostic approaches [16–19]. However, previous studies have concerned mainly in epitopes inducing cellular immunity, which is facilitated by MHC-I presentation (HLA-A, HLA-B, HLA-C). While those identifying potential epitopes restricted by the MHC-II molecules (HLA-DP, HLA-DQ, HLA-DR), which play roles in humoral responses against the parasite, are less common. Whereas both cellular and humoral immunities are important in host defense against *T. gondii* infection [20]. Hereby, this study aimed to identify potential epitopes restricted by HLA-DP, HLA-DQ, and HLA-DR on*T. gondii* proteins. The present study focused on the dense granule protein 5 (GRA5), which is known to be important in cell invasion, maintenance of parasitophorous vacuoles, and long-term survival of *T. gondii*, as well as highly immunogenic and expressed in all life stages of the parasite, either during acute or chronic phases of infection [21–28].The results of this study would be useful for the development of vaccines and diagnostic methods toward *T. gondii*.

2. Materials and Methods

2.1. Data resources

Twenty one complete GRA5 sequences of *T. gondii* identified as GT1 (n = 2), RH (n = 1), ME49 (n = 4), VEG (n = 2), MAS (n = 2), RUB (n = 2), FOU (n = 2), p89 (n = 2), VAND (n = 2), or GAB2-2007-GAL-DOM2 (n = 2) strains deposited in GenBank (http://www.ncbi.nlm.nih.gov/genbank/) until April 19th, 2015 were downloaded in fasta format. The accession numbers were as follows: KE387286 and AAQM03000223 (GT1 strain); L06091 (RH strain); NW_002234546, CM002039, KE138824, and XM_002369230 (ME49 strain); AAYL02000231 and KI544504 (VEG strain); KN044647 and AEXC02002287 (MAS strain); KN039988 and AFYV02001021 (RUB strain); AEYH02000464 and KN000236 (FOU strain); AEYI02000833 and KL998085 (p89 strain); AEYJ02000651 and KN042474 (VAND strain); AHZU02000792 and KN003125 (GAB2-2007-GAL-DOM2 strain).

2.2. Epitope prediction

The Immune Epitope Database (IEDB) (http://tools.immuneepitope.org/mhcii/) was used to predict the peptide binding of GRA5 sequences to HLA-DP, HLA-DQ, and HLA-DR molecules. The IEDB Recommended prediction was used, consisting of NN-align, SMM-align, CombLib, and/or NetMHCIIpan methods. Peptides with IC50 values <50 nM and percentile rank $\leq$ 1.0 were considered having high affinity. The ligation strength of GRA5 to defined HLA-DR molecules was also predicted using SYFPEITHI (http://www.syfpeithi.de/bin/MHCServer.dll/EpitopePrediction.htm). Peptides with SYFPEITHI score $\geq$ 15 were considerably having high binding. The sequence hydrophilicity was analyzed in DNAMAN version 8 software (Lynnon BioSoft, USA). Based on the results of these methods, the overlapped regions of the epitope prediction results which had good hydrophilicity were considered as potential MHC-II-restricted epitopes of GRA5.

3. Results and Discussion

GRA proteins have been suggested as promising antigens to induce host immunity and be targets of diagnostic approaches [16,29–33]. In particular, GRA5 has been used as an antigen in a vaccine study [34] and serological detection toward *T. gondii* with high specificity in both acute and chronic infections [35,36]. GRA5 has also been used as a strain-specific antigen to serotype *T. gondii* strains by enzyme-linked immunosorbent assay (ELISA) [37,38]. In the present study, 120 amino acid sequences of GRA5 of *T. gondii* GT1, RH, ME49, VEG, RUB, FOU, p89, VAND, and GAB2-2007-GAL-DOM2 strains were analyzed. The MAS isolates comprised 119 amino acids only due to a deletion in position 51 of the GRA5 sequence. GRA5 sequences in each strain were identical. GRA5 sequences of GT1 and RH strains, which were known as type I strain, were also identical with those of FOU and VAND strains (figure 1). Likewise, GRA5 sequences of VEG, RUB, p89, and GAB2-2007-GAL-DOM2 strains were also identical (figure 1).

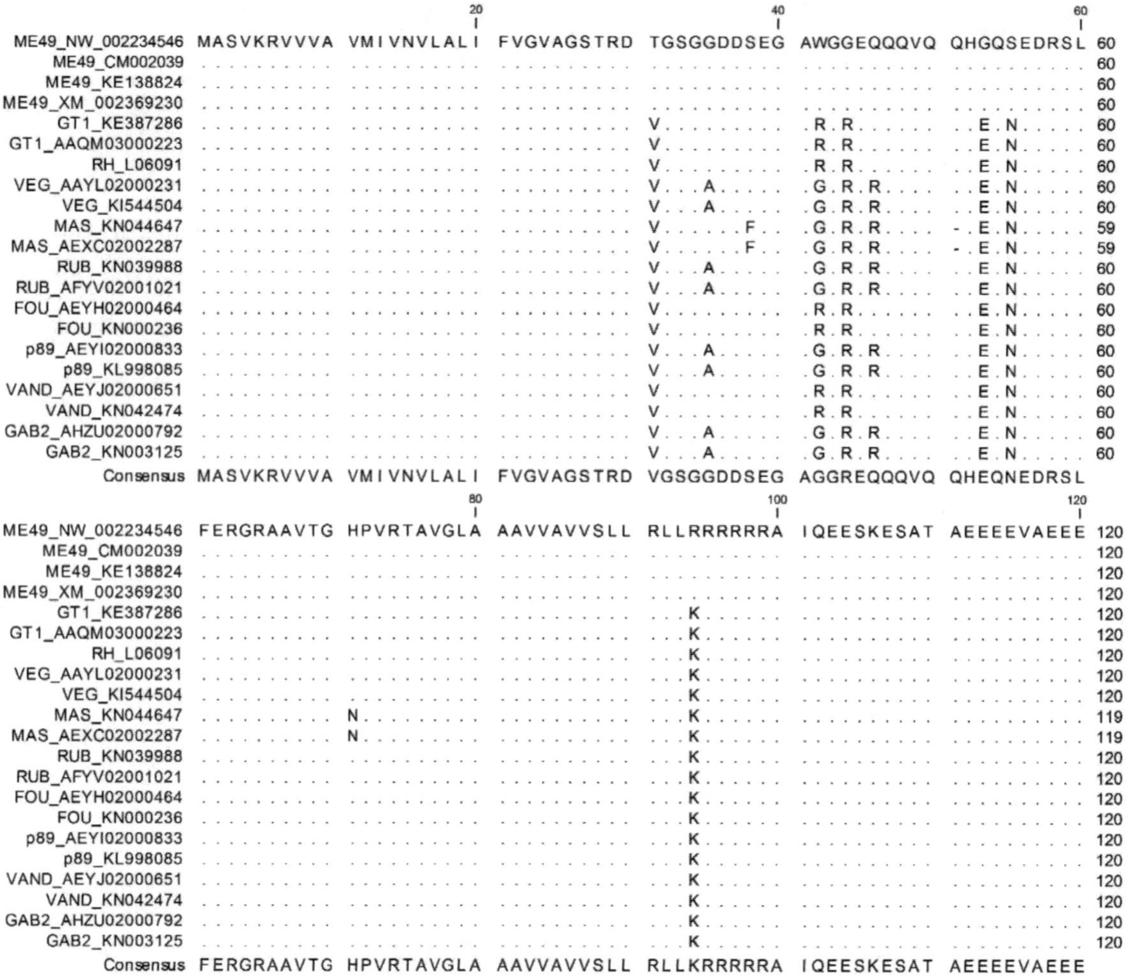

Figure 1. Alignment of GRA5 sequences from *T. gondii* GT1 (KE387286, AAQM03000223), RH (L06091), ME49 (NW_002234546, CM002039, KE138824, XM_002369230), VEG (AAYL02000231, KI544504), MAS (KN044647, AEXC02002287), RUB (KN039988, AFYV02001021), FOU (AEYH02000464, KN000236), p89 (AEYI02000833, KL998085), VAND (AEYJ02000651, KN042474), and GAB2-2007-GAL-DOM2 (AHZU02000792, KN003125) strains. Dots represent matching residues. Sequence numbering is in accordance with NW_002234546 isolate.

The induction of both $CD4^{+}$ and $CD8^{+}$ T cell responses is important in the formulation of successful vaccines against T. gondii [39]. The $CD8^{+}$ T cell response is initiated by MHC-I presentation, resulting in the death of infected cells [40,41]. While the CD4+ T cell response occurs with the presentation of MHC-II (HLA-DP, -DQ, -DR) molecules, resulting in cytokine production and stimulation of antibody production [40–42]. Therefore, identifying the antigens of *T. gondii* recognized by both MHC molecules is important. Bioinformatics has been widely used to predict epitopes, particularly for vaccine studies [29,43]. One of the indicators in predicting the existence of an epitope is the molecule's hydrophilicity; that is the tendency of a molecule to be exposed to the surface protein thereby potentially antigenic [44]. Here, a number of hydrophilic regions (hydrophilicity score > 0) in GRA5 of *T. gondii* strains were identified(table 1). Positions 54-57, 64, 93-100, 103-106, 111-120 of GRA5 sequences of all strains showed high hydrophilicity score (≥ 1).

Table 1. Hydrophilic positions in GRA5 of *T. gondii*.

Position	Hydrophilicity score in each *T. gondii* strain			
	GT1, RH, FOU, VAND	ME49	MAS	VEG, RUB, p89, GAB2-2007-GAL-DOM2
3	0.50	0.50	0.50	0.50
4	0.46	0.46	0.46	0.46
5	0.30	0.30	0.30	0.30
7	0.16	0.16	0.16	0.16
26	0.15	0.15	0.15	0.15
27	0.81	0.81	0.81	0.81
28-30	0.65	0.83	0.65	0.65
31	0.71	0.90	0.71	0.71
32-33	0.21	0.40	0.21	0.13
34	0.88	0.88	0.88	0.8
35	0.93	0.93	0.46	0.85
36-37	1.30	1.30	0.83	1.21
38	1.21	1.21	0.75	1.21
39	1.30	0.23	0.33	0.8
40	0.88	≤ 0	≤ 0	0.38
41-42	1.33	≤ 0	0.83	0.83
43	1.36	≤ 0	1.33	1.33
44	1.48	≤ 0	1.45	1.45
45	1.01	0.51	1.48	1.48
46	0.76	0.26	1.23	1.23
47	0.29	0.30	0.76	0.76
48	≤ 0	≤ 0	0.26	0.38
49	≤ 0	≤ 0	0.18	≤ 0
50	0.18	≤ 0	0.18	0.18
51	0.18	≤ 0	-	0.18
52	0.46	0.06	0.18	0.46
53	0.84	0.45	0.84	0.84
54	1.23	0.83	1.23	1.23
<u>55</u>	1.81	1.41	1.81	1.81
<u>56</u>	1.45	1.46	1.45	1.45
<u>57</u>	1.11	1.13	1.11	1.11
58-59	0.66	0.66	0.66	0.66
60	0.75	0.75	0.75	0.75
61	0.25	0.25	0.25	0.25
62	0.69	0.69	0.69	0.69
63	0.91	0.91	0.91	0.91
<u>64</u>	1.25	1.25	1.25	1.25
65	0.58	0.58	0.58	0.58

Table 1. Hydrophilic positions in GRA5 of *T. gondii*. (continued)

Position	Hydrophilicity score in each *T. gondii* strain			
	GT1, RH, FOU, VAND	ME49	MAS	VEG, RUB, p89, GAB2-2007-GAL-DOM2
66-67	0.01	0.01	0.01	0.01
92	0.60	0.60	0.60	0.60
93-94	1.40	1.40	1.40	1.40
95	2.20	2.20	2.20	2.20
96	3.00	3.00	3.00	3.00
97	2.41	2.41	2.41	2.41
98	1.61	1.61	1.61	1.61
99	1.14	1.14	1.14	1.14
100	1.06	1.06	1.06	1.06
101	0.98	0.98	0.98	0.98
102	0.53	0.53	0.53	0.53
103	1.11	1.11	1.11	1.11
104	1.83	1.83	1.83	1.83
105	1.85	1.85	1.85	1.85
106	1.35	1.35	1.35	1.35
107	0.86	0.86	0.86	0.86
108	0.73	0.73	0.73	0.73
109-110	0.65	0.65	0.65	0.65
111	1.01	1.01	1.01	1.01
112	1.51	1.51	1.51	1.51
113-117	1.33	1.33	1.33	1.33
118	1.10	1.10	1.10	1.10
119	1.75	1.75	1.75	1.75
120	2.50	2.50	2.50	2.50

[a]Underlined numbers are positioned in GRA5 sequences of all strains predicted with high hydrophilicity (score >1).

Epitopes in GRA5 were also predicted using online tools, including IEDB and SYFPEITHI databases. The results showed that there were 18, 18, 14, 19, 18, 19, 19, 19, 19, and 19 predicted epitopes with high antigenicity in GRA5 of *T. gondii* GT1, RH, ME49, VEG, MAS, RUB, FOU, p89, VAND, and GAB2-2007-GAL-DOM2 strains, respectively, as indicated by NN-align, SMM-align, NetMHCIIpan, and CombLib IC50 $<$ 50 and/or rank $\leq$ 1.0 in the IEDB-based analysis. All of these epitopes were restricted by HLA-DR molecules. Whereas SYFPEITHI tool predicted 142, 142, 145, 143, 147, 143, 143, 143, 143, and 143 HLA-DR-restricted epitopes with high binding (score $\geq$ 15) in GRA5 sequences of GT1, RH, ME49, VEG, MAS, RUB, FOU, p89, VAND, and GAB2-2007-GAL-DOM2 strains, respectively. A number of regions of the predicted results of the IEBD and SYFPEITHI analyzes were overlapped and conserved among *T. gondii* strains (table 2). Considering the epitope's affinity, ligation strength, and hydrophilicity, LRLLRRRRRRAIQEE sequence (aa 90-104) restricted by HLA-DRB1*0101, HLA-DRB1*0301 (DR17), and HLA-DRB1*0401 (DR4Dw4) was considered as the most potential MHC-II-restricted epitope in GRA5 of *T. gondii*.

Table 2. List of potential epitopes in GRA5 of *T. gondii* predicted using IEDB and SYFPEITHI.

Strain	Peptide	aa	HLA-DR	SMM-align IC_{50}/rank[a]	NN-align IC_{50}/rank[a]	Score[b]
GT1, RH, ME49, VEG, MAS, RUB, FOU, p89, VAND, GAB2-2007-GAL-DOM2	NVLALIFVGVAGSTR	15-29	B5*0101	19/0.1	4.3/0.6	-
			B1*1501 (DR2b)	-	-	24
	VLALIFVGVAGSTRD	16-30	B5*0101	19/0.1	3.8/0.5	-
			B1*0101	-	-	26
			B1*0401 (DR4Dw4)	-	-	20
			B1*1101	-	-	18
	LALIFVGVAGSTRDT	17-31	B5*0101	18/0.1	3.1/0.3	-
			B1*0101	-	-	16
			B1*1501 (DR2b)	-	-	18
	ALIFVGVAGSTRDTG	18-32	B5*0101	19/0.1	3.1/0.3	-
			B1*0101	-	-	25
			B1*0401 (DR4Dw4)	-	-	16
			B1*1101	-	-	16
			B1*1501 (DR2b)	-	-	16
	LIFVGVAGSTRDTGS	19-33	B5*0101	19/0.1	3.9/0.5	-
			B1*0101	-	-	16
			B1*0401 (DR4Dw4)	-	-	20
			B1*0701	-	-	22
	VVAVVSLLRLLRRRR	83-97	B1*1101	6/0.01	4.0/0.1	21
			B1*0101	-	-	17
			B1*0301 (DR17)	-	-	21
			B1*0401 (DR4Dw4)	-	-	25
			B1*1501 (DR2b)	-	-	20
	VAVVSLLRLLRRRRR	84-98	B1*1101	2/0.01	3.5/0.05	-
			B1*0101	-	-	18
			B1*1501 (DR2b)	-	-	20
	VVSLLRLLRRRRRRA	86-100	B1*1101	2/0.01	3.5/0.05	23
			B1*0101	-	-	17
			B1*0301 (DR17)	-	-	19
	VSLLRLLRRRRRRAI	87-101	B1*1101	2/0.01	4.1/0.1	23
			B1*0101	-	-	20
			B1*0301 (DR17)	-	-	19
			B1*0401 (DR4Dw4)	-	-	20
	LLRLLRRRRRRAIQE	89-103	B1*1101	6/0.01	5/0.2	23
			B1*0101	-	-	19
			B1*0301 (DR17)	-	-	18
			B1*0401 (DR4Dw4)	-	-	15
	LRLLRRRRRRAIQEE	90-104	B1*1101	7/0.01	8/0.9	15
			B1*0101	-	-	18
			B1*0301 (DR17)	-	-	18
			B1*0401 (DR4Dw4)	-	-	15
			B1*0701	-	-	16
GT1, RH, VEG, MAS, RUB, FOU, p89, VAND, GAB2-2007-GAL-DOM2	VAVVSLLRLLRRRRR	84-98	B5*0101	10/0.02	4.8/0.8	-
	VVSLLRLLRRRRRRA	86-100	B5*0101	9/0.02	4.0/0.5	-
	VSLLRLLRRRRRRAI	87-101	B5*0101	9/0.02	4.4/0.7	-
ME49	VVSLLRLLRRRRRRA	86-100	B1*0401 (DR4Dw4)	-	-	20

[a]IC_{50} values <50 nM and percentile rank ≤ 1.0 in the IEDB-based analysis are considered to have high affinity. The CombLib and NetMHCIIpan rank and IC_{50} values are not available for all peptides predicted.
[b]SYFPEITHI score ≥ 15 are predicted to have high binding.

A major limitation of this study is the limited number of sequences analyzed, reflecting the lack of *T. gondii* sequences in the database. Future works to report and analyze more *T. gondii* sequences are still needed in order to get a better understanding of the parasite. Overall, the results of this study are expected to be useful for studies concerning in developing *T. gondii* vaccine and diagnostic method. The predicted epitopes identified in this study would contribute knowledge about the immunobiology of *T. gondii* and help explain the difference in the immune response between *T. gondii* strains. As the limitation of bioinformatic analysis, the results of this study were based on prediction only. Hereafter, we suggest further studies to confirm the appearance of the predicted epitopes.

4. Conclusion

Considering the predicted epitope's affinity, ligation strength, and hydrophilicity, an epitope located in aa 90-104 (LRLLRRRRRRAIQEE) restricted by HLA-DRB1*0101, HLA-DRB1*0301 (DR17), and HLA-DRB1*0401 (DR4Dw4) is considerably the most potential MHC-II-restricted epitope in GRA5 of *T. gondii*.

Acknowledgement

This work was supported in part by a grant from APBN/DIPA UNS (No.339/UN27.11/PL/2015).

References

[1] Mercier C and Cesbron-Delauw M F 2015 *Trends Parasitol.***31** 60–71
[2] Weiss L M and Dubey J P 2009 *Int. J. Parasitol.***39** 895–901
[3] Denkers E Y 2003 *FEMS Immunol Med Microbiol.***39**193–203
[4] Lang C, Algner M, Beinert N, Gross U and Lüder C G 2006 *Microbes Infect.***8** 1994–2005
[5] Pereira-Chioccola V L, Vidal J E, and Su C 2009 *Future Microbiol.***4** 1363–79
[6] Remington J S, McLeod R, Wilson C B and Desmonts G 2011 Toxoplasmosis *Infectious Diseases of the Fetus and Newborn* (Philadelphia: Elsevier, 2011) pp 918–1041
[7] Prasetyo A A, Ariapramuda E R, Kindi E A, Dirgahayu P, Sari Y, Darmawan R, Hudiyono, Hartono and Kageyama S 2014 *Southeast Asian J Trop Med Public Health* **45** 1032–47
[8] Prasetyo A A, Sariyatun R, Reviono, Sari Y, Hudiyono, Haryati S, Adnan Z A, Hartono and Kageyama S 2015 *J ClinVirol* **70** 67–71
[9] Montoya J G and Liesenfeld O 2004 *Lancet.* **363** 1965–76
[10] Zhang N Z, Chen J, Wang M, Petersen E and Zhu X Q 2013 *Expert Rev Vaccines.* **12** 1287-99
[11] Bilgrami M and O'Keefe P 2014 *HandbClin Neurol.***121** 1321–44
[12] Tait E D and Hunter C A 2009 *Mem Inst Oswaldo Cruz.***104** 201–10
[13] Coombs G H and Müller S 2002 *Int J Parasitol.* **32** 497–508
[14] Innes E A 2010 *Expert Rev Vaccines.***9** 1117–9
[15] Zhang N Z, Huang S Y, Zhou D H, Xu Y, He J J and Zhu X Q 2014 *Genet Mol Res.* **13** 10669–77
[16] Wang Y, Wang G, Ou J, Yin H and Zhang D 2014 *Parasit Vectors.* **7** 474
[17] Wang Y, Wang G, Zhang D, Yin H and Wang M 2013 *ExpParasitol.* **135** 606–10
[18] Wang Y, Wang G, Zhang D, Yin H and Wang M 2013*Parasit Vectors.***6**1 25
[19] Wang Y and Yin H 2014 *Parasit Vectors.* **7** 180
[20] Lu G, Wang L, Zhou A, Han Y, Guo J, Song P, Zhou H, Cong H, Zhao Q and He S 2015 *Acta Trop.***146** 66–72
[21] Tilley M, Fichera M E, Jerome M E, Roos D S and White M W 1997 *Infect Immun.***65** 4598–605
[22] Charif H, Darcy F, Torpier G, Cesbron-Delauw M F and Capron A 1990 *ExpParasitol.***71** 114–24
[23] Achbarou A, Mercereau-Puijalon O, Sadak A, Fortier B, Leriche M A, Camus D and Dubremetz J F 1991 *Parasitology* **103 Pt 3** 321–9
[24] Dubremetz J F and Schwartzman J D 1993 *Res Immunol.***144** 31–3

[25] Cesbron-Delauw M F 1994 *Parasitol Today.* **10** 293–6
[26] Metsis A, Pettsersen E and Petersen E 1995 *ExpParasitol.***81** 472–9
[27] Carruthers V B, Moreno S N and Sibley L D 1999 *Biochem J.* **342** (Pt 2) 379–86
[28] Hughes H P A and van Knapen F 1982 *Int J Parasitol.* **12** 433–7
[29] Martin V, Supanitsky A, Echeverria P C, Litwin S, Tanos T, De Roodt A R, Guarnera E A and Angel S O 2004 *ClinDiagn Lab Immunol.***11** 704–10
[30] Ram H, Rao J R, Tewari A K, Banerjee P S and Sharma A K 2013 *Parasitol Res.***112** 2487–94
[31] Nigro M, Gutierrez A, Hoffer A M, Clemente M, Kaufer F, Carral L, Martin V, Guarnera E A and Angel S O 2003 *DiagnMicrobiol Infect Dis.***47** 609–13
[32] Vercammen M, Scorza T, Huygen K, De Braekeleer J, Diet R, Jacobs D, Saman E and Verschueren H 2000 *Infect Immun.***68** 38–45
[33] Cong H, Mui E J, Witola W H, Sidney J, Alexander J, Sette A, Maewal A, Bissati K E, Zhou Y, Suzuki Y, Lee D, Woods S, Sommerville C, Henriquez F, Roberts C W and McLeod R 2012 *Hum Immunol.***73** 1–10
[34] Igarashi M, Kano F, Tamekuni K, Machado R Z, Navarro I T, Vidotto O, Vidotto M C and Garcia J L 2008 *ExpParasitol.***118** 386–92
[35] Holec-Gąsior L and Kur J 2010 *ExpParasitol.***124** 272–8
[36] Ching X T, Lau Y L, Fong M Y, Nissapatorn V and Andiappan H 2014 *BioMed Res Int.* **2014** 690529
[37] Peyron F, Lobry J R, Musset K, Ferrandiz J, Gomez-Marin J E, Petersen E, Meroni V, Rausher B, Mercier C, Picot S, Cesbron-Delauw M -F 2006 Serotyping of *Toxoplasma gondii* in chronically infected pregnant women: predominance of type II in Europe and types I and III in Columbia (South America) *Microbes and Infections* **8** 2333-2340
[38] Morisset S, Peyron F, Lobry J_R, Garweg J, Ferrandiz J, Musset K, Gomez-Marin J_E, de la Torre A, Demar M, Carme B, Mercier C, Garin J_F and Cesbron-Delauw M_F 2008 *Microbes Infect.***10** 742–7
[39] Alexander J, Oseroff C, Dahlberg C, Qin M, Ishioka G, Beebe M, Fikes J, Newman M, Chesnut RW, Morton PA, Fok K, Appella E and Sette A 2002 *J Immunol.***168** 6189–98
[40] Da Silva S A, Mazini P S, Reis P G, Sell A M, Tsuneto L T, Peixoto P R and Visentainer J E 2009 *BMCInfect Dis.***9** 134
[41] Janeway Jr. C A, Travers P, Walport M and Shlomchik M J 2001 *Immunobiology* (New York: Garland Science)
[42] Kelly C D, Russo C M, Rubin B Y and Murray H W 1989 *ClinExpImmunol.***77** 397–402
[43] Bai Y, He S, Zhao G, Chen L, Shi N, Zhou H, Cong H, Zhao Q and Zhu X Q 2012 *Exp Parasitol.***132** 458–64
[44] Hopp T P and Woods K R 1981 *Proc Natl AcadSci U S A.***78** 3824–8

IOP Conf. Series: Materials Science and Engineering **107** (2016) 012053 doi:10.1088/1757-899X/107/1/012053

Virtual screening of Indonesian flavonoid as neuraminidase inhibitor of influenza a subtype H5N1

A A Parikesit[1], B Ardiansah[1], D M Handayani[1], U S F Tambunan[1*] and D Kerami[2]

[1]Bioinformatics Group, Department of Chemistry, Faculty of Mathematics and Science University of Indonesia Pondok Cina, Beji, Depok 16424 Indonesia
[2]Mathematics Computation Group, Department of Mathematics, Faculty of Mathematics and Science University of Indonesia Pondok Cina, Beji, Depok 16424 Indonesia

[*]E-mail: usman@ui.ac.id

Abstract. Highly Pathogenic Avian Influenza (HPAI) H5N1 poses a significant threat to animal and human health worldwide. The number of H5N1 infection in Indonesia is the highest during 2005-2013, with a mortality rate up to 83%. A mutation that occurred in H5N1 strain made it resistant to commercial antiviral agents such as oseltamivir and zanamivir, so the more potent antiviral agent is needed. In this study, virtual screening of Indonesian flavonoid as neuraminidase inhibitor of H5N1 was conducted. Total 491 flavonoid compound obtained from HerbalDB were screened. Molecular docking was performed using MOE 2008.10. This research resulted in Guajavin B as the best ligand.

1. Introduction

Influenza is an acute respiratory illness caused by influenza viruses with symptoms off ever, sore throat, chills, cough, headache, muscle aches, and fatigue [1]. The disease is classified as an infectious disease caused by a virus and can infect humans as well as animals. Avian influenza is a significant threat to poultry and human health in the world. Genetic diversity possessed by the influenza virus causing the disease becomes a pandemic, mainly due to the incorporation of the gene(gene reassortment) of avian influenza virus with an influenza virus that commonly occurs in humans [2].

The subtype H5N1 virus included in the Highly Pathogenic Avian Influenza (HPAI) with morbidity and mortality that reaches 100%. The H5N1 virus was first identified in the early 1900s in Italy [3]. However, the virus is of international concern since the H5N1 infection in humans that occurs the first time at 1997 in Hongkong and cause a significant number of mortality [2]. In Indonesia, the number of H5N1 infections in humans since the beginning of 2005 until December 2013 reached 195 cases with 163 deaths [4]. H5N1 infection showed clinical symptoms such as high fever (up to 38°C), respiratory disorders, limfofenia, abnormalities on chest radiographs, and in some cases diarrhea and vomiting [5].

One precaution against H5N1 infection is through vaccination. However, vaccination has a low efficiency in protecting the body when the influenza virus infected [6] or as a result, of antigenic drift and antigenic shift happens to the virus. Therefore, various antiviral drugs have been developed for the treatment of influenza. There are two classes of antiviral drugs that have been developed at this time, the class of adamantane and neuraminidase inhibitor [7]. Adamantane class inhibits the action of the

M2 ion channel. Amantadine and rimantadine were developed in 1960 and had been accepted in many countries [7]. Nevertheless, this class limited to influenza A virus only. Influenza virus resistance to adamantane group is also quite high. Therefore, the use of adamantane class has been limited by the Centers for Disease Control and Prevention (CDC) [8]. Since 2010, oseltamivir (Tamiflu®) and zanamivir (Relenza®), which includes a neuraminidase inhibitor class of antiviral drugs, were recommended by CDC and the World Health Organization (WHO) for the treatment of influenza A and B infections [9].

Nevertheless, the threat of the Highly Pathogenic Avian Influenza H5N1 notonly lies in a number of fatalities butal so because of high resistance to the commercial drugs. The H274Y mutation has been shown to occur inH5N1 influenza virus, causing virus H5N1 resistant to oseltamivir [10]. Therefore, the emergence of influenza viruses resistant to commercial antiviral medication has led to the need for the development of more potent antiviral drugs [11].

Various kinds of compounds have been developed neuraminidase inhibitor for the treatment of influenza virus infection. Flavonoids area a large group of compounds of natural ingredients that can be found either in higher plants as well as the low level of the plant, including algae. Flavonoids are known to have pharmacological character among anticancer, antibacterial, anti-inflammatory, and antiviral [12]. In the context of a drug likeness and ability to be a drug, a flavonoid class of compounds that has a low molecular mass with good bioavailability is considered feasible. Meanwhile, Indonesia is the country with the second highest biodiversity in the world after Brazil. Of the 40,000 species of plants that exist in the world, 30,000 of who are in the Indonesian archipelago, and among the30,000 species of plants in Indonesia, 9,600 of whom are known to have pharmacological activity [13-15]. Therefore, this study will carry out virtual screening natural materials of Indonesian flavonoid as neuraminidase inhibitors of influenza A virus subtype H5N1via molecular docking.

2. Materials and methods

2.1. Materials

This research was using computational study (*in silico*), by including personal computer intel core i7 with 4 GB of RAM, MOE 2008.10, ACDLabsChemSketch 12, Chem 3D Pro, VegaZZ, Jalview 2.8.0b.1, Toxtree 2.6.0, UCSF Chimera 1.9 and several online software from internet web servers[16-22]. Chemical structures in this research such as neuraminidase were obtained from online database NCBI, EMBL-EBI, PDB, and HerbalDB (http://herbaldb.farmasi.ui.ac.id/v3/) [13].

2.2. Method

*2.2.1. Preparation of Neuraminidase H5N1.*This step included searching of neuraminidase sequence, multiple sequences alignment, searching, and validation of three-dimensional structure of neuraminidase influenza A subtype H5N1. Structure validation was done by evaluating QMEAN4 and QMEAN-Z as a result of SWISS Modeling[23]. Structure validation was verified by Ramachandran plot analysis, which could be accessed through http://mordred.bioc.cam.ac.uk [24].

2.2.2. Preparation of Ligand. Three-dimensional structures of Indonesian Flavonoid was obtained from the herbalDB (http://herbaldb.farmasi.ui.ac.id/v3/), developed by Faculty of Pharmacy, University of Indonesia. These structures were then converted into Molecular Design Limited (MDL) Molfile (.mol) using VegaZZ.

2.2.3. Molecular Docking Simulation. There were stages of the molecular docking, namely preparation of neuraminidase and ligands for docking, and molecular docking process of protein-ligand [25].

2.2.4. Molecular Docking Analysis. Gibbs free energy from docking with MOE could be seen on an output in the format of MOE database (.mdb). Complex with the lowest free energy was then prepared for relocking step by retaining increase with a factor of 100. Moreover, LigX and MOE features could be used to see the interaction between ligand inhibitor with amino acid residue on target enzyme. It was possible to identify which functional group from the ligand that would bind to the amino acid on the active site of the enzyme.

3. Results and discussion

3.1. *Structure Validation of Three-Dimensional Neuraminidase.* Neuraminidase is an enzyme that plays an important role in the life cycle of the influenza virus. The sequence of amino acids that constitutes the enzyme neuraminidase can be searched on the site National Center for Biotechnology Information (NCBI) through https://www.ncbi.nlm.nih.gov/ URL. One of the features of the NCBI Influenza Virus Resource is providing the data from the NIAID Influenza Genome Sequencing Project and GenBank, along with devices for the annotation and analysis of influenza sequences. NCBI Influenza Virus Sequence Database also provides protein sequences and nucleotide sequences encoding regions of the influenza virus. The search of neuraminidase influenza A virus subtype H5N1 sequences was done by determining search settings as follows: type A, human host, country / area Indonesia, NA protein, and subtype H5N1. A total of 90 sequences of proteins were obtained, 58 of whom completed (c), 1 noncompleted (NC), and 31 putative (p). Sequences that can be used is the completed ones.

The main goal of Multiple Sequences Allignment (MSA) is determining the conserved regions pattern of the sequences. The conserved region shows the identical area between the sequences. MSA was done to get consensus sequences representing the entire sequence of the H5N1 influenza virus neuraminidase. MSA was done by using Clustal Omega, which can be accessed online through http://www.ebi.ac.uk/Tools/msa/clustalo/ [26]. MSA results were then analyzed using software Jalview. Jalview is a program for editing, visualization, and analysis of the MSA results. By using Jalview, conservation of the region, the quality, as well as consensus sequences, could be obtained. Neuraminidase consensus sequences were obtained through this software in FASTA format. BLAST (Basic Local Allignment Search Tool) was performed on consensus sequences to search the biggest homology with sequences in the GenBank database. It was done using NCBI BLAST which can be accessed online through http://blast.ncbi.nlm.nih.gov/Blast.cgi [27]. Similarity searching through NCBI-BLAST can be done by uploading files consensus sequences in FASTA format. Protein-protein BLAST (BLASTP) was utilized to search for the protein sequence of the inserted protein sequences. Results of similarity searching on the server will show the percentage of identity, the maximum score, and protein access codes which have similarities with the input sequence. BLAST results show the high level of sequence homology with the consensus sequence of the protein in the GenBank database, which is 99-100%. From the results of BLAST, the chosen sequence is the one with access number ABI49398.1 (neuraminidase [Influenza A virus (A / Indonesia / CDC739 / 2006 (H5N1))]) because it has the highest score (925) and 100% identical to the consensus sequence.

The modeling was done by using SWISS-MODEL, which can be accessed online through the sitehttp://swissmodel.expasy.org/. The comparative modeling or homology modeling is a reliable method for predicting the structure of the target and to obtain a three-dimensional structure of a homologous protein to use as a template [28]. Homology modeling was done in four stages, namely: identification of template structure, sequence alignment of targets and templates, modeling, and evaluation of the model results [23]. There are three modes that are commonly used in modeling, namely: automatic mode, alignment, and projects [23]. 3D modeling in this study was conducted in automatic mode, for identifying the residue amount between the target and template. The results can be trusted if the automatic mode showed that the similarity between common targets and templates is greater than 50% [23]. The template of 3ckz was obtained from the PDB. The results obtained through SWISS-MODEL modeling has a percentage of 97.14% identity to 3ckz template. Therefore, the modeled 3D structure can be trusted because it has a high percentage of identity. Neuraminidase structure modeling results using SWISS-MODEL can be seen in figure1.Based on the evaluation of residues in Ramachandran plot, there is only one aminoacid residue, glutamate(GLU), which is located in the outlier region, which is about 0.3% probability. It can be said that the structure of the generated 3D is good enough and can be used for the next stage of this research.

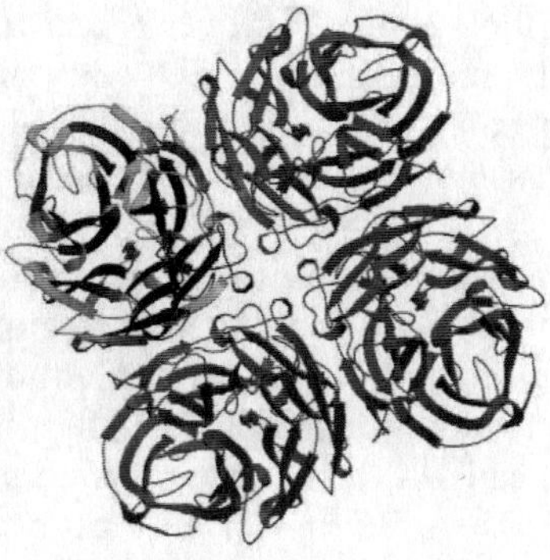

Figure 1. Valid structure of neuraminidase H5N1

Based on the evaluation of residues in Ramachandran plot, there is only one aminoacid residue, glutamate (GLU), which is located in the outlier region, which is about0.3%. It can be said that the generated 3D structure is good enough and can be used for the next stage of this research.

3.2. Visualization and Determination of Neuraminidase H5N1 Binding Site. The enzymes ABI49398.1 was visualized by using Gaussian Contact the Surface menu and maps of MOE2008.10. Visualization of ABI49398.1enzyme can be seen in figure2a. Red tape showing α-helical secondary structure, yellow indicates β-sheet structure, green indicates loop structure, while the blue shows the structure of the turn, an area slightly transparent visualization of the surface of the enzyme. It can be seen that the active side ABI49398.1 is relatively hydrophobic, but still has some hydrogen bonding region. The active enzymes and proteins are generally in the form of a hydrophobic cavity and involve side chain amino acid residues. They are making up the active site, and can be seen through the finder site features contained in MOE 2008.10. Site finder feature will perform calculations toward the active side of the 3D coordinates of receptor atoms. Visualization of the binding of the enzyme to the zanamivir standard was also done through the docking process. Standard zanamivir was obtained from PubChem with CID60855 access number. The obtained binding sites are Agr98, Glu208, Glu257, Arg273, Tyr324, Arg348 and Tyr382 (figure 2b).

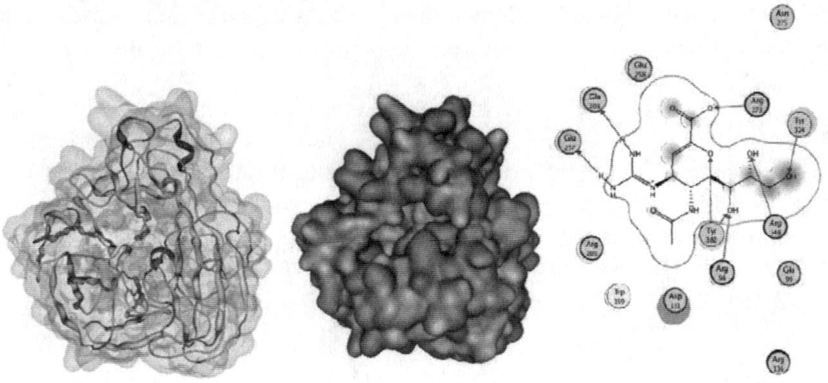

a b

Figure 2. (a) Visualization neuraminidase H5N1 using MOE and (b) binding site sequence ABI49398.1 with zanamivir standard

3.3. Molecular Docking Simulation and Analysis. Flavonoid ligand optimization was did on the MOE database viewer (dv). The first stage in the optimization of ligands is to wash the flavonoids and standard ligand. The goal of Wash treatment is to improve the structure of the ligand and improve the position of the hydrogen atom in the ligand. Then, partial charge determination was done by using a partial charge command. A selected forcefield for the determination of the partial load is MMFF94 that can be used to validate the position of the hydrogen Tomi the ligand [29]. MMFF94 is a force field that parameterized for small organic molecules. The same forcefield is used for energy minimization of ligands. Minimization process was carried out by the RMS gradient of 0.001kcal/mole. The ligands set was then stored in .Mdb format and ready for the simulation of molecular docking.

In the molecular docking, the enzyme is made rigid while the ligand is in a state of the freedom of movement to seek for a stable conformation. The utilized placement method is Triangle Matcher with the 1000 number of rounds. Triangle Matcher shows the random movement of the ligand in the active site of the enzyme to produce an optimal bond orientation. After ligand interacts with the enzyme, the affinity between the receptor and ligand will be evaluated using a scoring function by estimating the value of the binding energy [30-31]. The utilized scoring functions in molecular docking are London dG, which is the default parameter for MOE software. London dG scoring function will show the value of the Gibbs free binding energy of the enzyme conformation and ligand[32].

Molecular docking simulations were carried out to generate data in the form of Gibbs free energy of binding ($\Delta G_{binding}$). The obtained $\Delta G_{binding}$ value is the result of intermolecular hydrogen interactions, electrostatic, and weak Van der Waals forces that occurred between the ligand and the enzyme in a pose. This value is the initial parameters in conducting virtual screening. This is because the Gibbs free energy shows stability and strength of enzyme-ligand interactions. The lower the value of the Gibbs free energy, the enzyme is more stable and thermodynamically favored. From 491 flavonoids ligand screened at the beginning, there is 149 ligand that has a lower value ($\Delta G_{binding}$) than the standard. Initial screening was done by retaining parameter of 30. Then, redocking was done against 149 best ligands to increase retain parameter to 100, in order to get the best conformation of each ligand. In order to expose better data representation, the best 20 results from the redocking can be seen in table 1.Hence, the complete results of the docking for whole 491 ligands and redocking of 149 best ligands could be seen in this following supplementary material: http://staff.ui.ac.id/system/files/users/aditya.parikesit/material/supplementary_material_10jcc.pdf. The best ligand for ABI49398.1 is M0000338 (Guajavin B) with $\Delta G_{binding}$value of -67.2920 kcal/mol.

In general, the annotated interaction between guajavin B and neuraminidase is weak covalent ones (figure 3). Guajavin B is interacting with 10 amino acid residues on neuraminidase, namely: Glu99, Lys130, Asp131, Arg132, Trp159, Glu257, Arg273, Tyr324, Pro411, and Lys412 (figure 3). Guajavin B is included in the flavonol subclass of flavonoid. Flavanols hydroxyl substituent at position 3 and totally absent on double bond at C2-C3. This compound was formed from the reduction of flavanon [33].

Figure 3 shows that the ligand molecule is large enough to deter molecule entrance into the cavity of the enzyme. The formation of cation-aromatic interactions lies between the aromatic ring and Tyr324. Tyrosine will interact with aromatic ring due to delocalisation. Then, the hydrogen atoms in the side chains of Lys130, Arg132, Arg273, and Lys412 can form hydrogen interaction with oxygen atoms of hydroxyl groups on Guajavin B. Lysine is a positive polar (basic) amino acid due to the second primary amine group on the ε position of the aliphatic side chain.

Table 1. $\Delta G_{binding}$ of ligand

No	Ligand ID	$\Delta G_{binding}$ (kkal/mol)
1	**M00009338**	**-67,2920**
2	M00009213	-64,7416
3	M00005901	-63,8624
4	M00009105	-63,5440
5	M00014080	-63,4709
6	M00014082	-61,9593
7	M00013884	-59,5516
8	M00009106	-59,5235
9	M00014066	-58,6969
10	M00009235	-58,5701
11	M00005487	-58,4990
12	RMN00006834	-58,3978
13	RMN00006832	-57,1990
14	RMN00014803	-56,4708
15	RMN00006843	-56,4298
16	M00013746	-56,2798
17	M00014329	-55,9459
18	M00005900	-55,0116
19	M00006246	-54,9444
20	M00019135	-54,7079
21	Oseltamivir	-32,2472
22	Zanamivir	-41,6318

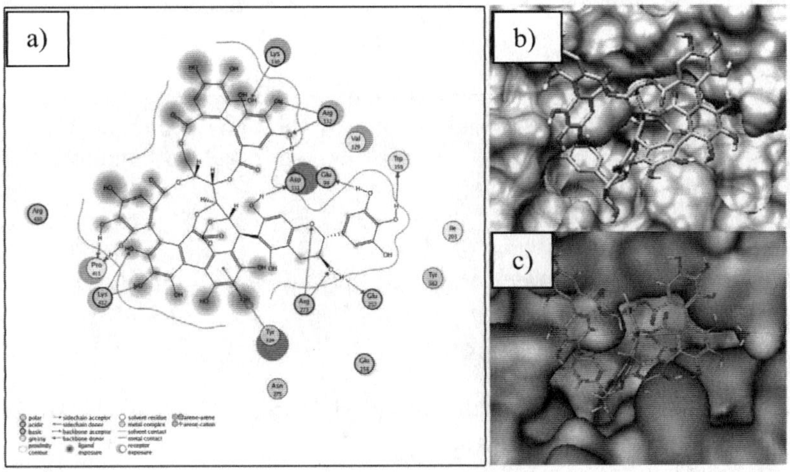

Figure 3: a) visualization of Guajavin B interaction with active site of neuraminidase, and the ligand entrance into the enzyme cavity based on (b) USCF Chimera 1.9 and (c) MOE 2008.10

While arginine is also positive polar (base) due to the guanidino group on the aliphatic side chain[34,35]. Residues of Glu99, Asp131, and Glu257 have formed hydrogen bonding interactions with the H atom of the carbonyl group in Guajavin B to act as an acceptor of H atoms. Glutamate and aspartate are polar amino acids and have negative charges because of the second carboxyl group on their side chains[34]. In addition, the residue of Pro411 is also forming hydrogen interactions with the H atom of the hydroxyl group of the ligand by using the nitrogen atom of the main chain of Pro411.

4. Conclusion

The predicted structure of neuraminidase of H5N1 has been found. Thus, based upon virtual screening of Indonesian natural products, the ligand M0000338 (Guajavin B) has been determined as the best possible lead compound with $\Delta G_{binding}$value of -67.2920 kcal/mol.

Acknowledgement

The authors would like to thanks, Hibah BOPTN Ditjen Dikti No. 0528/UN2.R12/HKP.05.00/2015 for providing research grant. Usman Sumo Friend Tambunan and Arli Aditya Parikesit supervised this research. Dwi Mustika and Bayu Ardiansyah worked on the technical details. Djati Kerami gave important suggestion to improve our pipeline. The proofreading was done by Ratih Dyah Puspitasari. Thanks also go to Dr. Arry Yanuar and his team, for providing access to Herbal DB database of Pharmacy Faculty, University of Indonesia. All authors were credited for writing this manuscript.

References

[1] Kong F 2009 Pilot clinical study on a proprietary elderly extract: efficacy in addressing influenza symptoms. *Online J Pharmacol Pharmacokinet* **5** 32–43

[2] Guan F, Poon LLM, Cheung C Y, Ellis T M, Lim W, Lipatov A S, Chan K H, Sturm-Ramirez K M, Cheung C L, Leung Y H C, Yuen K Y, Webster R G and Peiris J S M. 2004 H5N1 influenza: a protean pandemic threat. *Proc Natl Acad Sci U S A***101** (21) 8156–61

[3] Wang S-Q, Du Q-S and Chou K-C 2007 Study of drug resistance of chicken influenza A virus (H5N1) from homology-modeled 3D structures of neuraminidases. *Biochem Biophys Res Commun* **354**(3) 634–40

[4] WHO 2014 *Cumulative Number of Confirmed Human Cases for Avian Influenza A (H5N1) Reported to WHO.*;

[5] Chutinimitkul S, Suwannakarn K, Chieochansin T and Quynh L 2007 H5N1 Oseltamivir-resistance detection by real-time PCR using two high sensitivity labeled TaqMan probes. *J Virol Methods* **139**(1) 44–49

[6] Tomassini J, Selnick H, Davies M, Armstrong M E,Baldwin J, Bourgeois M, Hastings J, Hasuda D, Lewis J and Mcclements W 1994 Inhibition of Cap (m7GpppXm)-Dependent Endonuclease of Influenza Virus by 4-subsituted 2,4-dioxobutanoic Acid Compounds. *Antimicrob Agent Chemother* **38** 2827–2837.

[7] Samson M, Pizzorno A, Abed Y and Boivin G 2013 Influenza virus resistance to neuraminidase inhibitors. *Antiviral Res* **98** 174–185

[8] Behrens G, Gottschalk R, Guertler L, Harder T C, Hoffmann C, Kamps B S, Korsman S, Preiser W, Reyes-Teran G, Stoll M, Werner O and Zill G V 2006 *Influenza Report*. (Kamps BS, Hoffman C, Preiser W, eds.). Sevilla: Flying Publisher; Available at www.influenzareports.com.

[9] Rungrotmongkol T, Udommaneethanakit T, and Malaisree M 2009 Biophysical Chemistry How does each substituent functional group of oseltamivir lose its activity against virulent H5N1 in fluenza mutants? *Biophys Chem* **145** (1) 29–36

[10] Mckimm-breschkin JL, Selleck PW, Usman TB and Johnson MA 2007 Sensitivity of In fluenza A (H5N1) to Oseltamivir. *Emerg Infect Dis* **13** (9) 2005–2008

[11] Dao T, Tung B, Nguyen P, Thuong P, Yoo S and Kim E 2010 C-Methylated Flavonoids from Cleistocalyx operculatus and Their Inhibitory Effects on Novel Influenza A (H1N1) Neuraminidase. *J Nat Prod* **3423** 1636–1642

[12] Liu A, Wang H, Mingyuen S, Wang Y and Du G 2008 Bioorganic & Medicinal Chemistry Structure – activity relationship of flavonoids as influenza virus neuraminidase inhibitors and their in vitro antiviral activities. *Bioorg Med Chem.* **16** (15) 7141–7147

[13] Yanuar A, Mun'im A, Lagho A B A, Syahdi R R, Rahmat M and Suhartanto H 2011 Medicinal Plants Database and Three-Dimensional Structure of the Chemical Compounds from Medicinal Plants in Indonesia. *Int J Comput Sci* **8** (5) 180–183
[14] Syahdi R R, Mun'im A, Suhartanto H and Yanuar A 2012 Virtual screening of Indonesian herbal database as HIV-1 reverse transcriptase inhibitor. *Bioinformation* **8**(24) 1206–10
[15] Hayun A Y 2011 Virtual screening of 2,3-disubstituted-4(3H)-quinazolinones possessing benzenesulfonamide moiety for the COX-2 inhibitor. *Bioinformation* 7(5) 246 – 50.
[16] Vilar S, Cozza G and Moro S 2008 Medicinal chemistry and the molecular operating environment (MOE): application of QSAR and molecular docking to drug discovery. *Curr Top Med Chem* **8** (18) 1555–72
[17] Spessard GO. ACD Labs/LogP dB 3.5 and ChemSketch 3.51998 *J Chem Inf Model.*; **38** (6) 1250–1253
[18] Patlewicz G, Jeliazkova N, Safford R J, Worth A P and Aleksiev B 2008 An evaluation of the implementation of the Cramer classification scheme in the Toxtree software. *SAR QSAR Environ Res* **19** (5-6) 495–524
[19] Pettersen E F, Goddard T D, Huang C C, Couch G S, Greenblatt D M, Meng E C and Ferrin T E 2004 UCSF Chimera - A visualization system for exploratory research and analysis. *J Comput Chem* **25** (13) 1605–1612
[20] Hinchliffe A CS Chem3D Pro 3.5 and CS MOPAC Pro (Mac and Windows) UK.1997*Electron J Theor Chem*; **2** (1): 215–217
[21] Pedretti A, Villa L and Vistoli G 2004 VEGA – An open platform to develop chemo-bio-informatics applications, using plug-in architecture and script programming. *J Comput Aided Mol Des* 18(3) 167–173
[22] Clamp M, Cuff J, Searle S M and Barton G J 2004 The Jalview Java alignment editor. *Bioinformatics* **20**(3) 426–7
[23] Arnold K, Bordoli L, Kopp J and Schwede T 2006 The SWISS-MODEL workspace: a web-based environment for protein structure homology modeling. *Bioinformatics* **22**(2) 195–201
[24] Carugo O and Djinovic-Carugo K 2013 Half a century of Ramachandran plots. *Acta Crystallogr D Biol Crystallogr* **69** (Pt 8) 1333–41
[25] Parikesit A A and Tambunan K U S F 2013 Screening of Commercial Cyclic Peptides as Inhibitor Envelope Protein Dengue Virus (DENV) Through Molecular Docking and Molecular Dynamics. *Pakistan J Biol Sci* **16** (24) 1836–1848
[26] Thompson J D, Gibson T J and Higgins D G 2002 Multiple sequence alignment using ClustalW and ClustalX *Curr Protoc Bioinformatics* Chapter2: Unit 2.3
[27] Wheeler D L, Barrett T, Benson D A, Bryant S H, Canese K, Chetvernin V, Church D M, DiCuccio M, Edgar R, Federhen S, Geer L Y, Helmberg W, Kapustin Y, Kenton D L, Khovayko O, Lipman D J, Madden T L, Maglott D R, Ostell J, Pruitt K D, Schuler G D, Schriml L M, Sequeira E, Sherry S T, Sirotkin K, Souvorov A, Starchenko G, Suzek T O, Tatusov R, Tatusova T A, Wagner L and Yaschenko E 2007 Database resources of the National Center for Biotechnology Information. *Nucleic Acids Res* **35** D5–12
[28] Tambunan U S F and Parikesit A A 2013 In silico approach towards H5N1 virus protein and transcriptomics-based medication. *African J Biotechnol* **12** (21) 3110–3114
[29] Tambunan U S F, Parikesit A A, Dephinto Y and Sipahutar F R P 2014 Computational design of drug candidates for influenza A virus subtype H1N1 by inhibiting the viral neuraminidase-1 enzyme. *Acta Pharm* **64**(2) 157–72
[30] Kapetanovic I M 2008 Computer-aided drug discovery and development (CADDD): in silico-chemico-biological approach. *Chem Biol Interact* **171** (2) 165–76
[31] Stahl M and Rarey M 2001Detailed analysis of scoring functions for virtual screening. *J Med Chem.*;44:1035–1042. doi:10.1021/jm0003992.
[32] Manikrao A M and Khatale P N, Jawarkar R D and Vyas J V 2011 Presuming the Probable Anti-inflammatory Mechanism of Ursolic Acid : a plant-derived pentacyclic triterpenoid,

IOP Conf. Series: Materials Science and Engineering **107** (2016) 012053 doi:10.1088/1757-899X/107/1/012053

using molecular docking *Library (Lond)* **1**(2) 9–13

[33] Tanaka T, Ishida N, Ishimatsu M, Nonaka G and Nishioka I 1992Tannins and Related Compounds. CXVI. Six New Complex Tannins, Guajavins, Psidinins and Psiguavin from the Bark of Psidium guajava L. *Chem Pharm Bull (Tokyo)* **40** (8) 2092–2098

[34] IUPAC-IUB. IUPAC-IUB Joint Commission on Biochemical Nomenclature (JCBN) 1984 Nomenclature and symbolism for amino acids and peptides. Recommendations 1983. *Biochem J* 219(2) 345–73

[35] Silverman R B and Holladay M W 2014 *The Organic Chemistry of Drug Design and Drug Action*; doi:10.1016/B978-0-12-382030-3.00002-7.

In silico modification of Zn^{2+} binding group of suberoylanilide hydroxamic acid (SAHA) by organoselenium compounds as *Homo sapiens* class II HDAC inhibitor of cervical cancer

Usman Sumo Friend Tambunan[1*], Ridla Bakri[1], Arli Aditya Parikesit[1], Titin Ariyani[1], Ratih Dyah Puspitasari[1], Djati Kerami[2]
[1]Bioinformatics Group, Depatment of Chemistry, Faculty of Mathemathics and Science University of Indonesia, Jl. Salemba Raya No.4, Depok 16424, Indonesia
[2]Mathematics Computation Group, Department of Mathematics, Faculty of Mathemathics and Science University of Indonesia, Jl. Salemba Raya No.4, Depok 16424, Indonesia

[*]E-mail: usman@ui.ac.id

Abstract. Cervical cancer is the most common cancer in women, and ranks seventh of all cancers worldwide, with 529000 cases in 2008 and more than 85% cases occur in developing countries. One way to treat this cancer is through the inhibition of HDAC enzymes which play a strategic role in the regulation of gene expression. Suberoyl Anilide Hydroxamic Acid (SAHA) or Vorinostat is a drug which commercially available to treat the cancer, but still has some side effects. This research present *in silico* SAHA modification in Zinc Binding Group (ZBG) by organoselenium compound to get ligands which less side effect. From molecular docking simulation,and interaction analysis, there are five best ligands, namely CC27, HA27, HB28, IB25, and KA7. These five ligands have better binding affinity than the standards, and also have interaction with Zn^{2+} cofactor of inhibited HDAC enzymes. This research is expected to produce more potent HDAC inhibitor as novel drug for cervical cancer treatment.

1. Introduction
Cervical cancer occurs in the hollow area between the vagina and uterus, precisely in the cervical region [1]. This cancer is the second most common in women [2], as well as ranked seventh of all cancers in the world with as many as 529000 cases in 2008 and more than 85% of cases occur in developing countries. High-risk areas are found in East and West Africa, with a cumulative risk (0-74) of about 3.8%, South Africa (2.9%), South-Central Asia (2.6%), and Central Africa and the South Americas (2.5%) [3].

Cervical cancer is caused by human papilomavirus (HPV). HPV is from the family Papillomaviridae with the core material of double-stranded DNA and does not have a sheath (envelope) [4]. HPV enters the body through mucous membranes, does not circulate in the blood, but localized in the infected place, and cannot be grown in vitro [5].

A drug that used to suppress the growth of cervical cancer cells is an inhibitor of the enzyme histone deacetylase (HDAC). HDAC (EC.3.5.1) is an enzyme that acts as a catalyst for histone deacetylase [6]. It is a medium that can bind with oncogene transcription of genes with the aim of transforming the processes of cells into the media of the viral proliferation [7].

The most often used Inhibitors for class II HDAC activity is Vorinostat or suberoylanilide hydroxamic acid (SAHA). SAHA compound is having carbonyl and hydroxyl amine group. It binds zinc ion, Zn^{2+}, with aliphatic chains as a linker, a hydrophobic group at the other tail. SAHA is a drug that has been sold in the market, but has the side effect as inhibitor of osteoblast maturation [8]. It also raises a variety of symptoms, such as dizziness, diarrhea, vomiting, hyperglycemia, increased levels of protein in urine, to a shortage of platelets (trombocytopenia) [9]. With the side effects of vorinostat or SAHA, it is necessary to modify the group of these compounds to obtain drugs that have toxicity and lower side effects [10].

In a recent study, it was found that the element selenium has an important role in the body, it also acts as an anti-cancer. Organoseleium in the form of compounds of selenometionin and metilselenosistein are proven to be therapeutic compounds on colon cancer cases [11]. In other studies, the organoselenium metabolites can inhibit HDAC in the cases of prostate cancer [11].

An inhibitor of HDAC has at least three sides/regions, namely the attachment of the Zn^{2+} cofactor/the enzyme active site (Zn^{2+} chlating/binding region) [12], close to the hydrophobic site (hydrophobic cap) [13,14], and liaison (linker) [15] containing the connecting unit/CU with electronegative group.

This study will construct various ligand models derived from SAHA compounds with organoselenium modifications based on Zinc Binding Group (ZBG).

2. Material and Methods

2.1 Preparation of the 3D structure of class II HDAC Homo sapiens.

The sequence data of *Homo sapiens* Class II HDACs (4, 5, 6, 7, 9, 10) was prepared by dowloading them from the protein database at NCBI site (http://www.ncbi.nlm.nih.gov). Furthermore, the sequence was copied and stored with notepad in FASTA format. 3D structure of the *Homo sapiens* Class II HDACs was obtained from the Protein Data Bank (PDB) and stored in .pdb format. If there were proteins that were not available in the PDB, it was modeled by SWISS model and the data was obtained in .pdb format.

2.2 Sequence conservation of Homo sapiens Class II HDAC.

This alignment was done on the NCBI BLAST (Basic Local Alignment Search Tool) server and Clustal W at EMBL-EBI. Sequences of class II HDAC *Homo sapiens* enzymes were uploaded to the server, then performed the BLASTP (BLAST Protein).

2.3 Preparation of Homo Sapiens Class II HDAC enzyme.

HDACs sequence data were obtained and stored with .pdb format in the previous procedure, opened with the MOE (Molecular Operating Environment) 2008.10 software. Furthermore, the removal of amino acids chains, ligands/inhibitors, and unwanted solvents (usually H_2O) was conducted. Thus, 3D protonate menu was used on the compute section to add the polar hydrogen and led the charge of Zn^{2+}. The addition of polar hydrogen provides partial charge in the enzyme. The determination of the partial load was carried on the menu of compute mode that has been parameterized with AMBER99 forcefield. Then, the menu was utilized to minimize enzyme's energy to reach the lowest level. Enzymes are stored in .moe or .mdb format and used for molecular docking simulations.

2.3 Design and preparation of Homo sapiens Class II HDAC Ligand inhibitor.

Ligands were drawn in two dimensions using offline software program of ACDLabs ChemSketch 12.0. They are SAHA and its derivatives. Various SAHA modifications were utilized. The ligands were saved in MDL Molfile format. Ligands storage format was converted to MDL Mol using Vegazz software. All ligand then forwarded into MOE database. Ligand molecule was opened within .mol format using MOE software. Furthermore, the ligand was computed with "wash" menu. Ligand was adjusted with "partial charge" and optimized using MMFF94 forcefield.

2.4 Docking simulation.
Docking process was done by using the menu Compute-Simulation-Dock in MOE 2008.10 software. Docking results were stored in .mdb format. The selected area was the docking residues. The placement method was using a triangle matcher with repetition of energy readings every 1000000 position and other corresponding parameters contained in MOE 2008.10. The next stage of scoring function was using London dG, refinement by using forcefield. The first repetition was done in 30 times, while the latter only just showed one of the best from 30 repetitions.

2.5 Analysis of Molecular Docking.
Results were analyzed using molecular docking of MOE software. Results data is opened in the .mdb format. To make it easier to sort the results based on the $\Delta G_{binding}$, the data file was converted to .xls format by using the export menu. Ligand conformation that has the lowest binding energy was chosen to determine protein-ligand conformation from docking result. The selected enzyme-ligand complex was having a low binding energy value and the highest inhibition constants for further analysis.

2.6 Visualization of Molecular Interactions.
Visualization of 3D complex interactions with ligands enzyme was done through the rendering process. After going through the process, 3D structure of the enzyme can be illustrated through maps on the navigation of surface and compute menu. Then, the interacting ligand, cofactor, and the enzyme amino acid residues could be displayed. Interaction of amino acid residues in enzymes with ligands can be viewed by using the features of LigX on MOE. Initially, ligand was selected using the browser menu. Enzyme-ligand complex energy was minimized with the LigX Minimize menu, then clicked the Interaction menu. After that, an analysis of the interactions that occurred between amino acid residues in enzymes with ligand inhibitor was conducted.

3. Result and Discussion

3.1 Determination of class II HDAC Homo sapiens.
The results of the *Homo sapiens* Class II HDAC search in the NCBI sites are 91 sequences. The sequences were dowloaded in FASTA format, to conduct multiple sequnce alignment for each *Homo sapiens* Class II HDAC. The highest score was chosen as the modeling sequences. The obtained sequences of each enzyme HDAC class II are presented in Table 1.

Tabel 1. Sequences code of each enzyme HDAC class II *Homo sapiens*

Enzyme	Sequences code
HDAC4	P56524
HDAC5	Q9UQL6.2
HDAC6	Q9UBN7.2
HDAC7	Q8WUI4.2
HDAC9	Q9UKV0.2
HDAC10	Q969S8.1

From the modeling results, the enzyme HDAC5 and HDAC9 structures were quite valid, but HDAC10 showing results that are less valid. HDAC10 protein structure were taken in the absence of structural data that more comparable with the results HDAC10 modeling. After HDAC structure was obtained, then the determination of the active side of the structure was done.

The 3D structure of a protein can be accessed at the database of Research Collaboratory for Structural Bioinformatics Protein Data Bank (RCSB-PDB) through the website of http://rscb.org/pdb/.

IOP Conf. Series: Materials Science and Engineering **107** (2016) 012054 doi:10.1088/1757-899X/107/1/012054

3D structures of HDAC5, HDAC9, and HDAC10 were not found in the database RSCB-PDB, so it needs to conduct modeling with other software. FASTA sequences of HDAC5, HDAC9, and HDAC10 was served into its 3D models by using homology modeling software with online SWISS models (http://swissmodel.expasy.org/) (Table 2 and Figure 1).

Table 2. Results of 3D protein structure modeling class II HDAC Homo sapiens with SWISS Model

Protein	**QMEAN4** ***score***	***Identity***
HDAC5	-2.72	76.90%
HDAC9	-2.20	74.54%
HDAC10	-4.43	39.94%

After HDAC structure is obtained, then the active site of the structure was determined (Figure 2).

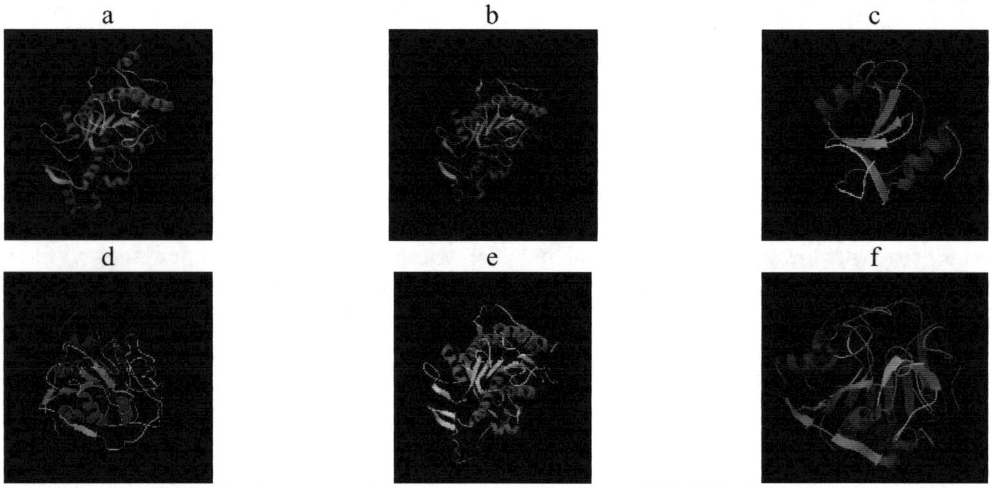

Figure 1. 3D structure of *Homo sapiens* class II HDAC: a.HDAC4, b.HDAC5, c. HDAC6, d. HDAC7, e. HDAC9, and f. HDAC10

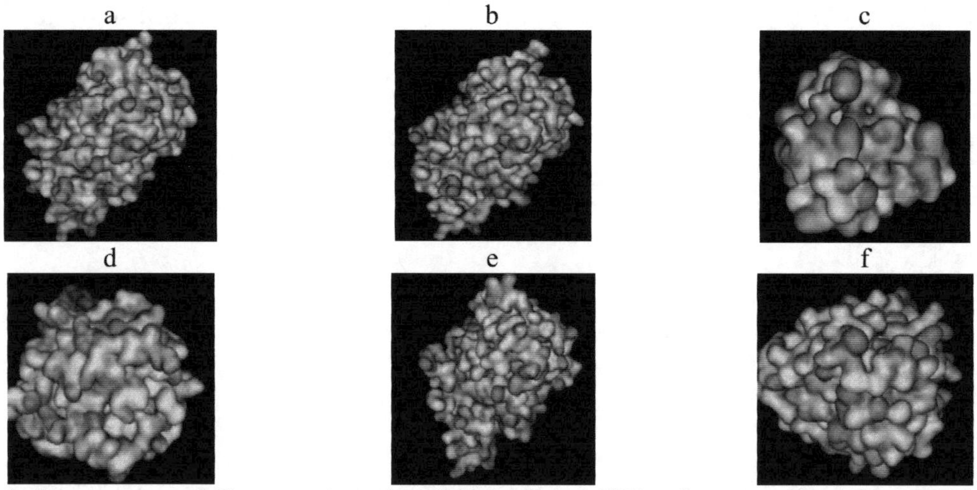

Figure 2. 3D structure of *Homo sapiens* class II HDAC in MOE *software* : a.HDAC4, b.HDAC5, c. HDAC6, d.HDAC7, e.HDAC9, and f.HDAC10
Description: purple: H-bonding, green: Hydrophobic region, and blue: Mild Polar

3.2 Determination of the Enzyme Active Side of class II HDAC Homo sapiens. Crystal protein contained in the PDB database is the protein inhibitor compound that has been shown to lower the catalytic activity. From this data, information can be found on the location of amino acid residues that become the catalytic site of enzyme. The PDB 3D structures of the HDAC4, HDAC6, and HDAC7 enzymes were found as well. Thus, the amino acid residues of the active site were obtained (Table 3). The structures of HDAC5, HDAC9, and HDAC10 were the result of homology modeling with SWISS Model.

The existence of Zn^{2+} cofactor became a key factor for inhibition of class II HDAC enzymes. It is designed so that the ligand inhibitor has a group that is able to interact or bind Zn^{2+} ions in order to inhibit the catalytic activity of the enzyme.

Table 3. location of class II HDAC *Homo sapiens* catalytic sites

Enzyme	Location of The Catalytic site
HDAC4	Ion Zn^{2+} with residu Asp196, His198, dan Asp290
HDAC5	Ion Zn^{2+} with residu Asp870, His872, dan Asp964
HDAC6	Ion Zn^{2+} with residu Cys5, His7, dan Cys78
HDAC7	Ion Zn^{2+} with residu Asp707, His709, dan Asp801
HDAC9	Ion Zn^{2+} with residu Asp820, His822, dan Asp914
HDAC10	Ion Zn^{2+} dengan residu Asp172, His174, dan Asp265

3.3 Design of Class II HDAC Homo sapiens Inhibitor. SAHA was modified to obtain a potential drug candidate with minimal side effects for the treatment of cervical cancer. Modifications were carried out on Zinc Binding Group (ZBG) by substituting organoselenium in the group, while maintained the structure of the CAP and the linker. Organoselenium compounds have been selected in previous studies with the same modification on ZBG selenium compounds to inhibit HDAC on lung cancer cell (Figure 3). In vitro studies proved that the result of this modification provides better performance than the standard compound. Therefore, modification was followed by substituting the variation of organoselenium to obtain the possibility of better results.

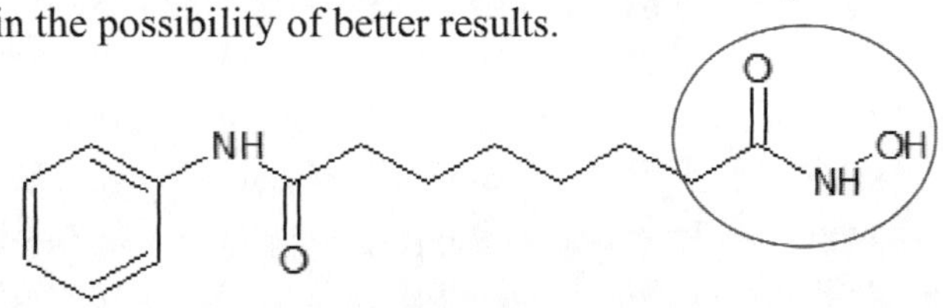

Figure 3. Modification of SAHA ligand. The area within blue circle is the Zinc Binding Group (ZBG).

Total ligands that have been successfully made are 1763, and 2 standard ligand compounds of SAHA and TSA.

3.4 Docking.
Molecular docking simulation process was iterated once on each of the six class II HDAC *Homo sapiens* enzymes with the same 1165 ligands. Docking process takes 18-50 hours, depending on the number of ligands and the computer processor specifications.

The selected 18 best ligands are HA27, UA28, MB25, UA30, A24, HB29, LA30, SA27, ZA28, UA30, KA7, HB28, CC27, A1A27, ZA25, EB25, IB25, and BA24. Six others were taken from the best ligand structure approach that has a good interaction with the six class II HDAC *Homo sapiens* enzymes and remain within the below standard criteria of $\Delta G_{binding}$ value. Those six ligands are B1B25, BB24, BC7, HB20, JA27, and LA26. The binding free energy ($\Delta G_{binding}$) of the best ligand with HDAC enzymes is presented in Table 4, while the inhibition constant is in Table 5, respectively.

Table 4. The binding free energy ($\Delta G_{binding}$) of the best ligand with HDAC enzymes

	$\Delta G_{binding}$ (kkal/mol)					
Ligand	**HDAC4**	**HDAC5**	**HDAC6**	**HDAC7**	**HDAC9**	**HDAC10**
SAHA	-29.420	-39.175	-12.562	-45.617	-25.340	-11.079
TSA	-53.710	-4.057	-1.032	-32.101	-17.760	27.276
A1A27	-26.044	-49.282	-14.273	-64.810	-30.377	-89.697
A24	-4.146	-59.975	-14.969	-59.968	-36.873	-10.939
B1B25	-42.219	-58.684	-1.628	-42.561	-42.004	-14.847
BA24	-39.919	-60.331	-17.748	-24.563	-34.751	63.416
BB24	-40.248	-58.882	-13.248	-31.459	-31.459	-82.618
BC7	-24.517	-42.294	-17.277	-44.085	-27.217	**-99.316**
CC27	-26.108	-36.836	-13.487	-58.605	-33.903	-12.392
EB25	-3.825	-59.831	-16.090	-3.794	-32.282	-13.166
HA27	-54.414	-58.829	-17.251	-41.632	-27.043	11.455
HB20	-25,480	-37.405	-18.655	-31.388	-26.977	-98.902
HB28	-32.076	-46.794	-2.436	-36.847	-45.414	11.430
HB29	-31.434	-51.708	**-26.456**	-0.375	-29.787	-38.785
IB25	-44.025	-60.535	-15.989	-33.678	-38.057	-12.633
JA27	-22.347	-60.990	-17.564	-46.531	-45.342	-1.163
KA7	-34.164	-36.715	-18.564	-38.382	-22.533	-98.975
LA26	-33.382	-40.031	-17.985	-33.393	-0.271	-98.266
LA30	-31.023	-43.625	-25.909	-43.167	-3.162	-19.152
MB25	-47.377	-57.618	-19.077	-52.978	-38.989	-88.693
SA27	-31.655	-4.733	-25.688	-53.560	-28.638	-88.903
UA28	**-57.837**	**-75.258**	-12.381	**-99.786**	**-63.495**	0.641
UA29	-28.383	-70.755	-15.962	-42.443	-23.721	-55.034
UA30	-46.946	-75.085	-14.086	-57.256	-2.646	-50.433
ZA25	-27.777	-58.900	-15.961	-17.130	-44.571	-14.146
ZA28	-40.935	-32.213	-25.553	-2.508	-25.874	-33.987

Note: The number in bold is the lowest free energy value

3.5 Interaction Analysis of HDAC with The Best Ligand.

After getting the value of $\Delta G_{binding}$ molecular docking results, the next step is to analyze the interaction between class II HDAC *Homo sapiens* enzymes with the ligand inhibitor. From the results of molecular docking database, the selected the file-browser menu would determine the ligand interactions. After the minimization of energy, visualization of the interaction between the enzyme and the ligand was obtained.

In HDAC4, the interaction was formed between Zn^{2+} cofactor ligand compounds with oxygen atoms at hydroxyl groups that bound to the amide group (Figure 4). The Zn^{2+} cofactor coordination complex bonding with oxygen atoms also bind 3 amino acids (His198, Asp196, and Asp290), which is the enzyme active site residues of HDAC4. Zn^{2+} coordination number in these bonds is four. Zn^{2+} electronic charges on oxygen have shown empty p orbitals (π acceptor). Then, the interaction between His872 side chains with a hydroxyl group attached to the Selenium was observed as well.

Table 5. Inhibition constant value of the best ligand with HDAC enzymes

	Inhibition constant (pKi)					
Ligand	**HDAC4**	**HDAC5**	**HDAC6**	**HDAC7**	**HDAC9**	**HDAC10**
SAHA	21.4463	28.5574	9.1573	33.2534	18.4721	8.0762
TSA	39.1529	2.9574	0.7523	23.4006	12.9465	-19.8834
A1A27	18.9853	35.9251	10.4046	47.2445	22.1439	65.3863
A24	3.223	43.7199	10.9119	43.7148	26.8793	7.9742
B1B25	30.7763	42.7788	1.1868	31.0257	30.6196	10.8230
BA24	29.0997	43.9794	12.9377	17.9057	25.3324	-46.2283
BB24	29.3395	42.9232	9.6574	22.9326	22.9326	60.2260
BC7	17.8721	30.8310	12.5944	32.1366	19.8404	**72.3983**
CC27	19.0319	26.8523	9.8316	42.7212	24.7142	9.0334
EB25	2.7883	43.6150	11.7291	2.7657	23.5326	9.5976
HA27	39.6661	42.8845	12.5754	30.3484	19.7135	-8.3503
HB20	18.5741	27.2671	13.5989	22.8809	19.6654	72.0965
HB28	23.3824	34.1114	1.7758	26.8603	33.1054	-8.3321
HB29	22.9144	37.6935	**19.2856**	0.2734	21.7138	28.2731
IB25	32.0929	44.1281	11.6555	24.5502	27.7424	9.2091
JA27	16.2903	44.4598	12.8036	33.9197	33.0529	0.8478
KA7	24.9045	26.7641	13.5326	27.9793	16.4259	72.1497
LA26	24.3344	29.1814	13.1105	24.3425	0.1976	71.6329
LA30	22.6148	31.8013	18.8869	31.4674	2.3050	13.9612
MB25	34.5364	42.0017	13.9065	38.6193	28.4218	64.6545
SA27	23.0755	3.4502	18.7258	39.0436	20.8762	64.8075
UA28	**42.1614**	**54.8608**	9.0254	**72.7409**	**46.2859**	-0.4673
UA29	20.6903	51.5782	11.6358	30.9396	17.2919	40.1181
UA30	34.2222	54.7346	10.2683	41.7379	1.9289	36.7641
ZA25	20.2486	42.9363	11.6351	12.4872	32.4909	10.3120
ZA28	29.8404	23.4823	18.6273	1.8283	18.8613	24.7755

Note: The number in bold is the highest inhibition constant value

From the visualization of HDAC5 interaction with ligand IB25, there is visible presence of a covalent bond between the Zn^{2+} cofactor with oxygen atoms in the carboxylic group of ZBG, also with the active site amino acid residues (His872, Asp870, and Asp964) (Figure 5). Then H in the amine backbone of Gly1005 was able to donate a proton to the oxygen atom on ZBG fragment and form hydrogen bonds.

The HDAC6 enzyme interaction with ligand CC27 shows that it does not bind Zn^{2+} cofactor, but provide hydrogen bonding interactions with residues Cys78 which became one of the amino acid residues in the enzyme active site (Figure 6). Secondary amine ligands were donating protons to the backbone chain of Cys78 residue. In the hydrophobic stamp or linker, there is no interaction between the ligand with the enzyme.

HDAC6 enzyme has a smaller structure than other class II HDAC enzymes. In addition, there are active sides on the outside, so it should be easier to give a pose that generate the interaction. However, there is also possibility of interactions with the standard SAHA and HDAC6. The Zn^{2+} on enzyme did not show any interaction with the standard compound.

Visualization of the interaction between the enzyme HDAC7 with ligand CC27 showed five covalent bond coordination between Zn^{2+} cofactor with three residues of the active sites (Asp707, His709, and Asp801) (Figure 7). His843 side chain residues showed interaction with the hydrophobic cap. Whereas His670 side chain interacts with the carbonyl of ZBG. Then, there is also a hydrogen bonding interaction between the H bonded to nitrogen with side chain residue of Asp707.

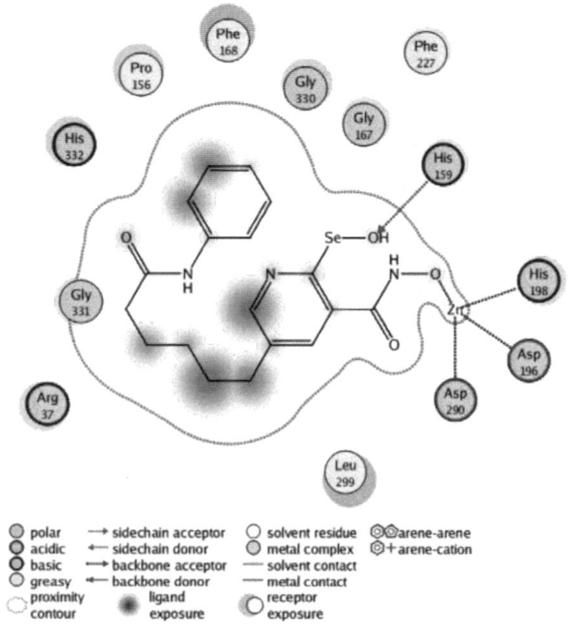

Figure 4. Visualization HDAC4 enzyme with ligand HA27

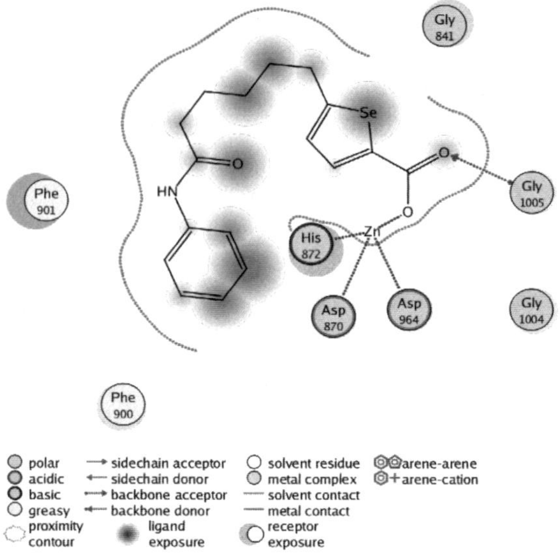

Figure 5. Visualization of HDAC5 enzyme interaction with ligand IB25

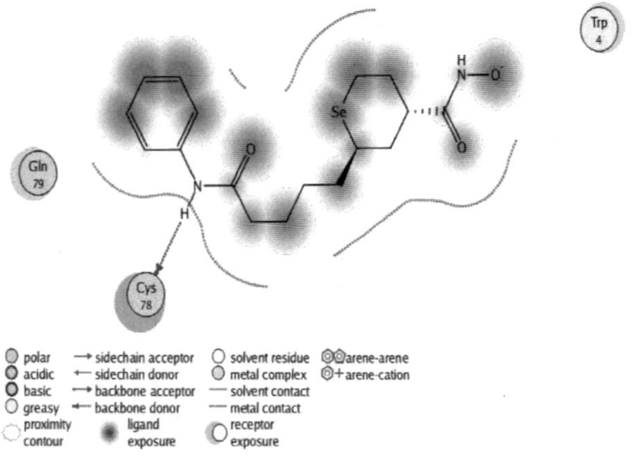

Figure 6. visualisation of enzyme HDAC6 interaction with ligandd CC27

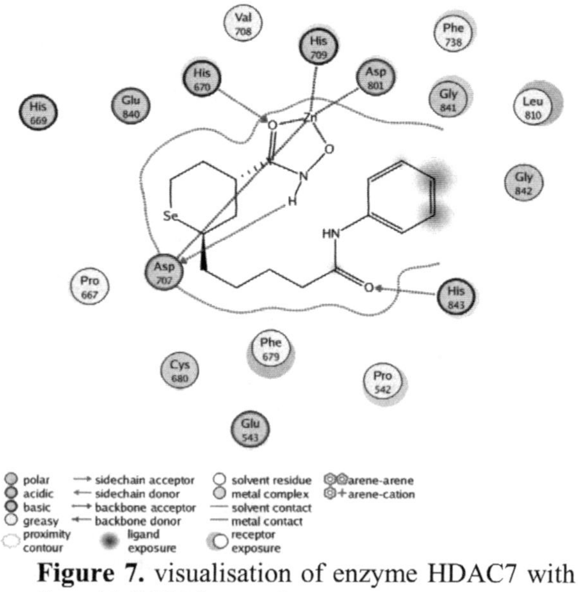

Figure 7. visualisation of enzyme HDAC7 with ligandd CC27 interaction

Next on the visualization of the interaction between the ligand HB28 and HDAC9 enzyme, the cofactor Zn^{2+} formed four bonds coordination with oxygen atoms at hydroxyl groups with three active site residues (Asp820, His822, and Asp914) (Figure 8). Then the active site of the side chain residues of Asp820 and Asp914 was also forming hydrogen bonding interactions with the hydroxyl groups on the H atom in ZBG, while the side chains of the active site residues His822 also forms a hydrogen bond interactions with the cation H in the aromatic group.

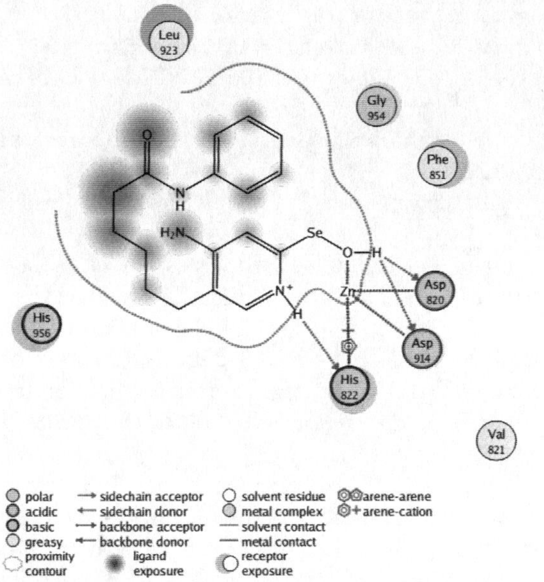

Figure 8. Visualisation of enzyme HDAC9 with ligandd HB28

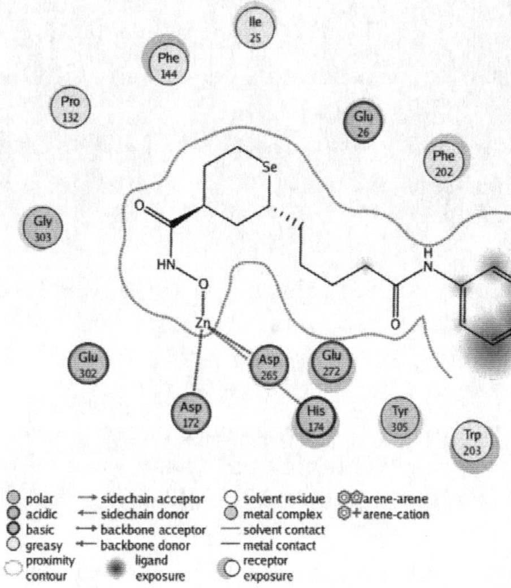

Figure 9. Visualisation of enzyme HDAC10 with ligandd KA7

At HDAC10, KA7 ligand formed coordination bond between Zn^{2+} cofactor with the active site residues (His172, His174, and Asp265). Thus, the oxygen atoms in the hydroxyl group were attached to an amide (Figure 9).

4. Conclusion

The designed compounds were organoselenium-modified version of SAHA at the ZBG. This modification resulted in 1763 ligands, and then advanced to the stage of molecular docking. From the analysis of $\Delta G_{binding}$ between the ligand inhibitor with a class II HDAC enzymes *Homo sapiens* as well as analysis and visualization of interactions, 24 best ligands were obtained. The interaction analysis between the class II HDAC *Homo sapiens* enzyme with the ligand inhibitor shows the covalent bond coordination between the Zn^{2+} cofactor and oxygen atom in the carboxylate groups on ZBG, also with amino acid residues active side.

Acknowledgement

The authors would like to thanks Hibah BOPTN Ditjen Dikti No. 0542/UN2.R12/HKP.05.00/2015 for providing research grant. Thanks also goes to Directorate of Research and Community Engagement-University of Indonesia (DRPM-UI) for the sponsorship in project "Pelaksanaan Hibah Penyelenggaraan Pelatihan Penulisan Artikel Untuk Jurnal Internasional Tahun Anggaran 2015", No.: 2774/UN2.R12/HKP.05.00/2015. Usman Sumo Friend Tambunan, Ridla Bakri, and Arli Aditya Parikesit supervised this research. Titin Aryani and Ratih Dyah Puspitasari worked on the technical details. Djati Kerami gave important suggestion to improve our pipeline. All authors were credited for writing this manuscript.

References

[1] Tambunan U S F, Bramantya N and Parikesit A A 2011 In silico modification of suberoylanilide hydroxamic acid (SAHA) as potential inhibitor for class II histone deacetylase (HDAC).*BMC Bioinformatics* **12** (Suppl 13) S23 doi:10.1186/1471-2105-12-S13-S23

[2] Naus M 2008 HPV vaccine revolutionizes prevention of cervical cancer *B C Med J.* **50**(5)

[3] Ferlay J, Shin H R, Bray F, Forman D, Mathers C and Parkin D M 2008 Estimates of worldwide burden of cancer in GLOBOCAN *Int J Cancer* 2010 **127**(12):2893–2917 doi:10.1002/ijc.25516

[4] Ciaran B J, Woodman S I C and L S Y 2007 The natural history of cervical HPV infection unresolved issues *Nat Rev Cancer* (doi:10.1038/nrc2050)

[5] Beskow A 2003 *Genetic risk factors for cervical carcinoma in situ.* Uppsala: Uppsala University

[6] Bouchain G and Delorme D 2003 Novel hydroxamate and anilide derivatives as potent histone deacetylase inhibitors: synthesis and antiproliferative evaluation *Curr Med Chem* **10** 2359–2372 doi:10.2174/0929867033456585

[7] Park J S, Kim E J, Kwon H J, Hwang E S, Namkoong S E and Um S J 2000 Inactivation of interferon regulatory factor-1 tumor suppressor protein by HPV E7 oncoprotein Implication for the E7-mediated immune evasion mechanism in cervical carcinogenesis *J Biol Chem.* **275** (10) 6764–6769 doi:10.1074/jbc.275.10.6764

[8] Tambunan U S F, Parikesit A A, Prasetia T and Kerami D 2013 In Silico Molecular Interaction Studies of Suberoylanilide Hydroxamic Acid and its Modified Compounds with Histones Deacetylase Class II *Homo sapiens* as Curative Measure towards Cervical Cancer In: *Engineering* Vol **5** Scientific Research Publishing 203–206. doi:10.4236/eng.2013.510B043

[9] Matsuoka H, Unami A, Fujimura T, Noto T, Takata Y, Yoshizawa K, Mori H, Aramori I and Mutoh S 2007 Mechanisms of HDAC inhibitor-induced thrombocytopenia. *Eur J Pharmacol* **571** (2-3):88–96. doi:10.1016/j.ejphar.2007.06.015

[10] Tambunan U S F and Wulandari E K 2010 Identification of a better Homo sapiens Class II HDAC inhibitor through binding energy calculations and descriptor analysis *BMC Bioinformatics* **11** Suppl 7:S16. doi:10.1186/1471-2105-11-S7-S16

[11] Lee J_I, Nian H, Cooper A_J, Sinha R, Dai J, Bisson W_H, Dashwood R_H and Pinto J_T 2009 Alpha-keto acid metabolites of naturally occurring organoselenium compounds as inhibitors of histone deacetylase in human prostate cancer cells *Cancer Prev Res (Phila)* **2** (7) 683–93 doi:10.1158/1940-6207.CAPR-09-0047

[12] Thangapandian S, John S, Lee Y, Arulalapperumal V and Lee K W 2012 Molecular Modeling Study on Tunnel Behavior in Different Histone Deacetylase Isoforms. *PLoS One.* **7** doi:10.1371/journal.pone.0049327

[13] Chen P C, Patil V, Guerrant W, Green P and Oyelere A K 2008 Synthesis and structure-activity relationship of histone deacetylase (HDAC) inhibitors with triazole-linked cap group. *Bioorg Med Chem.* **16** (9) 4839–53 doi:10.1016/j.bmc 03.050

[14] Huang D, Li X, Wei Y and Xiu Z 2012 A novel series of l-2-benzyloxycarbonylamino-8-(2-pyridyl)-disulfidyloctanoic acid derivatives as histone deacetylase inhibitors: design, synthesis and molecular modeling study *Eur J. Med. Chem.* **52** 111–22 doi:10.1016/j.ejmech.2012.03.009

[15] Rajak H, Kumar P, Parmar P, Thakur B S, Veerasamy R, Sharma P C, Sharma A K, Gupta A K and Dangi J S 2012 Appraisal of GABA and PABA as linker: design and synthesis of novel benzamide based histone deacetylase inhibitors *Eur J. Med. Chem.* **53** 390–7 doi:10.1016/j.ejmech.2012.03.058

IOP Conf. Series: Materials Science and Engineering **107** (2016) 012055 doi:10.1088/1757-899X/107/1/012055

Strain-specific variations in *Toxoplasma gondii* GRA1, GRA5, GRA6, GRA8, and GRA14

S Haryati[1,2,3], Y Sari[1,2,3], A A Prasetyo[1,2,4*] and R Sariyatun[1,2]

[1]A-IGIC (A-Infection, Genomic, Immunology & Cancer) Research Group, SebelasMaret University, Jl. Ir. Sutami 36A, Surakarta, 57126, Indonesia
[2]Center of Biotechnology and Biodiversity Research and Development, SebelasMaret University, Jl. Ir. Sutami 36A, Surakarta, 57126, Indonesia
[3]Department of Parasitology, Faculty of Medicine, SebelasMaret University, Jl. Ir. Sutami 36A, Surakarta, 57126, Indonesia
[4]Department of Microbiology, Faculty of Medicine, SebelasMaret University, Jl. Ir. Sutami 36A, Surakarta, 57126, Indonesia

*E-mail: afie.agp.la@gmail.com; afieagp@yahoo.com; afie@staff.uns.ac.id

Abstract. Diagnosis and identification of the genetic group of *T. gondii* (*Toxoplasma gondii*) are important to control better the *T. gondii* infection, particularly in immunocompromised people as HIV patients. This study aimed to identify strain-specific variations in *T. gondii* GRA1, GRA5, GRA6, GRA8, and GRA14 in order to help design such diagnostic tool to detect and characterize the parasite. Forty-three *T. gondii* GRA1, GRA5, GRA6, GRA8, and GRA14 sequences deposited in GenBank were aligned. A number of positions in the gene sequences were highly conserved. All GRA sequences had strain-specific positions, however, only GRA1, GRA5, and GRA6, which contained specific variations for each *T. gondii* lineage. In conclusion, *T. gondii* GRA1, GRA5, GRA6, GRA8, and GRA14 are predicted to contain highly conserved regions and positions with strain-specific variation, which might be useful for the design of diagnostic tools detecting and distinguishing T. gondiistrains.

1. Introduction

T. gondii (*Toxoplasma gondii*) is a protozoan parasite known to infect warm-blooded animals, including humans [1]. In human, *T. Gondii* usually persists as a long-lasting infection with a prevalence of 30-80% worldwide and even higher in tropical areas, as Indonesia [2–8].Nevertheless, *T.gondii* can cause a life-threatening condition in immuno-compromised individuals [9,10], miscarriage and fetal malformations in pregnant women, as well as great economic losses in livestock industries due to abortion in farm animals [10–12]. Therefore, early detection of *T. gondii* infection is crucial to prevent the unfavorable medical and economic impacts.

Identification of a potential target gene or protein is an important step in developing a diagnostic method to detect a pathological agent. In *T. gondii*, the dense granular (GRA) proteins are known as highly immunogenic and highly expressed proteins both in tachyzoite and bradyzoite, either during acute or chronic infection [13,14].Hence, GRA proteins have been suggested as potential targets to diagnose *T. gondii* infection [15–19], and hereby continuous study of the proteins are essential. Another important aspect of the diagnosis of *T. gondii* infection is the identification of *T. gondii* strain, as it allows the degree of the virulence to be determined [18]. Characterization of *T. gondii* strain is also important to understand the population biology of *T. gondii* and identify the infection source [18].

Therefore, this study aimed to investigate strain-specific nucleotides and amino acids in GRA1, GRA5, GRA6, GRA8, and GRA18 sequences, as well as conserved regions in the GRA genes. The results of this study would be useful for the development of either molecular or serological diagnostic method which is able to detect *T. gondii* and characterize its strain.

2. Materials and Methods

2.1. Data resources

In total,forty-three *T. gondii* sequences deposited in GenBank (http://www.ncbi.nlm.nih.gov/genbank/) were analyzed, which consisted of GRA1 (n = 9), GRA5 (n = 9), GRA6 (n = 9), GRA8 (n = 9), and GRA14 (n = 7). The *T. gondii* sequences comprised type I (GT1 and RH), II (ME49), and III (VEG) strains. The accession numbers of the sequences were presented in table 1.

Table 1. Accession numbers of *T. gondii* sequences analyzed in the study.

Gene	*T. gondii* strains			
	GT1	RH	ME49	VEG
GRA1	AAQM03000013, KE387274	EU983103	NW_002234537, KE138831, XM_002365660, CM002043	KI544492, AAYL02000011
GRA5	AAQM03000223, KE387286	L06091	NW_002234546, KE138824, XM_002369230, CM002039	KI544504, AAYL02000231
GRA6	AAQM03000202, KE387284	JN649063	NW_002234575, KE138836, XM_002371898, JX044213	KI544503, AAYL02000197
GRA8	AAQM03000244, KE387289	DQ835668	NW_002234547, KE138817, XM_002369485, CM002037	KI544505, AAYL02000250
GRA14	AAQM03000183, KE387283	FJ015061	KE138826, CM002040	KI544502, AAYL02000189

2.2. Multiple alignments and identification of strain-specific variations

Multiple alignments of GRA1, GRA5, GRA6, GRA8, and GRA14 nucleotide and amino acid sequences were built using ClustalW integrated into MEGA6 software [20,21]. Alignment results were checked manually in CLC Sequence Viewer version 7.6 (www.clcbio.com). The aligned sequences were directed to the identification of sites with high conservation and strain-specific nucleotides and amino acids in GRA1, GRA5, GRA6, GRA8, and GRA14.

3. Results and Discussion

The alignment results showed that some isolates did not contain complete GRA1 (nucleotide 1-67; amino acid 1-23) and GRA8 sequences (nucleotide 1-69 and 789-810; amino acid 1-23 and 264-269) (numbering was in accordance with NW_002234575, an ME49 isolate), hereby these positions were not analyzed. Numerous positions in GRA1, GRA5, GRA6, GRA8, and GRA14 of all *T. gondii* strains presented same nucleotides and amino acids (table 2). This highly conserved nucleotide and amino acid sequences would be useful for the design of molecular and serological detections, respectively, targeting *T. gondii* GRA genes/proteins.

A number of positions in GRA1, GRA5, GRA6, GRA8, and GRA14 were specific to certain *T. gondii* strains (table 3 and table 4). GRA1, GRA5, and GRA6 contained positions with specific nucleotides and amino acids for the three types of *T. gondii* lineages. Hereby, GRA1, GRA5, and GRA6 were suggested as the most potential target for such diagnostic method to detect and characterize the strains of T. gondii.

Table 2. Highly conserved positions in GRA sequences of *T. gondii*.

		Position
Nucleotide	GRA1	68, 70-123, 125-179, 182-248, 384-390, 392-420, 422-429, 431-476, 478-482, 485-623, 625-680, 682-687, 689-706
	GRA5	1-90, 93-103, 105-123, 125-129, 131-136, 138-157, 159-163, 165-280, 282-290, 292-363
	GRA6	1-18, 20-48, 50-83, 85-139, 141-148, 150-281, 283-313, 315-538, 540-553, 555-573, 575-591, 593-612, 613, 615-619, 621-636, 638-649, 651, 653-661, 663-670, 672-675
	GRA8	70-302, 304-337, 339-551, 553-637, 639-642, 644-650, 654-660, 662-668, 671-675, 677-687, 689-690, 692-669, 702-708, 710-717, 719-738, 740-745, 747-768, 770-771, 773-776, 778-788
	GRA14	1-126, 128-967, 969-1065, 1068-1091, 1093-1227
Amino acid	GRA1	24-41, 43-59, 61-85, 87-95, 97-98, 100-113, 115-116, 118-162, 164-190
	GRA5	1-30, 32-34, 36-41, 43, 45, 47-52, 54, 56-93, 95-120
	GRA6	1-6, 8-16, 18-46, 48-49, 51-93, 95-104, 106-179, 181-184, 186-191, 193-197, 199-203, 206, 208-212, 214-216, 219-220, 222-223
	GRA8	24-112, 114-212, 214, 216-220, 222, 225, 227-229, 232-233, 235-236, 238-239, 241-246, 248, 250-256, 260-263
	GRA14	1-42, 44-322, 324-355, 357-363, 365-408

[a]Sequence numbering for GRA1, GRA5, GRA6, and GRA8 is referring NW_002234575 (ME49 strain) isolated while that for GRA14 is in accordance with KE138826 (ME49 strain) isolated.

Table 3. Positions in GRA predicted with strain-specific nucleotides.

Specificity	Gene	Position/nucleotide[a]
GT1 and RH	GRA1	124T, 391G
	GRA5	91A, 92C, 124A
	GRA6	49G, 282A, 620C
ME49	GRA1	180G, 181C, 430A, 477G, 484T, 624C, 681A, 688T
	GRA5	124T, 130G, 158G, 164G, 281G, 291G
	GRA6	19T, 149G, 314C, 554G, 614A, 650G, 652A, 662A, 671A
	GRA8	303A, 338G, 552G, 638C, 643G, 661A, 669T, 670G, 676C, 688A, 691C, 700G, 701C, 709A, 718G, 739T, 746C, 769G, 772T, 777A
	GRA14	127C, 968A, 1066T, 1067C, 1092T
VEG	GRA1	421G
	GRA5	104C, 124G, 137G
	GRA6	84T, 140A, 539C, 592C, 637G

[a]Sequence numbering for GRA1, GRA5, GRA6, and GRA8 is referring NW_002234575 (ME49 strain) isolated while that for GRA14 is in accordance with KE138826 (ME49 strain) isolated.

Table 4. Positions in GRA predicted with strain-specific amino acids.

Specificity	Protein	Position/amino acid[a]
GT1 and RH	GRA1	42F, 86A
	GRA5	42R
	GRA6	17V, 94E, 207A
	GRA14	43P, 323D, 356S, 364D
ME49	GRA1	60E, 99M, 114E, 117F, 163N
	GRA5	31T, 42W, 44G, 53G, 55S, 94R,
	GRA6	7Y, 50R, 105A, 185G, 205E, 217G, 218S, 221E, 224Fs
	GRA8	113G, 213A, 221T, 223N, 224V, 226L, 230I, 231L, 234A, 237T, 240A, 247L, 249T, 257D, 258F, 259Q
VEG	GRA1	96D
	GRA5	35A, 42G, 46R,
	GRA6	47D, 180P, 198R, 213A

[a]Sequence numbering for GRA1, GRA5, GRA6, and GRA8 is referring NW_002234575 (ME49 strain) isolated, while that for GRA14 is in accordance with KE138826 (ME49 strain) isolated.

The major limitation of this study is the low number of sequences analyzed, reflecting the limited number of *T. gondii* sequences available in the database. Future works are suggested to identify and analyze more *T. gondii* sequences in order to better understand the biology and genetics of the parasite.

4. Conclusion

T. gondii GRA1, GRA5, GRA6, GRA8, and GRA14 are predicted to contain highly conserved regions and positions with strain-specific variation.

Acknowledgement
This work was supported in part by grants from the Indonesian Directorate of Higher Education-APBN/DIPA UNS (No. 501/UN27.11/PN/2014, No. 853/UN27.11/PN/2014, and No. 339/UN27.11/PL/2015).

References

[1] Mercier C and Cesbron-Delauw M F 2015 *Trends Parasitol.* **31** 60–71
[2] Weiss L M and Dubey J P 2009 *Int. J. Parasitol.* **39** 895–901
[3] Denkers E Y 2003 *FEMS Immunol Med Microbiol.* **39** 193–203
[4] Lang C, Algner M, Beinert N, Gross U and Lüder C G 2006 *Microbes Infect.* **8** 1994–2005
[5] Pereira-Chioccola V L, Vidal J E and Su C 2009 *Future Microbiol.* **4** 1363–79
[6] Remington J S, McLeod R, Wilson C B and Desmonts G 2011 *Toxoplasmosis Infectious Diseases of the Fetus and Newborn* (Philadelphia: Elsevier) pp 918–1041
[7] Prasetyo A A, Ariapramuda E R, Kindi E A, Dirgahayu P, Sari Y, Darmawan R, Hudiyono, Hartono and Kageyama S 2014 *Southeast Asian J Trop Med Public Health* **45** 1032–47
[8] Prasetyo A A, Sariyatun R, Reviono, Sari Y, Hudiyono, Haryati S, Adnan Z A, Hartono and Kageyama S 2015 *J. Clin Virol* **7** 067–71
[9] Montoya J G and Liesenfeld O 2004 *Lancet.* **363** 1965–76
[10] Zhang N Z, Chen J, Wang M, Petersen E and Zhu X Q 2013 *Expert Rev Vaccines.* **12** 1287-99
[11] Bilgrami M and O'Keefe P 2014 *Handb Clin Neurol.* **121** 1321–44
[12] Tait E D and Hunter C A 2009 *Mem Inst Oswaldo Cruz.* **104** 201–10
[13] Cesbron-Delauw M F 1994 *Parasitol Today.* **10** 293–6
[14] Hughes H P and van Knapen F 1982 *Int. J. Parasitol.* **12** 433–7
[15] Ching X T, Lau Y L, Fong M Y, Nissapatorn V and Andiappan H 2014 *Biomed Res.* **2014** 690529
[16] Holec-Gasior L and Kur J 2010 *Exp. Parasitol.* **124** 272–8
[17] Peyron F, Lobry J R and Musset K 2006 *Microbes Infect.* **8** 2333–40
[18] Switaj K, Master A, Skrzypczak M and Zaborowski P 2005 *Clin. Microbiol. Infect.* **11** 170–6
[19] Wang Y, Wang G, Ou J, Yin H and Zhang D 2014 *Parasit Vectors.* **7** 474
[20] Tamura K, Stecher G, Peterson D, Filipski A and Kumar S 2013 *Mol. Biol. Evol.* **30** 2725–9
[21] Thompson J D, Higgins D G and Gibson T J 1994 *Nucleic Acids Res.* **22** 4673–80

Screening of Marine Actinomycetes from Segara Anakan for Natural Pigment and Hydrolytic Activities

A Asnani[1*], D Ryandini[2], Suwandri[1]

[1]Deprtment of Chemistry, Faculty of Mathematic and Sciences, Jendral Sudirman University, Karangwangkal, North Purwokerto, Banyumas Regency, Central Java 53122
[2]Department of Biology, Faculty of Biology, Jendral Sudirman University Unsoed, Puwokerto Jl. Dr. Soeparno No. 63, Central Java 53122

[*]E-mail: ari.asnani@unsoed.ac.id

Abstract. Marine actinomycetes have become sources of great interest to natural product chemistry due to their new chemical entities and bioactive metabolites. Since April 2010, we have screened actinobacteria from five sites that represent different ecosystems of Segara Anakan lagoon. In this present study we focus on specific isolates, K-2C which covers 1) actinomycetes identification based on morphology observation and 16S rRNA gene; 2) fermentation and isolation of pigment; 3) structure determination of pigment; and 4) hydrolytic enzymes characterization; Methodologies relevant to the studies were implemented accordingly. The results indicated that K-2C was likely *Streptomyces fradiae* strain RSU15, and the best fermentation medium should contain starch and casein with 21 days of incubation. The isolate has extracellular as well as intracellular pigments. Isolated pigments gave purple color with λmax of 529.00 nm. The pigment was structurally characterized. Interestingly, *Streptomyces* K-2C was able to produce potential hydrolytic enzymes such as amylase, cellulase, protease, lipase, urease, and nitrate reductase.

1. Introduction

Actinomycetes are known as a group of microorganisms that are most responsible for the production of crucial bioactive compounds, particularly antibiotics. The potency of actinomycetes as antibiotic producing bacteria has made these microbial groups become very important. However, the bioactive compounds produced by actinomycetes are not limited against bacteria, but also effective as anti-fungal agents, anti-cancer, and some of them are as anti-virus [1]. Until now, the search of actinomycetes producing new types of bioactive compounds continues.

The research exploration of marine actinomycetes is aimed to explore the indigenous species which are able to produce potentially bioactive compounds, such as pigments, enzymes, anti-bacterial and bioremediation agents. Samples of sediment/soil and water are obtained from the mangrove region as an area of mangrove forest is a source of organic matter of weathering and decomposition of litter that is very beneficial for heterotrophic microorganism such as actinomycetes. Asnani & Ryandini have reported 26 isolates of actinomycetes halotolerant of mangrove areas Segara Anakan that have potency as sources of anti-bacterial compounds. Based on the observation of the colony and cell morphology, actinomycetes isolates generally obtained have morphological properties of actinomycetes genera *Streptomyces* [2].

Certain actinomycetes show the various color of pigments in the hyphae, and various pigment diffused into the fermentation medium such as beige, white, red, green, and brown pigment. Gunasekaran & Poorniammal have reported that *Penicillium* sp capable of producing extracellular red pigment [3]. The red pigment with the molecular formula of anthraquinone produced by *Penicillium oxalicum* has now mass produced as a natural pigment. More recently, Mohanasrinivasan et al., [4] reported that *Streptomyces coelicolor* MSIS1 able to produce the extracellular blue pigment that turns red in acidic conditions. Following these findings, we continue exploring one specific isolate, namely K-2 because of its potency to produce color pigments as well as hydrolytic enzymes. Hence, this study was aimed to identify actinomycetes producing pigment, the isolation, and structure identification of pigment, and to evaluate the hydrolytic capability of isolates.

2. Materials and Methods

2.1. Sites of Sampling

The research was exploratory. Five different sites that represent diverse ecosystems of Segara Anakan Cilacap were selected for the present study. Sediment and water samples were taken in the area of mangrove vegetation with specifications Rhizophora area sp., grassy land, and open land areas. Each location was taken three sampling points. Soil and sludge samples were taken at a depth of soil about 5-8 cm and then put into sterile bottles.

2.2. Actinomycetes Isolation

1 g of sediment or 1 mL of water was mixed with 9 mL of sterile distilled water, shaken until homogeneous and become suspension with a dilution rate of 10^{-1}. From 10^{-1} dilution, the dilution was carried out with sterile seawater to obtain a suspension with a dilution rate of 10^{-2}, 10^{-3}, 10^{-4} and 10^{-5}. The suspension with 10^{-5} dilution was pipetted into 1 mL and cultured in casein nitrate medium in a sterile petri dish. Media for cultivation was Starch Casein Nitrate (SCN) agar (10 g soluble starch; 1 g Casein; 1 g KNO_3; 0.5 g K_2HPO_4; 0.5 g $MgSO_4.7H_2O$; 0.5 g NaCl; 0.01 g $FeSO_4.7H_2O$; 20 g agar; pH 6.8-7.0) supplemented with nystatin [2]. Incubation was carried out at 28 °C for 7-9 days, or until the actinomycetes colonies grew. The characteristics observed in actinomycetes colonies grow rounded colonies, slow growth, colony surface (could) had hyphae, hard as dusty or velvety. Usually, there was a central point in the middle of the colony and typical colored of the colony.

2.3. Colony Identification

2.3.1. Morphology Identification. The colony morphology and cell identification were done by preparing Henrici's Slide Culture (HSC). The observations of the colony were intended to the form, surface, consistency, edges, color, growth radial and concentric growth. The observation of the presence or absence of cells was directed against the substrate mycelium and aerial mycelium, the appearance of the substrate mycelium, aerial mycelium appearance, presence of spores and structure, the color of spores, as well as the pigment that diffuses into the medium. The observations of the hyphae characteristics were then compared with Bergey's Manual of Determinative Bacteriology.

2.3.2. Species Identification. Identification of actinomycetes species was based on *16S rRNA*. DNA isolation *used* Presto^{TM} Mini gDNA Bacteria Kit Geneaid Instruction Manual. DNA obtained was amplified with 27F and 1492R primers using KAPA Taq^{TM} Extra Host start ReadyMix with dye. DNA electrophoresis used 1% of agarose gel submerged in 1x TAE buffer. Species determination used BLAST (Basic Local Alignment Search Tool).

2.4. Pigment Production.

The pigment production consisted of four stages, namely a) inoculum preparation, b) fermentation medium preparation, c) the pigment production, d) pigment isolation, and e) pigment extraction.

Actinomycetes isolated was a streak on SCN medium and incubated for 8 days at a temperature of 30 °C; then it served as the parent culture. Pigment production was done by inoculating five plugs sized of 0.5 mm from parent culture to 100 ml of liquid medium. The fermentation process was done for 14 t0 21 days at room temperature. Media for fermentation was SCN broth (10 g soluble starch; 1 g Casein; 1 g KNO_3; 0.5 g K_2HPO_4; 0.5 g $MgSO_4.7H_2O$; 0.5 g NaCl; 0.01 g $FeSO_4.7H_2O$; pH 6.8-7.0) supplemented with nystatin. After incubation, the culture producing pigment was centrifuged at 3000 rpm for 30 minutes. The filtrate obtained was an extracellular pigment; while the mycelium obtained was extracted with ethanol to obtain the intracellular pigment [5].

2.5. *Structure Determination of Pigment*

Pigment extract obtained was purified with silica gel 60 GF254 followed by column chromatography eluted with chloroform: methanol (9: 1). The pure pigment was characterized by UV-Vis spectrophotometry (Shimadzu UV-1800), FTIR spectroscopy (Perkin Elmer Spectrum 100), and NMR Spectrometer (NMR Jeolin 500 MHz).

2.6. *Hydrolytic Enzymes Characterization*

The physiology of the isolates was observed by a series of biochemical tests following the known procedure [6]. Biochemical tests included hydrolytic, enzymatic tests such as starch hydrolysis, CMC hydrolysis, casein hydrolysis, lipid hydrolysis, urea hydrolysis, and nitrate reduction.

3. Results and Discussion

Sediment and water samples were collected from Klaces (K) and Ujung Alang (U) areas in Segara Anakan. Sampling time was during the day with the local temperature of 28 ° C and pH waters around 7-8 which corresponded to the growth of actinomycetes such as the temperature of 25-30 ° C and a pH of about neutral pH. Actinomycetes cultivation was carried out on an agar medium Starch-casein nitrate (SCN) supplemented with antifungal nystatin to inhibit mold growth. SCN is a highly sensitive and specific medium for the growth of actinomycetes. Usha et al. managed to isolate 15 isolates of actinomycetes origin by using medium SCN soil [7]. Khan et al. also succeeded in isolating two actinomycetes isolate originating from the sponge using SCN medium supplemented with antifungal and antibacterial [8].

The population of actinomycetes which had colored mycelium was generally obtained from the litter under the mangrove vegetation [2]. This indicates that the litter of mangrove vegetation litter is rich in organic matter because the result of weathering leaf or root exudates. As estuarine areas, Segara Anakan region experiences water movement that causes a typical condition. Only microorganisms with high tolerance to a constantly changing of oxygen, temperature, salinity and humidity are able to survive.

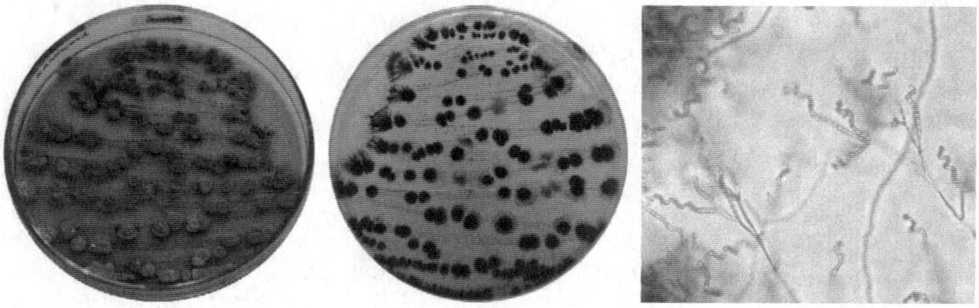

(a) (b) (c)

Figure 1. Colony observation of actinomycetes K-2C: (a) top (b) bottom, (c) mycelium

3.1. Colony Identification

Actinomycetes are a group of bacteria that forms unicellular mycelium that has the appearance of morphology such as mold. Unicellular mycelium is formed consisting of hyphae branched hyphae with a diameter of 0.5-0.8 μm. There are two types of mycelium, the substrate mycelium (vegetative) growth in the medium and aerial mycelium (aerial) growing on the surface of the medium. Aerial hyphae can be white, gray, violet, red, yellow, brown, green or any other color due to the pigmentation. In observation with a light microscope on a glass slide, aerial mycelium was refractive and bright, while the substrate mycelium is transparent and dark [1].

Based on morphology observation, the colony of K-2C has a size of 6 mm, red pigmentation to medium, irregular shape, powdery surface, convex elevation, margin undulate, gray aerial mycelium, and brown substrate mycelium; Morphological properties of K-2C were suggested to be similar with actinomycetes genera *Streptomyces*. Vegetative hyphae form a mycelium was branched. Aerial hyphae forming three or more mature spore chains, and a spiral-shaped sectional existed (figure 1).

Determination of the species of the isolate K-2C was done using 16S rRNA. The isolated DNA was amplified using PCR and sequenced in two-way forward and reverse with 27F and 1492R primers to get the base sequence. Determination of the base sequence of K-2C was carried out by pairwise alignment method in Bioedit program. The results of Bioedit program is as follows:

CATGCAGTCGACGATGAACCACCTTCGGGTGGGGATTAGTGGCGAACGGGTGAGTAACA
CGTGGGCAATCTGCCCTGCACTCTGGGACAAGCCCTGGAAACGGGGTCTAATACCGGATA
CTGACCTGCCAAGGCATCTTGGCGGGTCGAAAGCTCCGGCGGTGCAGGATGAGCCCGCG
GCCTATCAGCTTGTTGGTGAGGTAATGGCTCACCAAGGCGACGACGGGTAGCCGGCCTGA
GAGGGCGACCGGCCACACTGGGACTGAGACACGGCCCACACTCCTACGGGAGGCAGCAG
TGGGGAATATTGCACAATGGGCGCAAGCCTGATGCAGCGACGCCGCGTGAGGGATGACG
GCCTTCGGGTTGTAAACCTCTTTCAGCAGGGAAGAAGCGAAAGTGACGGTACCTGCAGA
AGAAGCGCCGGCTAACTACGTGCCAGCAGCCGCGGTAATACGTAGGGCGCAAGCGTTGT
CCGGAATTATTGGGCGTAAAGAGCTCGTAGGCGGCTTGTCGCGTCGGTTGTGAAAGCCCG
GGGCTTAACCCCGGGTCTGCAGTCGATACGGGCAGGCTAGAGTTCGGTAGGGGAGATCG
GAATTCCTGGTGTAGCGGTGAAATGCGCAGATATCANGAGGAACACCGGTGGCGAAGGC
GGATCTCTGGGCCGATACTGACGCTGAGGAGCGAAAGCGTGTGGAGCGAACAGGATTAG
ATACCCTGGTAGTCCACGCCGTAAACGGTGGGCACTAGGTGTGGGCAACATTCCACGTTG
TCCGTGCCGCAGCTAACGCATTAAGTGCCCCGCCTGGGGAGTACGGCCGCAAGGCTAAAC
TCAA

Total nucleotides	= 834 base pairs
Molecular weight	= 253152.00 Daltons, single stranded
Molecular	= 508419.00 Daltons, double stranded
Composition G+C	= 59.83%
Composition A+T	= 40.17%

Based on the BLAST analysis, isolate K-2C was identified to have similar sequences of 99% with a *Streptomyces fradiae* strain RSU15. Hangstrom et al., stated that isolates which have similarities sequences of the 16S rRNA gene could represent the same species while the sequence similarities between 93% -97% could represent the same identity at the level of genus, but different at the species level [9]. In other words, the difference between the 16S rRNA gene sequences of isolates stored in the data center if it is less than 3% showed the same species, while if more than 3% indicates a different species from the data center. From BLAST searches, K-2C taxonomy can be seen, as follows:

Domain: Bacteria
Phylum: Actinobacteria
Classic: Actinomycetes
Ordo: Actinomycetales
Familia: Streptomycetaceae
Genus: Streptomyces
Species: *Streptomyces fradiae* strain RSU15

3.2. Pigment of Actinomycetes K-2C

Aside from being able to produce antibiotics, actinomycetes are also able to produce the pigment that includes all rainbow colors. The bioactive pigments produced by microbes halotolerant has been reviewed. Those pigments have specific color and have an activity, such as anti-cancer, antibacterial, anti-inflammatory, antioxidant, as well as antibiotics. Asnani & Ryandini have also been reported marine actinomycetes expressing the color pigment. Observation of the pigment was observed on colony color and pigmentation to the substrate [5].

The microbes pigment can be isolated from three general stages, namely the cultivation of microbes, microbial fermentation to produce pigment, and pigment isolation. Measurement of the pigment concentration can be carried out using a UV spectrophotometer at the optimum wavelength (λ) for a particular color. For example, the yellow pigment can be observed on λ 390 nm, while the red pigment on λ 500 nm [10]. In this study, *Streptomyces* K-2 C produced purple pigment within 21 days of fermentation time, with optimum λ at 529.00 nm (figure 2). The wavelength in the range of 500 – 560 nm is the wavelength that can be absorbed by red and purple colors [11].

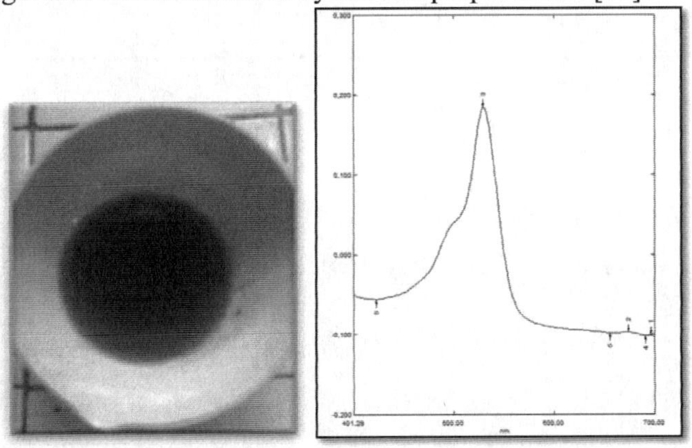

(a) (b)

Figure 2. Pigment production from actinomycetes K-2C: (a) purple pigment, (b) optimum λ at 529.00 nm

The highest production of purple pigment was obtained with fermentation with modified casein nitrate medium, and the lowest production was obtained from that of yeast extract-malt extract medium. Microbial fermentation medium is used for microbial growth as well as for the pigment production. Optimization of the fermentation medium can be done by varying sources of carbon and nitrogen sources. Palanichamy et.al., reported that antibiotics actinorhodin red-bluish can be isolated from *Streptomyces* sp in the fermentation medium using glycerol and mannose as a source of C; and Na-casein peptone as a source of N [12]. Optimization of fermentation medium in order for *Penicillium* sp to excrete extracellular red pigment was using soluble starch as a source of C and peptone as a source of N, at a temperature of 30 °C and pH 9.0. The optimum condition of fermentation medium was able to increase the production of red pigment up to sevenfold [3].

FTIR analysis was conducted to determine the functional groups of isolated pigment. Infrared spectra of purple pigment were shown in Figure 3 while the data of wave number and related functional groups was presented in table 1. Interpretation of IR spectra indicated that *Streptomyces* K-2C pigment contained a cyclic core, carbonyl, hydroxyl, alkyl (CH3 and CH2), non-conjugated one and secondary NH.

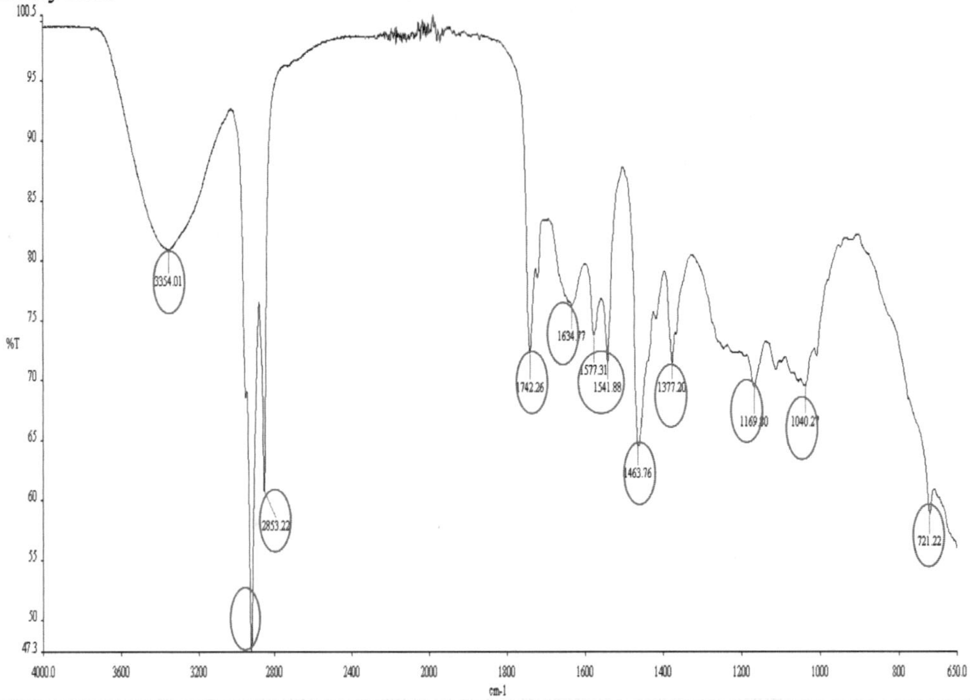

Figure 3. IR spectrum of pigment from *Streptomyces* K-2C

Table 1. FTIR data of purple pigment

Wavenumber (cm^{-1})	Frequency range (cm^{-1})	Correlated functional groups
2853.22	2850-3000	C-H alkane
1742.26	1650-1780	C=O ester
1634.77	1620-1680	C=C non conjugated
1577.31	1550-1640	N-H secondary
1463.76	1450-1470	$-CH_2-$
1377.20	1350-1370	$-CH_3$
1169.80	1000-1300	C-O ester
1040.27	990-1060	C-OH cyclic
721.22	650-1000	=C-H

1H-NMR spectra analysis was done to describe different types of hydrogen atoms in the pigment molecule. The information that can be obtained from the 1H-NMR spectrum is the chemical environment of hydrogen atoms, the number of hydrogen atoms in each environment and the cluster structures adjacent to each hydrogen atom [13]. 1H-NMR spectrum of the pigment Streptomyces K-2C is shown in Figure 4.

1H-NMR spectrum of the pigment Streptomyces K-2C showed a number of signals that appear in the chemical shift (δ) at 0.5 to 5.5 ppm. These data indicated aliphatic compounds. A chemical shift at 0.5 to 2.0 ppm was the signal for the H atoms bonded to C in a sp3 manner. Chemical shift at 0.87

ppm (3H) indicated the H atoms as $-CH_3$; δ = 1.29 ppm (2H) indicated the H atoms as $-CH_2$; and δ = 1.62 ppm (2H) indicated the H atoms as $-CH_2$ as well. Chemical shift at 2.34 ppm (2H) was a signal of H atom $-CH_2$ bonded to C heteroatoms. Chemical shift at 3.72 ppm (1H) indicated the H atoms as a form of –CH bonded to C heteroatoms. Chemical shift at 5.35 ppm (1H) is a signal to H bound to C double bond (-C = H). Based on the interpretation of the 1H-NMR spectrum, the purple pigment has H atoms in the form of methyl and methylene bonded to C Sp3; H form methylene attached to the C-heteroatom; Metin H form attached to the C-heteroatom; and H attached to C double bond.

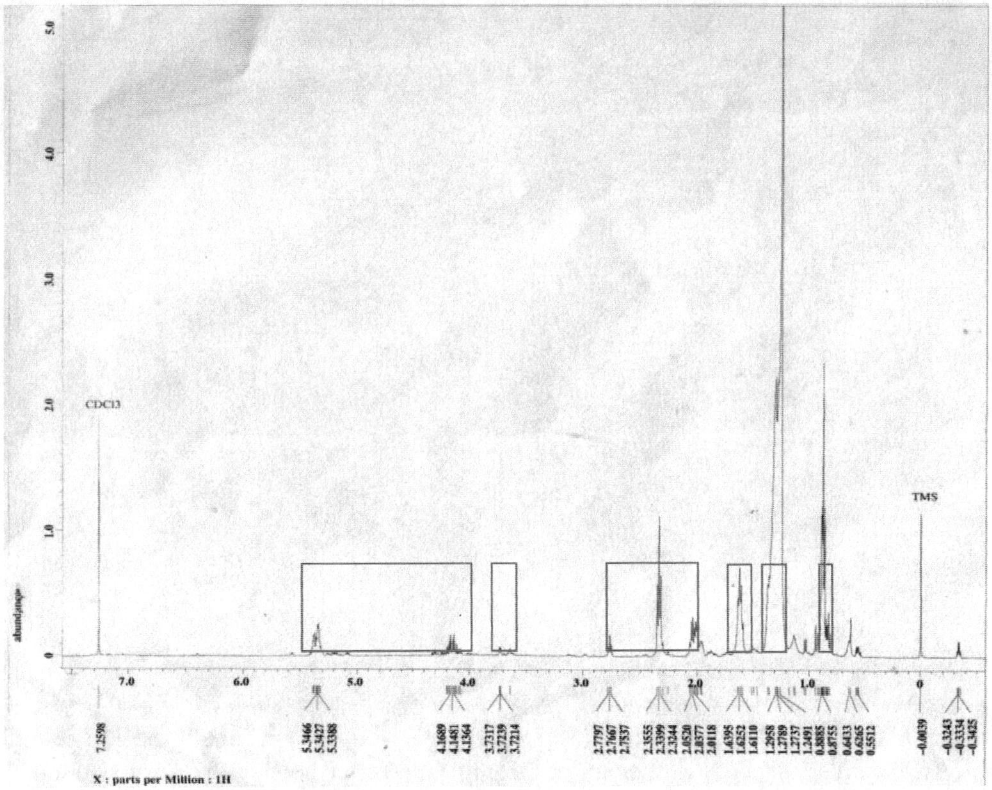

Figure 4. 1H-NMR spectrum of pigment from Streptomyces K-2C

3.3. Hydrolytic Enzyme Characteristics

The other actinomycetes potency is its ability to produce hydrolytic enzymes. The results from biochemical tests revealed the ability of *Streptomyces* K-2C in hydrolyzing starch, casein, lipids, urea, cellulose and reduced nitrate. These positive results indicated that the isolate were able to produce hydrolytic enzymes such as amylase, cellulase, protease, lipase, urease and nitrate reductase (table 2).

The further quantitative study indicated that the cellulolytic ability of K-2C was higher (cellulase index of 20.06 mm) compared to the amylolytic ability (amylase index of 6.26 mm) (figure 5). Indeed, the areas of mangrove forest are sources of organic matter of weathering and decomposition of litter that is very beneficial for heterotrophic microorganism such as actinomycetes. Thus, marine actinomycetes obtained from mangrove areas in Segara Anakan, such as *Streptomyces* K-2C, allegedly able to act as decomposers organic materials. In this regard, the hydrolytic ability of *Streptomyces* K-2C will have a promising prospect for industrial application.

Table 2. Biochemical Tests of *Streptomyces* K-2C

	Results
Enzyme Production:	
- Amylase	+
- Protease	+
- Lipase	+
- Cellulase	+
- Nitrate reductase	+
Carbohydrate Utilization	
- Galactose	-
- Sucrose	+
- Raffinose	+
- Mannitol	+
- Arabinose	+
- Xylose	-

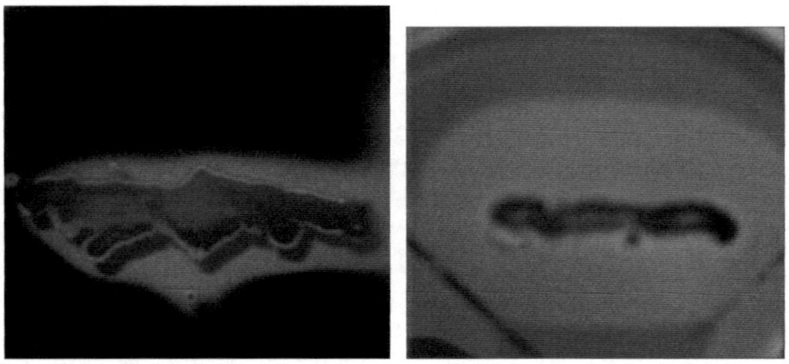

Starch hydrolysis (6,16mm) CMC hydrolysis (20,06mm)

Figure 5. Amylolytic and cellulolytic ability of Streptomyces K-2C

4. Conclusion

Marine actinomycetes K-2C was identified as *Streptomyces fradiae* strain RSU15. The isolate was able to produce purple pigment with λmax of 529.00 nm within 21 days of fermentation. Partial structure of the pigment was identified using FTIR and NMR. However further characterization is needed to have a complete pigment structure. Interestingly, the isolate was also able to produce potential hydrolytic enzymes namely amylase, cellulase, protease, lipase, urease, and nitrate reductase.

Acknowledgement

The authors are grateful to Indonesian Higher Education for providing research grant under *Strategies National* scheme in 2015. The authors also want to appreciate Nurhasanah for her work on pigment purification.

References

[1] Miyadoh S 1997 *Atlas of Actinomycetes*. Asakura Publishing Co. Ltd. Japan.

[2] Asnani A, and Ryandini D 2011 Screening of marine actinomycetes from Segara Anakan Indonesia for antimicrobial activity Proceeding ICONS 2011 Shaker Verlag Germany ISBN 978-3-8440-1403-7 ISSN 1434-5536.

[3] Gunasekaran S, and Poorniammal R 2008 Optimization of fermentation conditions for red pigment production from *Penicillium* sp. under submerged cultivation *Af. J. of Biotech.* **7 (12)** pp 1894-1898.

[4] Mohanasrinivasan V, SriramKalyan P, Nandi I, Subathradevi C, Selvarajan E, Suganthi V, and Jemimah N S 2013 Fermentative production of extracellular pigment from *Streptomyces coelicolor* MSIS1 *Res. J. Biotec.* **8 (4)** pp. 31-41.
[5] Asnani A, Ryandini D, and Suwandri 2014 Analysis of microbial pigment productivity in *Streptomyces* sp isolated from Segara Anakan. Joint Indonesia-UK Conference of Organic and Natural Product Chemistry, UGM, Yogyakarta.
[6] Asnani A, Ryandini D, and Suwandri 2014 Analisis potensi amilolitik dan selulolitik dari isolat aktinomisetes laut Seminar Nasional Percepatan Desa Berdikari Melalui Pemberdayaan Masyarakat Dan Inovasi Teknologi LPPM Unsoed Prosiding ISBN: 978-602-1643-13.
[7] Usha R, Ananthaselvi P, Venil C.K, and Palaniswamy M, 2010 Antimicrobial and antiangiogenesis activity of *Streptomyces parvulus* KUAP106 from mangrove soil *European J. of Biol. Sci.* **2(4)** 77-83.
[8] Khan S.T, Motohashi K, Kozonel, Takagi M, and Shin-ya K, 2010 Streptomyces associated with a marine sponge *Haliclona* sp; biosynthetic genes for secondary metabolites and product. *Environ. Microbiol.* **13(2)** 391-403.
[9] Hagstrom A, Pinhassi J, and Zweifel U.L, 2000 Biogeographical diversity among marine bacterioplankton. *Aquatic Microbial Ecology* ***21*(3)** 231-244.
[10] Kasim E, Astuti S, Nurhidayat N, 2005 Karakterisasi pigmen dan kadar lovastatin beberapa isolat *Monascus purpureus Biodiversitas* **6 (4)** hal. 245-247.
[11] Sahin N, 2011, Significance of absorption spectra for the chemotaxonomic characterization of pigmented bacteria *Turk. J. Biol.* **35** pp 167-175.
[12] Palanichamy V, Hundet A, Mitra B and Reddy N, 2011 Optimization of cultivation parameters for growth and pigment production by *Streptomyces* spp. isolated from marine sediment and rhizosphere soil *Int. J. of Plant and Animal Environm. Sci* **1 (3)** pp 158-170.
[13] Juliana V.A, Aisyah S, Mustapha I, 2010 Isolasi dan Karakterisasi Senyawa Turunan Terpenoid dari Fraksi n-Heksan *Momordica charantia L. J. Sains dan Tek. Kim.* ***1*(1)** hal. 88-93.

Characterization of active paper packaging incorporated with ginger pulp oleoresin

T Wiastuti, L U Khasanah*, Kawiji, W Atmaka, G J Manuhara, R Utami
Food Science and Technology, Sebelas Maret University, Jl. Ir. Sutami 36A Kentingan, Surakarta, Indonesia

*Email: liaumikhasanah@yahoo.co.id

Abstract. Utilization of ginger pulp waste from herbal medicine and instant drinks industry in Indonesia currently used for fertilizer and fuel, whereas the ginger pulp still contains high oleoresin. Active paper packaging were developed incorporated with ginger pulp oleoresin (0%, 2%, 4%, and 6% w/w). Physical (thickness, tensile strength, and folding endurance, moisture content), sensory characteristics and antimicrobial activity of the active paper were evaluated. Selected active paper then were chemically characterized (functional groups). The additional of ginger pulp oleoresin levels are reduced tensile strength, folding endurance and sensory characteristic (color, texture and overall) and increased antimicrobial activity. Due to physical, sensory characteristic and antimicrobial activity, active paper with 2% ginger pulp oleoresin incorporation was selected. Characteristics of selected paper were 9.93% of water content; 0.81 mm of thickness; 0.54 N / mm of tensile strength; 0.30 of folding endurance; 8.43 mm inhibits the growth of *Pseudomonas fluorescence* and 27.86 mm inhibits the growth of *Aspergillus niger* (antimicrobial activity) and neutral preference response for sensory properties. For chemical characteristic, selected paper had OH functional group of ginger in 3422.83 cm^{-1} of wave number and indicated contain red ginger active compounds.

1. Introduction

Ginger pulp waste was produced from herbal medicine and instant drinks industry or distillation stage. In Indonesia, ginger pulp waste currently used for fertilizer and fuel, whereas the ginger pulp still contains high oleoresin. Ginger pulp waste from UD Yogyakarta contains high oleoresin. Ginger pulp oleoresin contains Zingiber (37.13%). Zingiberene of ginger oleoresin decreased because of encapsulation (1.5%) [1]. An active compound from ginger pulp oleoresin also was affected by the extraction process. Extraction above 45°C, *Gingerol* decomposed into *Shogaol* that less quality than *Gingerol* [2].

Packaging technology widely developed. One of the innovative packaging technology development is active packaging. Active packaging can protect product quality and shelf life [3]. Active packaging was developed to modify active the internal environment. It has the active interaction with materials package and foodstuffs that could extend the shelf life. The systems of active packaging were O_2/CO_2 scavengers, CO_2/ethanol emitting systems, ethylene and moisture absorption systems, and antimicrobial/antioxidant releasing/containing systems [4]. Antimicrobial packaging made by incorporation and immobilization of antimicrobial agents or surface coating [5].

The paper packaging has bad barrier properties that unsuitable for prolong storage. To increase barrier properties of paper, the paper made by coating, laminating or adding by wax or resin [6]. The paper could be formed as active packaging because of the pore structure. Antimicrobial agent absorbed in pores that could increase the paper materials activity such as vapor and gas permeability, physical power, optical properties, surface properties and antimicrobial activity. The antimicrobial agent of packaging materials can diffuse in a food product (migration process) and extended product shelf life [7]. The active paper developed in several countries such as Japan, Italy, India and other [3].

The ginger has antimicrobial activity. Ginger oleoresin from Vietnam has high antimicrobial activity on Gram (+) bacteria and Gram (-) bacteria [8]. 0.5% of ginger essential oil incorporated into uwi starch edible film could inhibit *Escherichia coli* growth. Antibacterial compounds of ginger essential oil were gingerol, geraniol, zingiberene, zingerone, paradol and shogaol [9]. Antifungal activity of ginger oil was recorded 7.7-13.02 mm of zone inhibition for tested fungal strains. The antifungal compound of ginger was *gingerol*, *gingerdiol* and *shagelol*. [10, 11, 12]. For increasing the used of ginger waste, it processed into oleoresin that incorporated in the paper as an antimicrobial agent. Ginger pulp still contains high oleoresin that have an active compound for antimicrobial activity [1].

Several factors affected the active paper packaging characteristics such as chemical mechanism, storage and distribution condition, physical condition and mechanical properties of packages, sensory properties, and toxicity antimicrobial properties. This research purposed to characterized the physical (thickness, tensile strength, and folding endurance, moisture content), sensory characteristics and antimicrobial activity of active paper by adding ginger pulp oleoresin. Then selected active paper characterized the chemical characteristic by FTIR and compared with active paper no oleoresin.

2. Methodology

2.1. Ginger Pulp Oleoresin Preparation

The ginger pulp was air dried for 8 days until the water content decreased to 12-14%. Dried ginger pulp was milled and sieved to 30 mesh. The pulp was extracted by batch maceration and stirred at 500 rpm. 100-gram ginger pulp (30 mesh) and 600 ml ethanol 96% were mixed and stirred for 2 h at ambient condition, and settled for 12 h at ambient condition and filtered. The ginger pulp extract was evaporated by rotary evaporator RE-300 at 50 ^{o}C for oleoresin production [1].

2.2. Active Paper Preparation

15 g of piece filter paper (2 mm x 2 mm) was soaked for 24 h in 250 ml of distilled water. For pulping, the soaked paper and 250 ml distilled water blended in crusher-blender for 5 minutes. The mixture added with tapioca 4.5g/50 ml distilled water and crushed and blended for 5 minutes. 0.45 gram of chitosan in 100 ml acetic acid 1% added, crushed and blended for 5 minutes. Oleoresin emulsion prepared by mixed ginger pulp oleoresin (2%, 4%, and 6%) in 50 ml distilled water while added with tween 80 (1:1) until emulsion formed by stirred. Whereas one sample no added oleoresin used as a control. Pulp, chitosan, and oleoresin emulsion mixed and blended for 5 minutes. The last pulping process was poured into molding paper with screener (20 cm x 30 cm), trimmed, rolled and pressed into wet paper sheet formed. The sheet is dried for 48 h at ambient temperature (30-33 ^{o}C).

2.3. Physical Properties of Paper

The moisture content of the active paper was evaluated in the oven at 105 °C for 24 h. The thickness of the active paper was measured by Krisbow micrometer (0.01 mm of precision) [SNI 14-4977-1999]. Tensile strength was measured by Tensile Tester - Kao Tieh (Model KT-7010-A2). Whereas for analysis Tensile strength testing specimen strips were 15 mm wide and 200 mm length. Then both of tip (top and bottom) was clipped within 180 mm by a tensile tester. Run the motor until stop and the sample broken.

IOP Conf. Series: Materials Science and Engineering **107** (2016) 012057 doi:10.1088/1757-899X/107/1/012057

The tensile strength value can read directly at SNI 14-0437-1989. Folding endurance was measured by MIT Folding Endurance Tester. For preparation, the strips with a width of 15 ± 0.02 mm and a length of 150 mm are used for the test. The surfaces forming the slot in the folding head have a radius of 0.38 ± 0.02 mm. The clamp oscillates with 175 ± 25 double folds/min through an angle of 135 ± 2° in each direction. Whenever possible a tension of 0.5 kg_f shall be used. Record the number of double folds required to sever the specimen (SNI 0491:2009).

2.4. Antimicrobial Test

NA medium (Merck) were used for the growth *P. fluorescent* and PDA medium (Merck) was used for the growth of *A. niger*. *P. fluorescence* was inoculated on nutrient agar for 24 hours. Bacterial cells were ordered to obtain a cell concentration around 10^6 cells/ml. *A. niger* spores of 7-day cultures were harvested with sterile distilled water with 0.2% Tween 80 (v/v). The concentration of the conidia suspensions was determined using a hemocytometer with an optical microscope at 400 magnification, and 10^3 spores were inoculated in the center of the Petri dish for antimicrobial testing. Small paper discs (5 mm diameter) were used for the inhibition zone test. These discs were pre-sterilized non-thermally by keeping the ultraviolet lamp at a distance of 0.5 m for 5–6 h. Active papers containing a various concentration of ginger pulp oleoresin were tested in order to assess their antimicrobial activity. The discs were put on the lid of inoculated Petri dishes. The Petri dishes were incubated 24 h at 37 °C for the growth of *P. fluorescent* and 3 days at 37 °C for the growth of *A. niger*.

2.5. Sensory Properties

Sensory analysis performed by 30 untrained panelists used hedonic test included color, texture, flavor and overall.

2.6. Chemical Properties

The samples have been jarred and KBr-pelleted. The measurements, made on an IR spectrometer Nicolet/Avatar-360 plus device, have been processed with the Spectra Manager.

2.7. Statistical Analysis

SPSS (SPSS 16.0 for Windows) statistical software was used to calculate the analysis of variance (ANOVA) and Duncan Multiple Range Test (DMRT) was used for the comparison of means, with significance assigned at $P<0.05$.

3. Result And Discussion

The features of the package which make these materials suitable for packaging relate to appearance and performance. Appearance relates the visual impact as color, smoothness, and glossing. Whereas performance properties are related to the level of efficiency achieved during the manufacture of the pack, as stiffness, rigidity strength, tensile strength, wet strength, tear strength, fold endurance moisture content, air permeability, water absorbency, surface friction, surface tension, etc. [13].

Table 1. Physical Properties of Active Paper Packaging

Oleoresin Concentration	Water Content (%)	Thickness (mm)	Tensile Strength (N/mm)	Folding Endurance
0%	$11.11^{a} \pm 2.19$	$0.82^{a} \pm 0.03$	$1.66^{b} \pm 0.20$	0.48
2%	$9.93^{a} \pm 0.30$	$0.81^{a} \pm 0.00$	$0.54^{a} \pm 0.04$	0.30
4%	$11.60^{a} \pm 0.13$	$0.77^{a} \pm 0.05$	$0.41^{a} \pm 0.09$	0.12
6%	$11.86^{a} \pm 1.01$	$0.75^{a} \pm 0.01$	$0.56^{a} \pm 0.13$	0.30

Note: the numbers that are followed by the same letter in the same column and row is insignificantly different based on DMRT test. $\alpha = 0.05$.

Based on Table 1 the oleoresin variation concentration has no significant effect on the water content of active papers ($P_{value} > 0.05$). However, the water content decreased on incorporated 2% ginger pulp oleoresin (9.93%), then increased to oleoresin addition. The desired water content of the paper is between 4%-8% [14]. In this study, the water content of active papers more than was desired which are 9-11%. In preparation step, the active paper was added to chitosan and oleoresin emulsion that caused water content more than the reference. Chitosan was hygroscopic and have a greater capability to form hydrogen bonding [15, 16]. The presence of Tween 80 emulsifier in the chitosan film increased hygroscopicity due to its hydrophilic groups [16]. The oleoresin variation concentration has no significant effect on the thickness of active paper ($P_{value} > 0.05$). The thickness of active paper about 0.75-0.82 mm.

The tensile strength of active papers added with oleoresin were significantly lower than no oleoresin ($P\text{-}_{value} < 0.05$) (Table 1). Oregano oil enrichment on chitosan film also decreased the tensile strength [17]. The higher of oil content could be attributed to the discontinuities induced in the chitosan matrix by oil droplets which caused a loss of the film cohesion and mechanical resistance [18]. Furthermore, the increasing of water content caused decreasing of tensile strength. The moisture of paper-based packaging materials is mainly due to the development of hydrogen bonds between hydrophilic hydroxyl groups of cellulose fiber in the paper matrix and water molecules replacing the fiber-fiber interaction. Consequently, reducing the intermolecular interaction between fibers [19].

Folding endurance decreased during oleoresin addition (Table 1). Factor influenced the folding endurance were grams mature, temperature, water content, and structure of paper sheet. Fellers [20] reported that the increasing of grams mature caused the increasing of folding endurance. While, at the same grams mature, the increasing of thickness increased at high flexibility paper caused the increasing of folding endurance.

Table 2. Antimicrobial Activity of Active Paper Packaging

Oleoresin Concentration	Inhibition Zone (mm)	
	P. fluorescent	*A. niger*
0%	$6.73^{a} \pm 0.053$	$21.91^{ab} \pm 2.249$
2%	$8.43^{b} \pm 0.773$	$27.86^{b} \pm 3.182$
4%	$8.37^{b} \pm 0.212$	$16.55^{a} \pm 1.184$
6%	$8.57^{b} \pm 0.188$	$13.74^{a} \pm 6.227$

Note: the numbers that are followed by the same letter in the same column and row is insignificantly different based on DMRT test. $\alpha = 0.05$.

Active paper incorporated by ginger pulp oleoresin could inhibit bacterial and fungal growth (Table 2). Inhibition zone on *P. fluorescent* growth by the active paper were affected by the concentration of oleoresin ($P\text{-}_{value} < 0.05$) while those on *A. niger* were no affected ($P_{value} > 0.05$). Inhibition zone of *P.*

fluorescent about 6.73-8.57 mm (moderate response, inhibition zone 6-11 mm). Whereas inhibition zone of *A. niger* about 13.74-27.86 mm (strong inhibition, inhibition zone >11 mm). Elgayar et al [21], classified inhibition zone of tested essential oil were strongly active (inhibition zone, > 11 mm), moderately active (inhibition zone, 6-11 mm), and inactive (no inhibition zone, < 6). Inhibition zone of *P. fluorescence is* increased with the addition of oleoresin. Antimicrobial activity of control paper due to the nature of chitosan acetate. There were interactions between positively charged chitosan and negatively charged microbial cell wall leads to the leakage of the intracellular constituents [22].

Inhibition zone of *P. fluorescence* increased at 2% of ginger pulp oleoresin incorporation. In study Nurlaili [1], found 37.13% of zingiberene as antimicrobial. Antifungal activity of active paper about *A. niger* increased at 2% of ginger pulp oleoresin incorporation but decreased along oleoresin incorporated. Increasing of inhibition zone caused by the antimicrobial activity of chitosan enriched ginger pulp oleoresin incorporated in active paper. Oxygen compound of ginger was electronegative, and hydrocarbon compound (sesquiterpene or monoterpene) that have double bound so showed antimicrobial activity [23]. Whereas decreasing of inhibition zone caused by an emulsion of oleoresin incorporated strongly so active compound diffusion difficulty [24].

Table 3. Hedonic Test of Active Paper Packaging

Oleoresin Concentration	Parameter			
	Color	Flavor	Texture	Overall
0%	4.38^{c}	3.28^{a}	3.62^{b}	3.86^{b}
2%	3.86^{b}	3.41^{a}	3.00^{a}	3.41^{a}
4%	3.72^{ab}	3.41^{a}	3.07^{a}	3.52^{ab}
6%	3.41^{a}	3.41^{a}	2.93^{a}	3.34^{a}

Note: the numbers that are followed by the same letter in the same column and row is insignificantly different based on DMRT test. $\alpha = 0.05$.
Note of score:
1: really dislike, 2: dislike;3: neutral; 4: like, 5: really like.

An active paper by oleoresin incorporated decreased preference level except parameter of flavor (**Table 3**). The color of active paper with oleoresin preference level decreased caused dark brown of oleoresin made active paper turbidity than active paper no oleoresin. In parameter texture decreased caused higher of oil content in oleoresin loss of the film cohesion and mechanical resistance.

Table 4. Active Paper Vibration on Specific Wave Number (cm^{-1})

No.	Active paper (0%)	Active paper (2%)	Functional group
1.	3448.87	-	OH
2.	-	3422	OH Fenolic
3.	2902.03	2902.03	CH alkane and OH carbocyclic
4.	1633.78	1637.64	N-H (amine)
5.	1429.31	1431.24	C-C aromatic
6.	1372.41	1372.41	C-H alkane
7.	1319.37-1030.03	1319.37-1057.04	C-O dan C-N (aliphatic amine)
8.	895.97-661.61	897.9-665.47	C-H aromatic, N-H (amine primer and secondary)

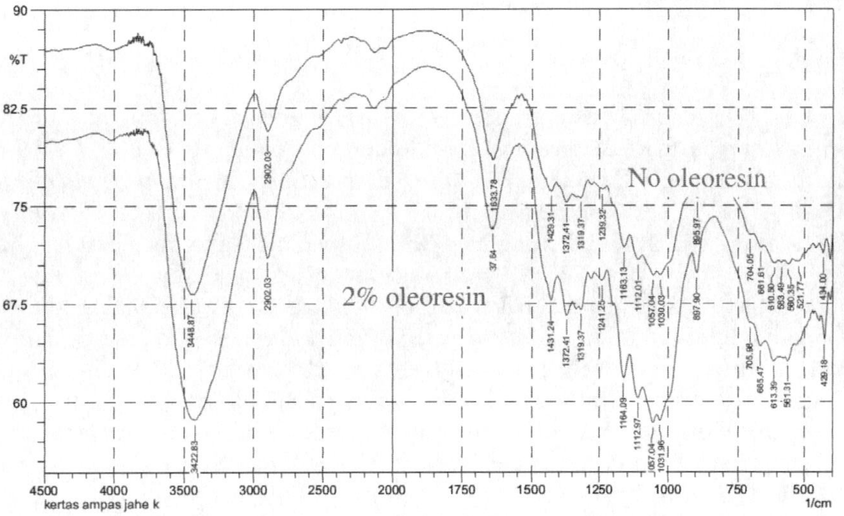

Figure 1 .FTIR of Active Paper Selected (2% oleoresin and no Oleoresin)

Figure 1 and Table 4 showed the peaks of cellulose spectrum, a derivative of cellulose and chitosan. Fibre of pine composed of cellulose found at 1160 cm^{-1}, 1316 cm^{-1}, 1370 cm^{-1} and 1424 cm^{-1} of wave number and xylan at 1734 cm^{-1}, 1460 cm^{-1} and 1240 cm^{-1} of wave number. Those wave number indicated as valency vibration of C-O, C-H, and C-C that related to cellulose and xylan [25]. Active paper has wave number 1163.13 and 1164.09 cm^{-1}; 1319.37 cm^{-1}; 1372.41 cm^{-1}; 1429.31 cm^{-1}, and 1240 and 1241.25 cm^{-1} were valency vibration to C-O, C-H and C-C that related to cellulose. In the study, also found acetyl group at (1319.37-1030.03) cm^{-1}, hydroxyl group at 3448.87 cm^{-1} and amine primer (1633.78 dan 1637.64) cm^{-1}. There was the structure of chitosan acetate. The ginger compound also found in active paper selected sample with wave number about 3379-3422 cm^{-1} and identified as valence vibrations corresponding to the OH groups of red ginger [26].

4. Conclusion

Active paper with ginger pulp oleoresin reduced tensile strength, folding endurance and sensory characteristic (color, texture and overall) and increased antibacterial activity but antifungal activity decreased. Due to physical, chemical, sensory characteristic and microbial activity, active paper with 2% ginger pulp oleoresin incorporation was selected. Characteristics of selected paper were 9.93% of water content; 0.81 mm of thickness; 0.54 N / mm tensile strength; 0.30 of folding endurance; 8.43 mm inhibits the growth of *P. fluorescent,* and 27.86 mm inhibits the growth of *A. niger* (antimicrobial activity) and neutral preference response for sensory properties. Chemical properties found O-H group of red ginger active compound.

Acknowledgements

This work was financially supported by a research project of PNBP No: 698/UN27/ PN/2015 from Sebelas Maret University, Indonesia.

References

[1] Nurlaili F A, Darmadji P and Pranoto Y 2014 Microencapsulation of Pulp Ginger (*Zingiber officinale var.Rubrum*) Oleoresin with Maltodextrin Coating *AGRITECH* **34** 1 p 23 27
[2] Gaedcke F and Bjorn F 2005 Ginger Extract Preparation *US Patent Application Publication* **US 2005/0031772** A1
[3] Brody A L, Strupinsky E R and Kline L R 2001 Active Packaging for Food Application *CRC Press* USA
[4] Balasubramanian A, Rosenberg L L E, Yam K and Chikindas M L 2009 Antimicrobial Packaging: Potential vs. Reality—A Review *Journal of Applied Packaging Research.* **3** 4 p 195
[5] Suppakul P, Miltz J, Sonneveld J, and Bigger S W 2003 Active Packaging Technologies with an Emphasis on Antimicrobial Packaging and its Applications *Journal of Food Science* **68** 2 p 411
[6] Opara U L and Mditshwa A 2013 A Review of The Role of Packaging in Securing Food System: Adding Value to Food Products and Reducing Losses and Waste *African Journal of Agricultural Research.* **8** 22 p 2622
[7] Rakchoy S, Suppakul P and Jiankarn T 2009 Antimicrobial Effects Of Vanillin Coated Solution for Coating Paperboard Intended for Packaging Bakery Products *As. J. Food Ag-Ind.* **2** 4 p 139
[8] Stoyanova A, Denkova Z, Nenov N, Slavehev A, Jirovetz L, Buchbauer G, Lien H N, Schimidt E and Geissler M 2006 $C_2H_2F_4$-SCFE-Oleoresins of Black Pepper (Piper Nigrum L.) and Ginger (Zingiber Officinale (L.) Rosc.) from Vietnam: Antimicrobial Testings Gas Chromatographic Analysis and Olfactoric Evaluation *Electron. J Environ Agric Food Chem* **5** 5 p 1617-1618
[9] Miksusanti H and Masril K I 2013 Antibacterial and Antioxidant of Uwi (*Dioscorea Alata L*) Starch Edible Film Incorporated with Ginger Essential Oil *International Journal of Bioscience Biochemistry and Biopharmaceutics* **3** 4 p 355
[10] Barman K L and Jha D K 2013 Comparative Chemical Constituents and Antimicrobial Activity of Normal and Organic Ginger Oils (*Zingiber officinale roscoe*) *International Journal of Applied Biology and Pharmaceutical Technology* **4** 1 p 263
[11] Supreetha Sharadadevi M, Sequira P S, Jithesh J, Shreyas T and Amit M 2011 Antifungal Activity of GingerExtract on *Candida Albicans*: An In-vitro Study *Journal of Dental Science and Research* **2** 2 p 20
[12] Sabra S M M, Al-Masoudi L M R, Al-Gehani S A H and Harbah A A O A 2014 Comparative Laboratory Study on Antimicrobial Effects of Fresh and Dry Ginger (*Zingiber officinale*) Taif KSA *IOSR-JESTFT* **8** 9 p 119
[13] Coles R, McDowell D and Kirwan M J 2003 Food Packaging Technology *Blackwell Publishing* Ltd. UK
[14] Li PY, Ramaswamy S and Bjegovic P 2003 Pre-emptive Control of Moisture Content in Paper Manufacturing Using Surrogate Measurements *Transactions of The Institute of Measurement and Control.* **25** 1 p 40
[15] Szymańska E and Winnicka K 2015 Stability of Chitosan—A Challenge for Pharmaceutical and Biomedical Applications Mar. Drugs. **2015** 13 p 1827
[16] Muresan A, Cerempai A, Dunca S, Muresan R and Butnaru R 2009 Aromatherapeutic Characteristics of Cotton Fabrics Treated With Rosemary Essential Oil. Cellulose Chem.Technol **43** 9-10 p 440
[17] Chi S 2004 Development and Characterization of Antimicrobial Food Coatings Based on Chitosan and Essential Oils *Thesis* University of Tennessee Knoxville

[18] Bonilla J, M Vargas L Atares and A Chiralt 2011 Physical properties of chitosan-basil essential oil edible films as affected by oil content and homogenization conditions Procedia Food Science **1** 2011 p 53
[19] Rhim J W 2010 Effect of Moisture Content on Tensile Properties of Paper-based Food Packaging Material *Food Sci. Biotechnol.* **19** 1 p 245-246
[20] Feller C, Iversen T, Lindstrom T, Nilsson T and Rigdhal M 1989 Ageing/Degradation of Paper *ISSN 0284-5636* Stockholm
[21] Elgayyar M, F A Draughon, D A Golden and J R Mount. 2001 Antimicrobial Activity of Essential Oils from Plants against Selected Pathogenic and Saprophytic Microorganisms *Journal of Food Protection* **64** 7 p 1019
[22] Huang L, Dai T, Xuan Y, Tegos G P and Hamblin M R 2011 Synergistic Combination of Chitosan Acetate with Nanoparticle Silver as a Topical Antimicrobial: Efficacy against Bacterial Burn Infections Antimicrob Agents Chemother **55** 7 p 3432-34338
[23] Sasidharan I and A N Menon 2010 Comparative Chemical Composition and Antimicrobial Activity Fresh and Dry Ginger Oill (*Zingiber officinale roscoe*) *Int. J Curr Pharm Res* **2** 4 p 42
[24] Maleki S, S M Seyyednejad, N M Damabi and H Motamedi 2008 Antibacterial Activity of The Fruits of Iranian Torilis leptophylla Against Some Clinical Pathogens *Pakistan Journal of Biological Sciences* **11** 9 p 1286-1289
[25] Fan M, D Dai and B Huang 2012 Fourier Transform-Material Analysis *Intech* Europe
[26] Purwakusumah E Dj, M Rafi, U D Syafitri, W Nurcholis dan M A Z Adzkiya 2014 Identification and Authentication of Jahe Merah Using Combination of FTIR Spectroscopy and Chemometrics *AGRITECH* **34** 1 p 83-84

IOP Conf. Series: Materials Science and Engineering **107** (2016) 012058 doi:10.1088/1757-899X/107/1/012058

Biological evaluation of certain substituted hydantoins and benzalhydantoins against microbes

Ika-Wiani Hidayat[1, 2*], Yee Yee Thu [2, 3], David St.C Black[2] and Roger W Read[2]
[1] Department of Chemistry, Padjadjaran University, Jl. Raya Bandung Sumedang Km. 21, Jatinangor 45363 Indonesia
[2] School of Chemistry, The University of New South Wales, UNSW Sydney NSW 2052, Australia
[3] Department of Botany, University of Yangon, Myanmar

[*]E-mail: i.wiani@unpad.ac.id; i.wiani@gmail.com

Abstract. Twenty-three synthetic (thio)hydantoins and benzalhydantoins were evaluated for antimicrobial activity against *Candida albicans*, *Malassezia furfur*, *Escherichia coli* and *Staphylococcus aureus*, by the paper disc diffusion method. 3-*n*-butyl-4'-nitrobenzalhydantoin showed very high activity against *E. coli* and high selectivity with respect to the other microorganisms, while 3-*n*-butyl-2'-bromo-4',5'-dimethoxybenzal hydantoin demonstrated very high selectivity in its activity against *M. furfur* and *S. aureus.* These compounds show the most promise as drug lead compounds.

1. Introduction

Infectious diseases are still a major cause of deaths, especially in developing countries, as new infectious diseases arise and an increasing number of multi-drug resistant strains of microbial pathogens emerge. The latter problem applies particularly through Gram positive bacteria, such as *S. aureus.* Despite remarkable advances over the past decades, the search for new drugs to combat and control this threat is ongoing [1-9]

Oxazolidinones have emerged as a promising new class of antimicrobial agents. The first of these to be authorised for use in Australia, linezolid **1**, was approved in 2003 for use in humans. Oxazolidinones operate through a unique mechanism that result in the absence of cross-resistance with other antibiotics. The mechanism involves early inhibition of protein synthesis by preventing the formation of the 70S initiation complex through binding to the 20S ribosomal RNA of the 50S bacterial ribosomal subunit.[9-17]

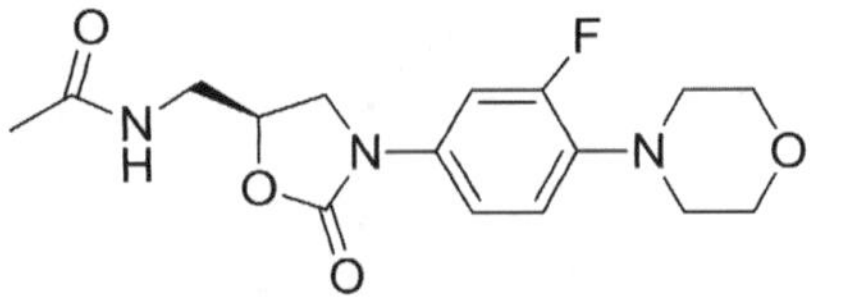

Linezolid **1**

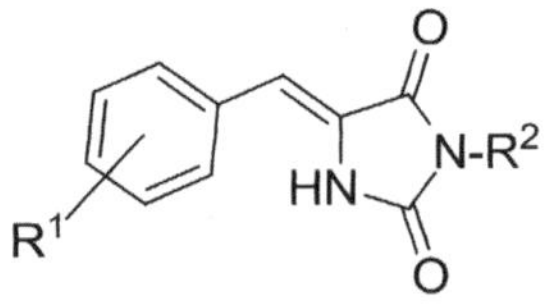

2 Linezolid mimic.

IOP Conf. Series: Materials Science and Engineering **107** (2016) 012058 doi:10.1088/1757-899X/107/1/012058

Our group has prepared a wide range of new, substituted benzalhydantoins (5-arylideneimidazolidine-2,4-diones) **2** as part of a synthetic study toward tandem approaches to the molecules. [18] It occurred to us that these can be viewed as spatial mimics of linezolid **1**. Like the oxazolidinones, 5-benzalhydantoins and 5-benzylhydantoins have attracted medicinal chemistry interest in their own rights because of their phenethylamine skeleton.[19] It was decided to evaluate a representative selection of hydantoin (imidazolidine-2,4-dione) precursors and final substituted benzalhydantoin compounds, in *in vitro* assays for antimicrobial activity.

2. Experimental

2.1. Procedure for Assessment of Antimicrobial Activity [20]

The culture medium for the fungi, *C. albicans* and *M. furfur*, was nutrient broth on agar medium 25 g/L at pH 7.0. The test organisms were inoculated in 10 mL assay broth in 50 mL Falcon™ tubes at 37 °C and incubated overnight. An aliquot (100 μL) of test organisms was then added to each of the assay plates and the broth spread on the agar. Paper discs (6 mm diameter size) impregnated with solutions of Amphotericin B, hydantoins **6-8**, thiohydantoins **3-5**, benzalhydantoins **9-22**, **24-25**, and benzylhydantoin alcohol **26** (20 μL of 1.0 mg/mL), were applied to the test plates and the plates were incubated for 24-36 h at 37 °C for *E. coli* and *S. aureus*. After the first 24 h the plates were examined and the diameter of any inhibition zone recorded. This process was repeated on all plates. A clear zone (inhibitory zone) surrounding the test disc indicated the presence of bioactivity in the compound (Table 1).

Meanwhile, the bacterial strains *E. coli* and *S. aureus* were maintained in malt extract agar medium 13 g/L at pH 7.0. A method similar to the bacterial test protocol was adopted, except that incubation was carried out at each step at 30 °C. The results are presented in table1.

3. Results and Discussions

In addition to the compounds mentioned above, it was thought worthwhile to study an example of an intermediate in the synthetic pathway, an hydroxybenzylhydantoin, as well as to study the effect of replacing the C(2)-carbonyl group oxygen with sulfur in both classes of compound. Thus, selected hydantoins **3-7**, benzalhydantoins **8-26**, were tested.

3.1. Antimicrobial Test Methodology

Two common strains of fungi, *Candida albicans* and *Malassezia furfur*, and the benchmark Gram negative and Gram positive bacteria, *Escherichia coli* and *Staphylococcus aureus*, respectively, were chosen as test organisms for compounds **3-26**, and the paper disc diffusion test method [20] was employed for their screening. Amphotericin B [21, 22] a well known commercial antifungal drug, was chosen as a reference substance for assessing the antimicrobial performance of these compounds and they were both tested against methanol as the control solvent.

The inhibition associated with each of the sample substances against each of the four test organisms was recorded in Table 1. Inhibition was recorded numerically in Table 1 as the diameter (width) of the inhibitory zone (clear zone). In addition, inhibition was accorded a ranking of "*", "**", or "***", corresponding to clear zones measured in the ranges 7–11 mm, 12–15 mm, and 16–20 mm, respectively.

3.2. General Observations

Eight compounds, namely hydantoin **8** and benzalhydantoins **12**, **17-21**, **24**, and **4**, had antimicrobial activity against all four test organisms (table 1). These compounds had in common the presence of halides (chlorine or bromine) at the 2-position and/or 3,4-dimethoxy substituents in the benzenoid ring of the arylmethylidene group.

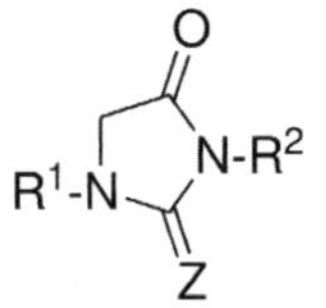

	Z	R¹	R²
6	O	H	Ph
7	O	H	allyl
8	O	H	*n*-Bu
3	S	H	Ph
4	S	Ac	Ph

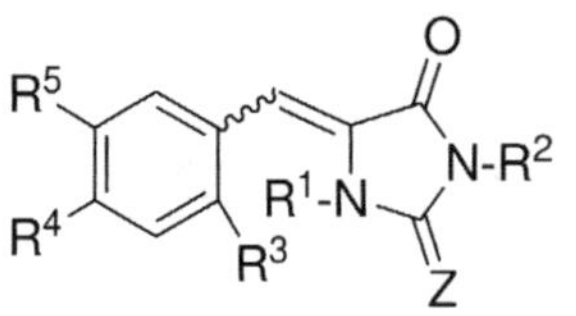

5, 9-25

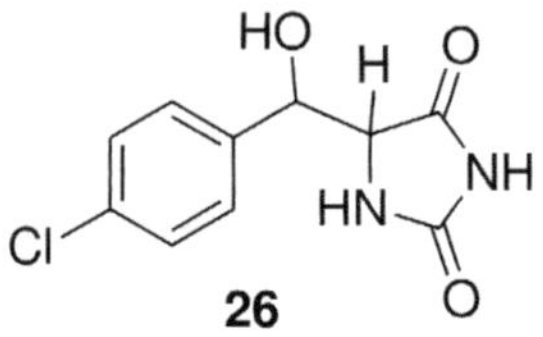

26

	Z	R¹	R²	R³	R⁴	R⁵
9	O	H	H	Cl	H	H
10	O	H	Ph	Cl	H	H
11	O	H	allyl	Cl	H	H
12	O	H	*n*-Bu	Cl	H	H
13	O	H	H	H	NO_2	H
14	O	H	Ph	H	NO_2	H
15	O	H	allyl	H	NO_2	H
16	O	H	*n*-Bu	H	NO_2	H
17	O	H	H	H	OMe	OMe
18	O	H	Ph	H	OMe	OMe
19	O	H	allyl	H	OMe	OMe
20	O	H	*n*-Bu	H	OMe	OMe
21	O	H	H	Br	OMe	OMe
22	O	H	Ph	Br	OMe	OMe
23	O	H	allyl	Br	OMe	OMe
24	O	H	*n*-Bu	Br	OMe	OMe
25	O	allyl	allyl	Br	OMe	OMe
5	S	H	Ph	H	$-OCH_2O-$	

Interestingly, the highest and broadest activity across all four microorganisms was demonstrated by 3,4-dimethoxybenezenemethylidene examples **17-21**. The introduction of an *ortho*-bromo substituent into these derivatives **21-25**, and in one example, replacement of the 3,4-dimethoxy substituents with a 3,4-methylenedioxy equivalent **5**, gave almost as much activity in many cases but with total loss of activity against some other species. This group included the most active substance, the *n*-butyl derivative **24**, and the least active compound, with N,N'-diallylation of the hydantoin ring.

Incorporation of such methoxy substituents therefore had a positive effect on activity but 2-halogenation and even replacement of the dimethoxy motif with the methylenedioxy motif has a significant impact on selectivity. In contrast, removal of the N(1)-H proton was detrimental and this suggested the positive involvement of an hydrogen bonding interaction associated with N(1).

3.3. Antifungal Activity

When the antifungal properties of the sample library members were assessed against *C. albicans* and *M. furfur*, the results in Table 1 revealed that the compounds generally had poor or no activity against *C. albicans* to the extent that all were of lower activity than the control, Amphotericin B. The most active compounds, benzalhydantoins **16**, **17**, and **20**, all exhibited inhibitory zones of 11 mm, which corresponded to the upper limits of the "*" (= weak activity) category, and there was no obvious correlation between the position and/or type of substituents and the activity.

Table 1. Inhibitory zones (clear zone, diameter in mm) of hydantoin derivatives against four test organisms [20] (Red coloured entries indicate results for compounds with at least one (***) level of microbial activity.)

	Fungi		Bacterial	
Compound Number	*C. Albicans*	*M. furfur*	*E. coli*	*S. aureus*
6	9 (*)	10 (*)	10 (*)	0
7	0	0	12 (**)	0
8	7 (*)	10 (*)	12 (**)	12 (**)
3[23]	0	0	10 (*)	0
4	7 (*)	10 (*)	8 (*)	17 (***)
9	9 (*)	10 (*)	10 (*)	0
10	0	0	10 (*)	14 (**)
11	7 (*)	16 (***)	12 (**)	14 (**)
12	0	16 (***)	14 (**)	14 (**)
13	0	12 (**)	13 (**)	12 (**)
14	0	7 (*)	13 (**)	12 (**)
15	11(*)	0	11(*)	13(*)
16	7 (*)	0	16 (***)	8 (*)
17	7 (*)	14 (**)	16 (***)	10 (*)
18	11 (*)	13 (**)	12 (**)	11 (*)
19	7 (*)	10 (*0	12 (**)	11 (*)
20	9 (*)	14 (**)	16 (***)	14 (**)
21	11 (*)	10 (*)	0	12 (**)
22	0	8 (*)	12 (**)	15 (**)
24	7 (*)	17 (***)	11 (*)	17 (***)
25	0	10 (*)	0	0
5[24,25]	7 (*)	10 (*)	0	15 (**)
26[26]	0	12 (**)	12 (**)	11 (*)
Amphotericin B	12 (**)	0	0	0
Control (MeOH)	0	0	0	0

Disc size: 6 mm (width), volume: 20 μL /disc (c = 1 mg/mL). Clear zone (inhibitory zone): * = 7–11 mm (weak activity); ** = 12–15 mm (high activity); *** = 16–20 mm (very high activity). Compound numbers are coded red in order to highlight the most active compounds.

In contrast, *M. furfur* was susceptible to eighteen of the twenty-three test substances, and of these eight showed "**" (= high activity) to "***" (= very high activity). The most active compounds, namely benzal hydantoins **11**, **12**, and **24**, had chlorine or bromine substituents in the benzenoid ring, and in one case a 3-allyl group and the other two cases a 3-butyl substituent in the hydantoin ring, with inhibitory zones within the range 16-17 mm (very high activity; ***).

There was therefore considerable variation in the bioactivity of the test compounds against fungi, but some examples of high activity were observed. As a result, it is suggested that compounds **11**, **12**, and **24** might serve as useful lead compounds for the development of drugs that are selective antifungal agents, particularly against *M. furfur*.

3.4. Antibacterial Activity

Further examination of the results in Table 1 indicated that bioactivity of the test compounds against bacteria appeared more common and higher than against fungi. Antibacterial activity against Gram-negative *E. coli* was only slightly more widespread (twenty bioactive out of twenty-three test compounds) than against Gram-positive *S. aureus* (eighteen bioactive), but the number that were observed to have "high" to "very high" activity were about the same. Almost all compounds showed activity against Gram-negative *E. coli*, except compounds **21**, **25** and **5**. Interestingly, compound **25** with allyl substituents in both N1 and N3 hydantoin ring positions was again inactive against Gram-positive *S. aureus* but compounds **21** and **5** showed high activity against Gram-positive *S. aureus*, with inhibitory zones within the range 12 - 15 mm (high activity; **).

The most active compounds against *E. coli* were compounds **16**, **17**, and **20** while those most active against *S. aureus* were compounds **24** and **4**. Compound **16**, was unusual in bearing a nitro group in the benzenoid ring, while all the other most active antibacterial compounds, namely **17**, **20**, **24** and **4**, bore two methoxy substituents. In addition, it was noted that of the most active antibacterials, compounds **16**, **20** and **24**, bore 3-*n*-butyl substituents.

Further reflection on the results in Table 1 revealed that compound **24** had quite modest levels of activity against *C. albicans* and *E. coli*, but was the only compound of those tested that showed the maximum level of bioactivity, "very high activity", against two or more microorganisms. Compound **24** therefore stands out as a major lead compound for further development in terms of drug development, especially as it demonstrates high selectivity in its biological activity, but also since it is active against a fungus and a Gram positive bacterium.

In a complementary sense, compound **16** appeared to be the only compound tested that showed very high activity against the Gram-negative bacterium, *E. coli*, as well as very high selectivity towards this microorganism (compounds **17** and **20** were more widespread inactivity).

In speculating on the most important features of compounds **16** and **24**, and indeed the other most bioactive compounds, it appeared that the 3-*n*-butyl substitution was important. If this meant that lipophilicity alone was important, then compounds **11**, **15**, **19**, **23**, and **25**, with their N-allyl substituents, should have shown promising bioactivity. Compound **23** was not available for testing, but compounds **11**, **15** and **19** did show parallel activity to the N-*n*-butyl analogues. Notably, compound **25**, was observed to have the least antimicrobial activity of all the benzal hydantoins and be one of the least active compounds overall of those tested. This reinforced the concept that a free N(1)-H group was needed for H-bonding to illicit high activity. Alternatively, the close proximity of the bromine and N(1)-allyl substituents in compound **25** might have created steric hindrance and/or a change in conformation of the molecule that interfered with biological activity. Whatever the cause, the poor bioactivity of compound **25** together with the promising activity of compounds **16** and **24** reveal the need for further structure-activity relationship studies on these compounds in the future.

4. Summary

A total of twenty three hydantoin and benzalhydantoin derivatives were investigated for antimicrobial activity against fungal (*C. albicans*, *M. furfur*) and bacterial (*E. coli* and *S. aureus*) microorganisms. A preliminary scan of the data showed that compounds had poor or no activity against *C. albicans*. All had lower activity than the control, Amphoterisin B, and within experimental error, all but two had lower activity against *C. albicans* than against the other three organisms. The two exceptions, benzalhydantoins **15** and **16**, were nitro derivatives, and of the set of nitro derivatives, **13-17**, these bore the most lipophilic groups at N(3). Curiously, those with N(3)-H and N(3)-phenyl substituents had similar activity in the bacterial assays but opposite selectivity towards *C. albicans* and *M. furfur*.

Eight compounds, hydantoins **8** and **4** and benzalhydantoins **11**, **17-20** and **24**, exhibited relatively strong activity against all four test organisms, although amongst these, compounds **17-20** showed the most consistent activity across all four organisms and compounds **24** and **4** showed the highest degree of selectivity. Compounds **17-20** all contained 3,4-dimethoxy substituents, which are commonly found in natural products and associated with biological activity. These substituents might have imparted a good degree of solubility and lipophilicity to the molecules. However, compounds **18-21** also had this motif and yet their bioactivity was very varied. Similarly, compound **5** had an equivalent 3,4-methylenedioxy substituent, and yet its bioactivity was also inconsistent against the four organisms.

It was concluded that compounds **21-24** and **5** might have had the conjugation of the 4'-methoxy substituent with the C(4)=O carbonyl group on the hydantoin segment of the more active molecules disrupted in the latter molecules. In particular, the large size of the bromine atom in compounds **21-25** might have caused a twist in the 5-methylidene bond on the hydantoin rings due to steric hindrance. For compound **5**, the combined effect of the C(2)=S thiono group, which would be less polarized that the equivalent C(2)=O carbonyl bond in the other examples, and the subsequent strengthening of the N(3)-C(4) with the opportunity for conjugation with the N(3)-phenyl substituent, might provide a different explanation for the supposed reduced conjugation between the 4'-methoxy group and the C(4)=O group.

Another general observation was that the relatively small, hydantoin derivatives **6**, **7** and **3** were active against the Gram negative bacterium, *E. coli* but ineffective against the Gram positive bacterium *S. aureus*. This might be explained by the more fragile cell wall of *E. coli*, which is supported by the observation that this was not the case for hydantoin analogues **6** and **4**, wherein the molecules were more lipophilic and able to penetrate the relatively thick wall of *S. aureus*.

Compounds **16** and **24** showed the most promise of all the candidates for further investigation because of their very high activity and specificity towards particular microorganisms.

Acknowledgements
Authors gratefully acknowledge small grant funding support from the Faculty of Science at The University of New South Wales, Sydney, Australia, and Direktorat Pendidikan Tinggi, Departemen Pendidikan Nasional, Republik Indonesia and Padjadjaran University, Bandung, Indonesia.

References

[1] Patrick G L 2009 *An Introduction to Medicinal Chemistry 4th Edition* (Oxford: Oxford University Press) pp 137 282 546-547
[2] Vicini P, Geronikaki A, Anastasia K, Incerti M and Zani F 2006 *Bioorg. Med. Chem.* **14** 3859-3864
[3] Tenover F C and McDonald L C 2005 *Curr. Opin. Infect. Dis.* **18** 300-305
[4] Pfeltz R F and Wilkinson B J 2004 *Curr. Drug Targets: Infect. Disord.* **4** 273-294
[5] Roberts M C 2004 *Curr. Drug Targets: Infect. Disord.* **4** 207-215
[6] Dessen A, Di Guilmi A M, Vernet T and Dideberg O 2001 *Curr. Drug Target: Infect. Disord.* **1** 63-77
[7] Muroi H, Nihei K, Tsujimoto K and Kubo I 2004 *Bioorg. Med. Chem.* **12** 583-587
[8] Wahab Khan M, Jahangiri Alam M, Rashid M A and Chowdhury R 2005 *Bioorg. Med. Chem.* **13** 4796-4805
[9] Torres-Viera C and Dembry L M 2004 *Curr. Opin. Infect. Dis.* **17** 541-547
[10] Anstead G M and Owens A D 2004 *Curr. Opin. Infect. Dis.* **17** 549-555
[11] Ford C W, Zurenko G E and Barbachyn M R 2001 *Curr. Drug Targets: Infect. Disord.* **1** 181-199
[12] Shinabarger D L, Marotti K R and Murray R W 1997 *Antimicrob. Agents Chemother.* **41** 2132–2136

IOP Conf. Series: Materials Science and Engineering **107** (2016) 012058 doi:10.1088/1757-899X/107/1/012058

[13] Swaney S M, Aoki H, Ganoza M C and Shinabarger D L 1998 *Antimicrob. Agents Chemother.* **42** 3251–3255
[14] Bulent B and Appelbaum P C 2004 *Int. J. Antimicrob. Agents* **23** 113–119
[15] Livermore D M 2003 *J. Antimicrob. Chemother* **51** (Suppl. 2) ii9–ii16
[16] Patel R P, Rouse M S, Piper K E and Steckelberg J M 1999 *Diagn. Microbiol. Infect. Dis.* **34** 119–122
[17] Blanc M, Cussac M, Boucherle A and Leclerc G 1992 *Eur. J. Med. Chem.* **27** 267-275
[18] Black D St C, Hidayat, I W and Read R W The University of New South Wales, unpublished
[19] Ward H P 1952 *J. Am. Chem. Soc.* **74** 4212–4213
[20] Landry M L 2007 Rhinoviruses. In *Manual of Clinical Microbiology* (7th ed.), Murray P R Editor-in-Chief, Baron E J, Pfaller M A, Tenover F C, Yolken R H, Eds, American Society for Microbiology Press: Washington DC 1999 pp. 983-989
[21] Gupta A K, Sauder D N and Shear N H 1994 *J. Am. Acad. Dermatol.* **30** 677–698
[22] Georgopapadakou N H and Walsh T J 1996 *Antimicrob. Agent Chemother.* **40** 279–291
[23] Finkbeiner H 1965 *J. Org. Chem.* **30** 3414–3419
[24] Wheeler H L and Hoffman C 1911 *Am. Chem. J.* **45** 368-383
[25] Cherouvrier J R, Carreaux F and Bazureau J P 2002 *Tetrahedron Lett.* **43** 3581-3584
[26] Eriksson A, Lespistoe M, Lundkvist M, Munck Af Rosenschoeld M and Zlatoidsky P 2004 Metalloproteinase Inhibitors. United States Patent US 2004/0147573 A1; July 29, 2004; 129 pp; see p 125

IOP Conf. Series: Materials Science and Engineering **107** (2016) 012059 doi:10.1088/1757-899X/107/1/012059

Ethyl-2-(3,5-Dihidroxyfenol): Phloroglucinol derivatives as potential anticancer material

Triana Kusumaningsih*, Maulidan Firdaus, Muhammad Widyo Wartono, Anif Nur Artanti, Desi Suci Handayani and Angga Eryanto Putro

Department of Chemistry, Faculty of Mathematics and Natural Sciences
Sebelas Maret University, Jl. Ir. Sutami 36 A Kentingan Surakarta 57126 Indonesia

*Email: triana_kusumaningsih@staff.uns.ac.id

Abstract. Ethyl-2-(3,5-dihidroxyfenol) based phloroglucinol compounds have been synthesized. Ethyl-2-(3,5-dihidroksifenol) were synthesized by reacting phloroglucinol with ethyl 2-chloro acetate in excess. Phloroglucinol reaction using 2-chloro ethyl acetate was carried out under reflux for 24 hours at a temperature of 56 °C. The reaction products were identified by a thin layer chromatography and were characterized by melting point test. Analysis of the structure of the products was obtained by FTIR spectrophotometer, ^{1}H-NMR, and ^{13}C-NMR. The result of the reaction between phloroglucinol and 2-chloro ethyl acetate was brownish black solid, and has a melting point of 191 °C. Based on the structural analysis by FTIR, ^{1}H-NMR and ^{13}C-NMR, the reaction product was a mixture of compounds which is ethyl 2- (3,5-dihidroksifenol) acetate, ethyl-2-(2,4,6-trioxocyclohexyl) acetate, and the rest of phloroglucinol which can not react.

1. Introduction

Phloroglucinol (1,3,5-trihidroxybenzene) and its derivatives have been used extensively in some commercial products. Phloroglucinol also widely used as a raw material or intermediate in the synthesis of medicinal substances, microbicides or in organic synthesis. Phloroglucinol also used as a dye for samples containing tannins on the analysis with a microscope and is also used as a dye in the leather industry, textile, and hair dyes. This compound is also used in the manufacture of explosives were stable against temperature and impacts, such as 1,3,4-triamino-2,4,6-trinitrobenzena [1].

Research on the synthesis of phloroglucinol compounds has been carried out by reacting with other compounds to form phloroglucinol derivatives to improve its bioactivity. The synthesis of a polyphenols based phloroglucinol aims to improve the bioactivity of phloroglucinol. It can be used as medicine because it was proven to prevent gastric ulcer in mice by decreasing oxidative stress in the stomach tissue. This study carried out the reaction between phloroglucinol reagent and ethyl-2-chloro acetate to form the derivatives of phloroglucinol [2].

Phloroglucinol compounds can be obtained from the isolation and synthesis. In order to enhance its bioactivity as a drug, the compounds can be prepared by derivatization of phloroglucinol. Phloroglucinol compounds have been produced through chemical synthesis. Phloroglucinol compound synthesis was done by modifying and adding a new group to improve its bioactivity as medicinal compounds. The

IOP Conf. Series: Materials Science and Engineering **107** (2016) 012059 doi:10.1088/1757-899X/107/1/012059

modifications of phloroglucinol compounds can be carried out by reacting it with an acyl phloroglucinol, phloroglucinol terpenes, glycosides phloroglucinol, halogen phloroglucinol, dimer phloroglucinol, trimer phloroglucinol, phlorotannin, and cyclic polyketides [3].

Phloroglucinol compound has two forms or tautomers, namely as the 1,3,5-trihidroxybenzene that has characters such as phenol, and 1,3,5-sikloheksanatrion (phloroglucin) which has a character such as ketones, in which the two tautomers are in equilibrium.

In the reaction that use a phloroglucinol reagent, i.e. 2-chloro ethyl acetate polyphenols, which reacts with phloroglucinol. The functional group can be in enol and keto forms. The reaction products were then to be analyzed by FTIR spectrophotometry, 1H-NMR and 13C-NMR to study their structure. In addition, the reaction product can be characterized by melting point analysis to ensure the product. Synthesis of phloroglucinol derivative with ethyl 2-chloroacetic has not been previously reported. The 2-chloro-ethyl acetate was chosen as a chemical reagent due to its availability and also inexpensive.

2. Methodology

2.1. Tools and Materials Research

The equipment used included infra-red (FT-IR, Prestige 8201 PC, KBr Pellet), Nuclear Magnetic Resonance (^{1}H-NMR, Agilent 400 MHz). The materials used are phloroglucinol (Sigma-Aldrich), ethyl 2-Chloro Acetate (E-Merck), calcium dichloride/$CaCl_2$ (E-Merck), Sodium Iodide/NaI (E-Merck), Potassium carbonate/K_2CO_3 (E-Merck), Sodium Sulfate/Na_2SO_4 (E-Merck), acetone p.An (E-Merck), chloroform (E-Merck), saturated sodium chloride, sodium hydroxide p.An (E. Merck), chloric acid p.An (E. Merck), dichloromethane (E. Merck), ethyl acetate (E. Merck), acetic acid (E. Merck), methanol (E. Merck), and distilled water.

2.2. Procedure

Into a 100 mL, three-necked round bottom flask equipped with a reflux condenser and a nitrogen pipe connector, phloroglucinol (0.25 mmol), ethyl-2-chloroacetic (0.2875 g), NaI (0.355 g) and K_2CO_3 (0.425 g) were dissolved in dry acetone (50 mL). The mixture was refluxed for 24 hours. After cooling, the excess of K_2CO_3 was filtered, and the solvent was evaporated. The residue obtained was dissolved in chloroform, and then was washed with 1 M HCl (3 times) and once with saturated NaCl. After drying with anhydrous Na_2SO_4, the solvent was evaporated, then recrystallized with a mixture of methanol-chloroform (1: 1). The resulted product was dried in a desiccator and analyzed by FT-IR, ^{1}H NMR, Scanning Electron Microscopy (SEM) and Surface Area Analyzer (SAA).

2.2.1. Determination of Melting Point. Test the melting point by using the tool Melting Point Apparatus SMP10 Stuart. Set the temperature observations were 0-200 °C.

2.2.2. Identification of the derived phloroglucinol compound. Phloroglucinol and phloroglucinol derivative were dissolved in the mixture of ethyl acetate: methanol at a ratio of 4:1 and then was performed in a Thin Layer Chromatography (TLC). The eluent used in the process was the mixture of dichloromethane, methanol, and acetic acid at a various ratio of 9:1:1; 6: 1:1; 1:1:1, respectively.

3. Results and Discussions

Phloroglucinol can be treated with several reagents, in order to obtain products include acyl phloroglucinol terpenes, glycosides phloroglucinol, halogen phloroglucinol, phloroglucinol dimer, trimer phloroglucinol, phlorotannin and cyclic polyketides. In this study, phloroglucinol was reacted with 2-chloro ethyl acetate.

The reaction of phloroglucinol derivative was carried out by reacting the compound with ethyl-2-chloro acetate. Phloroglucinol compounds were a white powder. Meanwhile, the ethyl-2-chloro acetate was a yellow colored liquid. The colour changed from white to brown-black. It indicates the production of a new compound. Phloroglucinol compound in the reaction process is capable of experiencing Tautomerizations. The reaction is described in Figure 1. The reaction phloroglucinol compounds can be in two different forms, which where keto and enol forms.

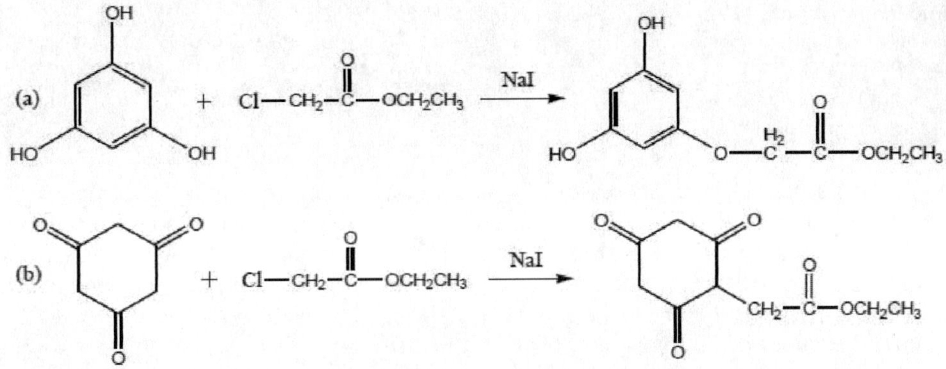

Figure 1. (a) scheme of phloroglucinol reaction with 2-chloro ethyl acetate in enol form (b) scheme of phloroglucinol reaction with 2-chloro ethyl acetate in keto form

3.1. Characterization of compound derivative phloroglucinol.

Characterization of the compounds was performed with melting point test resulting the 191 °C as the melting point. Meanwhile, the melting point of phloroglucinol compound is 217 °C. The change of melting point indicates that there was a reaction occurred between phloroglucinol compound, and ethyl-2-chloro acetate to produce new compounds.

3.2. Identification of Compounds Derived phloroglucinol.

The TLC identification was conducted by dissolved the product of the reaction in the mixture of ethyl acetate: methanol (4: 1). Meanwhile, the eluent was a mixture of dichloromethane: methanol: acetic acid in the ratio 9: 1: 1; 6: 1: 1; 1: 1: 1. The result of TLC identification are described in Figure 2.

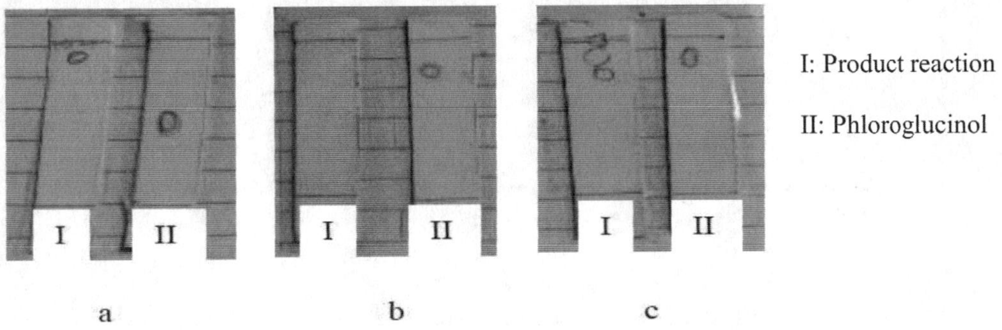

Figure 2. The TLC result of phloroglucinol and the product of reaction of phloroglucinol with the eluent of a) dichloromethane: methanol: acetic acid (9:1:1), b) dichloromethane: methanol: acetic acid (6:1:1), c) dichloromethane: methanol: acetic acid (1:1:1).

The analysis of TLC results is listed in Table 1.

Table 1. The result of TLC analysis on phloroglucinol and the product of reaction

Eluent	Phloroglucinol	Product reaction
dichloromethane : methanol : acetic acid (9:1:1)	1 spot, *Rf*: 0.58	1 spot, *Rf*: 0.92
dichloromethane : methanol : acetic acid (6:1:1)	1 spot, *Rf*: 0.86	no spot, *Rf*: 0
dichloromethane : methanol : acetic acid (1:1:1)	1 spot, *Rf*: 0.94	3 sopt *Rf*: 0.81 ; *Rf*: 0.94 ; *Rf*: 1.09
n-hexane : acetic acid (9:1) n-hexane : acetic acid (6:1) n-hexane : acetic acid (1:1)	No spot	No spot

Table 1 shows that the products reaction exists in three spots, it indicates that the compound is still a mixture, i.e., phloroglucinol, the residual of reaction products, and the reaction that proceeded to form the keto compounds of phloroglucinol and also in the enol form. Rf value is very characteristically for a particular compound in a particular eluent. The R_f of the product of the reaction in dichloromethane: methanol:acetic acid eluent is the highest, i.e. 0.94. It indicates low polarity. That is because the polar stationary phase. The more polar compounds will be stuck firmly in the stationary phase, resulting in a low R_f value.

3.3. Derivative Structure Analysis phloroglucinol

The FTIR spectra of the phloroglucinol compounds compared to the initial reaction is shown in Figure 3.

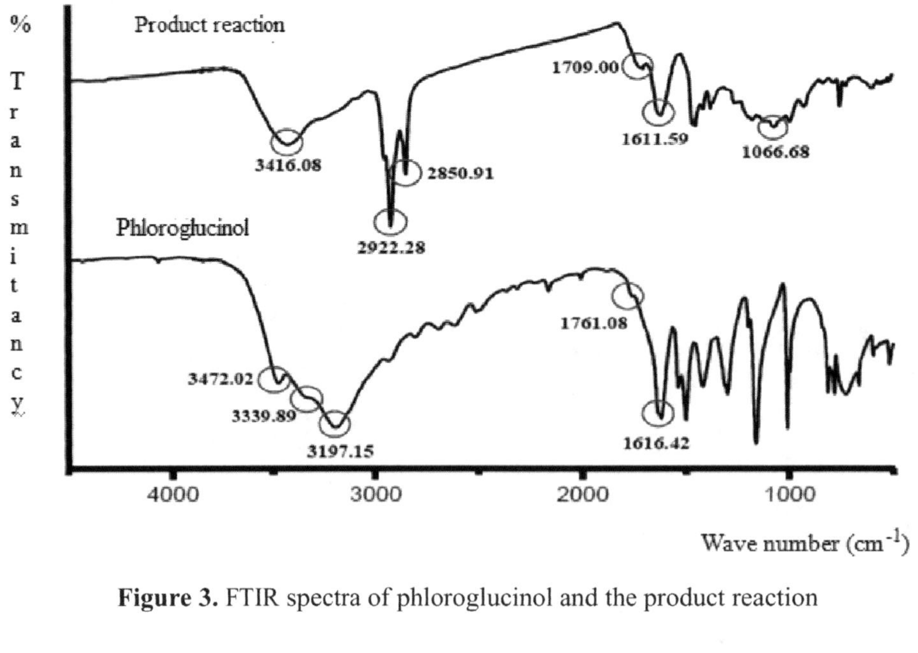

Figure 3. FTIR spectra of phloroglucinol and the product reaction

The FTIR spectra in Figure 3 can be presented in Table 2.

Table 2. Results of the analysis of the structure and its FTIR spectra

Wave number (cm^{-1})		
Phloroglucinol	**Product reaction**	**Functional group**
3472.02; 3339.89; 3197.15	3416.08	-OH
-	2922.28	C-H alkane
-	2850.91	C-H alkane
1761.08	1709.00	1,3,5substitute benzene
1616.42	1611.59	C=C cyclic
-	1066.68	aryl-O-CH_2

Table 2 shows that there is a new group that emerged in reaction to the compound produced from the substitution of the CH alkane of 2-chloro ethyl acetate that reacted with phloroglucinol compounds. There are two possible products which where keto and enol forms. These products were further analyzed by ^{1}H-NMR and ^{13}C-NMR spectroscopy.

3.3.1. Structural analysis using a spectrophotometer 1H-NMR. The reaction between phloroglucinol and 2-chloro ethyl acetate were analyzed using ^{1}H-NMR spectrophotometers. The spectrum is presented in Figure 4.

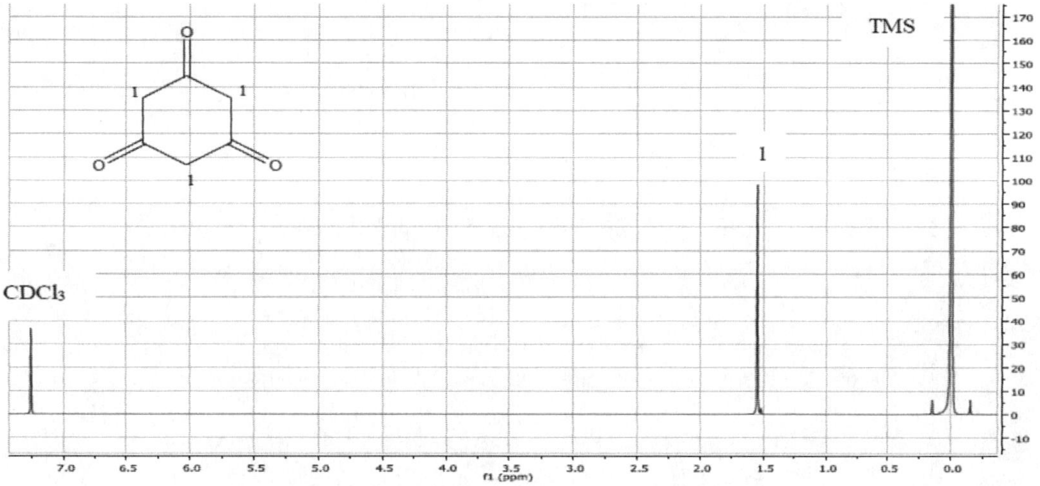

Figure 4. ^{1}H-NMR spectra of phloroglucinol compounds

Figure 4 shows that the structure of starting compound of phloroglucinol was changed. The structural change was described in Figure 1. The compounds undergo a process of phloroglucinol tautomerizations. Tautomerizations process depends on the several factors, including temperature, solvent, and pH. Phloroglucinol compounds can undergo tautomerization in polar protic solvents such as water; a single partner will be involved in a hydrogen bonding with solvents, making them less available to form hydrogen bonds with the enol form. It shows that the protic hydrogen atoms are bonded to an electronegative atom which in this case is oxygen [4].

Figure 4 shows that the analysis in the solvent $CDCl_3$, phloroglucinol are dominated in keto form. It because the phloroglucinol is in the solid phase and acidic conditions. Phloroglucinol compounds as well as curcumin. Curcumin in acidic conditions, neutral (with its pH at 25 °C was set as 7.0) and also in the solid phase are dominated by keto forms. Meanwhile, in alkaline condition is dominated by enol form [5]. Aromatic compounds which are in equilibrium with significant comparisons can be reduced into alkenes at a suitable condition. The product of reaction was also analyzed by ^{1}H-NMR to determine the structure. Based on the results of TLC identification, the compound has three spot which indicates that the reaction product is a mixture of compounds. The ^{1}H-NMR spectra of compounds as produced by phloroglucinol reaction are presented in Figure 5.

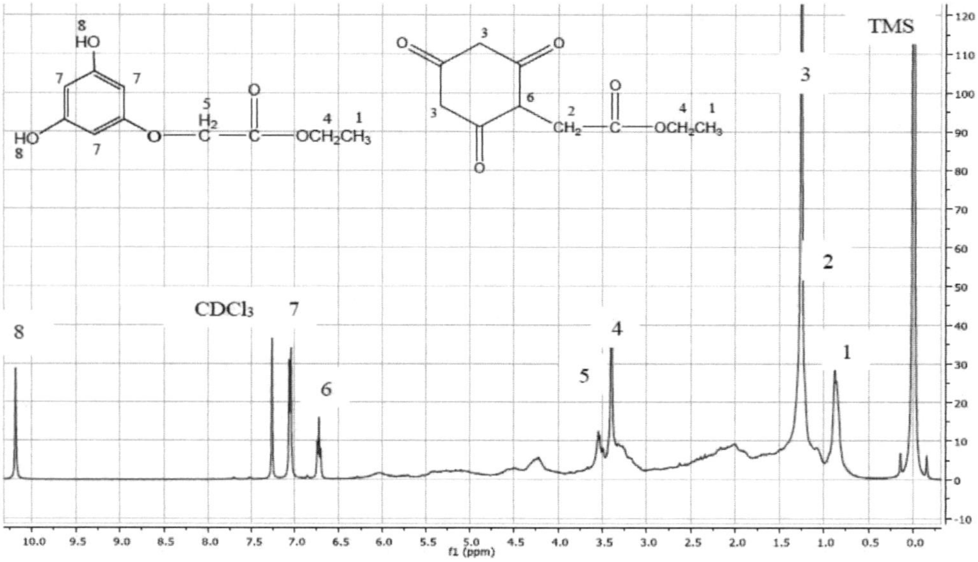

Figure 5. Spectra ^{1}H-NMR of the product of phloroglucinol reaction

Figure 5 shows 8 signals, indicating the existence of protons with 8 kinds of different environments. The interpretation of ^{1}H-NMR of data on the reaction products can be seen in Table 3.

Table 3. Interpretation of compound based on the 1H-NMR spectra

Signal	Chemical shift (ppm)	Integration	Functional group
1	0.83-0.89	1.27	Methyl (-CH3)
2	1.19	0.25	CH_2
3	1.25	2.17	CH_2
4	3.36-3.44	0.71	-O-CH_2
5	3.52-3.56	0.37	-O-CH_2-
6	6.7-6.74	0.34	-C-H
7	7.04-7.06	0.66	-C-H aromatic
8	10.18	0.38	Hidroxy (-OH)

The reaction scheme of phloroglusinol compounds in Figure 5, shows that the reaction product can be in the form of keto and enol. Two kinds of reactions produced enol from phloroglucinol which was reacted with 2-chloro ethyl acetate, or phloroglucinol in the form of keto compounds which reacted with the 2-chloro ethyl acetate compound. Integration of the reaction products showed a signal that appears on the spectra as two compounds of enol and keto form. Meanwhile, for the rest of phloroglucinol compound that was not reacted is appeared on signal number 3 with the integration of 2.17 (group H totaled 10, which belongs phloroglucinol 6H and 4H keto in the form of reaction products in the form of keto). Based on analysis of the 1H-NMR spectra in Figure 5, it is known that the product of these reactions was still a mixture of compounds.

3.3.2. The structural analysis by using ^{13}C-NMR. The product of reaction which are in enol form of phloroglucinol can be analyzed by using ^{13}C-NMR, and the result is in Figure 6.

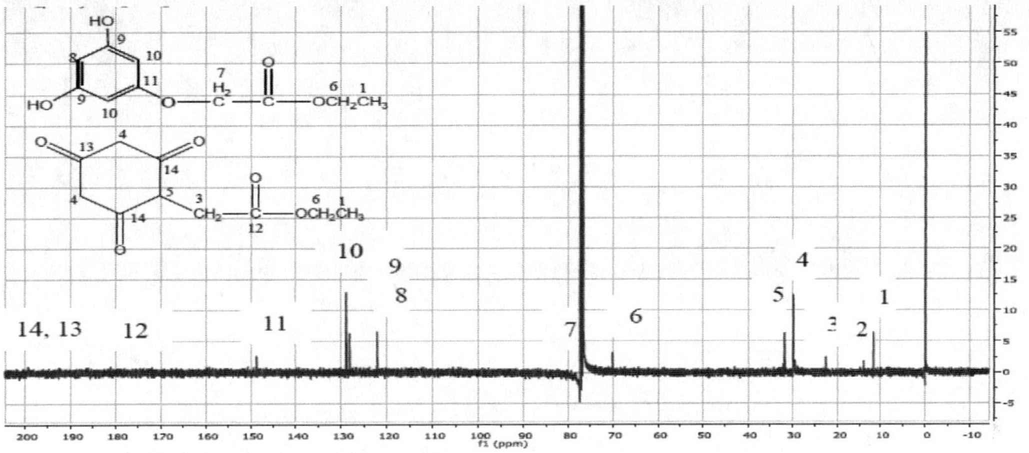

Figure 6. The ^{13}C-NMR spectra of reaction product of phloroglucinol

The interpretation of ^{13}C-NMR spectra is presented in Table 4.

Table 4. The interpretation of ^{13}C-NMR spectra

signal	Wave number (ppm)	Functional group
1	11.65	CH_3
2	14.13	CH_3
3	22.53	CH_2
4	29.53	CH_2
5	31.52	CH-R
6	70.15	-O-CH_2
7	77.18	-O-CH_2
8	122.225	C aromatic benzene
9	128.2	=C-O
10	128.95	C aromatic benzene
11	148.75	=C-O
12	181.0	-C=O
13	199.5	-C=O
14	206.2	-C=O

Table 4 shows the presence of a new compound. The results of ^{13}C-NMR analysis shows the presence of $-O-CH_2$, CH_3, $=CO$, C aromatic benzene groups. There are many other peaks that might come from phloroglucinol tautomeric keto compounds that react with 2-chloro ethyl acetate.

Structural analysis by using ^{1}H-NMR and ^{13}C-NMR, can be confirmed with the reaction mechanism between compounds in the form of enol phloroglucinol and 2-chloro ethyl acetate. The reaction mechanism is described in Figure 7. Meanwhile, the reaction mechanism between compounds phloroglucinol in the form of keto and 2-chloro ethyl acetate is described in Figure 8.

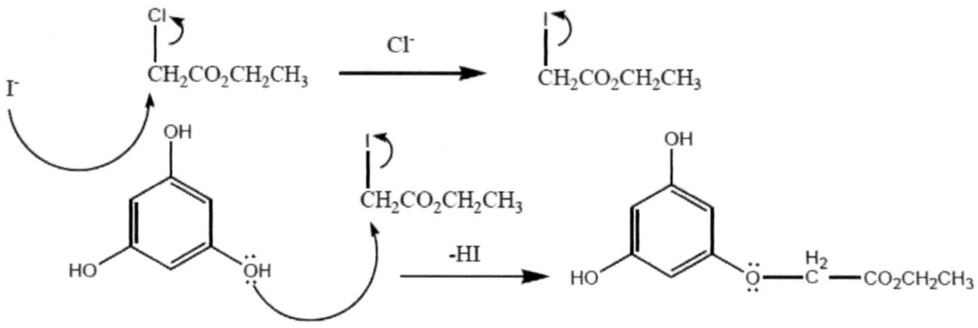

Figure 7. The reaction mechanism in the form of enol phlroglucinol

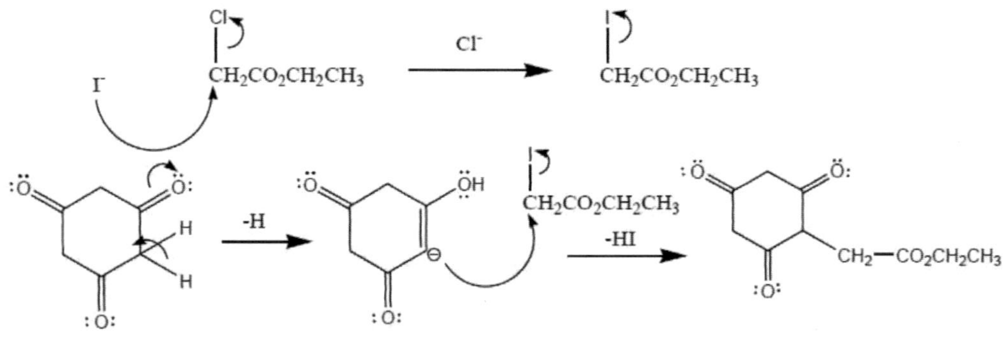

Figure 8. The reaction mechanism in the form of keto plroglucinol [6]

Reactions to phloroglucinol compound is strongly influenced by the leaving group on the alkyl halide. NaI use in this reaction, will change the ethyl-2-chloroacetic into ethyl-2-iodoasetat. Species I is a stronger nucleophile than Cl, therefore I able to repel Cl. As a leaving group I will more possibly move than Cl. The Cl substitution reaction by insitu I was occured to the hydroxy group of phloroglucinol. Furthermore, ploroglisinol that has hydroxyl group (OH), with O bearing a negative charge will bind the existing C atoms in the ethyl 2-iodoacetic which has positive charge. Then the I- as a leaving group will bind the existing H atoms in the hydroxy-owned phloroglucinol cluster.

Analysis using FTIR, ^{1}H-NMR and ^{13}C-NMR shows the structure of the reaction products formed phloroglucinol. The weak peak present in the FTIR, ^{1}H-NMR and ^{13}C-NMR analysis is a carbonyl group (-C=O) of ester compound that is present in the reaction products. Results of the reaction process of phloroglucinol compound with 2-chloro ethyl acetate according to the Figure 9 (phloroglucinol early in the enol form), that shows the substitution of hydroxy group at phloroglucinol with existing ester groups on 2-chloro ethyl acetate and it is described in Figure 10 (phloroglucinol early in the keto form).

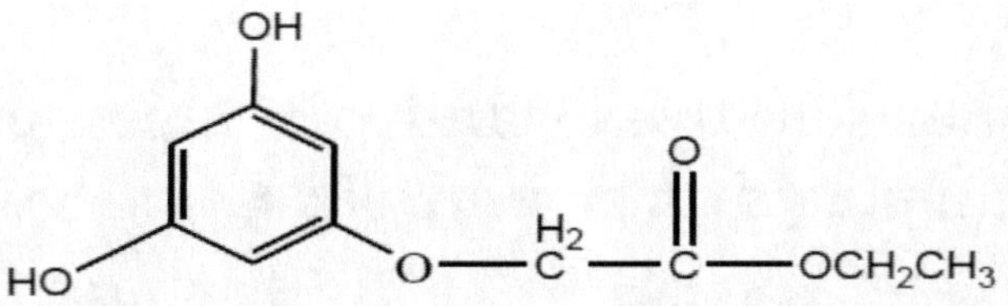

Figure 9. Structure of ethyl 2-(3,5- dihydroxyphenol) acetate

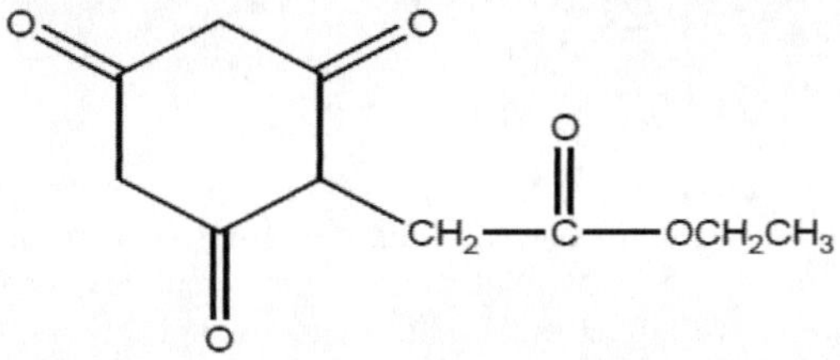

Figure 10. Structure of ethyl-2-(2,4,6-trioxocyclohexyl) acetate

4. Conclusions

The reaction between phloroglucinol and 2-chloro ethyl acetate produced a solid brownish black that has a melting point of 191 ^{o}C. The analysis using FTIR, ^{1}H-NMR, and ^{13}C-NMR confirm that the product of reaction is a mixture of compounds, i.e. phloroglucinol, ethyl 2-(3,5- dihydroxyphenol) acetate and ethyl-2-(2,4,6-trioxocyclohexyl) acetate.

References

[1] Frost J W 2007 Biosynthesis of Phloroglucinol and Preparation of 1,3-dihydroxybenzene therefrom United States Patent Application 20070178571

[2] Chaudhuri S R, Modak A, Bhaumik A and Swarnakar S 2011 Phloroglucinol Derivatives as Potential Antiulcer Compound that Inhibits Matrix metalloproteinase-9 *International Journal of Pharmaceutical Applications* **2**(4) 237-252

[3] Pal S I and Bharate S B 2006 Phloroglucinol Compounds of Natural Origin *Natural Product Reports* **23** (4) 558-591.

[4] Smith M B and March J 2006 March's Advanced Organic Chemistry: Reactions, Mechanisms, and Structure, Sixth Edition, Published Online: 18 Mei 2006, ISBN: 9780470084960, doi: 10.1002/0470084960

[5] Strimtakos A S and Sharma R A 2008 Comprehensive Invited Review Curcumin: Preventive and Therapeutic Properties in Laboratory Studies and Clinical Trials *Antioxid. Redox* **10** 511-546

[6] Kerry-Ann S 2010 Structure and Synthesis of Phloroglucinol Derivatives from Hypericum roeperianum. *Dissertation* University of KwaZulu-Natal South African

Acidity-controlled selective oxidation of alpha-pinene, isolated from Indonesian pine's turpentine oils (*pinus merkusii*)

Masruri[*], Mohamad Farid Rahman, and Bagus Nurkam Ramadhan

Chemistry Department, Brawijaya University, Jl. Veteran 65145 Malang, Indonesia

[*]E-mail: masruri@ub.ac.id

Abstract. Alpha-pinene was isolated in high purity from turpentine oil harvested from *Pinus merkusii* plantation. The recent investigation on selective oxidation of alpha-pinene using potassium permanganate was undertaken under acidic conditions. The result taught the selective oxidation of alpha-pinene in acidic using potassium permanganate lead to the formation of 2-(3-acetyl-2,2-dimethylcyclobutyl)acetaldehyde or pinon aldehyde. The study method applied reaction in various different buffer conditions i.e. pH 3, 4, 5, and 6, respectively, and each reaction product was monitored using TLC every hour. Product determination was undertaken on spectrometry basis such as infrared, ultra violet-visible, gas chromatography- and liquid chromatography-mass spectrometry.

1. Introduction

Selective oxidation of naturally abundant substrates is an invaluable strategy in order to get industrially important starting materials for synthesis, pharmaceuticals, and or agrichemicals. Alpha-pinene, isolated as a major volatile component from turpentine oils become focused our study. Recently, it has been reported as antibacterial [1], anti-inflammation [2], and insect attractant. Oxidation of its carbon-carbon double bond was also reported, such as provided its 1,2-amino-alcohol [3] and 1,2-diol [4-5]. This paper is disclosed recent study for its selective transformation provided pinon aldehyde. Similar papers have reported using a different strategy such as using catalyst ruthenium [6-9], analysis [10-11], palladium [12], metal-oxo [13], and potassium permanganate [14-15]. However, it was found low yield and selectivity were reported.

2. Materials and methods

2.1 Chemical and instrument

Chemicals for research is used and bought from the manufacturer or as mentioned, including potassium permanganate (Merck), ethyl acetate (SAP), n-hexane (SAP), magnesium sulfate anhydrate (Merck), sodium sulfate anhydrate (Merck), silica gel (column chromatography, Merck), chloroform (Smart Lab), acetone (Smart Lab), ethanol (LIPI), methanol (Smart Lab), and pre-coated TLC silica (Merck). Alpha-pinene was isolated and purified from turpentine oils (got from PT. Perhutani Anugerah Kimia). Procedure and analysis were performed following Masruri et al. (2007) [1]. Meanwhile, instrumentation for analysis the product such as an ultra violet-visible spectrophotometer (Shimadzu UV-1601), an infrared spectrophotometer (Shimadzu FTIR-8400S), gas chromatograph-mass spectrometer (Shimadzu QP2010S. Operational condition: capillary column Restex 30 m,

IOP Conf. Series: Materials Science and Engineering **107** (2016) 012060 doi:10.1088/1757-899X/107/1/012060

programed-temperature with increasing rate 10 °C per min, initial temperature 80 °C for 5 min, end temperature 250 °C, carrier gas nitrogen), and UV lamp for TLC spot detection.

2.2 General procedure

To a mixture of alpha-pinene (5.00 mL, 31.56 mmol) and potassium permanganate (7.48 g, 47.34 mmol) was added buffer sodium acetate solution (2,5 mL, 1.0 M) at 10 °C. This reaction mixture was stirred until reaction complete. Reaction's progress was monitored with spotting the sample on TLC plate, and the completed reaction was indicated as disappearing or consumed alpha-pinene during the reaction. Then, the reaction mixture was filtered, and the filtrate was extracted with ethyl acetate (3 x 10 mL). The combined ethyl acetate layers were dried under sodium sulfate anhydrate and concentrated by evaporation in a vacuum. Product purification was undertaken using flash column chromatography using ethyl acetate/n-hexane as eluent, and the purified product was isolated for further analysis using spectrometry methods (UV-Vis, FTIR, and GCMS).

2.3 Procedure for study pH reaction

The general procedure was followed with minor modification. The buffer sodium acetate solution was added in variation pH condition such as pH 3, 4, 5, and 6. Product separation, purification, and analysis were conducted using a similar procedure.

3. Result and discussion

3.1 alpha-Pinene from turpentine oils

Starting with the preparation of alpha-pinene from turpentine oil was the first step. Even though, the procedure already established for its purification, it was common founded that source of turpentine oil in different harvesting time and different manufacturer slightly provided different purity and composition [16].

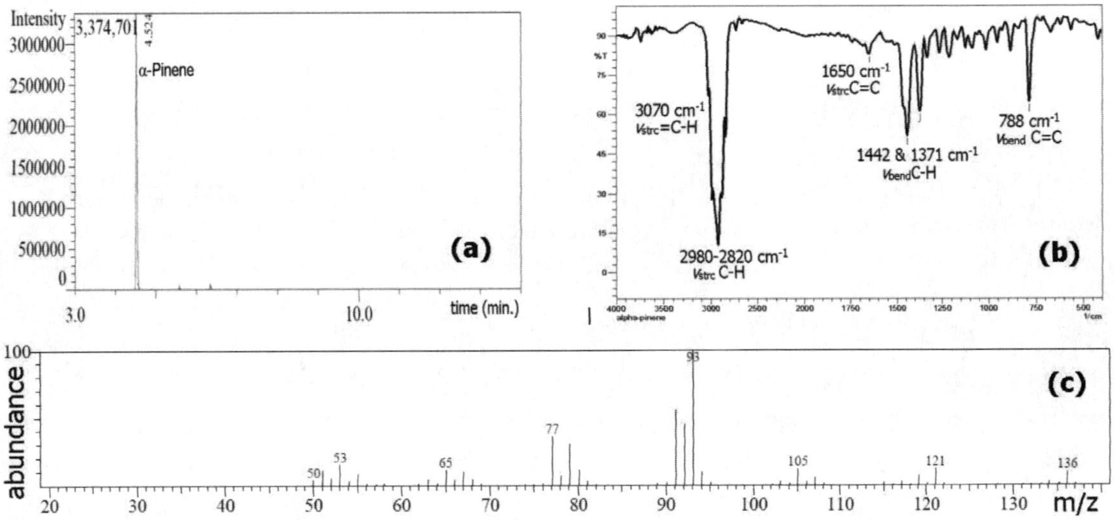

Figure 1. Characterisation of starting material α-pinene: (a) chromatogram, (b) FTIR spectra, and (c) mass spectra from GCMS

Purification of alpha-pinene from turpentine oils produced by PT Perhutani Anugerah Kimia, as starting material was undertaken using fractional distillation under reduced pressure [1] (Masruri et al., 2007). Alpha-pinene was afforded in first fraction as clear oil with 97% of purity (Figure 1b), density 0.8406 g/mL (29 °C), index refractive 1.4648 (26 °C), and boiling point 77-79 °C (40 mmHg). The structure is showed in scheme 1. Characterization of the alpha-pinene structure followed spectrometry data (Figure 1). Its infrared spectra significantly support the presence of functional groups contained (Figure 1a). Stretching vibration of =C-H was detected at 3070 cm^{-1}. This band was also correlated with the presence C=C stretching vibration at 1650 cm^{-1}, and its bending out of a plane in 788 cm^{-1}. The C=C alkene is the only functional group constructs its structure [1] (Masruri et al., 2007). Besides that, alkyl-group (CH_3, CH_2, and CH) was also detected. Symmetry and asymmetric stretching vibration found in between 2980 and 2820 cm^{-1} meanwhile bending vibration around 1442 and 1371 cm^{-1}. In addition, the mass spectra gave ion-molecule (m/z) 136.00 that correlates to the molecular weight of alpha-pinene with molecular formula $C_{10}H_{16}$ (Figure 1c). The exact calculation for $C_{10}H_{16}$ itself gave 136.1252 mass units. Moreover, the resulted spectra confirmed and agreed to that recorded in library spectra (Wiley7) by providing similarity index (SI) value 95% similarities to alpha-pinene.

3.2 Oxidation reaction

Previously, some papers reported oxidation of alpha-pinene using ozonolysis strategy provided low yield and selectivity of pinon aldehyde, pinonic acid, and some others oxidation products [10-11]. Another strategy was applied by employing ruthenium trichloride [6-9]. This catalyst is a precious metal and not easily accessed. Another metal such as osmium [5] and palladium [12] was also reported as a catalyst. This finding applied potassium permanganate to convert alpha-pinene. It was found the acidic condition lead the reaction formed pinon aldehyde as the major product. The range of pH study was conditioned at 3, 4, 5, and 6 respectively (scheme 1). Increasing the acidity of reaction mixture significantly improved the yield of pinon aldehyde. For example, reaction under buffer pH 6 provided pinon aldehyde in 17.46 %yield. It improved to 33.70% at pH 5, 38.16% at pH 4, and 77.68% at pH 3 (table 1). A different result was recorded that increasing of acidity did not affect the formation of pinonic acid. In average, it was isolated in between 4.20 and 8.98 %yield.

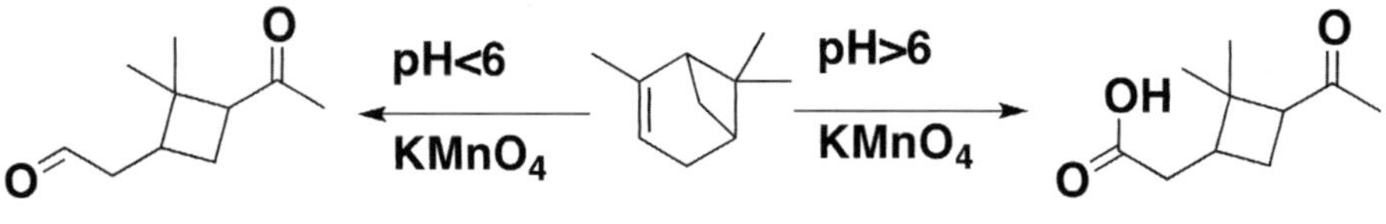

pinonaldehyde alpha-pinene pinonic acid

Figure 2. Oxidation reaction of alpha-pinene in acidic and basic condition using potassium permanganate

Table 1. Tabulation of oxidation of alpha-pinene in acidic condition

Entry	Reaction pH	Time (h)	Pinonaldehyde (yield, %)[b]	Pinonic acid (yield, %)[b]
1[a]	3	6	77.68	8.98
2[a]	4	6	38.16	7.25
3[a]	5	6	33.70	4.20
4[a]	6	6	17.46	5.61

[a]Reaction using buffer acetate and conducted at 10 °C
[b]Product yield were determined from GC chromatogram.

3.3 Product characterization

Pinonaldehyde was not a new compound. It has been reported and analyzed using gas chromatography-mass spectrometry of its deuterated compound. During the course, pinon aldehyde was characterized using infrared, ultraviolet-visible and mass spectra data (figure 3).

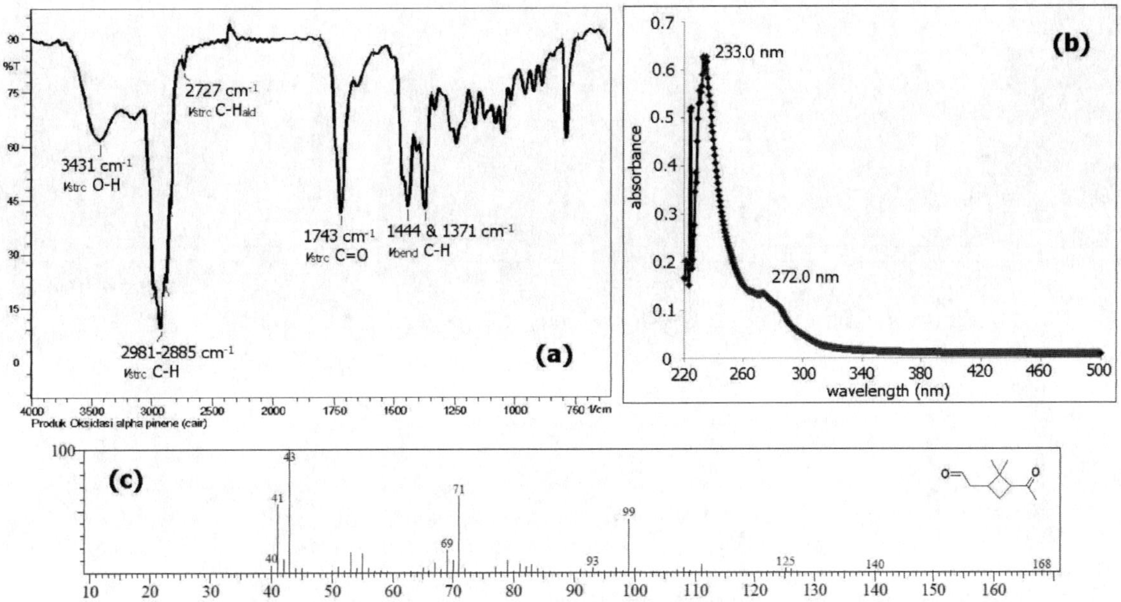

Figure 3. Spectra characteristic of pinon aldehyde: (a) infrared spectra, (b) ultra violet-visible spectra, and (c) mass spectra afforded from GCMS analysis.

The mass spectra provided a peak at m/z 168 for molecular ion (M^+). Fragmentation of this ion provided daughter ions such as m/z 140, 125, 99, 71, and 43 (base peak). Peak m/z 140 was predicted as the peak for decomposed of carbon monoxide of molecule ion. This fragmentation is characteristic for an aldehyde compound [17]. Another specific fragmentation was displacement of acylium group (CH_3CO) formed m/s 125 and m/z 43 (base peak). Moreover, characterization using ultraviolet-visible spectrophotometer gave two maximum wavelengths i.e. 233 and 272 nm. It was predicted 233 nm for $\pi \rightarrow \pi^*$ transition of C=C double bond, meanwhile 272 (low absorbance) as $n \rightarrow \pi^*$ transition on carbonyl group. In addition to this, the infrared spectra detected vibration band for C=O carbonyl of aldehyde (1743 cm^{-1}) and its C-H stretching (2727 cm^{-1}). The alkyl group was also detected their C-H stretching vibration in 1981-1885 cm^{-1} include the bending vibration in 1444 and1371 cm^{-1}. A minor O-H vibration still was detected, in 3431 cm^{-1}. In short, all the spectra support the existence of pinon aldehyde as the reaction product oxidation of alpha-pinene catalyzed by potassium permanganate under acidic condition.

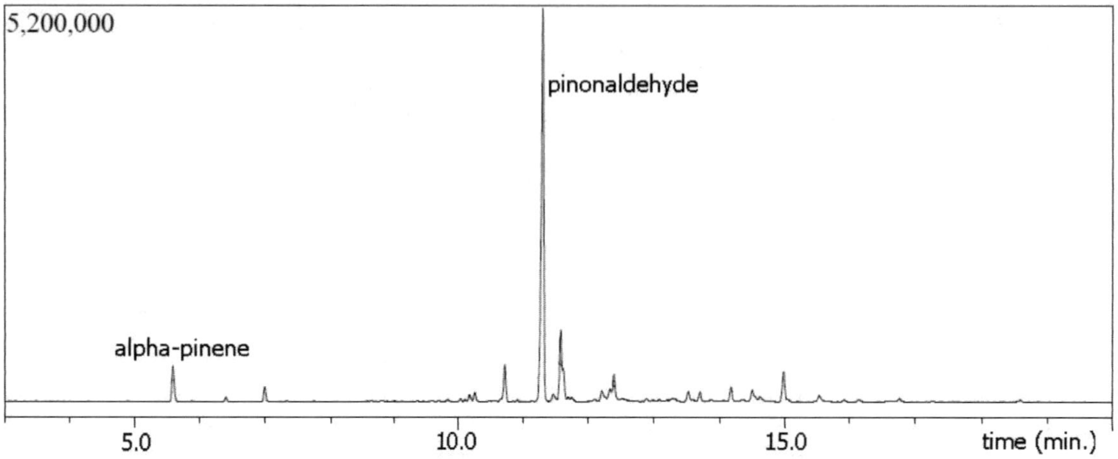

Figure 4. Chromatogram of pinon aldehyde afforded from GCMS analysis

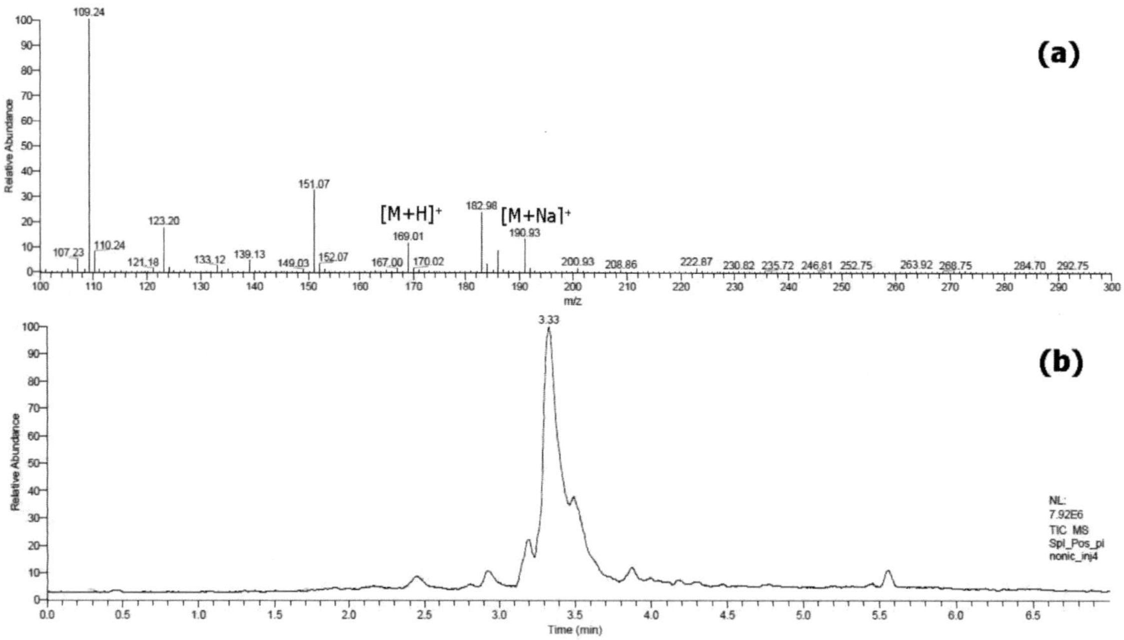

Figure 5. LCMS result of pinon aldehyde analysis; chromatogram (b) and its mass spectra (a)

Analysis using GCMS, in fact, was also afforded chromatogram (figure 4). Pinonaldehyde was detected as it in retention time 11.25 min. alpha-Pinene, starting material was detected as well at 5.62 min, including other minor products. At the same time, the product was also sent for LCMS analysis. The result was reported for it chromatogram and mass spectra (figure 5a-b). Mass spectra for pinon aldehyde were scanned at retention time 3.10-3.51 min (peak at 3.33 min). Pinonaldehyde was detected as its [M+Na] ion found at 190.99 and [M+H] detected at 169.01 mass units. Theoretical calculation for pinon aldehyde molecular formula $C_{10}H_{16}O_2$ was 168.12 mass units. Its mass plus sodium ($C_{10}H_{16}O_2Na$) gave 191.10 mass units while its protonated molecule ($C_{10}H_{17}O_2$) gave 169.12 mass units.

4. Conclusion

In summary, oxidation of alpha-pinene using potassium permanganate in acidic condition provided major product, pinon aldehyde. It chemoselectivity is improved by increasing the acidity of reaction conditions. Other products were also detected as pinonic acid. Its quantity slightly affected by the acidic condition.

Acknowledgement

Authors thanks to the Indonesian Ministry of Research, Technology, and Higher Education through Brawijaya University for the "Penelitian Unggulan Perguruan Tinggi (PUPT) research grant year 2015 to MSR and MFR. All authors have an equal contribution to preparation the paper; BNR is collecting data, MSR analyzing and writing the manuscript, and MFR evaluating the result.

References

[1] Masruri, Rahman M F and Prasodjo T I 2007 *Jurnal Ilmu-Ilmu Hayati (Life Sciences)*, **19** (1), 32-35.

[2] Khotimah H, Lyrawati D and Masruri 2006 Antiinflammatory effects of alpha-pinene extracted from Pinus mercury on levels of TNF-alpha signaling, iNOS, and apoptosis of neuronal cells, *Asian Symposium on Medicinal Plants, Spices, and Other Natural Products (ASOMP XII)*, 13-18 November 2006, Padang, Indonesia.

[3] Łączkowski K Z, Kmieciak A and Kozakiewicz A 2009 *Tetrahedron: Asym.*, **20** (13), 1487-1492.

[4] Erdik E, K [acaron] hya D and Daşkapan T 1998 *Synth. Comm.*, **28** (1), 1-7.

[5] Gomes M, and Antunes O A C 2001 *Catal. Comm.*, **2** (6), 225-227.

[6] Carlsen P H J, Katsuki T Martin, V S and Sharpless K B 1981 *J. Org. Chem.*, **46** (19), 3936-3938.

[7] Moglioni, Albertina G, García-Expósito E, Aguado G P, Parella T, Branchadell V, Moltrasio G Y and Ortuno R M 2000 *J. Org. Chem.*, **65**, 13, 3934-3940.

[8] Ziyat H, Ali M A, Karim A, Meliet C, Castanet Y and Mortreux A 2004 *Acta Chim. Slov.*, **51**, 223-230.

[9] Padala, Kishor and Jeganmohan M 2012 *Org. Let.*, **14**, 4, 1134-1137.

[10] Holloway F, Anderson H J and Rodin W 1955 *Ind. Eng. Chem.*, **47**, 10, 2111-2113.

[11] Fisher G S, Goldblatt L A and Stinson J S 1956 Converting alpha-pinene to mixtures of acids. In US Patents No US2750411 A: 1956

[12] Liu, Yan-Yun, Ren-Jie S, Cui-Yan W, Lu-Bin G, Ming H, Zhi-Qiang Wa, Ye-Xiang X and Jin-Heng L 2012 *Adv. Synth. Cat.*, **354**, 2/3, 347-353.

[13] Rubinstein, Amir, Jiménez-Lozanao P, Carbó J J, Poblet J M and Neumann R 2014 *J. Am. Chem. Soc.*, **136**, 31, 10941-10948.

[14] Tercio J, Ferreira B, Cruz W O, Vieira P C and Yonashiro M 1987 *J. Org. Chem.*, **52**, (16), 3698-3699.

[15] Masruri, Amini R W and Rahman M F 2015 Antibacterial activity of alpha-pinene derivative and its coordination compounds *Indonesian Journal of Chemistry* (Submitted, under reviewed)

[16] Amini R W, Masruri and Rahman M F 2014 *Jurnal Ilmu Kimia Universitas Brawijaya*, **1** (1), 147-153.

[17] Silverstein R M, Francis X, Webster and Kiemle D J 2005 Spectrometric Identification of Organic Compound, 7th edition, John Wiley & Sons Inc., USA

IOP Conf. Series: Materials Science and Engineering **107** (2016) 012061 doi:10.1088/1757-899X/107/1/012061

The co-pigmentation of anthocyanin isolated from mangosteen pericarp (*Garcinia Mangostana L.*) as Natural Dye for Dye-Sensitized Solar Cells (DSSC)

H Munawaroh, G Fadillah, L N M Z Saputri, Q A Hanif, R Hidayat, S Wahyuningsih*

Chemistry of Department, Faculty of Mathematics and Natural Sciences, Sebelas Maret University, Jl. Ir. Sutami, Kentingan, Surakarta 57126, Central Java, Indonesia

*E-mail: sayekti@mipa.uns.ac.id

Abstract. Study of color stability of anthocyanin from extract mangosteen pericarp (*Garcinia mangostana L.*) with co-pigmentation method has been conducted. Malic acid and ascorbic acid used as a co-pigment to stabilize the anthocyanin structure through formation of new binding between anthocyanin. Anthocyanin from mangosteen pericarp were isolated by several steps, including maceration, extraction, and Thin Layer Chromatography (TLC). The anthocyanin separation was conducted by TLC, while the identification of functional groups of those compound, were used FTIR (Fourier Transform Infrared Spectroscopy) for spectra analysis. Ultraviolet- visible absorption spectra have represented differences absorbance and color intensity in various pH. Copigmentation with malic acid and ascorbic acid in many composition and temperature were also well described. Meanwhile, anthocyanin-malic acid and anthocyanin-ascorbic acid have color retention higher than that of pure anthocyanin. Maximum color retention has been achieved at a ratio of 1:3 and 1:5 for ascorbic acid and malic acid, respectively. Therefore, the addition of ascorbic acid and malic acid as a co-pigment shows the ability to protect color retention of anthocyanin (mangosteen pericarp) from degradation process. The better efficiency of DSSC (η) have been achieved, whereas η of controlled anthocyanin, anthocyanin-ascorbic acid, and anthocyanin-malic acid were 0,1996%, 0,2922%, 0,3029%, respectively.

1. Introduction

Based on the manufacture of materials, there are two kinds of solar cells. The first solar cells are made of silicon. However, the fabrication of silicon solar cells is costly. While the new type of solar cells is dye-sensitized solar cells (DSSC), a device for the conversion of visible light into electricity based on the sensitization of wide band gap semiconductors [1]. The first DSSC was pioneered by Grätzel et al. in the early nineties, called Grätzel cells [1,2], these have attracted considerable attention due to their environmental friendliness and low cost [3]. The performance of the cell mainly depends on a dye used as a sensitizer. The important parameters of the efficiency of the cells are determined by the absorption spectrum of the dye and the anchorage of the dye to the surface of TiO_2 [4]. So far, the highest efficiency of DSSC by transition metal coordination compound (ruthenium polypridyl complexes) absorbed on nanocrystalline TiO_2 reached 8.7% [5]. However, ruthenium polypridyl complexes contain heavy metal which is undesirable for environment aspect and a limited amount of ruthenium hence their costly production [6]. Alternatively, natural dyes are being developed due to their availability, low cost [3], easy to prepare, nontoxic, and biodegradable [7].

By far, several of natural dyes that have been used as a sensitizer in DSSC system such as tannin [8], chlorophyll [9], anthocyanin [10], cyanine [11], and carotene [12]. Anthocyanin is very large group of red-blue pigments that can be found in flowers, fruits, fruit juice, leaves, stem, and root of the plants. One of them can be obtained from mangosteen (Garcinia Mangostana L.) pericarp. Anthocyanin in isolated from mangosteen pericarps has been studied before[13], but they have the weakness especially in the color stability and labile pigment due to very easily degraded. The efforts to enhance the stabilization of anthocyanin is co-pigmentation, an interaction between anthocyanin with other molecules such as metal (Al^{3+}, Fe^{3+}, Sn^{2+}, Cu^{2+}) and other organic molecules such as flavonoid compounds (flavonoids, flavanones, flavonols), and more. Their action of co-pigmentation tends to increase the color stability and protect from oxidation [14].

Based on the description above, this research focuses on co-pigmentation anthocyanin with malic acid and ascorbic acid. These extracted dyes were characterized by Thin Layer Chromatography (TLC), Infrared spectroscopy (IR), and UV-Vis absorption spectra. The photovoltaic properties of DSSC by purified product were studied.

2. Experimental

2.1. Sample preparation

Mangosteen pericarp was dried in an oven with temperature ± 40 °C. And then dried sample changed into a powder.

2.1.1. Sample maceration and extraction. Extraction of mangosteen pericarp powder used ethanol 96% and HCl 1M as a solvent with ratio 1:4 in 24 hours. Residue and filtrate were separated with vacuum filtration. Then filtrate was evaporated to remove the solvent with a rotary evaporator, yielded concentrated extract. Characterization concentrated extract was conducted with UV-vis spectrophotometry, Thin Layer Chromatography (TLC), and Fourier Transform Infrared (FTIR).

2.1.2. Characterization of anthocyanin extracts in various pH. Measurement of the color intensity of anthocyanin extract was performed on six pH conditions (1, 3, 5, 7, 9, 11). Each solution was measured at the maximum wavelength with a spectrophotometer at range 200-800 nm.

2.1.3. The color retention characterization of anthocyanin in co-pigmented with ascorbic acid. In this step, anthocyanin was diluted with ethanol than mixed with ascorbic acid in several ratio; 1:0, 1:1, 1:2, and1:3 (w/w). Then co-pigmented-anthocyanin was conducted in several of temperature (30, 40, 50, and 60 °C) to observe the effect of temperature in color stabilization. Then the result was characterized with UV-visible spectrophotometry in wavelength 520 nm.

2.1.4. The color retention characterization of anthocyanin co-pigmented with malic acid. In this step, anthocyanin was diluted with ethanol than mixed with ascorbic acid in several ratio; 1:5 and 1:10 (w/w). Then co-pigmented-anthocyanin was conducted in various of temperature (60 °C, 80 °C, 100 °C, 120 °C and 140 °C) to observe the effect of temperature in color stabilization. Then the result was characterized with UV-vis spectrophotometry in wavelength 520 nm.

2.2. Fabrication of DSSC

A fluorine-doped tin oxide (FTO) glass was used. The conductive glass substrate was first cleaned and coated with a layer of TiO_2 paste (0.5 g TiO_2 was added 2 mL ethanol and stirred until homogeny) by slip casting and dried at 400 °C for 30 min. After that, the glass substrate coated TiO_2 was immersed into anthocyanin dye overnight. To prepare the counter electrode, a platinum paste was deposited on the conductive substrate glass and dried at 400 °C for 30 min. The working electrode and counter electrode were assembled into a cell. A drop of electrolyte solution containing I-/I3-injected in the cell and the hole in the counter electrode was sealed using a sealing spacer. Therefore, DSSC arrangement was FTO|TiO_2|Anthocyanin (or co-pigmented-anthocyanin)| I-/I3-|Pt|FTO.

2.2.1. Characterization and measurement of photoelectrochemical. Solar energy conversion efficiency (the photo current-voltage (I-V) curve) was measured by Keithley multimeter under illumination. Based on I-V curve, the fill factor (FF) is defined as equation (1);

$$FF= (I_{max} \times V_{max})/(I_{sc} \times V_{oc}) \quad (1)$$

Where I_{max} and V_{max} are the photocurrent and photovoltage for maximum power output (P_{max}), I_{sc} and V_{oc} are the short-circuit photocurrent and open-circuit photovoltage, respectively. The overall energy conversion efficiency (η) is defined as equation (2)

$$\eta= (I_{sc} \times V_{oc} \times FF)/P_{in} \quad (2)$$

where P_{in} is the power of incident light.

3. Result and Discussion

3.1. Absorption of anthocyanin

Figure 1(a) shows the UV-Vis absorption spectrum of mangosteen extract. It was found that the absorption peak of mangosteen extract is about 520 nm. About 320, 340, and 450-520 nm indicate the presence of anthocyanin compounds. The present anthocyanin also depends on pH value. Figure 1(b) shows the absorption spectrum of anthocyanin shift to a greater wavelength in base condition or pH increase. Overall the spectrum shows a peak of anthocyanin at range 400-600 nm.

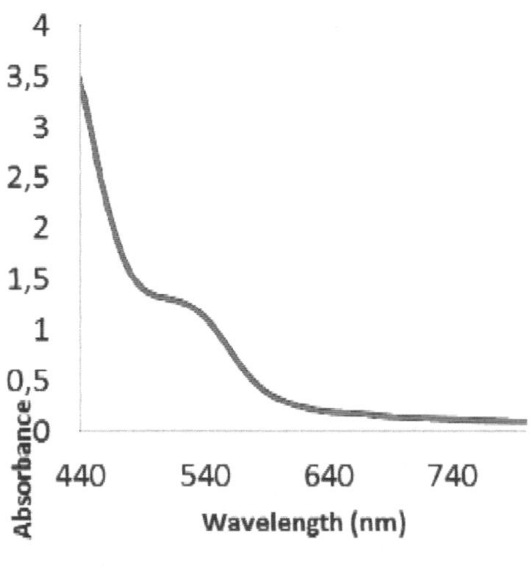

Figure 1. (a) Spectrum UV-Vis Anthocyanin.

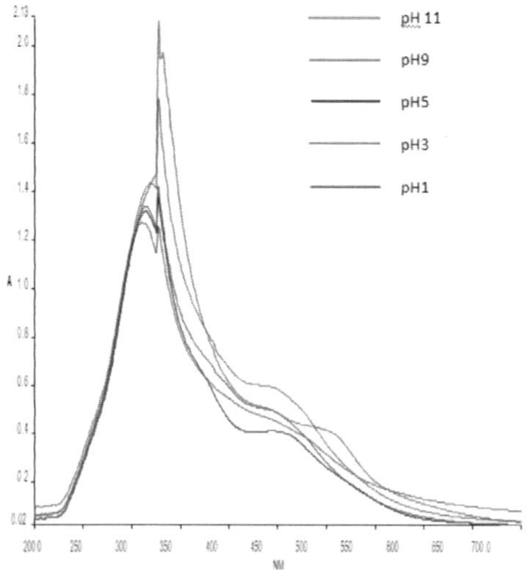

Figure 1. (b) Effect of pH.

Besides using UV-Vis, characterization of extract mangosteen was performed also with Thin Layer Chromatography (TLC) and infrared spectroscopy (IR). The purpose of using Thin Layer Chromatography (TLC) is to determine patterns of separation and purity of the compounds. Butanol:acetic acid: water used as eluent with a ratio of 4:1:5. The retention factor (Rf) obtained

0.255, this value related to the previous result that the Rf of anthocyanin approximately 0.1 to 0.4 [15]. While the characterization of anthocyanin using IR spectroscopy to determine the molecular structure of compounds (figure 2).

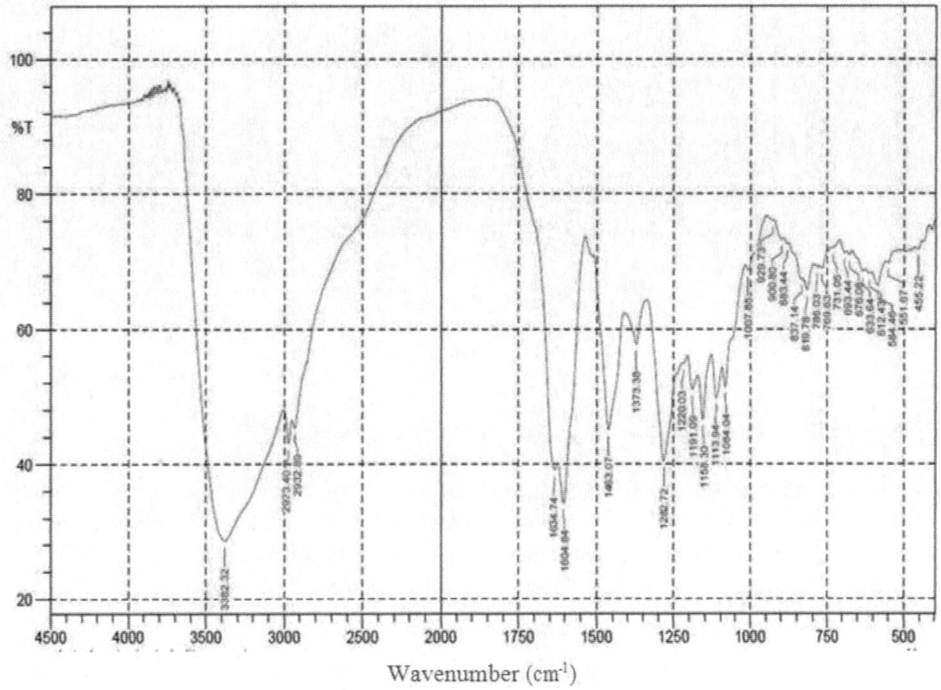

Figure 3. Spectra IR of Anthocyanin.

The FTIR spectra of these compounds show benzene groups vibration peaks between 455.22-929.7 cm^{-1}. The double bond C=C vibrations at 1100-1300 cm^{-1}. Two adjacent peaks at 1604-1634 cm^{-1} indicate carbonyl group (C=O). In addition, to the peak appearing at 2932-2937 cm^{-1} for C-H stretching. While the wide uptake 3362 cm^{-1} indicate the presence of OH group sand hydrogen bonding that occurs as intermolecular interaction of anthocyanin. All of these function groups establish the structure of anthocyanin that consist of benzene ring, double bond, a carbonyl group, C-H bond, and OH group.

3.2. Copigmentation anthocyanin: ascorbic acid

Figure 3 shows the effect of increasing temperature to color stability of anthocyanin-ascorbic acid co-pigmentation. Increasing absorbance of anthocyanin-ascorbic acid co-pigmentation followed by increasing mass ratio of ascorbic acid were added. The co-pigmentation of anthocyanin-ascorbic acid with ratio 1:3 have the greatest absorbance rather than ratio 1:1 and 1:2. However, there is an irregularity curve at ratio 1:2 caused by influence of heat that make the anthocyanin-ascorbic acid degraded quickly. High heating temperatures relatively can damage the structure of anthocyanin. Heating an anthocyanin-ascorbic acid cause an impact to their color stability due to degradation that may occur. Figure 5 represent that increasing temperature of heating anthocyanin-ascorbic acid was followed by degradation of color retention (color stability).

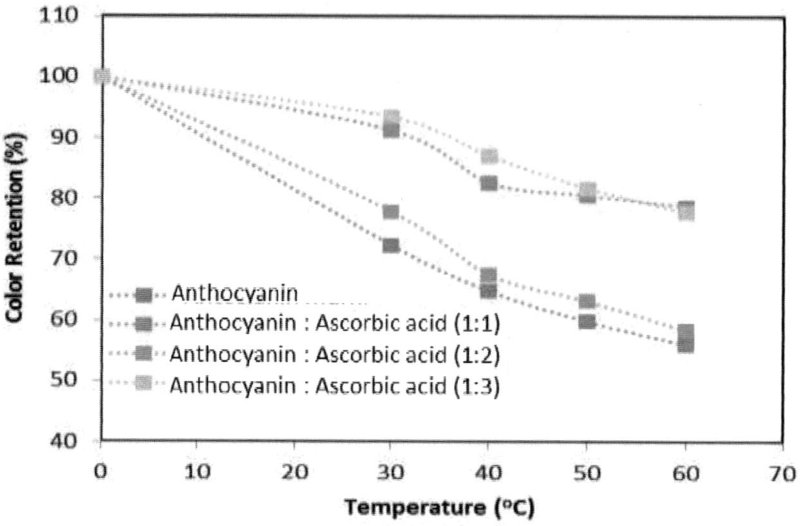

Figure 5. Color Retention of Anthocyanin-Ascorbic Acid.

3.3. Copigmentation anthocyanin: malic acid

Figure 4 shows the effect of increasing temperature to color stability of anthocyanin- malic acid co-pigmentation. The color stabilities anthocyanin-malic acid with ratio 1:0, 1:5, and 1:10 are 74.32; 92.89; and 98.64%, respectively. High heating temperatures relatively can damage the structure of anthocyanin. Heating an anthocyanin-malic acid cause an impact to their color stability due to degradation that may occur. Increasing temperature of heating anthocyanin-malic acid was followed by degradation of color retention (color stability).

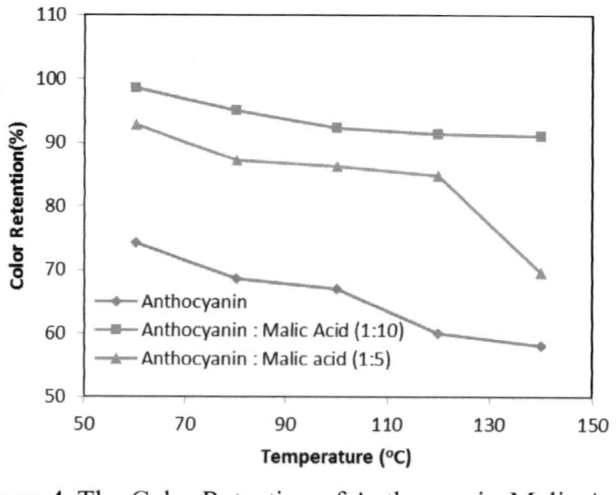

Figure 4. The Color Retention of Anthocyanin-Malic Acid.

3.4. Photoelectrochemical properties of DSSC with natural dye

Photovoltaic test of DSSC using co-pigmentation of natural dyes as sensitizers were performed by measuring the current-voltage (I-V) curve by solar simulator. The performance of DSSC was evaluated by energy conversion efficiency (η). The efficiency of anthocyanin, anthocyanin-ascorbic acid, and anthocyanin: malic acid are 0.199; 0.292; 0.303%, respectively. The result shows that co-pigmentation potentially increase the efficiency of DSSC from 46.40% to 51.75%. The polymerization that may occur between anthocyanin and co-pigment be able to stabilize the color of anthocyanin.

4. Conclusions

The anthocyanin was successfully extracted from mangosteen pericarp by maceration method. The characterization using TLC, UV-Vis, and FTIR shows the characteristic of anthocyanin, compared by the standard of anthocyanin. The efficiency of anthocyanin, anthocyanin-ascorbic acid, and anthocyanin-malic acid are 0.199; 0.292; 0.303%, respectively. The co-pigmentation with ascorbic acid and malic acid be able to increase the efficiency of DSSC from 46.40% to 51.75%.

References

[1] O' Regan B and Gratzel M 1991 A Low-Cost High Efficiency Solar Cell Based on Dye Sensitized Colloidal TiO_2 Films *Nature* **353**:737-740
[2] Maddu A, Zuhri M and Irmansyah 2010 *Penggunaan Ekstrak Antosianin Kol Merah sebagai Foto sentizer pada Sel Surya TiO_2 Nanokristal Tersentisasi Dye Bogor* Departemen FISIKA FMIPA Institut Pertanian Bogor
[3] Zhou H, Liqioing Wu, Yurong Gao and Tingli Ma 2011 Dye-Sensitized Solar Cells Using 20 Natural Dyes as Sensitizer *Photochem Photobiol* **219**:188-194
[4] Wongcharee K, Vissanu M and Sumaeth C 2007 Dye-Sensitized Solar Cell Using Natural Dyes Extracted from Rosella Blue Pea Flowers *Sol Ener Mat Sol C* **91**: 566-571
[5] Nazeeruddin M K, Klein C, Liska P and Gratzel M 2005 Synthesis of Novel Ruthenium Sensitizers And Their Application In Dye-Sensitized Solar Cells *Coord. Chem. Rev* **249**: 1460-1467
[6] Zhang G, Bala H, Cheng Y, Shi D, Lv X, Yu Q and Wang P 2009 High Efficiency and Stable Dye-Sensitized Solar Cells With an Organic Chromophore Featuring a Binary Phi-Conjugated Spacer *Chem Comm* 2198-2200
[7] Kamiloglu S, Ayca Ayfer P, Beraat O, John Y C and Esra C 2015 Color Retention, Anthocyanin Stability, And Antioxidant Capacity in Black Carrot (Daucuscarota) Jamsand Marmalades: Effect of Processing, Storage Conditions, And In Vitro Gastrointestinal Digestion *Roy Soc Ch* **13**: 1-10
[8] Espinosa R, Zumetta I and Santana J L 2005 Nanocrystalline TiO_2 Photosensitized With Natural Polymers With Enhanced Efficiency From 400 to 600 Nm *Sol Ener Mat Sol C* **85**: 359-369
[9] Kumara G R A, Kaneko S and Okuya M 2006 Shiso Leaf Pigments For Dye Sensitized Solid-State Solar Cell *Sol Energ Mat Sol C* **90**: 1220-1226
[10] Wongcharee, Meeyoo K V and Chavadej S 2007 *Dye-Sensitized Solar Cell Using Natural Dyes Extracted From Rosella and Blue Pea Flowers* Teknik Kimia: Universitas Teknologi Mahanakom
[11] Furukawa S, Lino H, Iwamoto T and Kukita K 2009 Characteristics Of Dye Sensitized Solar Cells Using Natural Dye *Thin solid films* **518**: 526-529
[12] Ortiz, Gomez I A, Maldonado V and Espadas A R P 2010 Dye Sensitized Solar Cells With Natural Dyes Extracted From Achiote Seeds *Sol Energ Mat Sol* **94**: 40-44
[13] Tanaka Y, Sasaki N and Ohmiya A 2008 Biosynthesis of plant pigments: anthocyanins betalains and carotenoids *The Plant journal* **54**: 733-749
[14] Boulton R 2001 The Copigmentation of Anthocyanins and Its Role In The Color Of Red Wine: A Critical Review *Am J Enol Viticul* **52** (2): 67-8
[15] Harborne J B 1987 *Metode Fitokimia: Penuntun Cara Modern Menganalisis Tumbuhan* Bandung: Penerbit ITB

Addition reaction of methyl cinnamate with 2-amino-4-nitrophenol

Venty Suryanti*, Fajar Rakhman Wibowo, Pranoto, Siti Robingatun Isnaeni, Meiyanti Ratna Kumala Sari and Sekar Handayani
Department of Chemistry, Faculty of Mathematics and Natural Sciences, Sebelas Maret University, Jl. Ir. Sutami 36A Surakarta, Central Java 57126 Indonesia

*Email: venty@mipa.uns.ac.id

Abstract. A novel compound which have one N-H fragment and nitrophenyl group has been designed and synthesized from cinnamaldehyde. The reaction was conducted in 3 step reactions to give the final product. Firstly, cinnamaldehyde was converted into cinnamic acid, which was then esterified with methyl alcohol to obtained methyl cinnamate. The last step was the addition reaction between methyl cinnamate and 2-amino-4-nitrophenol to give a cinnamaldehyde derivative, namely methyl-3-(2-hidroksi-5-nitrophenyl amino)-3-phenylpropanoate.

1. Introduction

Cinnamomum Verum J. Presl belongs to the Lauraceae family. Cinnamon is a good source of polyphenol compounds where over 300 volatile compounds were found in essential oils of cinnamon [1-2]. Oils and extracts from cinnamon have good antioxidant activity, which is ascribed to the presence of phenolic and polyphenolic compounds [3]. Cinnamaldehyde (3-phenyl-2-propanal) is the major component which present in cinnamon oil or cinnamon bark extract [4]. This compound having various substituents on the aromatic ring are useful for starting material in synthesizing of its derivatives.

Peralta-Dominguez *et al.* have established applications of organic molecules based on cinnamaldehyde for colorimetric chemosensors. A Schiff base derivative of cinnamaldehyde, 4-chloro-2-[(3-(4-(dimethylamino)phenyl)allyl-dene)amino]phenol has been synthesized and demonstrated to be a sensitive colorimetric Ni^{2+} sensor in aqueous solution, where the presence of Ni^{2+} into the sample solution produced a rapid color change from faded yellow to deep orange [15].

Suryanti *et al.* have demonstrated colorimetric anion sensors based on *N*-acetyl glyoxylic amides. These compounds were synthesized by one-step ring-opening reactions of *N*-acetylation with 4-nitrophenyl and 2,4-dinitrophenyl amines. These sensors contain two NH groups as anion binding sites and a nitrophenyl moiety as a signaling unit, where visible color changes occurred upon the addition of fluoride and cyanide anions. These anions formed hydrogen bonding interactions with the compounds and triggered deprotonation of those compounds [16]. Suryanti *et al.* also have reported the synthesized of another series of *N*-acetyl glyoxylic amides having N-H fragments [17]. In addition, series of *N*-octanoyl glyoxylic amides and oxalyl-bis-glycosylamines have been synthesized by ring opening of N-octanoyl isatins and bis-acylisatins, respectively, with a range of amines and amino acid alkyl esters [18-19]. These findings prompted us to synthesize novel compounds containing N-H fragment and nitrophenyl group. This present research was focused on the design and develop of target compound using cinnamaldehyde as starting material.

2. Materials and Methods

2.1. Chemistry

Reagents and solvents were purchased from commercial suppliers. All solvents employed were of analytical grade and used without further purification. Reactions were monitored by thin-layer chromatography (TLC) on TLC silica gel 60 F_{254}. Melting points were determined by an open tube capillary method using Stuart SMP 10 melting point apparatus and uncorrected. All spectrometric analyzes were carried out at the Faculty of Mathematics and Natural Sciences Laboratory Facilities, Sebelas Maret University. UV-Vis spectra were recorded on a UV-Vis Shimadzu Lambda 25 spectrophotometer at room temperature. IR spectra were recorded as KBr pellets on Shimadzu type FT-IR 8201 PC spectrophotometer. ^1H NMR dan ^{13}C NMR spectra were measured with an Agilent VNMR 400 MHz instrument at 25°C. Signals were recorded in parts per million (δ) downfield from TMS as an internal standard, and *J* values are given in Hz.

2.2. Procedures of compounds preparation

*2.2.1. Synthesize of cinnamic acid **2***

Cinnamaldehyde **1** (2.46 g), diethyl ether (100 mL), four drops of polysorbate 20 and water (100 mL) were placed and stirred in round bottom flask. CrO_3 (7.9 g) was added slowly and the reaction mixture was then stirred for 6 h. The mixture was cooled with ice in 4-5°C and then leaved at room temperature. The mixture was extracted with diethyl ether. The organic phase was collected, dried over MgSO4 and evaporated to dryness affording a white powder of 64.9% yield and m.p. 125°C. UV(EtOH): λ_{max} 201 (5715 $cm^{-1}M^{-1}$), 210 (4375), 267 (2086); IR (KBr): υ_{max} 3462, 3068, 3028, 2835, 1683, 1630, 1576, 1492, 1449, 1423, 1344, 1315, 1287, 1226, 1175, 933, 871, 765, 705; ^1H NMR (400 MHz, $CDCl_3$): δ 6.46 (dd, ArCHC$\underline{H}$, *J*= 16.0, 7.7 Hz, 1H), 7.26 (d, COOH, *J*=7.7 Hz, 1H), 7.39 (m, 3 x ArH, 3H), 7.56 (dd, 2 x ArH, *J*= 9.7 Hz, 2H), 7.80 (dd, ArC$\underline{H}$CH, *J*=5.9 Hz, 1H). ^{13}C NMR (100 MHz, $CDCl_3$): δ 117.2 (ArCH$\underline{C}$H), 128.3 (2 x ArC), 128.9 (2 x ArC), 130.9 (ArC), 134.0 (ArC), 146.9 (Ar$\underline{C}$HCH), 171.8 (COOH).

*2.2.2. Esterification of cinnamic acid **2** with methanol*

Cynamic acid (0.67 g) and methanol (10 mL) were placed in a round bottom flask. Two drops of saturated H_2SO_4 was added and the reaction mixture was heated under reflux for 6 hours. After cooling, the reaction mixture was evaporated to dryness affording a dark brown liquid of 42% yield and b.p. 33°C. UV(EtOH): λ_{max} 202 (7909 $cm^{-1}M^{-1}$), 211 (4672), 269 (2441); IR (KBr): υ_{max} 3413, 3028, 2927, 2854, 1715, 1634, 1578, 1493, 1450, 1314, 1279, 1203, 1171, 980, 868, 767, 701; ^1H NMR (400 MHz, $CDCl_3$): δ 3.81 (s, OCH_3, 3H), 6.44 (d, ArCHC$\underline{H}$, *J*=16 Hz, 1H), 7.38 (m, 3 x ArH, 3H), 7.52 (m, 2 x ArH, 2H); 7.68 (dd, ArC$\underline{H}$CH, *J*=5.9 Hz, 1H); ^{13}C NMR (100 MHz, $CDCl_3$): δ 51.7 ($\underline{C}OOCH_3$), 117.9 (ArCH$\underline{C}$H), 128.0 (2 x ArC), 128.9 (2 x ArC), 130.3 (ArC), 134.0 (ArC), 144.9 (Ar$\underline{C}$HCH), 171.8 ($\underline{C}OOCH_3$).

*2.2.3. Reaction of methyl cinnamate **3** and 2-amino-4-nitrophenol **4***

Methyl cinnamate (1 mmol) and 2-amino-4-nitrophenol (1.2 mmol) were reacted in 25 mL dichloromethane. The reaction mixture was refluxed for 48 h. After cooling, the reaction mixture was evaporated to dryness affording a yellowish brown gel of 68% yield and m.p. 81°C. UV(acetone): λ_{max} 328 (9091 $cm^{-1}M^{-1}$), 375 (10014); IR (KBr): υ_{max} 3531, 3375, 3306, 2952, 1709, 1693, 1632, 1525, 1499, 1451, 1336, 1294, 1204, 1172, 1079, 1039, 979, 876, 769, 746; ^1H NMR (400 MHz, CD_3OD): δ 3.61 (s, OCH_3, 3H), 6.50 (d, ArCHC$\underline{H}$, *J*=16 Hz, 2H), 6.74 (m, ArC$\underline{H}$CH, *J*=8.7 Hz, 1H), 7.36 (dd, 3 x ArH, *J*=10.9, 8.1 Hz, 3H), 7.51 (dd, *J*=10.9, 8.1 Hz, 1H), 7.57 (m, 3 x ArH, 3H), 7.66 (m, ArH, 1H); ^{13}C NMR (100 MHz, CD_3OD): δ 50.7 ($\underline{C}OOCH_3$), 109.6 (ArCH$\underline{C}H_2$), 109.7 (Ar$\underline{C}HCH_2$), 117.1,

127.8, 128.6, 130.1, 135.5, 134.2 (6 x ArC), 112.8, 114.7, 114.8, 140.9, 144.8, 151.1, (6 x $CHNO_2OH$), 167.7 ($\underline{C}OOCH_3$).

3. Results and Discussion

Scheme 1 summarizes the reaction steps occurred in this work, where the reaction was conducted in 3 step reactions to give the final product.

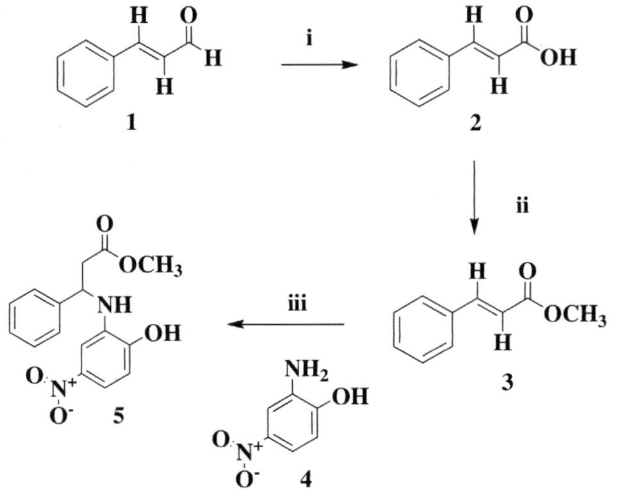

Scheme 1. *Reagents and conditions*: (i). $(C_2H_5)_2O$, H_2O, polysorbate 20, CrO_3, 6 h, r.t. (ii). CH_3OH *sat*. HCl, reflux, 6 h. (iii). CH_2Cl_2, reflux, 48 h.

*3.1. Synthesize of cinnamic acid **2***

Compound **2** was prepared by reacting of cinnamaldehyde **1** with chromium trioxide (CrO_3) in diethyl ether and water in the presence of polysorbate 20 as phase transfer catalyst. The chromium trioxide in water forms chromic acid (H_2CrO_4) which would then oxidized the cinnamaldehyde to cinnamic acid **2** as a white powder in 64.9% yield.

The ^1H NMR spectrum of compound **1** in $CDCl_3$ exhibited the singlet resonance for the aldehyde proton (-COH) at 9.69 ppm, whereas the ^{13}C NMR spectrum of the compound **2** in $CDCl_3$ exhibited the singlet resonance for the carboxylic proton (-COOH) at 7.26 ppm. In addition, proton shifts were observed in the ^1H NMR spectra of the compounds **1** and **2**. The resonances for –CH=CH- in the ^1H NMR spectrum of compound **1** were observed at 6.72 and 7.56 ppm, whereas in the ^1H NMR spectrum of compound **2** were observed at 6.46 and 7.80 ppm, respectively. The resonance for the carboxylic carbonyl (-COOH) of compound **2** was also confirmed by ^{13}C NMR spectroscopy spectrum which showed a peak at 171 ppm, whereas compound **1** exhibited a peak at 193 ppm for aldehyde carbonyl (-COH). Moreover, the resonances for –CH=CH- in the ^{13}C NMR spectrum of compound **1** were observed at 152 and 131 ppm, whereas in the ^1H NMR spectrum of compound **2** were observed at 117 and 146 ppm, respectively.

*3.2. Esterification of cinnamic acid **2** with methanol*

Compound **3** was synthesized in 42% yield by stirring of a mixture of compound **2** and methanol under reflux for 6 hours in the presence of saturated HCl. The analysis of compound **3** by ^1H NMR spectroscopy revealed the presence of new methyl protons at 3.81 ppm. The new resonance for this –CH_3 was also indicated by ^{13}C NMR spectroscopy spectrum which showed a peak at 51 ppm.

IOP Conf. Series: Materials Science and Engineering **107** (2016) 012062 doi:10.1088/1757-899X/107/1/012062

3.3. Reaction of methyl cinnamate 3 and 2-amino-4-nitrophenol 4

The reaction of compound **3** with 2-amino-4-nitrophenol **4** in dichloromethane under reflux for 48 hours resulted in the formation of compound **5** as a yellowish brown gel in 68% yield. Figure 1 and 2 showed the ^{1}H NMR and ^{13}C NMR spectra of compound **5**, respectively. The resonances for –CH=CH– in the ^{1}H NMR spectrum of compound **3** were observed at 6.44 and 7.68 ppm for one proton each. Conversely, the ^{1}H NMR spectrum of compound **5** exhibited peaks at 6.50 ppm for two protons and 7.80 ppm for one proton. In addition, resonances for –CH=CH– in the ^{13}C NMR spectrum of compound **3** were observed at 167 and 117 ppm. As shown in the ^{13}C NMR spectrum of compound **5**, these peaks were shifted to 169.6 and 109.7 ppm. Analysis of compound **5** by ^{1}H NMR spectroscopy revealed additional peak at 7.36 ppm for additional aromatic protons which integrated for three protons in total. Furthermore, additional six aromatic carbons were also observed in ^{13}C NMR spectrum of compound **5** at 112.8, 114.7, 114.8, 140.9, 144.8 and 151.1 ppm. These results indicated that the addition product of compound **5** was obtained in this reaction. Compound **5** contains one N-H fragment and nitrophenyl group as we expected which might be useful for the future application.

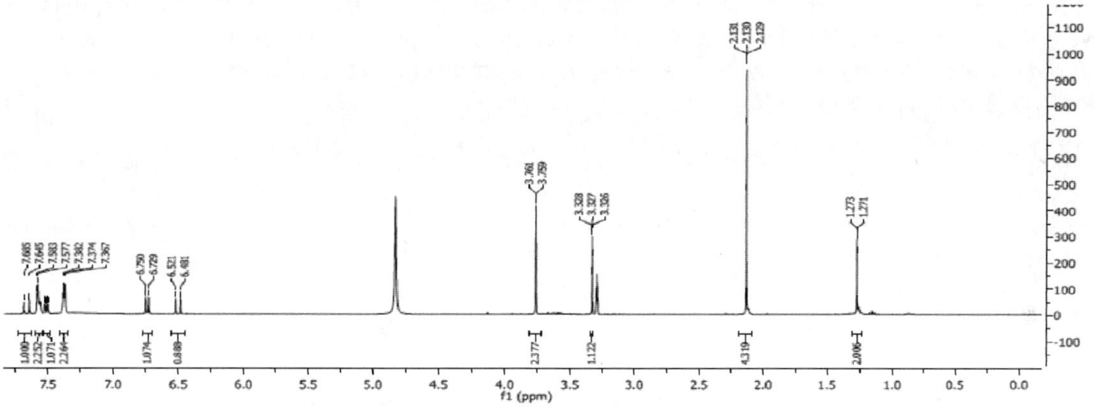

Figure 1. ^{1}H NMR spectrum of compound **5**.

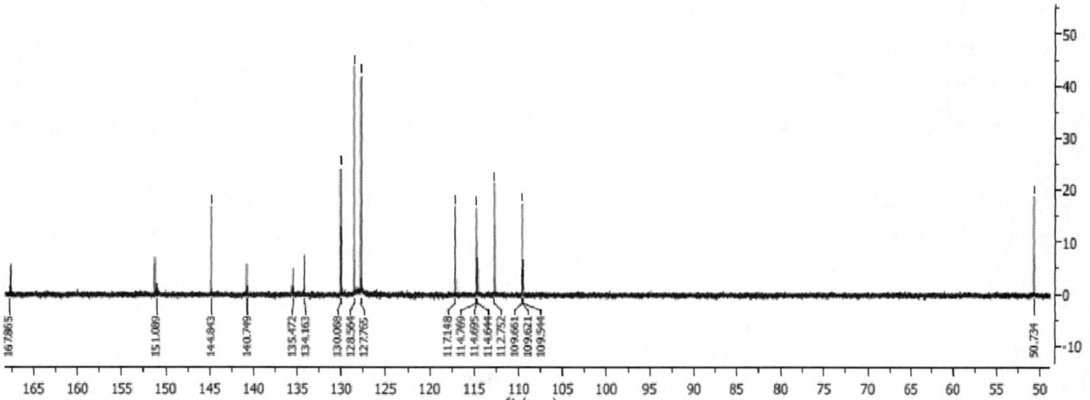

Figure 2. ^{13}C NMR spectrum of compound **5**.

In contrast to the above reaction, the reaction between methyl cinnamate **3** with 4-nitroaniline **6** and 2,4-dinitroaniline **7** were proved unsuccessful in producing desired products. The reaction between compound **3** and 4-nitroaniline **6** resulted in the formation of cinnamic acid **2**, whereas the starting material **3** and 2-4-dinitroaniline **7** were found by TLC to remain unreacted in the reaction of compound **3** and 2-4-dinitroaniline **7**. The difficulty encountered in the reaction of compound **3** with 4-nitroaniline **6** was unexpected as the previously described compound **5** was readily obtained from compounds **3** and **4** in a one-step reaction (Scheme 1). The difficulty in the reaction between compound **3** and 2,4 dinitroaniline **7** can possibly be explained by the desired product offering steric hindrance which could prevent the addition reaction from occurring.

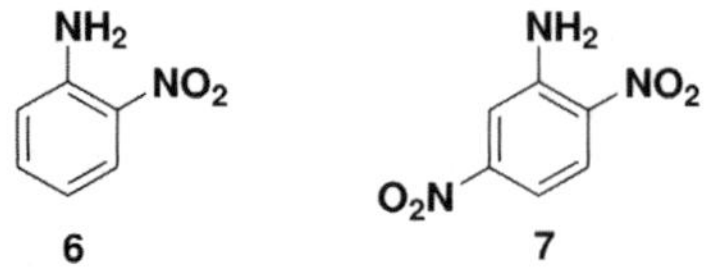

4. Conclusion

The addition reaction occurred between methyl cinnamate and 2-amino-4-nitrophenol. The reaction was conducted in 3 step reactions to give the final product from cinnamaldehyde as starting material. Firstly, cinnamaldehyde was converted into cinnamic acid, which was then esterified with methanol to have methyl cinnamate. The last step was the addition reaction between methyl cinnamate and 2-amino-4-nitrophenol to give methyl-3-(2-hidroksi-5-nitrophenyl amino)-3-phenylpropanoate as a yellowish brown gel in 68% yield.

Acknowledgments

The authors acknowledge their thanks to Sebelas Maret University for the financial support of this research in the form of Penelitian Unggulan Perguruan Tinggi (PUPT) Grant 2015 with contract number: 339/UN27.11/PL/2015.

References

[1] Seidemannt J 2005 Cinnamon-Lauraceae, Heidelberg: Springer-Verlag, 100-104
[2] Yan-Qua L, De-Xin K and Hong W 2013 *Ind.Crop. Prod.* **41** 269-278
[3] Singh G, Maurya S, Lampasona M P and Cantalan C A N 2007 *Food Chem. Toxicol.* 45 1650-1661
[4] Singh K and Mathela C S 2014 *Indian J. Chem.* **53B** 907-912
[5] Pontiki E, Hadjipavlou-Litina D, Litinas K and Geromichalos G 2014 *Molecules* **3** **19** 9655-9674
[6] Godoy M E, Rotelli A, Pelzer L and Tonn C E 2000 *Molecules* **5** 547-548
[7] Bairwa R, Kakwani M, Tiwari N R, Lalchandani J, Ray M K, Rajan M G R and Degani M S 2010 *Bioorg. Med. Chem. Lett.* **20** **5** 1623-1625
[8] Carvalho S A, da Silva E F, de Souza M V N, Lourenço M C S and Vicente F R 2008 *Bioorg. Med. Chem. Lett.* **18** **2** 538-541
[9] De P, Baltas M and Bedos-Belval F 2011 *Curr. Med. Chem.* **18** **11** 1672-1703
[10] De P, Yoya G K, Constant P, Bedos-Belval F, Duran H, Saffon N, Daffé M and Baltas M 2011 *J. Med. Chem.* **54** **5** 1449-1461
[11] Gainsborough, H 1928 *Lancet* **211** 5462 908-909
[12] Kakwani M K, Suryavanshi P, Ray M, Rajan M G R, Majee S, Samad A, Devarajan P and Degani M S 2011 *Bioorg. Med. Chem. Lett.* **21** **7** 1997-1999
[13] Yoya G K, Bedos-Belval F, Constant P, Duran H, Daffé M and Baltas M 2009 *Bioorg. Med. Chem. Lett.* **19** **2** 341-343
[14] Kankeaw U and Masong E 2015 *Int. J. Chem. Eng. App.* **6** **2** 91-94
[15] Peralta-Domínguez D, Rodríguez M, Ramos-Ortíz G, Maldonado J L, Meneses-Nava M A, Barbosa-García O, Santillan R and Farfán N 2015 *Sensors Actuator B* **207** 511–517

[16] Suryanti V, Bhadbhade M, Chawla H M, Howe E, Thordarson P, Black D Stc, Kumar N 2014 *Spectrochim. Acta A* **121** 662–669
[17] Suryanti V, Bhadbhade M, Bishop R, Black D StC and Kumar N 2012 *CrystEngComm* 14 7345-7354
[18] Suryanti V, Bhadbhade M, Bishop R, Black D StC and Kumar N 2013 *Tetrahedron* **69** 8446-8455
[19] Suryanti V, Condie G C, Bhadbhade M, Bishop R, Black D StC and Kumar N 2013 *Aust. J. Chem.* **67** 1270-1278

IOP Conf. Series: Materials Science and Engineering **107** (2016) 012063 doi:10.1088/1757-899X/107/1/012063

Isolation and identification of curcumin and bisacurone from rhizome extract of temu glenyeh (*Curcuma soloensis.* Val)

Rista A Vitasari, Fajar R Wibowo, Soerya D Marliyana and M Widyo Wartono*

Medicinal Chemistry Research Group, Chemistry Department, Faculty of Mathematic and Science, Sebelas Maret University, Jl. Ir. Sutami 36 A, Surakarta, Central Java, Indonesia 57126

*E-mail: widyo@staff.uns.ac.id

Abstract. Temu glenyeh (*Curcuma soloensis*. Val) is one of the medicinal plants that grow in Surakarta. This plant is similar with *C. longa* and *C. Xanthoriza*. Chemical constituents from an extract of the plant have never been studied. In this paper, we report the isolation of a terpenoid and curcumin from the rhizome of *C. soloensis*. The isolation was employed by soxhlet apparatus using acetone as solvent. The fractionation and purification of the compound from the acetone extracts were undertaken by vacuum liquid chromatography and flash chromatography. Identification of compounds used spectroscopy methods, such as FTIR, NMR (^{1}H NMR, ^{13}C NMR, COSY, HSQC and HMBC) and GC-MS. Isolated compounds were identified as curcumin (**1**) and bisacurone (**2**).

1. Introduction

Indonesian society has used various types of plants as medicine, and prevention of various diseases. Medicinal plants are generally based on inheritance or experience without knowing the exact chemical content. Many plants used as drugs in Indonesia are the member of Zingiberaceae family, such as ginger, turmeric, kencur, white ginger, temu ireng, temu glenyeh, temu lawak, lempuyang, and bengle. Plants of this family empirically use to treat stomach pains, carminative, diarrhea, rheumatism, asthma, inflammation, intestinal worms, scabies, and ulcers [1].

One of the medicinal plants, temu glenyeh or temu tis (*Curcuma soloensis* syn. *C. purpura scenas*) has the potential for drug plants. Chemical components of this plant have been identified but limited, only from essential oils of rhizome that contains terpenoids [2-5]. Essential oils of rhizome temu glenyeh have inhibitory against three strains of bacteria such as *Staphylococcus aureus*, *S. epidermis*, and *Streptococcus haemolyticus* [6]. While research on the n-hexane extract, extract of methylene chloride and ethyl acetate extracts of rhizome temu glenyeh, showed that extracts have antifungal activity against *Candida albicans*. Tested showed phytochemical compounds of the extract contain terpenoids. The highest antifungal activity possessed by the ethyl acetate extract, which also contain phenols compounds [7]. Temu glenyeh research largely focused on essential oils, while the isolation of compounds in the extract of temu glenyeh has not been studied. Therefore, in this study, we report the isolation and identification of curcumin (**1**) and bisacurone (**2**) from acetone extract of temu glenyeh rhizome.

2. Experimental detail

The rhizome of Temu glenyeh (*C. soloensis*) was collected from Boyolali district, Central Java and identified by staff Department of Biology, Universitas Sebelas Maret, Surakarta. Rhizome was crushed and dried to be a powder. Extraction of material used Soxhlet extractor apparatus and acetone as solvent. Isolation of compound from acetone extract used chromatography methods such as vacuum liquid chromatography (VLC) and flash chromatography. Stationary phase used silica gel G (VLC) and silica gel 60 (0.04 to 0.063 mm) 230-400 mesh ASTM (flash chromatography), all process were guided by thin layer chromatography (Si-gel f_{254}) and detected by UV 254 nm and cerium (IV) sulfate 2% as spray reagent. Spectrophotometer Fourier Transform Infra Red (FTIR) used Shimadzu Prestige-21, NMR spectroscopy by Agilent VNMR 400 MHz ($CDCl_3$ and TMS as standard), mass spectra by GC-MS QP-2010S SHIMADZU with AGILENT DB-1 column.

Dry powder rhizome of *Curcuma soloensis* (1.16 kg) was extracted using acetone (2.2 L) by Soxhlet apparatus. The filtrate then evaporated to resulting acetone extract (220 g) and fractionated using VLC and eluted with *n*-hexane : ethyl acetate ratio (9:1); (8:2) (2x); (7:3) (4x); (6:4) (4x); (5:5) (2x) and (4:6) (2x), (@150 mL) to obtain 17 fractions (figure 1), namely fraction A (0.533 g), B (0.976 g), C (0.501 g), D (0.522 g), E (0.822 g), F (0.276 g), G (0.212 g), H (0.183 g), I (0.131 g), J (0.202 g), K (0.385 g), L (0.403 g), M (0.85 g), N (0.937 g), O (0.695 g), P (0.397 g), and Q (0.752 g). Fraction M is bright orange powder, and analysis of the compound gives curcumin (**1**). Fraction P and Q then purified using flash chromatography using *n*-heksane: ethyl acetate ratio (7:3) (50 mL); (6:4) (150 mL); and (1:1) (100 mL), and gives 37 fraction and from fraction 29 – 37 gives compound **2** (108 mg). Compound **2** (bisacurone): viscouse oil, IR (KBr) v_{max} cm^{-1}: 3385, 2960, 2930, 1678, 1600; EI-MS (70 ev): 216 (M^+-$2H_2O$), 201, 173, 159, 145, 132, 119, 105, 91, 83(100), 65, 55, 39; ^1H and ^{13}C NMR see table 2.

3. Results and Discussion

Fraction M is bright orange powder, insoluble in water but soluble in acetone and ethanol. ^1H NMR spectra indicated similarities with previous research that identified as curcumin [8, 9]. Comparison of ^1H NMR spectra of isolated compound and curcumin is shown in table 1[8].

Table 1. Comparison of δH NMR of isolated compound and curcumin

No	δ H (ppm) (∑H, m, *J*)	
	Curcumin*	Isolated compound (1)
4, 4'	7.59 (2H, *d*, *J*=15.8)	7.60 (2H, *dd*, *J*=5.88; 15.56)
6, 6'	7.15 (2H, *d*, *J*=2)	7.12 (2H, *dd*, *J*=1.76; 8.08)
10, 10'	7.08 (2H, *dd* , *J*=1.8; 9)	7.05 (2H, *d*, *J*=1.6)
9, 9'	6.93 (2H, *d*, *J*=8.2)	6.92 (2H, *d*, *J*=8.2)
3, 3'	6.48 (2H, *d*, *J*=15.8)	6.48 (2H, *dd*, *J*=3.56; 15.72)
1	5.81 (1H, *s*)	5.86 (1H, *s*)
7, 7'- 2 OMe	3.97; 3.93 (6H, *s*, *s*)	3.95 (6H, *s*)

^{13}C NMR spectra of isolated 2 showed a peak of 15 carbon atoms that indicated a sesquiterpene. Carbonyl atom showed at the peak of 200.73 ppm. Signals of four carbons alkenes appear on the δC area from 123.94 to 155.78 ppm. The area of δC 48.72 to 16.93 ppm is carbons alkane comprising of two methine (CH), two methylenes (CH_2), and four methyls (CH_3), while the peak of δC 70.54 and 73.36 ppm indicated as two carbon atoms that binding to Oxygen (C-O).

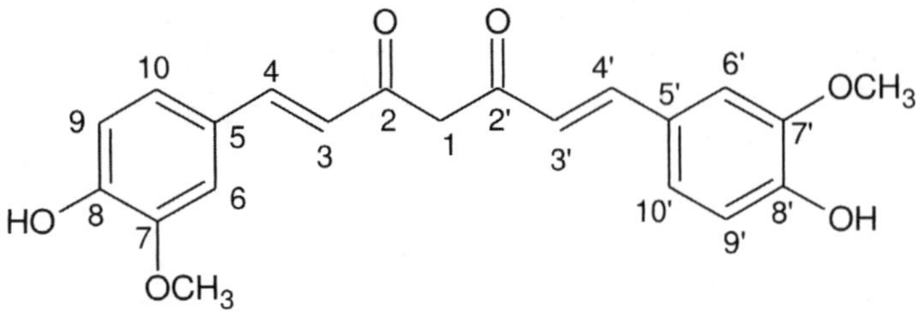

1

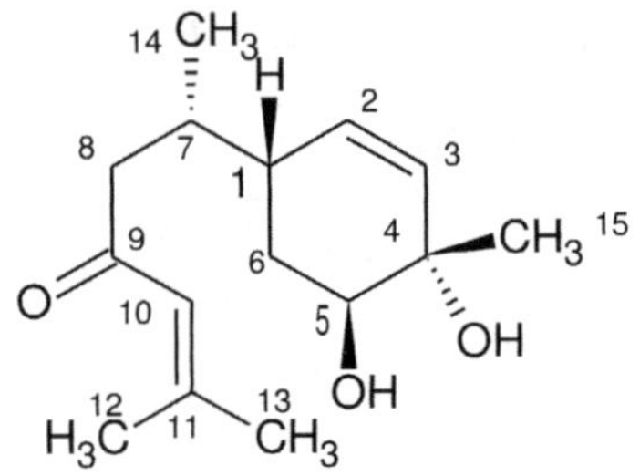

2

^{1}H NMR of compound **2** indicated 14 proton signals. Three signals derived from alkene proton (CH) at δ_H 5.64 (s, 1H), 5.64 (s, 1H) and 6.05 (s, 1H). This shows that there are no aromatic groups. Methine proton usually found on isoprene. It is supported by the presence of four methyls at 0.88 (d, 3H), 1.30 (s, 3H), 1.88 (s, 3H), and 2.12 (s, 3H). Another methine protons were shown at 2.19 (m, 1H), 2.29 (m, 1H) and 3.79 (dd, 1H, J = 3.32; 7.24 Hz). Proton signal methyl indicated at δ_H 0.88 (d, 3H, 6.6 Hz). For methylene protons themselves are shown in δ_H 1.72 (m), 1.82 (ddd, 1H, J = 3.36; 6.4; 13.8 Hz), 2.25 (m, 1H), 2.45 (*dd*, 1H, J = 4.96; 14.92 Hz).

The relationship between carbon with a proton in the bond, it can be seen from the HSQC spectra. There is three carbon signal that has no correlation with proton signals; it is possible a quaternary carbon. COSY spectra can be used to determine the correlation between proton-proton neighbors. The relationship between proton and carbon within 2-3 ties can be seen from the data HMBC. Correlation HSQC, HMBC, and COSY are shown in table 2.

Table 2. ^{1}H NMR (400 MHz), ^{13}C NMR (100 MHz, $CDCl_3$) and HSQC, HMBC, dan COSY Correlation of compound 2.

No. C	δC (ppm)	δH (ppm)(m, *J* Hz) (HSQC corr)	HMBC corr.	COSY
1	36.51	2.29 (1H, *m*)	C7, C14	H7, H2
2	132.00	5.64 (1H, *s*)	C1, C4	H1
3	132.30	5.64 (1H, *s*)	C1, C4	-
4	70.54	-	-	-
5	73.36	3.79 (1H, *dd, J*=7.24; 3.32)	-	-
6a		1.72 (1H, *m*)	C7, C1, C4, C5, C2, C3	H1, H5
6b	27.86	1.82 (1H, *ddd, J*=13.8; 6.4; 3.36)	-	H1, H5
7	33.12	2.19 (1H, *m*)	C1, C8	H14, H8b
8a		2.25 (1H, *m*)	C14, C7	
8b	48.72	2.45 (1H, *dd, J*=4.96; 14.92)	C14, C7, C9	H7
9	200.73	-	-	-
10	123.94	6.05 (1H, *s*)	C12, C13, C9	H13, H12
11	155.78	-	-	-
12	20.78	2.12 (3H, *s*)	C13, C10, C11	H10
13	27.70	1.88 (3H, *s*)	C12, C10, C11	H10
14	16.93	0.88 (3H, *d, J*=6.6)	C7, C1, C8	-
15	23.82	1.30 (3H, *s*)	C4, C5, C3	-

FTIR spectra (figure 1) of isolated compound 2 showed several functional groups, the presence of hydroxyl group (3385 cm^{-1}), Carbonyl (1678 cm^{-1}) and C-O at 1035 cm^{-1}. NMR spectra of isolated 2 suggested the compound have 15 C atoms, 24 H atoms and three O atoms with the molecular formula $C_{15}H_{24}O_3$ (MW = 252). However, mass spectra of compound 2 (figure 1) showed peak maximum at 216 that suggested molecular formula $C_{15}H_{20}O$. This fact indicated that compound **2** release two water molecules from chromatography heating process before ionization in the mass spectrometry (figure 2).

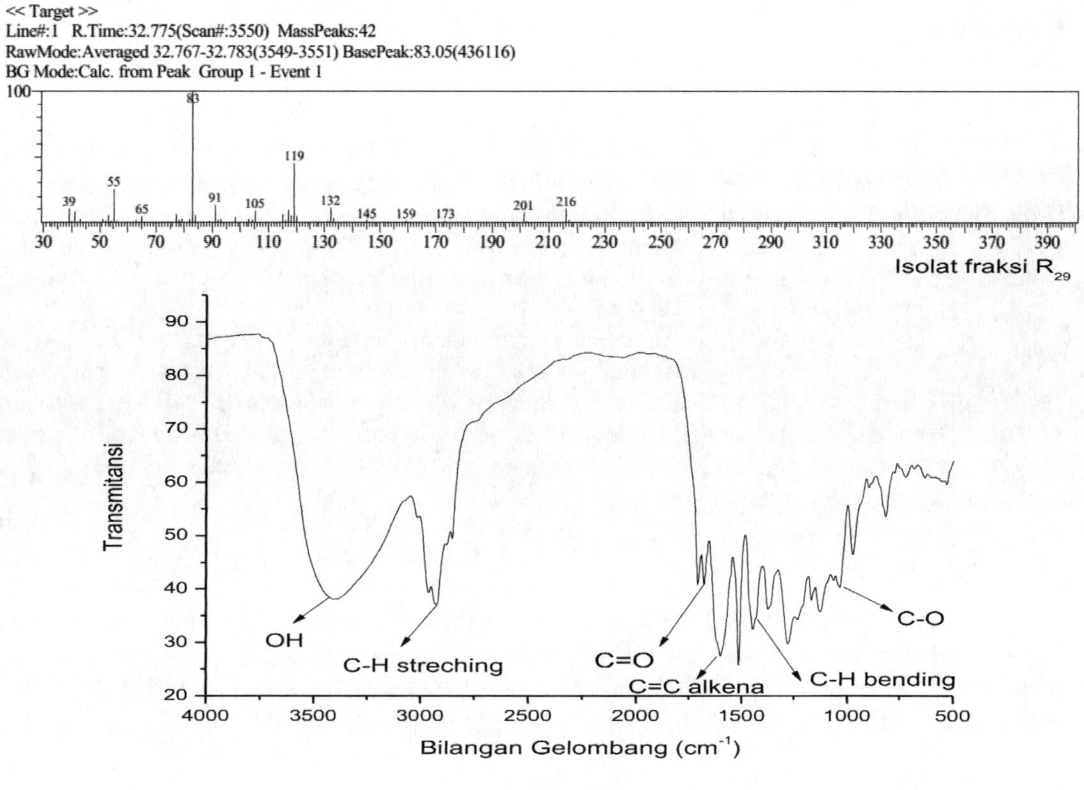

Figure 1. Mass spectra and FTIR spectra of compound **2.**

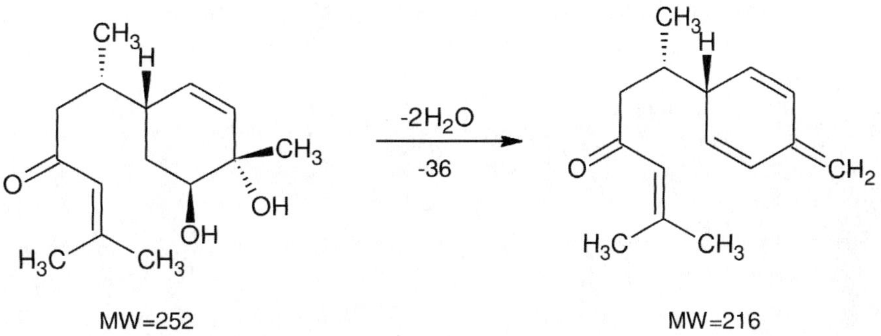

Figure 2. The reaction of bisacurone in the GC column under high temperature before ionization in the mass spectroscopy.

4. Conclusion

Based on data analysis FTIR, ^{1}H NMR, ^{13}C NMR, HSQC, HMBC, and COSY, suggested structure compounds **2** are derivatives of bisabolane terpenoids with molecular formula $C_{15}H_{24}O_3$ (MW = 252), A known compound that is bisacurone. Compound **2** previously has been isolated from the rhizome of the rhizome of *Curcuma xanthorhiza* [10] and *Curcuma longa* [11]. Comparison of the spectra suggested that the isolated compound **2** is bisacurone.

Acknowledgement

We thank LPPM UNS for providing us MRG grant for supporting this research.

References

[1] Heyne K 1987 *Tumbuhan Berguna Indonesia I* Penerjemah: Badan Penelitian dan Pengembangan Kehutanan Yayasan Sarana Wahajaya Jakarta

[2] Astuti P 2011 *Isolas, Identifikasi dan Uji Aktivitas Antibakteri Minyak Atsiri Rimpang Temu Glenyeh (Curcuma soloensis Val.) dari Wonogiri* Bachelor Thesis Universitas Sebelas Maret Surakarta

[3] Hong S L, Lee G S, Rahman S N S A, Hamdi O A A, Awang K, Nugroho N A, and Malek S N A 2011 Essential Oil Content of the Rhizome of *Curcuma purpurascens* Bl. (Temu Tis) and Its Antiproliferative Effect on Selected Human Carcinoma Cell Lines *The Scientific World Journal* **2014** pp.1-7

[4] Setyawan A D 2003 Keanekaragaman Kandungan Minyak Atsiri Rimpang Temu-temuan (Curcuma)/Diversity of essential oils constituent of Curcuma *Biofarmasi* **1** (2) pp. 44-49

[5] Shibuya H, Yoshihara M, Kitano E, Nagasawa M and Kitagawa I 1999 Qualitative and Quantitative Analysis of Essential Oil Constituents in Various Zedoariae Rhizoma (Gajutsu) by Gas Liquid Chromatography-Mass Spectrometry *Journal of the Pharmaceutical Society of Japan* **3** pp. 212-6

[6] Murningsih T, Rezeki S, Chairul H, Priyono S and Taufiq A 2000 Komposisi kimia dan Uji antibakteri minyak Atsiri rimpang Temu Glenyeh (*Curcuma soloensis* Val.) *Warta AKAB* **12** pp 37-45

[7] Suryanti V, Udin Z, and Harliana D 2006 Aktivitas Antijamur Ekstrak Rimpang Temu Glenyeh (*Curcuma soloensis Val.) Journal Alchemy* **5** pp. 31-38

[8] Rao E V, and Sudheer P 2011 Revisiting Curcumin Chemistry Part I: A New Strategy for The Synthesis of Curcuminoids, *Indian Journal of Pharmaceutical Sciences* **73** pp. 262-270

[9] Peng S F, Lee C Y, Hour M J, Tsai S C, Kuo D H, Chen F A, Shieh P C, and Yang J S 2014 Curcumin-loaded Nanoparticles Enhance Apoptotic Cell Death of U2OS Human Osteosarcoma Cells Through the Akt-Bad Signaling Pathway *Internstional Journal of Oncology* **44** pp. 238-246

[10] Uehara S, Yasuda I, Takeya K, and Itokawa H 1989 New Bisabolane Sesquiterpenoids from The Rhizomes of *Curcuma xanthorriza (Zingiberaceae) Chemistry Pharmacy Bulletin.* **37** pp. 237-240

[11] Oshiro M, Kuroyanagi M, and Ueno A 1990 Structures of Sesquiterpenes from Curcuma longa *Phytochemistry* **29** pp. 2201-2205

Physical characteristics of cinnamon oil microcapsule

R F Hermanto[1], L U Khasanah[1*], Kawiji[1], W Atmaka[1], G J Manuhara[1] and R Utami[1]

[1]Faculty of Food Science and Technology, Sebelas Maret University, Surakarta, Indonesia

*E-mail: liaumikhasanah@yahoo.co.id

Abstract. Cinnamon (*Cinnamomum burmanii*) oil products can be obtained from the bark by steam distillation. Essential oils are susceptible to high temperatures, oxidation, UV light, and humidity. Microencapsulation may change essential oils into powder, protect the sensitive core material and reduce the amount of flavor which lost during storage. In the microencapsulation, one of the important factors is the type of coating agent. The objective of this work was to characterize the cinnamon oil microcapsule. Ratio variations of coating agent maltodextrin and gum arabic were (1:0); (0:1); (1:1); (2:3). Physical characteristics such as water content, solubility, bulk density, surface oil, and microencapsulation efficiency of samples were investigated. Results showed that the ratio variations of the coating agent significantly affected the water content, bulk density, surface oil and microencapsulation efficiency but significantly affected the water solubility. Characteristics of selected microcapsule were 6.13% water content; 96.33% solubility; 0.46 g/cm^3 bulk density; 2.68% surface oil; 70.68% microencapsulation efficiency and microstructures were rather good.Introduction

1. Introduction

There are 54 species of cinnamon (*Cinnamomum sp.*) in the world, and 12 of them are located in Indonesia. *Cinnamomum Burman Nii* (from Indonesia) which also known as cassia vera, *Cinnamomum zeylanicum* (from Sri Lanka and Seychelles) and *Cinnamomum cassia* (from China) are types of cinnamon with the best demand in the world market. They are also the most famous spice plants. *Cinnamomum Burman Nii* type Blume or cassia vera is a traditional export product that is still controlled by Indonesia as the main exporting countries in the world [1]. In the international trade, cinnamon tree products are marketed in the form of bark, essential oils, and oleoresin. Volatile oil products can be obtained from the bark, twigs, branches, or leaves of the cinnamon tree by distillation. Steam distillation produced a higher yield. Previous research [2] reported that the yield of cinnamon essential oil by steam treatment was 0.66%.

Essential oil of cinnamon has a clear yellow to brownish color and cinnamaldehyde as the main component. Essential oil of cinnamon has been widely used as an ingredient in cosmetics, perfumes, foods and beverages flavor, used as antiseptic and antimicrobial. However, essential oils are susceptible to high temperature, oxidation, UV light, and humidity. Oxidative damage caused an unpleasant flavor and also

IOP Conf. Series: Materials Science and Engineering **107** (2016) 012064 doi:10.1088/1757-899X/107/1/012064

decreased both shelf life stability and sensory properties [3]. Microencapsulation is one of the techniques that can be used to protect the core material such as essential oil by changed the materials from a liquid into solid form. The products became easy to handle, and the core material can be protected from losing flavor [4].

In the microencapsulation process, coating agent affected the quality. Maltodextrin and gum arabic are a coating agents that often used in microencapsulation. Maltodextrin was able to protect the main compounds in ginger oil during the encapsulation process. Previous research reported the production of ginger oil microcapsule and ginger oleoresin microcapsule using maltodextrin as a coating agent. Sample with a ratio of essential oil and maltodextrin in 1:25 produced the best characteristics. However, due to low viscosity emulsion during the spray drying process made the skin layer weak and caused the core material to be less protected [5].

Gum arabic has unique properties compared to the other types of gum. With a concentration of 40-50%, gum arabic does not produce too high viscosity and can create a perfect stability. Gum arabic is an effective emulsifying agent due to protective colloid ability [6]. In soybean oil emulsification and microencapsulation using gum arabic study [7], the particle size ranged from 9-17 μm and the microencapsulation efficiency decreased from 100 to 48% by the addition of oil. Essential oil addition less than 50% would produce a good solubility in water powder. Maltodextrin and gum arabic in red palm oil microcapsules showed the best formulation with a ratio of maltodextrin and gum arabic and oil in 2: 3: 2. The water content of microcapsule was 0.92%, carotenoids 82.63 ppm and solubility in water 66.82% [8]. Therefore, a combination of coating agents such as maltodextrin and gum arabic could produce microcapsule with good characteristics. The objective of this work was to characterize the cinnamon oil microcapsule that contained maltodextrin and gum arabic as coating agents.

2. Method

2.1 Essential Cinnamon Oil Preparation

The bark of a cinnamon tree (Cinnamomum Burman Nii) from Bubakan, Girimarto, Wonogiri (Indonesia) was air dried for 5 days until the water content decreased less than 12%. Dried cinnamon bark was milled and sieved 10 mesh. The powder cinnamon was steam distillation for 4 hours.

2.2. Microemulsion

The microemulsion is made with a ratio of essential oils and coating as 1:25. The formula is then dissolved in distilled water with a ratio of coating material and distilled water as 1:20. Meanwhile, the variation ratio of coating material maltodextrin (DE 10-15 with Associated British Budi brand from Bratachem Chemical Store Yogyakarta) and gum arabic (powder Type 4687 from Bratachem Chemical Store Surakarta) is (1:0); (0:1); (1:1) and (2:3). Homogenization process is done using a homogenizer (Ultra-Turax) for 15 minutes at 8000 rpm.

2.3. Microencapsulation by spray drying

Microencapsulation is done by a spray dryer (Spray Dryer, Basic Lab Plant SD) with an inlet temperature of 109°C.

2.4. Emulsion characterization

2.4.1. Emulsion stability

Immediately after the emulsion preparation, 100 mL of samples were transferred gradually to 100 ml cylinders, sealed, stored at room temperature for 24 hours. After 24 hours, the volume of the upper phase measured. The stability was measured by % of separation and expressed as:

IOP Conf. Series: Materials Science and Engineering **107** (2016) 012064 doi:10.1088/1757-899X/107/1/012064

% Separation = [H1 / (H0) x 100%] where Ho represents the initial emulsion height and H1 is the phase upper height.

2.4.2. Emulsion droplet size
The droplet size distribution was measured using a microscope optic lab viewer. A small sample was added in the preparation and observed under the objective lens.

2.5. Powders analysis
2.5.1 Moisture Content
Analysis of water content was conducted using a thermogravimetric method by weighing the sample in the form of powder / material which has been refined as much as 1-2 grams in the weighing bottle. Then, dried the refined material in an oven with a temperature of 100-105°C for 3-5 hours, cooled in a desiccator and weighed. Pre-heat oven for 30 minutes longer cooled and weighed. This treatment was repeated until a constant weight [9].

2.5.2 Solubility
Solubility measurement was calculated based on the weight percentage of residues that could not go through a filter paper to the weight of sample used. Solubility analysis was performed by gravimetric method. Samples (0.75 g) was diluted in 100 ml of distilled water and filtered by vacuum filtration. Before the next step, the filter papers were dried in a 105°C oven for 30 minutes and weighed. After the filtration process, the filter papers along with the residues were dried in an oven at 105°C for approximately 3 hours, cooled in a desiccator for 15 minutes and weighed [10].

2.5.3. Surface oil and microencapsulation efficiency
Surface oil was calculated using the gravimetric method. Fifteen milliliters of hexane was added to 2 g of powder in a test tube, processed in the vortex for 2 minutes at room temperature. The mixture was filtered using Whatman No 1 filter paper and washed 3 times using 20 ml of hexane. After filtering and washing, both residue and filter paper were dried at 60°C until constant weight. The content of surface oil was calculated as a weight percentage difference of the powder, before and after the extraction and washing process. Microencapsulation efficiency (ME) was calculated using the formula: ME = [(Total oil - Surface oil) / Total oil x 100] [11].

2.5.4. Bulk density
Bulk density was determined by the method of tapping. 2 grams of powder was put in a 10 ml cylinder. A cylinder containing the powder knocked using a tap density to a constant volume. Bulk density was calculated by dividing the sample weight by volume [12].

2.5.5. Analysis by SEM
Microstructural observations on microcapsules made with SEM (JEOL JSM-type T-300). Samples were prepared by attaching little microcapsules on a specimen that has been coated with insulating conductive. Furthermore, the specimen coated with gold as filling material by Fine Coat Ion Sputter JFC models 1100. After sampling, the gold coated specimen was included in the column injector to be fired with electrons. The electron will be reflected back by the sample and is referred to as secondary electron. The secondary electron detector will show in the form of images [13].

2.6. Statistical Analysis
SPSS (SPSS 16.0 for Windows) statistical software was used to calculate the analysis of variance (ANOVA). Duncan Multiple Range Test (DMRT) was used for the comparison of means, with significance assigned at $P<0.05$.

3. Results and discussion

3.1. Emulsion characterization
Emulsion stability was ranged from 98.18%-100%. The addition of gum arabic as coating material resulted in higher emulsion stability. Gum arabic is an effective emulsion thickening agent because of its ability to protect the colloid. This type of thickener is also heated resistant to the process that uses heat. Gum arabic has unique properties such as high solubility and low viscosity. The other characteristics of gum arabic were also the ability to form texture and film, binder ability and also as an emulsifying agent because of its protein component [14]. A mixture of gum arabic as an emulsifying agent in anise oleoresin showed the result that coating material emulsion provides the highest stability (90.12%) and a mixture of gum arabic maltodextrin coating material provides emulsion stability of 75.39% [15].

Microencapsulation is a process using a relatively thin coating on a small particles of solids or liquid droplets and the dispersion liquid, where the particle size ranged between 1-5000 μm. In the process of spray drying, the emulsion droplet size or diameter ranges between 1-100 μm [16].

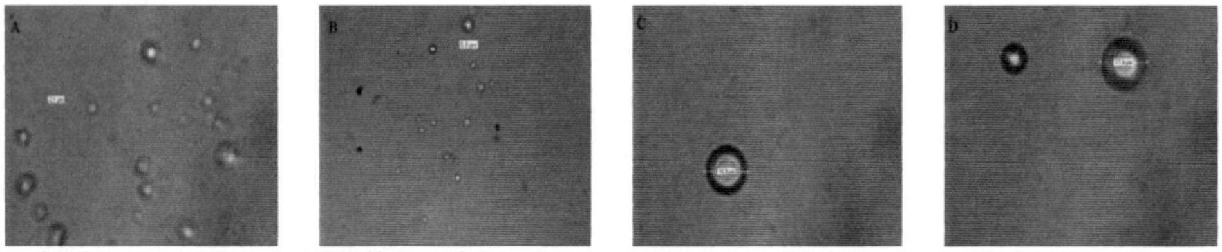

Figure 1. Cinnamon Essential Oil Emulsion (A) MD:GA 1:0, (B) MD:GA, (C) MD:GA 1:1, and (D) MD:GA 2: 3

The emulsion droplet size was ranging from 2.1 μm - 65.3 μm. The size of the microcapsules proved that the product will be made to microscopic size. Figure 1 showed that the core material essential oil of cinnamon is covered by the coating material. Homogenizing linseed oil with different wall materials can be combined to produce an emulsion with droplet diameter ranging from 0.6 to 26 μm [17].

3.2. Powder characterization

Table 1. Physical Properties of Microcapsule

Characteristic	Variation Ratio of Coating Agents (MD:GA)			
	(1:0)	(0:1)	(1:1)	(2:3)
Water Content (%)	5.691±0.14^a	7.715±0.61^b	6.502±0.44^a	6.134±0.24^a
Solubility (%)	97.002±1.42^a	91.393±6.55^a	97.467±0.39^a	96.333±0.69^a
Bulk density (g/ml)	0.426±0.001^a	0.491±0.011^c	0.453±0.006^b	0.456±0.007^b
Surface oil (%)	4.262±0.03^c	4.099±0.12^c	3.447±0.39^b	2.675±0.15^a
EM (%)	59.859±5.75^a	66.164±2.01^a	65.811±5.54^a	70.684±6.23^a

Water content is an important parameter that can affect the stability of the product during storage and distribution. The water content of the microcapsules was ranging between 5.691% - 7.715% (Table 1). The lowest water content (5.691%) was performed by the formula with a coating ratio of (MD: GA) (1:0), and the highest (7.715%) was showed by the formula with a coating ratio of (MD: GA) (0: 1). Those range of water content was a typical water content of the microcapsule products obtained from spray drying (2-6%) [18]. Microcapsules analysis with the same coating material showed a range of 5.9-8.1% [19]. The result indicated that the water content of cinnamon essential oil microcapsules was still in accordance with the typical water content in microcapsule products.

The addition of maltodextrin can reduce water content because maltodextrin has a low viscosity that made the solution could heat at high concentration, so water less absorbed and microcapsule were easily dried [20]. High viscosity indicates that the water trapped in the structure were high, so it will be difficult to be separated during the drying process. The increasing of maltodextrin has been used and the decreasing gum arabic has been used, decreased the viscosity. It can also be explained that if less water trapped inside the structure, it means that the water content was low. Previous research [8] reported that red palm oil microcapsule with the lowest water content (0.62%) was showed by the formula with a coating ratio of (MD: GA) and oil in (3:3): 2.

Solubility in water of microcapsule with gum arabic and maltodextrin as the coating material ranged from 94-98% [19]. The solubility value is determined by the ability of the coating material to soluble water. The solubility of cinnamon essential oil microcapsules ranged from 91.393% - 97.467% (Table 1). The highest solubility of MD: GA (1:1) coating material was equal to 97.467%. Maltodextrin is soluble in cold water perfectly so that it can release the flavor rapidly in certain applications [21]. Gum arabic also has a high solubility in water. So, a combination of both can produce microcapsule with high solubility in water [22]. The high solubility of maltodextrin will increase the solubility of microcapsule [23]. Compared to gum arabic that will form an irregular pattern of oil droplets that can affect the solubility, maltodextrin can improve solubility because it is more soluble in water. The formula of (MD:GA): Oil in (2:3):2 has a higher solubility (66.82%) due to the amount of maltodextrin (MD:GA):Oil(1:3):1. Besides, the size of droplet relatively more uniform or more stable when compared with the other formula, which is (MD: GA): Oil (3:3): 2 [8].

Bulk density analysis in the food industry is usually performed to help determine the capacity of the powder in packaging, storage and distribution process. Bulk density is the mass of the particle that is measured from a container per unit volume [23]. In this study, powder density ranged from 0.426 g/cm^3 - 0.491 g/cm^3 and increased gradually in line with the decline of maltodextrin (Table 1). Microcapsules with a coating ratio of MD: GA (1:0) showed the smallest bulk density (0.426 g/cm^3) while the other values are MD:GA (1:1) 0.453 g/cm^3; MD:GA (2:3) 0.456 g/cm^3. The largest bulk density (0.491 g/cm^3) performed by MD: GA (0:1). Small particle size would form a larger mass density as a result of a reduction in the cavities between the particles. High water levels will increase the weight of the material in the same container volume, and cause the increasing of bulk density [24].

Surface oil content is observed from the outer surface of the microcapsule wall. The existence of this oil is undesirable because it comes in contact with the surrounding environmental conditions so that it can be damaged. The percentage of surface oil is calculated by comparing the oil that is found in the surface with the oil that is purposely added so that it can also be seen from the total oil content.

The result showed that surface oil of microcapsules ranged from 2.675% - 4.262% (Table 1). The high addition of maltodextrin tends to increase the amount of oil encapsulated. Increasing the number of maltodextrins (a decrease in the amount of gum arabic) will increase the unencapsulated oil content. Maltodextrin is lipophobic so that it cannot bond with the oil molecules. In addition, the decreasing of gum arabic also reduced the oil coating ability of gum arabic as an emulsifier. Maltodextrin does not have lipophilic properties. Therefore, maltodextrin on lipid encapsulation by spray dryer method caused the emulsion stability and low oil retention, but the oil encapsulated has a resistance to oxidation [25]. Microcapsules with coating material combination of MD: GA (2:3) showed the lowest value of surface oil (2.675%). The higher oil on the surface (2.2%) resulted by GA compared to the combination of MD /GA (1.9%) [26].

Microencapsulation efficiency of cinnamon essential oil ranged between 59.841%-70.702% (Table 1). The highest efficiency in microcapsules was reached by a combination of coating material MD:GA (2:3). In the case of flavor encapsulation by spray drying, the retention of flavor depends on the ratio between maltodextrin and gum acacia with optimal retention at a ratio of 2/3 and 3/2 [27]. Besides, another study [26] showed that the GA coating material showed 90.5% efficiency while the coating material MD/GA generated microencapsulation efficiency of 91.4%. Increasing total content of maltodextrin in the emulsion coating material caused the decreasing of surface oil and increasing of microencapsulation efficiency although a recent study reported that increasing concentrations of total solids had little effect on the efficiency of microencapsulated powder [28]. Microcapsules with coating material MD:GA (1:0) showed the lowest microencapsulation efficiency (59.841%). Using various concentrations of maltodextrin and gum acacia in the microencapsulation of ethyl butyrate with the spray drying method, proved that the increasing concentration of maltodextrin (10-30%) was accompanied by an increase in the efficiency of the microencapsulation process [28].

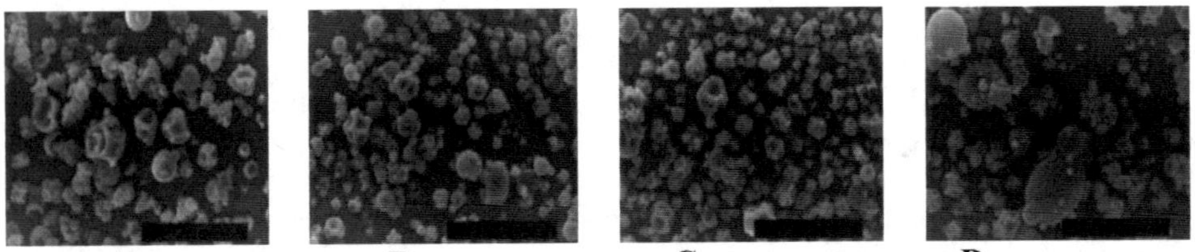

A B C D

Figure 2. Scanning Electron Microscopy (SEM) analysis of cinnamon essential oil microcapsules (A) MD:GA (1:0) with 1000x magnification; (B) MD:GA (0:1) with 1000x magnification; (C) MD:GA (1:1) with 1000x magnification; (D) MD:GA (2:3) with 1000x magnification

Figure 2 showed that in terms of size, although the microcapsules are relatively larger in figure D, but there is no significant different between the size of A, B and C microcapsules. The size of cinnamon essential oil microcapsule ranged from 1.92 - 30.8 μm. Particles of good quality, which size are less than 40 μm [29]. Characteristics of good microcapsules were thin wall thickness, large particle size and high capable of carrying components [30]. Microcapsules processed from spray drying which used polysaccharide as the coating material showed a typical surface. Those because the effects of coating agents composition, the process of atomization and drying, irregular wrinkles at the start of the drying process, and the effect of surface tension of the solution. Development of air or moisture in the dried particles can not also be perforated. The effect of this development is commonly called ballooning. Ballooning will affect the quality of the microcapsules. Ballooning was heavily influenced by the speed of drying and viscoelastic properties of the matrix coating material. Ballooning can also occur due to too high drying process pressure, thus made the hydrophobic interaction or also called as hydrophilic occurs between the carrier amphipathic, water, and substances that are encapsulated so that it would make it into the micelle. The small droplets of micelles formed (average diameter of 100 m) making it easily dried, forming a hard coating around the encapsulated substances [31].

4. Conclusion

The selected essential oil microcapsule was determined by looking at the results of the physical characteristics analysis. In bulk density, the best result was obtained by microcapsules with the coating material of MD:GA (0:1). In surface oil, the best results are microcapsules with coating material variation of MD:GA (2:3). While in microstructural parameter, good results are reported by microcapsules with coating material variations of MD: GA (0:1) and MD:GA (2:3) which also formed a spherical microstructure somewhat wrinkled. By some parameters from physical characteristics analysis, microcapsules with coating material variations of MD:GA (0:1) and MD:GA (2:3) have the same trend. However, oil surface was more important parameter due to an undesirable high oil surface could contact with the surrounding environmental conditions that can damage the microcapsule quality. Sample with the lowest content of surface oil was microcapsule with a coating material variation of MD:GA (2:3). In conclusion, this formulation was chosen as the best microcapsule. Characteristics of selected microcapsule were 6.13% water content; 96.33% solubility; 0.46 g/cm^3 bulk density; 2.68% surface oil; 70.68% microencapsulation efficiency and microstructures were rather good.

Acknowledgements

This work was financially supported by a research project of PNBP UNS No: 624/UN27.11/PL/2015 from Sebelas Maret University, Indonesia.

References

[1] Aprianto 2011 Cinnamon oleoresin extraction with ultrasonic in alcohol solvent *Chemical Engineering Magister Program of Diponegoro University* Semarang
[2] Khasanah L U, R Baskara K A, Rohula U, Dimas R A M and Mukhamad L 2013 Separation of cinnamon (Cinnamomum burmannii) essential oil compounds with fractionation distillation in various levels of distillation temperature *Procedia of Agribusiness National Seminar* November, 10th **2013**. p 353-359
[3] Velasco J, Dobarganes C, Marques-Ruiz G 2003 Variables affecting lipid oxidation in dried microencapsulated oils *Grass Activities* **54**: 304-314. DOI: 10.3989/gya 2003 V54.i3.246
[4] Soottitantawat A, H Yoshii, T Furuta, M Ohkawara, P Linko 2003 Microencapsulation by spray drying: influence of emulsion size on the retention of volatile compounds *Food engineering and physical properties* **68**. p 2256 - 2262
[5] Nurlaili F A 2011 Characterization of ginger essential oil microcapsule and ginger waste oleoresin (zingiber official var. rub rub) with maltodextrin coating *Post Harvest Technology Thesis of Agricultural Technology Faculty* Gadjah Mada University Yogyakarta
[6] Glicksman M 1983 *Food Hydrocolloids* Volume II CRC Press Inc Boca Raton Florida
[7] McNamee B F, E Dolores O, and Michael O'S 1998 Emulsification and microencapsulation properties of gum Arabic *J. Agric. Food Chem.* 1998 **46** 4551-4555
[8] Fasikhatun T 2010 Effect of Maltodextrin and Gum Arabic Concentration on The Characteristics of Red Palm Oil Microcapsule with Spray Drying Method *Agricultural Technology Thesis of Agricultural Technology Faculty* Bogor Agricultural University.
[9] AOAC. 1997. *Official Methods of Analysis*. Association of Official Analytical Chemistry.
[10] Fardiaz S 1992 *Food Microbiology 1* (Jakarta: PT Gramedia Pustaka Utama)
[11] Bae E K And S J Lee 2008 microencapsulation of avocado oil by spray drying using whey protein and maltodextrin. *Journal of Microencapsulation*, December 2008; **25**(8): 549–560

[12] Beristain, C I, H S Garcia and E J Vernon Carter 2001 Spray-dried Encapsulation of Cardamon (Elettaria cardamomum) Essential Oil with Mesquite (Prosopis juliflora) Gum. *Lebensm-Wiss u-Technol.*, **34**, 398-401

[13] Trindade, C S Favaro, A S Santana, E S Monterrey-Quintero, M A Trindade, F M Netto 2010 The use of spray drying technology to reduce the bitter taste of casein hydrolysate. *Food hydrocolloids*. Vol **24** 336-340

[14] Hui 1992 *Food additive* (Yogyakarta: UGM-Press)

[15] Chranioti C and Constantina T 2013 Arabic Gum Mixtures as Encapsulating Agents of Freeze-Dried Fennel Oleoresin Products. *Food Bioprocess Technol*

[16] Gharsallaoui A G R, Odile C, Andre´e V, and Re´mi S 2007 Applications of spray-drying in microencapsulation of food ingredients*: An overview. Food Research International* **40** (2007) 1107–1121

[17] Carneiro H C F, Renata V Tonon b, Carlos R F G, Miriam D H 2013 Encapsulation efficiency and oxidative stability of flaxseed oil microencapsulated by spray drying using different combinations of wall materials *Journal of Food Engineering* **115** (2013) 443–451

[18] Reineccius G A 2004 The spray drying of food flavors *Drying Technology* **22** (6):1289-1324

[19] Ahza A and Asep H S 1997 Microencapsulation of Peel Extract and Pulp of Lemon Fruit (Citrus aurantifolia swingle) and Its Application on Tea Bag. *Technology and Food Industry Bulletin* Vol. **VIII**. No. 2.

[20] Simanjuntak M 2007 *Optimation of Red Palm Oil Microcapsule Formulation with Maltodextrin, Gelatin, and Carboxymethyl Cellulose in Thin Layer Drying Process* Thesis Agricultural Technology Faculty (Bogor Agricultural University)

[21] Kenyon M and R J Anderson 1988 *Maltodextrin dan low-dextrose-equivalence corn syrup solids In Risch* Ed S J and G A Reineccius Flavour Encapsulation (Washington, D.C. : American Chemical Society) 7-10

[22] Setyaningsih D, Reni R, and Sugiyono 2007 Study of Vanilla Extract Microencapsulation *Journal of Agricultural Industry Technology* Vol. **19**(2), 64-70

[23] Srihari E, Farid S L , Rossa H, Helen W S 2010 *Effect of maltodextrin addition on powdered coconut milk manufacturing*. Chemical Engineering and Chemical Process Seminar, August 4th-5th **2010**. ISSN: 1411-4216

[24] Prabowo B 2010 Study of Physicochemical Characteristics of Yellow Millet Flour and Red Millet Flour. *Thesis. Post Harvest Technology* Sebelas Maret University Surakarta.

[25] Westing L and Rennecius F 1988 Shelf life of Storage Oil: Effect of Encapsulation by Spray drying, Extrusion, and Molecular Inclusion In *Flavor Encapsulation*; ACS Symposium Series **370** eds Risch, S J, Rennecius GA (Washington DC: *American Chemical. Society*)

[26] Gallardo, G, Leticia G, Vanina M, María C L, Dana B, Ramiro B, Ruth P I and Laura G H 2013 Microencapsulation of linseed oil by spray drying for functional food application *Food Research International* FRIN-04451

[27] Bhandari B R, E D Dumoulin, H M J Richard, I Noleau, A M Lembert 1992 Flavour Encapsulation by Spray Drying : Application to Citral and Linalyl Acetate *Journal of Food Science* Vol. **57**, No. 1, p. 217-221

[28] Domian E and Iwona W 2008 Microencapsulation of rapeseed oil based on the spray drying method. *polish journal of food and nutrition sciences*. Vol. **58**, No. 4, pp. 477-483

[29] Nedovic, Viktor Ana Kalusevica, Verica Manojlovicb, Steva Levica, Branko Bugarski 2011 *An overview of encapsulation technologies for food applications.* Procedia of Food Science **1** (2011) 1806 – 1815 11th International Congress on Engineering and Food (ICEF11)

[30] Finch C A 1993 *Industrial microencapsulation: polymers for encapsule walls* In Encapsulation and Controlled Release, Karsa, D R and Stephenson, R A (eds.) The Royal Society of Chemistry, Cambridge, pp. 1–17.

[31] Sheu T Y and M Rosenberg 1998 Microstructure of Microcapsules Consisting of Whey Proteins and Carbohydrates *Journal of Food Science* Volume **63**, No. 3, 1998.

IOP Conf. Series: Materials Science and Engineering **107** (2016) 012065 doi:10.1088/1757-899X/107/1/012065

O-methylation of natural phenolic compounds based on green chemistry using dimethyl carbonate

N I Prakoso[1*], P H Pangestu[2] and T D Wahyuningsih[2]
[1]Chemistry Department, Islamic University of Indonesia, Jl. Kaliurang Km. 14.5, Sleman, Daerah Istimewa Yogyakarta 55584, Indonesia
[2]Chemistry Department, Gadjah Mada University, Bulaksumur, Daerah Istimewa Yogyakarta 55281, Indonesia

[*]E-mail: nurcahyo.ip@uii.ac.id

Abstract. The alkyl aryl ether compounds, of which methyl eugenol and veratraldehyde are the simplest intermediates can be synthesized by reacting eugenol and vanillin with the green reagent dimethyl carbonate (DMC). The reaction was carried out under mild of temperature and pressure. Excellent yields and selective products were obtained (95-96%) after a few hours. In the end of the reaction, the catalysts (base and Phase Transfer Catalyst) can be recovered and regenerated.

1. Introduction

The Green Chemistry principle [1] is a concept that aims to prevent pollution at the source, starting raw materials, synthesis procedure, the design process and the finished product before potential pollutants. It is important to replace the reagents that are toxic and harmful to the reagents that are safe and do not produce byproducts such as inorganic salts.

It is known that methylation reagents such as dimethyl sulfate or methyl halides are hazardous compounds. Dimethyl carbonate (DMC) is an alternative reagent that can be used in the methylation reaction to replace dimethyl sulfate or methyl halides [2, 3, 4]. In preparation of fragrances, dyes, and pesticides, or antioxidants in oils and fats or stabilizers of plastics, alkyl aryl ether compounds are needed [5, 6, 7].

Methylation of phenol with DMC performed in an autoclave at 160 °C using a base catalyst has been patented by Iori and Romano [8]. Several kinds of catalysts such as amines and a tertiary phosphine, penta alkilguanidin and cesium carbonate were used by Barcello *et al.*, Lee and Shimizu [9, 10] in the reaction of phenolic compounds with dialkyl carbonates at temperatures between 120 to 200 °C with the autogenous pressure system. Bomben *et. al.* [11] reported on the use of polyethylene glycol as solid-gas phase transfer and potassium carbonate as a base at high temperature. In this research, selective methylation at the hydroxy group in high yield was selected. However, the drawback of the procedure is hard to use because of the high boiling point is needed.

Ouk *et al.* [12] reported the methylation reaction with milder conditions where the methylation with DMC performed on the system solid-liquid phase transfer and the reaction was carried out at a temperature of 90-100 °C at atmospheric pressure. This reaction was used potassium carbonate as a base and tertbutyl amoniumbromida (TBAB) as PTC to give the best results on the methylation of *p*-cresol after a reaction time of 5 hours with a yield of 99%.

Eugenol and vanillin are a natural phenolic compound which are relatively abundant in Indonesia. As phenolic compounds, eugenol and vanillin can be converted into other compounds that are more useful with through the methylation reaction of eugenol to methyl eugenol and vanillin to veratraldehyde [13,14]. Research on the methylation of eugenol and vanillin have been reported [15]. Methylation reaction performed using DMS and PTC N- cetyl-N, N, N-trimethylammonium bromide (CTAB) for 2 hours at 29 °C. The product of methyl eugenol and veratraldehyde obtained with good yield. The disadvantage of this method is the DMS reagent is toxic and dangerous. To overcome these problems, we have studied the reaction of DMC with eugenol and vanillin under solid/liquid phase transfer conditions.

2. Experimental Detail

All chemicals used were of analytical grade from Merck and Co. Inc. include vanillin, methyl tertiary butyl ether (MTBE), dimethyl carbonate (DMC), potassium carbonate (K_2CO_3), hydrochloric acid (HCl), N-cetyl-N, N, N-trimethylammonium bromide (CTAB), tetrabutylammonium bromide (TBAB), polysorbate 80, and sodium sulfate anhydrous (Na_2SO_4). The substrate with laboratory grade is eugenol with 98.89% of purity. The instrument used in this study include infrared spectrophotometer (FT-IR, Shimadzu Prestige-21), Gas Chromatograph - Mass Spectrometer (GC-MS, Shimadzu QP-2010s, Rastek RXi-5MS, FID) and Gas Chromatography (GC, Hewlett Packard 5890, HP 5, FID).

Synthesis of methyl eugenol and veratraldehyde were carried out by refluxing eugenol or vanillin, DMC, K_2CO_3 and PTC for 5 hours. At the end of the reaction, the base (K_2CO_3) was recovered by filtration and the PTC was separated by extraction with an aqueous hydrochloric acid (until pH 1) and methyl tertiary butyl ether (MTBE). The organic phase, containing products, were evaporated with a rotary evaporator and analyzed by gas chromatography. In order to optimize the yield of the reaction the PTC/DMC/reaction time ratio and catalyst system were modified. The PTC recovered in the aqueous phase can be regenerated [16].

3. Results and Discussion

3.1. Methylation by varying of phase transfer catalyst

The results of methylation of eugenol and vanillin with CTAB, TBAB and Tween 80 as PTC via GC analysis are shown in table 1.

Table 1. Effect of catalyst system on the yield

Substrate	DMC (mmol)	PTC	Catalyst (mmol)	Time (h)	Yield (%)	Purity (%)
Eugenol	16	CTAB	0.6	5	96.94	98.61
"	16	TBAB	0.6	5	92.65	94.24
"	16	Tween 80	0.6	5	trace	-
Vanillin	16	CTAB	0.6	5	19.10	19.51
"	16	TBAB	0.6	5	38.50	38.81
"	16	Tween 80	0.6	5	trace	-

By looking at the yield and purity were not significantly different between using CTAB and TBAB as PTC for O-methylation of eugenol (Table 1). If compared with O-methylation of vanillin, using TBAB give better yield than CTAB, 19.10% when using CTAB and raised to 38.50% of yield when using TBAB. Therefore, tetrabutylammonium bromide (TBAB) was selected as PTC and used in the subsequent reaction. Selection of TBAB referred to one of the principles of green chemistry, which is to reduce the waste generated.

3.2. Methylation by varying the amount of dimethyl carbonate

The results of methylation of eugenol and vanillin with a variation of DMC via GC analysis are shown in table 2.

Table 2. Effect of various amount of DMC on the yield

Substrate	PTC	Catalyst (mmol)	DMC (mmol)	Time (h)	Yield (%)	Purity (%)
Eugenol	TBAB	0.6	8	5	52.57	55.06
"	"	0.6	12	5	75.36	79.85
"	"	0.6	16	5	**91.59**	**94.24**
Vanillin	"	0.6	16	5	38.50	38.81
"	"	0.6	24	5	**48.20**	**48.43**
"	"	0.6	32	5	51.33	51.44

The results showed that the optimum concentration used in the methyl eugenol synthesis and veratraldehyde synthesis using DMC were 16 and 24 mmol. The use of DMC classified as excess. This is possible in addition to functioning as a methylation reagent, DMC also serves as a solvent. This fact is shown in table 2, where increasing concentrations of DMC further improve the yield and purity of product.

3.3. Methylation by varying the amount of phase transfer catalyst

The results of methylation of eugenol and vanillin with a variation of PTC's amounts via GC analysis are shown in table 3.

Table 3. Effect of various amount of PTC on the yield

Substrate	PTC	Catalyst (mmol)	DMC (mmol)	Time (h)	Yield (%)	Purity (%)
Eugenol	TBAB	0	16	5	0	0
"	"	0.2	16	5	59.73	62.91
"	"	0.4	16	5	**96.22**	**99.00**
"	"	0.6	16	5	91.59	94.24
Vanillin	"	0	24	5	0	0
"	"	0.2	24	5	trace	-
"	"	0.4	24	5	trace	-
"	"	0.6	24	5	**48.20**	**48.43**

The yield of methyl eugenol is not significantly different on the use of 0.4 mmol or 0.6 mmol of PTC. Therefore, referring to the principles of green chemistry is the efficient use of chemical reagents, the selected amount of the PTC is 0.4 mmol. This was not followed by a veratraldehyde synthesis were not obtained good results when the amount of the catalyst is under 0.6 mmol.

3.4. Methylation by varying the reaction time

When the reaction time was increased from 2 hours to 3 hours, the yield of methyl eugenol increased significantly almost twice, being 96.16% with 99.37% of purity. At the vanillin, there is a significant difference that the reaction time takes longer than eugenol, almost twice. Therefore, it is determined that the optimal reaction time for the methylation reaction using dimethyl carbonate are 3 hours for yielded methyl eugenol and 10 hours for yielded veratraldehyde. The optimum reaction time via GC analysis shown in table 4. In comparison, the reaction time of veratraldehyde synthesis is longer than methyl eugenol synthesis.

Table 4. Effect of various reaction time on the yield

Substrate	PTC	Catalyst (mmol)	DMC (mmol)	Time (h)	Yield (%)	Purity (%)
Eugenol	TBAB	0.6	16	1	21.75	24.98
"	"	0.6	16	2	54.91	59.97
"	"	0.6	16	3	**96.16**	**99.37**
"	"	0.6	16	4	92.38	95.61
"	"	0.6	16	5	91.59	94.24
		0.4	**16**	**3**	**96.12**	**99.33**
Vanillin	"	0.6	24	5	48.20	48.43
"	"	0.6	16	10	85.47	86.51
"	"	**0.6**	**24**	**10**	**95.23**	**95.41**

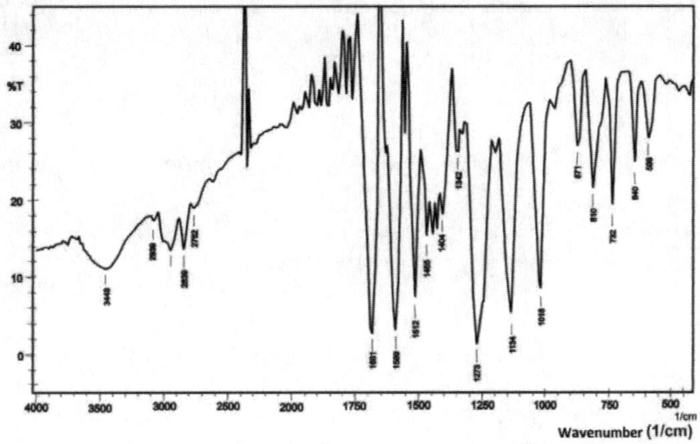

(a)

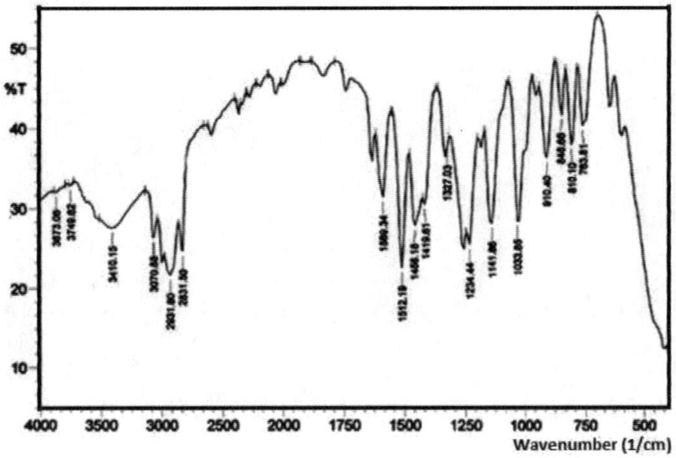

(b)

Figure 1. IR-spectra for veratraldehyde (a) and methyl eugenol (b)

Structure elucidation of methylation products from optimum methods were performed by means of IR (figure 1), GC (figure 2) and MS (figure 3). For methyl eugenol, IR (neat) : 3070, 2931, 2831, 1512, 1589, 1458 and 1033 cm^{-1}. The molecular formula was determined to be $C_{11}H_{14}O_2$ based on MS data (M^+, calculated for being 178). For veratraldehyde, IR (neat) : 3194, 2954, 2846, 2731, 1689, 1600,

1400, and 1300 cm^{-1}. The molecular formula was determined to be $C_9H_{10}O_3$ based on MS data (M^+, calculated for being 166).

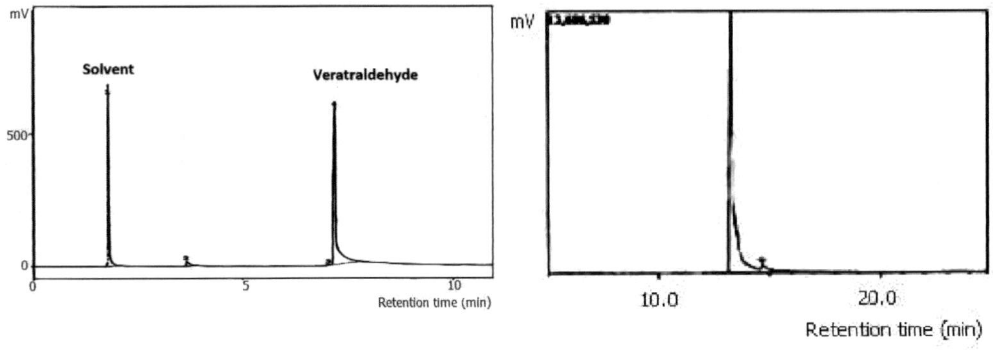

(a) (b)

Figure 2 Chromatogram for veratraldehyde (a) and methyl eugenol (b)

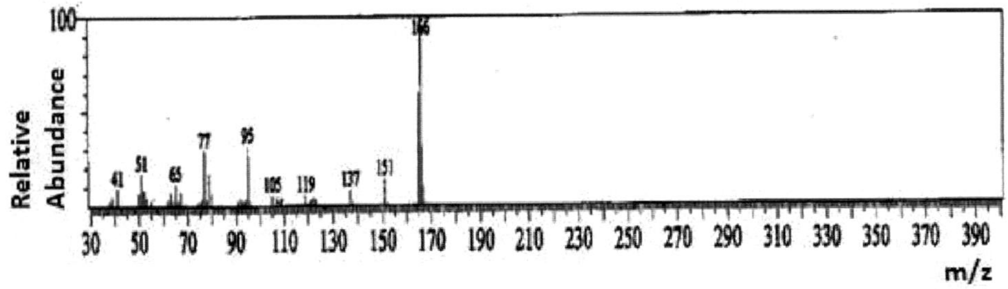

(a)

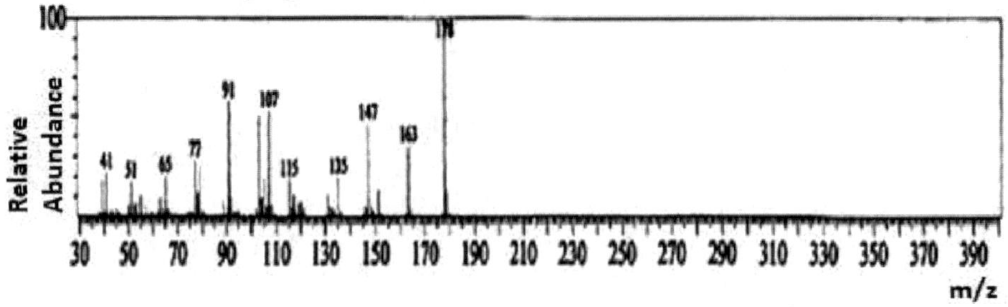

(b)

Figure 3 Mass spectra for veratraldehyde (a) and methyl eugenol (b)

Other research by Ouk *et. al.* [12] reported methylation of p-cresol with DMC takes 5 hours. Shieh *et. al.* [17] reported methylation of *p*-chlorophenol with DMC takes 16 hours (table 5). The different reactivity between phenolic compounds can be explained by the inductive effect of aldehyde group at the vanillin, halide group at *p*-chlorophenol, an alkyl group at *p*-cresol and allyl group in eugenol (Figure. 4).

Table 5. Effect of functional groups on the methylation reaction with DMC

Substrate	Functional Group Properties	Reaction time (hours)
Eugenol[a]	Electron Donating	3
p-Kresol[b]	Electron Donating	5
Vanillin[c]	Electron Withdrawing	10
p-chloro-phenol[d]	Electron Withdrawing	16

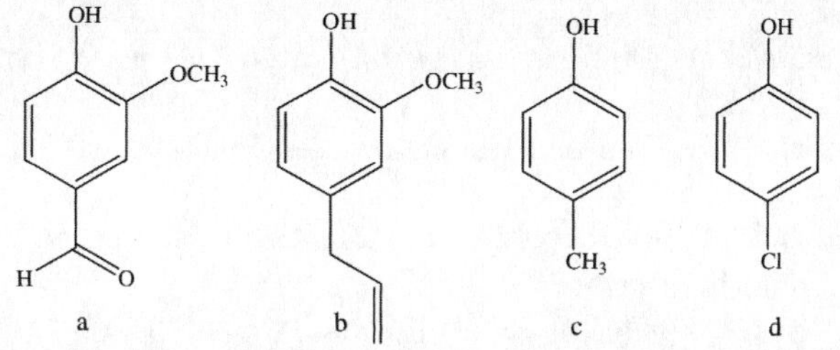

Figure 4 (a) Vanillin; (b) Eugenol; (c) *p*-Cresol; (d) *p*-Chloro-Phenol

Aldehyde group at the vanillin and halide group at *p*-chlorophenol are electron withdrawing groups while the allyl group on eugenol and alkyl groups on *p*-cresol are electron donating groups. The existence of electron withdrawing groups will cause electron density at the OH-group (at vanillin and *p*-chlorophenol) will be reduced, so the ability of nucleophiles would be reduced. It makes the reaction will be more difficult and takes a long time. The opposite occurs in eugenol and *p*-cresol, where the electron donating groups will increase the electron density of OH groups. The ability of nucleophile will increase, the reaction will easily occur and takes a short time. The possible mechanism of this reaction is shown in figure 5.

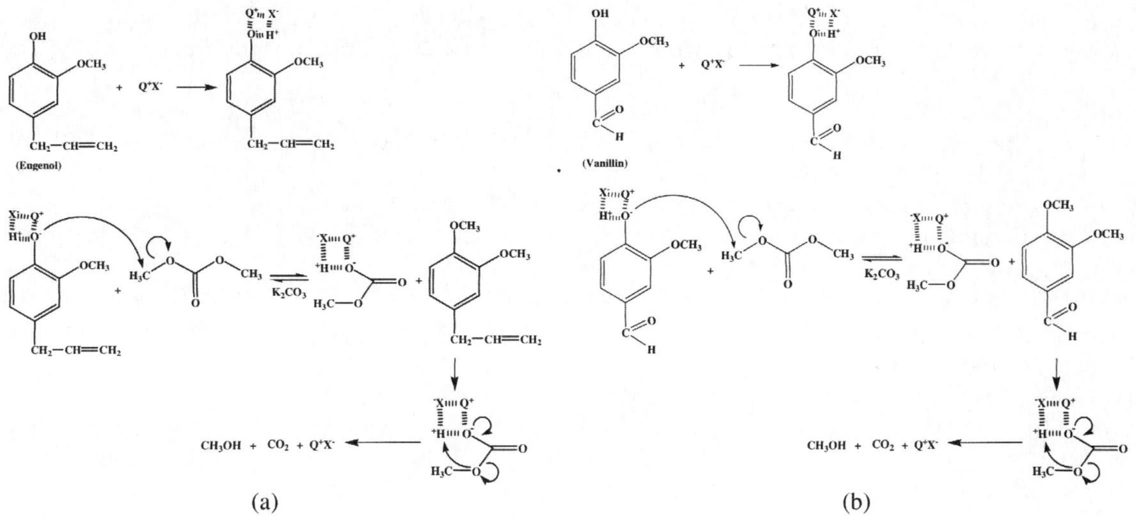

(a) (b)

Figure 5. The possible mechanism of methylation of Eugenol (a) and Vanillin (b) using DMC and PTC.

IOP Conf. Series: Materials Science and Engineering **107** (2016) 012065 doi:10.1088/1757-899X/107/1/012065

4. Conclusion

In conclusion, we have presented optimum conditions of the methylation reaction of eugenol and vanillin by using DMC as the environmental friendly reagent. The reaction was conducted under reflux conditions at 90 °C and obtained in 95–96% yield. Furthermore, the base and PTC which act as catalysts can be recovered and regenerated easily.

References

[1] Anastas P and Warner J C 1998 *Green Chemistry: Theory and Practice* (New York: Oxford University Press)

[2] Clark J and Macquarrie D 2002 *Handbook of Green Chemistry and Technology* (London: Blackwell Science Ltd)

[3] Kidwai M and Mohan R 2005 Green Chemistry: An Innovative Technology *Found. Chem.* **7** 269–287

[4] Sharma S K, Chaudhary A and Singh R V 2008 Gray Chemistry Versus Green Chemistry: Challenges and Opportunities *Rasayan J. Chem.* **1** 68-92

[5] Ono Y 1996 Dimethyl Carbonate for Environmentally Benign Reactions *Pure and Appl. Chem.* **68** 367-375

[6] Memoli S, Selva M and Tundo P 2001 Dimethylcarbonate for eco-friendly methylation reactions *Chemosphere* **43** 115–121

[7] Dorothea G 1991 *Phenol derivatives. Ullmann's Encyclopedia of Industrial Chemistry* Barbara E, Stephen H, Gail S Eds VCH Weinheim **A19**

[8] Iori G and Romano U 1985 Process for Synthesizing Allyl Carbonates of Polyhydric Alcohols and Their Derivatives *Patent US* 4 508 656

[9] Barcello G, Grenouillat D, Senet J P and Sennyey G 1990 Penta alkylguanidines as Etherification and Esterification Catalysts *Tetrahedron* **46** 1839-1848

[10] Lee Y and Shimizu I 1998 Convinient O-Methylation of Phenols with Dimethyl Carbonate *Synletters* **10** 1063

[11] Bomben A, Selva M, Tundo P and Valli L 1999 A Continuos-flow O-Methylation of Phenols with Dimethyl Carbonate in a Continuous Fed Stirred Tank Reactor *Ind. Eng. Chem. Res.* **38** 2075-2079

[12] Ouk S, Thiebaud S, Borredon E, Legars P and Lecomte L 2002 O-Methylation of Phenolic Compounds with Under Solid/Liquid Phase Transfer System *Tetrahedron Lett.* **43** 2661-2663

[13] Guenther E 1990 *The Essential Oils* **II** (Jakarta: Universitas Indonesia Press)

[14] Sastrohamidjojo H 1981 *A Study of Some Indonesian Essential Oil* Desertasi FMIPA UGM Yogyakarta

[15] Zuliatun S 1995 *Study the Use of Polysorbate 80 (Tween 80) and N-Cetyl-N, N, N-trimethylammonium Bromide as Phase Transfer Catalysts in Methyleugenol Synthesis* Doctoral Thesis FMIPA-UGM Yogyakarta

[16] Berris B C 1991 Phase Transfer Catalyst Recovery *Patent US* **5** 030 757

[17] Shieh W C, Dell S and Repic O 2001 1,8-Diazabicyclo[5.4.0]undec-7-ene (DBU) and Microwave-Accelerated Green Chemistry in Methylation of Phenols, Indoles, and Benzimidazoles with Dimethyl Carbonate *Org. Lett.* **3** 4279-4281

The effect of polyethylene glycol on shellac stability

Khairuddin[1*], Edi Pramono[2], Suryadi Budi Utomo[3], Viki Wulandari[1], A'an Zahrotul W[1], Francis Clegg[4]
[1]Department of Physics, Sebelas Maret University, Jl. Ir. Sutami 36A Surakarta 57126 Indonesia
[2]Deaprtment of Chemistry, Sebelas Maret University, Jl. Ir. Sutami 36A Surakarta 57126 Indonesia
[3]Department of Chemistry Education, Sebelas Maret University, Jl. Ir. Sutami 36A, Surakarta 57126 Indonesia
[4]Materials and Engineering Research Institute, Sheffield Hallam University, Sheffield, S1 1WB, U.K

[*]E-mail: khairuddin_70@yahoo.com

Abstract. The effect of polyethylene glycol (PEG) having amolecular weight of 1000 and 2000 on shellac stability has been investigated in this research. The shellac was shellac wax free, and the solvent was ethanol 96 %.Shellac films were prepared by solventevaporationmethod.The stability of shellac was investigated using insoluble solid test, Fourier Transform Infra Red (FTIR), Thermogravimetry Analyzer (TGA), and Water Vapour Transmission Rate (WVTR). The results showed that stability of shellac decreased after heating at 125°C for 10,30,90,and 180 minutes, and storing for 1 month at 27 °C and 85 relative humidity (RH). PEG improved the stability, and the most stable effect was achieved through PEG1000.

1. Introduction

Shellac is a natural polymer produced by insects in the forest in tropical countries. It wasnon-toxigenic, renewable and naturally decomposable so its products could meet sustainable and environmental protection requirements. It had been applied in food and medicine coating, metallurgy, military,etc. The attractive of shellac use relied on its; high adhesive strength, great electrical insulation property, waterproofness, moisture and corrosion resistance, ultraviolet screening, oil resistance, enteric nature (acid resistance) and good dye-fixing.

However, shellac also has disadvantages relating to its stability. Over time its molecular groups could undergo intramolecular chemical bonding (polymerization) which led to changes in its properties, it becomes more brittle and less soluble. Shellac underwent polymerization/aging by either heating or exposure to humidity [1]. Since most of the acids contained more than one hydroxyl group and some more than one carboxyl group it was believed that this polymerization was a result of self-esterification of the material as shown in the proposed mechanism by Limmatpavirat [2] in Figure 1. This polymerization manifested itself in the so-called blocking of the material as the individual shellac flakes stuck together. Hence, the stability of shellac could be improved by protecting these functional groups [2,3]. The amount of polymerization in shellac correlated to the percentage of insoluble solid.

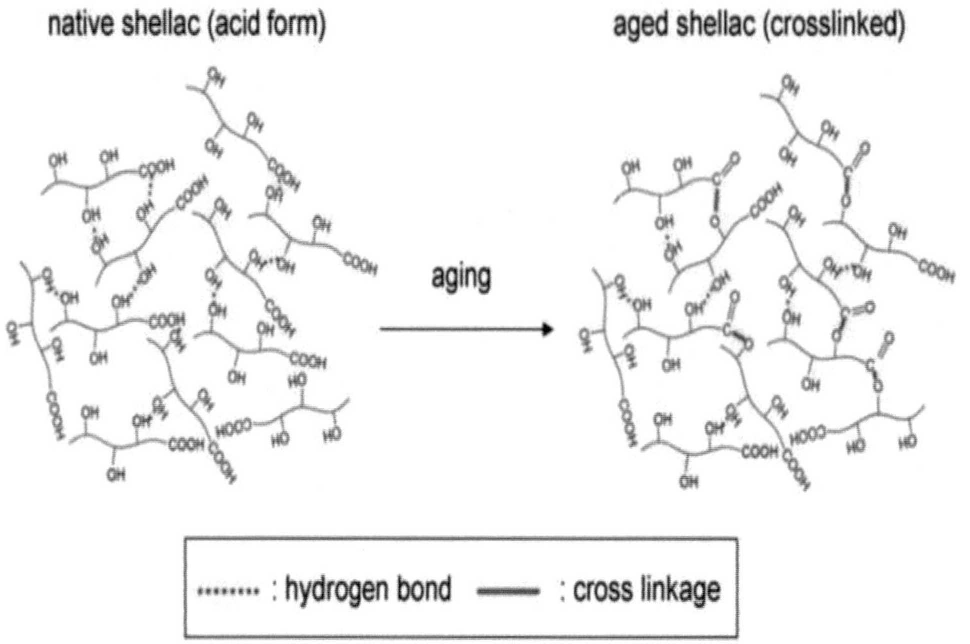

Figure 1. Proposed mechanism of polymerization in shellac [2].

Several attempts have been done to address this issue. Alkali treatment of shellac and thus composite salt formation (by carboxylate ions) improved flexibility and stability but worsen water barrier properties [4]. A comparison between native and hydrolysed shellacs prepared as films in free acid and ammonium salt forms by using ethanol and ammonium hydroxide solution, respectively, did show that alkali treatment increased acid value, solubility at pH 7 and strain, but decreased ester value, water vapour permeability coefficient and stress. The composite salt formation also led to the stabilisation of shellac [3-6], which was believed to occur through the protection of polymerisable groups and stopped the formation of esters through the reversal of the salt formation mechanism (a known aging effect). This can extend the shelf-life of films from 3 to over 12 months when stored at 40 °C and 75%RH. Although composite salt formation provided flexibility to shellac films and coatings, plasticizers were still required to overcome their brittleness. Although this could lead to poorer water barrier properties, plasticizers also served to protect shellac from polymerisation and thus stabilized their properties, this was believed to occur through a hydrogen bonding mechanism which protected the reactive groups. Anan *et al.* [7] showed that PEG with Mw of 400 was better than that of 200 and 4000 and reasoned it to the balance between a number of hydrogen bonding environments and molecular volume. An investigation to look at Mw 1000 and 2000 is deemed necessary as these may provide further stabilization. The high Mw are less likely to migrate or leach from polymerisable sites and thus, lead to higher stable films, however too high an MW will lead to poor plasticising effects. Other plasticizers such as triacetin or diethyl phthalate provided less stability.

2. Experimental

2.1. Materials

Shellac wax free was supplied by Sigma-Aldrich UK. Alcohol 96 % was supplied by Merck Germany. Plasticizer polyethylene glycol was supplied by Sigma-Aldrich UK.

2.2. Preparation of shellac and plasticised shellac films
All films presented were prepared by solutioncasting. Shellac films were made by dissolving 1.25 gr of shellac in 12.5 ml ethanol (96%) while PEG/shellac films (10 wt%) were prepared by mixing and stirring shellac solution and PEG solution. The shellac solutions were made by dissolving 1.125 gr of shellac in 10 ml ethanol and stirred for 2 hours while PEG solutions were made by dissolving 0.125 gr of PEG in 2.5 ml ethanol and stirred for 2 hoursusing a magnetic bar/stirrer at room temperature. The solution was then cast into Petri dish, then dried at 50 °C for 7 hours before being analysed.

2.3. Characterization

2.3.1. Insoluble solid. Polymerisation of shellac was accelerated by heating at 125°C for 10, 30, 90, 180 and 360 minutes. The heated shellac film was dissolved in ethanol (96%) for 3 hours. The solution was then filtered using a metal filter (150 mesh). The insoluble solid was dried at room temperature until a constant weight was obtained and the percentage calculated.

2.3.2. Fourier Transform Infra Red (FTIR). A Thermo Nicolet NEXUS spectrometer along with an MCT (mercury cadmium telluride) detector cooled by liquid nitrogen was used for all the FTIR analysis. The spectrometer was coupled with a single reflection diamond ATR cell (GrasebySpecac, U.K). This setup has the trade name "Golden Gate". The film was placed on top of the crystal diamond and the spectra of the films taken. The wavelength of interest ranged from 700 to 4000 cm^{-1} and was collected with aresolution of 4 cm^{-1} and 64 scans. Omnic software was used to collect and analyse the spectra.

2.3.3. ThermogravimetryAnalyzer (TGA). All TGA measurements were performed on a Mettler Toledo TG50 ThermogravimetricAnalyser using an alumina crucible. Data was recorded at a heating rate of 20 °C/min between 35 and 800 °C after an initial isothermal stage where the sample was kept at 35 °C for 15 minutes. The analysis was performed under a nitrogen atmosphere with a flow rate of 10 ml/min.

2.3.4. Water vapour transmission rate (WVTR). An experimental procedure to measure the WVTR was conducted using aluminium Payne cups (supplied by Sheen Instruments) with exposed surface area of 10 cm^2 and containing silica gel desiccant (8 g in each cup), which is closed by the material to be tested and dried overnight in an oven at 100 °C. The cups were placed in a controlled atmosphere (humidity chamber). The cups were weighed at suitable time intervals and the WVTR were determined from the increase in mass when this increase had become proportional to the time interval. The WVTR result was calculated from the equation (1) [8].

$$\text{WVTR} = (240 \times m) / (S \times t) \quad (1)$$

where is the total duration in hours, m is the increase in mass in milligrams of the assembly during the time t, and S is surface area exposed by the Payne cups. Shellac and PEG/shellac films were equilibrated in the humidity oven at 23 °C and 85 % relative humidity (RH) for 24 hours before performing the WVTR measurement. The relative humidities of 85% were achieved by placing saturated potassium chloride salt solutions, in a humidity chamber [9].

3. Result and Discussion
Figure 2 shows the insoluble solid of both native shellac (no PEG presence in shellac) and shellac plasticized by PEG of different molecular weight after heating at 125 °C for adifferent time, i.e.,, 10, 30, 90, 180, 360 minutes. It shows that the insoluble solid increases with increasing heating time are indicating that the degree of polymerization increases with increasing heating time for both native shellac and plasticized shellacs. When samples were heated, the energy will break the carboxyl bond

to create ester. The longer the heating, the higher amount of ester produced. These results suggest that plasticizer polyethylene glycol improved thestability of shellac althoughit did not stop polymerization of shellac.The O-H group of PEG may protect the carboxyl group of shellac resulting bonding in PEG/shellac is stronger than that of shellac. A similar trend of theinsoluble solid test was also reported by Anan [7] using PEG200, 4000, and 4000. It also showed that PEG1000 gave the most stable effect on protecting of shellac compared to PEG2000.The amount of carboxyl group that can be protected by PEG1000 may be higher than that of PEG2000 as PEG2000 has longer molecular chains.

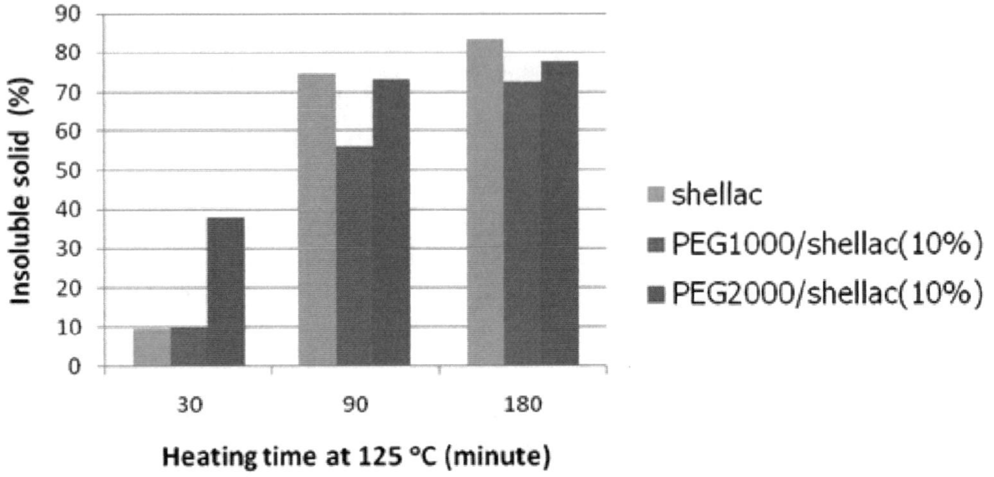

Figure 2. Effect of PEG on shellac stability via insoluble

Identification of shellac polymerization using FTIR has not been reported so far in the literature. According to proposed mechanism in Figure 1, polymerization of shellac is thus assigned by the increase of peak intensity bands of C=O stretching of ester 1709 cm^{-1} and the decrease of as O-H stretching of thehydroxyl group at 3400 cm^{-1}. The intensity of the band at 2942 cm^{-1} occurs due to the C-H stretching vibration [10] and does not change with achange in thedegree of shellac polymerization. Therefore, the intensity ratios of the band of O-H $cm^{-1}$to C-H, and the intensity ratios of the band of C=O to C-H from all films then correlates to the polymerization in shellac.

Typical FTIR spectrum of shellac, PEG1000/shellac (10 wt%), and PEG2000/shellac (10 wt%) after heating is shown in Figure 3 giving the information of peak intensity bands of C=O, O-H, and C-H due to heating. The peak intensities of O-H to C-H and C=O to C-H were rationed, and the results are given in Figure 4 and Figure 5, respectively. It showed that the peak intensity ratios of O-H to C-Hdecreased while the peak intensity ratio of C=O to C-H increased indicating that polymerization occurred in shellac and plasticized shellac supporting insoluble solid test in Figure 1. The rate of peak intensity ratio of O-H to C-H and C=O to C-H is the smallest in PEG1000/shellac, followed by PEG2000/shellac, and shellac.

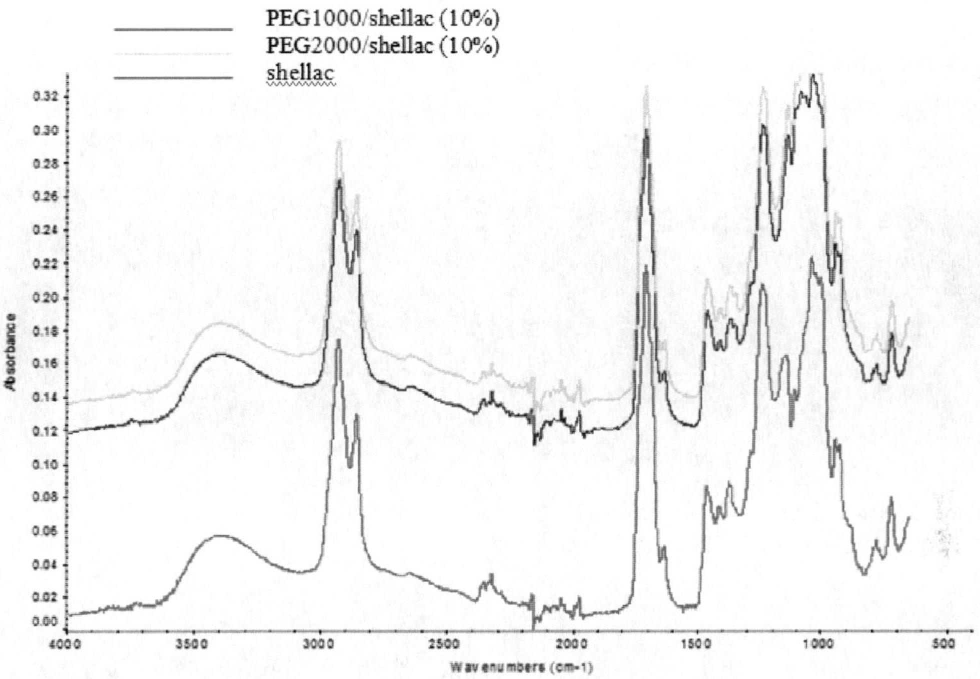

Figure 3. FTIR spectrum of shellac and PEG shellac system heated at 125 °C for 30 minutes.

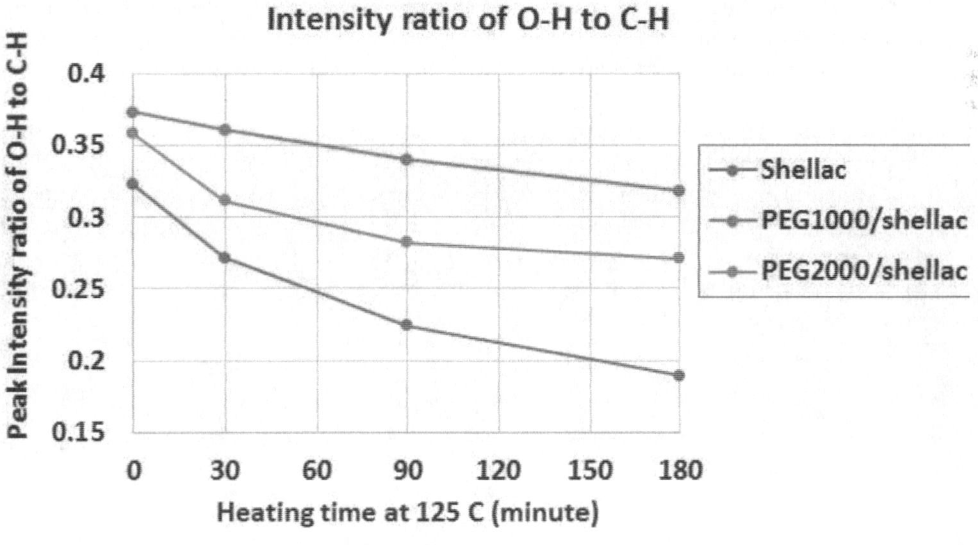

Figure 4. Peak intensity band of O-H of shellac, PEG1000/shellac, and PEG2000/shellac after heating at 125 °C for 0, 30, 90, and 180 minutes

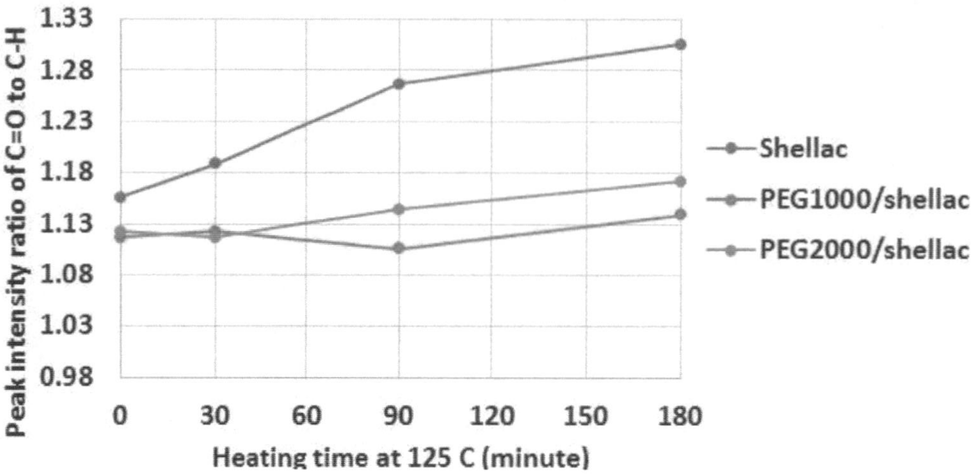

Figure 5. Peak intensity band of O-H of shellac, PEG1000/shellac, and PEG2000/shellac after heating at 125 °C for 0, 30, 90, and 180 minutes

The thermograms collected from the shellac, PEG1000/shellac, and PEG2000/shellac is presented in Figure 6. In general, there are 3 areas of weight loss. The first weight losses observed in the range 35-150 °C relates to water or solvent present in the sample. The second weight losses observed in the range 250 – 475°C relates to structural decomposition of the shellac. Above 500°C, relatively little weight is lost which relates to remained compound or carbon. The TGA data shows that adding PEG improved thermal degradation of shellac. PEG1000 increased thermal degradation temperature of shellac by 20 °C compared to only 5 °C by PEG2000.

The water vapour transmission rate (WVTR) of shellac, PEG1000/shellac, and PEG2000/shellac collected at day 0 (samples were prepared), and after 1 month is given in Figure 7. Lower WVTR correlates to good water vapour barrier properties, and vice versa. It shows that adding PEG reduces water vapour barrier properties of shellac. PEG may break hydrogen bond between molecular chains of shellac, in addition to hydrophilic properties of PEG. Generally, PEG reduces water vapour barrier properties of thepolymer [10]. The decrease of water barrier properties was higher due to PEG2000 than that of PEG1000. Furthermore, the water vapour barrier properties decreased after one-month storage in 85 %RH indicating either polymerization or penetration of water vapour into shellac chains reduced the water barrier properties of shellac.

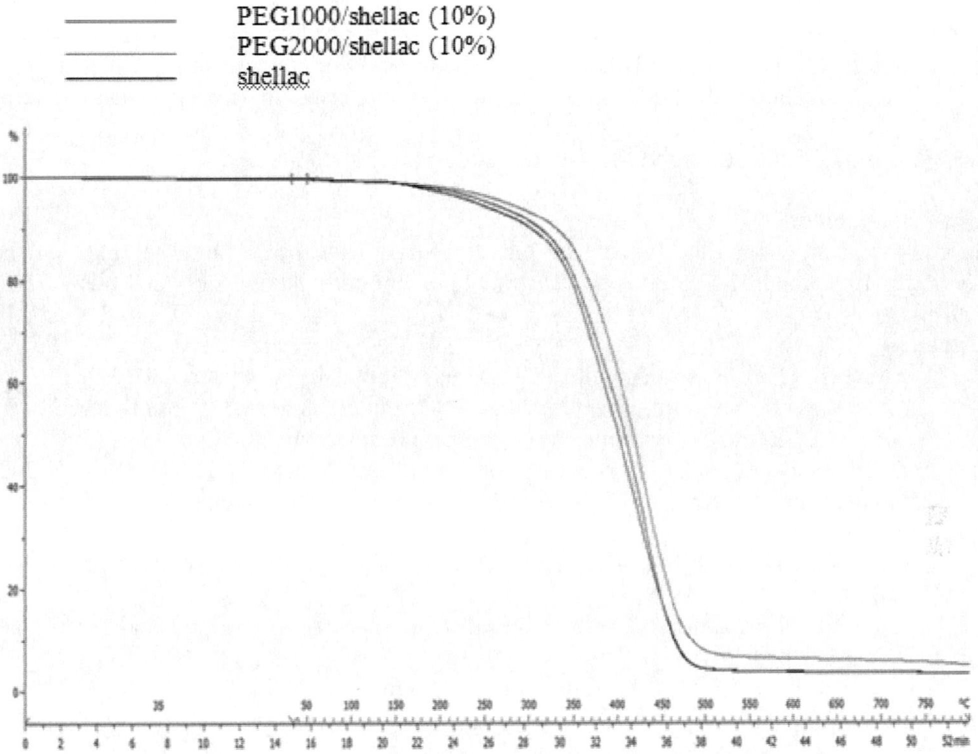

Figure 6. Thermograms of shellac, PEG1000/shellac, and PEG2000/shellac

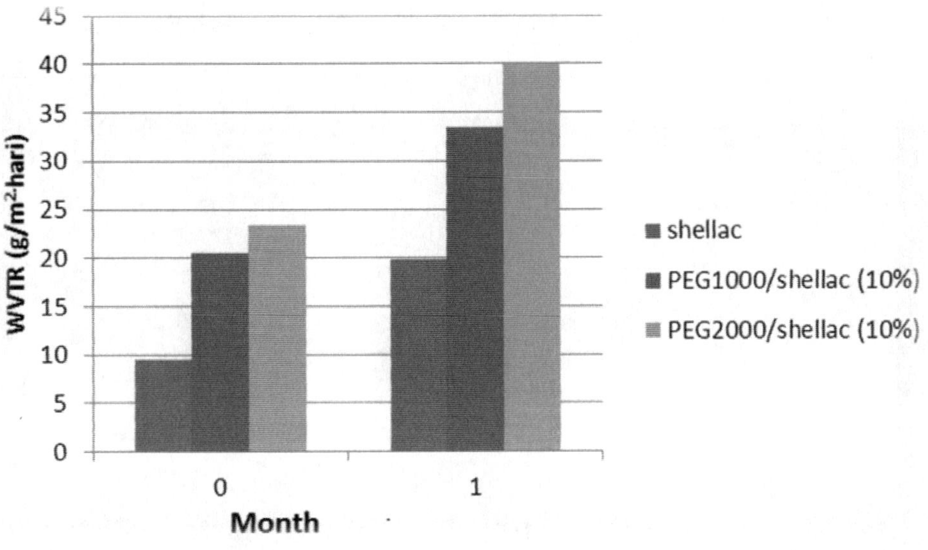

Figure 7. Water vapour transmission rate of shellac, PEG1000/shellac, and PEG2000/shellac

4. Conclusion

It has been shown that polymerization in shellac and plasticized shellac occurred by heating the samples at 125 °C. The percentage of the polymerization is increased with heating time as shown by theinsoluble solid test. The polymerization was slowed down by PEG, and the most stable effect was given by PEG1000. The trend of polymerization was confirmed by FTIR as shown by the decrease of the peak intensity ratios of theO-H band to C-H band, and the increase of intensity ratio of C=O band to C-H band. Adding PEG1000 and PEG2000 also increased thermal degradation temperature of shellac up to 20 °C and 5 °C, respectively. The water vapour barrier properties of shellac decreased both after adding PEG and stored for 1 month at 85 % RH at 27 °C.

Acknowledgement
The authors acknowledge the Indonesian Government for awarding "International collaboration and scientific publication 2014" grants with contract number 6563/UN27.16/PN/2014 dated 10 June 2014.

References

[1] Oesman, Z, Ph.D Thesis, Johannes Guttenberg University, Germany, 2012
[2] Limmatvapirat, S., Limmatvapirat, C., Puttipipatkhachorn,S., Nunthanid,J., Anan, M.L. European Journal of Pharmaceutics and Biopharmaceutics 67, 690–698, 2007
[3] Qussi,B&Suess,W.G. Drug Development & Industrial Pharmacy 32 (4), 403-412,2006.
[4] Limmatvapirat, S., Nuntahid,J., Anan,M.L, Pharmaceutical Development & Technology 10, 41–46, 2005.
[5] Specht, F., Saugestad,M., Waaler, T., Miller, B.W., Pharmaceutical Technology 10, 20–28, 1998.
[6] Anan, M.L., Nunthanid,J., Limmatvapirat, S.Journal of Agricultural and Food Chemistry. 58, 12934-12940, 2010.
[7] British Standard (BS 2782-8:Method 820A). Method of testing Plastics part 8, 1996.
[8] Equilibrium relative humidty at http://www.omega.com/temperature/z/pdf/z103.pdf
[9] Derry J, Master Thesis, University of Oslo, Norwegia, 2012
[10] Khairuddin,Ph.D Thesis, Sheffield Hallam University, UK, 2012

IOP Conf. Series: Materials Science and Engineering **107** (2016) 012067 doi:10.1088/1757-899X/107/1/012067

The langmuir isotherm adsorption equation: The monolayer approach

Patiha*, E Heraldy, Y Hidayat, M Firdaus

Department of Chemistry, Faculty of Mathematic and Natural Sciences, Sebelas Maret University, Jl. Ir. Sutami 36A, Surakarta, 57126, Jawa Tengah, Indonesia

*Email: patiha31@yahoo.co.id

Abstract. Langmuir adsorption isotherm equation is generally derived using the kinetics approach and based on some assumptions. These assumptions are usually not accompanied by an adequate explanation related with the mechanisms and the determination techniques of its adsorption constants. This study presents a simple technique on the determination of the isotherm equation based on the main assumption so called the monolayer assumption. Furthermore, a more precise technique for determining the adsorption constants is also described. This theoretical study is supported with statistical analysis on the experimental data adapted from the literature. Two theories based on adsorption mechanism and based on rate were introduced. Regardless of its limitations, the results show that the technique for the determination is valid.

1. Introduction

Adsorption is the adhesion of an adsorbate such as a fluid, liquid, or gas, by creating a thin layer or film on the surface of an adsorbent whether it is a solid or liquid. The bonding between adsorbate-adsorbent could be physical or chemical bonding. The chemical bonding provides stronger bonding than physical bonding, and the film might be single or multiple layers. Adsorbate can be separated from the adsorbent and the process is called desorption.

There are some factors influencing the adsorption process, but, at the basic level, the most widely studied is the influence of temperature and could be assumed as a monolayer. The adsorption process is usually studied at a given temperature and recognized as an adsorption isotherm. One of them is the Langmuir adsorption isotherm.

Langmuir equation is derived based on the rate of adsorption and desorption [1,2]. If S_o is the maximum available sites at the adsorbent surface, S the amount occupied, $S_1 = S_o - S$ is vacant, and C is the concentration of the solution, then

$$\text{the rate of adsorption} = k_1 S_1 C$$
$$= k_1 [So - S] C \quad (1)$$

The rate of desorption is proportional the number of occupied sites S:

$$\text{rate of desorption} = k_2 S \quad (2)$$

IOP Conf. Series: Materials Science and Engineering **107** (2016) 012067 doi:10.1088/1757-899X/107/1/012067

k_1 and k_2, are adsorption and desorption rate constant respectively.

At equilibrium, the rate of adsorption = rate of desorption, or Equation (1) = Equation (2), or

$$k_1 [S_o - S] C = k_2 S \tag{3}$$

$$k_1 S_o C = [k_2 + k_1 C] S$$

$$S = \frac{k_1 S_o C}{k_2 + k_1 C}$$

$$S = \frac{S_o C}{k_2 / k_1 + C} \tag{4}$$

If $S_o = K_{ads}$ and $k = (k_2 / k_1)$ then

$$S = \frac{K_{ads} C}{k + C} \tag{5}$$

K_{ads} and k are constants with values depend upon the nature of the adsorbent and adsorbate.

There are 3 problems arises with regard to eq. (5). The first is how to relate it with the monolayer assumption. In order to be meaningful, eq. (5) should be derived from the assumption. Second, adsorption phenomena are irreversible processes. Thus, it needs a different reaction mechanism, not just based on the assumption that, at equilibrium, the rate of adsorption = rate of desorption. Third, the curve (of S versus C) of eq. (5) is a hyperbolic equation. It is similar with reactions catalyzed by enzyme [3]. Accordingly, it is difficult to determine the value of K_{ads} and k directly from eq. (5). By taking the reciprocal of both sides of eq. (5), however, it is converted into its linear form

$$\frac{1}{S} = \frac{1}{K_{ads}} + \frac{k}{K_{ads}}\left(\frac{1}{C}\right) \tag{6}$$

K_{ads} value is obtained from the intercept and k from the slope of the curve of $(1/S)$ vs. $(1/C)$.

It is interesting to compare eq. (6) with the equation obtained from eq. (5) on the condition of C is relatively higher and lower than k.

If C is higher than k and so $C+k \approx C$, eq. (5) becomes

$$S = K_{ads} \tag{7}$$

The reciprocal of eq. [7] is

$$\frac{1}{S} = \frac{1}{K_{ads}} \tag{8}$$

The equation (8) is different with eq. (6).

Inserting eq. (7) into eq. (6) gives

$$\frac{1}{K_{ads}} = \frac{1}{K_{ads}} + \frac{k}{K_{ads}}\left(\frac{1}{C}\right)$$

or

$$k = 0 \tag{9}$$

That is to say, eq. (6) does not apply if C is relatively higher than k.

If C is lower than k and so $C+k \approx k$, eq. (5) will become

$$S = \frac{K_{ads}C}{k} \tag{10}$$

The reciprocal of eq. (10) is

$$\frac{1}{S} = \frac{k}{K_{ads}}\left(\frac{1}{C}\right) \tag{11}$$

which is also differ with eq. (6)

By assumption, $S_o = K_{ads}$ and so, if all surface occupied (or $S = S_o$), then $S = K_{ads}$ and eq. (10) [or eq. (11)] becomes

$$k = C \tag{12}$$

That is to say, eq. (6) also does not apply if C is relatively lower than k. So, it can be concluded that, theoretically, eq. (6) is ambiguous and therefore, it can not be used to determine, both K_{ads} and k.

Based on the previously discussion, there are three objectives of this study i.e. to derive the Langmuir adsorption isotherm equation which is based on the monolayer assumption, to derive the Langmuir adsorption isotherm equation which is based on the monolayer assumption and irreversible mechanism, and to find an easier and reliable technique for the determination of K_{ads} and k.

2. Methodology

This research is a literature study. The first goal is to drive eq. (5) which is based on the monolayer assumption. The second goal is to drive eq. (5) which is based on the monolayer assumption and the reaction mechanism involving the product (irreversible process). The third goal is to find an easier and reliable technique for the determination of K_{ads} and k. The validity is based on the comparison of the results obtained from of eq. (7) and eq. (10) and with obtained from eq. (6).

This study uses data, not from adsorption experiment, but taken from the bromination of acetone [4], $(CH_3)_2CO + Br_2 \rightarrow CH_3COCH_2Br + Br^- + H^+$. This reaction has the rate law of

$$-d[P]/dt = \frac{(k_1k_2k_3)/k_{-1}[A][H^+][Br_2]}{k_{-2}[H^+]+k_3[Br_2]} \tag{13}$$

If in the beginning, $[Br_2]_o$, is relatively higher than $k_{-2}[H^+]$, at the early-stage, the reaction is zero-order with respect to $[Br_2]$ but at the near-end is first-order. During the reaction process, $[Br_2]$ is decreased while $[H^+]$ increased (released from acetone). But, if in the beginning, $[Br_2]_o$ is relatively lower than $k_{-2}[H^+]$, for the same reason, the reaction is always first-order.

As seen in Table 1. and Table 2., the data are not presented in concentration unit but in absorbance of Br_2 in the solution, measured at $\lambda = 400$ nm and room temperature. It is merely done for the easier use of Equation (6), Equation (7) and Equation (10). Data in Table 1 is obtained from the experiment where, $[H^+]_o$ is relatively low (and so, $[Br_2]$ will be higher than $k_{-2}[H^+]$). Data in Table 2 is obtained from the experiment where, $[H^+]_o$ is relatively high (and so $[Br_2]$ is also lower than $k_{-2}[H^+]$).

Table 1. Absorbance, A, of Br_2 in Bromination of Acetone: 10 mL Acetone 4.0 M, 5.0 mL HCl 1.0 M, 25 mL H_2O, 10 mL Br_2 0.05 M [5]

	Early stage		Near end	
No.	***t / s***	**A**	***t / s***	**A**
1.	10	0.483	590	0.073
2.	20	0.473	600	0.066
3.	30	0.463	610	0.059
4.	40	0.454	620	0.053
5.	50	0.444	630	0.048
6.	60	0.434	640	0.043

Table 2. Absorbance, A, of Br_2 in Bromination of Acetone: 10 mL Acetone 4.0 M, 7.5 mL HCl 1.0 M, 22.5 mL H_2O, 10 mL Br_2 0.05 M [5]

	Early stage		Near end	
No.	***t / s***	**A**	***t / s***	**A**
1.	10	0.493	330	0.103
2.	20	0.483	340	0.091
3.	30	0.473	350	0.081
4.	40	0.464	360	0.072
5.	50	0.454	370	0.064
6.	60	0.445	380	0.057

Then, to make it more similar to the experimental technique on adsorption, each observation which was taken at the same time interval and the observation resulting at the end of one-time interval being considered as the initial value for a new time interval. Each direct observation is taken as C_o and the following as C. The difference between each pair is taken as S (thus, the amount of S is always taken at the same time–interval). The setting of the data from both tables is presented in Table 3. and Table 4.

Table 3. Setting the data of Table 1.

No.	*t/s*	Early-stage			Near-end			
		C_o	C	$C_o - C = S$	*t/s*	C_o	C	$C_o - C = S$
1.	10	0.483	0.473	0.010	590	0.073	0.066	0.007
2.	20	0.473	0.463	0.010	600	0.066	0.059	0.007
3.	30	0.463	0.454	0.009	610	0.059	0.053	0.006
4.	40	0.454	0.444	0.010	620	0.053	0.048	0.005
5.	50	0.444	0.434	0.010	630	0.048	0.043	0.005
6.	60	0.434			640	0.043		

The value of K_{ads} and k is determined using eq. (6) for the conventional and eq. (7) and eq. (10) for the new method.

Table 4. Setting the data of Table 2.

No.	t/s	Early-stage			t/s	Near-end		
		C_o	C	$C_o - C = S$		C_o	C	$C_o - C = S$
1.	10	0.493	0.483	0.010	330	0.103	0.091	0.012
2.	20	0.483	0.473	0.010	340	0.091	0.081	0.010
3.	30	0.473	0.464	0.009	350	0.081	0.072	0.009
4.	40	0.464	0.454	0.010	360	0.072	0.064	0.008
5.	50	0.454	0.445	0.009	370	0.064	0.057	0.007
6.	60	0.445			380	0.057		

Based on equation (7), the process is of the zero- order with respect to *C*: the value of K_{ads} is obtained from the intercept of the linear curve of *S* versus *C*. This value is then used to find the value of *k*, by inserting it into the value of the slope of the linear curve of *S* versus *C* using eq. (10) (the process is of the first-order with respect to *C*). The results are then compared with that obtained from equation (6).

The process is said to be zero-order if the value of the slope close to 0.00 and first-order if the value of the intercept close to 0.00.

3. Discussions And Implications

3.1. Discussions

3.1.1. Monolayer theory. There are 5 kinds of interaction in adsorption process: between adsorbent-adsorbate, adsorbate-adsorbate, adsorbate-solvent, adsorbent-solvent, and solvent- solvent. In order the adsorption process to occur, the adsorbent-adsorbate interaction must be the strongest. The monolayer is formed if the strength of the bond between adsorbate-adsorbent molecules is stronger than adsorbate-adsorbate, adsorbate-solvent, adsorbent-solvent, and solvent- solvent. The position of the first adsorbed molecules will act as the initiator to the position of the next adsorbed molecule. The second molecule will be tied adjacent to the first molecule, and because of strong adsorbent-adsorbate interaction, besides and not above. Then, the reaction mechanism can be described as equation (14).

$$S_f + C \xrightarrow{k_{add\text{-}1}} S_{f-1}C_1^*,\ S_{f-1}C_1^* + C \xrightarrow{k_{kads\text{-}2}} S_{f-2}C_2^*, \ldots,\ S_{f-n}C_n^* + C \xrightarrow{k_{ads\text{-}n}} S_nC_n^* \quad (14)$$

S_f, and S_n are the fraction of S_l and of S_o respectively while $S_nC_n^*$ is the intermediate. Of course, a lot of initiators which be formed but each will increase gradually. But, these intermediates do not accumulate to any significant extent during the reaction. They are in the steady-state condition. Then, there are two possible termination mechanisms: the equilibrium and the irreversible mechanism approach (intermediate $S_n\ C_n^*$ turned into the product). The first approach would lead to the first goal and the other to the second.

3.1.2. First goal. By steady-state approximation,. The addition of these equations gives equation (15).

$$dS_1C_{tot}^* / dt = 0 \quad (15)$$

where $S_1C_{tot}^* = \sum_{n=0}^{\infty} S_nC_n^*$.

In each step, one intermediate is consumed and one is produced (by undergoing further adsorption or back to the original by desorption). Hence, these steps do not affect $S_1C^*_{tot}$ and $dS_1C^*_{tot}/dt$. Therefore, it is only the rate at equilibrium that need to be considered.

To make it easier, it is assumed, the intermediate reactivities are independent of size and all have the same rate constant and for adsorption, $k_{ads-1} = k_{ads-2} = \ldots\ldots, = k_1$; and for desorption, $k_{des-1} = k_{des-2} = \ldots\ldots, = k_2$. These assumptions are not exaggerated, because in a reversible reaction, the constants k_1 (and k_2) always has the same value even if determined at different concentration [6].

At equilibrium,

$$k_1 [\, S_o - S] \,\mathrm{C} = k_2 S \tag{3}$$

which, on solving, gives eq. (6)

3.1.3. Second goal. Under the second approach, the mechanism of the reaction is assumed to take place as eq. (16)

$$S + C \underset{k_{-1}}{\overset{k_1}{\longleftrightarrow}} [SC^*] \xrightarrow{k_2} P \tag{16}$$

$[SC^*]$ is an intermediate and P is the product.

The first step is reversible that stay close to equilibrium and the second is irreversible. The latter is the rate-determining step or

$$dP/dt = -dC/dt = k_2[SC^*] \tag{17}$$

Applying the steady state approximation on $[SC]^*$ gives

$$\begin{aligned} d[SC^*]/dt &= 0 = k_1 SC - k_{-1}[SC^*] - k_2[SC^*] \\ k_1 SC &= k_1[SC^*] - k_2[SC^*] \end{aligned} \tag{18}$$

If the maximum available sites are S_o, then $S = S_o - [SC]^*$ and eq. [18] becomes

$$\begin{aligned} k_1(S_o - [SC^*])C &= k_1[SC^*] + k_2[SC^*] \\ k_1 S_o C &= (k_1 + k_2)[SC^* + k_1 C[SC^*] \\ [SC^*] &= \frac{k_1 S_o C}{(k_1 + k_2) + k_1 C} \end{aligned} \tag{19}$$

Substitution of eq. [19] into eq. [17] gives

$$\begin{aligned} -dC/dt &= \frac{k_1 k_2 S_o C}{(k_1 + k_2) + k_1 C} \\ -dC/dt &= \frac{k_2 S_o C}{(k_{-1} + k_2)/k_1 + C} \end{aligned} \tag{20}$$

If

$$(k_{-1} + k_2)/k_1 = k \tag{21}$$

And

$k_2 S_o = K_{ads}$ (22)

then eq. [15] becomes

$$-dC/dt = \frac{K_{ads}C}{k+C} \quad (23)$$

Because the rate is proportional to the amount adsorbed at a certain time, then eq. (23) becomes

$$S = \frac{K_{ads}C}{k+C} \quad (6)$$

3.1.4. Third goal. The results of the calculation are presented in Table 5. and Table 6.

Table 5. Calculation Results of Table 3.

Method	**New**			**Equation (3)**		
Data Source	***A***	***B***	***r***	***A***	***B***	***r***
Early-stage	0.0100	-0.0004	0.0146	106.5711	-1.9708	-0.0297
Near-end	0.0004	0.1040	0.9404	9.3918	8.4730	0,9411

Based on Table 5., the new method confirms that at the early stage, the process is zero-order (slope B = -0.0004; the curve is linear with x-axis) and at the near-end is first-order (intercept A = 0.0004). The eq. (6) is not valid at the early-stage (negative slope) but seems valid at the near-end. However, the curve (rate versus concentration) of the first-order reaction cross the x-axis at x = 0 and the y-axis at y = 0.

Table 6. Calculation Results of Table 4.

Method	**New**			**Equation (6)**		
Data Source	***A***	***B***	***r***	***A***	***B***	***r***
Early-stage	0,0000	0.0206	0.5651	-1.6749	49.1769	0,5650
Near-end	-0.0012	0.1419	0,9937	-10.9633	8.7668	0.9969

Based on Table 6., the new method, confirms that at the early-stage (intercept A = 0.0000) and at the near-end (intercept A = -0.0012), the process is always first-order. There are two slopes, but it is the slope at the near-end that should be used. This slope has higher *r* (= 0,9997).

From a kinetic point of view, in eq. (5) the reaction is always first-order to *So*, either if *C* is much lower or higher than *k*. This means, doubling *So* will double the rate (and so as K_{ads}). This theory is very much in accordance with the fact found from the value of the slopes obtained from Table 5. and Tabel 6. The slope from Table 6. is, somewhat, 1.5 times higher than that from Table 5. This is because, the concentration of HCl used in Table 6. is also 1.5 times higher. Then, if the concentration of HCl is used for calculation, approximately, the same value of *k* will be obtained by using eq. (10). This is not the case in the use of eq. (6); both give the same slope value. So, as the conclusion, the value of K_{ads} and *k* must determine separately, using eq. (7) and eq. (10).

3.2. Implications

There are two cases that will be discussed further as the possible implication of this study. First, literature sources state that Freundlich adsorption isotherm [7], can be derived from the Langmuir equation [8]. It is also implicit on the above discussion; the Freundlich equation is the equation of Langmuir on the condition of C is relatively lower than k. The problem is, the b constant in the Freundlich equation,

$$S = KC^b \tag{23}$$

still need a further discussion, especially when related to reaction mechanism concept.

In many textbooks Freundlich isotherm is presented as an empirical equation with no or limited explanation on the physical significance of both K and or b [9], and [10] [11]. Eq. (21) can be readily derived by kinetics approach [12]. However, it is not yet fully associated with the concept of reaction order and also not associated with the monolayer assumption. The reaction orders are usually integers or half-integers (1/2, 3/2, ...), positive or negative [13]. The value of b in the adsorption of soluble organic nitrogen (SON) by granular activated carbon (GAC) is 0.746 [14], glycolic acid and acetic acid by Amberlite IRA-67 are 0.124 and 0.3174 respectively [15], and Procion Red MX-5B by filter paper is 0.57 [16]. In general, the parameter b is dimensionless and typically has a value of $b < 1$. The parameter is a measure of heterogeneity of a system and a more homogeneous system would have a b value approaching unity a more heterogeneous system would have a b value approaching zero [17]. The fractal Langmuir equation has also been introduced [18], but still not clearly related to the reaction order concept and the monolayer assumption.

The next problem is the technique of experiment for the determination of the constants. As has been discussed above, the values of the constants are determined from S, using the concentrations data obtained at equilibrium condition. In facts, sometimes, equilibrium condition need too long times. However, $S = C_o - C$ and rate are proportional to the amount adsorbed at a certain time. Moreover, the rate constant of reversible reaction could be determined by initial rate method and will give the same value at any stages of a reaction. There is no need to wait for the equilibrium achieved [19].

The experiment can also be done from an experiment on the condition of C is higher and lower than k. The observation is just done at the very beginning of the reaction on the condition of the constant time interval. This is done by using the physical properties resulting at the end of the one-time interval is being considered as the initial value for a new time interval. Because the amount absorbed is directly proportional to the rate, the constants can be determined by way of rate concept. Basically, there are 2 methods, differential and integral. The differential method gives the true order directly but not the exact value of the rate constant. The integral method gives the exact value. It is also easier to use because using C data directly [20].

4. Conclusion and Recommendation

4.1. Conclusion

We present a simple technique on the determination of the isotherm equation based on the monolayer assumption and a product-based mechanism. In addition, precise technique for the adsorption constants determination is also explained. K_{ads} and k should be determined separately i.e. K_{ads} from the intercept of the linear plot of S versus (relatively high) C and k from the calculated value of K_{ads} and the slope of the linear plot of S versus (relatively low) C. The results show that this technique is more valid.

4.2. Recommendation

We proposed two suggestions. First is to derive the fractal Langmuir isotherm adsorption equation from monolayer assumption and relate it with the fractal Freundlich isotherm equation. Second is to determine K_{ads}, k, and b by way of initial rate or better integral method.

IOP Conf. Series: Materials Science and Engineering **107** (2016) 012067 doi:10.1088/1757-899X/107/1/012067

References

[1] Langmuir I 1916 Part I The Research Laboratory of The General Electric Company 2221
[2] Langmuir I 1918 Part II The Research Laboratory of The General Electric Company 1848
[3] Espenson J H 1995 *Chemical Kinetics and Reaction Mechanisms* 2nd Edition (New York McGraw-Hill Inc) p 34 -35
[4] Farrington D 1970 *Experimental Physical Chemistry* 7th Ed (Tokyo Kogakusha Company Ltd) p 152-155
[5] Patiha 2013 *Kajian Krirtis terhadap Persamaan-persamaan dan Teknik untuk Penentuan Tetapan Michaelis-Menten* Penelitian Fundamental Surakarta FMIPA UNS
[6] Patiha 2010 *Pendekatan Kinetika terhadap Reaksi Kesetimbangan* Penelitian Fundamental. Surakarta FMIPA UNS
[7] Freundlich HMF 1906 Uber die Adsorption in Lösungen *Z. Phys. Chem.* **57(A)** 385–470
[8] Adamson A W 1982 *Physical Chemistry of Surfaces* 4th Ed (New York John Wiley & Sons Inc) p 373
[9] Essington M E 2004 *Soil and Water Chemistry An integrative approach* (Boca Raton F. L) p 343
[10] Spark D L 2003 *Environmental Soil Chemistry* 2nd ed (San Diego) p 151
[11] Sime R J 1990 *Physical Chemistry Methods Techniques and Experiments* (Holtt Rinehartand Winston) p 529
[12] Skopp J 2009 Derivation of the Freundlich Adsoption Isotherm from Kinetics *J Chem. Educ.* **86 (11)** 1341-1343
[13] Levine I N 2009 *Physical Chemistry* 6th Ed (New York McGraw-Hill Inc) p 516 – 517
[14] Sawyer C N, P L McCarty and G N Parking 2003 *Chemistry for Environmental Engineering and Science* 5th Ed (New York McGraw-Hill Companies Inc) p 100-105
[15] Uslu H, Inci I and Bayazit S S 2010 Adsorption Equilibrium Data for Acetic Acid ang Glycolic Acid onto Amberlite IRA-67 *J. Chem. Eng. Data.* **55** 1295-1299
[16] Girado G and Ayllo J A 2011 A Simple Adsorption Experiment *J. Chem. Educ.* **88** 624-628
[17] Gregory T, Karns C L, Shimizu K D 2005 A Critical Examination of the Use of Freundlich Isotherm in Characterizing Molecularly Imprinted Polymers (MIPS) *Analytica Chimica*
[18] Sposito G 1980 Soil *Sci Soc Am J* **44** 652-654
[19] Patiha 2013 Penentuan Tetapan Laju Reaksi Balik dan Tetapan Kesetimbangan dengan Pendekatan Reaksi Searah dan Hukum Laju Reaksi Maju *Alchemy* **9** 145-155
[20] Laidler K J 1987 *Chemical Kinetics* 3rd Edition (New York Harper Collins Publisher Inc).

IOP Conf. Series: Materials Science and Engineering **107** (2016) 012068 doi:10.1088/1757-899X/107/1/012068

Theoretical study of adsorption and dissociation of NH_3 on pentanuclear Fe(111) surface

G Purwiandono[1*], Triyono[2] and K Wijaya[2]

[1]Chemistry Department, Islamic University of Indonesia, Yogyakarta-55584, Indonesia

[2]Chemistry Department, Gadjah Mada University, Yogyakarta-55281, Indonesia

[*]E-mail: gani_js@yahoo.com

Abstract. Theoretical study regarding the adsorption and dissociation of NH_3 molecule on Fe(111) surface has been carried out. The method used was DFT-B3LYP, and the basis sets used were ECP and 6-311G**. This research aimed at giving the theoretical understanding of adsorption and dissociation of NH_3 molecule on Fe(111) surface. The adsorption and dissociation were studied based on the energetic parameter, bond length, electron population, vibration and orbital interaction. The result of theoretical calculation revealed that the on top position is the interaction position with the minimum energy for the adsorption of NH_3 molecule on Fe(111) surface. The analysis of electron population of the two composing molecular orbitals indicated that the transfer of electron has an important role in the mechanism of adsorption. The visualization of molecular orbital showed that the transfer of electron occurred in the electron pairs of the interacting orbital. Relating to the adsorption mechanism and the overlapping of interaction orbitals, NH_3 molecule provides the Highest Occupied Molecular Orbital (HOMO) as the interaction pair for the Lowest Unoccupied Molecular Orbital (LUMO) on Fe surface. The insignificant difference between the oxidation state of N and H atoms as well as the higher hemolytic N-H bond dissociation energy (compared to heterolytic dissociation) cause the adsorption mechanism of NH_3 on Fe(111) model to occur through homolytic-dissociative chemisorption.

1. Introduction

Iron (Fe) is known as the heterogenic catalyst in NH_3 (ammonia) synthesis through Haber-Bosch process. Initially, the iron oxide form used as a catalyst in NH_3 synthesis is $Fe_2O_3{\cdot}nH_2O$. It is known that the factor affecting the catalytic activity of Fe in NH_3 synthesis is the magnetic properties/magnetic moment of the surface which is represented by the crystallographic surface field of Fe [1].

Theoretically, by using ab initio DFT (Density Functional Theory) method, the calculation regarding the catalytic activity of Fe and its mechanism can be carried out. The calculation mechanism is performed by observing the structural model and the electronic properties of the crystal as well as the effects of magnetic properties on the occurring reaction.

Blonski and Kiejna [2] have carried out the theoretical study using ab initio method on the structure, electronic and magnetic properties of various types of single crystal Fe. It is found that single crystal Fe(111) has the highest surface energy as well as the highest magnetic properties among each type of Fe crystal which consists of 4 – 10 composing atoms. Satoh [3] has conducted the theoretical calculation for the catalytic activity of Fe(110) and Fe(111) as the adsorbent of NH_3 molecule. The optimum position for NH_3 adsorption on Fe surface is ‘on top’ position.

The model of NH_3 molecule could be adsorbed on Fe surface with 3 possible positions: on top of the apical atom, on top of the bridge site and on the three-fold site, known as on top, bridge and hollow positions. Adsorption of a molecule or an atom on the metal surface is dependent on the type and the crystallographic surface field of metal as well as the type of adsorbed matter [4].

Ammonia adsorption on Fe surface can be explained by the bonding formation between electron pair of NH_3 and Fe orbital, forming a donor-acceptor bonding. Confirmation of donor-acceptor bonding formation is carried out by observing the bond length of NH_3 molecular model when adsorbed on Fe surface as well as the magnetic and vibration properties of Fe-NH_3 interaction. The bond length will change and will be expressed as the bond-length correction of the addition of bonding with Fe [5]. GGA method with GENECP-6-311G** combination basis sets has a validation for $Fe_{(10)}$ system, giving the result which is similar to the experimental data. Therefore, the basis sets can be used in the calculation of energy and the structure of Fe-NH_3 complex [3].

2. Methods

The research was conducted in the Austrian-Indonesian Centre (AIC) for Computational Chemistry in Universitas Gadjah Mada, Yogyakarta. The hardware used consists of a computer with the specification as follows: Intel Dual Core 2,66 GHz processor with 4 GB Random Access Memory (RAM) and the software consist of GaussView version 5.08, Gaussian 09 [6], GaussSum version 2.2 [7], and Chemcraft 1.6 [8]. Windows platform was used for formulating the data of bond length and cluster geometry of the molecule under study (input) as well as the visualization of the result (output).

3. Results and Discussion

3.1. Adsorption Model of NH_3 Molecule on Fe(111)

Validation analysis of B3LYP/ECP basis sets in the calculation of Fe(111) field model was shown in Table 1. Spencer et al. [12] have given the evaluation of experimental data of crystal lattice constant of various metals. The crystal lattice constant of Fe(111) was 2.86Å. Doan [10] and Kittel [11] have given the evaluation of experimental data of magnetic moment and binding energy of each atom in various crystals. The magnetic moment of Fe(111) crystal was 2.22μ_B, and the binding energy of each atom was 18 kcal.mol^{-1}. Data resulted from computational calculation showed that the evaluation of crystal lattice constant, magnetic moment and lattice bonding energy are similar to the experimental data.

Table 1. Parameters of Fe(111) system model.

	Lattice constant (Å)	Magnetic moment (μ_B)	Bonding energy of each atom (kcal·mol^{-1})
B3LYP/ECP	2.86	2.22	18.54243
Experimental data	2.86 [12]	2.22 [10,11]	18 [10,11]

Figure 1 showed that there is a shifting of electron density from NH_3 model molecule to Fe(111) surface. NH_3 molecule used its lone pair electrons to interact with Fe(111) surface. According to Figure 1, it can be known that the adsorption of NH_3 molecule in various interacting positions on Fe(111) surface occurred due to the shifting of electron density from NH_3 molecule to Fe(111) surface.

Fielicke et al. [12] generated the equation to evaluate the adsorption energy of the metal-ligand complex system as follows: $\Delta E_{M\text{-}X} = E_{M\text{-}X} - E_M - E_X$
in which, $E_{M\text{-}X}$ is the total energy of metal-ligand complex, E_M is the metal energy and E_X is the ligand energy. It is known that the adsorption energy is the binding energy between metal and ligand, therefore, the lower the adsorption energy; the more stable the metal-ligand complex.

Adsorption energy of on top position is -20.37 kcal·mol^{-1}. The interaction of NH_3 molecule and Fe(111) surface in on top position is an ideal interacting position. On top position has the lowest interaction distance which enables the low coverage. The low coverage is preferred in the adsorption process.

Evaluation of Fe bond length in Fe(111) model and N atom in NH_3 model showed that on top position provided the most dominant interaction sites of all positions. The decrease of interaction distance of Fe_5NH_3 from on top position to bridge and hollow caused the increase of high coverage in the catalytic reaction. This could decrease the active site of the catalyst and increase the residence time of the adsorbate. Thus, the optimum distance of adsorption process in the catalyst is needed [9].

IOP Conf. Series: Materials Science and Engineering **107** (2016) 012068 doi:10.1088/1757-899X/107/1/012068

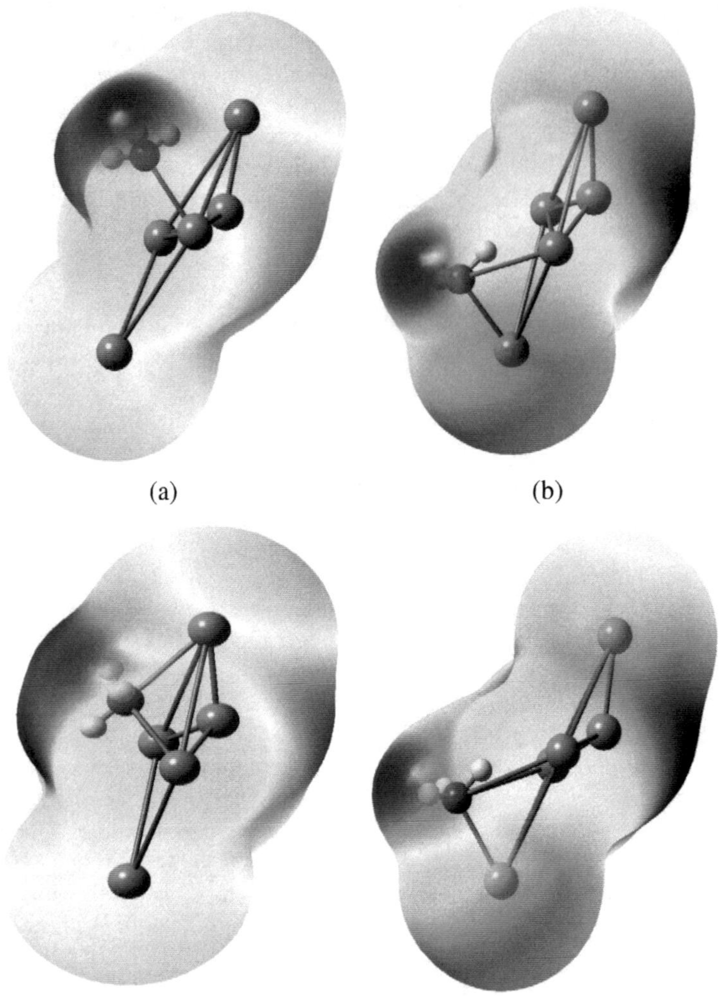

(a) (b) (c) (d)

Figure 1. Interaction model of NH_3 adsorption on Fe(111) surface: (a) on top, (b) bridge, (c) hollow 1 and (d) hollow 2.

Table 2. The distance and energy of NH_3 adsorption on Fe(111) surface.

Adsorption position	The optimum bond distances of Fe-NH_3 (Å)	Adsorption energy (kcal.mol^{-1})
on top	2.40	- 20.37
bridge	2.82	- 13.76
hollow 1	3.24	- 8.43
hollow 2	3.18	- 9.39

The decrease of interaction bond length up to the optimum distance eased the transfer of an electron from NH_3 molecule to Fe(111) surface, thus, increasing the adsorption interaction energy. The lowest interaction distance between Fe and N in on top position makes the electron transfer in this position to be the highest of all other interacting positions.

3.2. Electron Population Analysis

Once the evaluation of electron calculation on Fe(111) surface and NH_3 molecule, as well as the interaction area between those two molecules, are conducted, electron changes related to the interaction could be known and could be elaborated as given in Table 3.

IOP Conf. Series: Materials Science and Engineering **107** (2016) 012068 doi:10.1088/1757-899X/107/1/012068

Table 3. Electron population change in the adsorption of Fe_5-NH_3 complex, given in electron per volume (m^3).

Electron density	on top	bridge	hollow 1	hollow 2
1. Fe surface	1.007	0.275	0.021	0.024
2. Electron transfer from Fe surface to NH_3 molecule	0.018	0.013	0.001	0.002
a. new bond formation	0.018	0.013	0.001	0.002
b. transfer to NH_3	0.000	0.000	0.000	0.000
3. NH_3 molecule	0.010	-0.001	-0.001	-0.001
4. Electron transfer from NH_3 molecule to Fe surface	0.212	0.090	0.060	0.057
a. new bond formation	0.126	0.064	0.043	0.042
b. transfer to Fe surface	0.085	0.026	0.016	0.015

Electron population analysis using electron population approach could not explain the effect of Fe(111) surface on its composing atoms in electron transfer. The evaluation of electron population showed that electron transfer is the main factor in the adsorption of NH_3 molecule on Fe(111) surface.

3.3. Orbital Interaction

An initial study of electron transfer is based on the orbital interaction of Fe(111)-NH_3 complex system from the occupied molecular orbital of NH_3 to the empty molecular orbital of Fe(111). It is known that the electron transfer with a reversed direction is not significant as shown in Table 3. From the result of structural calculation, 4 pairs of occupied-unoccupied orbital were obtained. The electron transfer from NH_3 molecule to Fe(111) surface was represented by 4 orbital pairs for either alfa or beta spins.

Table 4 showed the charges of the orbital interactions which are suitable for the transfer of an electron from NH_3 molecule to Fe(111) surface at an optimum distance in various interacting positions. The value of each pair of interaction orbital demonstrated the electron transfer and electron population that participated in the formation of a new bond.

Table 4. Orbital interaction from NH_3 molecule to Fe(111) surface, given in electron per volume (m^3).

Adsorption position	Spin	Pair 1	Pair 2	Pair 3	Pair 4
on top	A	0.202	0.031	0.027	0.008
	B	0.184	0.021	0.019	0.018
bridge	A	0.197	0.022	0.017	0.012
	B	0.177	0.027	0.022	0.019
hollow 1	A	0.108	0.005	0.004	0.001
	B	0.105	0.005	0.004	0.001
hollow 2	A	0.107	0.005	0.004	0.001
	B	0.106	0.005	0.004	0.001

From Table 4, it is known that the interacting orbital pair 1 participated most significantly in the transfer of an electron from NH_3 molecule to Fe(111) surface in each position. Electron population which is dominant in the formation of the new bond between NH_3 and Fe(111) surface is pair 1. Orbital interaction demonstrated that a small part of electron population shifted from NH_3 molecule to Fe(111) surface and to the area between the two molecules. This shows that the electron population shifting from NH_3 molecule was caused not only by the electron transfer but also affected by the function of Fe orbital in forming the interaction with the orbital function of NH_3 molecule. The orbital interaction also theoretically explains that hollow interacting position is insignificant in the bonding formation between NH_3 molecule and Fe(111) surface.

The adsorption strength is dependent on how far the orbital interaction in the adsorption process is. In the adsorption model, the stronger orbital interaction in the initial stage of adsorption results in the more effective orbital interaction on the surface. The local interaction orbital can also strengthen the further orbital interaction, thus, enables NH_3 molecule to achieve the optimum distance in the structural calculation.

3.4. Homolytic and Heterolytic Dissociative Adsorption of N-H in NH_3 Molecule

Figure 2 (a and b) showed the homolytic and heterolytic dissociative adsorption of Fe-N and Fe-H in Fe_5NH_3 complex. The result if geometric calculation showed that the dissociation of Fe-N and Fe-H occurred homolytically by forming the N atom radical for Fe-N and by forming H atom radical for Fe-H. The calculation showed that NH_3 heterolytic dissociation did not occur due to the insignificant difference of the oxidation states of N and H atoms. In addition, the homolytic dissociation energy of N-H bond is more negative than the heterolytic one. Therefore, the adsorption of NH_3 molecule on Fe(111) surface occurred through homolytic-dissociative chemisorption.

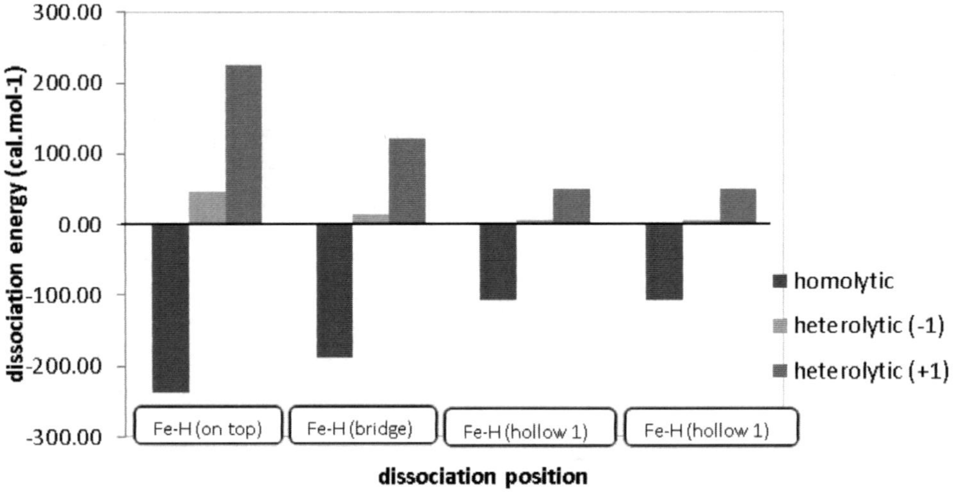

(a)

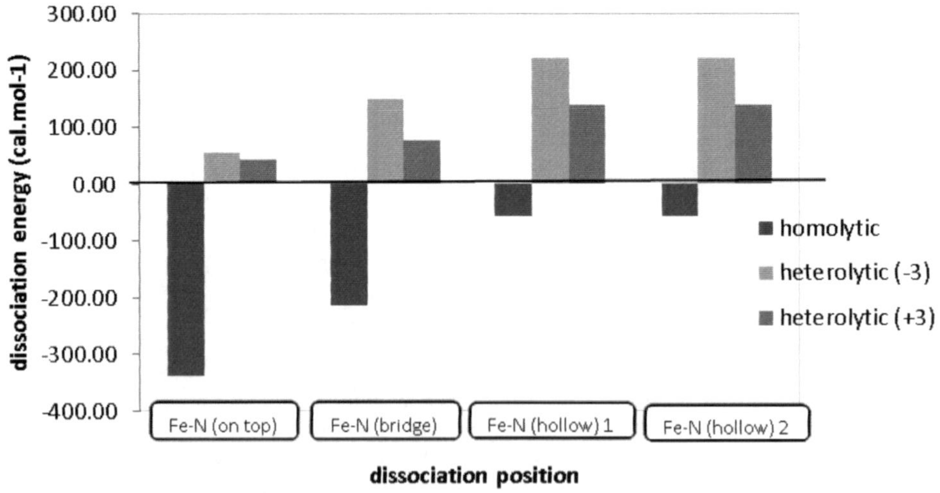

(b)

Figure 2. Homolytic and heterolytic dissociations of (a) Fe-H and (b) Fe-N on Fe(111) model.

Figure 2 (a) and (b) showed that based on the total dissociation energy, the dissociation of Fe-N on Fe_5-NH_3 complex occurred homolytically. The homolytic dissociation of Fe-N and Fe-H was preferable in on top position, followed by bridge and hollow positions.

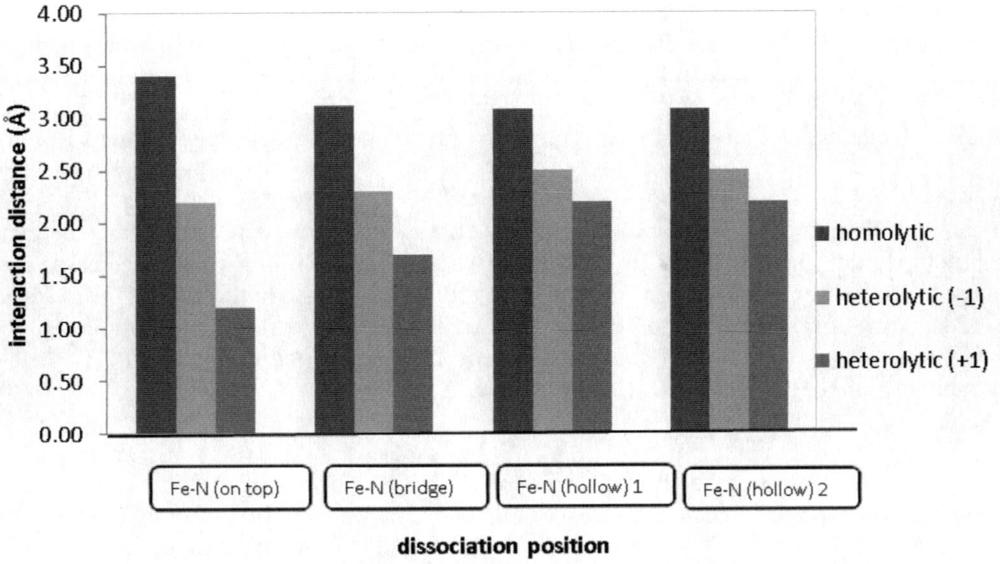

(a)

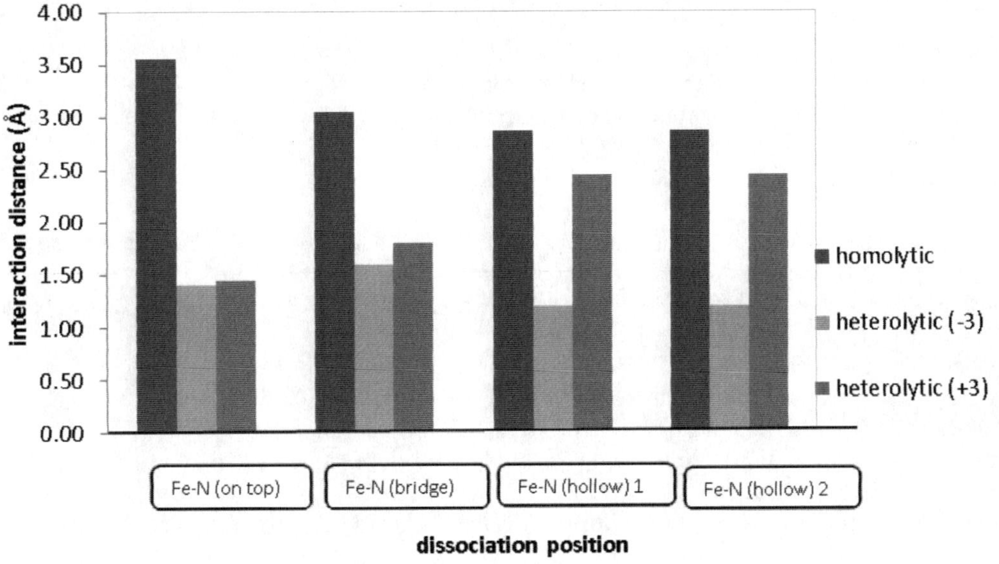

(b)

Figure 3. Homolytic and heterolytic distance of (a) Fe-H and (b) Fe-N bond dissociations on Fe(111) model.

The easiness of the dissociation of N and H atoms is affected by the difference of oxidation state and the interaction distance that affects the energy of interaction model of Fe-N and Fe-H in Fe_5-NH_3 complex. Figure 3 (a) and (b) gave both homolytic and heterolytic bond dissociation distance of Fe-H and Fe-N in Fe(111) model. The formed interaction distance is the representation of bond interaction energy. The greater the interaction distance causes the electron transfer to occur insignificantly, lowering the interaction energy and making the dissociation process become easier. The low

IOP Conf. Series: Materials Science and Engineering **107** (2016) 012068 doi:10.1088/1757-899X/107/1/012068

interaction distance will facilitate the electron transfer and will make Fe-N and Fe-H bonds stronger, thus, the dissociation process becomes difficult to occur.

The evaluation of the value of the dissociation energy stated the easiness of bond breaking process both symmetrically and asymmetrically. The breaking in which one electron resulted from the interaction is attracted to each interacting atom (symmetry breaking) is easier to occur than the breaking in which two interaction electrons are attracted only to one atom (asymmetry breaking).

4. Conclusions

The most stable position of NH_3 adsorption on Fe(111) surface is on top position with the energy of -90.56 kcal·mol^{-1}. Bond length of Fe-N in Fe_5-NH_3 in various adsorption positions decreases in the order: hollow > bridge > on top. Electron transfer in on top position is the highest of all positions. The vibrational frequency of N-H bond in NH_3 molecule decreases in the order: hollow > bridge > on top. The low vibrational frequency of Fe-N caused the strong NH_3 adsorption on Fe_5 field (111) and the dissociation process of N-H in NH_3 become easy to occur.

References

[1] Spencer A N D, Schoonmaker R C and Somorjai G 1982 Iron Single Crystals as Ammonia Synthesis Catalysts: Effect of Surface Structure on Catalyst Activity *J. Catal.* **74** 129

[2] Blonski P and Kiejna A 2006 Structural Electronic and Magnetic Properties of bcc Iron Surfaces *J. Surf. Sci.* **601** 123–133

[3] Satoh S, Fujimoto H and Kobayashi H 2006 Theoretical Study of NH_3 Adsorption on Fe (110) and Fe (111) Surfaces *J. Phys. Chem. B* **110** 4846-4852

[4] Somorjai G A 1994 *Introduction to Surface Chemistry and Catalysis* John Wiley and Sons Inc. New York

[5] Ohashi N, Yoshikawa K, Endou A, Takami S, Kubo M and Miyamoto A 2001 Adsorption properties of SO_2 on Ultrafine Precious Metal Particles Studied Using Density Functional Calculation *Appl. Surf. Sci.* **177** 180-188

[6] Frisch M J, Trucks G W, and Schlegel HB 2009 *Gaussian 09, Revision A.02* Gaussian Inc. Wallingford CT

[7] O'Boyle N M, Tenderholt A L and Langner K M 2008 a Library for Package-Independent Computational Chemistry Algorithms *J. Comput. Chem.* **29** 839-845

[8] Zhurko G A and Zhurko D A 2009 *Chemcraft: Graphical Program for Working with Quantum Chemistry Computations 1.6.* chemcraftprog.com

[9] Heyden B E and Bradshaw A M 1983 The Adsorption of CO on Pt (111) Studied by Infrared Reflection – Absorption Spectroscopy *J. Surf. Sci.* **125**, 787–802

[10] Doan G E 1953 *The Principles of Physical Metallurgy* 3rd ed. McGraw-Hill New York

[11] Kittel C 1996 *Introduction to Solid State Physics* 7th ed. John Wiley & Sons Inc. New York

[12] Fielicke A, Gruene P, Meijer G and Rayner D M 2009 The Adsorption of CO on Transition Metal Clusters: A Case Study of Cluster Surface Chemistry *J. Surf. Sci.* **603** 1427–1433

IOP Conf. Series: Materials Science and Engineering **107** (2016) 012069 doi:10.1088/1757-899X/107/1/012069

The effects of sintering temperature on dielectric constant of Barium Titanate ($BaTiO_3$)

DianisaKhoirum Sandi[1], Agus Supriyanto[1], and Anif Jamaluddin[2], Yofentina Iriani[1]*

[1]Physics Department, Faculty of Mathematics and Natural Science, Sebelas Maret University, Jl.Ir.Sutami 36 A Kentingan, Surakarta 57126
[2]Physics Department, Faculty of Teacher Training and Education, Sebelas Maret University, Jl.Ir. Sutami 36 A Kentingan, Surakarta 57127

*E-mail:yopen_2005@yahoo.com

Abstract.Barium Titanate (BT) has been synthesized using solid-state reaction method. Raw materials are Barium Carbonate ($BaCO_3$) and Titanium Dioxide (TiO_2). These materials are mixed for 6 h and sintered at a temperature of 1000°C, 1100°C, and 1200°C for 2 h. The sintering temperature was varied to investigate its effects on microstructure and dielectric constant of BT. The XRD patterns showed that BT becomes homogenous, with the large lattice parameter as the increase of sintering temperature. The crystal structure of BT is tetragonal. The crystalline size and crystallinity of BT at a sintering temperature of 1000°C are 37 nm and 97%. Those values for BT at a sintering temperature of 1100°C are 38 nm and 96%. At a sintering temperature of 1200°C, the values are 41 nm and 97%. The dielectric constant of BT at a sintering temperature of 1000°C, 1100°C, and 1200°C are 148, 163, and 185, respectively. It can be concluded that sintering temperature affects microstructure and dielectric constant of BT. High sintering temperature produces a high dielectric constant of BT. It indicates that crystalline size increases.

1. Introduction

The ferroelectric material is a material which undergoes spontaneous polarization without subjected to an electric field [1-2]. In its development, ferroelectric materials become more important as main materials in electronic devices [3]. Barium Titanate (BT) or BT is one of the ferroelectric materials with a high dielectric constant which is most commonly used in the electronic industries. It used to the main materials manufacture of electronic components such as capacitors, multilayer capacitor (MMLc), and energy storage devices [1,4,5].

Currie temperature (T_C) of BT is 120°C. When the condition is under T_C, BT materials are in ferroelectric phase and dielectric constant increases with increasing temperature, while the condition is above T_C, these materials are in paraelectric phase which dielectric constant getting down as temperature increases [1,6]. Ferroelectric phase has a tetragonal crystal, and paraelectric phase has cube crystal [2,7]. The change of BT microstructure affects properties of dielectric while the change of structural of BT is affected by the change of temperature on its fabrication [7].

Several methods which are used to synthesize BT are a sol-gel method, hydrothermal, co-precipitation and solid state reaction method. Compared with other methods, solid state reaction is easy, simple, and low cost. Gomez *et al.*, 2000; Micleа*et al*, 2007; Pavlovic*et al*, 2008; Kim *et al.*, 2009;

Othman *et al.*, 2014 using solid state reaction method to fabricate BaTiO3 [8-12]. Their researchers focus on mixing process using ball mill or milling with mixing time variation. The sintering temperature is done under a temperature of 1000°C approximately. XRD patterns of their fabrication showed that there are still peaks of raw materials ($BaCO_3$ and TiO_3) with high intensity.

In this paper, BT materials were synthesized using solid-state reaction method. BT was sintered at temperatures of 1000°C, 1100°C, and 1200°C. Its aims are to determine the effects of sintering temperatures on microstructure and dielectric constant of BT.

2. Experiment Method

Barium titanate ceramics has been synthesized using solid-state reaction method. Raw materials of $BaCO_3$ and TiO_2 were weighed according to the Equation (1).

$$BaCO_3 + TiO_2 \rightarrow BaTiO_3 + CO_2 \quad (1)$$

Raw materials were mixed for 6 h in a mortar using the mechanical technique, pressed to form bulk, and sintered for 2 h at a temperature of 1000°C, 1100°C, and 1200°C.

Samples characterization used X-Ray Diffraction (XRD) instrument byBruker D8 Advance with λ_{Cu} is 1.54056 Å to determine its microstructure. Crystalline size and crystallinity were found using Equation (2) and Equation (3).

$$D = \frac{k\lambda}{B\cos\theta} \quad (2)$$

$$Crystallinity = \frac{I_{maxpeak} - I_{minpeak}}{I_{maxpeak}} \quad (3)$$

D is the crystalline size, β is FWHM, λ is the wavelength of the X-ray source, θ is the angle of diffraction peaks, and k is Scherer constant. Scherer constant for cube crystal is 0.94, whereas the other one is 1 [13,14].

Samples also were characterized using LCR-800 Series instrument byGwinstek.The calculation of dielectric constant is according to Equation (4).

$$K = \frac{Cd}{\varepsilon_0 A} \quad (4)$$

K is dielectric constant of BT, C is capacitance, d is thickness of dielectric material (thickness of bulk BT), A is surface area of bulk BT, and ε_0 is permittivity of vacuum (8.85 x 10^{-12} F / m) [15].

3. Results and Discussion

The result of characterization using XRD is shown by Figure1. XRD patterns were matched with International Commission Data Diffraction (ICDD)database and them How the peaks correspond to the peaks of BT with database number #752116. The peaks with the mark (*) are $BaCO_3$ with database number#5225. They indicate that mixture of BT was still not homogeneous, or raw materials have not reacted perfectly. Increasing temperature does not affect the diffraction angle but affects the peaks of intensity. The peaks of $BaCO_3$ become lower which indicating the higher temperature made BT more homogenous. Due to high sintering temperature provides high energy for raw materials to react (through diffusion), then the mixture becomes more homogenous.

XRD patterns wereanalyzedusingGeneralStructureAnalysis System software (GSAS) withRietveldmethodtoensurethatthe peaks ofXRD pattern is BT.The results of the refinement process using GSAS software shown in Figure 2. The black graph is a graph of data obtained from XRD (observed data) that was incorporated into the GSAS software. The red graph is a graph from ICDD data (data calculation). Blue line is the difference between observed data and calculation data. The green line is the background. And the orange line is the Chi-square (χ^2). If the data calculation closer to or in agreement with the observed data, Chi-Square line (χ^2) will be straight, and the value will be smaller or close to 1.

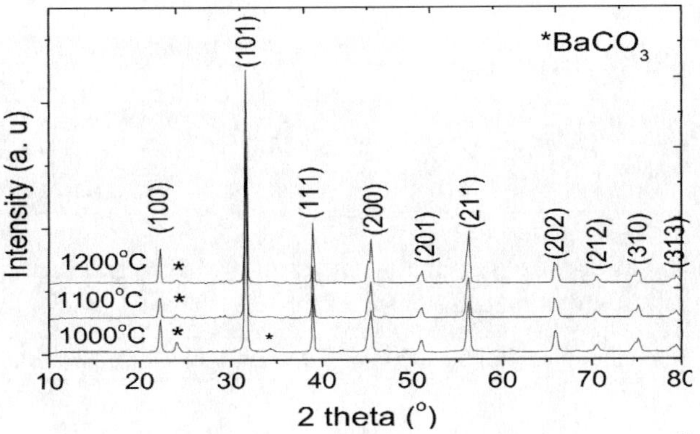

Figure 1. XRD patterns of BTatsinteringtemperatures of1000°C, 1100°C, and 1200°C

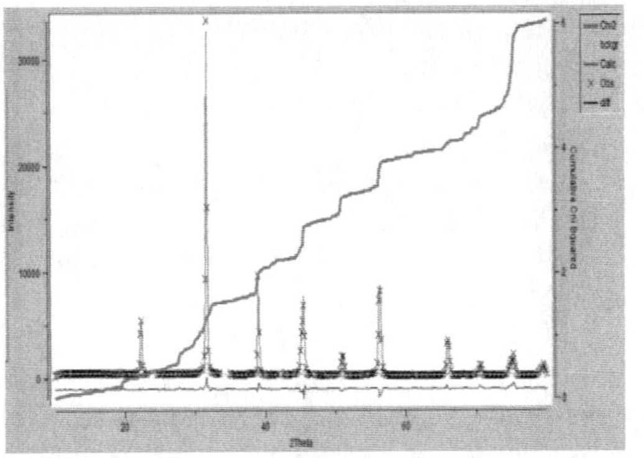

(a)

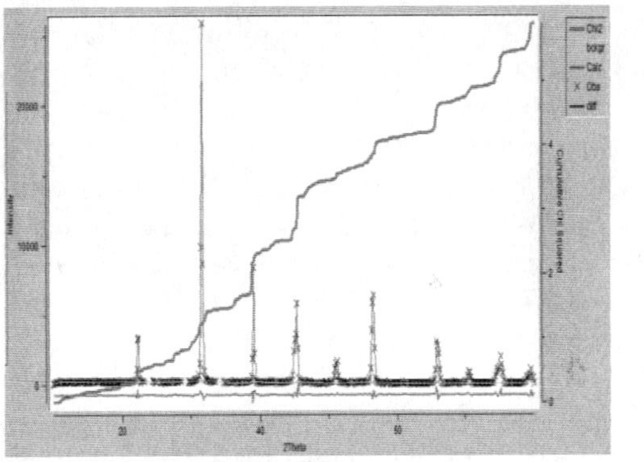

(b)

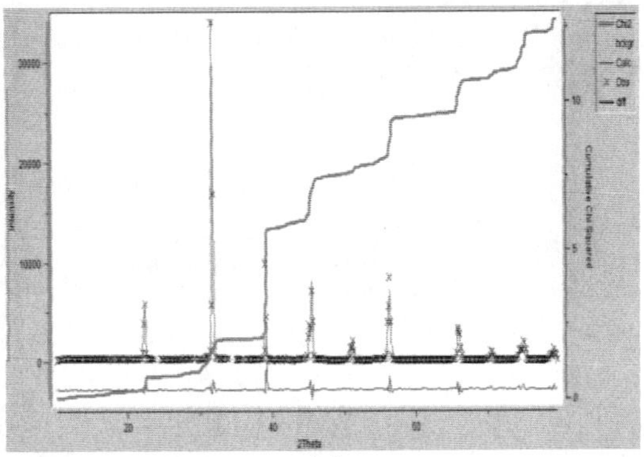

(c)

Figure 2. The results of refinement of XRD patterns of BT using GSAS software at sintering temperature of (a) 1000°C (b) 1100°C (c) 1200°C

The results of refinement process show that the initial allegation is true, that those peaks belong to BT with tetragonal crystal structure which a =b≠c and α=β=γ=90°. Chi-square (χ^2) of refinement results of BT samples at sintering temperature of 1000°C, 1100°C, and 1200°C, are 6.065; 5.910; and 12.880, respectively.

Table 1.BTlattice parameters from the results of refinement process using GSAS software

Sintering temperature (°C)	Lattice parameter (Å)	
	a=b	C
1000	3.983	4.009
1100	3.987	4.015
1200	3.978	4.024

Table 2. Crystalline size at plane orientation of (1 0 1) and crystallinity of BT with variation of sintering temperature

Sintering temperature (°C)	Crystalline size (nm)	Crystallinity (%).
1000	37	97
1100	38	96
1200	41	97

The results of refinement process confirm lattice parameter of BT. Table 1 shows the lattice parameter of BT, which is influenced by sintering temperature. The lattice parameters of BT tend increasingly longer when the sintering temperature increased.

The peaks of XRD pattern can be used to calculate FWHM values. Then, it used to calculate the crystallite size and crystallinity of BT crystal using Eq. 2 and Eq. 3. Table 2 shows the effect of sintering temperature on the crystalline size and crystallinity. The crystalline size of BT increases as the higher sintering temperature. Sintering temperature gives energy to the atoms diffuse and agglomerates. That energy becomes bigger when the sintering temperature increased, and then more atoms agglomerate, thus, the crystalline size becomes larger [16]. That is why BT's lattice parameter tends increasingly longer.

The regularity of the atomic arrangement of materials is expressed by crystallinity. Crystallinity is influenced by sintering temperature. The higher sintering temperature causes the energy that is received by atoms to become larger (diffusion rate becomes faster) so that the bond of atoms become strong and regular. The high value of crystallinity shows that atoms bonding are robust and regular. The crystallinity values of BT shown in Table 2.

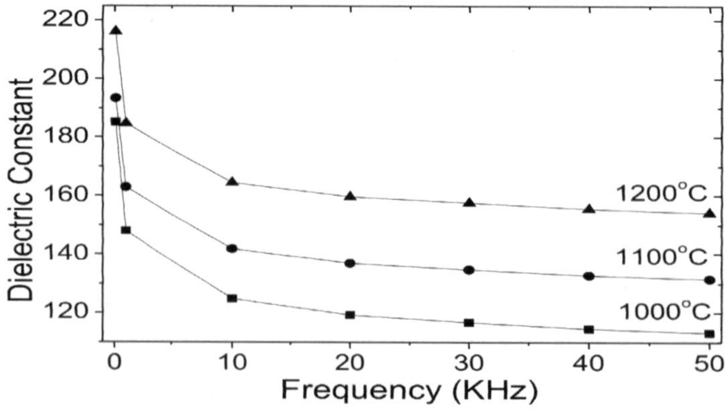

Figure 3. The effect of frequency on dielectric constant of with sintering temperature variation

IOP Conf. Series: Materials Science and Engineering **107** (2016) 012069 doi:10.1088/1757-899X/107/1/012069

Capacitance was obtained from characterization using LCR meter instrument. It used to calculate dielectric constant using Eq. 4. Frequency was varied at 0.1 KHz to 50 KHz to find out its effect on dielectric constant. Figure 3 is the graph of the dielectric constant of BT to the change of frequency with variation sintering temperature. The frequency affects capacitance; the higher frequency makes capacitance become lower so that dielectric constant is so. The dielectric constant of ferroelectric materials will be lower as increasing of frequency but tends to be more stable [1].

Increasing of sintering temperature causes the dielectric constant of BT increases. Dielectric constant values of BT, which are measured at a frequency of 1 KHz for samples at a sintering temperature of 1000°C, 1100°C, and 1200°C are 148, 163, and 185, respectively. In ferroelectric materials crystals, there are domains. Domain is the area which has polarization with the same direction. Ferroelectric domains become larger as the larger crystalline size, thus, the polarization becomes stronger. It indicates that dielectric constant of BT increases.

4. Conclusions

BTwas synthesized using solid-state reaction method at a sintering temperature of 1000°C, 1100°C, and 1200°C. The sintering temperature affects microstructure and dielectric constant of BT. The higher sintering temperature made a change in lattice parameter. The crystallinity of BT crystals is high for each sintering temperature. The higher sintering temperature made dielectric constant of BT becomes higher because it is considered that the crystalline size becomes larger.

5. Acknowledgements

Graduate Grant contract in 2015 with the number 624 / UN27.11 / PL / 2015 that has provided research grants. So with these funds, the writer can do this research.

6. References

[1] Vijatovic M M, Bobic J D, Stojanovic B D2008*Science of Sintering***40**235-244.
[2] Yadav A K, Gautam C R, and Mishra A2014*Journal of Advanced Ceramics***3(2)** 137–146
[3] Kongtaweelert S, Sinclair D C, and Panichphant S 2006 *Current Applied Physics***6** 474–477
[4] Kholodkova A,Danchevskaya M, and FionovA 2012*Nanocon2012***10**
[5] Modani U S, and Jagrawal G 2012 *International Journal of Recent Technology and Engineering (IJRTE)***1**65-72
[6] Kosanovic D, Obradovi N, Zivojinovic J, Filipovic C, Maricic A, Pavlovic V, Tang Y, and Ristic M 2012 *Science of Sintering* **44** 47-55.
[7] Sonia 2011Synthesis & Characterization of Modified Barium Titanate Ferroelectrics by Modified Solid State Reaction and Microwave Sintering Routes*Disertation* National Institute of Technology Rourkela, India.
[8] Gomes C, Benitez C, and Ramirez B 2000 *Ceramics International***26** 271-277
[9] Miclea C, Tanasoiu C, Spanulescu I, Miclea CF, Gheorghiu A, Amarande L, Cioangher M and Miclea CT 2007 *Roumanian Journal of Information Science and Technology***10** 335-345
[10] Pavlovic VP, Stojanovic BD, Pavlovic VB, Marinkovic Z, Zivkovic L and Ristic 2008 *Science of Sintering***40** 21-26
[11] Kim HT, Kim JH, Jung WS and Yoon DY 2009 *Journal of Ceramics Processing Research***10** 753-757
[12] Othman K, Hassan A, Abdelal O and Elshazly E 2014 International *Journal of Scientific &Enginering Research***5** 1460-1464
[13] MaharsiR, Jamaludin A, Iriani Y 2014*JurnalFisika Indonesia***XVIII** 52
[14] Suryanarayana C and Grant Norton M1998 *X-Ray Diffraction (XRD): A practical approach*(Washington: Washington State University)
[15] Serway R A, and Jewet J W 2004 *Physics for Scientist and Engineers*(United States of America : Thomson Brooks/Cole)

IOP Conf. Series: Materials Science and Engineering **107** (2016) 012069 doi:10.1088/1757-899X/107/1/012069

[16] Callister WD 2007 *Materials science and engineering* 7th*ed.* Department of Metallurgical Engineering, University of Utah. U.S.A.(New Jersey: John Wiley & Sons, Inc)

Institute of Physics
Dirac House, Temple Back
Bristol BS1 6BE UK

ISSN: 1757-8981
ISBN 978-1-5108-2084-5

CURRAN ASSOCIATES INC.
proceedings.com